# Environmental Stressors in Health and Disease

**OXIDATIVE STRESS AND DISEASE**

Series Editors

LESTER PACKER, PH.D.
ENRIQUE CADENAS, M.D., PH.D.
University of Southern California School of Pharmacy
Los Angeles, California

1. Oxidative Stress in Cancer, AIDS, and Neurodegenerative Diseases, *edited by Luc Montagnier, René Olivier, and Catherine Pasquier*
2. Understanding the Process of Aging: The Roles of Mitochondria, Free Radicals, and Antioxidants, *edited by Enrique Cadenas and Lester Packer*
3. Redox Regulation of Cell Signaling and Its Clinical Application, *edited by Lester Packer and Junji Yodoi*
4. Antioxidants in Diabetes Management, *edited by Lester Packer, Peter Rösen, Hans J. Tritschler, George L. King, and Angelo Azzi*
5. Free Radicals in Brain Pathophysiology, *edited by Giuseppe Poli, Enrique Cadenas, and Lester Packer*
6. Nutraceuticals in Health and Disease Prevention, *edited by Klaus Krämer, Peter-Paul Hoppe, and Lester Packer*
7. Environmental Stressors in Health and Disease, *edited by Jürgen Fuchs and Lester Packer*

*Additional Volumes in Preparation*

Handbook of Antioxidants, Second Edition, Revised and Expanded, *edited by Enrique Cadenas and Lester Packer*

*Related Volumes*

Vitamin E in Health and Disease: Biochemistry and Clinical Applications, *edited by Lester Packer and Jürgen Fuchs*

Vitamin A in Health and Disease, *edited by Rune Blomhoff*

Free Radicals and Oxidation Phenomena in Biological Systems, *edited by Marcel Roberfroid and Pedro Buc Calderon*

Biothiols in Health and Disease, *edited by Lester Packer and Enrique Cadenas*

Handbook of Antioxidants, *edited by Enrique Cadenas and Lester Packer*

Handbook of Synthetic Antioxidants, *edited by Lester Packer and Enrique Cadenas*

Vitamin C in Health and Disease, *edited by Lester Packer and Jürgen Fuchs*

Lipoic Acid in Health and Disease, *edited by Jürgen Fuchs, Lester Packer, and Guido Zimmer*

Flavonoids in Health and Disease, *edited by Catherine Rice-Evans and Lester Packer*

# Environmental Stressors in Health and Disease

edited by

JÜRGEN FUCHS
*J. W. Goethe University*
*Frankfurt, Germany*

LESTER PACKER
*University of Southern California School of Pharmacy*
*Los Angeles, California*

MARCEL DEKKER, INC. NEW YORK • BASEL

**ISBN: 0-8247-0530-0**

This book is printed on acid-free paper.

**Headquarters**
Marcel Dekker, Inc.
270 Madison Avenue, New York, NY 10016
tel: 212-696-9000; fax: 212-685-4540

**Eastern Hemisphere Distribution**
Marcel Dekker AG
Hutgasse 4, Postfach 812, CH-4001 Basel, Switzerland
tel: 41-61-261-8482; fax: 41-61-261-8896

**World Wide Web**
http://www.dekker.com

The publisher offers discounts on this book when ordered in bulk quantities. For more information, write to Special Sales/Professional Marketing at the headquarters address above.



Current printing (last digit):
10 9 8 7 6 5 4 3 2 1

**PRINTED IN THE UNITED STATES OF AMERICA**

# Series Introduction

Oxygen is a dangerous friend. Overwhelming evidence indicates that oxidative stress can lead to cell and tissue injury. However, the same free radicals that are generated during oxidative stress are produced during normal metabolism and thus are involved in both human health and disease.

Free radicals are molecules with an odd number of electrons. The odd, or unpaired, electron is highly reactive as it seeks to pair with another free electron.
Free radicals are generated during oxidative metabolism and energy production in the body.
Free radicals are involved in:
- Enzyme-catalyzed reactions
- Electron transport in mitochondria
- Signal transduction and gene expression
- Activation of nuclear transcription factors
- Oxidative damage to molecules, cells, and tissues
- Antimicrobial action of neutrophils and macrophages
- Aging and disease

Normal metabolism is dependent on oxygen, a free radical. Through evolution, oxygen was chosen as the terminal electron acceptor for respiration. The two unpaired electrons of oxygen spin in the same direction; thus, oxygen is a biradical, but is not a very dangerous free radical. Other oxygen-derived free radical species, such as superoxide or hydroxyl radicals, formed during metabolism or by ionizing radiation are stronger oxidants and are therefore more dangerous.

In addition to research on the biological effects of these reactive oxygen species, research on reactive nitrogen species has been gathering momentum. NO, or nitrogen monoxide (nitric oxide), is a free radical generated by NO synthase (NOS). This enzyme modulates physiological responses such as vasodilation or signaling in the brain. However, during inflammation, synthesis of NOS (iNOS) is induced. This iNOS can result in the overproduction of NO, causing damage. More worrisome, however, is the fact that excess NO can react with superoxide to produce the very toxic product peroxynitrite. Oxidation of lipids, proteins, and DNA can result, thereby increasing the likelihood of tissue injury.

Both reactive oxygen and nitrogen species are involved in normal cell regulation in which oxidants and redox status are important in signal transduction. Oxidative stress is increasingly seen as a major upstream component in the signaling cascade involved in inflammatory responses, stimulating adhesion molecule and chemoattractant production. Hydrogen peroxide, which breaks down to produce hydroxyl radicals, can also activate NF-κB, a transcription factor involved in stimulating inflammatory responses. Excess production of these reactive species is toxic, exerting cytostatic effects, causing membrane damage, and activating pathways of cell death (apoptosis and/or necrosis).

Virtually all diseases thus far examined involve free radicals. In most cases, free radicals are secondary to the disease process, but in some instances free radicals are causal. Thus, there is a delicate balance between oxidants and antioxidants in health and disease. Their proper balance is essential for ensuring healthy aging.

The term *oxidative stress* indicates that the antioxidant status of cells and tissues is altered by exposure to oxidants. The redox status is thus dependent on the degree to which a cell's components are in the oxidized state. In general, the reducing environment inside cells helps to prevent oxidative damage. In this reducing environment, disulfide bonds (S—S) do not spontaneously form because sulfhydryl groups kept in the reduced state (SH) prevent protein misfolding or aggregation. This reducing environment is maintained by oxidative metabolism and by the action of antioxidant enzymes and substances, such as glutathione, thioredoxin, vitamins E and C, and enzymes such as superoxide dismutase (SOD), catalase, and the selenium-dependent glutathione and thioredoxin hydroperoxidases, which serve to remove reactive oxygen species.

Changes in the redox status and depletion of antioxidants occur during oxidative stress. The thiol redox status is a useful index of oxidative stress mainly because metabolism and NADPH-dependent enzymes maintain cell glutathione (GSH) almost completely in its reduced state. Oxidized glutathione (glutathione disulfide, GSSG) accumulates under conditions of oxidant exposure, and this changes the ratio of oxidized to reduced glutathione; an increased ratio indicates oxidative stress. Many tissues contain large amounts of glutathione, 2–4 mM in erythrocytes or neural tissue and up to 8 mM in hepatic tissue. Reactive oxygen and nitrogen species can directly react with glutathione to lower the levels of this substance, the cell's primary preventative antioxidant.

Current hypotheses favor the idea that lowering oxidative stress can have a clinical benefit. Free radicals can be overproduced or the natural antioxidant system defenses weakened, first resulting in oxidative stress, and then leading to oxidative injury and disease. Examples of this process include heart disease and cancer. Oxidation of human low-density lipoproteins is considered the first step in the progression and eventual development of atherosclerosis, leading to cardiovascular disease. Oxidative DNA damage initiates carcinogenesis.

Compelling support for the involvement of free radicals in disease development comes from epidemiological studies showing that an enhanced antioxidant status is associated with reduced risk of several diseases. Vitamin E and prevention of cardiovascular disease is a notable example. Elevated antioxidant status is also associated with decreased incidence of cataracts and cancer, and some recent reports have suggested an inverse correlation between antioxidant status and occurrence of rheumatoid arthritis and diabetes mellitus. Indeed, the number of indications in which antioxidants may be useful in the prevention and/or the treatment of disease is increasing.

Oxidative stress, rather than being the primary cause of disease, is more often a secondary complication in many disorders. Oxidative stress diseases include inflammatory bowel diseases, retinal ischemia, cardiovascular disease and restenosis, AIDS, ARDS, and neurodegenerative diseases such as stroke, Parkinson's disease, and Alzheimer's disease. Such indications may prove amenable to antioxidant treatment because there is a clear involvement of oxidative injury in these disorders.

In this new series of books, the importance of oxidative stress in diseases associated with organ systems of the body will be highlighted by exploring the scientific evidence and the medical applications of this knowledge. The series will also highlight the major natural antioxidant enzymes and antioxidant substances such as vitamins E, A, and C, flavonoids, polyphenols, carotenoids, lipoic acid, and other nutrients present in food and beverages.

Oxidative stress is an underlying factor in health and disease. More and more evidence is accumulating that a proper balance between oxidants and antioxidants is involved in maintaining health and longevity and that altering this balance in favor of oxidants may result in pathological responses causing functional disorders and disease. This series is intended for researchers in thc basic biomedical sciences and clinicians. The potential for healthy aging and disease prevention necessitates gaining further knowledge about how oxidants and antioxidants affect biological systems.

Oxidative stress arising from exposure to environmental stressors such as irradiation, chemicals, and atmospheric pollutants is a major source of pathophysiological change leading to aging and disease initiation and progression. In particular, such responses would be involved in disorders to the exposed parts of the body such as the skin, eyes, and respiratory tract.

This volume highlights the role of the environment in oxidative processes involved in health and disease.

*Lester Packer*
*Enrique Cadenas*

# Preface

Industrial, agricultural, pharmacological, and lifestyle applications have increased the kinds and amounts of xenobiotic agents to which humans are exposed. A generally accepted toxicity hypothesis is that some of these agents mediate cellular injury by virtue of their ability to cause oxidative stress and subsequently trigger oxidative injury in susceptible biomolecules and/or to modulate redox-sensitive elements, such as signal transduction pathways and gene expression. Environmental pollution with prooxidant xenobiotics may have an impact on the incidence of diseases of the lung, the skin, the eye, and the immune system, which are the major targets directly exposed to environmental stressors. The stress imposed on human health by environmental oxidants may be reduced by increased consumption of micronutrients and antioxidants. The objective of this book is to review issues of molecular and cellular biology as well as clinical aspects of environmental oxidants and their effects on target organs, with the aim of helping the reader comprehend the complexities of free radical aspects of environmental medicine while at the same time providing useful clinical information on common illnesses and tissue injury.

Part I focuses on the basic pathology induced by prooxidant environmental xenobiotics, such as inflammation, immune response, signal transduction, regulation of gene expression, and carcinogenesis. Parts II to IV focus on the target organs of environmental oxidants—the lung, skin, and eye, and the immune system. Chapters cover (1) the toxic effects of cigarette smoke, diesel exhaust particles, asbestos, silica, and oxidizing atmospheric gases on the lung; (2) the action of ultraviolet radiation, environmental and occupational prooxidant xenobiotics in skin function and disease; and (3) aspects of eye injury mediated through oxidative stress, in particular the oxidant pathogenesis of cataract formation, anterior segment damage, and macular degeneration.

We believe that this volume will provide an important and timely contribution to the biomedical community.

*Jürgen Fuchs*
*Lester Packer*

# Contents

## II. Target Organs of Environmental Oxidants: Lung



## III. Target Organs of Environmental Oxidants: Skin



## IV. Target Organs of Environmental Oxidants: Eye

# Contributors

**Paul S. Bernstein, M.D., Ph.D.** Moran Eye Center, University of Utah School of Medicine, Salt Lake City, Utah

**David R. Blake** Royal Hospital for Rheumatic Diseases, Bath, England

**Vince Castranova** National Institute for Occupational Safety and Health, Morgantown, West Virginia

**Fei Chen** National Institute for Occupational Safety and Health, Morgantown, West Virginia

**Ching K. Chow, Ph.D.** Graduate Center for Nutritional Sciences, University of Kentucky, Lexington, Kentucky

**Carroll E. Cross, M.D.** Department of Medicine and Physiology, University of California, Davis, California

**Claire A. Davies** Bone and Joint Research Unit, St. Bartholomew's and the Royal London School of Medicine and Dentistry, London, England

**Min Ding** National Institute for Occupational Safety and Health, Morgantown, West Virginia

**Jürgen Fuchs, Ph.D., M.D.** Department of Dermatology, J. W. Goethe University, Frankfurt, Germany

**Keith Green** Department of Ophthalmology, Medical College of Georgia, Augusta, Georgia

**Barry Halliwell, D.Sc.** Department of Biochemistry, National University of Singapore, Singapore

**Andrea Hubbard, Ph.D.** University of Connecticut School of Pharmacy, Storrs-Mansfield, Connecticut

**Takamichi Ichinose** Ohita University of Nursing and Health Sciences, Ohita-Notsuhara, Japan

**Lisa M. Kamendulis** Department of Pharmacology and Toxicology, Indiana University, Indianapolis, Indiana

**Santosh K. Katiyar** Department of Dermatology, Case Western Reserve University, Cleveland, Ohio

**Nikita B. Katz, M.D., C. Phil.** Institute for Natural Resources, Berkeley, California

**Kaye H. Kilburn, M.D.** Department of Medicine, University of Southern California, Keck School of Medicine, Los Angeles, California

**Yong-Chul Kim** Institute for Virus Research, Kyoto University, Kyoto, Japan

**James E. Klaunig, Ph.D.** Department of Pharmacology and Toxicology, Indiana University, Indianapolis, Indiana

**Sunye Kwack, M.D.** Department of Medicine, University of California, Davis, California

**Yong-Won Kwon** Institute for Virus Research, Kyoto University, Kyoto, Japan

**Lee Ann Laurent-Applegate, Ph.D.** Department of Obstetrics, University Hospital, Lausanne, Switzerland

**Samuel Louie, M.D.** Department of Medicine, University of California, Davis, California

**Michael I. Luster, Ph.D.** Toxicology and Molecular Biology Branch, National Institute for Occupational Safety and Health, Morgantown, West Virginia

**Hiroshi Masutani** Department of Biological Responses, Institute for Virus Research, Kyoto University, Kyoto, Japan

**Kenneth P. Mitton, B.Sc. (Hon.), Ph.D.** Kellogg Eye Center, University of Michigan, Ann Arbor, Michigan

**Brooke T. Mossman, Ph.D.** Department of Pathology, University of Vermont College of Medicine, Burlington, Vermont

**Hasan Mukhtar** Department of Dermatology, Case Western Reserve University, Cleveland, Ohio

**Hajime Nakamura** Institute for Virus Research, Kyoto University, Kyoto, Japan

**Akira Nishiyama** Institute for Virus Research, Kyoto University, Kyoto, Japan

**Lester Packer, Ph.D.** Department of Molecular Pharmacology and Toxicology, University of Southern California School of Pharmacy, Los Angeles, California

**Maurizio Podda** Department of Dermatology, J. W. Goethe University, Frankfurt, Germany

**Sharanya Reddy, Ph.D.** Department of Medicine, University of California, Davis, California

**Masaru Sagai, Ph.D.** Faculty of Health Sciences, Aomori University for Health and Welfare, Aomori City, Japan

**Stefan Schwarzkopf** University Hospital, Lausanne, Switzerland

**Xianglin Shi, Ph.D.** Pathology and Physiology Research Branch, National Institute for Occupational Safety and Health, Morgantown, West Virginia

**Arti Shukla, Ph.D.** Department of Pathology, University of Vermont College of Medicine, Burlington, Vermont

**Petia P. Simeonova** National Institute for Occupational Safety and Health, Morgantown, West Virginia

**Allen Taylor, Ph.D.** Jean Mayer USDA Human Nutrition Research Center on Aging, Tufts University, Boston, Massachusetts

**Cynthia R. Timblin, Ph.D.** Department of Pathology, University of Vermont College of Medicine, Burlington, Vermont

**John R. Trevithick, Ph.D.** Department of Biochemistry, Faculty of Medicine and Dentistry, University of Western Ontario, London, Ontario, Canada

**Val Vallyathan** National Institute for Occupational Safety and Health, Morgantown, West Virginia

**Albert van der Vliet, Ph.D.** Department of Medicine, University of California, Davis, California

**Stefan U. Weber** Department of Molecular and Cell Biology, University of California, Berkeley, California

**Paul G. Winyard** St. Bartholomew's and the Royal London School of Medicine and Dentistry, London, England

**Pat Wong, Ph.D.** Department of Medicine, University of California, Davis, California

**Junji Yodoi, M.D., Ph.D.** Department of Biological Responses, Institute for Virus Research, Kyoto University, Kyoto, Japan

**Thomas Zollner** Department of Dermatology, J. W. Goethe University, Frankfurt, Germany

# Environmental Stressors in Health and Disease

# 1

# Free Radical Reactions in Human Disease

**BARRY HALLIWELL**

*National University of Singapore, Singapore*

## I. INTRODUCTION

The biomedical literature continues to resound with suggestions that "free radicals" and other "reactive species" are involved in different human diseases. They have been implicated in over 100 disorders, ranging from scleroderma and hemorrhagic shock to cardiomyopathy and cystic fibrosis to gastrointestinal ischemia, AIDS, hearing loss (1–7) and even male pattern baldness (8). The various chapters that follow provide further illustrations. This wide range of disorders implies that free radicals are not something esoteric, but that their increased formation accompanies tissue injury in most, if not all, human diseases (9). Reasons for this are summarized in Figure 1. Sometimes free radicals make a significant contribution to disease pathology; at other times they may not, as Figure 2 illustrates. A major task of researchers in this field is to distinguish between these scenarios so as to be able to create useful therapies: there is no therapeutic advantage to inhibiting free radical damage when it is merely an accompaniment to tissue injury. As Table 1 summarizes, demonstrating that free radicals are important in any disease involves much more than a mere demonstration of their formation in increased amounts. The same comment applies to nitric oxide, cytokines, leukotrienes, and any of the other potential mediators of tissue injury.

A clear understanding of some basic definitions and principles is necessary to guide us in evaluating the role of free radicals in human disease and in designing antioxidants for therapeutic use. This chapter provides those definitions and then considers how best to evaluate the role of free radicals and other reactive species in disease and how to evaluate the effects of therapeutic antioxidants in vivo.

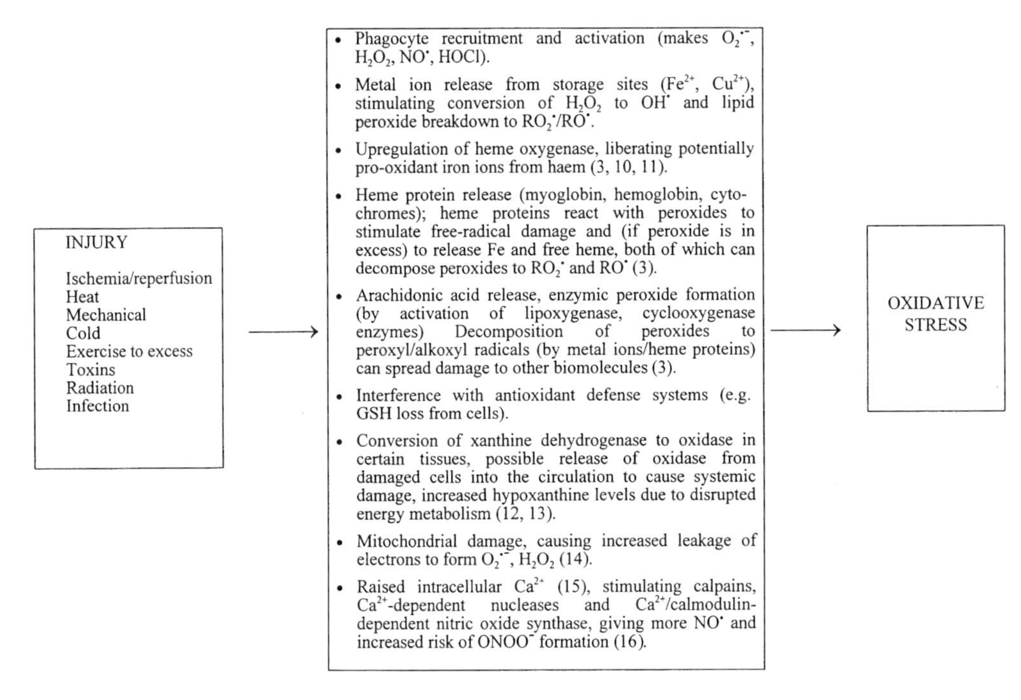


**Figure 1** Some of the reasons why tissue injury (by any mechanism) causes oxidative stress.

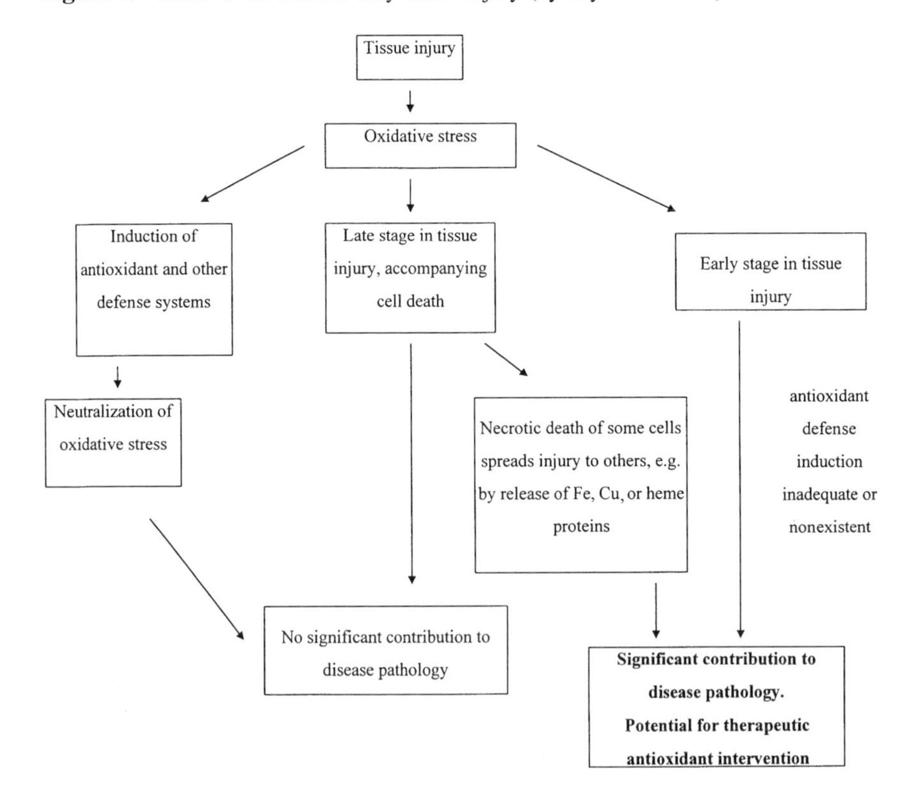


**Figure 2** Free radicals: cause and consequence.

**Table 1** Criteria for Implicating Reactive Species (RS) as a Significant Mechanism of Tissue Injury in Human Disease

1. The RS should always be present at the site of injury.
2. Its time course of formation should be consistent with the time course of tissue injury, preceding or accompanying it.
3. Direct application of the RS to the tissue at concentrations within the range found in vivo should reproduce most or all of the damage observed.
4. Removing the RS or inhibiting its formation should diminish the injury to an extent related to the degree of removal of the RS or inhibition of its formation.

These criteria apply equally well to other agents (e.g., nitric oxide, prostaglandins, leukotrienes, and cytokines). All of these agents have been implicated in multiple diseases, but their importance has been proven only infrequently.

## II. BASIC DEFINITIONS

### A. Free Radical

Electrons in atoms and molecules move within regions of space known as orbitals. Each orbital holds a maximum of two electrons, since no two electrons can have the same four quantum numbers. A free radical is defined as *any species capable of independent existence that contains one or more unpaired electrons*, an unpaired electron being one that is alone in an orbital. Examples of free radicals are atomic hydrogen ($H^{\bullet}$) and atomic sodium ($Na^{\bullet}$). Others include the oxygen-centered radicals such as superoxide ($O_2^{\bullet -}$) and hydroxyl ($OH^{\bullet}$), the sulfur-centered radicals such as thiyl ($RS^{\bullet}$), and carbon-centred radicals such as trichloromethyl ($CCl_3^{\bullet}$). Nitric oxide ($NO^{\bullet}$) and nitrogen dioxide ($NO_2^{\bullet}$) are free radicals in which the unpaired electron is delocalized between different atoms. A superscript dot is used to designate a free radical.

Most biological molecules are nonradicals, containing only paired electrons. A free radical can be made by loss of one electron from a nonradical (X),

$$X \rightarrow X^{\bullet +} + e^-,$$

by a gain of one electron,

$$X + e^- \rightarrow X^{\bullet -},$$

by homolytic fission of a covalent bond (splitting of the electron pair forming the bond so as to leave one electron on each of the originally bonded atoms),

$$X - Y \rightarrow X^{\bullet} + Y^{\bullet},$$

or by abstraction of a hydrogen atom. Since $H^{\bullet}$ has one electron, its removal must leave behind an unpaired electron on the atom to which the $H^{\bullet}$ was covalently bonded. Some examples,

$$CH_4 + Cl^{\bullet} \rightarrow CH_3^{\bullet} + HCl,$$

$$\text{Lipid-H} + OH^{\bullet} \rightarrow \text{lipid}^{\bullet} + H_2O,$$

$$RSH + RO_2^{\bullet} \rightarrow RS^{\bullet} + RO_2H.$$

Antioxidants are usually not free radicals; thus when an antioxidant is acting as a free radical scavenger, an antioxidant-derived radical will be created (radicals beget radicals).

Similarly, if a free radical loses or gains a single electron, it usually ceases to be a free radical. Thus, for nitric oxide,

$$NO^{\bullet} \rightarrow e^{-} + NO^{+}\text{(nitrosyl cation)},$$

$$NO^{\bullet} + e^{-} \rightarrow NO^{-}\text{(nitroxyl anion)}.$$

When free radicals react together, there can also be a net loss of radicals. Radical–radical termination reactions are often beneficial by disposing of reactive radicals. However, this is not always true; sometimes the products are damaging. For example, reaction of $O_2^{\bullet-}$ and $NO^{\bullet}$ generates peroxynitrite, $ONOO^-$ (16),

$$O_2^{\bullet-} + NO^{\bullet} \rightarrow ONOO^-$$

and self-reactions of peroxyl radicals can generate singlet $O_2$ (17).

## B. Reactive Species

The term reactive oxygen species (ROS), often used in the free radical field, is a collective one that includes not only oxygen-centered radicals such as $O_2^{\bullet-}$ and $OH^{\bullet}$, but also some nonradical derivatives of oxygen, such as hydrogen peroxide ($H_2O_2$), singlet oxygen $^1\Delta g$, and hypochlorous acid (HOCl) (Table 2). A similar term, reactive nitrogen species, has diffused into the literature over the past 3 years. The term reactive chlorine species has recently made an appearance (Table 2).

## C. Oxidative Stress

The term oxidative stress is widely used in the literature, but is often vaguely defined. In essence, it refers to a serious imbalance between production of reactive species and antioxidant defense. Sies introduced the term in 1985, and defined it in 1991 (19) as *a disturbance in the prooxidant–antioxidant balance in favor of the former, leading to potential damage.*

Oxidative stress can result from:

1. *Diminished levels of antioxidants* (e.g., mutations affecting the activities of antioxidant defense enzymes such as CuZnSOD, MnSOD, and glutathione peroxidase, or toxins that deplete antioxidant defenses). For example, many xenobiotics are metabolized by conjugation with GSH; high doses can deplete GSH and cause oxidative stress even if the xenobiotic is not itself a generator of reactive species (3). Deficiencies in dietary antioxidants and other essential dietary constituents can also lead to oxidative stress (1,3).

2. *Increased production of reactive species* [(e.g., by exposure of cells or organisms to elevated levels of $O_2$), the presence of toxins that are themselves reactive species (e.g., $NO_2^{\bullet}$) or are metabolized to generate reactive species (e.g., paraquat), or excessive activation of "natural" systems producing such species (e.g., inappropriate activation of phagocytic cells in chronic inflammatory diseases, such as rheumatoid arthritis and ulcerative colitis)]. Mechanism 2 is usually assumed to be more relevant as a source of oxidative stress in human diseases and is frequently the target of attempted therapeutic intervention, but rarely is much attention paid to the antioxidant nutritional status of sick patients, which can often be very poor (20,21).

Oxidative stress can result in:

1. *Adaptation* of the cell or organism by upregulation of defense systems, which

**Table 2** Definition and Nomenclature of Reactive Species

| Radicals | Nonradicals |
|---|---|
| **Reactive Oxygen Species (ROS)** | |
| Superoxide, $O_2^{\bullet-}$ | Hydrogen peroxide, $H_2O_2$ |
| Hydroxyl, $OH^{\bullet}$ | Hypobromous acid, HOBr[a] |
| Hydroperoxyl, $HO_2^{\bullet}$ | Hypochlorous acid, HOCl[b] |
| | Ozone $O_3$ |
| Lipid peroxyl, $LO_2^{\bullet}$ | Singlet oxygen ($O_2{}^1\Delta g$) |
| Lipid alkoxyl, $LO^{\bullet}$ | Lipid peroxides, LOOH |
| | Peroxynitrite, $ONOO^-$ |
| **Reactive Chlorine Species (RCS)** | |
| Atomic chlorine, $Cl^{\bullet}$ | Hypochlorous acid, HOCl[b] |
| | Nitryl (nitronium) chloride $NO_2Cl$[c] |
| | Chloramines |
| | Chlorine gas ($Cl_2$) |
| **Reactive Nitrogen Species (RNS)** | |
| Nitric oxide, $NO^{\bullet}$ | Nitrous acid, $HNO_2$ |
| Nitrogen dioxide, $NO_2^{\bullet}$ | Nitrosyl cation, $NO^+$ |
| | Nitroxyl anion, $NO^-$ |
| | Dinitrogen tetroxide, $N_2O_4$ |
| | Dinitrogen trioxide, $N_2O_3$ |
| | Peroxynitrite, $ONOO^-$ |
| | Nitronium (nitryl) cation, $NO_2^+$ |
| | Alkyl peroxynitrites, ROONO |
| | Nitryl (nitronium) chloride, $NO_2Cl$[c] |

Reactive oxygen species (ROS) is a collective term that includes both oxygen radicals and certain nonradicals that are oxidizing agents and/or are easily converted into radicals (HOCl, HOBr, $O_3$, $ONOO^-$, $^1O_2$, $H_2O_2$). RNS is also a collective term including nitric oxide and nitrogen dioxide radicals, as well as such nonradicals as $HNO_2$ and $N_2O_4$. $ONOO^-$ is often included in both categories. ''Reactive'' is not always an appropriate term: $H_2O_2$, $NO^{\bullet}$, and $O_2^{\bullet-}$ react quickly with only a few molecules, whereas $OH^{\bullet}$ reacts quickly with almost everything. $RO_2^{\bullet}$, $RO^{\bullet}$, HOCl, HOBr, $NO_2^{\bullet}$, $ONOO^-$, and $O_3$ have intermediate reactivities.

[a] HOBr could also be regarded as a ''reactive bromine species.''

[b] HOCl is often included as a ROS.

[c] $NO_2Cl$ is a chlorinating and nitrating species produced by reaction of HOCl with $NO_2^-$ (18).

may (a) completely protect against damage; (b) protect against damage to some extent, but not completely; and (c) ''overprotect,'' (e.g., the cell is then resistant to higher levels of oxidative stress imposed subsequently).

As an example of (b), if adult rats are gradually acclimatized to elevated $O_2$, they can tolerate pure $O_2$ for much longer than control rats, apparently due to increased synthesis of antioxidant defense enzymes and of GSH in the lung. Although the damage is slowed, it is not prevented (22). As an example of (c), treatment of *E. coli* with low levels of $H_2O_2$ increases transcription of genes regulated by the oxyR protein and renders the bacteria resistant to higher $H_2O_2$ levels (23). Examples of type (c) adaptation in animals are rarer, but one may be provided by ischemic preconditioning (24). A brief period of ischemia in pig hearts led to depression of contractile function, and administration of antioxidants

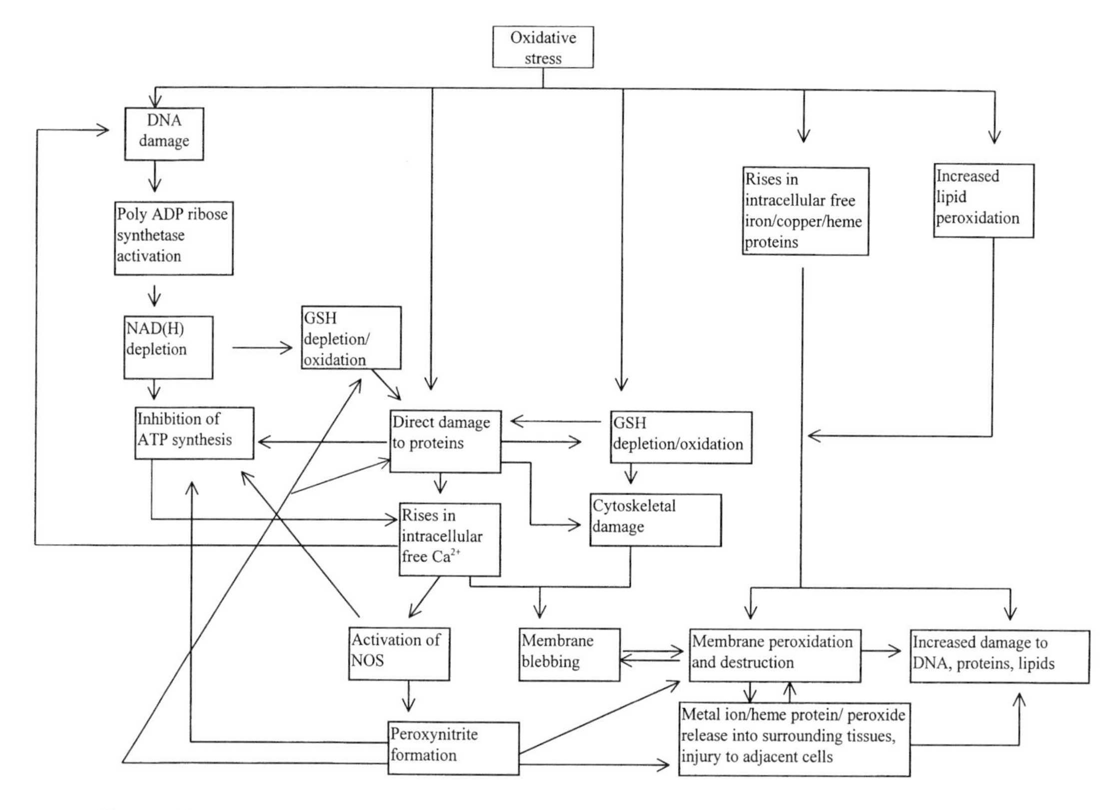


**Figure 3** Interacting and overlapping methods of cell injury by oxidative stress.

offered protection against this "myocardial stunning." However, repeated periods of ischemia led to quicker return of contractile function, but this adaptive response was blocked in the antioxidant-treated animals. Hence, reactive species produced by ischemia/reperfusion were initially damaging, but also led to a response protective against subsequent insult.

2. *Tissue injury.* This can involve damage to any or all molecular targets: lipids, DNA, carbohydrates, and proteins (including lipoproteins and nucleoproteins). When measurements of such damage are made after injury has started, it may not be clear which biomolecules were attacked first, since injury mechanisms are interrelated in complex ways (Fig. 3). Indeed, depending on the cell or tissue under study, the primary molecular target of oxidative stress may vary (3,25). DNA is an important early target of damage when $H_2O_2$ is added to many mammalian cells (26), and often increased DNA damage (measured as strand breakage or formation of modified DNA bases) occurs before *detectable* lipid peroxidation or oxidative protein damage. The word *detectable* is emphasized because such conclusions are obviously dependent on the assays used to measure oxidative damage. For example, measurement of protein carbonyls (27) would not detect oxidative protein damage occurring by oxidation of essential -SH groups on membrane ion transporters or the conversion of methionine to its sulfoxide.

Figure 3 may seem complex, but it is an oversimplification. In particular, it omits the effects of reactive species on signal transduction systems. Overactivation of such systems by mild oxidation, or inhibition of them by excessive oxidation, may also contribute to cell injury (28–30). Effects on the transport/levels of ions other than $Ca^{2+}$, iron, and copper (31) have also been omitted.

3. *Cell death* from multiple mechanisms, such as the rupture of membrane blebs (32). Excessive activation of poly(ADP ribose) polymerase (PARP) can so deplete intracellular $NAD^+$ and NADH levels that the cell cannot make ATP and dies (3). Peroxynitrite toxicity involves PARP activation in several cell types (33).

Cell death can occur by essentially two mechanisms, necrosis and apoptosis; both can result from oxidative stress, among other causes (32). In necrotic cell death, the cell swells and ruptures, releasing its contents into the surrounding area and affecting adjacent cells. Contents released can include antioxidants such as catalase or GSH, and prooxidants such as lipid peroxides, heme proteins, and ions of such transition metals as copper and iron. Hence, even if a cell dies as a result of mechanisms other than oxidative stress, necrotic cell death can impose oxidative stress on the surrounding tissues (Figs. 2, 3). In apoptosis, the cell's own intrinsic ''suicide mechanism'' is activated; apoptosing cells do not release their contents and so apoptosis does not, in general, cause disruption to surrounding cells.

The role of reactive species in apoptosis is confusing. Oxidative stress often induces apoptosis (30) in many cell types (e.g., by causing mitochondrial damage leading to cytochrome c release). In addition, apoptosis induced by other mechanisms is often accompanied by increased production of reactive species (34,35). Yet high levels of reactive species (including $H_2O_2$ and probably $ONOO^-$ and HOCl) can halt apoptosis by oxidizing active-site thiol groups that are essential to the catalytic activity of the caspases (30). Indeed, some transformed cell lines appear to produce reactive species in order to prevent apoptosis (36).

## D. Antioxidant

Like the term oxidative stress, ''antioxidant'' is a vague word, often used but rarely defined. One can make almost any chemical exert antioxidant effects in vitro by choosing appropriate assay conditions (37). To reflect the fact that calling something an antioxidant means little without specifying the assay methodology used, Halliwell and Gutteridge (38) defined an antioxidant as *any substance that, when present at low concentrations compared with those of an oxidizable substrate, significantly delays or prevents oxidation of that substrate*. The term ''oxidizable substrate'' includes every type of molecule found in vivo. This definition emphasizes the importance of the damage target studied and the source of reactive species used when antioxidant action is examined in vitro.

When reactive species are generated in vivo, many antioxidants come into play (1,3,6,19,39,40). Their relative importance depends upon which species is generated, how it is generated, where it is generated, and what target of damage is measured. For example, when human blood plasma is exposed to nitrogen dioxide ($NO_2^{\bullet}$) or to ozone ($O_3$), urate acts as an ''antioxidant'' protective agent (41). By contrast, urate appears to provide little protection against damage to plasma constituents by hypochlorous acid (42). Similarly, if the source of oxidative stress is kept the same but a different damage target is measured, different answers can result. For example, gas-phase cigarette smoke causes lipid peroxidation in plasma, which is prevented by ascorbate (43). However, ascorbate does not protect against damage to plasma proteins by cigarette smoke, as measured by the carbonyl assay (44). Hence, putative therapeutic antioxidants should be examined for their effects against a wide range of molecular targets in vitro and in vivo: it is perfectly possible to have, for example, an antioxidant inhibitor of lipid peroxidation that has no effect on, or even aggravates, oxidative damage to DNA or proteins. Diethylstilbestrol may be one such compound—antioxidant toward lipids (45) but prooxidant toward DNA (46).

## III. MEASURING OXIDATIVE DAMAGE IN VIVO: THE KEY TO EVALUATION OF THE ROLE OF REACTIVE SPECIES AND THE EFFECTS OF THERAPY

Table 1 lists the major criteria needed to establish that reactive species (or any other putative tissue-injurious agents) are important mediators of disease pathology. To meet these criteria, it is necessary to be able to measure reactive species and the damage they cause. Such measurements are also important in clinical trials of therapeutic agents. Suppose, for example, that company X develops an antioxidant to treat a particular disease, but finds that it has no therapeutic benefit in clinical trials. Some possible explanations are:

1. Not enough antioxidant was administered to give sufficient levels to act at the appropriate site.
2. The antioxidant is directed against the wrong molecular target (e.g., an inhibitor of lipid peroxidation would not be expected to protect where the significant molecular damage is occurring to proteins or DNA).
3. Reactive species make no significant contribution to the disease pathology.

Before reaching the unfortunate conclusion in (3), possibilities (1) and (2) need to be ruled out by appropriate measurements of oxidative damage. Consider now a happier situation for the company, in which the drug is clinically effective. This could mean that either

1. Reactive species do make a significant contribution to disease pathology.
2. The antioxidant has other beneficial pharmacological effects unrelated to its action on reactive species. For example, many chain-breaking antioxidant inhibitors of lipid peroxidation also inhibit lipoxygenase enzymes (47), important in biosynthesis of leukotrienes. Again, measurements of oxidative damage should accompany clinical tests of the effects of antioxidants to show whether or not they affected reactive species to an extent correlated with the clinical improvement (Table 1).

In most human diseases, oxidative stress is a consequence and not a primary cause of the disease process. Tissue damage by infection, trauma, toxins, abnormally low or high temperatures, and other causes usually leads to formation of increased amounts of "injury mediators," such as prostaglandins, leukotrienes, $NO^{\bullet}$, interleukins, and cytokines, such as TNFs and reactive oxygen/nitrogen/chlorine species. Do these various agents make a significant contribution to the disease pathology, or is their formation of little or no consequence (Fig. 2)? The answer probably differs for each agent and for different diseases. The following chapters in this book show us that, although oxidative stress may usually be a secondary event, it nevertheless does play an important role in many human diseases. The challenge is to develop effective therapeutic antioxidants, demonstrate their benefit to patients, and prove that they are working by an antioxidant mechanism. The remainder of this chapter will be devoted to this key question of methodology.

## A. Basic Principles

Assessment of the role of reactive species and the effects of therapy upon damage caused by them could involve

1. Measurement of the levels of reactive species, either directly (e.g., by ESR or aromatic hydroxylation) or indirectly (e.g., by ESR spin-trapping).
2. Measurement of the damage that they cause (i.e., oxidative damage).

Methods currently available for the direct measurement of reactive species are of limited applicability to humans (3) and so most clinical studies focus on the measurement of end-products of damage. This is to some extent logical, since it is the damage caused by reactive species that is important rather than the total amount generated. For example, if highly reactive hydroxyl radicals are generated within a cell, many of them may react with unimportant targets. The small fraction that reacts with DNA (to cause strand breakage and formation of mutagenic base or sugar oxidation products), with key proteins, or with lipids (initiating lipid peroxidation) may be the important fraction.

## B. Oxidative DNA Damage

One obvious problem in studying damage to DNA by reactive species is the limited availability of human tissues from which to obtain it. Most studies are performed on DNA isolated from lymphocytes or total white cells from human blood, and it is assumed that changes here are reflected in other tissues. Sperm, buccal cells, placenta, and biopsies of muscle, skin, colon, and other tissues are other potential sources of DNA.

A question of even greater importance is whether we can measure oxidative DNA damage accurately. For most of the methods currently employed, DNA must first be isolated from tissues. Then it must be hydrolyzed and the hydrolysate prepared for analysis of oxidized bases, usually by high-performance liquid chromatography (HPLC) or mass spectrometry (MS) methods. Isolation, hydrolysis, and analysis steps all have the potential to cause artifactual further oxidation of DNA, raising the apparent level of DNA damage and invalidating the measurement. Nevertheless, several laboratories are developing analytical protocols that appear to minimize or prevent such artifacts (48).

A third important question is, which oxidized DNA bases should be measured? Since its introduction by Kasai and his colleagues (reviewed in Ref. 49) and aided by the development of a simple electrochemical detection method linked to HPLC (50), the measurement of 8-hydroxy-2′-deoxyguanosine (8OHdG) has become the most widely used technique for assessing oxidative DNA damage. Reasons for this include the availability of a sensitive assay, the fact that 8OHdG is formed in DNA by several reactive oxygen species (e.g., $OH^{\bullet}$ and $^1O_2$), and the importance of this lesion in vivo as reflected by its known mutagenicity in inducing GC $\rightarrow$ TA transversions, as well as the multiple mechanisms that appear to have evolved to remove 8OHdG from DNA, and prevent its incorporation into DNA from the nucleotide precursor pool (49,51).

There are drawbacks, however, to relying entirely on 8OHdG as a biomarker of oxidative DNA damage. 8OHdG levels are not a quantitative marker of oxidative damage to DNA; 8OHdG is only a minor product of attack on DNA by reactive nitrogen (e.g., $ONOO^-$, $HNO_2$) and chlorine (e.g., HOCl, $NO_2Cl$) species (reviewed in Ref. 26). Other products of attack on guanine residues in DNA by reactive oxygen species include Fapy-guanosine (52). The ratio of 8OHdG to Fapy-guanosine is affected by the redox state of the cell (i.e., it is decreased at low $O_2$ concentrations) and by the presence of transition metal ions (52). Hence, in principle, a more advantageous method is to measure a wide range of DNA base damage products by mass spectrometry, which allows for rigorous identification of products (26,52). Mass spectrometry can also measure products of attack on DNA by reactive nitrogen and chlorine species. It is usually coupled with gas chromatography, although liquid chromatography/mass spectrometry techniques for measurement of DNA base oxidation products are undergoing rapid development. However, both DNA bases and their oxidation products can sometimes be modified artifactually by the hydrolysis and derivatization procedures used in conventional gas chromatography/mass spectrometry (53). Fortunately, techniques that appear to minimize these problems now exist (26,48,53).

Levels of 8OHdG and other modified bases, as measured in DNA isolated from human tissues, represent the dynamic equilibrium between rates of oxidative DNA damage and rates of repair of that damage. It follows that levels of oxidized bases can change not only because of changes in the rate of DNA damage, but also because of alterations in repair rate. Fortunately, 8OHdG is excreted in human urine and its measurement has been used as an assessment of ''whole-body'' DNA damage (54). Estimates of this excretion rate can be combined with measurements of 8OHdG in cellular DNA to study the question of rates of repair and rates of damage. For example, administration of 2-nitropropane to rats led to elevated tissue levels of 8OHdG. Levels of 8OHdG subsequently decreased, accompanied by an increase in urinary 8OHdG excretion (54). Cigarette smoking raises 8OHdG levels in human white-cell DNA (55) as well as 8OHdG excretion rates (54). Therefore, both nitropropane and constituents of cigarette smoke appear to produce elevations in the rate of oxidative DNA damage in vivo, and the rate of repair (as indicated by 8OHdG excretion) also increases. Measurements of only urinary excretion rates of 8OHdG should be interpreted with caution. For example, an agent that increases 8OHdG excretion rates might be interpreted as ''bad'' (thought to increase DNA damage), but it might in fact be ''good'' (if it stimulated repair and therefore decreased steady-state 8OHdG levels in DNA). The level of 8OHdG in urine is thought to be unaffected by diet and this compound is not thought to be metabolized in humans (54), although more detailed studies on both of these points are probably required. In addition, it is possible that some or all of the 8OHdG excreted in human urine may arise not from DNA, but from oxidation of deoxyguanosine triphosphate (dGTP) in the DNA precursor pool (51). If so, it means that 8OHdG excretion rates are not a quantitative index of oxidative damage to guanine residues in DNA. As in the case of measurements of steady-state levels of oxidative DNA damage, the measurement of additional products in urine might reveal unsuspected events. For example, in patients treated with adriamycin, levels of 8OHdG excretion did not change, but there was a significant rise in 5-(hydroxymethyl) uracil in the urine (56).

### C. Lipid Peroxidation

Developing accurate methods for measuring lipid peroxidation has been a task even more complex than that of developing accurate methods for measuring DNA damage. There is a growing belief that the most valuable biomarker of lipid peroxidation in the human body may be the $F_2$-isoprostanes, which are specific products arising from the peroxidation of arachidonic acid residues in lipids (57,58), although more work on developing assay methodology and standardizing it between laboratories needs to be done (58). Isoprostanes can be accurately and sensitively measured by mass spectrometric techniques (57–59), so that the levels present in human plasma and urine can be measured accurately. Isoprostane levels are elevated in smokers and in diseases associated with oxidative stress and are diminished by the administration of antioxidants such as vitamin E (57–64). Isoprostanes appear to turn over rapidly, being both metabolized and excreted (57,65). Detection of isoprostanes and their metabolites in urine seems to be a useful biomarker of whole-body lipid peroxidation (57–59).

Different families of isoprostanes arise from peroxidation of eicosapentaenoic acid residues ($F_3$-isoprostanes) and docosahexaenoic acid residues ($F_4$-isoprostanes, sometimes called neuroprostanes) (66–68). These families can be distinguished from $F_2$-isoprostanes in the mass spectrometer and measuring them may be helpful in determining the rates at which different polyunsaturated fatty acid (PUFA) residues peroxidize in vivo. It is well known that increasing the number of double bonds in a PUFA increases its susceptibility to oxidation in vitro (3), but whether highly unsaturated PUFAs undergo oxidation at a faster rate in vivo remains to be established. However, increased levels of $F_4$-isoprostanes in the brain cortex of patients with Alzheimer's disease have been reported (68,69). If isoprostanes (especially $F_4$-isoprostanes) escape into the CSF (70), or even into the blood, they may turn out to be an important biomarker of lipid peroxidation in the brain. Such a biomarker would be useful to assess the effect of antioxidant therapy on oxidative damage to the nervous system in patients suffering from neurodegenerative diseases.

### D. Oxidative Protein Damage

Oxidative damage to proteins may be important in vivo both in its own right (affecting the function of receptors, enzymes, transport proteins, etc., and perhaps generating new antigens that can provoke immune responses) (71,72), and because it can contribute to secondary damage to other biomolecules (e.g., inactivation of DNA repair enzymes and loss of fidelity of DNA polymerases in replicating DNA).

The chemical reactions resulting from attack of reactive species upon proteins are more complex than for DNA, since there are 20 amino acid residues rather than 4 DNA bases, and each amino acid can give rise to multiple products upon oxidation. Free radical attack on proteins can generate amino acid radicals, which react with $O_2$ to give peroxyl radicals and then protein peroxides, which can decompose in complex ways, promoted by heat or transition metal ions (73).

Several assays for damage to specific amino acid residues in proteins have been developed, including measurements of dihydroxyphenylalanine (produced by tyrosine hydroxylation), γ-glutamyl and 2-amino-adipic semialdehydes, valine hydroxides (produced from valine hydroperoxides), tryptophan hydroxylation and ring-opening products, 2-oxohistidine, dityrosine, and ortho- and meta-tyrosines, products of attack of $OH^{\bullet}$ upon phenylalanine (73–79). The levels of any one (or, preferably, of more than one) of these products in proteins could in principle be used to assess the balance between oxidative protein damage and the repair (or hydrolytic removal) of damaged proteins. The products most exploited to date have been the hydroxylated phenylalanines and bityrosine. For example, levels of ortho-tyrosine and dityrosine in human lens proteins have been reported in relation to age (76) and dityrosine has been measured in human atherosclerotic lesions (79). Levels of 2-amino-adipic semialdehyde in plasma were reported to be elevated in human subjects supplemented with fruit juices (80).

### E. The Carbonyl Assay

Oxidative protein damage has been measured most often in human or animal studies by using the carbonyl assay, a ''general'' assay of oxidative protein damage (81). The carbonyl assay is based on the ability of several reactive species to attack amino acid residues in proteins (particularly histidine, arginine, lysine, and proline) to produce carbonyl functions that can be measured after reaction with 2,4-dinitrophenylhydrazine (27). The carbonyl assay has become widely used and many laboratories have developed individual protocols for it (reviewed in Refs. 81–83). However, there is considerable variation in the ''baseline'' levels of protein carbonyls in certain tissues, depending on how the assay is performed (82). By contrast, most groups seem to obtain broadly comparable values for protein carbonyls in human plasma, of $<1$ nmol/mg protein, so plasma protein carbonyls should in principle be a useful marker of oxidative protein damage for nutritional studies. However, Davies and Dean (73) suggested that carbonyl levels are an overestimate of steady-state levels of oxidative protein damage.

More research is needed to identify the molecular nature of the carbonyls detected in tissues and body fluids (i.e., which amino acid residues have been damaged and on what proteins). Western blotting assays based on the use of anti-DNPH antibodies have been developed for this purpose (83,84). It must be borne in mind that protein carbonyls can also be generated by the covalent binding of certain aldehyde end-products of lipid peroxidation (such as HNE and MDA) to proteins (85) and by glycation reactions (86).

## IV. CONCLUSION

Our ability to assess oxidative damage in human tissues and body fluids by meaningful and validated assays is improving rapidly, and the increasing cooperation between laboratories in exchanging samples (48,87,88) is contributing to this improvement. Progress in understanding the effects of treatment on oxidative damage in human disease will be greatly facilitated when biomarkers of oxidative damage are routinely included in clinical trials of therapeutic agents.

## REFERENCES

1. Frei H, ed. Natural Antioxidants in Human Health and Disease. San Diego: Academic Press, 1994.
2. Halliwell B, Cross CE, Gutteridge JMC. Free radicals, antioxidants and human disease. Where are we now? J Lab Clin Med 1992; 119:598–620.
3. Halliwell B, Gutteridge JMC. Free Radicals in Biology and Medicine, 3rd ed. Oxford: Clarendon Press, 1999.
4. Olanow CW, Jenner P, Youdim M, eds. Neurodegeneration and Neuroprotection in Parkinson's Disease. London: Academic Press, 1996.
5. Richter C, Gogvadze V, Laffranchi R, Schlapbach R, Schweizer M, Suter M, Walter P, Yaffee M. Oxidants in mitochondria: from physiology to diseases. Biochim Biophys Acta 1995; 1271: 67–74.
6. Sies H, ed. Antioxidants in Disease, Mechanisms and Therapy. London: Academic Press, 1996.
7. Jacono AA, Hu B, Kopke RD, Henderson D, Van De Water TR, Steinman HM. Changes in cochlear antioxidant enzyme activity after sound conditioning and noise exposure in the chinchilla. Hearing Res 1998; 117:31–38.

8. Giralt M, Cervello I, Nogues MR, Puerto AM, Ortin F, Argany N, Mallol J. Glutathione, glutathione S-transferase and reactive oxygen species of human scalp sebaceous glands in male pattern baldness. J Invest Dermatol 1996; 107:154–158.
9. Halliwell B, Gutteridge JMC. Lipid peroxidation, oxygen radicals, cell damage and antioxidant therapy. Lancet 1984; i:1396–1398.
10. Lamb NJ, Quinlan GJ, Mumby S, Evans TW, Gutteridge JMC. Haem oxygenase shows prooxidant activity in microsomal and cellular systems: implications for the release of low-molecular-mass iron. Biochem J 1999; 344:153–158.
11. Vercellotti GM, Balla G, Balla J, Nath K, Eaton JW, Jacob S. Heme and the vasculature: an oxidative hazard that includes antioxidant defenses in the endothelium. Artifi Cell Blood Subst and Immobil Biotechnol 1994; 22:207–221.
12. Yokoyama Y, Beckman JS, Beckman TK, Wheat JK, Cash TG, Freeman BA, Parks DA. Circulating xanthine oxidase: potential mediator of ischemic injury. Am J Physiol 1990; 258: G564–G570.
13. Houston M, Estevez A, Chumley P, Aslan M, Marklund S, Parks DA, Freeman BA. Binding of xanthine oxidase to vascular endothelium. Kinetic characterization and oxidative impairment of nitric oxide-dependent signaling. J Biol Chem 1999; 274:4985–4994.
14. Ambrosio G, Zweier JL, Duilio C, Kuppusamy P, Santoro G, Elia PP, Tritto I, Cirillo P, Condorelli M, Chiarello M. Evidence that mitochondrial respiration is a source of potentially toxic oxygen free radicals in intact rabbit hearts subjected to ischemia and reflow. J Biol Chem 1993; 268:18532–18541.
15. Orrenius S, McConkey DJ, Bellomo G, Nicotera P. Role of $Ca^{2+}$ in toxic cell killing. Trends Pharmacol Sci 1989; 10:281–285.
16. Beckman JS, Chen J, Ischiropoulos H, Crow JP. Oxidative effects of peroxynitrite. Meth Enzymol 1994; 233:229–240.
17. Foote CS, Clennan EL. Properties and reactions of singlet dioxygen. In: Foote CS, Valentine JS, Greenberg A, Liebman JF, eds. Active Oxygen in Chemistry. London: Blackie, 1995:105–140.
18. Eiserich JP, Cross CE, Jones DA, Halliwell B, Van der Vliet A. Formation of nitrating and chlorinating species by reaction of nitrite with hypochlorous acid. J Biol Chem 1996; 271: 19199–19208.
19. Sies H, ed. Oxidative Stress: Oxidants and Antioxidants. New York: Academic Press, 1991.
20. Scorah CJ, Downing C, Piripitsi A, Gallivan L, Al-Hazaa AH, Sanderson MJ, Bodenham A. Total vitamin C, ascorbic acid, and dehydroascorbic acid concentrations in plasma of critically ill patients. Am J Clin Nutr 1996; 63:760–765.
21. Williams A, Riise GC, Anderson BA, Kjellström C, Schersten H, Kelly FJ. Compromised antioxidant status and persistent oxidative stress in lung transplant recipients. Free Rad Res 1999; 30:383–393.
22. Kinnula VL, Crapo JD, Raivio KO. Generation and disposal of reactive oxygen metabolites in the lung. Lab Invest 1995; 73:3–19.
23. Kullik I, Stevens J, Toledano MB, Storz G. Mutational analysis of the redox-sensitive transcriptional regulator oxyR: regions important for DNA binding and multimerization. J Bact 1995; 177:1285–1291.
24. Sun JZ, Tang XL, Park SW, Qui Y, Turrens JF, Bolli R. Evidence for an essential role of reactive oxygen species in the genesis of late preconditioning against myocardial stunning in conscious pigs. J Clin Invest 1996; 97:562–568.
25. Hyslop PA, Hinshaw DB, Halsey WA, Jr., Schraufstatter IU, Sauerhaber RD, Spragg RG, Jackson JH, Cochrane CG. Mechanisms of oxidant-mediated cell injury. The glycolytic and mitochondrial pathways of ADP phosphorylation are major intracellular targets inactivated by hydrogen peroxide. J Biol Chem 1998; 263:1665–1675.
26. Halliwell B. Can oxidative DNA damage be used as a biomarker of cancer risk in humans? Free Rad Res 1998; 29:469–486.

27. Amici A, Levine RL, Tsai L, Stadtman ER. Conversion of amino acid residues in proteins and amino acid homopolymers to carbonyl derivatives by metal-catalyzed oxidation reactions. J Biol Chem 1989; 264:3341–3346.
28. Burdon RH. Superoxide and hydrogen peroxide in relation to mammalian cell proliferation. Free Rad Biol Med 1995; 18:775–794.
29. Finkel T. Oxygen radicals and signaling. Curr Opin Cell Biol 1998; 10:248–253.
30. Hampton MB, Orrenius S. Redox regulation of apoptotic cell death. Biofactors 1998; 8:1–5.
31. Bychkov R, Pieper K, Reid C, Milosheva M, Bychkov E, Luft FE, Haller H. Hydrogen peroxide, potassium currents and membrane potential in human endothelial cells. Circulation 1999; 99:1719–1725.
32. Nicotera P, Orrenius S. Molecular mechanisms of toxic cell death: an overview. Meth Toxicol 1994; 1B:23–28.
33. Scott GS, Jakeman LB, Stokes BT, Szabo C. Peroxynitrite production and activation of poly (adenosine diphosphate-ribose) synthetase in spinal cord injury. Ann Neurol 1999; 45:120–124.
34. Tsujimoto Y. Role of bcl-2 family proteins in apoptosis: apoptosomes or mitochondria? Genes Cells 1998; 3:697–707.
35. Jabs T. Reactive oxygen intermediates as mediators of programmed cell death in plants and animals. Biochem Pharmacol 1999; 57:231–245.
36. Clement MV, Pervaiz S. Reactive oxygen intermediates regulate cellular response to apoptotic stimuli: an hypothesis. Free Rad Res 1999; 30:247–252.
37. Halliwell B. Food-derived antioxidants. Evaluating their importance in food and in vivo. Food Sci Agr Chem 1999; 1:67–109.
38. Halliwell B, Gutteridge JMC. Free Radicals in Biology and Medicine, 2nd ed. Oxford: Clarendon Press, 1989.
39. Packer L, ed. Oxygen radicals in biological systems. Parts C and D. Meth Enzymol 1994; 233:1–711; 234:1–704.
40. Packer L, ed. Oxidants and Antioxidants. Part B. Meth Enzymol 1999; 300:3–481.
41. Halliwell B, Cross CE. Oxygen-derived species: their relation to human disease and environmental stress. Environ Health Perspect 1994; 102 (suppl 10):5–12.
42. Hu ML, Louie S, Cross CE, Motchnik P, Halliwell B. Antioxidant protection against hypochlorous acid in human plasma. J Lab Clin Med 1992; 121:257–262.
43. Frei B, Forte TM, Ames BN, Cross CE. Gas phase oxidants of cigarette smoke induce lipid peroxidation and changes in lipoprotein properties in human blood plasma. Biochem J 1991; 277:133–138.
44. Reznick AZ, Cross CE, Hu M, Suzuki YJ, Khwaja S, Safadi A, Motchnik PA, Packer L, Halliwell B. Modification of plasma proteins by cigarette smoke as measured by protein carbonyl formation. Biochem J 1992; 286:607–611.
45. Wiseman H, Halliwell B. Carcinogenic antioxidants: Diethylstiboestrol, hexoestrol and 17 alpha-ethynyl-oestradiol. FEBS Lett 1993; 322:159–163.
46. Roy D, Liehr JG. Elevated 8-hydroxydeoxyguanosine levels in DNA of diethylstilboestrol-treated Syrian hamsters: covalent DNA damage by free radicals generated by redox cycling of diethylstilboestrol. Cancer Res 1991; 51:3882–3885.
47. Laughton MJ, Evans PJ, Moroney MA, Hoult JRS, Halliwell B. Inhibition of mammalian 5-lipoxygenase and cyclo-oxygenase by flavonoids and phenolic dietary additives. Biochem Pharmacol 1991; 42:1673–1681.
48. Special Issue. DNA damage: measurement and mechanism. Free Rad Res 1998; 29:461–624.
49. Kasai H. Analysis of a form of oxidative DNA damage, 8-hydroxy-2′-deoxyguanosine, as a marker of cellular oxidative stress during carcinogenesis. Mutat Res 1997; 387:146–163.
50. Floyd RA, Watson JJ, Wong PK, Altmiller DH, Rickard RC. Hydroxy free radical adduct of deoxyguanosine: sensitive detection and mechanisms of formation. Free Rad Res Commun 1986; 1:163–172.

51. Mo JY, Maki H, Sekiguchi M. Hydrolytic elimination of a mutagenic nucleotide, 8-oxodGTP, by human 18-kiloDalton protein: sanitization of nucleotide pool. Proc Natl Acad Sci USA 1992; 89:1021–11025.
52. Dizdaroglu M. Facts about the artifacts in the measurement of oxidative DNA base damage by gas chromatography-mass spectrometry. Free Rad Res 1998; 29:551–563.
53. Cadet J, d'Ham C, Douki T, Pouget JP, Ravanat JL, Sauvaigo S. Facts and artifacts in the measurement of oxidative base damage to DNA. Free Rad Res 1998; 29:541–550.
54. Loft S, Deng X-S, Tuo J, Wellejus A, Sorensen M, Poulsen HE. Experimental study of oxidative DNA damage. Free Rad Res 1998; 29:525–539.
55. Kiyosawa H, Suko M, Okudaira H, Murata K, Miyamoto T, Chung MH, Kasai H, Nishimura S. Cigarette smoking induces formation of 8-hydroxydeoxyguanosine, one of the oxidative DNA damages, in human peripheral leukocytes. Free Rad Res Commun 1990; 11:23–27.
56. Faure H, Mousseau M, Cadet J, Guimier C, Tripier M, Hida H, Favier A. Urine 8-oxo-7,8-dihydro-2′-deoxyguanosine versus 5-(hydroxymethyl) uracil as DNA oxidation marker in adriamycin-treated patients. Free Rad Res 1998; 28:377–382.
57. Roberts LJ, II, Morrow JD. The generation and actions of isoprostanes. Biochim Biophys Acta 1997; 1345:121–135.
58. Lawson JA, Rokach J, Fitzgerald GA. Isoprostanes: formation, analysis and use as indices of lipid peroxidation in vivo. J Biol Chem 1999; 247:24441–24444.
59. Li H, Lawson JA, Reilly M, Adiyaman M, Hwang SW, Rokach J, Fitzgerald GA. Quantitative high performance liquid chromatography/tandem mass spectrometric analysis of the four classes of $F_2$-isoprostanes in human urine. Proc Natl Acad Sci USA 1999; 96:181–186.
60. Pratico G, Iuliano L, Mauriello A, Spagnoli L, Lawson JA, Rokach J, Maclouf J, Violi F, Fitzgerald GA. Localization of distinct $F_2$-isoprostanes in human atherosclerotic lesions. J Clin Invest 1997; 100:2028–2034.
61. Practico D, Tangirala R, Rader DJ, Rokach J, Fitzgerald GA. Vitamin E suppresses isoprostane generation in vivo and reduces atherosclerosis in Apo E-deficient mice. Nat Med 1998; 4: 1189–1192.
62. Davi G, Allessandrini P, Mezzetti A, Minotti G, Bucciarelli T, Constantini F, Cipollone F, Bon GB, Ciabattoni G, Patrono C. In vivo formation of 8 epi prostaglandin $F_{2\alpha}$ is increased in hypercholesterolaemia. Arterioscler Thromb Vasc Biol 1997; 17:3230–3235.
63. Mallat Z, Philip I, Lebret M, Chatel D, Maclouf J, Tedgui A. Elevated levels of 8-iso-prostaglandin $F_{2\alpha}$ in pericardial fluid of patients with heart failure. Circulation 1998; 97:1536–1539.
64. Morrow JD, Frei B, Longmire AW, Gaziano JM, Lynch SM, Syhr Y, Strauss WE, Oates JE, Roberts LJ, II. Increase in circulating products of lipid peroxidation ($F_2$-isoprostanes) in smokers. N Engl J Med 1995; 332:1198–1203.
65. Basu S. Metabolism of 8-isoprostaglandin $F_{2\alpha}$. FEBS Lett 1998; 428:32–36.
66. Nourooz-Zadeh J, Halliwell B, Anggard EE. Evidence for the formation of $F_3$-isoprostanes during peroxidation of eicosapentaenoic acid. Biochem Biophys Res Commun 1997; 236:467–472.
67. Nourooz-Zadeh J, Liu EHC, Anggard EE, Halliwell B. $F_4$-isoprostanes: a novel class of prostanoids formed during peroxidation of docosahexaenoic acid. Biochem Biophys Res Commun 1998; 242:338–344.
68. Roberts JL II, Montine TJ, Markesbery WR, Tapper AR, Hardy P, Chemtob S, Dettbarn WD, Morrow JD. Formation of isoprostane-like compounds (neuroprostanes) in vivo from docosahexaenoic acid. J Biol Chem 1998; 273:13605–13612.
69. Nourooz-Zadeh J, Liu EHC, Yhlen B, Anggard EE, Halliwell B. $F_4$-isoprostanes as a specific marker of docosahexaenoic acid peroxidation in Alzheimer's disease. J Neurochem 1999; 72: 734–740.
70. Montine TJ, Beal MF, Cudkowicz ME, O'Donnell BS, Margolin RA, McFarland L, Bachrach AF, Zackert WE, Roberts LJ, Morrow JD. Increased CSF $F_2$-isoprostane concentration in probable AD. Neurology 1999; 52:562–565.

71. Halliwell B. Biochemical mechanisms accounting for the toxic action of oxygen on living organisms: the key role of superoxide dismutase. Cell Biol Int Rep 1978; 2:113–128.
72. Casciola-Rosen L, Wigley F, Rosen A. Scleroderma autoantigens are uniquely fragmented by metal-catalysed oxidation reactions: implications for pathogenesis. J Exp Med 1997; 185:71–79.
73. Davies MJ, Dean RT. Radical-Mediated Protein Oxidation. From Chemistry to Medicine. Oxford: Oxford University Press, 1997.
74. Daneshvar B, Frandsen H, Autrup H, Dragsted LO. γ-Glutamyl semialdehyde and 2-amino-adipic semialdehyde: biomarkers of oxidative damage to protein. Biomarkers 1997; 2:117–123.
75. Uchida K, Kawakishi S. 2-Oxo-histidine as a novel biological marker for oxidatively modified proteins. FEBS Lett 1993; 332:208–210.
76. Wells-Knecht MC, Huggins TG, Dyer DG, Thorpe SR, Baynes JW. Oxidized amino acids in lens proteins with age. Measurement of o-tyrosine and dityrosine in the aging human lens. J Biol Chem 1993; 268:12348–12352.
77. Kaur H, Halliwell B. Aromatic hydroxylation of phenylalanine as an assay for hydroxyl radicals. Measurement of hydroxyl radical formation from ozone and in blood from premature babies using improved HPLC methodology. Anal Biochem 1994; 220:11–15.
78. Giulivi C, Davies KJA. Dityrosine and tyrosine oxidation products are endogenous markers for the selective proteolysis of oxidatively modified red blood cell hemoglobin by (the 19S) proteasome. J Biol Chem 1993; 268:8752–8759.
79. Leeuwenburgh C, Rassmussen JE, Hsu FF, Muller DM, Pennathur S, Heinecke JW. Mass spectrometric quantitation of markers for protein oxidation by tyrosyl radical, copper, and hydroxyl radical in LDL isolated from human atherosclerotic plaques. J Biol Chem 1997; 272: 3520–3526.
80. Young JF, Nielsen SE, Haralsdottir J, Daneshvar B, Lauridsen ST, Knuthsen P, Crozier A, Sandstrom B, Dragsted LO. Effect of fruit juice intake on urinary quercetin excretion and biomarkers of antioxidant status. Am J Clin Nutr 1999; 69:87–94.
81. Levine RL, Garland D, Oliver CN, Amici A, Climent I, Lenz AG, Ahn BW, Shaltiel S, Stadtman ER. Determination of carbonyl content in oxidatively modified protein. Meth Enzymol 1990; 186:464–487.
82. Evans P, Lyras L, Halliwell B. Measurement of protein carbonyls in human brain tissue. Meth Enzymol 1999; 300:145–156.
83. Levine RL, Williams JA, Stadtman ER, Shacter E. Carbonyl assays for determination of oxidatively modified proteins. Meth Enzymol 1994; 233:346–357.
84. Keller J, Halmes NC, Hinson JA, Pumford NR. Immuno-chemical detection of oxidized proteins. Chem Res Toxicol 1993; 6:430–433.
85. Esterbauer H, Schaur RG, Zollner H. Chemistry and biochemistry of 4-hydroxynonenal, malonaldehyde and related aldehydes. Free Rad Biol Med 1991; 11:81–128.
86. Liggins J, Furth AJ. Role of protein-bound carbonyl groups in the formation of advanced glycation end-products. Biochem Biophys Acta 1997; 1361:123–130.
87. Lunec J, Herbert KE, Jones GDD, Dickinson L, Evans M, Mistry N, Chauhan D, Capper G, Zheng Q. Development of a quality control material for the measurement of 8-oxo-7,8-dihydro-2′-deoxyguanosine, an *in vivo* marker of oxidative stress, and a comparison of results from different laboratories. Free Rad Res, in press.
88. ESCODD. Comparison of different methods of measuring 8-hydroxylated guanine as a marker of oxidative DNA damage. Free Rad Res 2000; 32:333–341.

# 2

# Radicals and Inflammation: Mediators and Modulators

**CLAIRE A. DAVIES**

*St. Bartholomew's and the Royal London School of Medicine and Dentistry, London, England*

**DAVID R. BLAKE**

*Royal Hospital for Rheumatic Diseases, Bath, England*

**PAUL G. WINYARD**

*St. Bartholomew's and the Royal London School of Medicine and Dentistry, London, England*

## I. INTRODUCTION

Inflammation is the local physiological response to tissue injury, which can be initiated by a microbial infection, physical agent, irritant, or corrosive chemical. It is usually categorized into acute and chronic inflammation. Acute inflammation is the initial response to tissue injury, where chronic inflammation is the subsequent reaction that may occur if the stimulus cannot be removed. Free radicals have been shown to be involved in both physiological processes (1). However, before giving a detailed discussion of the involvement of free radicals in inflammation, an outline of inflammation is described.

## II. INFLAMMATION

The physical characteristics of inflammation were described by Celsus as redness, swelling, heat, and pain. Loss of function, a consequence of inflammation, was later added to the initial list by Virchow (2). These physical symptoms are brought about by the increased blood flow to the tissues, increased vascular permeability, and the formation of a cellular exudate. The release of mediators such as bradykinin and serotonin also induces pain.

## A. Acute Inflammation

### *1. Vascular Events*

Tissue damage or the invasion of a pathogen leads to the activation of the coagulation system, the fibrinolytic system, and the kinin system (3). In turn, this indirectly activates complement via its alternative pathway. Complement peptides aid the acute response by opsonizing the invading microorganism, attracting polymorphonuclear (PMN) leukocytes to the area, and facilitating the degranulation of mast cells and the lysis of some bacteria by the lytic pathway (4).

Bradykinin is produced due to the activation of the kinin system. Its potent vasodilator action causes increased vascular permeability of the local endothelium. In vascular endothelium and fibroblasts, bradykinin has been shown to activate phospholipase $A_2$ ($PLA_2$) and liberate arachidonate (5). The formation of arachidonate facilitates the production of prostaglandins (PT), thromboxanes, and leukotrienes (LT) (3), which increases vascular permeability and attracts phagocytes to the area. Histamine (released from mast cells), bradykinin, and nitric oxide ($NO^•$) increase permeability of the postcapillary venules allowing the entry of phagocytes into the inflamed area and the initiation of the immune response via the lymphatics.

### *2. Cellular Events*

Polymorphonuclear leukocytes are attracted to the affected area by cytokines, leukotrienes, and complement fractions, C3a and C5a. Leukocytes then adhere to the vascular walls of the vessels adjacent to the inflamed area. Adherence of leukocytes to the endothelium is dependent on the receptor-ligand binding between the selectins on the endothelial surface and the lectins expressed on the polymorphs. Neutrophils are activated by proinflammatory mediators such as IL-8, which are released from activated endothelial cells. Lectins on the cell surface of the neutrophils are shed and replaced by integrins that mediate the firm attachment to, and penetration of, the endothelial layer. The migration of these cells is also caused by pathogens releasing chemotaxins [e.g., formyl peptides or substances produced locally like C5a or leukotriene $B_4$ ($LTB_4$)] (Figure 1 summarizes acute inflammation; however, further information can be found in References 6 and 7.)

### *3. Killing Mechanisms*

The invading pathogen can be removed either by the formation of the C56789 complement complex or by activated phagocytes. Removal of the pathogen by phagocytosis is performed in two ways: oxygen dependent and oxygen independent. The latter involves the release of lysozyme, lactoferrin, and cationic peptides. Oxygen-dependent killing requires the production of free radicals such as superoxide ($O_2^{•-}$) and the hydroxyl radical ($^•OH$).

## B. Chronic Inflammation

If the stimulus or causative agent is persistent, the inflammation becomes chronic in nature. Chronic inflammation is characterized by the infiltration of macrophages and lymphocytes into the inflamed area. Macrophages phagocytose and remove dead tissue cells and polymorphs but can also damage healthy cells via the production of free radicals such as $NO^•$ and $O_2^{•-}$. Macrophages and fibroblasts produce cytokines such as interleukin-1 (IL-1) and tumor necrosis factor-α (TNF-α), and radicals such as $NO^•$, which mediate cell proliferation and the inflammatory process. T-cells are involved in the removal of the antigen in

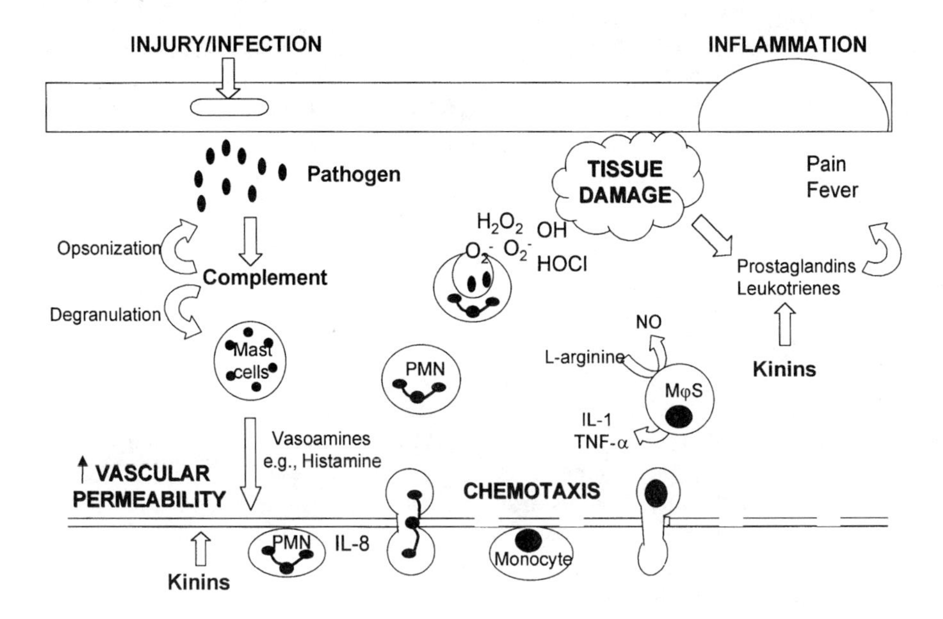


**Figure 1** Summary of the acute inflammatory responses. Invasion of a pathogen leads to the activation of various systems including the kinin and complement systems. Components from these systems increase the vascular permeability, attract PMNs and monocytes to the area, and cause the release of prostaglandins and leukotrienes. The incoming phagocytes then ingest and kill the bacteria by releasing an array of free radicals and enzymes. Cytokines such as IL-8, IL-1, and TNF-α are also involved in the regulation of inflammatory response. Excessive production of free radicals and cytokines is known to induce tissue damage.

a more specific way. Antigen presenting cells (APC) are thought to uptake and process the antigen via their class II major histocompatibility complexes. The antigen is presented to the T-cell by APCs and then specific antibodies can be produced by B-cells to remove the offending organism (8). The formation of a granuloma from the mass of immune cells attracted to the infected area is also a hallmark of chronic inflammation.

Rheumatoid arthritis (RA) is an example of a chronic inflammatory joint disease, in which the exogenous antigen is unknown. It has been suggested that the inflammation seen in RA is initiated by an antigen-induced interaction between APCs and T-cells (9). Currently, the perpetuation of the inflammatory process by this pathway appears to be under question. Several groups have suggested that the inflammatory process may be independent of T-cells (10) and that it may be driven by synovial fibroblasts and macrophages (11,12). It is clear that the proinflammatory cytokines such as IL-1 and TNF-α secreted by macrophages play an important role in the disease (13–15). In the inflamed joint, the lining increases to five to ten cell layers, mainly due to the accumulation of synovial fibroblasts (16). Cytokines and growth factors are thought to stimulate synovial fibroblasts to change into aggressive, proliferative, and invasive cells that express oncogenes and secrete connective tissue degrading enzymes such as collagenase, gelatinase, and stromelysin (also known as matrix metalloproteinases). In the joint, these activated synovial fibroblasts form a tissue called the pannus along with blood vessels and mononuclear cells (17,18). The pannus tissue erodes adjacent cartilage and bone via the production of matrix metalloproteinases (MMPs). As well as this proteinase-induced damage, a considerable amount of tissue destruction is thought to be caused by the generation of free radicals and their reactive intermediates (some of these mechanisms will be described later).

## III. RADICAL PRODUCTION IN INFLAMMATION

A free radical can be defined as a species capable of independent existence that contains one or more unpaired electrons. A free radical is usually denoted by the superscript dot (e.g., $^{\bullet}OH$). The unpaired electron(s) present in the molecule or atom of the free radical makes it very unstable and hence very reactive. In mammalian systems, free radicals usually exist at low steady-state concentrations in vivo. An increase in free radical production is a necessary part of the inflammatory response. However, if the inflammation is not resolved, free radicals are thought to cause tissue damage. The production of free radicals at inflammatory sites will be described here, along with their regulative and damaging effects in the inflammatory process.

### A. Reactive Oxygen Species

#### *1. Superoxide*

A one-electron reduction of oxygen gives rise to $O_2^{\bullet-}$. In biological systems, $O_2^{\bullet-}$ is generated as a result of many physiological or pathophysiological processes such as the production of energy by the mitochondrial electron transport chain (19), the metabolism of certain substances by cytochrome P450s (20), and by xanthine oxidase in ischemia-reperfusion injury (21). However, in terms of the inflammatory response, the production of $O_2^{\bullet-}$ by phagocytes (such as macrophages and neutrophils) in the respiratory burst is often the most important. The importance of the respiratory burst in killing certain microorganisms is highlighted by the susceptibility of patients with chronic granulomatous disease to infection (22).

At rest, neutrophils and macrophages use very little oxygen. However, once they become activated, their oxygen consumption increases markedly (23). This increase in oxygen consumption is known as the respiratory burst. The oxygen is utilized by an enzyme called nicotinamide adenine dinucleotide phosphate, reduced form (NAPDH) oxidase, which catalyzes the production of $O_2^{\bullet-}$ by the one-electron reduction of oxygen. NADPH is used as an electron donor [Eq. 1].

$$2O_2 + NADPH \rightarrow 2O_2^{\bullet-} + NADP^+ + H^+. \quad [1]$$

NADPH oxidase consists of the cytosol subunits $p47^{PHOX}$, $p40^{PHOX}$, $p67^{PHOX}$ and Rac2 and the membrane-bound subunits, Rap 1A and cytochrome $b_{558}$, $p22^{PHOX}$ and $gp91^{PHOX}$ (24). Receptors on the cell surface of the phagocyte bind antigen and certain complement components, triggering an array of signal transduction pathways. These cell signals cause the phosphorylation of $p47^{PHOX}$, which is essential for the activation of the enzyme (25). However, the phosphorylation of the $p40^{PHOX}$ and $p67^{PHOX}$ may also be important in converting the enzyme from its inactive to its active state (26,27). The cytosolic components then migrate to the membrane where they bind to cytochrome $b_{558}$ to assemble the active en-

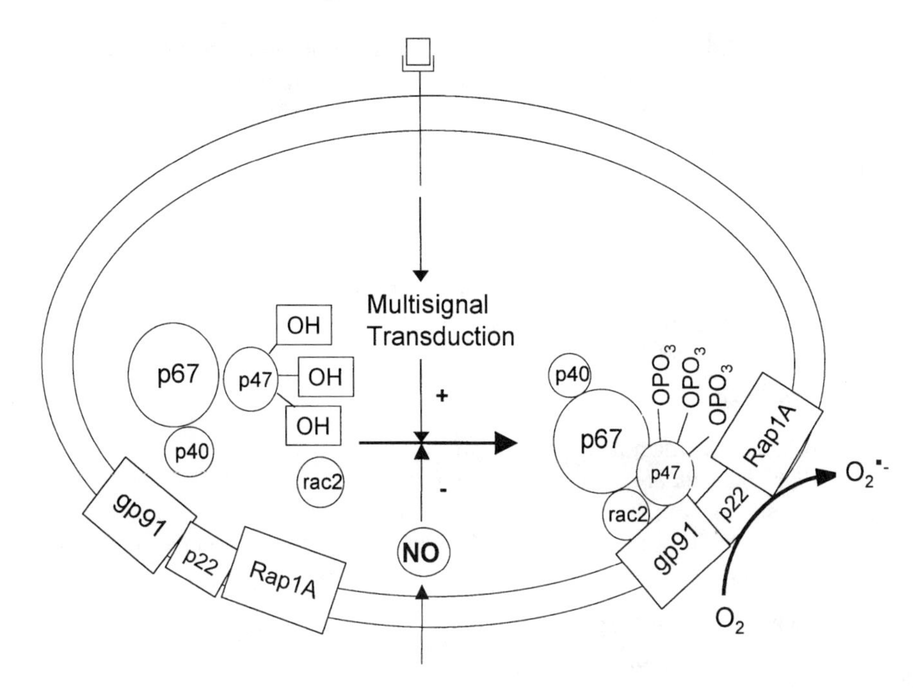


**Figure 2** Activation of NADPH oxidase. Receptors on the cell surface of neutrophils recognize and bind their specific ligands. In turn, this leads to the activation of signal transduction pathways, which controls the assembly of NADPH oxidase. Phosphorylation of p47 causes the cytosolic components to migrate to the membrane where they bind cytochrome $b_{588}$ to form the activated complex. NADPH oxidase is then able to catalyze the conversion of $O_2$ to $O_2^{\bullet -}$. NO is thought to inhibit the formation of the oxidase.

zyme complex (Fig. 2). NADPH oxidase produces $O_2^{\bullet -}$ by transferring electrons from oxygen by means of its electron carrying flavin and heme groups (24). The initial stimulus dictates whether the $O_2^{\bullet -}$ is produced phagosomally or extracellularly.

Sulfhydryl groups are important for the function of the oxidase and several naturally occurring sulfhydryl blockers have been identified. For example, the aldehyde 4-hydroxynonenal has been shown to inhibit the oxidase by the blockade of the −SH groups (28). This may be physiologically significant, as this aldehyde is a major product of lipid peroxidation and hence will be present in tissues with oxidative stress. Nitric oxide ($NO^{\bullet}$) and nitrosothiols (RSNO) also inhibit NADPH oxidase (29,30). However, they act by preventing the translocation of the cytosolic components of the oxidase. It is thought that both $NO^{\bullet}$ and RSNO may act by reacting with the cysteine-SH groups to form RSNOs on the components of the oxidase.

### *2. Hydrogen Peroxide*

Hydrogen peroxide ($H_2O_2$) is not a radical, as it does not process an unpaired electron. However, $H_2O_2$ is well known as a reactive oxygen species (ROS). Hydrogen peroxide is produced when $O_2^{\bullet -}$ dismutates. Although this reaction favors acidic conditions, nonenzymatic generation of $H_2O_2$ is also known to occur at neutral pH. However, the enzyme superoxide dismutase (SOD) facilitates the formation of $H_2O_2$ by catalyzing the dismutation of $O_2^{\bullet -}$ at neutral or alkaline pH [Eq. 2] (31). $H_2O_2$ has been shown to be produced extracellularly and within the phagosome. In 1994, Karnovsky (32) showed that $H_2O_2$ was generated in the phagosome by demonstrating the peroxidase-dependent oxidation of 3,3-diaminobenzidine and the precipitation of cerium perhydroxide.

$$2O_2^{\bullet -} + 2H^+ \xrightarrow{\text{SOD}} H_2O_2 + O_2. \quad [2]$$

$O_2^{\bullet -}$ is mildly cytotoxic and $H_2O_2$ is only bactericidal or cytotoxic at very high concentrations. Hence the production of $O_2^{\bullet -}$ and $H_2O_2$ do not account for the cytotoxicity of the respiratory burst. It is generally accepted that these species are intermediates for other, more toxic species. The nature of these species is questionable and their production in vivo is even more controversial.

### 3. *Hydroxyl Radical*

$^{\bullet}OH$ has a short half-life, which makes this powerful oxidant extremely reactive. Consequently, when $^{\bullet}OH$ radicals are generated in vivo, they will react with molecules at or within a few nanometers of their site of production. Cellular generation of $^{\bullet}OH$ can occur by three routes: from $H_2O_2$ in the Fenton reaction [Eq. 3], by the iron-catalyzed reaction of $O_2^{\bullet -}$ with $H_2O_2$ [Eq. 4], and by the decomposition of peroxynitrite [Eq. 5]:

$$Fe^{2+} + H_2O_2 \rightarrow Fe^{3+} + {}^{\bullet}OH + OH^-, \quad [3]$$

$$O_2^{\bullet -} + H_2O_2 \xrightarrow{\text{Iron(salt)}} O_2 + {}^{\bullet}OH + OH^-, \quad [4]$$

$$ONOOH \rightarrow {}^{\bullet}OH + NO_2. \quad [5]$$

Tissue damage and bactericidal killing by neutrophils has been attributed to $^{\bullet}OH$. The fact that $^{\bullet}OH$ is cytotoxic was supported by evidence showing that the addition of iron chelators and $^{\bullet}OH$ scavengers protected against tissue injury (33). It was also demonstrated that the incorporation of SOD into the phagosome prevented microbial killing (34). However, more recent data have shown that neutrophils do not contain the metal catalyst required for the Fenton reaction (35). Hence, it is unlikely that hydroxyl radicals are produced within the phagosome. In biological systems, transition metals such as copper or iron are also normally sequestered in complexes such as ceruloplasmin and transferrin, respectively. Despite this, neutrophils have been shown to produce hydroxyl radicals in the presence of proteolytically degraded transferrin (36,37). Furthermore, $O_2^{\bullet -}$ and $NO^{\bullet}$ are thought to be capable of liberating iron from ferritin, hence stimulating the formation of $^{\bullet}OH$ [see Eq. 9] (38,39). Therefore, the extracellular production of $OH^{\bullet}$ will occur when the target cells release ''free'' iron that can catalyze the Fenton reaction.

### 4. *Production of ROS in Inflammatory Diseases*

The literature is full of evidence for the production of ROS in tissues from patients with a wide range of inflammatory diseases. Despite this, the majority of these reports have used methods to quantitate secondary products from ROS reactions such as malondialdehyde (40) and 8-oxo-7-hydrodeoxyguanosine (8-oxodG) (these products of tissue damage will be discussed in more detail in Sec. VI). However, the formation of $O_2^{\bullet -}$, $H_2O_2$, and $^{\bullet}OH$ has been demonstrated in several types of inflammatory cells. For instance, Carter et al. (41) detected the production of $O_2^{\bullet -}$ and $H_2O_2$ in stimulated endothelial cells using flow cytometry, and Britigan et al. (42) detected $O_2^{\bullet -}$ and $OH^{\bullet}$ generation in stimulated neutrophils by electron paramagnetic resonance (EPR) spectrometry.

The measurement of SOD activity or expression is often used to implicate $O_2^{\bullet -}$ in physiological and pathophysiological processes. Sumii et al. (43) measured SOD activity in the synovial fluid from patients with RA, osteoarthritis (OA), and control subjects. They concluded that $O_2^{\bullet-}$ generation was increased in patients with RA and OA when compared to controls, as SOD activity was higher in these groups of patients. Mazzetti et al. (44) confirmed these results when they demonstrated that the concentration of copper/zinc SOD was increased in patients with RA when compared to controls. Furthermore, Pronai et al. (45) demonstrated that peripheral blood polymorphonuclear cells (PMNs) from RA patients generate more $O_2^{\bullet-}$ than PMNs isolated from healthy controls.

#### *5. ROS and Inflammation*

The production of ROS influences many aspects of the inflammatory process (e.g., the breakdown and removal of pathogens, chemotaxis of phagocytes to the inflamed area, and regulation of certain parts of inflammation). One example described below is the effect of ROS on inflammatory cell adhesion to the vascular endothelium. Others include the inhibition of prostacyclin (46) and prostaglandin $E_2$ ($PGE_2$) synthesis (47), maintenance of myeloperoxidase in its active form, activation of apoptosis (48), and the activation of transcription factors such as activator protein-1 (AP-1) and nuclear factor-kappa B (NF-κB) (which will be discussed later).

*Cell Adhesion.* Interestingly, ROS such as $H_2O_2$ and hydroperoxides have been shown to enhance the adhesion of PMNs to the endothelial cell surface (49). McIntyre and colleagues reported that high concentrations of $H_2O_2$ may stimulate an influx of extracellular calcium ions, which in turn would activate $PLA_2$ and, ultimately, platelet activating factor (PAF) synthesis (49). Alternatively, PAF synthesis could be activated by oxidants that disrupt the cell membrane leading to an influx of calcium ions. Furthermore, the synthesis of $LTB_4$ can be enhanced by oxidants via iron-catalyzed reactions. Activated PAF and $LTB_4$ can upregulate directly and indirectly the synthesis and expression of several PMN adhesion molecules. For instance, PAF enhances the expression of CD11/CD18 on PMNs and P-selectin on the cell surface of endothelial cells (49,50). Indirectly, PAF can induce the synthesis of intracellular adhesion molecule-1 (ICAM-1), P-selectin, and E-selectin by activating the transcription factor NF-κB (the direct activation of NF-κB by ROS will be described later). Siebenlist and colleagues (51) have reported that $LTB_4$ enhances the expression of certain adhesion molecules (e.g., ICAM-1 on endothelial cells via the activation of NF-κB) (51,52).

The antiadhesion properties of $NO^{\bullet}$ are probably due to its rapid reaction with $O_2^{\bullet-}$, resulting in the formation of peroxynitrite. However, $NO^{\bullet}$ may cause the same effect by inactivating iron or heme proteins that catalyze oxidative reactions [Eqs. 8 and 9] or by limiting the activation of NF-κB by increasing the synthesis and stability of IκB (53). Furthermore, nitric oxide synthase (NOS) inhibitors have been used to demonstrate that $NO^{\bullet}$ decreases the production of $LTB_4$ and PAF, thus limiting leukocyte migration and adhesion (54).

### B. Reactive Chlorine Species

#### *1. Hypochlorous Acid*

The $O_2^{\bullet-}$ and hydrogen peroxide formed during the respiratory burst are used to generate hypochlorous acid (HOCl). Although HOCl is not a radical, it is very strong oxidant. HOCl is responsible for killing many types of bacteria. Its bactericidal targets include iron sulfur proteins, membrane transport proteins (55), adenosine triphosphate (ATP)-

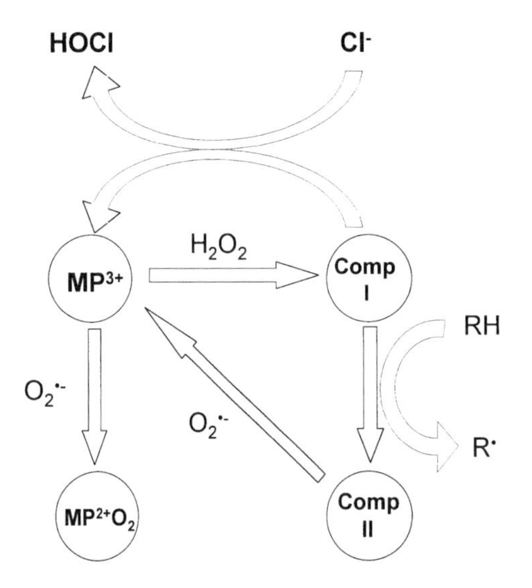


**Figure 3** HOCl production. Ferric myeloperoxidase ($MP^{3+}$) reacts with $H_2O_2$ to form the compound I. Compound I oxidizes chloride ions, by transfer of two electrons, to produce HOCl. Superoxide recycles the native enzyme by converting compound II back to $MP^{3+}$ thus preventing the accumulation of compound II. Compound I and Compound II are also involved in the oxidation of numerous organic substrates (RH).

generating systems (56), and the origin replication site for deoxyribonucleic acid (DNA) synthesis (57).

HOCl is produced by a heme oxidase called myeloperoxidase, which is released from the azurophilic cytoplasmic granules in neutrophils (58). Recent evidence has suggested that HOCl is produced phagosomally (59). Myeloperoxidase uses hydrogen peroxide and chloride ions as substrates to catalyze the production of HOCl (Fig. 3). $H_2O_2$ is produced from the dismutation of $O_2^{\bullet-}$ as described previously (Sec. III.A.2), and the cytoplasmic chloride ions are pumped out of activated neutrophils. HOCl formation accounts for 20 to 70% $H_2O_2$ liberated by inflammatory cells (60,61).

Ferric myeloperoxidase ($MP^{3+}$) reacts with $H_2O_2$ to form the redox intermediate compound I. Compound I oxidizes chloride ions by a single two-electron transfer to produce hypochlorous acid. Myeloperoxidase also oxidizes numerous organic substrates by a two-electron transfers involving compound I and compound II enzyme intermediates (62). Myeloperoxidase is not totally reliant on chloride ions and can use other substrates, such as iodide, bromide, thiocyanate, and nitrite (63–65). However, poor peroxidase substrates trap the enzyme as compound II. If compound II accumulates, hypohalous acid production is inhibited unless $O_2^{\bullet-}$ recycles the native enzyme. $H_2O_2$ itself can also readily inactivate myeloperoxidase (66,67). At neutral pH, $O_2^{\bullet-}$ has been shown to prevent the reversible (68) and irreversible inhibition (69) and therefore enhances production of HOCl (Fig. 3).

Both $NO^{\bullet}$ and peroxynitrite are known to react with myeloperoxidase. $NO^{\bullet}$ reacts with ferrous myeloperoxidase, which is likely to cause the inhibition of the enzyme (70). Peroxynitrous acid has been shown to react rapidly with $MP^{3+}$ to convert the enzyme to compound II (71). $NO_2$ is produced when nitrite interacts with compound II to regenerate the native enzyme. $NO_2$ can then go on to nitrate proteins (e.g., it is known to convert tyrosyl residues to 3-nitrotyrosine) (72).

### 2. *Secondary Oxidants*

HOCl reacts with a variety of amines to form relatively stable chloramines (73) (e.g., taurine is converted to taurine chloramine (Tau Cl) (74). Chloramines are long-lived oxidants, but they are generally less toxic to cellular targets than HOCl. Within phagosomes, it is possible that chloramines would be toxic in high concentrations. However, extracellularly, toxic concentrations are unlikely to be reached (75). Free amino acids, extracellular and intracellular proteins, and macromolecular structures on bacteria are targets for chlorination (73). Recently, electron spin resonance spectrometry has been used in combination with spin trapping to identify a myeloperoxidase-dependent pathway for the production of ${}^{\bullet}OH$ by isolated neutrophils (76). Hampton et al. (77) suggested that ${}^{\bullet}OH$ is formed as a result of a reaction between HOCl and $O_2^{\bullet -}$. However, it is unclear whether sufficient ${}^{\bullet}OH$ is produced to play a significant role in cytotoxicity.

Singlet oxygen is an excited state of oxygen, which is capable of reacting with a variety of biomolecules including membrane lipids to initiate peroxidation. Theoretically, singlet oxygen could be produced by a reaction between hydrogen peroxide and HOCl. However, the evidence for singlet oxygen production by neutrophils is inconclusive.

### 3. *Reactive Chlorine Species and Inflammation*

Reactive chlorine species (RCS) are strong oxidants that are essential for the breakdown and removal of certain pathogens. At high concentrations, HOCl causes cell lysis and is capable of oxidizing $\alpha$1-antitrypsin (78) and activating neutrophil collagenase (79). In addition, chlorination of proteins by HOCl makes them more susceptible to degradation by proteinases (80). For example, exposure of fibronectin to HOCl causes alterations in both the primary and tertiary structure of the protein and increases the possibility of degradation by elastase. Interestingly, both HOCl and TauCl inhibit the generation of macrophage inflammatory mediators such as $NO^{\bullet}$, TNF-$\alpha$, and $PGE_2$ (81). Furthermore, HOCl is thought to have an effect on other aspects of the inflammatory process such as endothelial permeability and leukocyte adherence to microvascular endothelium (82).

## C. Reactive Nitrogen Species

### 1. *Nitric Oxide*

$NO^{\bullet}$ is an inorganic radical that exists in the gaseous state at room temperature. Initially, it was thought of as an air pollutant rather than a biological signaling molecule. In 1980, the relaxation of smooth muscle by acetylcholine was shown to be dependent on a substance named endothelium-derived growth factor (EDRF). The chemical nature of EDRF was not discovered until 1987, when Furchgott (83) and Ignarro et al. (84) independently pointed out that EDRF and $NO^{\bullet}$ had very similar biological properties. In the same year, Palmer and coworkers, using a bioassay and a chemiluminescence assay simultaneously, demonstrated that EDRF was in fact $NO^{\bullet}$ (85).

$NO^{\bullet}$ is produced as part of the inflammatory response from activated cells such as macrophages. It has cytotoxic properties against tumor cells, parasitic fungi, intracellular bacteria, protozoans, and helminths (86,87). In addition, $NO^{\bullet}$ has been implicated in the killing of bacteria by neutrophils (88). The mechanism of cytotoxicity is unclear, although $NO^{\bullet}$ is thought to damage mitochondrial complexes I and II, mitochondrial aconitase, glyceraldehyde-3-phosphate dehydrogenase, ribonucleotide reductase, and DNA (89). Acute inhibition of mitochondrial cytochrome oxidase or bacterial oxidase may assist in killing the pathogen.

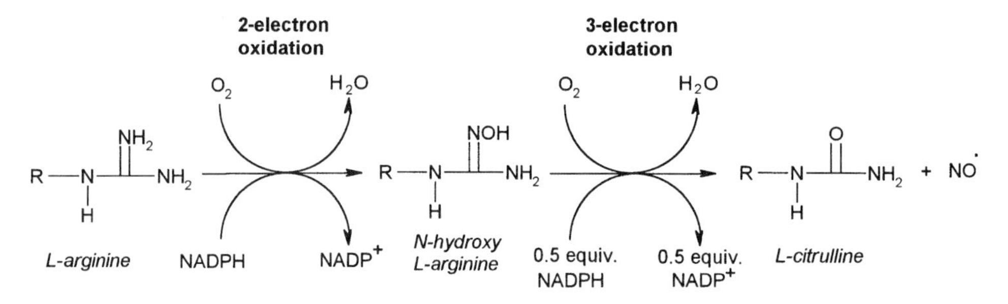


**Figure 4** NO production. L-arginine undergoes a 2-electron oxidation to form N-hydroxy-L-arginine. N-hydroxy-L-arginine undergoes a further 3-electron oxidation to form L-citrulline and NO. $O_2$ is required for the reaction, along with NADPH, which acts as an electron donor.

NO· is produced by a group of enzymes called the nitric oxide synthases (NOS). NOS catalyzes the conversion of L-arginine and oxygen to L-citrulline and NO·. Figure 4 shows a more detailed picture, where L-arginine is converted to N-hydroxy-L-arginine, which involves a two-electron oxidation supported by one molecule of NADPH. The conversion of this intermediate to L-citrulline and NO· requires the insertion of a second molecule of oxygen and involves an additional three-electron oxidation (90). This reaction also requires the presence of flavin adenine dinucleotide (FAD), flavin mononucleotide (FMN), and tetrahydrobiopterin ($BH_4$) as cofactors (91). There are three isoforms of NOS: (1) neuronal NOS (nNOS), which was first identified in neuronal cells of the central and peripheral nervous systems, but has now been shown to be expressed in other cell types such as skeletal muscle; (2) endothelial NOS (eNOS) first identified in endothelial cells; and (3) inducible NOS (iNOS), which was first detected in murine macrophages (92). eNOS and nNOS are both constitutively expressed, whereas the expression of iNOS is inducible and can be activated by cytokines (such as IL-1 and TNF-$\alpha$), endotoxins, and lipopolysaccharide (LPS) (93). In terms of inflammation, iNOS, which is expressed in macrophages (94), neutrophils (95), and other cell types, is probably the most important isoform. Recently, it was shown that xanthine oxidase can produce NO· under hypoxic conditions from nitrite and nitrate due to its reductase activity (96,97). Xanthine oxidase converts nitrite to NO· by accepting an electron from nicotinamide adenine dinucleotide, reduced (NADH), which is then converted to nicotinamide adenine dinucleotide ($NAD^+$). Zhang et al. (96) showed the production of NO· by xanthine oxidase in inflamed synovial tissue. However, whether this method of NO· production is viable in vivo is yet to be determined.

### 2. *Secondary Reactive Nitrogen Species*

NO· reacts with a variety of biomolecules both directly and indirectly. Metal complexes such as those found in hemoglobin and cytochrome P-450 (see Sec. III.C.4) and free radicals such as peroxyl radicals are known to react with NO· directly. In addition, NO· can undergo nitration, oxidation, and nitrosation reactions to give rise to secondary products known as reactive nitrogen species (RNS) (98). These reactions of NO· and their effects on the inflammatory process are discussed in further detail.

*Nitrite and Nitrate.* Oxidation of NO· leads to the formation of a multitude of nitrogen oxides, including nitrite. In whole blood, nitrite is then converted to nitrate by reacting with oxyhemoglobin [Eq. 6]. Nitrate can also be formed directly when NO· and oxyhemoglobin react together to form nitrate and methemoglobin [Eq. 7] (90). The concentration of $NO^{\bullet}$ in vivo is dependent on this reaction, as it acts as a detoxification mechanism and metabolic route for $NO^{\bullet}$. Its success is largely due to the affinity of $NO^{\bullet}$ for oxyhemoglobin and the high concentrations of oxyhemoglobin found in blood.

$$NO_2^- + 2HbO_2 + 2H^+ \rightarrow 2Hb^+ + O_2 + H_2O + NO_3^-, \qquad [6]$$

$$Hb(Fe - O_2) + NO \rightarrow \text{met Hb(Fe(III))} + NO_3^-. \qquad [7]$$

In addition, nitrite can be produced when $NO^{\bullet}$ scavenges deleterious metallo-oxo species formed, for example, by the action of hydrogen peroxide on iron (II) or (III) (99). Here, $NO^{\bullet}$ is acting as an antioxidant, in which $NO^{\bullet}$ is protecting tissues from peroxide-mediated damage [Eqs. 8 and 9] (100).

$$Fe^{2+} + H_2O_2 \rightarrow Fe^{4+} = O + H_2O, \qquad [8]$$

$$Fe^{4+} = O + NO \rightarrow Fe^{3+} + NO_2^-. \qquad [9]$$

*Nitrosothiols.* S-Nitrosothiols (RSNO) are known to be formed in vivo by the $NO^{\bullet}$-dependent S-nitrosation of thiol-containing proteins and peptides such as albumin and hemoglobin (101,102). The S-nitrosylation of thiols accelerates their oxidation and increases their reactivity in reactions with various functional groups. However, RSNOs are considerably more stable than $NO^{\bullet}$. Therefore, it is only logical to suggest that RSNOs may provide a way of regulating the bioavailability of $NO^{\bullet}$ and/or serve to increase its range of action. For instance, nitrosothiols are relied on in the blood for vascular relaxation because free $NO^{\bullet}$ cannot coexist with hemogoblin (103). S-nitrosothiols are also likely to be involved in inflammation via the host defense mechanisms, as they have potent antimicrobial properties whereas $NO^{\bullet}$ does not (104). Furthermore, it has been proposed that RSNO groups in proteins are important in the metabolism of $NO^{\bullet}$ and in the regulation of cellular functions such as the transport and targeting of the $NO^{\bullet}$ group to specific thiol regulatory effector sites, including enzymes and signaling proteins. In addition, RSNOs perform functions similar to $NO^{\bullet}$ such as vasodilatation and inhibition of platelet aggregation.

RSNOs may act as antioxidants; it has been suggested that they are resistant to decomposition by ROS and thus may prevent the diffusion controlled reaction between $NO^{\bullet}$ and $O_2^{\bullet-}$ to form peroxynitrite (101). More recently, these findings have been contradicted by an investigation performed by Jourd'heil et al. (105), who demonstrated that GSNO is readily decomposed by $O_2^{\bullet-}$. However, RSNOs are acting as antioxidants in both scenarios by limiting the amount of substrates available.

*Peroxynitrite.* Peroxynitrite is formed from a reaction between $NO^{\bullet}$ and $O_2^{\bullet-}$ which is a diffusion-controlled reaction [Eq. 10]. $NO^{\bullet}$ has a higher affinity for $O_2^{\bullet-}$ than SOD (106). However, peroxynitrite formation is limited in normal tissue because the concentration of $NO^{\bullet}$ produced is not high enough to compete with SOD (107). In pathological conditions, macrophages and other cell types are capable of producing large amounts of $NO^{\bullet}$ via iNOS, which is activated by bacterial endotoxins and cytokines released in inflammation (93). In some cases, the quantity of $NO^{\bullet}$ produced can be 100- to 1000-fold greater than normally formed by endothelial cells for signal transduction (107). These elevated concentrations of $NO^{\bullet}$ allow the formation of peroxynitrite to take place as $NO^{\bullet}$ competes with endogenous SOD for $O_2^{\bullet-}$. Episodes of inflammation also lead to the increased production of $O_2^{\bullet-}$ from phagocytic cells, which leads to the production of large amounts of peroxynitrite. In fact, the formation of peroxynitrite has been demonstrated in macrophages (108), human neutrophils (109), and cultured endothelial cells (110).

$$NO + O_2^{\bullet -} \rightarrow {}^{-}OONO. \quad [10]$$

Peroxynitrite is a potent oxidant, which can oxidize many biological components including lipids, sulfhydryls, and iron sulfur centers and zinc fingers (111,112). In the presence of a metal catalyst such as iron (III), peroxynitrite can form a species similar to the nitronium ion [Eq. 11].

$$Fe^{3+} + {}^{-}OONO \rightarrow Fe^{2+} - O^{-} - NO_2^{+}. \quad [11]$$

This intermediate readily nitrates tyrosine residues on proteins, resulting in the formation of 3-nitrotyrosine residues (113). The nitration of these tyrosine residues can cause the protein to become antigenic and can prevent phosphorylation of certain receptors central to signal transduction (114).

### *3. RNS Production in Inflammatory Diseases*

There is an array of evidence implicating the production of $NO^{\bullet}$ in inflammatory diseases such as rheumatoid arthritis, osteoarthritis, and glomerulonephritis. In the majority of these studies, iNOS expression has been measured as an index of $NO^{\bullet}$ production. Many different inflammatory cells have been shown to express iNOS [e.g., chondrocytes (115), synoviocytes (116), macrophages (117), and neutrophils (118)]. However, expression of iNOS does not necessarily mean that the cells are producing large amounts of $NO^{\bullet}$. Essentially in inflammation it appears that iNOS needs to be induced by the elevated levels of proinflammatory cytokines such as IL-1, TNF-$\alpha$, and interferon-$\gamma$ (IFN-$\gamma$), which are characteristic of arthritis (119). Furthermore, the control of $NO^{\bullet}$ production is also dependent on anti-inflammatory cytokines such as IL-4 and IL-10, which have been reported to inhibit iNOS expression in macrophages. The regulation of iNOS has proved to be more complicated, as the expression of iNOS in a variety of cell types is different. For instance, IL-1$\beta$ and TNF-$\alpha$ readily induce iNOS in murine macrophages; however, these cytokines do not affect the expression of iNOS in human leukocytes (120).

In terms of arthritic diseases, the importance of $NO^{\bullet}$ has been suggested by the fact that inhibition of iNOS has been shown to suppress arthritis in several animal models (116,121). In addition, urinary nitrate excretion and iNOS expression have been reported to be substantially increased during joint inflammation in both adjuvant-induced and collagen-induced models of arthritis in rats (122) (increased $NO^{\bullet}$ synthesis is thought to be reflected by the urinary excretion of nitrate). This observation was confirmed in patients with RA when increases in the excretion of urinary nitrate were demonstrated (119,123). Furthermore, the treatment of RA patients with prednisolone has been shown to reduce urinary nitrate levels (124). Commonly, the measurement of nitrite and nitrate is used as another index of $NO^{\bullet}$ production. For example, Farrell et al. (125) implicated $NO^{\bullet}$ in arthritis by detecting increased nitrite concentrations in the serum and synovial fluid from patients with RA and OA when compared to controls. Ueki et al. (126) confirmed these results in RA patients but also demonstrated that serum nitrite levels correlated significantly with clinical manifestations of RA, such as early morning stiffness, the number of tender or swollen joints, and C-reactive protein. In addition, they showed that serum nitrite levels correlated significantly with serum TNF-$\alpha$ and IL-6 levels, thus suggesting that increased endogenous $NO^{\bullet}$ synthesis reflects abnormalities of immunoregulation in the joints of RA patients.

Although peroxynitrite and nitrosothiols are thought to accumulate in biological fluids during immune activation and inflammation, there is very little evidence to confirm these suggestions. Kaur et al. (127) demonstrated that nitrotyrosine (a marker of peroxynitrite production) was elevated in serum and synovial fluid samples from patients with rheumatoid arthritis. They also observed that no detectable nitrotyrosine was detected in serum and synovial fluid samples taken from normal volunteers, patients with osteoarthritis, and patients with early-stage rheumatoid arthritis. S-nitrosoproteins have also been shown to reach high concentrations in the synovial fluid of RA patients when compared to the serum (128). This suggests, therefore, that nitrosoproteins are formed within in the joint from the NO$^•$ generation by inflammatory cells.

Unfortunately, the majority of evidence for the production of NO$^•$ in inflammatory conditions has arisen from the indirect measurement of NO$^•$. This is due mainly to the short half-life of this radical, which in tissues and blood is approximately 10 s and 0.46 ms, respectively (90). Its half-life in whole blood is greatly reduced due the reaction of NO$^•$ with hemoglobin. One of the possible products from this reaction is nitrosylhemoglobin, which can be measured by electron paramagnetic resonance spectrometry. Presently, the measurement of this species remains one of the few ways of directly measuring NO$^•$ in whole blood. Weinberg and coworkers (129) have used this method to measure NO$^•$ in MRL-*lpr*/*lpr* autoimmune mice which developed manifestations of autoimmunity, including arthritis, vasculitis, and glomerulonephritis. Whole blood taken from the MRL-*lpr*/*lpr* autoimmune mice contained increased levels of nitrosylhemoglobin when compared to controls, demonstrating in situ NO$^•$ formation in this model. In principle, the measurement of nitrosylhemoglobin in human inflammatory conditions such as rheumatoid arthritis and systemic lupus erythematosus (SLE) should be possible, although it has not yet been reported. The reasons for this either lie within the limitations of the method or the need to take into account the ethical considerations while dealing with human patients.

Elevated NO$^•$ production has been reported in a multitude of diseases, and therefore increased NO$^•$ formation cannot be viewed as a specific marker for any particular disease state. Nevertheless, it is clear that excessive levels of NO$^•$ can lead to tissue injury and contribute to the progression of many diseases. Indeed, animal studies have shown that, by inhibiting NO$^•$ production, diseases such as SLE (130) and arthritis (121,131) can be attenuated.

### *4. RNS and Inflammatory Diseases*

*Nitric Oxide, Prostaglandins, and Leukotrienes.* Prostaglandins and leukotrienes are produced during inflammation by the enzymes cyclooxygenase (COX) and lipoxygenase, respectively. Cyclooxygenase and lipoxygenase are enzymes belonging to the cytochrome P450 family. NO$^•$ is thought to interact with this family of enzymes in two ways: (1) reversible inhibition, where NO$^•$ binds to the heme center to prevent oxygen binding, thus inhibiting catalysis and (2) irreversible inhibition mediated by RNS formed by the autoxidation of NO$^•$.

The current evidence suggesting a role for NO$^•$ in prostaglandin synthesis contains some conflicting results. The effect of NO$^•$ on COX is different in various cell types. For instance, NO$^•$ has been shown to increase the production of prostaglandins in RAW 264.7 macrophages (132), bovine cartilage, chrondrocytes (133), and rat mesangial cells (134). In contrast, NO$^•$ has been shown to be responsible for a decrease in prostaglandin synthesis in cultured murine macrophages (135) and LPS-stimulated peritoneal macrophages (136). Furthermore, the addition of a NOS inhibitor has been shown to increase the production of prostaglandin, $PGE_2$ (115,137). Initially, it was thought that NO$^•$ might regulate COX by interacting with its iron catalytic site. However, there is no evidence to suggest a direct interaction at in vivo concentrations of NO$^•$. Alternatively, NO$^•$ may act at a cellular level of COX expression, which could explain the conflicting evidence for different cell types. Nevertheless, the effect of NO$^•$ on COX appears to be concentration dependent (e.g., in chrondrocytes there is evidence that low amounts of NO$^•$ increase $PGE_2$ production while high amounts inhibit $PGE_2$ production) (137). It is also possible that NO$^•$ may elicit its effect through peroxynitrite, which has also been implicated in the activation of COX.

The generation of leukotrienes by lipoxygenase is inhibited by NO$^•$; however, the mechanism is unclear. Experimental studies have demonstrated that NO$^•$ inhibits the formation of lipoxygenase-derived products (138,139). Maccarone et al. (139) suggested that large fluxes of NO$^•$ inactivated the enzyme by the reduction of its $Fe^{3+}$ to $Fe^{2+}$, whereas Rubbo et al. (140) demonstrated that leukotriene production was inhibited by NO$^•$ via a mechanism involving lipid radical scavenging.

*NO$^•$ and Lipid Peroxidation.* NO$^•$ is thought to terminate lipid peroxidation by reacting with peroxyl radicals as shown in Eq. [12] (141).

$$LOO^• + NO^• \rightarrow LOONO. \quad [12]$$

Oxidation of low-density lipoproteins is also prevented by this mechanism, which may be important in diseases such as atherosclerosis. Additionally, NO$^•$ has been shown to protect against oxidation of organic hydroperoxides found in lipid membranes. For example, the NONOate, PAPA/NO protects against the production of *tert*-butyl hydroperoxide or cumene hydroperoxide (142). Rubbo and coworkers confirmed that NO$^•$ acts as an antioxidant when they demonstrated that NO$^•$ suppressed xanthine oxidase-mediated lipid peroxidation (140). Hydroperoxides formed during oxidative stress react with heme proteins to form hypervalent metalloproteins. These hypervalent metalloproteins are capable of inducing lipid peroxidation and can decompose to release intracellular iron. NO$^•$ exerts its antioxidant effect by reacting with these metalloproteins to restore them to their ferric form [Eqs. 8 and 9]. In turn, this prevents the release of intracellular iron and limits oxidative damage (99,100).

*Other Effects of NO$^•$ on Inflammation.* The inflammatory process can be influenced by NO$^•$ in a multitude of different ways. Some of these have been covered previously; however, NO$^•$ has many other effects. For instance, the inhibition of the tissue inhibitor of metalloproteinase-1 (TIMP-1) via peroxynitrite (143) and the inhibition of mast cell degranulation. Furthermore, several groups have shown that NO$^•$ may attenuate its own synthesis, thus acting as a negative feedback factor regulating NO$^•$ levels (144,145). Constitutive isoforms of NOS appear to be more susceptible to inhibition than iNOS, which may explain why higher fluxes of NO$^•$ can be produced by iNOS. Recently, it has become apparent that NO$^•$ may regulate secretion of proinflammatory cytokines as part of a positive feedback mechanism. For example, NO$^•$ may increase the catabolic activity of IL-1 by suppressing the synthesis of its receptor antagonist, IL-1ra. Additionally, McInnes et al. (146) illustrated that the addition of a NO$^•$ donor causes the induction of TNF production in cultured synoviocytes and macrophages.

Finally, NO$^•$ can also inhibit platelet and neutrophil adhesion to endothelial monolayers as well as causing the inhibition of leukocyte $O_2^{•-}$ and HOCl production. These effects have been covered in more detail in Section III.C.5, which deals with $O_2^{•-}$ production and its involvement in inflammation.

### *5. NO• and Arthritis*

In inflammatory diseases such as rheumatoid arthritis, NO• has the effects described above but it is also involved in articular damage. For instance, it has been shown to inhibit collagen and proteoglycan synthesis and activate metalloproteinases.

*NO•, Collagen, and Proteoglycans.* Cartilage is made up of chondrocytes, which are responsible for synthesizing and maintaining the extracellular matrix, which forms hyaline cartilage. NO• affects the synthesis and degradation of the two proteins, proteoglycan and collagen, that are found in cartilage.

NO• has been shown to suppress proteoglycan synthesis possibly via IL-1 (147). In contrast, NO• is thought to afford protection from the degradation of the proteoglycans. Evidence suggesting this role for NO• in arthritis was provided by Stefanovic-Racic et al. (148) who demonstrated that proteoglycan degradation was enhanced when a NOS inhibitor was added to IL-1-stimulated bovine and rabbit cartilage.

Prolyl hydroxylase is an enzyme that crosslinks collagen chains to provide tensile strength. NO• may affect collagen metabolism by inhibiting this enzyme (149). Consequently, this results in the accumulation of underhydroxylated and underannealed collagen α-chains, which are proteolyzed. Again, IL-1 has been shown to be involved in this NO•-mediated process. IL-1 has been demonstrated to suppress collagen synthesis in rabbit articular chrondrocytes at two levels: a pretranslational level, determined by Col2A1 messenger ribonucleic acid (mRNA) expression that is insensitive to NO•, and a translational or posttranslational level that is NO•-dependent (149).

*NO• and Proteinases.* Matrix metalloproteinases contribute to the degradation of cartilage and bone in arthritis. NO• is thought to influence the synthesis of these MMPs. However, whether NO• stimulates or suppresses MMP production is debatable. For instance, Amin et al. (115) demonstrated that inhibition of NOS by N-methyl-L-aspartic acid (L-NMA) enhanced both $PGE_2$ and MMP-3 production in OA affected cartilage. In contrast, other groups have shown that NO• activates MMPs in bovine cartilage (150), rat mesangial cells (151), and chrondrocytes (152). Furthermore, Stadler et al. (137) demonstrated that NO• had little effect on the induction of MMPs in chrondrocytes exposed to IL-1.

Tamura et al. (151) proposed that basic fibroblast growth factor (bFGF; a heparin binding factor from the extracellular matrix) was released as a result of matrix degradation mediated by NO• in rabbit chrondrocytes. They also reported that the increased production of MMPs and bFGF could both be blocked by NO• inhibition. Consequently, they hypothesized that bFGF release through NO• could stimulate cell proliferation and thus angiogenesis in the synovium of arthritic patients.

## D. Antioxidants

Evolution has provided our bodies with a well-balanced mechanism to neutralize free radicals. These specialized, protective components are known collectively as the "antioxidant defense system." Antioxidants control and regulate the concentration of free radicals, thereby preventing tissue damage and unwanted physiological effects. In the body, there is a diverse range of antioxidants that even include free radicals such as NO. Antioxidant defense systems are traditionally split into two groups—primary and secondary antioxidants. Primary antioxidants, such as vitamin E, glutathione, and SOD, interact directly with $O_2^{\bullet -}$, or other radicals, whereas secondary antioxidants, including proteolytic enzymes and

DNA repair enzymes, scavenge free radicals and other reactive species. Metal chelators such as transferrin and ceruloplasmin have an indirect antioxidant effect because they act by scavenging transition metals required to catalyze the production of secondary radicals such as $^{\bullet}OH$ (see Refs. 153, 154 for further information). Biologically, the most important antioxidants are glutathione, glutathione peroxidase, vitamins E and C, SOD, catalase, and urate. These antioxidants are strategically positioned within cells to provide maximal protection [e.g., manganese SOD, catalase, and glutathione are found in the mitochondria where free radical leakage occurs frequently (155,156)].

Antioxidants thus provide protection to tissues and organelles in inflammatory diseases. However, excessive free radical production causes the depletion of these antioxidant pools, leading to an imbalance in antioxidant/oxidant ratio. For example, our group has previously demonstrated that vitamin E and vitamin C levels are depleted in RA (157,158). Consequently, the depletion of these protective compounds contributes to the tissue destruction seen in inflammatory diseases.

## IV. RADICALS, ISCHEMIA/REPERFUSION INJURY, AND INFLAMMATORY DISEASES

The occlusion of a blood vessel causes the surrounding tissues to become ischemic due to the lack of oxygenated blood supply. Ischemia is a symptom of many diseases, including myocardial infarction, thrombotic stroke, embolic vascular occlusions, angina pectoris, cardiac surgery, and organ transplantation. Lack of oxygen causes damage and cell death within the tissue. In response to this tissue damage, the inflammatory process is activated, leading to complement activation, increased endothelial permeability, $O_2^{\bullet -}$ generation, cytokine release, and activation of platelets with the release of serotonin and adenosine diphosphate (ADP).

Although ischemia is harmful to the surrounding tissues, reperfusion has also been shown to cause significant damage (159). Paradoxically, it is important to reoxygenate quickly in order to preserve the tissue, but this can cause damage via reperfusion injury. Hearse (160) suggested the concept of reperfusion injury; prior to that, it was assumed that the increased cell death on the reintroduction of oxygen to the area was due to the death of cells that had been irreversibly injured during the ischemic phase (161), in the human heart.

An early indication that molecular oxygen was involved in reperfusion injury was shown by Hearse et al. (162), who demonstrated that the reperfusion of an ischemic heart with an oxygenated solution enhanced injury, whereas a hypoxic solution did not increase myocardial injury. Since then, in vitro and in vivo studies have shown that the reoxygenation of ischemic myocardium is accompanied by the production of ROS (163,164). The major source of free radicals in ischemia reperfusion injury has not yet been confirmed. However, it is commonly believed that $O_2^{\bullet -}$ may be formed by the enzyme xanthine oxidase, which is located in the vascular endothelium. Normally xanthine oxidase/dehydrogenase oxidizes xanthine to uric acid using NAD as an acceptor [Eq. 13],

$$\text{Xanthine} + H_2O + NAD^+ \rightarrow \text{Uric Acid} + NADH + H^+. \qquad [13]$$

In ischemic tissues, oxidative phosphorylation is interrupted preventing the synthesis of energy-rich phosphates including ATP and creatine phosphate. Depletion of ATP during ischemia has two consequences: dysfunction of ATP-dependent membrane ion pumps, with consequent rise in both mitochondrial and cytosolic calcium levels, and an increase in the concentration of hypoxanthine (165). The increases in intracellular calcium concentrations activate a calcium-dependent protease, which in turn converts xanthine dehydrogenase to xanthine oxidase (166). Ischemia, tumor necrosis factor, IL-1 and IL-3, neutrophil elastase and C5, and the chemotactic peptide N-formyl-Met-Leu-Phe can activate this protease (167,168). The conversion of xanthine dehydrogenase to xanthine oxidase enables this enzyme to convert molecular oxygen to $O_2^{\bullet-}$ during the reperfusion of ischemic tissues (169) [Eq. 14].

$$\text{Xanthine} + H_2O + O_2 \rightarrow \text{Uric Acid} + 2O_2^{\bullet-} + H^+. \qquad [14]$$

Xanthine oxidase can use either xanthine or hypoxanthine as its substrate in the production of $O_2^{\bullet-}$. The breakdown of adenine nucleotides such as adenosine monophosphate (AMP) leads to the production of xanthine and hypoxanthine via adenosine and inosine (Fig. 5). The formation of $O_2^{\bullet-}$ in this way is also dependent on the reintroduction of molecular oxygen; hence, $O_2^{\bullet-}$ is not produced until the vessel is reperfused (169).

The $O_2^{\bullet-}$ formed is not responsible for the tissue damage caused in the ischemic area. Generally, this is thought to be a result of the formation of $H_2O_2$ and $^\bullet OH$. As mentioned previously, the generation of $^\bullet OH$ requires the presence of iron (II). Although iron binding is well developed in the body during ischemia, this may be lost when partially damaged tissues release iron due to decompartmentalization or disorganization.

Experimental evidence for the involvement of xanthine oxidase in ischemia-reperfusion injury was suggested when allopurinol, an inhibitor of the enzyme, was re-

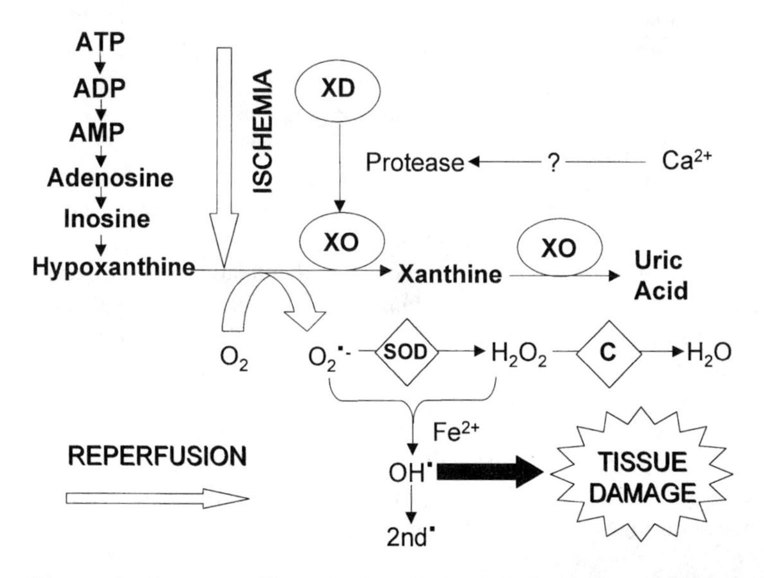


**Figure 5** Summary of free radical production in ischemic-reperfusion injury. During the ischemic phase, hypoxanthine accumulates in the tissue due to the breakdown of ATP products. Xanthine dehydrogenase (XD) is also converted in xanthine oxidase (XO) by the action of a protease. Reperfusion of the vessel enables xanthine oxidase to convert oxygen into superoxide. Superoxide dismutase (SOD) catalyzes the dismutation of superoxide to hydrogen peroxide, which can be converted into water by another antioxidant enzyme, catalase (C). In the presence of catalytic iron, superoxide and hydrogen peroxide can form hydroxyl radicals and other secondary radicals (2nd) which can cause tissue damage.

ported to protect reperfused tissues in canine models (170). However, both xanthine oxidase and xanthine dehydrogenase are present in considerable amounts in canine models whereas they are virtually absent in humans (165). Obviously, this casts doubt on whether this is the main route of radical production in ischemia-reperfusion injury in humans. It is possible that during ischemia-reperfusion free radicals may be generated by the cyclooxygenase and lipoxygenase pathways or by the mitochondrial transport chain.

Some investigators hold free radicals responsible for the increased levels of intracellular calcium seen in ischemia-reperfusion injury. Consequently, these large calcium influxes cause cellular effects that accelerate ischemia-reperfusion injury (e.g., the activation of phospholipases, which leads to the generation of arachidonic acid metabolites and ATP depletion (171). Radicals are thought to promote calcium overload in three ways: (1) interfering with calcium transport either at the sarcoplasmic reticulum or cell membrane; (2) impairing mitochrondrial oxidative phosphorylation and directly inhibiting glyceraldehyde-3-phosphate dehydrogenase; and (3) irreversibly inhibiting anaerobic glycolysis (172,173). Furthermore, it appears that calcium overload may enhance free radical generation by the activation of xanthine oxidase through calcium-dependent proteinases or by causing the leakage of free radicals from the electron transport chain.

### A. Ischemia-Reperfusion and RA

In rheumatoid arthritis, the joint becomes swollen due to the accumulation of synovial fluid and inflammatory exudate. In turn, this increases the resting intra-articular pressure within the rheumatoid joint, which is significantly increased in comparison to the normal joint (174). Unlike the normal joint, which maintains a subatmospheric pressure during exercise, the rheumatoid joint reaches intra-articular pressures of 200 mmHg, which is in excess of the synovial capillary perfusion pressure of 30–60 mmHg. The blood flow becomes impeded by the occlusion of the capillary bed and the synovium becomes ischemic (175). At this point, the partial pressure of oxygen ($pO_2$) has been shown to decline (176). Reperfusion of the synovial membrane occurs on the cessation of exercise. Consequently, the $O_2$ concentration increases and xanthine oxidase utilizes xanthine and hypoxanthine produced during the ischemic phase to generate $O_2^{\bullet-}$.

Experimental evidence has shown that both normal and rheumatoid synovia exhibit xanthine oxidase/dehydrogenase activity (177). Immunohistochemical studies have confirmed the presence of xanthine oxidase in synovial endothelial cells. Furthermore, in vitro studies using electron paramagnetic resonance spectroscopy have demonstrated the production of free radicals from the rheumatoid synovium, which can be suppressed by the xanthine oxidase inhibitor, allopurinol (178). The presence of iron-saturated ferritin in rheumatoid synoviocytes also implies that $^{\bullet}OH$ could be generated via the Fenton reaction. Potentially, the ''free iron'' needed to catalyze the Fenton reaction may be liberated from ferritin by $O_2^{\bullet-}$ produced during reperfusion (179).

## V. CELL SIGNALING AND GENE ACTIVATION

In the inflammatory process, free radicals are thought to mediate some of their effects via redox regulated transcription factors such as AP-1 and NF-κB. For instance, the cellular adhesion of neutrophils and monocytes is partly enhanced by ROS through the activation of NF-κB. Furthermore, the redox regulation of these transcription factors may play an important role in the generation of growth factors and cytokines. Consequently, it has been suggested that free radicals indirectly regulate cellular functions of the inflammatory process, such as T-cell proliferation and differentiation via the production of IL-2 (180).

### A. Activator Protein-1

AP-1 is a transcription factor that is redox regulated. The AP-1 binding site is also known as the tetradeconyl phorbol acetate responsive element and is found in various genes, including those encoding human collagenase, stromelysin, transforming growth factor-α and -β, IL-2, and tissue inhibitor of metalloproteinases-1 (181).

AP-1 is a dimeric complex that consists of two oncogenic proteins, Fos and Jun. In stimulated cells, the production of these proteins is regulated by the intracellular redox status. For instance, several groups have shown that c-fos and c-jun mRNA levels are strongly induced by $H_2O_2$ and ultraviolet (UV) light in osteoblasts, fibroblasts, and epithelial cells (98,182,183). Oxidants are thought to upregulate the translation of Fos mRNA by activating the binding of serum responsive factor (SRF) to the serum responsive element (SRE) (182).

AP-1 proteins consist of either a Fos-Jun heterodimer or a Jun-Jun homodimer. The transcriptional activation of effector genes occurs when either of these complexes binds the AP-1 binding site (184). The binding of the Fos-Jun complex to DNA has been shown to be modulated by the oxidation-reduction of a single cysteine present in the DNA binding domain of each subunit (Fig. 6) (185,186). Oxidation of this residue by diamide or glutathione disulfide (GSSG) inhibits DNA binding, whereas reduction by NADPH, dithiothreitol (DTT), or GSH enhances DNA binding (187). Free radicals such as nitric oxide have been shown to inhibit AP-1 by reacting with the critical cysteine groups (188). Furthermore, $H_2O_2$ has only been shown to moderately enhance DNA binding whereas antioxidants, including pyrrolidine dithiocarbamate (PDTC), butylated hydroxyanisole (BHA), and thioredoxin have been demonstrated to strongly activate the DNA binding and transactivation of AP-1 (189,190). Consequently, AP-1 has been classified as a secondary antioxidant response factor.

Furthermore, the nuclear protein known as Ref-1 may mediate the redox regulation of AP-1. The mechanism by which Ref-1 facilitates DNA binding is not thought to involve its binding to the Fos/Jun complex or to the DNA. Instead, it is hypothesized that Ref-1 reduces the crucial cysteine groups in the Fos and Jun proteins (Fig. 6) (191). The stimulatory effect of Ref-1 can be diminished under oxidative conditions due to the oxidation of a functional cysteine group (185). However, upon the addition of thioredoxin, an enzyme that catalyzes the reduction of cysteine residues, the stimulatory activity of Ref-1 was restored and AP-1 binding resumed (191).

### B. NF-κB

NF-κB is a transcription factor involved in controlling of the expression of several genes associated with the inflammatory process. Its target genes include cytokines and growth factors, cytokine receptors, stress proteins, cellular adhesion molecules, and immunoregulatory molecules such as MHC class I and II (192). Furthermore, the activation of NF-κB has been shown to stimulate the transcription of the iNOS gene leading to increased $NO^{\bullet}$ production (193). A negative feedback loop seems to exist here, as $NO^{\bullet}$ has been shown to inhibit NF-κB activation by enhancing the stability or synthesis of IκB, the inhibitory subunit of NF-κB (53).

NF-κB is a dimeric protein that can consist of variable subunits; however, classically

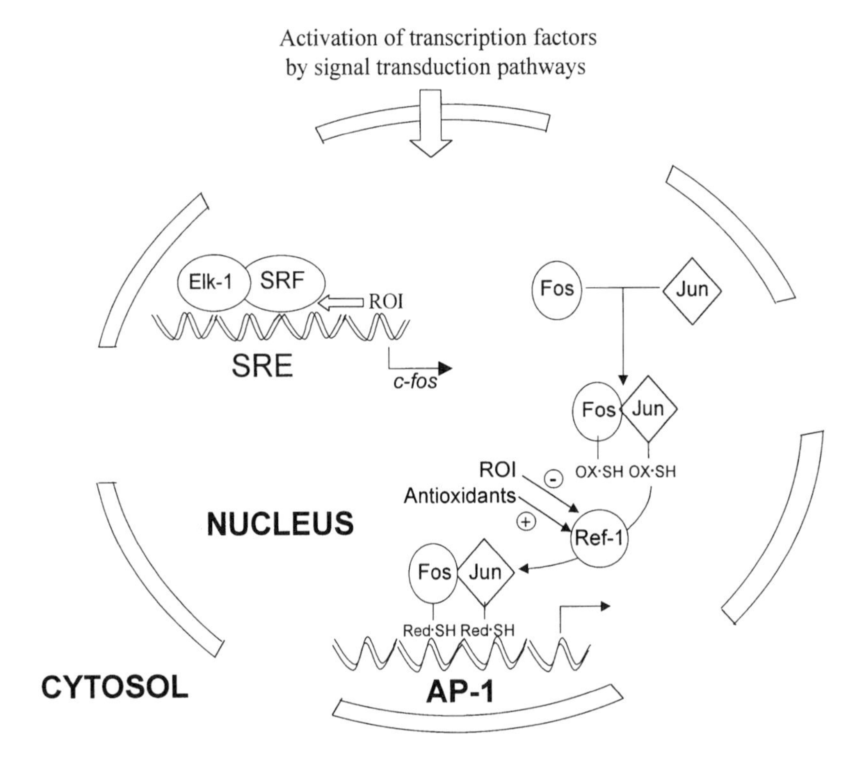


**Figure 6** Cell signaling pathways induce the transcription of the c-fos and c-jun genes and lead to the translation of the Fos and Jun proteins. ROI are thought to upregulate the translation of Fos mRNA by activating the binding of serum response factor to the serum response element. The binding of the Fos/Jun complex to the DNA binding site is induced by the reduction of a crucial cysteine group present on each subunit. The protein, Ref-1, is responsible for reduction of these cysteine groups and thus the activation of DNA binding. ROI are thought to participate in the regulation of Ref-1 by inactivating the protein, whereas antioxidants have been suggested to have the opposite effect. Abbreviations: Reduced cysteine (Red SH) and oxidized cysteine (Ox SH).

it is formed from a p50 subunit and a p65 (Rel A) subunit. In the cytoplasm of resting cells, the NF-κB complex is bound to its inhibitory protein, IκB. Upon stimulation of the cells, IκB dissociates from the NF-κB complex, revealing the nuclear location sequences in p65 and p50. The free NF-κB complex then translocates to the nucleus, where it regulates gene transcription through its binding to the kappa B motif of the target gene (194). An array of agents has been shown to stimulate NF-κB activation, including TNF-α, IL-1, phorbol esters, viruses, LPS, calcium ionophores, cyclohexamide, and ionizing radiation. Furthermore, the activation of NF-κB is redox-controlled and various oxidants have been implicated in its activation, such as $H_2O_2$ (195) and oxidized low-density lipoprotein (LDL) (196). Chen et al. (197) were the first to identify a large kinase that phosphorylates IκB at serine residues, S36 and S32. More recently, several groups have identified an IκB kinase (IKK) complex, which contains two kinase molecules, IKKα and IKKβ. It is unclear whether the latter kinase is the same as the one studied by Chen et al. (197). ROS have been shown to be important in phosphorylation of IκB, which is necessary for the dissociation of IκB from NF-κB, although the precise mechanism by which ROS controls

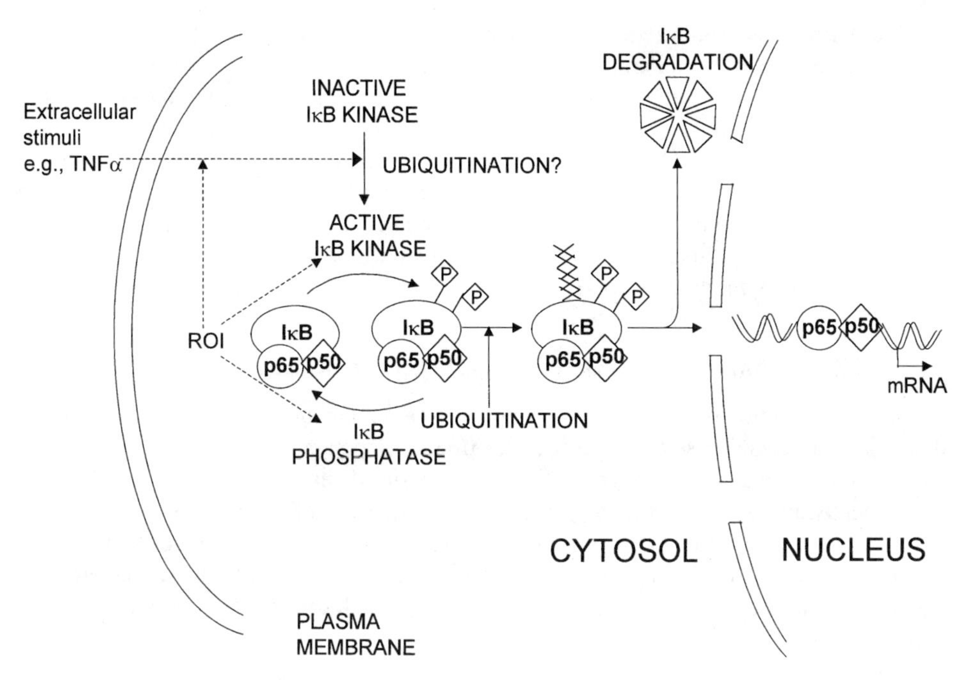


**Figure 7** Extracellular stimuli activate IκB kinase via specific signaling pathways. Activated IκB kinase phosphorylates IκB, the inhibitory subunit of NF-κB. In turn, this causes the ubiquitination of IκB and its consequent proteolytic degradation. The free NF-κB complex then translocates to the nucleus, where it regulates gene transcription through its binding to the kappa B motif of the target gene. ROS may participate in the activation of NF-κB by activating IκB kinase (directly or through signaling pathways), or possibly via the suppression of phosphatase activity.

phosphorylation is unclear (198). ROS may participate in the activation of NF-κB by activating IκB kinase directly or through signaling pathways, or possibly through the suppression of phosphatase activity (Fig. 7). However, ROS do not appear to be involved in the ubiquitination of IκB, which is essential for the proteolytic degradation of IκB and thus the translocation of NF-κB to the nucleus.

## C. AP-1 and NF-κB in Inflammatory Diseases

AP-1 and NF-κB have been implicated in a variety of disease states including rheumatoid arthritis, osteoarthritis, and ischemia-reperfusion injury. In arthritis, they have been used to show that the DNA binding activities of AP-1 and NF-κB were significantly higher in RA than OA synoval tissue (199). Asahara et al. (200), confirmed this when they indicated that the DNA binding activity of AP-1 was increased in synovial tissues from RA patients and that AP-1 activation correlated with disease activity. Additionally, the expression of c-fos and c-jun has been shown to be elevated in RA more than OA (199). Kinne and colleagues (201) detected c-fos/c-jun expression in the cells present within the lining layer and diffuse infiltrates of the synovial membrane. The expression of NF-κB was also elevated in both RA and OA when compared to controls (199). Our studies using immunohistochemical staining have demonstrated that the activated form of NF-κB is present in both vascular endothelial cells and macrophage-like cells in the rheumatoid synovuim (202).

In terms of ischemia-reperfusion injury, the DNA binding of AP-1 has been shown to increase during ischemia and decrease after reperfusion in a rat model of myocardial ischemia (203). Shimizu et al. (204) confirmed these findings when they discovered that AP-1 and NF-κB binding was elevated in areas of myocardial infarction in rats. Furthermore, the mRNA levels of Ref-1 become elevated during the early stages of hypoxia and these increased levels persist until the cells have been reperfused (205). These findings are consistent with the concept that the reperfusion of an ischemic tissue generates oxidizing conditions (via the production of free radicals) that would inactivate Ref-1 and in turn decrease the DNA binding of AP-1.

## VI. TISSUE DAMAGE CAUSED BY RADICALS IN DISEASE

Excessive production of free radicals is known to cause damage to DNA, lipids, carbohydrates, and proteins. These reactions between tissue components and free radicals generate stable products of tissue damage. Indicators of tissue damage are generally stable and consequently they are frequently measured to implicate free radicals in many disease states (although for the purpose of this section, the examples used will relate to the measurement of these markers in RA). Disease progression is not only affected by tissue damage but can be influenced by the products of tissue damage, which can regulate the inflammatory process.

### A. DNA Damage

Free radical production in inflammation is known to cause DNA damage by fragmenting DNA chains. For example, 8-hydroxydeoxyguanosine (a product of ROS-mediated DNA fragmentation) is significantly elevated in lymphocyte DNA from RA patients when compared to healthy controls (206,207). This damage activates p53, which has been shown to accumulate in cells exposed to radiation, UV light, and free radicals (208,209). Furthermore, the expression of p53 in synoval tissue was also found to be significantly higher in RA patients when compared to OA patients and controls (210).

Activation of p53 allows time for DNA repair by arresting cell growth. If the DNA damage is extensive and cannot be repaired, the cells may be killed by apoptosis. Largely this ensures that the damaged DNA is not replicated and that mutations do not materialize. However, mutated cells can survive especially if apoptosis is impaired. As a result, p53 mutations have been detected in inflammatory diseases (e.g., p53 mutations have been found in RA synoval tissue but not in control synovial tissue) (211).

Overall free-radical-induced DNA damage may lead to p53 mutations. These mutations could potentially cause proinflammatory changes such as an excessive production of metalloproteinases and cytokines in the RA synovium. In fact, Tak et al. (210) have suggested that the transformed cells found in the synovium may contribute to the invasion of the pannus and joint destruction.

### B. Lipid Damage

It is well recognized that free radicals attack fatty acids present in cell membranes including linoleic acid, arachidonic acid, and other polyunsaturated fatty acids. The presence of redox active metals such as copper and iron can facilitate the initiation of lipid peroxidation. Overall this leads to the production of secondary radicals, such as peroxyl and alkoxyl radicals, conjugated dienes, and lipid hydroperoxides. These products are often used as markers of lipid peroxidation; for instance, Merry et al. (212) demonstrated that lipid peroxidation was elevated in a group of exercised RA patients by measuring conjugated dienes. Degradation of lipid hydroperoxides generates toxic aldehydes, most commonly malondialdehyde and 4-hydroxynonenal. Malondialdehyde and other lipid peroxidation products react with thiobarbituric acid to form thiobarbituric acid reactive species (TBARs), which can be measured by UV spectrophotometry. In fact, this method has been used to demonstrate an increase in TBARs in exercised RA patients when compared to controls (213). In addition, TBARs and iron concentration have been shown to correlate with disease activity in RA (214). Other studies have also shown that 4-hydroxynonenal is present in rheumatoid synovial fluid and serum and that the levels correlate with disease activity (215).

These findings become relevant when they are related to the effects that lipid peroxidation products have on the inflammatory process. For example, free radical attack of lipid membranes causes membrane disorganization and loss of membrane fluidity. Furthermore, cytotoxic aldehydes are thought to affect inflammatory process by blocking macrophage activity, affecting chemotaxis and enzyme activity, inhibiting protein synthesis, and T-cell recognition of antigens (216). Furthermore, 4-hydroxynonenal is biologically active and can cause severe cell dysfunction both biochemically and genetically. It also acts as a chemotaxin for PMNs at picomolar concentrations and can inhibit cell proliferation (217). Thus, these products of lipid peroxidation may act as proinflammatory and anti-inflammatory regulators of inflammatory diseases.

### C. Carbohydrate Damage

Synovial fluid contains a polysaccharide known as hyaluronic acid, which forms the central axis of the proteoglycans and maintains the viscosity of synovial fluid. In the rheumatoid joint, conditions of oxidative stress cause the breakdown of the glycosidic linkages and fragmentation of hyaluronic acid. Consequently, this leads to reduced viscosity of the synovial fluid and destabilization of the connective tissue. In vitro studies performed by Grootveld et al. (218) defined several of the intermediates and end products of radical-induced hyaluronate fragmentation. Identical products were then found to be elevated in RA patients' synovial fluid after exercise (219). In this case, free radical damage to hyaluronic acid is thought to be ˙OH mediated mainly due to the presence of formate in the synovial fluid. More recently, glycosidic bond breakages and subsequent formate generation have been demonstrated in cartilage exposed to free radicals (220).

In addition, free radicals are known to react with a number of sugars, including glucose, mannitol, and deoxy sugars. Schiller et al. (220) have demonstrated that exposure of glucose solutions to oxygen free radicals leads to the formation of formate and malondialdehyde, suggesting ˙OH-mediated damage. Furthermore, this group has demonstrated that hypochlorous acid attacks the *N*-acetyl groups of the polymeric carbohydrates of cartilage, like hyaluronic acid and *N*-acetylglucosamine (221). Increased levels of acetate and *N*-acetyl groups (breakdown products of *N*-acetylglucosamine) were detected in RA synovial fluid and these levels correlated with myeloperoxidase activity, therefore suggesting that HOCl is also involved in cartilage destruction (221).

### D. Protein Damage

In biological systems, free radical damage to proteins is irreversible and often leads to protein unfolding. Peroxyl and hydroxyl radicals formed during inflammation can cause amino acid oxidation, decarboxylation, deamination, and structural damage such as polypeptide chain scission and cross-linking. Alterations to certain proteins can have an influence on the inflammatory process. For example, peroxynitrite is known to deposit nitrate tyrosine residues on proteins, which prevents the phosphorylation of these residues and thereby affects the cellular regulation and signal transduction pathways of inflammation (222). In addition, free-radical-induced protein damage can make proteins more susceptible to degradation by proteinases (e.g., TauCl modifies proteins by transchlorination and facilitates their degradation by proteinases) (80).

Many proteins are known to be susceptible to free radical damage. Immunoglobin G (IgG) and α-1-antitrypsin are good examples that are relevant to inflammatory diseases like RA.

### 1. *IgG*

Rheumatoid factor is an autoantibody directed against ''self'' IgG, which is regularly found in the sera of patients with autoimmune diseases. The antigenicity of ''self'' IgG is caused by an alteration in the immunoglobin structure. In vitro studies have demonstrated that exposure of IgG to radicals causes protein denaturation and aggregation, suggesting that free radicals may be involved in IgG damage in vivo (223). In fact, our group has demonstrated that allergic inflammation could be maintained by radical-altered IgG in the rat air pouch model (224). In ROS-induced inflammation, a decrease in the galactosylation of IgG correlated with the peak in oxidative damage, suggesting that ROS may be acting by destroying the galactose moiety of IgG (225,226). However, oxidation of cysteine, tryptophan, tyrosine, and lysine amino residues found in IgG can cause the formation of autofluorescent forms of monomeric and polymeric IgG. Furthermore, elevated levels of fluorescent IgG have been detected in RA patients after exercise (227).

A range of proinflammatory actions appears to result from the formation of ROS-altered IgG. For example, Swaak and colleagues (228) demonstrated that ROS-altered IgG has an increased ability to bind the first component of the complement cascade. In addition, the production of complexes consisting of rheumatoid factor and fluorescent IgG have been shown to stimulate the production of $O_2^{\cdot-}$ from PMNs (229).

### 2. *α-Antitrypsin*

In inflammation, the elastase secreted by activated PMNs can cause a considerable degree of tissue damage. In vivo, inhibitors such as α-antitrypsin are present that control elastase activity. However, in inflammatory diseases such as RA, α-antitrypsin can become inactivated by proteolytic cleavage by metalloproteinases or by free-radical-mediated oxidation (230,231). Both metalloproteinases and free radicals inactivate α-antitrypsin by targeting the functional methionine residue at its active site (231,232). In rheumatoid synovial fluid, products of enzyme cleavage and oxidation have been detected, indicating that this is occurring in the rheumatoid joint (232). In terms of joint damage, the imbalance caused by α-antitrypsin inactivation would result in an increase in elastase activity, and therefore elevate tissue destruction seen at the inflammatory site.

## VII. CONCLUSION

Free radicals modulate cellular mechanisms which are central to aspects of the inflammatory process such as the adhesion of leukocytes, prostaglandin synthesis, and the production of inflammatory mediators such as TNF-α and IL-1. In some cases, they indirectly exert their effects by activating transcription factors, while in other cases they directly interact with the cellular component to control its activity. The mechanisms regulating the in vivo concentration of radicals also seem, in part, to be controlled by the radicals themselves. For example, $NO^{\bullet}$ inhibits $O_2^{\bullet-}$ production by inhibiting NADPH oxidase. In the case of $NO^{\bullet}$, it appears that its concentration is critical in determining whether it acts as a signaling molecule and antioxidant, or a species responsible for causing tissue damage. Although the body's antioxidant defense system provides protection against radical-induced tissue damage, antioxidants often become depleted in tissues that have become chronically inflamed or ischemic. Consequently, this imbalance in the antioxidant/oxidant ratio leads to an excess of free radicals, which mediates damage to lipids, proteins, carbohydrates, and DNA. In RA, the products of tissue damage further activate the inflammatory response and elevate tissue and bone destruction.

Over the last 20 years, our knowledge of free radicals and their involvement in inflammatory diseases has grown rapidly. Hopefully, this knowledge and future findings will lead to the identification of regulatory steps that can be targeted by new and existing antioxidant drugs that will control chronic inflammatory diseases like RA. However, the way the body normally controls the fine balance between the concentration of radicals and antioxidants will be difficult to reestablish by traditional pharmacological approaches to disease therapy.

## REFERENCES

1. Winyard PG, Blake DR, Evans CH. Free radicals and inflammation, 1st ed. Berlin: Birkhauser, 1999.
2. Underwood JCE. Pathology, 1st ed. Edinburgh: Churchill Livingstone, 1992.
3. Rang HP, Dale MM, Ritter JM. Pharmacology, 4th ed. Edinburgh: Churchill Livingstone, 1999.
4. White M. Mediators of inflammation and the inflammatory process. J All Clin Immunol 1999; 103(3 Pt 2):S378–81.
5. Kaya H, Hong SL. Bradykinin-induced activation of phospholipase A2 is independent of the activation of polyphosphoinositide-hydrolyzing phospholipase C. Adv Prostagland Thrombox Leukotr Res 1989; 19:580–583.
6. Gallin JI, Snyderman R. Inflammation: Basic principles and clinical correlates, 3rd ed. Philadelphia: Lippincott Williams & Wilkins, 1999.
7. Trowbridge HO. Inflammation: a review of the Process, 5th ed. Chicago: Quintessence, 1997.
8. Roitt IM, Brostoff J, Male DK. Immunology, 3rd ed. London: Gower Medical Publishers, 1993.
9. Panayi GS. T-cell-dependent pathways in rheumatoid arthritis. Cur Opin Rheumatol 1997; 9(3):236–240.
10. Fox DA. The role of T cells in the immunopathogenesis of rheumatoid arthritis: new perspectives. Arthritis Rheum 1997; 40(4):598–609.
11. Zvaifler NJ, Firestein GS. Pannus and pannocytes. Alternative models of joint destruction in rheumatoid arthritis. Arthritis Rheum 1994; 37(6):783–789.
12. Boots AM, Wimmers-Bertens AJ, Rijnders AW. Antigen-presenting capacity of rheumatoid synovial fibroblasts. Immunology 1994; 82(2):268–274.
13. Proudman SM, Cleland LG, Mayrhofer G. Effects of tumor necrosis factor-alpha, interleukin 1beta, and activated peripheral blood mononuclear cells on the expression of adhesion molecules and recruitment of leukocytes in rheumatoid synovial xenografts in SCID mice. J Rheumatol 1999; 26(9):1877–1889.
14. Mino T, Sugiyama E, Taki H, Kuroda A, Yamashita N, Maruyama M, Kobayashi M.

Interleukin-1-alpha and tumor necrosis factor-alpha synergistically stimulate prostaglandin E2-dependent production of interleukin-11 in rheumatoid synovial fibroblasts. Arthritis Rheum 1998; 41(11):2004–2013.
15. Maini RN, Elliott M, Brennan FM, Williams RO, Feldmann M. TNF blockade in rheumatoid arthritis: implications for therapy and pathogenesis. APMIS 1997; 105(4):257–263.
16. Broker BM, Edwards JC, Fanger MW, Lydyard PM. The prevalence and distribution of macrophages bearing Fc gamma R I, Fc gamma R II, and Fc gamma R III in synovium. Scand J Rheumatol 1990; 19(2):123–135.
17. Remmers EF, Sano H, Wilder RL. Platelet-derived growth factors and heparin-binding (fibroblast) growth factors in the synovial tissue pathology of rheumatoid arthritis. Sem Arth Rheumatol 1991; 21(3):191–199.
18. Gay S, Gay RE, Koopman WJ. Molecular and cellular mechanisms of joint destruction in rheumatoid arthritis: two cellular mechanisms explain joint destruction? Ann Rheum Dis 1993; 52 (suppl 1):S39–47.
19. Li Y, Zhu H, Trush MA. Detection of mitochondria-derived reactive oxygen species production by the chemilumigenic probes lucigenin and luminol. Biochim Biophys Acta 1999; 1428(1):1–12.
20. Dai Y, Rashba-Step J, Cederbaum AI. Stable expression of human cytochrome P4502E1 in HepG2 cells: characterization of catalytic activities and production of reactive oxygen intermediates. Biochemistry 1993; 32(27):6928–6937.
21. McCord JM. Oxygen-derived free radicals in postischemic tissue injury. N Engl J Med 1985; 312(3):159–163.
22. Thrasher AJ, Keep NH, Wientjes F, Segal AW. Chronic granulomatous disease. Biochim Biophys Acta 1994; 1227(1–2):1–24.
23. Gabig TG, Bearman SI, Babior BM. Effects of oxygen tension and pH on the respiratory burst of human neutrophils. Blood 1979; 53(6):1133–1139.
24. Babior BM. NADPH oxidase: an update. Blood 1999; 93(5):1464–1476.
25. Huang J, Kleinberg ME. Activation of the phagocyte NADPH oxidase protein p47(phox). Phosphorylation controls SH3 domain-dependent binding to p22(phox). J Biol Chem 1999; 274(28):19731–19737.
26. Fuchs A, Bouin AP, Rabilloud T, Vignais PV. The 40-kDa component of the phagocyte NADPH oxidase (p40phox) is phosphorylated during activation in differentiated HL60 cells. Eur J Biochem 1997; 249(2):531–539.
27. Benna JE, Dang PM, Gaudry M, Fay M, Morel F, Hakim J, Gougerot-Pocidalo MA. Phosphorylation of the respiratory burst oxidase subunit p67(phox) during human neutrophil activation. Regulation by protein kinase C-dependent and independent pathways. J Biol Chem 1997; 272(27):17204–17208.
28. Siems WG, Capuozzo E, Verginelli D, Salerno C, Crifo C, Grune T. Inhibition of NADPH oxidase-mediated superoxide radical formation in PMA-stimulated human neutrophils by 4-hydroxynonenal—binding to -SH and -NH2 groups. Free Rad Res 1997; 27(4):353– 358.
29. Fujii H, Ichimori K, Hoshiai K, Nakazawa H. Nitric oxide inactivates NADPH oxidase in pig neutrophils by inhibiting its assembling process. J Biol Chem 1997; 272(52):32773–32778.
30. Ding J, Knaus UG, Lian JP, Bokoch GM, Badwey JA. The renaturable 69- and 63-kDa protein kinases that undergo rapid activation in chemoattractant-stimulated guinea pig neutrophils are p21-activated kinases. J Biol Chem 1996; 271(40):24869–24873.
31. Chapple IL. Reactive oxygen species and antioxidants in inflammatory diseases. J Clin Period 1997; 24(5):287–296.
32. Karnovsky MJ. Cytochemistry and reactive oxygen species—a retrospective. Histochemistry 1994; 102:15–27.
33. Tauber AI, Babior BM. Evidence for hydroxyl radical production by human neutrophils. J Clin Invest 1977; 60(2):374–379.

34. Johnston RB Jr, Keele BB Jr, Misra HP, Lehmeyer JE, Webb LS, Baehner RL, Rajagopalan KV. The role of superoxide anion generation in phagocytic bactericidal activity. Studies with normal and chronic granulomatous disease leukocytes. J Clin Invest 1975; 55(6):1357–1372.
35. Cohen MS, Britigan BE, Chai YS, Pou S, Roeder TL, Rosen GM. Phagocyte-derived free radicals stimulated by ingestion of iron-rich Staphylococcus aureus: a spin-trapping study. J Infect Dis 1991; 163(4):819–824.
36. Britigan BE, Edeker BL. Pseudomonas and neutrophil products modify transferrin and lactoferrin to create conditions that favor hydroxyl radical formation. J Clin Invest 1991; 88(4): 1092–1102.
37. Klebanoff SJ, Waltersdorph AM. Prooxidant activity of transferrin and lactoferrin. J Exp Med 1990; 172(5):1293–1303.
38. Reif DW, Simmons RD. Nitric oxide mediates iron release from ferritin. Arch Biochem Biophys 1990; 283(2):537–541.
39. Bolann BJ, Ulvik RJ. Release of iron from ferritin by xanthine oxidase. Role of the superoxide radical. Biochem J 1987; 243(1):55–59.
40. Gambhir JK, Lali P, Jain AK. Correlation between blood antioxidant levels and lipid peroxidation in rheumatoid arthritis. Clin Biochem 1997; 30(4):351–355.
41. Carter WO, Narayanan PK, Robinson JP. Intracellular hydrogen peroxide and superoxide anion detection in endothelial cells. J Leukotr Biol 1994; 55(2):253–258.
42. Britigan BE, Rosen GM, Thompson BY, Chai Y, Cohen MS. Stimulated human neutrophils limit iron-catalyzed hydroxyl radical formation as detected by spin-trapping techniques. J Biol Chem 1986; 261(36):17026–17032.
43. Sumii H, Inoue H, Onoue J, Mori A, Oda T, Tsubokura T. Superoxide dismutase activity in arthropathy: its role and measurement in the joints. Hiroshima J Med Sci 1996; 45(2): 51–55.
44. Mazzetti I, Grigolo B, Borzi RM, Meliconi R, Facchini A. Serum copper/zinc superoxide dismutase levels in patients with rheumatoid arthritis. Int J Clin Lab Res 1996; 26(4):245–249.
45. Pronai L, Ichikawa Y, Ichimori K, Nakazawa H, Arimori S. Association of enhanced superoxide generation by neutrophils with low superoxide scavenging activity of the peripheral blood, joint fluid, and their leukocyte components in rheumatoid arthritis: effects of slow-acting anti-rheumatic drugs and disease activity. Clin Exp Rheumatol 1991; 9(2):149–155.
46. Davidge ST, Baker PN, Laughlin MK, Roberts JM. Nitric oxide produced by endothelial cells increases production of eicosanoids through activation of prostaglandin H synthase. Circ Res 1995; 77(2):274–283.
47. Inoue T, Fukuo K, Morimoto S, Koh E, Ogihara T. Nitric oxide mediates interleukin-1-induced prostaglandin E2 production by vascular smooth muscle cells. Biochem Biophys Res Commun 1993; 194(1):420–424.
48. Buttke TM, Sandstrom PA. Oxidative stress as a mediator of apoptosis. Immunol Today 1994; 15(1):7–10.
49. McIntyre TM, Patel KD, Zimmerman GA, Prescott SM. Oxygen radical mediated leukocyte adherence. In: Granger DN, Schmid-Schonbein GW, eds. Physiology and Pathophysiology of Leukocyte Adhesion. New York: Oxford University Press, 1995:261–277.
50. Bienvenu K, Russell J, Granger DN. Platelet-activating factor promotes shear rate-dependent leukocyte adhesion in postcapillary venules. J Lipid Med 1993; 8(2):95–103.
51. Siebenlist U, Franzoso G, Brown K. Structure, regulation and function of NF-kappa B. Ann Rev Cell Biol 1994; 10:405–455.
52. Collins T, Read MA, Neish AS, Whitley MZ, Thanos D, Maniatis T. Transcriptional regulation of endothelial cell adhesion molecules: NF-kappa B and cytokine-inducible enhancers. FASEB J 1995; 9(10):899–909.
53. Peng HB, Libby P, Liao JK. Induction and stabilization of I kappa B alpha by nitric oxide mediates inhibition of NF-kappa B. J Biol Chem 1995; 270(23):14214–14219.

54. Kurose I, Wolf R, Grisham MB, Aw TY, Specian RD, Granger DN. Microvascular responses to inhibition of nitric oxide production. Role of active oxidants. Circ Res 1995; 76(1):30–39.
55. Albrich JM, Gilbaugh JH 3d, Callahan KB, Hurst JK. Effects of the putative neutrophil-generated toxin, hypochlorous acid, on membrane permeability and transport systems of Escherichia coli. J Clin Invest 1986; 78(1):177–184.
56. Barrette WC Jr, Hannum DM, Wheeler WD, Hurst JK. General mechanism for the bacterial toxicity of hypochlorous acid: abolition of ATP production. Biochemistry 1989; 28(23): 9172–9178.
57. Rosen H, Orman J, Rakita RM, Michel BR, VanDevanter DR. Loss of DNA-membrane interactions and cessation of DNA synthesis in myeloperoxidase-treated Escherichia coli. Proc Natl Acad Sci USA 1990; 87(24):10048–10052.
58. Bainton DF, Ullyot JL, Farquhar MG. The development of neutrophilic polymorphonuclear leukocytes in human bone marrow. J Exp Med 1971; 134(4):907–34.
59. Hazen SL, Hsu FF, Mueller DM, Crowley JR, Heinecke JW. Human neutrophils employ chlorine gas as an oxidant during phagocytosis. J Clin Invest 1996; 98(6):1283–1289.
60. Foote CS, Goyne TE, Lehrer RI. Assessment of chlorination by human neutrophils. Nature 1983; 301(5902):715–716.
61. Weiss SJ, Klein R, Slivka A, Wei M. Chlorination of taurine by human neutrophils. Evidence for hypochlorous acid generation. J Clin Invest 1982; 70(3):598–607.
62. Kettle AJ, Winterbourn CC. Myeloperoxidase: a key regulator of neutrophil oxidant production. Redox Rep 1997; 3(1):3–15.
63. Thomas EL, Fishman M. Oxidation of chloride and thiocyanate by isolated leukocytes. J Biol Chem 1986; 261(21):9694–9702.
64. van Dalen CJ, Whitehouse MW, Winterbourn CC, Kettle AJ. Thiocyanate and chloride as competing substrates for myeloperoxidase. Biochem J 1997; 327 (Pt 2):487–492.
65. Eiserich JP, Hristova M, Cross CE, Jones AD, Freeman BA, Halliwell B, van der Vliet A. Formation of nitric oxide-derived inflammatory oxidants by myeloperoxidase in neutrophils. Nature 1998; 391(6665):393–397.
66. Edwards SW, Nurcombe HL, Hart CA. Oxidative inactivation of myeloperoxidase released from human neutrophils. Biochem J 1987; 245(3):925–928.
67. Naskalski JW. Myeloperoxidase inactivation in the course of catalysis of chlorination of taurine. Biochim Biophys Acta 1977; 485(2):291–300.
68. Kettle AJ, Gedye CA, Winterbourn CC. Superoxide is an antagonist of antiinflammatory drugs that inhibit hypochlorous acid production by myeloperoxidase. Biochem Pharmacol 1993; 45(10):2003–2010.
69. Kettle AJ, Winterbourn CC. Influence of superoxide on myeloperoxidase kinetics measured with a hydrogen peroxide electrode. Biochem J 1989; 263(3):823–828.
70. Bolscher BG, Wever R. The nitrosyl compounds of ferrous animal haloperoxidases. Biochim Biophys Acta 1984; 791(1):75–81.
71. Floris R, Piersma SR, Yang G, Jones P, Wever R. Interaction of myeloperoxidase with peroxynitrite. A comparison with lactoperoxidase, horseradish peroxidase and catalase. Eur J Biochem 1993; 215(3):767–775.
72. Shibata H, Kono Y, Yamashita S, Sawa Y, Ochiai H, Tanaka K. Degradation of chlorophyll by nitrogen dioxide generated from nitrite by the peroxidase reaction. Biochim Biophys Acta 1995, 1230:45–50.
73. Thomas EL. Myeloperoxidase-hydrogen peroxide-chloride antimicrobial system: effect of exogenous amines on antibacterial action against Escherichia coli. Infect Immunol 1979; 25(1):110–116.
74. Marcinkiewicz J, Chain BM, Olszowska E, Olszowski S, Zgliczynski JM, Enhancement of immunogenic properties of ovalbumin as a result of its chlorination. Int J Biochem 1991; 23(12):1393–1395.

75. Smith JA. Neutrophils, host defense, and inflammation: a double-edged sword. J Leukotr Biol 1994; 56(6):672–686.
76. Ramos CL, Pou S, Britigan BE, Cohen MS, Rosen GM. Spin trapping evidence for myeloperoxidase-dependent hydroxyl radical formation by human neutrophils and monocytes. J Biol Chem 1992; 267(12):8307–8312.
77. Hampton MB, Kettle AJ, Winterbourn CC. Inside the neutrophil phagosome: oxidants, myeloperoxidase, and bacterial killing. Blood 1998; 92(9):3007–3017.
78. Weiss SJ. Tissue destruction by neutrophils. N Engl J Med 1989; 320(6):365–376.
79. Suomalainen K, Sorsa T, Lindy O, Saari H, Konttinen YT, Uitto VJ. Hypochlorous acid induced activation of human neutrophil and gingival crevicular fluid collagenase can be inhibited by ascorbate. Scand J Den Res 1991; 99(5):397–405.
80. Zgliczynski JM, Stelmaszynska T, Domanski J, Ostrowski W. Chloramines as intermediates of oxidation reaction of amino acids by myeloperoxidase. Biochim Biophys Acta 1971; 235(3):419–424.
81. Marcinkiewicz J, Grabowska A, Bereta J, Stelmaszynska T. Taurine chloramine, a product of activated neutrophils, inhibits in vitro the generation of nitric oxide and other macrophage inflammatory mediators. J Leukotr Biol 1995; 58(6):667–674.
82. Tatsumi T, Fliss H. Hypochlorous acid and chloramines increase endothelial permeability: possible involvement of cellular zinc. Am J Phys 1994; 267(4 Pt 2):H1597–1607.
83. Furchgott RF. An historical survey and prospects of research on EDRF. Nippon Heikatsukin Gakkai Zasshi 1987; 23(6):435–440.
84. Ignarro LJ, Byrns RE, Buga GM, Wood KS. Endothelium-derived relaxing factor from pulmonary artery and vein possesses pharmacologic and chemical properties identical to those of nitric oxide radical. Circ Res 1987; 61(6):866–879.
85. Palmer RM, Ferrige AG, Moncada S. Nitric oxide release accounts for the biological activity of endothelium-derived relaxing factor. Nature 1987; 327(6122):524–526.
86. Hibbs JB Jr, Taintor RR, Vavrin Z, Rachlin EM. Nitric oxide: a cytotoxic activated macrophage effector molecule. Biochem Biophys Res Commun 1988; 157(1):87–94.
87. Nathan CF, Hibbs JB Jr. Role of nitric oxide synthesis in macrophage antimicrobial activity. Curr Opin Immunol 1991; 3(1):65–70.
88. Malawista SE, Montgomery RR, van Blaricom G. Evidence for reactive nitrogen intermediates in killing of staphylococci by human neutrophil cytoplasts. A new microbicidal pathway for polymorphonuclear leukocytes. J Clin Invest 1992; 90(2):631–636.
89. Kroncke KD, Fehsel K, Kolb-Bachofen V. Nitric oxide: cytotoxicity versus cytoprotection—how, why, when, and where? Nitric Oxide 1997; 1(2):107–120.
90. Feelisch M, Stampler JS. Methods of Nitric Oxide Research, 1st ed. New York: Wiley, 1996.
91. Klatt P, Schmidt K, Werner ER, Mayer B. Determination of nitric oxide synthase cofactors: heme, FAD, FMN, and tetrahydrobiopterin. Meth Enzymol 1996; 268:358–365.
92. Kiechle FL, Malinski T. Nitric oxide. Biochemistry, pathophysiology, and detection. Am J Clin Pathol 1993; 100(5):567–575.
93. Moncada S. The L-arginine: nitric oxide pathway. Acta Phys Scand 1992; 145(3):201–227.
94. Reiling N, Ulmer AJ, Duchrow M, Ernst M, Flad HD, Hauschildt S. Nitric oxide synthase: mRNA expression of different isoforms in human monocytes/macrophages. Eur J Immunol 1994; 24(8):1941–1944.
95. Amin AR, Attur M, Vyas P, Leszczynska-Piziak J, Levartovsky D, Rediske J, Clancy RM, Vora KA, Abramson SB. Expression of nitric oxide synthase in human peripheral blood mononuclear cells and neutrophils. J Inflamm 1995; 47(4):190–205.
96. Zhang Z, Naughton D, Winyard PG, Benjamin N, Blake DR, Symons MC. Generation of nitric oxide by a nitrite reductase activity of xanthine oxidase: a potential pathway for nitric oxide formation in the absence of nitric oxide synthase activity. Biochem Biophys Res Commun 1998; 249(3):767–772.
97. Millar TM, Stevens CR, Benjamin N, Eisenthal R, Harrison R, Blake DR. Xanthine oxidore-

ductase catalyses the reduction of nitrates and nitrite to nitric oxide under hypoxic conditions. FEBS Lett 1998; 427(2):225–228.
98. Wink DA, Mitchell JB. Chemical biology of nitric oxide: Insights into regulatory, cytotoxic, and cytoprotective mechanisms of nitric oxide. Free Rad Biol Med 1998; 25(4–5):434–456.
99. Gorbunov NV, Osipov AN, Day BW, Zayas-Rivera B, Kagan VE, Elsayed NM. Reduction of ferrylmyoglobin and ferrylhemoglobin by nitric oxide: a protective mechanism against ferryl hemoprotein-induced oxidations. Biochemistry 1995; 34(20):6689–6699.
100. Wink DA, Hanbauer I, Laval F, Cook JA, Krishna MC, Mitchell JB. Nitric oxide protects against the cytotoxic effects of reactive oxygen species. Ann NY Acad Sci 1994; 738:265–278.
101. Stamler JS, Singel DJ, Loscalzo J. Biochemistry of nitric oxide and its redox-activated forms. Science 1992; 258(5090):1898–1902.
102. Jia L, Bonaventura C, Bonaventura J, Stamler JS. S-nitrosohaemoglobin: a dynamic activity of blood involved in vascular control. Nature 1996; 380(6571):221–226.
103. Hou Y, Guo Z, Li J, Wang PG. Seleno compounds and glutathione peroxidase catalyzed decomposition of S-nitrosothiols. Biochem Biophys Res Commun 1996; 228(1):88–93.
104. De Groote MA, Granger D, Xu Y, Campbell G, Prince R, Fang FC. Genetic and redox determinants of nitric oxide cytotoxicity in a Salmonella typhimurium model. Proc Natl Acad Sci USA 1995; 92(14):6399–6403.
105. Jourd'heuil D, Mai CT, Laroux FS, Wink DA, Grisham MB. The reaction of S-nitrosoglutathione with superoxide. Biochem Biophys Res Commun 1998; 246(3):525–530.
106. Carmichael AJ, Steel-Goodwin L, Gray B, Arroyo CM. Reactions of active oxygen and nitrogen species studied by EPR and spin trapping. Free Rad Res Commun 1993; 19(suppl 1):S1–16.
107. Beckman J, Tsai J. Reactions and diffusion of nitric oxide and peroxynitrite. Biochemist 1994; 8–10.
108. Ischiropoulos H, Zhu L, Beckman JS. Peroxynitrite formation from macrophage-derived nitric oxide. Arch Biochem Biophys 1992; 298(2):446–451.
109. Carreras MC, Pargament GA, Catz SD, Poderoso JJ, Boveris A. Kinetics of nitric oxide and hydrogen peroxide production and formation of peroxynitrite during the respiratory burst of human neutrophils. FEBS Lett 1994; 341(1):65–68.
110. Kooy NW, Royall JA, Ischiropoulos H, Beckman JS. Peroxynitrite-mediated oxidation of dihydrorhodamine 123. Free Rad Biol Med 1994; 16(2):149–156.
111. O'Donnell VB, Eiserich JP, Chumley PH, Jablonsky MJ, Krishna NR, Kirk M, Barnes S, Darley-Usmar VM, Freeman BA. Nitration of unsaturated fatty acids by nitric oxide-derived reactive nitrogen species peroxynitrite, nitrous acid, nitrogen dioxide, and nitronium ion. Chem Res Tox 1999; 12(1):83–92.
112. Radi R, Beckman JS, Bush KM, Freeman BA. Peroxynitrite oxidation of sulfhydryls. The cytotoxic potential of superoxide and nitric oxide. J Biol Chem 1991; 266(7):4244–4250.
113. Herce-Pagliai C, Kotecha S, Shuker DE. Analytical methods for 3-nitrotyrosine as a marker of exposure to reactive nitrogen species: a review. Nitric Oxide 1998; 2(5):324–336.
114. Li X, De Sarno P, Song L, Beckman JS, Jope RS. Peroxynitrite modulates tyrosine phosphorylation and phosphoinositide signalling in human neuroblastoma SH-SY5Y cells: attenuated effects in human 1321N1 astrocytoma cells. Biochem J 1998; 331 (Pt2):599–606.
115. Amin AR, Attur M, Patel RN, Thakker GD, Marshall PJ, Rediske J, Stuchin SA, Patel IR, Abramson SB. Superinduction of cyclooxygenase-2 activity in human osteoarthritis-affected cartilage. Influence of nitric oxide. J Clin Invest 1997; 99(6):1231–1237.
116. Stefanovic-Racic M, Stadler J, Georgescu HI, Evans CH. Nitric oxide synthesis and its regulation by rabbit synoviocytes. J Rheumatol 1994; 21(10):1892–1898.
117. Yui Y, Hattori R, Kosuga K, Eizawa H, Hiki K, Kawai C. Purification of nitric oxide synthase from rat macrophages. J Biol Chem 1991; 266(19):12544–12547.

118. McCall TB, Boughton-Smith NK, Palmer RM, Whittle BJ, Moncada S. Synthesis of nitric oxide from L-arginine by neutrophils. Release and interaction with superoxide anion. Biochem J 1989; 261(1):293–296.
119. Grabowski PS, Macpherson H, Ralston SH. Nitric oxide production in cells derived from the human joint. Br J Rheumatol 1996; 35(3):207–212.
120. Rabinovitch A, Suarez-Pinzon WL, Sorensen O, Bleackley RC. Inducible nitric oxide synthase (iNOS) in pancreatic islets of nonobese diabetic mice: identification of iNOS-expressing cells and relationships to cytokines expressed in the islets. Endocrinology 1996; 137(5):2093–2099.
121. McCartney-Francis N, Allen JB, Mizel DE, Albina JE, Xie QW, Nathan CF, Wahl SM. Suppression of arthritis by an inhibitor of nitric oxide synthase. J Exp Med 1993; 178(2): 749–754.
122. Cannon GW, Openshaw SJ, Hibbs JB Jr, Hoidal JR, Huecksteadt TP, Griffiths MM. Nitric oxide production during adjuvant-induced and collagen-induced arthritis. Arthritis Rheum 1996; 39(10):1677–1684.
123. Stichtenoth DO, Gutzki FM, Tsikas D, Selve N, Bode-Boger SM, Boger RH, Frolich JC. Increased urinary nitrate excretion in rats with adjuvant arthritis. Ann Rheum Dis 1994; 53(8):547–549.
124. Stichtenoth DO, Fauler J, Zeidler H, Frolich JC. Urinary nitrate excretion is increased in patients with rheumatoid arthritis and reduced by prednisolone. Ann Rheum Dis 1995; 54(10):820–824.
125. Farrell AJ, Blake DR, Palmer RM, Moncada S. Increased concentrations of nitrite in synovial fluid and serum samples suggest increased nitric oxide synthesis in rheumatic diseases. Ann Rheum Dis 1992; 51(11):1219–1222.
126. Ueki Y, Miyake S, Tominaga Y, Eguchi K. Increased nitric oxide levels in patients with rheumatoid arthritis. J Rheumatol 1996; 23(2):230–236.
127. Kaur H, Halliwell B. Evidence for nitric oxide-mediated oxidative damage in chronic inflammation. Nitrotyrosine in serum and synovial fluid from rheumatoid patients. FEBS Lett 1994; 350(1):9–12.
128. Hilliquin P, Borderie D, Hernvann A, Menkes CJ, Ekindjian OG. Nitric oxide as S-nitrosoproteins in rheumatoid arthritis. Arthritis Rheum 1997; 40(8):1512–1517.
129. Weinberg JB, Gilkeson GS, Mason RP, Chamulitrat W. Nitrosylation of blood hemoglobin and renal nonheme proteins in autoimmune MRL-lpr/lpr mice. Free Rad Biol Med 1998; 24(1):191–196.
130. Oates JC, Ruiz P, Alexander A, Pippen AM, Gilkeson GS. Effect of late modulation of nitric oxide production on murine lupus. Clin Immun Immunopathol 1997; 83(1):86–92.
131. Stefanovic-Racic M, Stadler J, Evans CH. Nitric oxide and arthritis. Arthritis Rheum 1993; 36(8):1036–1044.
132. Salvemini D, Masferrer JL. Interactions of nitric oxide with cyclooxygenase: in vitro, ex vivo, and in vivo studies. Meth Enzymol 1996; 269:12–25.
133. Manfield L, Jang D, Murrell GA. Nitric oxide enhances cyclooxygenase activity in articular cartilage. Inflamm Res 1996; 45(5):254–258.
134. Tetsuka T, Baier LD, Morrison AR. Antioxidants inhibit interleukin-1-induced cyclooxygenase and nitric-oxide synthase expression in rat mesangial cells. Evidence for post-transcriptional regulation. J Biol Chem 1996; 271(20):11689–11693.
135. Swierkosz TA, Mitchell JA, Warner TD, Botting RM, Vane JR. Co-induction of nitric oxide synthase and cyclo-oxygenase: interactions between nitric oxide and prostanoids. Br J Pharmacol 1995; 114(7):1335–1342.
136. Habib A, Bernard C, Lebret M, Creminon C, Esposito B, Tedgui A, Maclouf J. Regulation of the expression of cyclooxygenase-2 by nitric oxide in rat peritoneal macrophages. J Immunol 1997; 158(8):3845–51.
137. Stadler J, Stefanovic-Racic M, Billiar TR, Curran RD, McIntyre LA, Georgescu HI, Simmons

RL, Evans CH. Articular chondrocytes synthesize nitric oxide in response to cytokines and lipopolysaccharide. J Immunol 1991; 147(11):3915–3920.
138. Kanner J, Harel S, Granit R. Nitric oxide, an inhibitor of lipid oxidation by lipoxygenase, cyclooxygenase and hemoglobin. Lipids 1992; 27(1):46–49.
139. Maccarrone M, Corasaniti MT, Guerrieri P, Nistico G, Finazzi Agro A. Nitric oxide-donor compounds inhibit lipoxygenase activity. Biochem Biophys Res Commun 1996; 219(1):128–133.
140. Rubbo H, Parthasarathy S, Barnes S, Kirk M, Kalyanaraman B, Freeman BA. Nitric oxide inhibition of lipoxygenase-dependent liposome and low-density lipoprotein oxidation: termination of radical chain propagation reactions and formation of nitrogen-containing oxidized lipid derivatives. Arch Biochem Biophys 1995; 324(1):15–25.
141. Padmaja S, Huie RE. The reaction of nitric oxide with organic peroxyl radicals. Biochem Biophys Res Commun 1993; 195(2):539–544.
142. Wink DA, Cook JA, Krishna MC, Hanbauer I, DeGraff W, Gamson J, Mitchell JB. Nitric oxide protects against alkyl peroxide-mediated cytotoxicity: further insights into the role nitric oxide plays in oxidative stress. Arch Biochem Biophys 1995; 319(2):402–407.
143. Frears ER, Zhang Z, Blake DR, O'Connell JP, Winyard PG. Inactivation of tissue inhibitor of metalloproteinase-1 by peroxynitrite. FEBS Lett 1996; 381(1–2):21–24.
144. Abu-Soud HM, Loftus M, Stuehr DJ. Subunit dissociation and unfolding of macrophage NO synthase: relationship between enzyme structure, prosthetic group binding, and catalytic function. Biochemistry 1995; 34(35):11167–11175.
145. Hurshman AR, Marletta MA. Nitric oxide complexes of inducible nitric oxide synthase: spectral characterization and effect on catalytic activity. Biochemistry 1995; 34(16):5627–5634.
146. McInnes IB, Leung BP, Field M, Wei XQ, Huang FP, Sturrock RD, Kinninmonth A, Weidner I, Mumford R, Liew FY. Production of nitric oxide in the synovial membrane of rheumatoid and osteoarthritis patients. J Exp Med 1996; 184(4):1519–1524.
147. Van de Loo FA, Arntz OJ, Van den Berg WB. Effect of interleukin 1 and leukaemia inhibitory factor on chondrocyte metabolism in articular cartilage from normal and interleukin-6-deficient mice: role of nitric oxide and IL-6 in the suppression of proteoglycan synthesis. Cytokine 1997; 9(7):453–462.
148. Stefanovic-Racic M, Morales TI, Taskiran D, McIntyre LA, Evans CH. The role of nitric oxide in proteoglycan turnover by bovine articular cartilage organ cultures. J Immunol 1996; 156(3):1213–1220.
149. Cao M, Westerhausen-Larson A, Niyibizi C, Kavalkovich K, Georgescu HI, Rizzo CF, Hebda PA, Stefanovic-Racic M, Evans CH. Nitric oxide inhibits the synthesis of type-II collagen without altering Col2A1 mRNA abundance: prolyl hydroxylase as a possible target. Biochem J 1997; 324 (Pt 1):305–310.
150. Murrell GA, Jang D, Williams RJ. Nitric oxide activates metalloprotease enzymes in articular cartilage. Biochem Biophys Res Commun 1995; 206(1):15–21.
151. Tamura T, Nakanishi T, Kimura Y, Hattori T, Sasaki K, Norimatsu H, Takahashi K, Takigawa M. Nitric oxide mediates interleukin-1-induced matrix degradation and basic fibroblast growth factor release in cultured rabbit articular chondrocytes: a possible mechanism of pathological neovascularization in arthritis. Endocrinology 1996; 137(9):3729–3737.
152. Trachtman H, Futterweit S, Garg P, Reddy K, Singhal PC. Nitric oxide stimulates the activity of a 72-kDa neutral matrix metalloproteinase in cultured rat mesangial cells. Biochem Biophys Res Commun 1996; 218(3):704–708.
153. Blake DR, Winyard PG. Immunopharmacology of Free Radical Species, 1st ed. San Diego: Academic Press, 1995.
154. Halliwell B, Gutteridge JMC. Free Radicals in Biology and Medicine, 3rd ed. Oxford: Oxford University Press, 1999.
155. Ji LL, Stratman FW, Lardy HA. Antioxidant enzyme systems in rat liver and skeletal muscle.

Influences of selenium deficiency, chronic training, and acute exercise. Arch Biochem Biophys 1988; 263(1):150–160.
156. Lawrence RA, Burk RF. Glutathione peroxidase activity in selenium-deficient rat liver. Biochem Biophys Res Commun 1976; 71(4):952–958.
157. Lunec J, Blake DR. The determination of dehydroascorbic acid and ascorbic acid in the serum and synovial fluid of patients with rheumatoid arthritis (RA). Free Rad Res Commun 1985; 1(1):31–39.
158. Fairburn K, Grootveld M, Ward RJ, Abiuka C, Kus M, Williams RB, Winyard PG, Blake DR. Alpha-tocopherol, lipids and lipoproteins in knee-joint synovial fluid and serum from patients with inflammatory joint disease. Clin Sci 1992; 83(6):657–664.
159. Levine RL. Ischemia: from acidosis to oxidation. FASEB J 1993; 7(13):1242–1246.
160. Hearse DJ. Reperfusion of the ischemic myocardium. J Mol Cell Cardiol 1977; 9(8):605–616.
161. Jennings RB, Schaper J, Hill ML, Steenbergen C Jr, Reimer KA. Effect of reperfusion late in the phase of reversible ischemic injury. Changes in cell volume, electrolytes, metabolites, and ultrastructure. Circ Res 1985; 56(2):262–278.
162. Hearse DJ, Humphrey SM, Nayler WG, Slade A, Border D. Ultrastructural damage associated with reoxygenation of the anoxic myocardium. J Mol Cell Cardiol 1975; 7(5):315–324.
163. Zweier JL. Measurement of superoxide-derived free radicals in the reperfused heart. Evidence for a free radical mechanism of reperfusion injury. J Biol Chem 1988; 263(3):1353–1357.
164. Kramer JH, Arroyo CM, Dickens BF, Weglicki WB. Spin-trapping evidence that graded myocardial ischemia alters post-ischemic superoxide production. Free Rad Biol Med 1987; 3(2):153–159.
165. Kilgore KS, Lucchesi BR. Reperfusion injury after myocardial infarction: the role of free radicals and the inflammatory response. Clin Biochem 1993; 26(5):359–370.
166. Schaffer SW, Roy RS, McMcord JM. Possible role for calmodulin in calcium paradox-induced heart failure. Eur Heart J 1983; 4(suppl H):81–87.
167. Friedl HP, Till GO, Ryan US, Ward PA. Mediator-induced activation of xanthine oxidase in endothelial cells. FASEB J 1989; 3(13):2512–2518.
168. Bulkley GB. Endothelial xanthine oxidase: a radical transducer of inflammatory signals for reticuloendothelial activation. Br J Surg 1993; 80(6):684–686.
169. Maxwell SR, Lip GY. Free radicals and antioxidants in cardiovascular disease. Br J Clin Pharmacol 1997; 44(4):307–317.
170. Werns SW, Shea MJ, Mitso SE, Dysko RC, Fantone JC, Schork MA, Abrams GD, Pitt B, Lucchesi BR. Reduction of the size of infarction by allopurinol in the ischemic-reperfused canine heart. Circulation 1986; 73(3):518–524.
171. Becker LC, Ambrosio G. Myocardial consequences of reperfusion. Prog Card Dis 1987; 30(1):23–44.
172. Goldhaber JI, Weiss JN. Oxygen free radicals and cardiac reperfusion abnormalities. Hypertension 1992; 20(1):118–127.
173. Stewart JR, Blackwell WH, Crute SL, Loughlin V, Hess ML, Greenfield LJ. Prevention of myocardial ischaemia/reperfusion injury with oxygen free-radical scavengers. Surg Forum 1982; 33:317–320.
174. Jayson MI, Dixon AS. Intra-articular pressure in rheumatoid arthritis of the knee. 3. Pressure changes during joint use. Ann Rheum Dis 1970; 29(4):401–408.
175. Blake DR, Merry P, Unsworth J, Kidd BL, Outhwaite JM, Ballard R, Morris CJ, Gray L, Lunec J. Hypoxic-reperfusion injury in the inflamed human joint. Lancet 1989; 1(8633):289–293.
176. Unsworth J, Outhwaite J, Blake DR, Morris CJ, Freeman J, Lunec J. Dynamic studies of the relationship between intra-articular pressure, synovial fluid oxygen tension, and lipid

peroxidation in the inflamed knee: an example of reperfusion injury. Ann Clin Biochem 1988; 25:370–377.
177. Allen RE, Outhwaite JM, Morris CJ, Blake DR. Xanthine oxidoreductase is present in human synovium. Ann Rheum Dis 1987; 46(11):843–845.
178. Allen RE, Blake DR, Nazhat NB, Jones P. Superoxide radical generation by inflamed human synovium after hypoxia. Lancet 1989; 2(8657):282–283.
179. Bolann BJ, Ulvik RJ. Release of iron from ferritin by xanthine oxidase. Role of the superoxide radical. Biochem J 1987; 243(1):55–59.
180. Arai KI, Lee F, Miyajima A, Miyatake S, Arai N, Yokota T. Cytokines: coordinators of immune and inflammatory responses. Ann Rev Biochem 1990; 59:783–836.
181. Karin M. Signal transduction and gene control. Curr Opin Cell Biol 1991; 3(3):467–473.
182. Nose K, Shibanuma M, Kikuchi K, Kageyama H, Sakiyama S, Kuroki T. Transcriptional activation of early-response genes by hydrogen peroxide in a mouse osteoblastic cell line. Eur J Biochem 1991; 201(1):99–106.
183. Janssen YM, Matalon S, Mossman BT. Differential induction of c-fos, c-jun, and apoptosis in lung epithelial cells exposed to ROS or RNS. Am J Phys 1997; 273(4 Pt 1):L789–796.
184. Aruoma OI, Halliwell B. Molecular Biology of Free Radicals in Human Diseases, 1st ed. London: OICA International, 1998.
185. Abate C, Patel L, Rauscher 3d FJ, Curran T. Redox regulation of fos and jun DNA-binding activity in vitro. Science 1990; 249(4973):1157–1161.
186. Walker LJ, Robson CN, Black E, Gillespie D, Hickson ID. Identification of residues in the human DNA repair enzyme HAP1 (Ref-1) that are essential for redox regulation of Jun DNA binding. Mol Cell Biol 1993; 13(9):5370–5376.
187. Sun Y, Oberley LW. Redox regulation of transcriptional activators. Free Rad Biol Med 1996; 21(3):335–348.
188. Conant K, Ahmed U, Schwartz JP, Major EO. IFN-gamma inhibits AP-1 binding activity in human brain-derived cells through a nitric oxide dependent mechanism. J Neuroimmunol 1988; 88(1–2):39–44.
189. Meyer M, Schreck R, Baeuerle PA. $H_2O_2$ and antioxidants have opposite effects on activation of NF-kappa B and AP-1 in intact cells: AP-1 as secondary antioxidant-responsive factor. EMBO J 1993; 12(5):2005–2015.
190. Holmgren A. Thioredoxin and glutaredoxin systems. J Biol Chem 1989; 264(24):13963–13966.
191. Xanthoudakis S, Miao G, Wang F, Pan YC, Curran T. Redox activation of Fos-Jun DNA binding activity is mediated by a DNA repair enzyme. EMBO J 1992; 11(9):3323–3335.
192. May MJ, Ghosh S. Signal transduction through NF-kappa B. Immunol Today 1998; 19(2): 80–88.
193. Nunokawa Y, Oikawa S, Tanaka S. Human inducible nitric oxide synthase gene is transcriptionally regulated by nuclear factor-kappaB dependent mechanism. Biochem Biophys Res Commun 1996; 223(2):347–52.
194. Gilston V, Blake DR, Winyard PG. Inflammatory mediators, free radicals and gene transcription. In: Winyard PG, Blake DR, Evans CH, eds. Free Radicals and Inflammation. Basel: Birkhauser, 1999.
195. Baeuerle PA. The inducible transcription activator NF-kappa B: regulation by distinct protein subunits. Biochim Biophys Acta 1991; 1072(1):63–80.
196. Liao F, Andalibi A, deBeer FC, Fogelman AM, Lusis AJ. Genetic control of inflammatory gene induction and NF-kappa B-like transcription factor activation in response to an atherogenic diet in mice. J Clin Invest 1993; 91(6):2572–2579.
197. Chen ZJ, Parent L, Maniatis T. Site-specific phosphorylation of I-kappa B-alpha by a novel ubiquitination-dependent protein kinase activity. Cell 1996; 84(6):853–862.
198. Traenckner EB, Wilk S, Baeuerle PA. A proteasome inhibitor prevents activation of NF-kappa B and stabilizes a newly phosphorylated form of I kappa B-alpha that is still bound to NF-kappa B. EMBO J 1994; 13(22):5433–5441.

199. Han Z, Boyle DL, Manning AM, Firestein GS. AP-1 and NF-kappa B regulation in rheumatoid arthritis and murine collagen-induced arthritis. Autoimmunity 1998; 28(4):197–208.
200. Asahara H, Fujisawa K, Kobata T, Hasunuma T, Maeda T, Asanuma M, Ogawa N, Inoue H, Sumida T, Nishioka K. Direct evidence of high DNA binding activity of transcription factor AP-1 in rheumatoid arthritis synovium. Arthritis Rheum 1997; 40(5):912–918.
201. Kinne RW, Boehm S, Iftner T, Aigner T, Vornehm S, Weseloh G, Bravo R, Emmrich F, Kroczek RA. Synovial fibroblast-like cells strongly express jun-B and C-fos proto-oncogenes in rheumatoid- and osteoarthritis. Scand J Rheumatol 1995; 101(suppl):121–125.
202. Marok R, Winyard PG, Coumbe A, Kus ML, Gaffney K, Blades S, Mapp PI, Morris CJ, Blake DR, Kaltschmidt C, Baeuerle PA. Activation of the transcription factor nuclear factor-kappaB in human inflamed synovial tissue. Arthritis Rheum 1996; 39(4):583–591.
203. Omura T, Yoshiyama M, Shimada T, Shimizu N, Kim S, Iwao H, Takeuchi K, Yoshikawa J. Activation of mitogen-activated protein kinases in in vivo ischemia/reperfused myocardium in rats. J Mol Cell Cardiol 1999; 31(6):1269–1279.
204. Shimizu N, Yoshiyama M, Omura T, Hanatani A, Kim S, Takeuchi K, Iwao H, Yoshikawa J. Activation of mitogen-activated protein kinases and activator protein-1 in myocardial infarction in rats. Cardiol Res 1998; 38(1):116–124.
205. Yao KS, Xanthoudakis S, Curran T, O'Dwyer PJ. Activation of AP-1 and of a nuclear redox factor, Ref-1, in the response of HT29 colon cancer cells to hypoxia. Mol Cell Biol 1994; 14(9):5997–6003.
206. Bashir S, Harris G, Denman MA, Blake DR, Winyard PG. Oxidative DNA damage and cellular sensitivity to oxidative stress in human autoimmune disease. Ann Rheum Dis 1993; 52(9):659–666.
207. Harris G, Bashir S, Winyard PG. 7,8-Dihydro-8-oxo-2′-deoxyguanosine present in DNA is not simply an artefact of isolation. Carcinogenesis 1994; 15(2):411–413.
208. Sionov RV, Haupt Y. Apoptosis by p53: mechanisms, regulation, and clinical implications. Springer Semin Immunopathology 1998; 19(3):345–362.
209. Messmer UK, Ankarcrona M, Nicotera P, Brune B. p53 expression in nitric oxide-induced apoptosis. FEBS Lett 1994; 355(1):23–26.
210. Tak PP, Smeets TJ, Boyle DL, Kraan MC, Shi Y, Zhuang S, Zvaifler NJ, Breedveld FC, Firestein GS. p53 overexpression in synovial tissue from patients with early and longstanding rheumatoid arthritis compared with patients with reactive arthritis and osteoarthritis. Arthritis Rheum 1999; 42(5):948–953.
211. Firestein GS, Echeverri F, Yeo M, Zvaifler NJ, Green DR. Somatic mutations in the p53 tumor suppressor gene in rheumatoid arthritis synovium. Proc Natl Acad Sci USA 1997; 94(20):10895–10900.
212. Merry P, Grootveld M, Lunec J, Blake DR. Oxidative damage to lipids within the inflamed human joint provides evidence of radical-mediated hypoxic-reperfusion injury. Am J Clin Nutr 1991; 53(1 suppl):362S–369S.
213. Satoh K. Serum lipid peroxide in cerebrovascular disorders determined by a new colorimetric method. Clin Chim Acta 1978; 90(1):37–43.
214. Rowley D, Gutteridge JM, Blake DR, Farr M, B Halliwell. Lipid peroxidation in rheumatoid arthritis: thiobarbituric acid-reactive material and catalytic iron salts in synovial fluid from rheumatoid patients. Clin Sci 1984; 66(6):691–695.
215. Selley ML, Bourne DJ, Bartlett MR, Tymms KE, Brook AS, Duffield AM, Ardlie NG. Occurrence of (E)-4-hydroxy-2-nonenal in plasma and synovial fluid of patients with rheumatoid arthritis and osteoarthritis. Ann Rheum Dis 1992; 51(4):481–484.
216. Rhodes J. Erythrocyte rosettes provide an analogue for Schiff base formation in specific T cell activation. J Immunol 1990; 145(2):463–469.
217. Esterbauer H, Zollner H, Schaur RJ. Hydroxyalkenals: cytotoxic products of lipid peroxidation. Atlas Sci Biochem 1988; 1:311–319.
218. Grootveld M, Henderson EB, Farrell A, Blake DR, Parkes HG, Haycock P. Oxidative damage to hyaluronate and glucose in synovial fluid during exercise of the inflamed rheumatoid joint.

Detection of abnormal low-molecular-mass metabolites by proton-n.m.r. spectroscopy. Biochem J 1991; 273(Pt 2):459–467.
219. Henderson EB, Grootveld M, Farrell A, Smith EC, Thompson PW, Blake DR. A pathological role for damaged hyaluronan in synovitis. Ann Rheum Dis 1991; 50(3):196–200.
220. Schiller J, Arnhold J, Schwinn J, Sprinz H, Brede O, Arnold K. Reactivity of cartilage and selected carbohydrates with hydroxyl radicals: an NMR study to detect degradation products. Free Rad Res 1998; 28(2):215–228.
221. Schiller J, Arnhold J, Grunder W, Arnold K. The action of hypochlorous acid on polymeric components of cartilage. Biol Chem Hoppe-Seyler 1994; 375(3):167–172.
222. Stadtman ER, Berlett BS. Reactive oxygen-mediated protein oxidation in aging and disease. Drug Met Rev 1998; 30(2):225–243.
223. Lunec J, Blake DR, McCleary SJ, Brailsford S, Bacon PA. Self-perpetuating mechanisms of immunoglobulin G aggregation in rheumatoid inflammation. J Clin Invest 1985; 76(6): 2084–2090.
224. Lunec J, Brailsford S, Hewitt SD, Morris CJ, Blake DR. Free radicals: Are they possible mediators for IgG denaturation and immune complex formation in rheumatoid arthritis. Int J Immunother 1987; 3:39–43.
225. Griffiths HR, Lunec J. The effects of oxygen free radicals on the carbohydrate moiety of IgG. FEBS Lett 1989; 245(1–2):95–99.
226. Griffiths HR, Dowling EJ, Sahinoglu T, Blake DR, Parnham M, Lunec J. The selective protection afforded by ebselen against lipid peroxidation in an ROS-dependent model of inflammation. Agents Actions 1992; 36(1–2):107–111.
227. Blake DR, Merry P, Unsworth J, Kidd BL, Outhwaite JM, Ballard R, Morris CJ, Gray L, Lunec J. Hypoxic-reperfusion injury in the inflamed human joint. Lancet 1989; 1(8633):289–293.
228. Swaak AJ, Kleinveld HA, Kloster JF, Hack CE. Possible role of free radical altered IgG in the etiopathogenesis of rheumatoid arthritis. Rheumatol Int 1989; 9(1):1–6.
229. Henderson EB, Winyard PG, Grootveld M, Blake DR. Pathophysiology of reperfusion injury in human joints. In: Das DK, ed. Hypoxic Reperfusion in Disease. Boca Raton: CRC Press, 1993:430–469.
230. Winyard PG, Zhang Z, Chidwick K, Blake DR, Carrell RW, Murphy G. Proteolytic inactivation of human alpha-1-antitrypsin by human stromelysin. FEBS Lett 1991; 279(1):91–94.
231. Chidwick K, Winyard PG, Zhang Z, Farrell AJ, Blake DR. Inactivation of the elastase inhibitory activity of alpha-1-antitrypsin in fresh samples of synovial fluid from patients with rheumatoid arthritis. Ann Rheum Dis 1991; 50(12):915–916.
232. Zhang Z, Winyard PG, Chidwick K, Farrell A, Pemberton P, Carrell RW, Blake DR. Increased proteolytic cleavage of alpha-1-antitrypsin (alpha-1-proteinase inhibitor) in knee-joint synovial fluid from patients with rheumatoid arthritis. Biochem Soc Trans 1990; 18(5): 898–899.

# 3

# Autoimmunity Caused by Oxidizing Foreign Compounds

**KAYE H. KILBURN**

*University of Southern California, Keck School of Medicine, Los Angeles, California*

Seven years ago I examined this topic, focusing on chemical inducers of human connective tissue diseases (once called "collagen diseases") that are accompanied by circulating autoantibodies (1). Several reviews are helpful (2–4), especially that from the European advisory subgroup on toxicology (2). Human autoimmune diseases are often organ related as the antibodies are to nucleic acids, DNA, mitochondria, myelin basic protein, brush border, thyroid antigen, thyroid stimulating hormone, and components of platelets and spermatozoa. Environmental factors include drugs, metals, and industrial chemicals and particles (1,5). One unifying concept is that oxidation by hydroxyl radicals, singlet oxygen, aldehydes, and peroxide enzymes peroxidizes lipids in membranes and thiols and alters methylation and protein binding sites on DNA.

## I. UNIFYING CONCEPT

The best solution to overwhelming complexity and a deluge of information is to simplify concepts and arrange them in a plausible order. Major decisions concern what to keep and what to ignore. It is well known that blood vessels are a site for antibody deposition. New ideas, key concepts, and new interpretations are needed to stimulate hypotheses about systemic lupus erythematosus (SLE), for example. Angiogenesis may be critical to connective tissue disease (6). Blood vessels are probably released for endometrial stimulation during the estrous phase of the menstrual cycle. I propose that angiogenesis has a key role in connective tissue disease. Angiogenesis-inhibiting factors prevent female reproduction (7), and thus are contraceptive. Would they be therapeutic for the connective tissue diseases including lupus?

## II. HISTORICAL REVIEW

Despite 150 years of investigations (8), systemic lupus erythematosus and related connective tissue diseases (scleroderma, periarteritis nodosa, and dermatomyositis) are regarded as diseases of unknown cause. In the nineteenth century, lupus was limited to the skin (discoid). As it gained attention Kaposi recognized the progression and discerned acute or subacute disease that could ''endanger and destroy life'' (9). This recognition of visceral lesions and systemic signs ushered in a new era and a causal connection to tuberculosis was proposed, which was subsequently disproved. In 1900 Sir William Oster found the presence of visceral lesions in 29 patients (18 were males, 12 of whom were aged 3 to 12 years; 17 had arthralgia; and 14 nephritis) (10).

Early in the twentieth century, many more patients were diagnosed with SLE. There was a female preponderance of up to 10:1, a few childhood instances, and clear separation of discoid (skin limited) and systemic or disseminated (visceral) forms. In addition, renal involvement, endocarditis (11), and central nervous system signs of neurological cerebritis and mental alteration were recognized (12). Failure of tear and salivary glands was also observed (Sjögren's syndrome). The false (biological) positive test for syphilis seen in 1909–1910 has since 1983 been attributed to autoantibodies to cardiolipin, a phospholipid, which caused bleeding and thrombosis and is different from the complement fixation factor of syphilis. Wire loop lesions in the glomeruli of the kidney were noted by 1924 and endocardial lesions confirmed in 1930 (11).

In 1943, the characteristic lupus erythematosus (LE) cell with purple-staining cytoplasmic globules as seen by Hargraves, who published his results in 1949 (13). This ''tart cell'' is seen more frequently in specimens of bone marrow from LE patients and enhanced if they are allowed to incubate after removal. A serum factor in gamma globulin produced these changes in non-LE marrow cells, leading to the conclusion that the globules were phagocytosed material and could be found in many cells in the tissues of LE patients.

In 1957, the demonstration of a factor in serum that reacts with DNA made antibody measurements replace bone marrow examination (14,15). This antibody reacted with native double-strand DNA and with the denatured single-strand DNA. Other antibodies were described (16). In 1963, the ''lupus band test'' used fluorescent microscopy of skin biopsy specimens to demonstrate immunoglobulin deposited at the dermoepithelial junction; this reaction is seen in clinically uninvolved patients as well as those with rheumatoid arthritis (17).

The frequency of SLE sky-rocketed after 1950. Before then, only 1 to 3 patients per year were diagnosed at several large academic medical centers. In Sweden, the SLE incidence went from 1:100,000 population in 1938–1939 and 1948–1949 to 4.8:100,000 per year in 1954–1955 (18). Steroid treatment began in 1950, and increased life expectancy from a 52% 4-year survival in 1949–1953 to 87% 10-year survival; however, for those patients with renal involvement it was less (65%) (19). Neonatal LE in infants of women with LE or those who developed it shortly after parturition was associated with complete heart block in infants and myocardial hematoxylin bodies (20,21). Perhaps this is a continuation of heart involvement from Libman-Sachs endocarditis (11), which decreased after introduction of steroid therapy.

The historical summary finds that the causes of SLE began acting in the twentieth century, that women are targets 8 to 10 times as frequently as men, and that serum factors (autoantibodies) react in skin, kidney, heart, brain, and other vascular beds to stimulate collagen deposition. An array of related lesions have been described, including scleroderma, dermatomyositis, and mixed connective tissue disease. The emphasis on blood vessels since 1947 (22) suggests mechanisms of angiogenesis as a focusing concept (6). Since 1950, it has been thought that many chemicals—initially therapeutic drugs but now broadened to include metals, industrial chemicals, plastics, and silica—can induce lupus (2,5). Despite some differences in the disease manifestations of SLE of unknown origin, this clue cannot be ignored. As mechanisms of the pathogenesis have emerged, another theory has also been proposed, that oxidative stress produces lipid peroxidation shown by plasma 8 epi-$PGF_{2\alpha}$ (23) and stimulates the antiphospholipid antibodies. This thesis would accept the observations that silica crystals ($SiO_2$) initiate connective tissue diseases (24).

The fact that elicited anti-antibodies (AA) (immunoglobulins) are deposited in the kidneys, heart, and skin and that the AA titers increase before clinical exacerbations are detected suggests their causal role. Steroid therapy to suppress AAs avoids the exacerbations (25).

## III. OXIDATION PROCESSES

It has been speculated that oxidation of many different cell components stimulates autoantibodies (Table 1). Another approach to chemical causation is to look for associations between exposures to foreign compounds (e.g., silica, L-tryptophan, vinylchloride, and trichloroethylene). The wave of reviews at the end of the 1980s has not been repeated. Many accounts of individual patients have appeared as case reports, book chapters, and editorials (26–29) about drug-induced lupus syndromes. A standard topic a decade ago, it continues to show new associations with increased numbers of offending agents. In addition, therapeutic administration of biological agents, such as IFN-α and IFN-β has been associated with antibodies to gastric parietal cells, thyroid antigen, and thyroglobulin (30). Parvovirus, a foreign chemical that grows, has induced a lupus syndrome as has Epstein-Barr virus (31). New observations expand the definitions, with increases in numbers and variety of inducer substances and broadening the concept of autoimmune pathogenesis.

Questionnaire data on almost 4000 people exposed to over 20 environmental polluting chemicals showed that the clinical symptoms of lupus were common in these subjects as compared to unexposed subjects (Table 2). We used the American Rheumatism Association (ARA) questionnaires for SLE 1971/1981 (32), which have usually been used to survey hospital patients and populations. Our inquiries from those who were environmentally exposed to those unexposed showed polychlorinated byphenyls (PCBs) (groups A,B,C) elevated most scores significantly, as did trichloroethylene (TCE) (group D) and chlorine combined with creosol (group E). A water exposure to PCB and TCE combined (group F) and air exposures to arsenic and air (group G) to arsenic plus alkaline fluorine and chlorine air (group H), and to permethrin, a pyrethroid insecticide (group I) were associated with fewer differences.

The individual symptoms/signs of hair loss, mouth sores, pleurisy, and sun sensitivity were more common than were proteinuria, anemia, numbness, and rheumatism. Seizures were too infrequent to evaluate in these groups. Interpretation of the meaning of the questions may be a problem in such medically unsophisticated, environmentally exposed populations.

There are two possibilities. One is that these symptoms are common in chemically exposed people, as are the 35 other symptoms we surveyed by questionnaires, and thus

**Table 1** Survey of Human Autoimmune Diseases[a]

| Diseases | Self-antigens (as defined by the autoantibodies involved) |
|---|---|
| Autoimmune chronic active hepatitis, virus negative | Membrane and microsomes of liver cells including P-450 cytochrome isoenzymes |
| Autoimmune hemolytic anemia | Membrane components of erythrocytes |
| Bullous pemphigoid | Basement membrane of skin |
| Goodpasture's syndrome (glomerulonephritis and alveolitis with linear immunoglobulin deposits along the glomerular and alveolar basement membranes) | Components of the glomerular basement membrane (GBM) and alveolar BM |
| Guillain-Barre syndrome | Myelin and other components of the sheets of peripheral nerves |
| Hashimoto's thyroiditis | Cytoplasmic or microsomal thyroid antigen, thyroglobulin |
| Idiopathic leukocytopenia | Membrane components of leukocytes |
| Idiopathic thrombocytopenia | Membrane components of platelets |
| Male infertility (certain cases) | Spermatozoa |
| Myasthenia gravis | Acetylcholine receptor at the neuromuscular synapsis |
| Pemphigus vulgaris | Desmosomes linking epithelial cells of the skin |
| Pernicious anemia | Intrinsic factor (produced by parietal cells for absorption of vitamin $B_{12}$) |
| Primary Addison's disease | Microsomal antigens in the adrenal cortex |
| Progressive systemic sclerosis (scleroderma) | Various antigens in cell nuclei, especially nucleoli |
| SLE | Various nuclear antigens, especially double-stranded DNA; antigens on leukocytes and erythrocytes |
| Thyrotoxicosis | TSH receptors |
| Wegener's granulomatosis (inflammatory disease of veins and arteries, especially in the lung and kidneys) | Alkaline phosphatase-like material on endothelial cells and neutrophils (see Ref. 50) |

*Source*: Adapted from Reference 2.

[a] Diseases in which pathogenic autoimmune reactions are certain or likely because the self antigens involved have been relatively well defined.

are nonspecific. If they are nonspecific, then inferences about possible SLE are difficult. However, if these symptoms are specific (central to connective tissue responses), then they are a step toward recognition of a connective tissue disease before or below the level of clinical disease. If elevated ANA titers are added to four or five or more symptoms as a confirming criterion (32), recognition still remains difficult. The timing of ANA sampling must be defined and delimited as levels decrease progressively after acute elevations from Spanish toxic oil and probably after chronic exposures as were measured in the Phoenix trichloroethylene-exposed population. This raises the question of which and how many autoantibodies should be measured, and, most importantly, what is the level or threshold of ANA needed for labeling a disorder as autoimmune. Does it require that

**Table 2** Responses to Lupus Questionnaire by Groups of Chemical-Exposed People

| Group: | | A | B | C | D | E | F | G | H | I |
|---|---|---|---|---|---|---|---|---|---|---|
| *n*: | Total | 38 | 99 | 154 | 236 | 99 | 117 | 43 | 75 | 32 |
| Rheumatism | 6 | + | + | + | + | + | + | 0 | 0 | 0 |
| Raynaud's disease | 6 | + | + | + | + | 0 | 0 | + | + | 0 |
| Mouth sores | 8 | + | + | + | + | + | 0 | + | + | + |
| Anemia | 4 | + | + | + | 0 | 0 | 0 | 0 | 0 | + |
| Malar Rash | 7 | + | + | + | + | + | + | + | 0 | 0 |
| Sun sensitivity | 8 | + | + | + | + | + | + | + | 0 | + |
| Pleurisy | 8 | + | + | + | + | + | + | + | 0 | + |
| Proteinuria | 6 | + | 0 | + | + | + | 0 | 0 | + | + |
| Hair loss | 9 | + | + | + | + | + | + | + | + | + |
| Seizures | 0 | 0 | 0 | 0 | 0 | 0 | 0 | 0 | 0 | 0 |
| Total | | 9 | 8 | 9 | 8 | 7 | 5 | 6 | 4 | 6 |

*n* = Number of subjects studied; A = PCBs, incinerator medical waste; B = PCB, pumps natural gas pipeline; C = TCE, jet turbine repair; D = TCE, environmental microchip manufacturing; E = Chlorine-creosol, environmental spill; F = PCB–TCE, aluminum casting; G = Arsenic, mine cleanup; H = Arsenic, trioxide environmental; I = Permethrin, aircraft cabins.

autoantibodies be elevated for each individual or for the population? If so, which ones should be measured?

Finally, how does one apply criteria developed for individuals to populations? What cautions must one apply to do this? Perhaps the question should be turned around to ask how clinical diagnoses be made in practice on those selected by epidemiological criteria? These are logical questions that should be answered before discussing the role of autoantibodies. Are they simply markers for a process or do they play a role in the pathogenesis? In one instance (e.g., a mouse experimental model for autoimmune myocarditis), antibodies to myosin were localized within the cardiac matrix (33). The antifibrillarin antibodies, the $NH_2$ terminal, and COOH terminal portions overlap nuclear protein of the Epstein-Barr virus and the capsid protein from herpes simplex virus, suggesting again a broader, less specific, mode of active antibody formation that may relate to chemicals of which viruses are a special (living) class (34).

Last, studies on beryllium disease suggest that a specific genetic marker HLA DDB1 produced when glutamate is present in position 69 instead of lysine may provide a genetic marker for people susceptible to chemically induced lupus erythematosus (35).

## IV. DRUG- AND FOOD-INDUCED LUPUS

### A. Drugs

A definition of an oxidized foreign compound is broad, including oxygen and ozone, and extending to many drugs and chemicals in the workplace and environment that can elicit typical symptoms such as arthralgia and elevated ANA. Withdrawal of the offending drugs relieves symptoms and decreases ANA titers. Drugs causing lupus (Table 3) (4,36) have one of three features: (1) they have an amine group, $NH_2^-$, and arylamine or hydrazine function; (2) a SH, OSH sulfhydryl or thiono-sulfur group; or (3) are hydantoins that may form phenols, OHs.

ANA titers may be elevated in 20 to 50% of patients on chronic treatment with

**Table 3** Chemical Used as Drug Associated with Lupus

| | | | |
|---|---|---|---|
| *Drugs definitively associated with drug-related lupus* | | | |
| Chlorpromazine | | | |
| Hydralazine | | | |
| Isoniazid | | | |
| Methyldopa | | | |
| Minocycline | | | |
| Procainamide | | | |
| Quinidine | | | |
| *Other drugs associated with drug-related lupus* | | | |
| Acebutolol | Estrogens | Minoxidil | Prophythiouracil |
| Acecainide | Ethosuximide | Nalidixic acid | Propranolol |
| Allopurinol | Ethylphenacemide | Nitrofurantoin | Psoralen[a] |
| Aminoglutethimide | Gold salts | Nomifensine[a] | Pyrathiazine |
| Amoproxan | Griseofulvin | Oxyphenisatin | Pyrithoxine |
| Anthiomaline | Guanoxan | Oxyprenolol | Quinine |
| Anti-tumor necrosis factor-α | Ibuprofen | Para-amino-salicylic acid | Reserpine |
| Atenolol | Interferon-α | Penicillamine | Spironolactone |
| Benoxaprofen | Interferon-γ | Penicillin | Streptomycin |
| Bleomycin | Interleukin-2 | | |
| Captopril | Labetalol | Perazine | Sulindac |
| Carbamazepine | Leuprolide acetate | Perphenazine | Sulfadimethoxine |
| Chlorprothixene | Levadopa | Phenelzine | Sulfamethoxy-pyridazine |
| Chlorthalidone | Levomeprazone | Phenopyrazone[a] | Sulfasalazine |
| Cimetidine | Lithium carbonate | Phenylbutazone[a] | Tetracyclines |
| Cinnarizine | Lovastatin | Phenylethylace-tylurea[a] | Tetrazine |
| Clonidine | Mephenytoin | Phenytoin | Thionamide[a] |
| Danazol | Methimazole | Practolol[a] | Thioridazide |
| Diclofenac | Methlysergide[a] | Prazosin | Timolol eyedrops |
| 1-2-dimethyl-3 hydroxy-pyride-4-1 | Methylthiouracil[a] | Primidone | Tolazamide |
| Diphenylhydantoin | Metoprolol | Prindolol | Tolmetin |
| Disopyramide | Metrizamide | Promethazine | Trimethadione |
| Enalapril | | Propafenone | |

*Source*: Adapted from Reference 36.
[a] No longer made.

drugs that induce lupus, such as chlorpromazine, but only 1 or 2% show a lupus syndrome (2,28,36). Cytokines, interleukins, and antitissue necrosis factor can elicit ANA and anti-dsDNA antibodies and cause arthropathies in some patients; a few develop the lupus syndrome (26,30,35).

## B. Foods

Foods and dietary factors have numerous probable relationships such as protein malnutrition and some that suggest intriguing associations during an era of widespread public experiments with herbs (37).

Feeding the amino acid canavanine from derived alfalfa sprouts, which is an analogue of L-arginine, produces an SLE picture in experimental animals. Macaques fed alfalfa seeds developed antibodies to dsDNA (38), which persisted for 2 years. Canavanine inhibits mitogenic responses of human peripheral blood T-cells, especially CD8 plus T-cells, increased IgG production, and anti-DNA activity. Antibodies to nuclear antigens, double-stranded DNA (38), and red blood cells cause anemia and the deposition of immunoglobulin and complement in the kidney and skin (38,39).

Effects on immune responses generally include suppression, which produces immunodeficiency, are found from other dietary agents (36). Protein energy malnutrition is most common in children from developing countries producing thymic atrophy, decreased spleen weight, decreased numbers of T-cells with increased null cells, impaired lymphokine production, and delayed cutaneous hypersensitivity. Deficiency of vitamin A, of pyridoxine, vitamin C, vitamin E, and zinc reduce immune responses, while vitamin D inhibits these reductions (37).

Ozone is another environmental source of oxidizing potential that is a byproduct of petroleum burning particularly as gasoline in motor vehicles (5). Ozone decreases the numbers of alveolar macrophages residing in the lung, probably by lysis of cells, and decreases cell mobility and phagocytic activity and CD4-CD8 T-cell ratios (40,41). Recently, aldehydes hexanal and 2-nonenal were identified in lung lavage fluids after ozone.

Ultraviolet light also mediates immunosuppression, perhaps by releasing urocanic acid (42) and vitamin D (43), which is a powerful immunosuppressant that may be prevented by retinoids and carotenoids.

## V. MODELS OF AUTOIMMUNE DISEASE

Idiotypic (id) interactions, defined by Bigazzi (4,27) as those regulated by autoantigen-specific suppressor T-cells, occur due to the diversity of the variable (V) regions of immunoglobulins that are immunogenic and cause formation of anti-id antibodies. They may be initiated by xenobiotics or by their own diversity but connect by natural autoantibodies when self/non-self-discrimination fails (44). Current research on how the picture of self is produced (45), a new attractor-like inflammation, may arouse the pathogenic mechanism of graft-versus-host type (44).

Animal experiments have focused on graft-versus-host disease (GVHD) models or their surrogates, emphasizing the popliteal lymph node (PLN) assay in mice to assay chemicals for their potential for autoimmune disorder (2). If the model is accepted, this bioassay can compare autoimmune potential after substituting the side chains and prosthetic groups of molecules like the anticonvulsant hydantoins, succinimides, butyrolactones and the potential agents of Spanish oil syndrome, the imidazalidinethiones (46,47).

A human cell model system, lymphocyte transformation, may serve as a screening test for chemicals using human lymphocytes from a sensitive individual incubated with the chemical. This assay is a useful screen but is insufficient to prove an autoimmune pathogenesis. When results are negative, metabolites may be tested after preincubation of the chemical with hepatic microsomes (P-450) to convert the chemical to an intermediate that is active in vivo (48). Similarly, scratch tests may be useful in screening. Besides the PLN assay in mice, other animal models have proved useful (e.g., cats develop autoantibodies and a positive direct antiglobulin test when given 6-propothiouracil for 4 to 8 weeks) (49).

## VI. CHEMICALS ASSOCIATED WITH CONNECTIVE TISSUE DISEASES

In the section that follows, chemicals that had been reported to cause collagen-like diseases will be reviewed and compared (Table 4).

### A. Heavy Metals

Gold, cadmium, and mercury can stimulate AAs in animal models (50), but gold (50,51) and mercury (51–53) are also capable of causing thrombocytopenia (51) and glomerulonephritis in human subjects (54,55) and antoimmune disease in rats (53). Mercury elicits an anti-nucleolar autoantibody that targets fibrillarin, a 34-kDa protein found in nucleolar ribonucleoprotein particles (29). This is thought to depend on mercury interacting with cysteine in the fibrillarin sequence, altering the physicochemical part of the autoantigen. In turn, this is consistent with many drugs that cause lupus interaction with cellular thiols, reducing intercellular bridges or breaking protein cross-links (5). This mechanism, the interaction with cellular thiols, would tie together induction of LE by mercury, vinyl chloride, iodide, and toxic oil (5).

Other metals are implicated, cadmium in antoimmune kidney disease (56) and chromium and lithium in lupus (57,58). Lithium causes insulin-dependent diabetes and autoimmune thyroid (59) and kidney disease (60). Zinc has been linked to multiple sclerosis (61,62) and molybdenum to ANA-negative hypersensitivity (63). Beryllium elicits lymphocyte proliferation in those genetically marked (34).

### B. Vinyl Chloride

Vinyl chloride (VC) ($C_2H_3Cl$) was reported in 1974 by Lange et al. (64) to produce a scleroderma-like illness characterized by multisystemic involvement of collagen tissue with pulmonary fibrosis, skin sclerosis, fibrosis of liver and spleen, capillary disturbances, thrombocytopenia, paresthesia, and angiosarcoma of the liver (Table 1). The most extensive study was initiated clinically by Ward et al. (65), who found that 58 of 320 (18%) exposed workers had this scleroderma-like syndrome, 19 were moderately affected, and 9 were severely disabled. Eight of the latter but none of the moderately disabled group had ANA antibodies. A follow-up of 44 of the 53 available from the 58 workers by Black et al. (66) found 21 had severe and 23 had mild vinyl chloride scleroderma. HLA DR5 phenotype was of similar frequency in the industrial workers who developed the systemic sclerosis syndrome with vinyl chloride (VC) and in 50 classic idiopathic scleroderma patients but neither anti-centromere or anti-Scl 70 antibodies were found in the VC group.

### C. Trichloroethylene

Occupational exposure to trichloroethylene (TCE) that caused scleroderma was described in 1957 in a 24-year-old woman who used TCE to remove grease from aluminum plates (67). Since then, solvents, especially TCE, have shown this association, as reported in six of seven patients in Australia (68,69) and in seven of nine patients in Japan (70). Epoxides are formed in the metabolism of TCE, which appear to attack endothelial cells to produce vasculitis. In one fatal case (71), a 19-year-old TCE-exposed dry cleaner died after a few weeks of Raynaud's phenomena, nailfold hemorrhages, muscle weakness and aching, swollen fingers, and impotence. ANA was elevated and liver function showed elevated

**Table 4** Possible Relationships Between Human Sclerotic and Lupus-like Diseases and Chemical Exposures

| Chemical | References | Observation |
|---|---|---|
| Gold | Hall (50) | Gold and penicillamine nephropathy |
| | Kotsy, et al. (51) | Thrombocytopenia with auronofin Rx |
| Mercury | Pilletier, et al. (52) | Renal disease |
| | Bigazzi (53) | Autoimmune disease in rats, mice, rabbits. Fibrillarin-membranous glomerulopathy |
| | Cardenas, et al. (54) | 50 workers, more enzymes and antigen leakage with more Hg excreta |
| | Schrallhammer-Benkler, et al. (55) | Lichenoid drug eruption and ANA |
| | Pollard, et al. (29) | $HgCl_2$-induced ANOA in mice, cell death after Hg enters cell, aberrant fibrillarin migration—cysteines interact Hg |
| Lithium | Hassmann and McGregor (59) | Autoimmune thyroid disease |
| | Santella, et al. (60) | Focal glomerulosclerosis from lithium bicarbonate |
| Chromium | Kilburn and Warshaw (58) | Accompanied trichloroethylene and chlorinated solvents—symptoms, ANA |
| | Tsankov, et al. (57) | Induced pemphigus from workers with chromium |
| Cadmium | Bernard, et al. (56) | Autoimmune kidney disease |
| Zinc | Stein, et al. (61) | Multiple sclerous cluster with zinc |
| | Schiffer, et al. (62) | Zinc on experimental allergic encephalomyelitis |
| Molybdenum | Federmann, et al. (63) | 24-year-old woman implanted with metal plates had fever, ANA negative SLE: patch test positive, lymphocyte transformation showed delayed hypersensitivity |
| Beryllium | Richeldi, et al. (34) | Genetic marker, lymphocyte proliferation test |
| Occupational vinyl chloride | Lange, et al. (64) | Skin sclerosis |
| | Ward, et al. (65) | Lung fibrosis |
| | Black, et al. (66) | Nervous system paresthesia |
| | | Vessels—capillary inflammation, intimal fibrosis |
| | | Thrombocytopenia |
| | | Symptoms—fatigue, cold burning pain, emotional instability, loss of libido, impotence |
| | | Autoantibodies—not detected |

**Table 4** Continued

| Chemical | References | Observation |
|---|---|---|
| Tetrachloroethylene | Sparrow (72) | 19-year-old male dry cleaner, 4 years, elevated ANA titers, systemic sclerosis |
| TCE | Reinl (67) | 24-year-old woman, degreasing—scleroderma |
| Solvents | Yamakage and Ishikawa (70) | 7/9 patients in Japan—Raynaud's phenomenon, sclerosis; 6 had lung fibrosis |
| Solvents, toluene, xylene, white spirit | Walder (68,69) | 6/7 solvent workers in Australia; added 5 in 1983—scleroderma |
| Carbon tetrachloride & TCE | Saihan, et al. (74) | 43-year-old male, neuropathy, Raynaud's phenomenon, sclerosis |
| Organic solvents | Sverdrup (75) | Scleroderma in 8/9 manufacturing workers |
| TCE | Lockey, et al. (71) | 47-year-old female, fatal scleroderma 6 months after 2.5 h dermal exposure to TCE |
| TCE organic solvents | Nietert, et al. (73) | Odds ratio 3.3, Anti-scl-70AA |
| | Vojdani, et al. (76) | Increased autoantibody in computer manufacturing workers |
| Environmental TCE and solvents (water exposure) | Byers, et al. (78) | ANA in 10/23 family members of leukemia patients |
| TCE and solvents | Kilburn and Warshaw (58) | Increased symptoms of SLE and increased ANA titers versus controls |
| TCE and solvents | Kilburn (personal observation) | Antibodies to smooth muscle and myelin (basic protein) |
| Solvents—chemical wastes | Kardestuncer and Frumkin (79) | SLE 3/300 in Georgia community |
| Polymerizing epoxy resins [bis (4-amino-3-methyl-cyclohexyl methane)] | Yamakage, et al. (80) | 6/233 workers had skin sclerosis and muscle weakness |
| Spanish oil (syndrome) | Tabuenca (81) | PVIZT exposure: pulmonary edema, GVH disease, PLN assay |
| | Alonso-Ruiz, et al. (82) | |
| L-Tryptophan | Silver, et al. (83) | 9 patients with edema, pruritis, paresthesia, myalgia, scleroderma, eosinophilic fasciitis; 2 with ANA |
| | Belongia, et al. (84) | 63 patients with fatigue, muscle tenderness, cramps, arthralgias, rashes, and dyspnea |

| | | |
|---|---|---|
| Hydantoins, succinides and butyrolactones | Kammuller, et al. (46,47) | Grand mal anticonvulsants PLM assay, petite mal anticonvulsant PLN assay, anticonvulsant PLN assay |
| Silica | Gunther and Schuchardt (87), Pernis and Paronetto (86), Pearson (85) | 25 × increased risk of scleroderma from occupational silica exposure often simulataneous with silicosis; $SiO_2$ acts as adjuvant |
| | Rodnan, et al. (88) | 43% with PSS heavily exposed to silica |
| | Ziskind, et al. (89) | Excess SLE in silicosis |
| | Sanchez-Roman, et al. (24) | 32 of 50 subjects exposed to silica scouring powder had connective tissue disease |
| | Masson, et al. (90) | 8, 7 miners mineral dust with ANA |
| | Conrad, et al. (91) | SLE in excess in uranium miners |
| | Koeger, et al. (92) | 24 of 764 CTD patients were silica exposed (16 miners) PSS>SA>SLE>DM |
| Silicone | Barker, et al. (94) | Leakage from gel implants demonstrated in vitro |
| | Van Nunen, et al. (97) | SLE classic mixed connective tissue disease—Raynaud's rheumatoid arthritis with Sjögren's syndrome |
| | Fock, et al. (96) | Injected: SLE, scleroderma, idiopathic thrombocytopenic purpura |
| | Kumagai, et al. (93) | Report 18 review 28; of 24 with definite disease, some had been injected with paraffin |
| | Spiera (95) | 4.4% of 113 scleroderma patients had breast implant versus 0.3% of 286 rheumatoid arthritis patients |
| | Varga, et al. (99) | 8 patients had progressive systemic sclerosis (PSS), 5 SLE, 6 rheumatoid, 4 had PSS after silicone gel implants |
| | Shoaib and Patten (personal observation) | Arthritis in 22, human adjuvant disease or PSS Rh factor in 21 ANA titers increased in 8 after SGI in 17, SG saline in 4 and one silicone injection |
| | Press, et al. (98) | 11 of 24 referred silicone breast implants has connective tissue disease 10 had ANA>1:320 |
| | Gabriel, et al. (101) | 5:749 breast implant women followed 7.8 years, 10:1498 community controls |
| | Bridges, et al. (100) | 156 Silicone breast implants, 29 had a connective tissue disease, 14 scleroderma |
| | Hennekens, et al. (102) | 10,830 silicone breast implants 1.24 RR connective tissue disease |
| | Teuber and Gershwin (103) | 2 children of mothers with silicone breast implants, myalgias and +ANA |
| | Levine and Ilowite (104) | 6/8 breast fed children of SBI mothers had scleroderma like esophageal dysfunction |

gamma glutyl transpeptidase and bromsulphthalin retention, but there were no circulating immune complex, no LE cells, and no smooth muscle or antimitochondrial antibodies.

Tetrachloroethylene has produced systemic sclerosis and elevated ANA (72) as has TCE with organic solvents (73–75). Workers in computer manufacturing also had abnormal antibodies (76). These chlorinated alkene solvents resemble vinyl chloride and are metabolized via expoxides to trichloroacetic acid and trichloroethanol. If this metabolism is intracellular, perinuclear epoxides could adduct to DNA (71). When it occurs near neurofilaments of axons, binding to the neurofilaments interferes with their nutritional function, including transport of protein, causing periodic axonal swelling, an increase in axonal neurofilaments, and secondary demyelination (77).

### D. Environmental Exposure to Trichloroethylene and Solvents

Effects of prior well water exposures to TCE, perchloroethylene and 1,2 transdichloroethylene were examined by measuring five autoantibodies in the sera of 23 family members of leukemia patients at Woburn, Massachusetts. Autoantibodies were found in 48% (11/23), of which 10/23 were antinuclear (78). Observations of elevated ANA coupled with increased frequencies of the 11 American Rheumatology Association (ARA) symptoms for SLE in men and women chronically exposed to TCE 1,1,1-trichloroethene, other volatile organic chemicals (VOCs), and chromium in Tucson, Arizona, strengthened the suggestion that environmental chemical exposure may induce lupus that resembles the drug-induced lupus syndrome (58). Recently, another study that addressed mainly human neurobehavioral effects of environmental chemicals with exposure again focused on TCE, showed autoantibodies to smooth muscle and immunoglobulin G (IgG), and IgM autoantibodies to myelin basic protein (79). Both studies demonstrated impaired neurophysiological and neuropsychological performance and affective disturbances shown by elevated Profile of Mood States (POMS) scores and general unwellness with greatly elevated neurological and respiratory (irritative) complaints. Solvents and chemical manufacturing and waste were associated with excessive SLE in Georgia (79). In a South Carolina study of workers with PSS, organic solvent exposure increased risk of scleroderma by three times, which was highest for TCE (73).

### E. Epoxy Resins

Six of 233 workers using polymerizing epoxy resins had skin sclerosis (80). The causal agent is thought to be bis-(4-amino-3-methycyclothexyl methane) entering through the lungs and rapidly producing severe sclerosis of the skin and muscle weakness. This amine is related to amine chemicals used as therapeutic agents: procainamide: d-penicillamine, chlorpromazine, and isoniazid.

### F. Spanish Toxic Oil

Spanish toxic oil syndrome was named in 1981 for an epidemic of over 2600 illnesses, including 200 patients with scleroderma, in subjects who had consumed rape seed oil deliberately adulterated with aniline. Over 100 patients died. Important features were fever, respiratory distress with pulmonary edema and pleural effects, which was responsible for most deaths, exanthems, flushed cheeks, generalized lymphadenopathy, hepatosplenomegaly, and neurological signs of cerebral edema (81,82). ANA titers were elevated in 35 to 80% of patients during the first 8 months of the toxic oil scleroderma-like disease, but dropped after exposure ceased so that most were normal after 3 years (82). This temporal sequence resembles the behavior of ANA titers in drug-induced lupus (DLE). A metabolite of phenyl thiourea from Spanish oil, 1-phenyl 5-vinyl imidazolidinethione (PV12T), produced a graft-versus-host-like syndrome when experimentally injected in animals and produced a positive popliteal lymph node assay in mice (46,47).

## G. Tryptophan

In 1989 and 1990, a curious syndrome of scleroderma, pruritis, edema, myalgia, and eosinophilic fascitis was observed in subjects who had taken tryptophan, particularly in Minnesota (83), New Mexico, and South Carolina (84). In nine South Carolina patients, involvement of skin, muscle, nerve, and pulmonary tissue led to skin biopsies that were typical for scleroderma. Two had ANA titers. Activation of indoleamine 2,3-dioxgenase was postulated together with impairment of the hypothalamic pituitary adrenal axis. The 63 Minnesota patients (83) had fatigue, muscle tenderness, cramps, and arthralgias, two-thirds had rashes, and one-half were dyspneic and had difficulty climbing stairs. An etiological link was made to a production product or contaminant that eluded as a peak during high pressure liquid chromatography and was associated with reduced powdered carbon in a purification step of the fermentation product by one manufacturer of tryptophan.

## H. Silica, Paraffin, and Silicones

Silica particles have a well-recognized capacity in human subjects to produce adjuvant-like responses or immune stimulation (85,86). A 25-fold increase in scleroderma was seen in German workers with silicosis or silica exposure (87) and 43% of men diagnosed with progressive systemic sclerosis in Pittsburgh had prolonged and heavy silica particle exposure (88). SLE has also been observed in excess in silicosis (89). A French study of six slate miners, a gravel worker, and dental technician found high ANA titers, no Scl-70 antibodies, arthritis, pericarditis, and pleurisy (90). The most instructive series of 50 Spanish workers making scouring powder with a high 70 to 90% silica content from a work force of 300 showed 32 had connective tissue disease, including six with Sjögren's syndrome, five with systemic sclerosis, three with SLE, five with an overlap syndrome; 13 were poorly defined; and 36 of 50 had ANA. HLA-$DR_3$ was not significantly increased in tissue typing (24).

Uranium miners heavily exposed to quartz dust in Germany had 28 definite SLE based on four American Rheumatism Association criteria, all had ANA, and 15 probably had SLE with two or three ARA criteria, giving an estimated SLE prevalence of 93/100,000. They also showed antibodies, but less arthritis and photosensitivity (91). Twenty-four, 3% of 764 patients in a French rheumatory clinic, were silica exposed (most as miners and sandblasters) and had more scleroderma and ANA (92).

## I. Injection of Paraffin and Silicones

Implantation of foreign chemicals for breast augmentation provided another wave of iatrogenic connective tissue disease labeled by some as human adjuvant disease (93). Paraffin may behave as an immune adjuvant, particularly as an emulsion of mineral oil in water stabilized with a surfactant (usually lanolin and glycoproteins) from myobacterial or other microorganisms (85). Such adjuvant or immunostimulatory activity has been ascribed not only to complete Freund's adjuvant (85) in an oil vehicle but to silica (86), procainamide, hydralazine, d-penicillamine, tienilic acid, and trinitrobenzene sulfonic acid (48).

Scleroderma or human adjuvant illness in women who had breast augmentation by injection of paraffin or silicones or later silicone gel implants has been reported repeatedly since 1972 and finally led to FDA restrictions on the latter devices in 1992. Their leakage was demonstrated in vitro 15 years ago (94).

## J. Silicone Gel Implants

Of 19 implanted women, 5 had SLE, 8 progressive scleroderma, and 6 had rheumatoid arthritis (95). Of 113 scleroderma patients, 4.4% had a breast implant while in contrast only 0.3% of 286 patients with rheumatoid arthritis had had implants (95). Thrombocytopenic purpura have also been reported (96). SLE, mixed connective tissue disease, rheumatoid arthritis, and Sjögren's syndrome were also described (97). Arthritis followed silicone gel implants or injection and rheumatoid factor was elevated in 21 of 22 patients and ANA in 8 of 22 patients (B.O. Shoaib and B.M. Patten, personal communication). They also found central nervous system manifestations in many of these 22 patients, including sensorimotor neuropathy in 16, ALS in 2, MS in 2, myositis in 1, and myasthenia gravis in 1.

Twenty-four female patients with clinical syndromes similar to autoimmune disease were referred to Scripps Research Institute (La Jolla, CA) from July 1989 to January 1992 (98). Eleven women met criteria for defined diseases, seven had scleroderma, others had lupus erythematosus, rheumatoid arthritis, or overlapping autoimmune diseases. Ten of 11 had elevated ANA titers greater than 1:320 in 9 with patterns by indirect immunofluorescence, immunodiffusion, Western blot analysis, and immunoprecipitation similar to idiopathic forms of disease. Trauma to the breast accelerated the disease by reducing the average onset of symptoms from 8.4 to 2.8 years after implantation. Of the other 13 women, 4 had myositis, 4 had fibromyalgia, 4 had chronic fatigue syndrome, and 1 had arthralgias. Seven had ANA titers of 1:80 or greater.

Four women presented with progressive scleroderma 6 to 15 years after cosmetic mammoplasty, and had chronic inflammation at sites of elastomer leakage (99). Twenty-four of 156 women with silicone breast implants had connective tissue disease, 14 had scleroderma (100). At the Mayo Clinic, in 1994, 5 of 749 women with breast implants, and 10 of 1418 community controls had the same frequency of connective tissue disease (101). A larger study of women health professionals reported 10,830 had breast implants and 11,850 had reported connective tissue diseases between 1962 and 1991. The relative risk of connective tissue disease in the breast implant group was elevated to 1.24 95%CI 0.08 − 1.41 (102). Transfer to infants transplacentally or by breast feeding probably occurred in two female children who developed myalgias and +ANA (103). Six of eight other breast-fed children had scleroderma-like esophageal dysfunction (104).

The pathogenesis is unclear, but silicone definitely leaks into tissue and free silica particles are found in lymph nodes. Implant gels are composed of polydimethylsiloxane 69%, silica 30%, and catalyst 1% in a silica elastomere shell. In the United States alone, an estimated 2 million women have received such implants. In light of the known effects of silica and mineral oil adjuvants (86,87), such a large reservoir of silica in the body may induce continual immunostimulation. Fortunately, many patients recover from the connective tissue manifestation after the implants are removed, but neurotoxic effects are more lasting (B.O. Shoaib and B.M. Patten, personal communication) and therapy is difficult (97).

## K. Computer Manufacturing

The 289 exposed workers had been exposed for 10 years or more to chemicals in computer manufacturing, including phthalic anhydride, formaldehyde, isocyanate, trimillitic anhydride, and aliphatic and aromatic hydrocarbons. They were compared to 110 unexposed people. The exposed had significantly higher levels of ANA, rheumatoid factor, myelin basic protein IgG, IgM, IgA, and IgG, and IgM and IgA immune complexes. They also had headaches, mental status changes, and peripheral neuropathies (76).

## L. Hair Dyes

A population study in Georgia showed association between connective tissue disease and hair dye use (105). Using the Johns Hopkins lupus patient collection, a case control study found no association (106).

## M. Collagen Dermal Implants

Bovine dermal collagen is used by subcutaneous dermal injection to change the contours of the human face (107). Nine patients developed dermatomyositis or polymyositis on average 6.4 months after being injected with the bovine collagen, eight had delayed-type hypersensitivity, and five of six tested had serum antibodies to collagen (108).

# VII. TYPES OF DISEASE

## A. Graft-Versus-Host Disease

Allogenic bone marrow transplantation in human subjects for leukemia has caused severe and incapacitating graft-versus-host disease (GVHD). Because of its clinical similarities to VC- and TCE-induced autoimmune disease (109,110), it is included as a foreign chemical disease. GVHD resembles collagen vascular diseases, particularly scleroderma with dry eyes, pulmonary insufficiency, and wasting (111). Chronic GVHD is characterized by elevated eosinophilia, circulating autoantibodies, hypergammaglobulinemia, and plasmacytosis of lymph nodes and viscera. Malar rashes are common, and, after 1 year, thickened hidebound skin and alopecia are characteristic.

Eight of 20 patients were reported to have five or more of the ARA criteria for lupus, 11 of 17 had circulating autoantibodies that were ANA in eight and mitochrondial in three. Small-vessel intimal lesions were common. An in vitro active cytotoxic agent against endothelial cells (112) and a non-HLA alloantigen from endothelial cells in scleroderma serum have been implicated in renal allograph rejections (113). Chemical modified membrane determinants on B-cells or macrophages plus MAC class II molecules are needed to produce GVHD autologous target cells. B-cells, hematopoietic cells, or dendritic cells are altered to non-self by chemicals. Then autologous T-cells recognize and react against these altered cells (114).

In humans, prime candidate chemicals to do this are hydralazine, diphenylhydantoin (DPH), d-penicillamine, nitrofurantoin, mercuric chloride, Epstein-Barr virus, rubella virus, and cytomegalovirus. Models for this sequence typically carry certain major histocompability alleles (e.g., nonirradiated F1 mice in which the immunostimulatory or scleroderma pathology may develop). Alternately, the other major pathological response is suppressive or hypoplastic manifested by pancytopenia, aplastic anemia, thymic hypoplasia, and hypogammaglobulinemia. Graft-versus-host disease has been modeled by the popliteal lymph node response to injected chemicals in mouse hindlimbs (115).

This GVHD model provides a means to examine chemical induction of scleroderma-like disease as by DPH and similar anticonvulsant drugs (114) as lymph node proliferative activity (PLN). The PLN assay in mice (115) responds to chemical modification of the B-cell surface molecule and MHC class II molecules in mice provide a conceptual framework and an assay system for chemicals, which has tied together once disparate evidence (46).

## B. Organ-Specific Autoantibodies and Diseases

Less evidence has been found thus far of organ-specific autoantibodies induced by chemicals than for syndromes despite acceptance of several important diseases as having an autoimmune pathogenesis (Table 4).

### 1. *Hepatitis*

The drugs most often associated with autoimmune serologic reactions in 157 cases from five hepatologists were clometacin, fenofibrate, papaverine, and tienilic acid (116). Halothane and tienilic acid (tricryorafen, a uricosuria diuretic) produce immunoallergic hepatitis in humans with circulating antibodies against cell organelles. One protein, cytochrome (P-450-8) from human adult liver microsomes is recognized in most sera from patients with anti-liver/kidney microsome (anti-$LKM_2$) antibodies (48). These antibodies specifically inhibit the hydroxylation of tienilic acid by human liver microsomes. This suggests that autoantibodies form when the enzyme (P-450-8), present in endoplasmic reticulum, is alkylated by a reactive metabolite and migrates to the surface of the hepatocyte membrane. The modified protein is recognized by the cells reacting to the part of the molecule derived from the reactive metabolite. When this scenario is followed, the strategic borderline between self and non-self alluded to by Ward (117) has been crossed (3). Thus autoantibodies formed in hepatitis induced by administering tienilic acid catalyze the metabolic oxidation of the drug. It remains to be demonstrated whether anti-$LKM_2$ antibodies are responsible for the hepatitis in the one affected of 10,000 patients treated with tienilic acid. The recognition of self versus non-self is critical in the pathogenesis of autoimmune disease and the concept of autoimmunity. Therefore, observations that cytochrome P-450-8 elicits autoantibodies after conversion by a metabolite it initiated are important (114).

### 2. *Renal Disease*

Rapidly progressive glomerulonephritis has been seen with and without pulmonary hemorrhage (Goodpasture syndrome) associated with antibodies to glomerular basement membrane both bound and circulating (118). Detailed interviews with eight patients from a group of 13 with Goodpasture's antiglomerular basement membrane antibody mediated glomerulonephritis (anti-GBM) (autoimmune glomerulonephritis) found six of eight had extensive exposure to industrial solvents often as a heated vapor or mist (119). Degreasing and painting and paint stripping with heated solvents were frequent. One man fueled jet aircraft. Linearly deposited IgG was demonstrated along GBM by direct immunofluorescent microscopy in five patients and three of four had circulating antibodies. Goodpasture's syndrome has been induced in rabbits by instilling gasoline intratracheally and basement membrane antibodies were found in edematous lungs as well as in the kidneys (120).

Another example is the mercury-induced glomerulopathy in (PVG/C) rats which is associated with antinuclear antibodies against non-histone nucleoprotein and vasculitis (29,121). Both general T-cell reactivity to phytohemagglutinin and suppressor cell reactivity to concanavalin A were decreased in the mercury-diseased rats that are comparable to human drug-induced autoimmune disease. It seems logical to extend the search for chemical causes to myasthenia gravis, Guillain Barre syndrome, and autoimmune thyroiditis (122). The specific autoantibodies to acetylcholine receptor protein and to thyroglobulin are known, methods are well developed, and the frequency of these human diseases provides opportunities to search meticulously for association with chemicals.

### 3. *Systemic Vasculitis*

In some patients with systemic vasculitis, circulating autoantibodies to neutrophil cytoplasmic antigens have been detected using a solid-phase radioimmunoassay and their titers correlated with disease activity (123). Neurophil alkaline phosphatase has a component inserted in the cell membrane that might be recognized in systemic vasculitis and retains its enzymatic activity, which appears to be important in tissue injury (6). As young women apparently need one or more angiogenesis factors to replace the shed endometrium, observations that those with SLE and have low androgen levels and worsen during oral estrogen therapy (124,125) suggest examining these interactions. Also SLE exacerbations increased body mass index and fat mass independent of corticosteroids (126), again suggesting angiogenesis.

### 4. *Neurological Diseases*

Myasthenia gravis is characterized by weak muscles that fatigue quickly in use and is the prototypic antireceptor autoimmune disease with antibodies against the nicotinic acetylcholine receptor (AChR) (127,128) (Table 4). Anti-AChR autoantibodies are controlled by thymic T-cells that arise from impaired self-recognition (128). Levels of anti-AChR above 1.1 U/L are found in 91% of patients and the thymus is enlarged. D-Penicillamine, a four-carbon fragment of penicillin G with amino and thio groups is used to treat Wilson's disease because it chelates copper. It has also been used to treat rheumatoid arthritis, where it is thought to modulate T-lymphocyte function, immunosuppress the disease, and reduce IgM levels (129). In about 35% of patients with Wilson's disease and rheumatoid arthritis treated with penicillamine, autoimmune diseases develop including progressive systemic sclerosis, systemic lupus erythematosus, myasthenia gravis, polymyositis, and membranous glomerulonephritis (130). Between 0.4 and 1% of rheumatoid arthritis patients treated for several months develop myasthenia gravis and antibodies to AChR, frequently with antibodies to straited muscle (129). A mechanism that may induce autoimmune responses in myasthenia gravis is gold forming a mercaptide with cell surface thiol groups and D-penicillamine forming a mixed disulfide with these same thiol groups (127). In this regard, a possible insight has been provided in lupus psychosis by finding that antiribosomal p protein antibodies occur which are not found in lupus without psychosis, in other psychotics, or in normal controls (131). As CNS complaints occur in half of lupus patients, this is a promising lead as to its pathogenesis.

Amyotropic lateral sclerosis (ALS) is a devastating disease of unknown cause. When the spinal cords and motor cortices of patients with ALS were examined for IgG using immunohistochemical methods, motor neurons showed a granular staining pattern characteristic of binding to rough endoplasmic reticulation. A proportion of pyramidal cells were also stained. No staining was noted in control human tissues. Degenerating horn cells and microglia also stained for HLA-DR (132). Incubation of single mammalian skeletal muscle fibers with IgG from ALS reduced peak $Ca^{2+}$ current in the dehydropyridine-sensitive $Ca^{2+}$. Charge movement and effects were lost when IgG was boiled or absorbed with skeletal tubular membranes (133).

The findings that lupus psychosis was associated with a special antibody, that myasthenia gravis was mediated by an antibody to the acetylcholine receptor, and that ALS showed IgG binding to motor neurons invite speculation that multiple sclerosis, Parkinson's, ALS, and dementias of the Alzheimer's disease type may be due to chemical-induced autoimmunity (134). On one hand, at least three myelin-associated proteins—myelin basic protein, proteolipid protein, and GM-1 ganglioside—elicit autoantibodies. On the other hand, environmental chemicals used as food—cycad fruit in Guam and Japan (135) and ground (chicking) peas (136,137) in India—produce characteristic illnesses, chemically induced neurotoxicity that resembles spontaneous diseases. An ALS Parkinson's disease analogue is produced by a cycad-derived amino acid B, N-methylamino L-alanine (135) and a corticomotor neuronal deficit is produced by the amino acid B, N-oxyalylamino L-alanine from the chicking pea (*Lathyrus sativa*) (136,137). Whether immune mechanisms have a role or autoantibodies are present in these models is not known.

Overlap of chemically induced autoantibody production (ANA) and chemical neurotoxicity occurred in Tucson residents exposed via well water to TCE and other VOCs plus chromium (58). Although association does not imply a causal link, in a pilot study of Phoenix residents exposed to an even wider assortment of chemicals, especially VOCs, we found neurobehavioral impairments and antimyelin basic protein antibodies to both IgG and IgM (K. Kilburn, unpublished).

Perhaps these studies of cohorts exposed to low doses of environmental chemicals, who show autoimmune responses (ANA or anti-SM or anti-myelin antibodies) with lupus erythematosus or scleroderma, answer Bigazzi's question, ''Are there chemicals whose administration results in irreversible or progressive autoimmune disease?'' (4).

Parallel investigation of autoimmunity in the organ-specific disorders could be useful in establishing whether autoantibodies are elicited and, if so, are they involved in pathogenesis or simply an epiphenomenon. Epileptic patients treated with diphenylhydantoin have antibodies (ANA) (109) and so do schizophrenic patients treated with chloropromazine (138), but neither show additional CNS disease during drug treatment (139). This demonstrates that chemicals may induce autoantibodies without associated disease. However, a proper prospective study of children started on DPH and followed for many years for autoantibodies and associated diseases has not been reported. Therefore, more data and careful analysis must precede acceptance of each autoantibody-associated disease.

## VIII. MECHANISMS

Vinyl chloride was postulated to be incorporated by the liver into amino acid synthesis of a structurally abnormal protein that would be antigenically foreign (65). DNA is a poor immunogen (140) so that a plausible first step is to find a mechanism for increasing its immunogenicity. DNA adducts to epoxides formed during metabolism of chloroalkenes (77). DNA epoxide adducts stimulate autoantibodies and may behave as do formaldehyde serum albumin conjugates (141), hydralazine (142), hydralazine human serum albumin conjugates (143), DNA photo oxidation products (144), and acrylamide. If exposure to TCE is chronic, sufficient antigen may be available to form immune complexes. Concommitantly small chemical molecules, especially epoxides, may attach to B-cell surface molecules and thus arouse T-cells to react against them, as in Kammuler's model (46).

Two chemicals that cause drug-induced lupus—procainamide and hydralazine—bind to and alter DNA (142). Hydralazine can complex with soluble nucleoprotein and change its physical properties without changing its antigenicity while procainamide binds to single-stranded and native DNA (138). Rabbits immunized with hydralazine conjugated to human serum albumin developed rising titers of antibodies to the drug and to single-stranded and native DNA. The importance of the free amine group is shown by studies of *N*-acetylprocainamide, which lacks a free amine group and does not induce ANA or SLE (138). After 55 years of experiencing ''drug-induced lupus'' (145), the lesson is clear that connective tissue diseases can be induced by chemicals used as drugs. A list that shows the number of these associations (Table 3) is from a recent review (35).

### A. Genetic Factors

Just as the linkage of trichloroethylene to antinuclear and other autoantibodies and typical lupus symptoms proclaims associations that may have causal significance, the possibility that widely spread environmental contaminants, such as polychlorinated biphenyls, especially dioxins, in concert with genetic predisposition (perhaps as manifested by the MHC antigens HLA-DR5 for scleroderma and HLA-DR3 for SLE, may explain seemingly peculiar selection and variable disease manifestations within population groups. It is clear that genetic factors influence the risk of chemical-induced SLE or scleroderma acting via the HLA, major histocompatibility, acetylation, and cytochrome P-450 pathway (116). For example, impaired sulfoxidation of carbocysteine, which is similar to d-penicillamine, was associated with toxicity in patients with rheumatoid arthritis. In subjects with excessive oxidation, HLA-DR3 was an independent risk factor (146). Eight mechanisms postulated for autoimmunity due to chemicals, mostly from animal experiments (4), have included complexing with autoantigen (hydralazine), release of autoantigen (gold, cadmium) cross-reaction with autoantigen (hydralazine), immunogen, or hapten (penicillamine), inhibition of T-suppressor cells (methyldopa, practobol, procainamide, and mercury), stimulation of T-helper cells (procainamide, beta-blockers, phenytoin, and mercury), stimulation of B-cells (mercury, beta-blockers and phenytoin, penicillamine, iodine), and stimulation of macrophages (penicillamine, propylthiouracil, and iodine).

## IX. CONCLUSIONS

Rheumatoid arthritis, lupus erythematosus, scleroderma, and dermatomyositis can be induced by chemicals that also raise autoantibodies to altered body components. Rheumatoid arthritis has been induced by silica, by paraffin injection and silicones, and, in many instances, autoantibodies have been observed (Table 4). Although many of the chemicals that induce lupus have been drugs (147), TCE and other solvents and silicones in breast implants clearly extend and broaden these observations. Scleroderma, in contrast, has been linked more often to industrial chemicals especially chlorinated and aromatic solvents (75), although bleomycin has been implicated (148).

Polymyositis-dermatomyositis is another disease with a rising incidence. It has tripled from 1963 to 1982 in Allegheny County, PA, with a female-to-male ratio of 5:1 in the childbearing years (149). Antibodies to a saline extractable nuclear antigen (JO-1) were associated with interstitial lung disease (150). This circles us back to the silicosis-SLE association and again raises causal questions about angiogenesis (6). Perhaps the link is oxidative stress (23), although the intermediate steps are still elusive. Certainly the lists of causal agents for these ''connective tissue diseases'' overlap remarkably (151).

1. Mechanisms are diverse.
2. Structure function relationships are not identifiable.
3. Chemically induced delayed hypersensitivity has not been explored but in hepatitis effector T-cell reactions are not accompanied by autoantibody production.
4. Genetic predisposition appears to be essential and may be either immunogenic (MHC phenotype) or chemical (acetylator phenotype, sulfoxidizer, or aromatic hydrocarbon receptor).

## X. FUTURE DIRECTIONS

We need more cause-directed searches for chemical initiators in patients with autoimmune disease, and inclusion of autoantibody measurements in more epidemiological studies of subjects exposed occupationally or environmentally to chemicals. The flare in SLE without raising serologic titers (152) suggests finding the responsible agents. Among the communication problems are better definitions of relapses (25,153) and flares (152) and the relationship of it or these to oxidants (23). More hypothesis testing of chemical agents in animal models would blend chemical induction against variable genetic endowments.

Of further specific importance are explorations on interrelationships between SLE and the sclerosis disorders and CNS disorders building on the elegant models of antiacetylcholine receptors in myasthenia gravis and Guillain Barre syndrome, and extending to dementia and the amyotrophic lateral sclerosis–Parkinson's disease–dementia syndrome (134,135). There is enough evidence for functional CNS impairment in autoimmune CNS

**Table 5** Phenomena To Explain

1. SLE prevalence went from rare, less than 1:100,000 to 4–10 per 100,000 from 1950 to 2000. SLE is a female disease with an 8:1 ratio.
2. Females release more angiogenic factors.
3. New blood vessels lack vasoregulatory factors.
4. Cycling may lead to errors in vasoregulation.
5. Certain environmental chemicals raise SLE prevalence 10 times or more, silica dust, procainamide, and hydrazine.
6. Oxidizing agents increase prevalence.
7. Inhalation and lung processing of particles stimulate PMN leukocytes to release $O_3^+$ intracellarly via myeloperoxidase.
8. Curiously, many antirheumatic agents inhibit angiogenesis and these same agents cause lupus: gold, penicillamine, sulfasalazine, methotrexate, and nonsterodial anti-inflammatory agents.
9. Would angiogenesis-inhibiting factors modify the joint and serous surface manifestations of SLE?
10. Two other processes and loci are important: endothelial growth and depositions of immunoglobulins as in glomeruli, cerebral vessels, and the cardiac endothelium.

disease (61,110,111), particularly in GVHD, as in the lupus psychosis antiribosomal P-protein work (131) to justify further investigations. Certainly finding elevated anticardiolipin antibodies in SLE patients with epilepsy in Taiwan (154) suggests another link to vasculitis. Such studies should be coupled with a diligent thoughtful search for evidence of chronic exposure to chemicals in each patient's environment.

This review has suggested certain phenomena that need to be explained (Table 5). Studies in the last 7 years make the point that lupus continues to fascinate physicians. It even has its own journal. An association of lupus and multiple myeloma has been found in Nogales, AZ (155), associated with chemical pollution of air and water from ''border'' industries. Some connections between phenomena as associations should be explored, such as ''oxidizing influences'' that increase vascular reactivity and angiogenesis factors and may disturb synovium, and other serous surfaces may integrate some of the amazing features of connective tissue disease in the twenty-first century.

Finally, if this chapter has made connections requiring leaps of faith, it should be understood that the author is a student of immunology, not a practitioner. Perhaps considerable data should be ignored and larger leaps will prove to be beneficial, since this will provoke thought and stimulate exploration of these diseases. The association of connective tissue disease with autoimmune activation does not have an invariable pathogenic connection. Thus we should move away from the ''idiopathic'' label to search for chemical causes of these diseases.

## REFERENCES


1. Kilburn KH, Warshaw RH. Chemical-induced autoimmunity. In: Dean JH, Luster MI, Munson AE, Kimber I, eds. Immunotoxicology and Immunopharmacology, 2nd ed. New York: Raven Press, Ltd., 1994:523–538.
2. Gleichmann E, Kimber I, Purchase IFH. Immunotoxicology: suppressive and stimulatory effects of drugs and environmental chemicals on the immune system. Arch Toxicol 1989; 63:257–273.
3. Haustein UF, Ziegler V. Environmentally induced systemic sclerosis like disorders. Int J Dermatol 1985; 24:147–151.
4. Bigazzi PE. Autoimmunity induced by chemicals. Clin Toxicol 1988; 26:125–156.
5. Yoshida S, Gershwin ME. Autoimmunity and selected environmental factors of disease induction. Sem Arthr Rheum 1993; 22:399–419.
6. Walsh DA. Angiogenesis and arthritis. Rheumatology 1999; 38:103–112.
7. Klauber N, Rohan RM, Flynn E, D'Amato RJ. Critical components of the female reproductive pathway are suppressed by the angiogenesis inhibitor AGM-1470. Nature Med 1997; 4:443–446.
8. Benedek TG. Historical background of discoid and systemic lupus erythematosus. In: Dubois' Lupus Erythematosus, 5th ed. Baltimore, MD: Williams & Wilkins, 1997:3–16.
9. Kaposi M. Neue beitrage zur kenntnis des lupus erythematosus. Arch Dermat Syph 1872; 4:36–78.
10. Osler W. The visceral lesions of the erythema group. Br J Dermatol 1900; 12:227–245.
11. Libman E, Sachs B. A hitherto undescribed form of valvular and mural endocarditis. Arch Intern Med 1924; 33:701–737.
12. Jarcho S. Lupus erythematosus associated with visceral vascular lesions. Bull J Hopkins Hosp 1936; 59:262–270.
13. Hargraves MM. Production in vitro of the LE cell phenomenon: use of normal bone marrow elements and blood plasma from patients with acute disseminated lupus erythematosus. Proc Staff Mayo Clin 1949; 24:234–237.

14. Ceppelini R, Polli E, Celada F. A DNA-reacting factor in serum of a patient with lupus erythematosus diffusus. Proc Soc Exp Biol Med 1957; 96:572–574.
15. Robbins WC, Holman HR, Deicher HR, Kunkel HG. Complement fixation with cell nuclei and DNA in lupus erythematosus. Proc Soc Exp Biol Med 1957; 96:575–579.
16. Koffler D, Carr RI, Agnello V, Fiezi T, Kunkel HG. Antibodies to polynucleotides: distribution in human serum. Science 1969; 166:1648–1649.
17. Burnham TK, Neblett TR, Fine G. The application of the fluorescent antibody technic to the investigation of lupus erythematosus and various dermatoses. J Invest Dermatol 1963; 41: 451–456.
18. Svanborg A, Solvell L. Incidence of disseminated lupus erythematosus. JAMA 1957; 165: 1126–1128.
19. Albert DA, Hadler NH, Ropes MW. Does corticosteroid therapy affect the survival of patients with systemic lupus erythematosus? Arthritis Rheum 1979; 22:945–953.
20. Hogg GR. Congenital acute lupus erythematosus associated with subendocardial fibroelastosis. Am J Clin Pathol 1957; 28:648–654.
21. McCue CM, Mantakas ME, Tingelstad JB, Ruddy S. Congenital heart block in newborns of mothers with connective tissue disease. Circulation 1977; 56:82–90.
22. Rich AR. Hypersensitivity in disease, with special reference to periarteritis nodosa, rheumatic fever, disseminated lupus erythematosus and rheumatoid arthritis. Harvery Lect 1947; 42: 106–147.
23. Ames PRJ, Alves J, Murat I, Isenberg DA, Nourooz-Zadeh J. Oxidative stress in systemic lupus erythematosus and allied conditions with vascular involvement. Rheumatology 1999; 38:529–534.
24. Sanchez-Roman J, Wichmann I, Salaberri J, Varela JM, Nunez-Roldan A. Multiple clinical and biological autoimmune manifestations in 50 workers after occupational exposure to silica. Ann Rheum Dis 1993; 52:534–538.
25. Bootsma H, Spronk P, Derksen R, deBoer G, Wolters-Dicke H, Hermans J, Limburg P, Gmelig-Meyling F, Kater L, Kallenberg C. Prevention of relapses in systemic lupus erythematosus. Lancet 1995; 345:1595–1599.
26. Love LA. New environmental agents associated with lupus-like disorders. Lupus 1994; 3: 467–471.
27. Bigazzi PE. Autoimmunity and heavy metals. Lupus 1994; 3:449–453.
28. Yung RL, Richardson BC. Pathophysiology of drug-induced lupus. In: Lahita RG, ed. Systemic Lupus Erythematosus, 3rd ed. San Diego: Academic Press, 1999:909–928.
29. Pollard KM, Lee DK, Casiano CA, Bluthner M, Johnston MM, Tan EM. The autoimmunity-induced xenobiotic mercury interacts with the autoantigen fibrillarin and modifies its molecular and antigenic properties. J Immunol 1997; 158:3521–3528.
30. Kausman D, Isenberg DA. Role of the biologics in autoimmunity. Lupus 1994; 3:461–466.
31. Nesher G, Osborn TG, Moore TL. Parvovirus infection mimicking systemic lupus erythematosus. Semin Arthr Rheum 1995; 24:297–303.
32. Levin RE, Weinstein A, Peterson M, Testa MA, Rothfield NF. A comparison of the sensitivity of the 1971 and 1982 American Rheumatism Association criteria for the Classification of systemic lupus erythematosus. Arthr Rheum 1984; 27:530–538.
33. Liao L, Sindhwani R, Rojkind M, et al. Antibody-medicated autoimmune myocarditis depends on genetically determined target organ sensitivity. J Exp Med 1995; 181:1123–1131.
34. Kasturi KN, Hatakeyama A, Spiera H, Bona CA. Antifibrillarin autoantibodies present in systemic sclerosis and other connective tissue diseases interact with similar epitopes. J Exp Med 1995; 181:1027–1036.
35. Richeldi L, Sorrentino R, Saltini C. DPBI glutamate 69: a genetic marker of beryllium diseases. Science 1993; 262:242–244.
36. Mongey AB, Hess EV. Drug and environmental lupus: clinical manifestations and differ-

ences. In: Lahita RG, ed. Systemic Lupus Erythematosus, 3rd ed. San Diego: Academic Press, 1999:929–943.
37. Delafuente JC. Nutrients and immune responses. Rheum Dis Clin North Am 1991; 17:203–212.
38. Malinow MR, Bardana EJ, Pirofsky B, Craig S. Systemic lupus erythematosus like syndrome in monkeys fed alfalfa sprouts: role of a nonprotein amino acid. Science 1982; 216:415–417.
39. Montanaro A, Bardana EJ Jr. Dietary amino acid-induced systemic lupus erythematosus. Rheum Dis Clin North Am 1991; 17:323–332.
40. Wright ES, Dziedzic D, Wheeler CS. Minireview. Cellular, biochemical and functional effects of ozone: New research and perspectives on ozone health effects. Toxicol Lett 1989; 51:125–145.
41. Goodman JW, Peter-Fizaine FE, Shinpock SG, et al. Immunologic and hematologic consequences in mice of exposure to ozone. J Environ Pathol Toxicol Oncol 1989; 9:243–252.
42. Noonan FP, De Fabo EC. Immunosuppression by ultraviolet B radiation: initiation by urocanic acid. Immunol Today 1992; 13:250–254.
43. Lemire JM. Immunomodulatory role of 1,25-dihydroxyvitamin D3. J Cell Biochem 1992; 49:26–31.
44. Huetz F, Jacquemart F, Rossi CP, et al. Autoimmunity: The moving boundaries between physiology and pathology. J Autoimmun 1988; 1:11–22.
45. Vertosick FT, Kelly RH. Immune network theory: A role for parallel distributed processing? Immunology 1989; 66:1–7.
46. Kammuller ME, Penninks AH, deBakker JM, Thomas C, Bloksma N, Seinen W. An experimental approach to chemically induced systemic (auto) immune alterations: The Spanish toxic oil syndrome as an example. In: Fowler BA, ed. Mechanisms of Cell Injury: Implications for Human Health. New York: John Wiley and Sons, 1987:175–192.
47. Kammuller ME, Bloksma N, Seinen W. Chemical induced autoimmune reactions and Spanish toxic oil syndrome: focus on hydantoins and related compounds. Clin Toxicol 1988; 26: 157–174.
48. Beaune PH, Dansette PM, Mansuy D, et al. Human antiendoplasmic reticulum autoantibodies appearing in a drug induced hepatitis are direct against a human liver cytochrome P-450 that hydroxylates the drug. Proc Natl Acad Sci 1987; 84:551–555.
49. Aucoin DP, Peterson ME, Hurvitz AI, Drayer DE, Lahita NG, Quimby TW, Reidenberg MM. Propylthiouracil induced immune mediated disease in the cat. J Pharmacol Exper Ther 1985; 234:13–18.
50. Hall CL. The natural course of gold and penicillamine nephropathy: a long-term study of 54 patients. Adv Exp Med Biol 1989; 252:247–256.
51. Kotsy MP, Hench PK, Tani P, McMulian R. Thrombocytopenia associated with auronofin therapy: evidence of a gold-dependent immunologic mechanism. Am J Hematol 1989; 30: 236–239.
52. Pilletier L, Bellon B, Tournade H. In: Bona CA, Kaushik AK, eds. Chemical-Induced Autoimmunity in Molecular Immunology of Self Reactivity. New York: Marcel Dekker, Inc., 1992:315–323.
53. Bigazzi PE. Lessons from animal models: the scope of mercury-induced autoimmunity. Clin Immunol Immunopathol 1992; 65:81–84.
54. Cardenas A, Roels H, Bernard AM, et al. Markers of early renal changes induced by industrial pollutants. I. Application to workers exposed to mercury vapour. Br J Ind Med 1993; 50: 17–27.
55. Schrallhammer-Benkler K, Ring J, Przybilla B, Landthaler M. Acute mercury intoxication with licheniod drug eruption followed by mercury contact allergy and development of antinuclear antibodies. Acta Dermatol Venereol (Stockholm) 1992; 72:294–296.
56. Bernard AM, Roels HR, Foidart JM, Lauwerys RL. Search for antilaminim antibodies in the

serum of workers exposed to cadmium, mercury vapour or lead. Int Arch Occup Environ Health 1987; 59:303–309.
57. Tsankov N, Stransky L, Kostowa M, Mitrowa T, Obreschkowa E. Induced pemphigus caused by occupational contact with Basochrom. Dermat Beruf Umwelt. Occup Environ Dermat 1990; 38:91–93.
58. Kilburn KH, Warshaw RH. Prevalence of symptoms of systemic lupus erythematosus (SLE) and of fluorescent antinuclear antibodies associated with chronic exposure to trichloroethylene and other chemicals in well water. Environ Res 1992; 57:1–9.
59. Hassman RA, McGregor AM. Lithium and autoimmune thyroid disease. Lithium Ther Monogr 1988; 2:134–146.
60. Santella RN, Rimmer JM, MacPherson BR. Focal segmental glomerulosclerosis in patients receiving lithium carbonate. Am J Med 1988; 84:951–954.
61. Stein EC, Schiffer RB, Hall WJ, Young N. Multiple sclerosis and the workplace: report of an industry-based cluster. Neurology 1987; 37:1672–1677.
62. Schiffer RB, Herndon RM, Eskin T. Effects of altered dietary trace metals upon experimental allergic encephalomyelitis. Neurotoxicol 1990; 11:443–450.
63. Federmann M, Morell B, Graetz G, Wyss M, Elsner P, von Thiessen R, Wuthrich B, Grob D. Hypersensitivity to molybdenum as a possible trigger of ANA-negative systemic lupus erythematosus. Ann Rheum Dis 1994; 53:403–405.
64. Lange CE, Juhe S, Veltman G. Uber das auftreten von angiosarkomen der leber bei zwei arbeitern der PVC herstellenden industrie. Dtsch Med Wschr 1974; 99:1598–1599.
65. Ward AM, Udnoon S, Watkins J, Walker AE, Darke CS. Immunological mechanisms in the pathogenesis of vinyl chloride disease. Br Med J 1976; 1:936–938.
66. Black CM, Welsh KI, Walker AE, Bernstein RM, Catoggio LJ, McGregor AR, Jones JKL. Genetic susceptibility to scleroderma like syndrome induced by vinyl chloride. Lancet 1976; Jan:53–55.
67. Reinl W. Sklerodermic durch trichlorethylene in working? Z Arch Arteits 1957; 7:58–60.
68. Walder BK. Solvents and scleroderma. Lancet 1965; 2:436.
69. Walder BK. Do solvents cause scleroderma? Int J Dermatol 1983; 22:157–158.
70. Yamakage A, Ishikawa H. Generalized morphea like scleroderma occurring in people exposed to organic solvents. Dermatological 1982; 165:186–193.
71. Lockey JE, Kelly CR, Cannon GW, Colby TV, Aldrich V, Livingston GK. Progressive systemic sclerosis associated with exposure to trichloroethylene. J Occup Med 1987; 29:493–496.
72. Sparrow GP. A connective tissue disorder similar to vinyl chloride disease in a patient exposed to perchlorethylene. Clin Exper Dermatol 1977; 2:17–22.
73. Nietert PJ, Sutherland SE, Silver RM, Pandey JP, Knapp RG, Hoel DG, Dosemeci M. Is occupational organic solvent exposure a risk factor for scleroderma? Arthr Rheum 1998; 41(6):1111–1118.
74. Saihan EM, Burton JL, Heaton KW. A new syndrome with pigmentation, scleroderma, gynaecomastia, Raynaud's phenomenon and peripheral neuropathy. Br J Dermatol 1978; 99: 437–440.
75. Sverdrup B. Do workers in the manufacturing industry run an increased risk of getting scleroderma? (letter) Int J Dermatol 1984; 23:629.
76. Vojdani A, Ghoneum M, Brautbar N. Immune alteration associated with exposure to toxic chemicals. Toxicol Industr Health 1992; 8:239–254.
77. Savolainen H. Some aspects of the mechanisms by which industrial solvents produce neurotoxic effects. Chem Biol Interactions 1977; 18:1–10.
78. Byers VS, Levin AS, Ozonoff DM, Baldwin RW. Association between clinical symptoms and lymphocyte abnormalities in a population with chronic domestic exposure to industrial solvent contaminated domestic water supply and a high incidence of leukaemia. Cancer Immunol Immunother 1988; 27:77–81.

79. Kardestuncer T, Frumkin H. Systemic lupus erythematosus in relation to environmental pollution: an investigation in an African-American community in North Georgia. Arch Environ Health 1997; 52:85–90.
80. Yamakage A, Ishikawa H, Saito Y, Hattori A. Occupational scleroderma-like disorder occurring in men engaged in the polymerization of epoxy resins. Dermatologica 1980; 161: 33–44.
81. Tabuenca JM. Toxic allergic syndrome caused by ingestion of rapeseed oil denatured with aniline. Lancet 1981; Sept:567–568.
82. Alonso-Ruiz A, Zea-Mendoze AC, Salazar-Vallinas JM, Rocamora-Ripoll A, Beltran-Gutierrez J. Toxic oil syndrome: a syndrome with features overlapping those of various forms of scleroderma. Semin Arthr Rheum 1986; 15:200–212.
83. Silver RM, Heyes MP, Maize JC, Quearry B, Vionnet-Fausset M, Sternberg EM. Scleroderma, fascitis and eosinophilia associated with the ingestion of tryptophan. N Engl J Med 1990; 322:874–881.
84. Belongia EA, Hedberg CN, Gleich GJ, White KE, Mayeno AN, Loegeriny DA, Dunnette SL, Pirie PL, McDonald KL, Osterholm MT. An investigation of the cause of eosinophilia myalgia syndrome associated with tryptophan use. N Engl J Med 1990; 323:357–365.
85. Pearson CM. Development of arthritis, periarthritis, periostitis in rats given adjuvants. Proc Soc Exp Biol Med 1973; 143:95–99.
86. Pernis B, Paronetto F. Adjuvant effect of silica (tridymite) on antibody production. P.S.E.B.M. 1962; 110:390–392.
87. Gunther G, Schuchardt E. Silikose und progressive. Skleroderm Dtsch Med Ochenschr 1970; 95:467–468.
88. Rodnan GP, Benedek TG, Medsger TA, Gammarata RJ. The association of progressive systemic sclerosis (scleroderma) with coal miners' pneumoconiosis and other forms of silicosis. Ann Intern Med 1967; 66:323–334.
89. Ziskind M, Jones RN, Weill H. Silicosis. ARRD 1967; 113:643–664.
90. Masson C, Audran M, Pascaretti C, Chevailler A, Subra JF, Tuchais E, Kahn MF. Silica-associated systemic erythematosus lupus or mineral dust lupus? Lupus 1997; 6:1–3.
91. Conrad K, Mehlhorn J, Luthke K, Dorner T, Frank KH. Systemic lupus erythematosus after heavy exposure to quartz dust in uranium mines: clinical and serological characteristics. Lupus 1996; 5(1):62–69.
92. Koeger AC, Lang T, Alcaix D, Milleron B, Rozenberg S, Chaibi P, Arnaud J, Mayaud C, Camus JP, Bourgeois P. Silica-associated connective tissue disease. A study of 24 cases. Medicine 1995; 74(5):221–237.
93. Kumagai Y, Shiokawa Y, Medsger TA, Rodnan GP. Clinical spectrum of connective tissue disease after cosmetic surgery. Arth Rheum 1984; 27:1–12.
94. Barker DE, Retsky MI, Schultz S. Bleeding of silicone from bag gel breast implants, and its clinical relation to fibrous capsule reaction. Plast Reconstr Surg 1978; 61:836–841.
95. Spiera H. Scleroderma after silicone augmentation mammoplasty. J Am Med Assoc 1988; 260:236–238.
96. Fock KM, Feng PH, Tey BH. Autoimmune disease developing after augmentation mammoplasty: Report of 3 cases. J Rheum 1984; 11:98–100.
97. Van Nunen SA, Gatenby PA, Basten A. Post mammoplasty connective tissue disease. Arthr Rheum 1982; 25:694–696.
98. Press RL, Peebles CL, Kumagai Y, Ochs RL, Tan EM. Antinuclear autoantibodies in women with silicone breast implants. Lancet 1992; 340:1304–1306.
99. Varga J, Schumacher HR, Jimenez SA. Systemic sclerosis after augmentation mammoplasty with silicone implants. Ann Intern Med 1989; 111:377–383.
100. Bridges AJ, Conley C, Wang G, Burns DE, Vasey FB. A clinical and immunologic evaluation of women with silicone breast implants and symptoms of rheumatic disease. Ann Intern Med 1993; 118:929–936.

101. Gabriel SE, O'Fallon WM, Kurland LT, Beard CM, Woods JE, Melton LJ III. Risk of connective-tissue diseases and other disorders after breast implantation. N Engl J Med 1994; 330:1697–1702.
102. Hennekens CH, Lee IM, Cook NR, Hebert PR, Karlson EW, LaMotte F, Manson JE, Buring JE. Self-reported breast implants and connective-tissue diseases in female health professionals. JAMA 1996; 275:616–621.
103. Teuber SS, Gershwin ME. Autoantibodies and clinical rheumatic complaints in two children of women with silicone gel breast implants. Int Arch Allergy Immunol 1994; 103:105–108.
104. Levine JJ, Ilowite NT. Scleroderma like esophageal disease in children breast-fed by mothers with silicone breast implants. JAMA 1994; 271:213–216.
105. Freni-Titulaer LWJ, Kelley DB, Grow AG, McKinley TW, Arnett FC, Hochberg MC. Connective tissue disease in southeastern Georgia: A case-control study of etiologic factors. Am J Epidemiol 1989; 130:404–409.
106. Petri M, Allbritton J. Hair product use in systemic lupus erythematosus. A case-control study. Arthr Rheum 1992; 35:625–629.
107. Yarborough JM. Review the treatment of soft tissue defects with injectable collagen. Am J Med Sci 1985; 290:28–31.
108. Cukier J, Beauchamp RA, Spindler JS, Spindler S, Lorenzo C, Trentham DE. Association between bovine collagen dermal implants and a dermatomyositis or a polymyositis-like syndrome. Ann Intern Med 1993; 118:920–928.
109. Gleichmann E, Gleichmann H. Graft versus host reaction: a pathogenetic principle for the development of drug allergy, autoimmunity and malignant lymphoma in non chemiric individuals. Hypothesis. Z Krebsforsch 1976; 85:91–109.
110. Gleichmann E, Pals ST, Rolink AG, Radaszkiewicz T, Gleichmann H. Graft versus host reactions: clues to the etiopathology of a spectrum of immunological diseases. Immunol Today 1984; 5:324–332.
111. Shulman HM, Sullivan KM, Weiden PL, McDonald GB, Striker GE, Sale GE, Hackman R, Tsoi MS, Stort R, Thomas ED. Chronic graft versus host syndrome in man. Am J Med 1980; 69:204–216.
112. Kahaleh MB, Sherer GK, LeRoy EC. Endothelial injury in scleroderma. J Exp Med 1979; 149:1326–1339.
113. Moraes JR, Stastny P. A new antigen system expressed in human endothelial cells. J Clin Invest 1977; 60:449–454.
114. Gleichmann H, Pals ST, Radaszkiewicz T. T cell dependent B cell proliferation and activation induced by the drug diphenylhydantoin in mice. Hematol Oncol 1983; 1:165–176.
115. Ford WL, Burr W, Simonson M. A lymph node weight assay for the graft versus host activity of rat lymph cells. Transplantation 1970; 10:258–266.
116. Homberg JC, Andre C, Abuaf N. A new anti liver kidney microsome antibody (anti LKM2) in tienilic acid induced nephritis. Clin Exp Immunol 1984; 55:561–570.
117. Ward AM. Evidence of an immune complex disorder in vinyl chloride workers. Proc R Soc Med 1976; 69:289–291.
118. Lerner RA, Glassock RJ, Dixon FJ. The role of anti glomerular basement membrane antibody in the pathogenesis of human glomerulonephritis. J Exp Med 1967; 126:989–1004.
119. Beirne GJ, Brennan JT. Glomerulonephritis associated with hydrocarbon solvents. Arch Environ Health 1972; 25:365–369.
120. Yamamoto T, Wilson CB. Binding of anti basement membrane antibody to alveolar basement membrane after intratracheal gasoline instillation in rats. Am J Pathol 1987; 126:497–505.
121. Weening JJ, Hoedemaeker J, Bakker WW. Immunoregulation and anti nuclear antibodies in mercury induced glomerulopathy in the rat. Clin Exp Immunol 1981; 45:64–71.
122. Patterson PY, Day ED. Neuroimmunologic disease: experimental and clinical aspects. In:

Dixon FJ, Fisher DW, eds. The Biology of Immunologic Disease. Sunderland, MA: Sinauer Association, Inc., 1983.

123. Lockwood CM, Bakes D, Jones S, Whitaker KB, Moss DW, Savage CO. Association of alkaline phosphatase with an autoantigen recognised by circulating anti neutrophil antibodies in systemic vasculitis. Lancet 1987; 1:716–720.
124. Jungers P, Dougados M, Pelissier C, Kuttenn F, Tron F, Lesavre P, Bach JF. Influence of oral contraceptive therapy on the activity of systemic lupus erythematosus. Arthr Rheum 1982; 25:618–623.
125. Jungers P, Nahoul K, Pelissier C, Dougados M, Tron F, Bach JF. Low plasma androgens in women with active or quiescent systemic lupus erythematosus. Arthr Rheum 1982; 25: 454–457.
126. Kipen Y, Briganti EM, Strauss BJG, Littlejohn GO, Morand EF. Three year follow-up of body composition changes in pre-menopausal women with systemic lupus erythematosus. Rheumatology 1999; 38:59–65.
127. Smiley JD, Moore SE. Southwestern internal medicine conference: Molecular mechanisms of autoimmunity. Am J Med Sci 1988; 295:478–496.
128. Stobo JD. Autoimmune antireceptor diseases. In: Biology of Immunologic Disease. Dixon FJ, Fisher DW, eds. Sunderland, MA: Sinauer Assoc., Inc. 1983.
129. Jaffe IA. Penicillamine treatment of rheumatoid arthritis: Rationale, pattern of clinical response, and clinical pharmacology and toxicology. Proc NY Acad Med 1976; 354:11–24.
130. Bacon PA, Tribe CR, MacKenzie JC, Verrier Jones J, Cumming RH, Amer B. Penicillamine nephropathy in rheumatoid arthritis. Q J Med 1976; 180:661–684.
131. Bonfa E, Glombek SJ, Kaufman LD, Skelly S, Weissbach H, Brot N, Elkon KB. Association between lupus psychosis and antiribosomal P protein antibodies. N Engl J Med 1987; 317: 265–271.
132. Engelhardt JI, Appel SH. IgG reactivity in the spinal cord and motor cortex in amyotrophic lateral sclerosis. Arch Neurol 1990; 47:1210–1216.
133. Delbono O, Garcia J, Appel SH, Stefani E. IgG from amyotrophic lateral sclerosis affects tubular calcium channels of skeletal muscle. Am J Physiol 1991; 260:C1347–1351.
134. Benzing WC, Jufson EJ, Jennes L, Armstrong DM. Reduction of neurotensin immunoreactivity in the amygdala in Alzheimer's disease. Brain Res 1990; 537:298–302.
135. Spencer PS, Nunn PB, Hugo J, Ludolph AC, Ross SM, Roy DN, Robertson RC. Guam amyotrophic lateral sclerosis Parkinson dementia linked to a plant excitant neurotoxin. Science 1987; 237:517–522.
136. Kissler A. Lathyismus monatschrift. Psychiatr Neurol 1947; 113:345–375.
137. Ludolph AC, Hugon J, Dwivedi MP, Schaumburg HH, Spencer PS. Studies on the aetiology and pathogenesis of motor neuron diseases. Brain 1987; 110:149–165.
138. Schoen RJ, Trentham DE. Drug induced lupus: an adjuvant disease? Am J Med 1981; 71: 5–8.
139. Weinstein A. Drug induced systemic lupus erythematosus. Prog Clin Immunol 1980; 4:1–21.
140. Friou GJ. Double stranded DNA: an antigen of unique significance. J Lab Clin Med 1978; 91:545–549.
141. Patterson R, Dykewicz MS, Evans R 3d, Grammer LC, Greenberger PA, Harris KE, Lawrence ID, Pruzansky JJ, Roberts M, Shaughnessy MA, et al. IgG antibody against formaldehyde human serum proteins: a comparison with other IgG antibodies against inhalant proteins and reactive chemicals. J Allergy Clin Immunol 1989; 84:359–366.
142. Eldredge NT, Robertson WVB, Miller JJM. The interaction of lupus inducing drugs with deoxyribonucleic acid. Clin Immunol Immunopathol 1974; 3:263–271.
143. Yamauchi Y, Litwin A, Adams L, Zimmer H, Hess EV. Induction of antibodies to nuclear antigens in rabbits by immunization with hydralazine human serum albumin conjugates. J Clin Invest 1975; 56:958–969.

144. Blomgren SE, Vaughan JH. The immunogenicity of photo oxidized DNA and of the photoproduct of DNA and procainamide hydrochloride. Arthr Rheum 1968; 11:470.
145. Hoffman BJ. Sensitivity to sulfdiazine resembling acute disseminated lupus erythematosus. Arch Dermatol Syph 1945; 51:190–192.
146. Emery P, Panayi GS, Huston G, et al. D penicillamine induced toxicity in rheumatoid arthritis: the role of sulphoxidation status and HLA DR3. J Rheumatol 1984; 11:626–632.
147. Fritzer MJ. Drugs recently associated with lupus syndromes. Lupus 1994; 3:455–459.
148. Finch WR, Rodnan GP, Buckingham RB, Prince RK, Winkelstein A. Bleomycin-induced scleroderma. J Rheumatol 1980; 7:651–659.
149. Oddis CV, Conte CG, Steen VD, Medsger TA Jr. Incidence of polymyositis-dermatomyositis: a 20 year study of hospital diagnosed cases in Allegheny County, PA 1963–1982. J Rheumatol 1990; 17:1329–1334.
150. Hochberg MC, Feldman D, Stevens MB, Arnett FC, Reichlin M. Antibody to jo-1 in polymyositis/dermatomyositis: association with interstitial pulmonary disease. J Rheumatol 1984; 11:663–665.
151. Plotz PH, Dalakas M, Leff RL, Love LA, Miller FW, Cronin ME. Current concepts in the idiopathic inflammatory myopathies: polymyositis, dermatomyositis, and related disorders. Ann Intern Med 1989; 111:143–157.
152. Petri M, Genovese M, Engle E, Hochberg M. Definition, incidence, and clinical description of flare in systemic lupus erythematosus. Arthr Rheum 1991; 34:937–944.
153. Guzman J, Cardiel MH, Arce-Salinas A, Sanchez-Guerrero J, Alarcon-Segovia D. Measurement of disease activity in systemic lupus erythematosus. Prospective validation of 3 clinical indices. J Rheumatol 1992; 19:1551–1558.
154. Liou HH, Wang CR, Chen CJ, Chen RC, Chuang CY, Chiang IP, Tsai MC. Elevated levels of anticardiolipin antibodies and epilepsy in lupus patients. Lupus 1996; 5:307–312.
155. Christensen B. Burning Border health issues. Environ Health Persp 1995; 103:542.

# 4

# Role of Oxidative Stress in Chemical Carcinogenesis

**JAMES E. KLAUNIG and LISA M. KAMENDULIS**

*Indiana University, Indianapolis, Indiana*

## I. OVERVIEW OF CHEMICAL CARCINOGENESIS

A cause-and-effect and dose-responsive relationship has been established between exposure to chemical carcinogens and subsequent development of neoplasia in animals. Chemically induced neoplasia is a multistep process involving damage to the genome initially followed by clonal expansion of the DNA-damaged cell eventually leading to a neoplasm. Chemical carcinogens have been shown to impact at all of the stages of this tumorigenesis process. It has become apparent that chemical and physical agents that induce cancer may do so through several different cellular and molecular mechanisms. Williams and Weisberger (1), recognizing the apparent differences by which chemicals participate in the carcinogenesis process, coined the phrases ''genotoxic'' and ''epigenetic'' (nongenotoxic) in describing activities of chemicals and physical agents that induced cancer.

While the specific definitions of genotoxic and nongenotoxic will vary somewhat, there are generally agreed upon attributes for placement in these categories (Table 1). Genotoxic agents usually refer to chemicals that directly damage genomic DNA, which in turn can result in mutation. Chemicals in this category are usually activated in the target cell and produce a dose-response increase in neoplasm formation. Although a threshold dose is apparent for most genotoxic agents, for regulatory purposes genotoxic carcinogens are frequently assumed to be without threshold, and exposure to one molecule of a genotoxic carcinogen results in neoplastic development. Most genotoxic agents are mutagenic in eukaryotic and prokaryotic in vitro bioassay models. Frequently, these compounds invoke their tumorigenic response at or close to the site of application or administration of the compound. A second category includes genotoxic agents that require metabolism or activation by cellular metabolic pathways to produce their effect. These chemicals fre-

**Table 1** Attributes of Genotoxic and Nongenotoxic Carcinogens

| |
|---|
| Genotoxic Carcinogens |
| Mutagenic |
| DNA reactivity |
| Tumorigenicity is dose response |
| No threshold (?) |
| Can be complete carcinogens |
| Irreversible |
| Usually not strain- or species-specific |
| Function at initiation and progression stages of cancer process |
| Multiorgan tumorigenicity seen |
| Nongenotoxic Carcinogens |
| Nondirectly DNA reactive |
| Nonmutagenic |
| Exhibit threshold |
| Usually exhibit strain-, species-, and tissue specificity |
| Function at the tumor-promotion stage of the cancer process |
| Reversible |

quently invoke their effect in target organs that are able to metabolize and activate the procarcinogenic form of the chemical to its ultimate DNA interactive form. Based on this requirement for activation, procarcinogenic genotoxic chemicals usually are tissue-specific. As noted above, genotoxic agents eventually result in modification to cellular DNA. This DNA reactivity results in mutation in the target cell.

As noted above, a separate category of carcinogenic compounds has been identified that appear to function through a non-DNA-reactive mechanism or possibly through indirect mechanisms. While much less is known about the exact mode of action of nongenotoxic carcinogens, these agents neither induce mutation in short-term eukaryotic and prokaryotic mutation assays nor induce direct DNA damage in the target organ. These agents modulate cell growth and cell death and exhibit dose-response relationships between exposure and tumor formation. While the exact mechanism(s) of action for neoplastic cell formation has not been established, changes in gene expression and cell growth parameters appear paramount in their mode of action. Nongenotoxic compounds frequently require chronic treatment and exhibit a threshold dose to elicit carcinogenicity. Many nongenotoxic carcinogens appear to function during the promotion stage of the cancer process.

Three distinct stages (initiation, promotion, and progression) of the carcinogenesis process have been identified based on experimental data (Fig. 1). Initiation involves the formation of a mutated, preneoplastic cell from a genotoxic event. The formation of the preneoplastic, initiated cell is an irreversible, but dose-dependent process. Promotion involves the selective clonal expansion of the initiated cell through an increase in cell growth and/or a decrease in apoptosis in the target cell population. The events of this stage are dose dependent and reversible upon removal of the tumor promotion stimulus. Progression, the third stage, involves cellular and molecular changes that occur from the preneoplastic to the neoplastic state. This stage is irreversible, involves genetic instability, changes in nuclear ploidy, and disruption of chromosome integrity.

Although a number of chemical carcinogens that function through nongenotoxic mechanisms have been identified, no definitive mechanism(s) of action has been thor-

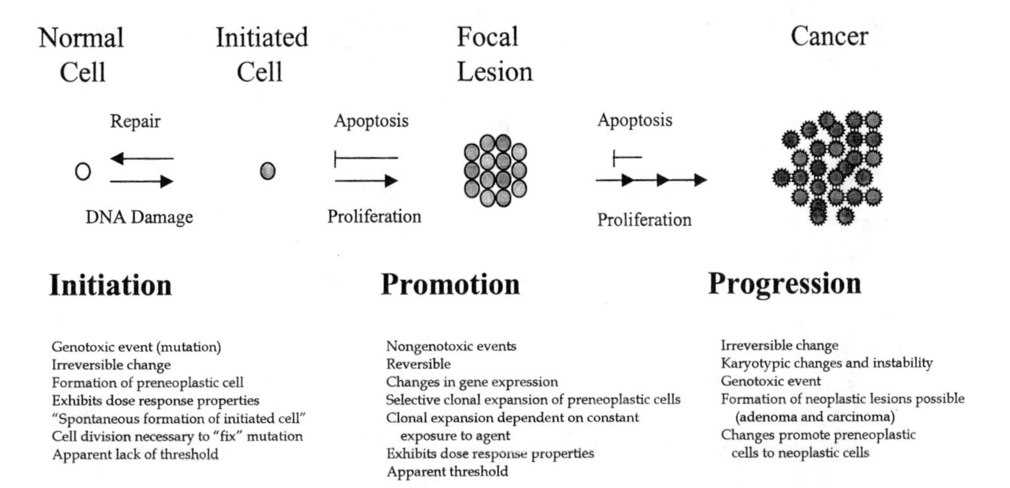


**Figure 1** Multistage process of chemical carcinogenesis. The induction of hepatic cancer by chemical carcinogens is a multistage process involving the selective promotion of an initiated cell to form a focal lesion. This clonal expansion may occur through either an increase in cell proliferation and/or a decrease in lesion apoptosis. Further cell growth may eventually lead to the progression of the focal lesion to form a neoplastic lesion.

oughly defined for this class of compounds. One common effect of most nongenotoxic carcinogens and tumor promoters is the induction of cell proliferation in the target tissue. The increase in cell proliferation occurs either through an increase in DNA synthesis and subsequent mitosis and/or a depression of apoptosis. The observed cell proliferation elicited by nongenotoxic carcinogens may arise from interaction with cellular receptors, modulation of growth factors and disruption of cell growth regulation, regenerative growth secondary to necrosis, and/or inhibition of apoptosis (2). Chemical compounds that induce cell proliferation can be divided into those that induce cytolethality with a resultant compensatory hyperplasia and those that function through mitogenic mechanisms (3). While receptor-mediated changes have been reported with selective members of mitogenic compounds, for the most part, the mechanism of induction of DNA synthesis and cell proliferation has not been thoroughly defined.

Increased replicative DNA synthesis and subsequent cell division have been linked to each of the stages of the carcinogenesis process (3–5). With compound-related induction of DNA synthesis, two possible mechanisms have been proposed for the induction of cancer. In one, an increase in DNA synthesis and mitosis by the nongenotoxic agent may induce mutations in dividing cells through misrepair. This, coupled with continual cell division pressure, may serve to ''fix'' heritable mutations in a cell resulting in an initiated, preneoplastic cell that through the pressure of additional and sustained cell proliferation clonally expands into a preneoplastic lesion and eventually a neoplasm. Alternately, the induction of DNA synthesis and mitosis by nongenotoxic agents may serve to allow the selective clonal growth of already ''spontaneously initiated cells'' (6–8). In this latter case, the nongenotoxic agents would serve to promote the expansion of initiated or preneoplastic cells as part of the tumor promotion phase of the multistep carcinogenesis process (9–11). The carcinogenicity of many compounds, including phenobarbital, cyproterone acetate, and ethinylestradiol, appear to act through a cell proliferation mechanism that is dependent on the continued administration of the compound (12–15). For both of these scenarios, a sustained and chronic induction of DNA synthesis and cell proliferation is required for maintenance and clonal expansion of preexisting preneoplastic cells and eventual neoplastic development. In concert with the selective stimulation of cell proliferation seen in preneoplastic cells, some nongenotoxic agents appear to have ''mito-inhibitory'' effects on normal cells (noninitiated cells). This mito-suppression in the liver is characterized by a reduction in basal/normal hepatocellular turnover and/or capacity to be stimulated to proliferate in vivo or in vitro (9,16–18). Additionally, recent studies in the liver have ascribed a role for the liver macrophage, the Kupffer cell, in the modulation of nongenotoxic carcinogen-induced cell proliferation via the release of TNF-α (19).

In maintaining cell number within a tissue, a dynamic equilibrium exists between mitosis and apoptosis, such that these processes are balanced. Apoptosis is a normal mechanistic process in which cells with severe damage to DNA or other critical cellular macromolecules, or cells with a low apoptotic threshold, are selectively removed from the tissue. With respect to carcinogenesis, it is generally accepted that the cancer process is a result of an imbalance between cell growth and death. Thus, the apoptotic process serves as a cellular mechanism to counter aberrant proliferation. The inhibition of apoptosis has been linked to tumor promotion in vivo (20,21). This inhibition of apoptosis has also been shown in vitro by chemical agents that induce reactive oxygen species, including the tumor promoter 12-o-tetradecanoylphorbol-13-acetate (TPA), and ionizing radiation (22). Several tumor promoters appear to act by invoking a resistance to apoptosis, possibly through the induction of enzymatic and nonenzymatic antioxidants in the initiated cells, thus allowing for clonal expansion of altered cells. Consistent with the concept that liver tumor promoters produce effects that are reversible upon removal of the compound, withdrawal of liver tumor promoters results in an enhanced rate of apoptosis that ultimately restores the liver to its original cell number (23). The increase in apoptosis often seen concomitantly with increases in cell proliferation by liver tumor promoters has been proposed as a protective mechanism that functions to delete genetically damaged cells prior to cell replication. Collectively, these results are consistent with the concept that apoptosis may protect against carcinogenesis by deleting altered and potentially mutagenic cells and support the theory that inhibition of apoptosis is an important aspect of tumor promotion induced by chemical agents.

## II. REACTIVE OXYGEN SPECIES IN CHEMICAL CARCINOGENESIS

Oxidative stress is classically defined as an imbalance between prooxidants and antioxidants in favor of the former, resulting in an overall increase in cellular levels of reactive oxygen species (24). Within cellular tissues, reactive oxygen species that include superoxide anion, hydrogen peroxide, and the hydroxyl radical are produced under normal physiological conditions (e.g., phagocytosis) and aerobic metabolism. An estimated 5% of the molecular oxygen consumed by the body is converted to reactive oxygen species through these normal metabolic processes (25). In addition, the body burden of reactive oxygen species can be increased by pathological conditions, inflammation, ischemia/reperfusion, xenobiotic metabolism, and lifestyle factors (diet, environmental tobacco smoke, alcohol) and thus contribute to the induction of oxidative stress. The existence of reactive oxygen species provides the potential to damage or alter cell macromolecules through interaction with DNA, protein, and lipids. Etiologically, oxidative stress has been linked to the pathogenesis of many age-related and chronic diseases including cancer (26–29).

The mechanism for the carcinogenic activity elicited by nongenotoxic carcinogens that induce oxidative stress remains unresolved. Oxidative stress induced by nongenotoxic carcinogens may interfere with physiological processes of cells. Oxidative DNA damage may be critical in carcinogenesis if it results in extensive enough changes to induce gene mutations. However, the induction of oxidative stress may also regulate gene expression either indirectly, through alteration of methylation status in the genome, or directly, through activation of gene transcriptional pathways.

Experimentally, a relationship between induction of oxidative stress and neoplastic development by carcinogenic chemicals has been demonstrated (30). Cellular damage from reactive oxygen species under normal physiological conditions is prevented by antioxidant systems, both enzymatic (i.e., catalase, superoxide dismutase, glutathione peroxidase) and nonenzymatic (i.e., vitamin E, vitamin C, glutathione) (31). Additionally, DNA repair enzymes exist which function to remove altered bases produced by oxidative mediated reactions. Several oxidative DNA repair enzymes have been identified from mammalian cells that function to remove oxidized bases from DNA and nucleotide pools (32–35). Paradoxically, the efficiency of repair may be enhanced following exposure to reactive oxygen species as expression of many DNA repair enzymes is upregulated following oxidative stress (36–37).

Carcinogens that increase reactive oxygen species (if left unbalanced by detoxification/antioxidant processes) will result in oxidative DNA or lipid damage. In relation to cancer formation and development, oxidative DNA adduct formation is considered a potentially relevant mechanism (38). Oxidative modifications to DNA have been estimated to reach $10^5$ per cell per day (39). Current methods have revealed more than 100 forms of oxidative DNA damage that result from interaction of reactive oxygen species with DNA (40–42). The types of damage encountered include modification of the purine, pyrimidine, and deoxyribose moieties of DNA, as well as the introduction of single- and double-stranded breaks in DNA and DNA cross-links. Hydroxylation of guanine at the C8 position results in the formation of 8-hydroxydeoxyguanosine (OH8dG). This DNA adduct appears to be the most predominant oxidative modification to DNA and has been shown to be mutagenic (43–46). The mutagenicity of OH8dG may occur due to the lack of base-paring specificity observed when OH8dG is present in DNA (47). In vitro it has been shown that either dC or dA could be incorporated opposite OH8dG (43–46). In addition, reactive oxygen species can react with dGTP in the nucleotide pool to form OH8dG (47b). During DNA replication, OH8dG has been postulated to be incorporated into DNA from the oxidized nucleotide pool (opposite dC or dA on the template strand) by DNA polymerases, resulting in A:T to C:G transversions during the next round of scheduled DNA replication (48,49). In addition, studies examining cellular transformation have shown that OH8dG induces a dose-related increase in cellular transformation that could be blocked by the antioxidant β-carotene providing further evidence that OH8dG formation is related to the carcinogenic process (50).

Aside from OH8dG, other oxidative-derived DNA adducts appear important in chemical carcinogenesis. Radical-mediated damage to cellular biomembranes results in the formation of malondialdehyde, a by-product of lipid degradation. Via the interaction of malondialdehyde with DNA, an adduct is formed that is thought to play a role in the cancer process (51,52). Copper, a transient metal ion, is involved with the stabilization of chromosome structure via association at guanine:cytosine base pair (G:C) sites in DNA (53). In addition, copper mediates the formation of the hydroxyl radical from $H_2O_2$ by the Haber–Weiss reaction (54). Due to the proximity of the copper ion to guanine and cytosine residues in DNA, the formed hydroxyl radicals attack the nearby guanine or cytosine bases. Specific gene mutations have been ascribed to this reaction, including the induction of G:C pair mutations seen in the p53 gene in tumor tissue and may relate to the cancer process (55–57).

## III. OXIDATIVE STRESS GENERATION BY CARCINOGENS

A number of xenobiotics produce liver tumors in mice but are considered of little or doubtful significance to humans. The mechanism(s) for the species selectivity, while realized, is unknown. One possible mechanism that has been linked to the cancer process is xenobiotic metabolism. Xenobiotic metabolism may directly result in formation of reactive intermediates that can bind to DNA and result in base changes and subsequent miscoding of DNA, the compound or metabolite may induce DNA transcription or bind to cellular receptors and initiate signal transduction cascades. Any of the above may result in activation of oncogenes, induction of DNA replication, mitosis, and cell proliferation and thus contribute to the cancer process. In addition to these potential direct actions, a further series of indirect events associated with metabolism may be involved in carcinogenesis. Central to these actions is the production of reactive oxygen radicals. Environmental agents including nongenotoxic carcinogens can directly generate or indirectly induce the generation of reactive oxygen species in cells (58). Many epigenetic agents, including peroxisome proliferators, chlorinated compounds, radiation, metal ion, barbiturates, and phorbol esters have shown the induction of oxidative stress in vitro and in vivo (30).

Metabolic activation and production of reactive oxygen species by cytochrome P450 is associated with many chronic diseases, including chemical carcinogenesis, and has been related to DNA damage, activation of protein kinase C, oncogenes, hyperplasia, and the metastatic process (59). Reactive oxygen species can be produced from many possible sources during metabolism—through redox cycling in the presence of molecular oxygen, through peroxidase-catalyzed single-electron drug oxidations, and through futile cycling of cytochromes P450. Compounds that are metabolized by and induce cytochrome P450 enzymes may produce reactive oxygen species (superoxide anion and hydrogen peroxide) through futile cycling (60). This has been proposed as a possible contributor for the observed induction of oxidative stress by a number of chlorinated and nonchlorinated compounds (e.g., dieldrin, TCDD, lindane, phenobarbital) that also induce the activities of cytochrome P450 isoenzymes in rodent liver (30). P450 futile cycling is especially seen with cytochrome P450 2E1. This enzyme is involved in the oxygenation of difficult-to-oxidize substrates, including ethanol, and thus is believed to generate reactive oxygen species near the site of substrate oxidation (61,62). Stabilization of 2E1 protein, an event known to occur following exposure to many substrates of the 2E1 enzyme, leads to a prolonged burst of reactive oxygen species production that has been associated with tissue necrosis, mutation, and malignancy.

Induction of cytochrome P450 has been suggested as a predictive marker of potential chemical carcinogenesis, perhaps due to an increase in reactive oxygen species. A common feature of chemicals that induce mouse liver tumors or promote the growth of previously initiated rat liver foci is that they induce cytochrome P450 2B (63). Phenobarbital, the liver tumor promoter, induces and is metabolically oxygenated by P450 2B, but with difficulty, such that the catalytic cycle is uncoupled and superoxide anion is released (64). Thus, the correlation between P450 2B induction and tumor promotion following phenobarbital exposure may be related to futile cycling and induction of reactive oxygen species (63). Other chemicals such as the peroxisome proliferators are potent inducers of the cytochrome P4504 family. It has been proposed that additional $H_2O_2$ will leak out of newly formed peroxisomes, producing an increase in the redox shift toward oxidation in the cell (65–67). This hypothesis has not been proven and recent studies have suggested that while an increase in oxidative DNA damage is seen with peroxisome proliferators, this increase does not correlate with the relative peroxisome proliferating activity of the compound (68). These latter findings do not preclude a role for oxidative stress in the cancer induction by peroxisome proliferators but may remove the linkage between the peroxisome organelle and the cancer process.

P450 oxidation of xenobiotics is higher in the mouse than in rats or humans (69). The ability to metabolize the carcinogen 7,12-dimethylbenzanthrecene is inversely correlated with longevity and body weight among several species (70). This again suggests a role for metabolism-dependent reactive oxygen species generation in chronic disease. Regardless of their genesis during metabolism, reactive oxygen species may contribute to the cancer process via the conversion of superoxide anion and hydrogen peroxide to the hydroxyl radical in the presence of iron (Fenton and Haber–Weiss reaction) to yield DNA damage. The oxidative DNA damage may then result in the production of single- or double-stranded DNA breaks, base modifications, or rearrangements. Alternately, reactive oxygen species may modulate gene expression and result in altered growth regulation (71–74).

## IV. OXIDATIVE STRESS AND CELL GROWTH REGULATION

A role for reactive oxygen species and oxidative stress has been proposed for both the stimulation of cell proliferation and the stimulation of cell deletion by apoptosis (75–78). The mechanisms for the involvement of oxidative stress in the stimulation of the cell proliferation and apoptotic processes are not known, but clearly do not involve a single universal mechanism. The effects of reactive oxygen species and oxidative stress within cells appear to be cell-type-specific and depend upon the source and intercellular concentration of reactive oxygen species. Thus, the involvement of reactive oxygen species in cell growth regulation is complex and dependent on a number of cellular and biochemical parameters.

Concerning cell proliferation, the induction of oxidative stress has particularly profound effects on the tumor promotion stage of carcinogenesis (79). Both $H_2O_2$ and superoxide anion have been shown to induce mitogenesis in several mammalian cell types (77,80). Likewise, reactive oxygen species, generated endogenously in cells, either via endogenous metabolism or xenobiotic metabolism, stimulate cell proliferation. Furthermore, by lowering endogenous levels of reactive oxygen species, supplementation of cells with antioxidants such as superoxide dismutase, catalase, β-carotene, and various flavonoids, inhibit cell proliferation in vitro (81,82). The induction of cell proliferation appears to occur only at exposure to low concentrations or transient exposure to reactive oxygen species (83,84).

Oxidative stress also appears to be a mediator of apoptosis. In the apoptosis process, high concentrations of reactive oxygen species trigger an apoptotic signaling pathway the leads to cell deletion (85). Through depletion of intercellular glutathione, the major regulatory protein for maintaining cellular redox balance, or through excess production of reactive oxygen species, a growing number of endogenous and exogenous compounds are being identified that induce apoptosis through reactive oxidative stress-mediated processes. Endogenous molecules and processes (arachidonic acid metabolism prostaglandins, and lipid hydroperoxides), redox cycling compounds (quinones, adriamycin), and growth regulatory factors (transforming growth factor–β and tumor necrosis factor–α) induce apoptosis via the generation of reactive oxygen species (86–92). Also, antioxidants such as *N*-acetyl cysteine, glutathione, and dithiothreitol inhibit the apoptotic process further supporting the link between reactive oxygen species induction and apoptosis (93–95).

## V. OXIDATIVE STRESS AND GENE EXPRESSION

Reactive oxygen species also appear to play an important role in several signaling pathways through the activation of the gene transcription factors NF-κB and AP-1 (96). The activation of these transcription factors is involved in both cell proliferation and apoptosis. Apparently, the relative concentration of reactive oxygen species formed dictates whether a mitogenic or an apoptotic pathway will ensue. The intracellular concentration of reactive oxygen species produced appears to influence the selective activation of these transcription factors. Thus, the genes transcribed may partially explain the observation that both cell deletion and division can occur following exposure to reactive oxygen species or reactive oxygen species–generating compounds.

Methylation status of cellular DNA influences gene expression. Reactive oxygen species, in turn, have been shown to modify methylation patterns in cells. Post-DNA synthetic methylation of the 5 position on cytosine [5-methylcytosine (5mC)] is the only naturally occurring modification to DNA in higher eukaryotes. In vertebrate organisms, approximately 4% of the cytosine residues in the genome are modified to 5mC (97). It is generally accepted that 5mC in DNA can affect cellular processes including development and differentiation; however, it is strongly evidenced that the presence of 5mC is involved in cancer development (98).

In tumor cells, the methylation pattern of various genes is often altered such that both hypermethylation and hypomethylation become apparent (99). Also, during carcinogenesis, widespread hypomethylation of the genome as well as hypermethylation in areas that are normally unmethylated is seen (99–101). The degree of methylation within a gene inversely correlates with the expression of that gene (102,103). Hypermethylation of genes is associated with decreased gene expression or gene silencing. Important to the cancer process, tumor suppressor genes are known to be hypermethylated and subsequently inactivated. It has been suggested that carcinogens may cause errors in methylation and lead to the formation of cells with increased tumorigenic potential (104). Conversely, hypomethylation can lead to enhanced gene expression and has been associated with increased mutation rates (105). Oncogenes have been shown to become hypomethylated and their expression amplified (99,106). Exposure of rats to hepatocarcinogens or a methyl-deficient diet (which itself is know to induce tumor formation) have been shown to decrease hepatic levels of S-adenosylmethionine (SAM) (107). A decrease in this cofactor would therefore promote hypomethylation and subsequent expression of oncognes. In rats fed a diet deficient in choline or methyl donor groups, hypomethylation of c-myc, c-fos, and c-H-ras protooncogenes was seen and was associated with hepatocarcinogenesis (108). Also consistent with the role of methylation on carcinogenesis process, administration of SAM has been shown to inhibit hepatocarcinogenesis. In addition to changes in gene expression, methylation also regulates chromosome stability. Changes in DNA methylation patterns can alter chromatin structure and result in deletions, inversions, and loss of chromosomes (109).

Several chemical carcinogens are known to modify DNA methylation, methyltransferase activity, and chromosomal structure. Of particular importance, the formation of oxidative DNA lesions has been linked to changes in DNA methylation profiles and the carcinogenesis process. Oxidative DNA damage can interfere with the ability of methyltransferases to interact with DNA, thus resulting in a generalized hypomethylation of cytosine residues at CpG sites. The formation of OH8dG in DNA by reaction of the hydroxyl radical or singlet oxygen with DNA can lead to hypomethylation of DNA since the presence of OH8dG in CpCpGpGp sequences inhibits the methylation of adjacent C residues. Additionally, OH8dG formation can interfere with the normal function of DNA methyltransferase and thus alter DNA methylation status (110). Thus, oxidative DNA damage may be an important contributor to the carcinogenesis process affected by the loss of DNA methylation, allowing the expression of normally quiescent genes. Also, the abnormal methylation pattern observed in cells transformed by chemical oxidants may contribute to an overall aberrant gene expression and promote the tumor process.

Although DNA damage with resulting mutation may be produced by reactive oxygen species generators, the induction of oxidative stress by nongenotoxic carcinogens may also directly modify gene expression. Reactive oxygen species formed in the mitochondria and in the cytosol are important determinants of the redox state of protein cysteine residues and constitute a regulatory mechanism of protein conformation and function (111). Reactive oxygen species–dependent redox cycling of thiol groups on cysteine residues is also important for the maintenance of protein–protein and protein–DNA interactions that regulate biological events in signal transduction pathways. Glutathione is an important regulator of the redox state of thiol groups. An increase in intracellular GSSG concentration can arise from the breakdown of $H_2O_2$ by glutathione peroxidase (112). Because of the relatively low concentration of GSSG compared with GSH, a minor elevation in the oxidation of GSH to GSSG, via increased reactive oxygen species burden or $H_2O_2$ metabolism, can result in a significant elevation in intercellular GSSG levels. This cellular redox shift may be significant enough to promote oxidation of sulfhydryl groups on cysteine residues and alter protein conformation and/or function. Redox cycling of cysteine residues is one of several oxidant-dependent mechanisms that regulate the activity of many transcription factors involved in cell proliferation and differentiation (113).

The induction of oxidative stress by xenobiotics appears to function to alter gene expression through signaling pathways including cAMP-mediated cascades; a calcium-calmodulin activated pathway; and intracellular signal transducers such as nitric oxide (114). This modification of gene expression, as noted above, may result in either cell proliferation or selective cell death apoptosis or necrosis (115–117). Calcium is a signaling factor responsible for regulating a wide range of cellular processes (118) including cell proliferation, differentiation, and apoptosis (119). Increased cellular levels of reactive oxygen species have been reported to release calcium from intracellular store in different cell lines and result in the activation of kinases, such as protein kinase C (PKC) (120–122). The activation of PKC was linked to the toxic mechanism of TCDD, chlorinated hydrocarbons, and TPA (123–126). The effects of oxidative stress on calcium and kinase activity also appear to be related to activation of transcription factors. Transcription factors are small molecular weight proteins that bind with promoter regions of genes and regulate the transcription of genes involved in development, growth, and aging of cells (127). Of particular importance, the transcription factors, nuclear factor-κB (NF-κB) and activator protein-1 (AP-1), are well studied for their regulation by reactive oxygen species (113,128–133)

As noted above, NF-κB plays a central role in the regulation of many genes involved in cellular defense mechanisms, pathogen defenses, immunological responses, and expression of cytokines and extracellular matrix-associated proteins. Active NF-κB exists as a dimer of two proteins: a p50/p52 protein of the NF-κB −1/−2 family, and a p65 protein, a member of the REL/RELA/RELB family (134). Inactive NF-κB complex is formed by two p50 homodimers or a p50/p65 heterodimer bound to an Iκ-B protein (135,136). The NF-κB/κ-B complex resides in the cytoplasm of cells. The activation of the NF-κB occurs in response to extracellular stimuli, physical stress, metal ion, asbestos, alcohol, and chemical treatment (137–141). In addition, pathogenic and inflammatory stimuli activate NF-κB (lipopolysaccharide, UVC ionizing radiation, interleukin-1, and tumor necrosis factor. Reactive oxygen species derived from mitochondrial respiration have been implicated as the second messengers involved in activation of NF-κB by tumor necrosis factor and interleukin-1 (129,142,143). Further substantiating the importance of reactive oxygen species on the induction of transcription factors are studies demonstrating that the activation of NF-κB by nearly all stimuli can be blocked by antioxidants, including L-cysteine, NAC, thiols, and vitamin E (144,145). NF-κB activation appears to be selectively mediated by peroxides as activation of NF-κB was observed only following exposure to $H_2O_2$ or butylperoxide, and not superoxide or hydroxyl radicals (146). Likewise, NF-κB activity was increased in cells that overexpressed superoxide dismutase and decreased in cells overexpressing catalase (147).

AP-1 represents a family of transcription factors that specifically bind to and transactivate a *cis*-acting transcriptional control DNA element. AP-1 exists as either a heterodimer of the protein products of individual members of the FOS and JUN gene families, or as a homodimer of JUN proteins (148,149). AP-1 controls the expression of many genes including those encoding extracellular matrix proteins (collagenase, stromolysin), cell cycle proteins (cyclin D), and various growth factors (TGF-1β, cytokines) (150). Expression of c-JUN and c-FOS can be induced by a variety of compounds including nongenotoxic and tumor-promoting compounds (carbon tetrachloride, phenobarbital, TPA, TCDD, cadmium, alcohol, ionizing radiation, asbestos), many of which generate reactive oxygen species (138–140,151–163). ROS generated by ionizing radiation have been shown to induce c-FOS and c-JUN expression by a mechanism that can be inhibited by NAC (164–166) and that involves activation of MAP kinases (165).

Both pro- and antioxidants affect the activation of AP-1 and c-fos (133). While the activation of AP-1 is related to cell proliferation and apoptosis, whether a different consequence results following activation of AP-1 by prooxidants or antioxidants remains unclear. Through regulation of gene transcription factors, and disruption of signal transduction pathways (144,167–172), reactive oxygen species are intimately involved in the maintenance of concerted networks of gene expression that may relate to neoplastic development.

## VI. OXIDATIVE STRESS AND GAP JUNCTIONAL MODIFICATION

In multicellular organisms, the intercellular exchange of cellular factors from one cell to a neighboring cell is mediated through either extracellular signal molecules (i.e., hormones) or between adjacent cells through gap junctional intercellular communication (173,174). Gap junctions are comprised of plaques of transmembrane channels composed of connexin hexamers forming connexin hemichannels (175). The hemichannels of one cell connect with hemichannels on neighboring cells to form a transmembrane conduit between the two adjacent cells. This conduit allows for the exchange of ions and low-molecular-weight water-soluble materials (less than 1000 daltons) between the cells. Growth regulatory signal-transducing substances, including calcium, cAMP, and inositol triphosphate, substances involved in the regulation of the cell cycle, cell growth, and cell death, are able to pass through the gap junction (176–178). Therefore, through gap junctions, a steady level of low-molecular-weight messenger molecules is maintained among cell populations.

Gap junctional intercellular communication has been shown to be modulated during the cancer process (179). Cell lines established from tumors of both animal and human tissues exhibit decreased gap junctional intercellular communication. In fact, during multistage rat hepatocarcinogenesis, gap junctional intercellular communication has been shown to progressively decrease. Additionally, liver carcinomas have also displayed reduced ability to communicate with neighboring cells (180,181). Growth control in GJIC-deficient tumorigenic cells is restored by transfection with connexin genes (182,183). Blockage of cell-to-cell communication has particularly been associated with the tumor promotion stage of chemical carcinogenesis, and has been suggested as a mechanism of action for the tumor promotion process (183–189). Many tumor-promoting compounds demonstrate a tissue- and species-specific inhibition of gap junctional intercellular communication in vivo and in vitro following exposure (190). A correlation between the ability of a compound to block cell-to-cell communication in cultured cells and its ability to induce rodent tumors through nongenotoxic mechanisms has also been demonstrated (190,191). Additionally, the inhibition of gap junctional intercellular communication observed following treatment with tumor-promoting compounds correlates with species and strain sensitivity of the chemical (184,192,193).

The induction of oxidative stress has been shown to modulate cell-to-cell communication. Hydrogen peroxide ($H_2O_2$), an established tumor promoter (194), inhibits GJIC that is reversible and involves a glutathione-dependent signal transduction pathway (195). In addition to the direct action of active oxygen species on GJIC, many chemicals that produce reactive oxygen species have been shown to inhibit GJIC in a variety of cells in culture. The phorbol ester and tumor promoter 12-o-tetradecanoylphorbol-13-acetate (TPA) has been shown to produce $H_2O_2$ in murine epidermal keratinocytes and is known to block intercellular communication (196). Similarly, DDT and paraquat, nongenotoxic liver carcinogens, selectively block GJIC in isolated mouse hepatocytes (184,197). Additionally, carbon tetrachloride inhibition of GJIC in rat hepatocytes occurs through an oxidative stress–mediated process (198). Further support for the involvement of reactive oxygen species in the inhibition of gap junctional intercellular communication by liver tumor promoters comes from studies examining antitumor-promoting compounds and/or cancer chemopreventive agents that function as antioxidants. Vitamin E and the polyphenolic fraction of green tea, EGCG, have been reported to abolish the inhibition of GJIC in mouse hepatocytes following treatment with phenobarbital and DDT (197,199). Thus, one function of antioxidants may be to maintain expression of gap junctions in cells. Additionally, alteration in cellular oxidative stress status appears to affect the assembly of gap junctions in neoplastic nodules (200,201).

The mechanism for the involvement of gap junctional intercellular communication in the carcinogenesis process may relate to the fact that the gap junction mediates the passage of both positive and negative growth regulatory molecules between neighboring cells. Blockage of gap junctional intercellular communication between normal and preneoplastic cells creates an environment in which preneoplastic cells are isolated from growth controlling factors of normal surrounding cells (185). Inhibition of gap junctional intercellular communication would therefore be expected to be involved in decreased regulation of homeostatic growth control, allowing the preneoplastic cell to escape the growth control of normal surrounding cells resulting in clonal expansion.

## VII. CONCLUSION

The induction of cancer by chemicals is a multistep, multievent process that involves DNA damage, the formation of a preneoplastic initiated cell, and the subsequent clonal proliferation of the initiated cell to the neoplastic state. Experimental evidence has attributed a role for oxidative stress and reactive oxygen species in this multistage process. Reactive oxygen species can be generated from endogenous (mitochondria, P450 cycling, peroxisomes, and imbalance of antioxidant systems) and exogenous sources (radiation and xenobiotics) by chemical carcinogens. Following generation, if not detoxified, reactive oxygen species have been shown to produce DNA damage through both direct (OH8dG formation) and indirect (lipid peroxidation products) resulting in cell mutation and/or changes in gene expression. In addition to direct DNA changes, changes in cell membrane continuity (modulation of cell-to-cell interaction through gap junctions), modification of gene expression, and changes in second messengers may occur. These cellular events can modify the cell cycle/apoptosis status of the target cell resulting in clonal growth. The oxidative stress mode of action of chemical carcinogens can be modified and prevented by reducing the amount and type of reactive oxygen species produced and/or increasing the antioxidant defense of the cell providing a chemopreventive means for cancer reduction. The induction of oxidative stress and the resulting cellular modification and damage thus participates in the cancer process.

## REFERENCES

1. Williams GM, Weisburger JH. Carcinogen risk assessment. Science 1983; 221:6.
2. Bursch W, Lauer B, Timmermann-Trosiener I, Barthel G, Schuppler J, Schulte-Hermann R. Controlled cell death (apoptosis) of normal and putative preneoplastic cells in rat liver following withdrawal of tumor promoters. Carcinogenesis 1984; 5:453–458.
3. Butterworth BE. Consideration of both genotoxic and nongenotoxic mechanisms in predicting carcinogenic potential. Mut Res 1990; 239:117–132.
4. Pitot HC, Goldsworthy T, Moran S. The natural history of carcinogenesis: implications of experimental carcinogenesis in the genesis of human cancer. J Supramolec Struct Cell Biochem 1981; 17:133–46.
5. Marsman DS, Cattley RC, Conway JC, Popp JA. Relationship of hepatic peroxisome proliferation and replicative DNA synthesis to the hepatocarcinogenicity of the peroxisome proliferators di(2-ethylhexyl)phthalate and [4-chloro-6-(2,3-xylidino)-2-pyrimidinylthio]-acetic acid (Wy-14643) in rats. Carcinogenesis 1988; 48:6739–6744.
6. Stott WT, Reita RH, Schumann AM, Watanabe PG. Genetic and nongenetic events in neoplasia. Food Cosmet Toxicol 1988; 19:567–576.
7. Ames BN, Gold LS. Too many rodent carcinogens: mitogenesis increases mutagenesis. Science 1990; 249:970–971.

8. Columbano A, Rajalakshmi S, Sarma DS. Requirement of cell proliferation for the initiation of liver carcinogenesis as assayed by three different procedures. Cancer Res 1981; 41:2079–83.
9. Farber E. Clonal adaption during carcinogenesis. Biochem Pharmacol 1990; 39:1837–1846.
10. Slaga TJ, Butler AP. Cellular and biochemical changes during multistage skin tumor promotion. Princess Takamatsu Symp 1983; 14:291–301.
11. Slaga TJ. Multistage skin carcinogenesis: a useful model for the study of the chemoprevention of cancer. Acta Pharmacol Toxicol 1984; 55(suppl 2):107–124.
12. Schulte-Hermann R, Schupple J, Ohde G, Timmermann-Trosiene I. Effect of tumor promoters on proliferation of putative preneoplastic cells in rat liver. Carcinogen Comprehens Surv 1982; 7:99–105.
13. Schulte-Hermann R, Schupple J, Timmermann-Trosiener I, Ohde G, Bursch W, Berger H. The role of growth of normal and preneoplastic cell populations for tumor promotion in rat liver. Environ Health Persp 1983; 50:185–94.
14. Schulte-Hermann R. Initiation and promotion in hepatocarcinogenesis. Arch Toxicol 1987; 60:179–81.
15. Kolaja KL, Bunting KA, Klaunig JE. Inhibition of tumor promotion and hepatocellular growth by dietary restriction in mice. Carcinogenesis 1996; 17:1657–1664.
16. Yager JD, Zurlo J, Sewall CH, Lucier GW, He H. Growth stimulation followed by growth inhibition in livers of female rats treated with ethinyl estradiol. Carcinogenesis 1995; 15: 2117–2123.
17. Eckl PM, Meyer SA, Whitcombe WR, Jirtle L. Phenobarbital reduces EGF receptors and the ability of physiological concentrations of calcium to suppress hepatocyte proliferation. Carcinogenesis 1988; 9:479–483.
18. Tsai W-H, Zarnegar R, Michalopoulos GK. Long-term treatment with hepatic tumor promoter inhibits mitogenic responses of hepatocytes to acidic fibroblast growth factor and hepatocyte growth factor. Cancer Lett 1991; 59:1–3–108.
19. Rose ML, Rusyn I, Bojes HK, Germolec DR, Luster M, Thurman RG. Role of Kupffer cells in peroxisome proliferator-induced hepatocyte proliferation. Drug Metab Rev 1999; 31:87–116.
20. Schulte-Hermann R, Timmermann-Trosiner I, Barthel G. DNA synthesis, apoptosis, and phenotypic expression as determinants of growth of altered foci in rat liver during phenobarbital promotion. Cancer Res 1990; 50:5127–5135.
21. Schulte-Hermann R, Grasl-Kraupp B, Bursch W. Tumor development and apoptosis. Int Arch Allergy Immunol 1994; 105:363–367.
22. Tomei LD, Kanter P, Wenner CE. Inhibiiton of radiation-induced apoptosis in vitro by tumor promoters. Biochem Biophys Res Commun 1988; 155:324–331.
23. Columbano A, Endoh T, Denda A, Noguchi O, Nakae D, Hasgawa K, Ledda-Cloumbano GM, Zedda AI, Konshi Y. Effects of cell proliferation and cell death (apoptosi and necrosis) on the early stages of rat hepatocarcinogenesis. Carcinogenesis 1996; 17:395–400.
24. Sies H. Oxidative stress: Introductory remarks. In: Sies H, ed. Oxidative Stress. New York: Academic Press, 1985:1.
25. Barber DA, Harris SR. Oxygen free radicals and antioxidants: A review. Am Pharm 1994; NS34:26–35.
26. Guyton KZ, Kensler TW. Oxidative mechanisms in carcinogenesis. Br Med Bull 1993; 49: 523–544.
27. Trush MA, Kensler TW. An overview of the relationship between oxidative stress and chemical carcinogenesis. Free Rad Biol Med 1991; 10:201–209.
28. Vuillaume M. Reduced oxygen species, mutation, induction and cancer initiation. Mut Res 1987; 186:43–72.
29. Witz G. Active oxygen species as factors in multistage carcinogenesis. Proc Soc Exp Biol Med 1991; 198:675–682.

30. Klaunig JE, Xu Y, Bachowski S, Jiang JZ. Free-radical oxygen-induced changes in chemical carcinogenesis. In: Wallace KB, ed. Free Radical Toxicology. Boston: Taylor and Francis, 1997:375–400.
31. Hochstein P, Atallah AS. The nature of oxidants and antioxidant system in the inhibition of mutation and cancer. Mut Res 1988; 202:363–375.
32. Tchou J, Grollman AP. Repair of DNA containing the oxidatively-damaged base, 8-oxoguanine. Mut Res 1993; 299:277–287.
33. Rosenquist TA, Zharkov DO, Grollman AP. Cloning and characterization of a mammalian 8-oxyguanine DNA glycosylase. Proc Natl Acat Sci USA 1997; 94:7429–7434.
34. Sakumi K, Furuichi M, Tsuzuki T, Kakuma T, Kawabata S, Maki H, Sekiguchi M. Cloning and expression of cDNA for a human enzyme that hydrolyzes 8-oxo-dGTP, a mutagenic substrate for DNA synthesis. J Biol Chem 1993; 268:23524–23530.
35. Slupska MM, Baikalov C, Luther WM, Chiang J-H, Wei Y-F, Miller JH. Cloning and sequencing a human homolog (hMYH) of the *Escherichia coli* MutY gene whose function is required for the repair of oxidative DNA damage. J Bacteriol 1996; 178:3885–3892.
36. Wani G, Milo GE, D'Ambrosio SM. Enhanced expression of the 8-oxo-7,8-dihydrodeoxyguanosine triphosphatase gene in human breast tumor cells. Cancer Lett 1998; 125:123–130.
37. Bases R, Franklin WA, Moy T, Mendez F. Enhanced excision repair activity in mammalian cell after ionizing radiation. Int J Radiat Biol 1992; 62:427–441.
38. Ames BN, Gold LS, Willett WC. The causes and prevention of cancer. Proc Natl Acad Sci USA 1995; 92:5258–5265.
39. Fraga CG, Shigenaga MK, Park JW, Degan P, Ames BN. Oxidative damage to DNA during aging: 8-hydroxy-2′-deoxyguanosine in rat organ DNA and urine. Proc Natl Acad Sci USA 1990; 87:4533–4537.
40. von Sonntag C. New aspects in the free-radical chemistry of pyrimidine nucleobases. Free Rad Res Comm 1987; 2:217–24.
41. Dizdaroglu M. Oxidative damage to DNA in mammalian chromatin. Mut Res 1992; 275: 331–342.
42. Demple B, Harrison L. Repair of oxidative damage to DNA: enzymology and biology. Ann Rev Biochem 1994; 63:915–48.
43. Shibutani S. Insertion of specific bases during DNA synthesis past the oxidation damaged base 8-oxodG. Nature 1991; 349:431–4343.
44. Wood ML, Diadaroglu M, Gajewski E, Essigmann JM. Mechanistic studies of ionizing radiation and oxidative mutagenesis: genetic effects of single 8-hydroxyguanine (7-hydro-8-oxoguanine) residue inserted at a unique site in a viral genome. Biochemistry 1990; 29:7024–7032.
45. Moriya M, Du C, Bodepudi V, Johnson F, Takeshita M, Grollman A. Site-specific mutagenesis using a gapped duplex vector: a study of translesion synthesis past 8-oxodeoxyguanosine in *E coli*. Mut Res 1991; 254:281–288.
46. Kamiat H. C-Ha-ras containing 8-hydroxyguanine at codon 12 induces point mutations at the modified and adjacent positions. Cancer Res 1992; 52:3484–3485.
47. Akiyama M, Maki H, Sekiguchi M. A specific role of MutT protein: to prevent dG:dA mispairing in DNA replication. Proc Natl Acad Sci USA 1989; 86:3949–3952.
48. Cheng KC, Cahill DS, Kasai H, Nishimura S, Loeb LA. 8-hydroxyguanine, an abundant form of oxidative DNA damage, causes G→T and A→C substitutions. J Biochem 1992;267: 166–172.
49. Maki H, Sekiguchi M. MutM protein specifically hydrolyses a potent mutagenic substrate for DNA synthesis. Nature 1992; 355:273–275.
50. Zhang H, Kamendulis LM, Xu Y, Klaunig JE. The role of 8-hydroxy-2′-deoxyguanosine in morphological transformation of Syrian hamster embryo (SHE) cells. Toxicol Sci 2000; 56: 303–312.
51. Floyd RA. 8-hydroxy-2′-deoxyguanosine in carcinogenesis. Carcinogenesis 1990; 11:1447–1450.

52. Marnett LJ. Lipid peroxidation-DNA damage by malondialdehyde. Mut Res 1999; 424:83–95.
53. Pezzano H, Podo F. Structure of binary complexes of mono- and polynucleotides with metal ions of the first transition group. Chem Rev 1980; 80:365–399.
54. Halliwell B. Mechanism involved in the generation of free radicals. Path Biol 1996; 44:6–13.
55. Hollstein M, Sidransry D, Vogelstein B. P53 mutations in human cancer. Science 1991; 253: 49–53.
56. Bressac B, Kew M, Wands J, Ozturk M. Selective G to T mutations of p53 gene in hepatocellular carcinoma from southern Africa. Nature 1991; 350(6317):429–431.
57. Lasky T, Silbergeld E. p53 mutations associated with breast, colorectal, liver, lung, and ovarian cancers. Environ Health Persp 1996; 104:1324–1331.
58. Rice-Evans C, Burdon R. Free radical-lipid interactions and their pathological consequences. Prog Lipid Res 1993; 32:71–110.
59. Parke DV. The cytochromes P450 and mechanisms of chemical carcinogenesis. Environ Health Persp 1994; 102:852–853.
60. Parke DV, Sapota A. Chemical toxicity and reactive oxygen species. Int J Occup Med Environ Health 1996; 9:331–340.
61. Eksrom G, Ingleman-Sundberg M. Rat liver microsomal NADPH-supported oxidase activity and lipid peroxidation dependent on ethanol inducible cytochrome P-450. Biochem Pharmacol 1989; 38:1313–1319.
62. Kuklielka E, Cederbaum AI. The effect of chronic ethanol on consumption on NADH- and NADPH-dependent generation of reactive oxygen intermediates by isolated rat liver nuclei. Alcohol Alcoholism 1992; 27:233–239.
63. Rice JM, Diwan BA, Hu H, Ward JM, Nims RW, Lubet RA. Enhancement of hepatocarcinogenesis and induction of specific cytochrome P450-dependent monooxygenase activities by the barbiturates allobarbital, aprobarbital, pentobarbital, secobarbital, and 5-phenyl- and 5-ethylbarbituric acids. Carcinogenesis 1994; 15:395–402.
64. Ingelman-Sundberg M, Hagbjork AL. On the significance of cytochrome P450-dependent hydroxyl radical-mediated oxygenation mechanism. Xenobiotica 1982; 12:673–686.
65. Rao MS, Reddy JK. An overview of peroxisome proliferator-induced hepatocarcinogenesis. Environ Health Perspect 1991; 93:205–209.
66. Tamura H. Long-term effects of peroxisome proliferators on the balance between hydrogen peroxide-generating and scavenging capacities in the liver of Fischer-344 rats. Toxicol 1990; 63:199–213.
67. Wade N, Marsman DS, Popp JA. Dose related effects of hepatocarcinogen Wy 14,623 on peroxisomes and cell replication. Fundament Appl Toxicol 1992; 18:149–154.
68. Rose ML, Rivera CA, Bradford BU, Graves LM, Cattley RC, Schoonhoven R, Swenberg JA, Thurman RG. Kupffer cell oxidant production is central to the mechanism of peroxisome proliferators. Carcinogenesis 1999; 20(1):27–33.
69. Lornez J, Glatt HR, Fleishman R, Ferlinz R, Oesch F. Drug metabolism in man and its relationship to that in three rodent species: Monooxygenase, epoxide hydrolase, and glutathione-S-transferase activities in subcellular fractions of lung and liver. Biochem Med 1984; 32:43–56.
70. Schwartz AG, Moore CJ. Inverse correlation between species life-span and capacity of cultured fibroblasts to metabolize polycyclic hydrocarbon carcinogens. Fed Proc 1979; 38: 1989–1992.
71. Hsie AW, Recio L, Katz DS, Lee CQ, Wagner M, Schenley RL. Evidence for reactive oxygen species inducing mutation in mammalian cells. Proc Natl Acad Sci USA 1986; 83:9616–9620.
72. Troll W, Weisner R. The role of oxygen radicals as a possible mechanism of tumor promotion. Ann Rev Toxicol 1985; 25:509–528.
73. Kensler TW. Free radicals in tumour promotion. Adv Free Rad Biol Med 1986; 2:347–387.

74. Kensler TW, Trush MA. The role of oxygen radicals in tumour promotion. Environ Mutagen 1984; 6:593–616.
75. Burdon RH. Superoxide and hydrogen peroxide in relation to mammalian cell proliferation. Free Rad Biol Med 1995; 18:775–794.
76. Buttke TM, Sandstrom S. Oxidative stress as a mediator of apoptosis. Immunol Today 1994; 15:7–10.
77. Burdon RH, Rice-Evans C. Free radicals and the regulation of mammalian cell proliferation. Free Rad Res Commun 1989; 6:345–358.
78. Slater AF, Stefan C, Nobel I, van den Dobbelsteen DJ, Orrenius S. Signalling mechanisms and oxidative stress in apoptosis. Toxicol Lett 1995; 82–83:149–53.
79. Cerutti PA. Prooxidant states and tumor promotion. Science 1985; 227:375–381.
80. D'Souza RJ, Phillips EM, Jone PW, Strange RC, Aber GM. of hydrogen peroxide with interleukin-6 and platelet-derived growth factor in determining mesangial cell growth: effect of repeated oxidant stress Clin Sci 1993; 86:747–751.
81. Burdon RH, Gill V. Cellularly generated active oxygen species and HeLa cell proliferation. Free Rad Res Commun 1993; 19:203–213.
82. Alliangana DM. Effects of beta-carotene, flavonoid quercitin and quinacrine on cell proliferation and lipid peroxidation breakdown products in BHK-21 cells. East Afr Med J 1996; 73: 752–757.
83. Fiorani M, Cantoni O, Tasinto A, Boscoboinik D, Azzi A. Hydrogen peroxide- and fetal bovine serum-induced DNA synthesis in vascular smooth muscle cells: positive and negative regulation by protein kinase C isoforms. Biochem Biophys Acta 1995; 1269:98–104.
84. Yang M, Nazhat NE, Jiang X, Kelsey SM, Blake DR, Newland AC, Morris CJ. Adriamycin stimulates proliferation of human lymphoblastic leukaemic cells via a mechanism of hydrogen peroxide ($H_2O_2$) production. Br J Haematol 1996; 95:339–344.
85. Dypbukt JM, Ankarcrona M, Burkitt M, Sjöholm A, Ström K, Orrenius S, Nicotera O. Different prooxidant levels stimulate growth, trigger apoptosis, or produce necrosis of insulin-secreting RINm5F cells. The role of intracellular polyamines J Biol Chem 1994; 269:30553–30560.
86. Sandstom PA, Roberts PA, Folks TM, Buttke TM. HIV gene expression enhances T cell susceptibility to hydrogen peroxide-induced apoptosis. AIDS Res Hum Retrovir 1993; 9: 1107–1113.
87. Sandstrom PA, Mannie MD, Buttke TM. Inhibition of activation-induced death in T cell hybridomas by thiol antioxidants: oxidative stress as a mediator of apoptosis. J Lekocyte Biol 1994a; 221–226.
88. Kim HS, Lee JH, Kim IK. Intracellular glutathione level modulates the induction of apoptosis by delta 12-prostaglandin. J Prostagland 1996; 51:413–425.
89. Aoshima H, Satoh T, Sakai J, Yamada M, Enokido Y, Ekeuchi T, H Hatanaka. Generation of free radicals during lipid hydroperoxide-triggered apoptosis in PC12h cells. Biochem Biophys Acta 1997; 1345:35–42.
90. Hamilton RF, Li L, Felder TB, Hollihan Q. Bleomycin induces apoptosis in human alveolar macrophages Am J Physiol 1995; 269:L318–L325.
91. Sanchez A, Albarez AM, Benito M, Fabregat I. Apoptosis induced by transforming growth factor-beta in fetal hepatocyte primary cultures: involvement of reactive oxygen intermediates J Biol Chem 1996; 271:7416–7422.
92. Cossarizza A, Francheshi C, Monti D, Solvioli S, Bellesia E, Rivabene L, Rainaldi G, Tinara A, Malorni W. Protective effect of N-acetylcysteine in tumor necrosis factor-alpha-induced apoptosis in U937 cells: the role of mitochondria. Exp Cell Res 1995; 220:232–240.
93. Sandstrom PM, Tebbey PW, Van Cleave L, Buttke TM. Lipid hydroperoxides induce apoptosis in T cells displaying a HIV-associated glutathione peroxidase deficiency. J Biol Chem 1994; 269:798–801.
94. Gabby M, Tauber M, Porat S, Simatov R. Selective role of glutathione in protecting human neuronal cells from dopamine-induced apoptosis. Neuropharmacology 1996; 35:571–578.

95. Abello PA, Fidler SA, Buchman TG. Thiol reducing agents modulate induced apoptosis in porcine endothelial cells. Shock 1994; 2:79–83.
96. Manna SK, Zhang HJ, Yan T, Oberley LW, Aggarwal BB. Overexpression of manganese superoxide dismutase suppresses tumor necrosis factor-induced apoptosis and activation of nuclear transcription factor-β and activated protein-1. Biol Chem 1998; 273:13245–13254.
97. Ehrlich M, Gama SM, Huang LH, Midgett RM, Kuo KC, McCune RA, Gehrke C. Amount and distribution of 5-methylcytosine in human DNA from different types of tissues and cells. Nucl Acid Res 1982; 10:2709–2721.
98. Liard PW, Jaenisch R. The role of DNA methylation in cancer genetics and epigenetics. Ann Rev Genet 1997; 30:441–464.
99. Counts JL, Goodman JI. Hypomethylation of DNA: a nongenotoxic mechanism involved in tumor promotion. Toxicol Lett 1995; 82/83:663–672.
100. Baylin SB, Makos M, Wu JJ, Yen RW, de Bustros A, Vertino P, Nelkin BD. Abnormal patterns of DNA methylation in human neoplasia: potential consequences for tumor progression. Cancer Cells 1991; 3:383–390.
101. Counts SL, Sarmiento JI, Harbison ML, Downing JC, McClain RM, Goodman JI. Cell proliferation and global methylation status after phenobarbital and/or choline-devoid, methionine-deficient diet administration. Carcinogenesis 1996; 17:1251–1257.
102. Costello JF, Futscher BW, Tano K, Graunke DM, Peiper RO. Graded methylation in the promoter and body of the O6-methylguanine DNA methyltransferase (MGMT) gene correlates with MGMT expression in human glioma cells. J Biol Chem 1994; 269:17228–17237.
103. Boyes J, Bird A. DNA methylation inhibits transcription directly via a methyl-CpG binding protein. Cell 1991; 64:1123–1134.
104. Boehm TL, Drahovsky D. Alteration of enzymatic methylation of DNA cytosines by chemical carcinogens: a mechanism involved in the initiation of carcinogenesis. J Natl Cancer Inst 1983; 71:429–33.
105. Chen RZ, Peterson U, Beard C, Jackson-Grusby L, Jaenisch R. DNA Hypomethylation leads to elevated mutation rate. Nature 1998; 395:89–93.
106. Belinsky SA, Nikula KJ, Palmisano WA, Michels R, Saccomanno G, Gabrielson E, Baylin SB, Herman JG. Aberrant methylation of $p16^{INK4a}$ is an early event in lung cancer and a potential biomarker for early exposure. Proc Natl Acad Sci USA 1998; 95:11891–11896.
107. Griffin S, Karran P. Incision at DNA G:T mispairs by extracts of mammalian cells occurs preferentially at cytosine methylation sites and is not targeted by a separate GT binding reaction. Biochemistry 1993; 32:13032–13039.
108. Wainfen E, Poirier LA. Ethyl groups in carcinogenesis: effects on DNA methylation and gene expression. Cancer Res 1992; 52:2071–2077.
109. Pogribny IP, Basnakian AG, Miller BJ, Lopatina NG, Poirier LA, James SJ. Breaks in genomic DNA and within the p53 gene are associated with hypomethylation in livers of folate/methyl-deficient rats. Cancer Res 1995; 55:1894–1901.
110. Turk PW, Laayoun A, Smith SS, Weitzman SA. DNA adduct 8-hydroxy-2′-deoxyguanosine (8-hydroxyguanosine) affects function of human DNA methyltransferase. Carcinogenesis 1995; 16:1253–1256.
111. Ziegler DM. Role of reversible oxidation-reduction of enzyme thiols-disulfides in metabolic regulation. Annu Rev Biochem 1985; 54:305–329.
112. Cotgreave IA, Moldéus P, Orrenius S. Host biochemical defense mechanisms against prooxidants. Annu Rev Pharmacol Toxicol 1988; 28:189–212.
113. Xanthoudakis S, Miao GC, Curran T. The redox and DNA-repair activities of Ref-1 are encoded by nonoverlapping domains. Proc Natl Acad Sci USA 1994; 91:23–27.
114. Kerr LD. Signal transduction: the nuclear target. Curr Opin Cell Biol 1992; 4:496–501.
115. Timblin CR, Janssen YMW, Mossman BT. Free-radical-mediated alterations of gene expression by xenobiotics. In: Wallace KB, ed. Free Radical Toxicology. Boston: Taylor and Francis, 1997:325–349.

116. Kass GEN. Free-radical-induced changes in cell signal transduction. In: Wallace KB, ed. Free Radical Toxicology. Boston: Taylor and Francis, 1997:349–374.
117. Klaunig JE, Ruch RJ. Strain and species effects on the inhibition of hepatocyte intercellular communication by liver tumor promoters. Cancer Lett 1987; 36:161–168.
118. Berridge MJ. The biology and medicine of calcium signaling. Molec Cell Endocrinol 1994; 98(2):119–124.
119. Whitfield JF. Calcium signals and cancer. Crit Rev Oncogen 1992; 3:55–90.
120. Larsson R, Cerutti P. Translocation and enhancement of phosphotransferase activity of protein kinase C following exposure in mouse epidermal cells to oxidants. Cancer Res 1989; 49:5627–5632.
121. Brawn MK. Oxidant-induced activation of protein kinase C in UC11MG cells. Free Rad Res 1995; 22:23–37.
122. Dreher D, Junod AF. Role of oxygen free radicals in cancer development. Eur J Cancer 1996; 32A:30–38.
123. Zorn NE. Alterations in splenocyte protein kinase C (PKC) activity by 2,3,7,8-tetrachlorodibenzo-p-dioxin in vivo. Toxicol Lett 1995; 78:93–100.
124. Kass GEN, Duddy SK, Orrenius S. Alteration in splenocyte protein kinase C (PKC) activity by 2,3,7,8-tetrachlorodibenzo-p-pdioxin in vivo. Toxicol Lett 1989; 78:93–100.
125. Jirtle RL, Hankins GR, Reisenbichler H, Boyer IJ. Regulation of mannose-6-phosphate/insulin-like growth factor II receptors and transforming growth factor–β during liver tumor promotion with phenobarbital. Carcinogenesis 1994; 15:1473–1478.
126. Brockenbrough JS. Reversible and phorbol ester-specific defect of protein kinase C translocation in hepatocytes isolated from phenobarbital-treated rats. Cancer Res 1991; 51(1):130–136.
127. Vellanoweth RL. Biology of disease transcription factors in development, growth, and aging. Lab invest 1994; 70:784–799.
128. Devary Y, Gottlieb RA, Lau LF, Karin M. Rapid and preferential activation of the c-jun gene during the mammalian UV response. Mol Cell Biol 1991; 11:2804–2811.
129. Devary Y, Gottlieb RA, Smeal T, Karin M. The mammalian ultraviolet response is triggered by activation of Src tyrosine kinases. Cell 1992; 71:1081–1091.
130. Okuno H, Akahori A, Sato H, Xanthoudakis S, Curran T. Escape from redox regulation enhances the transforming activity of Fos. Oncogene 1993; 8:695–701.
131. Toledano MB, Leonard WJ. Modulation of transcription factor NF-kappa B binding activity by oxidation-reduction in vitro. Proc Natl Acad Sci USA 1991; 88:4328–4332.
132. Schreck R, Albermann K, Baeuerle PA. Nuclear factor kappa B: an oxidative stress-responsive transcription factor of eukaryotic cells. Free Rad Res Commun 1992; 17:221–237.
133. Müller JM, Cahill MA, Rupee RA, Baeuerle PA, Nordheim A. Antioxidants as well as oxidants activate c-fos via Ras-dependent activation of extracellular-signal-regulated kinase 2 and Elk-1. Eur J Biochem 1997; 244:45–52.
134. Nabel GJ, Verma IM. Proposed NF-κB/IB family nomenclature. Genes Dev 1993; 7:2063.
135. Baeuerle PA, Henkel T. Function and activation of NF-κB in the immune system. Annu Rev Immunol 1994; 12:141–179.
136. Piette J, Piret B, Bonizzi G, Schoonbroodt S, Merville M-P. Multiple redox regulation in NF-B transcription factor activation. Biol Chem 1997; 378:1237–1245.
137. Martin RD, Schmid JA, Hofer-Warbinek R. The NF-kB/Rel family of transcription factors in oncogenic transformation and apoptosis. Mut Res 1999; 437:231–243.
138. Hart BA. Characterization of cadmium-induced apoptosis in rat lung epithelial cells: evidence for the participation of oxidant stress. Toxicology 1999; 133:43–58.
139. Lu SC. Effect of ethanol and high-fat feeding on hepatic gamma-glutamylcysteine. Hepatology 1999; 30:209–214.
140. Radler-Pohl A. The activation and activity control of AP-1 (fos/jun). Ann NY Acad Sci 1993; 684:127–148.

141. Gilmour PS. Free radical activity of industrial fibers: role of iron in oxidative stress and activation of transcription factors. Environ Health Perspect 1997; 105(suppl 5):1313–1317.
142. Schulze-Osthoff K, Los M, Baeuerle PA. Redox signalling by transcription factors NF-κB an AP-1 in lymphocytes. Biochem Pharmacol 1995; 50:735–741.
143. Mohan N, Meltz MM. Induction of nuclear factor B after low-dose ionizing radiation involves a reactive oxygen intermediate signaling pathway. Rad Res 1990; 140:97–104.
144. Schulze-Osthoff K, Bauer MK, Vogt M, Wesselborg S. Oxidative stress and signal transduction. Int J Vitam Nutr Res 1997; 67:336–342.
145. Staal FJT, Roederer M, Herzenberg LA. Intracellular thiols regulate activation of nuclear factor kB and transcription of human immunodeficiency virus. Proc Natl Acad Sci USA 1990; 87:9943–9947.
146. Schreck R, Rieber P, Baeuerle PA. Reactive oxygen intermediates as apparently widely used messagers in the activation of the NF-kB transcription factor and HIV-1. EMBO J 1991; 10: 2247–2258.
147. Schmidt KN, Armstad P, Cerutti P, Baeuerle PA. The roles of hydrogen peroxide and superoxide as messengers in the activation of transcription factor NF-κB. Biol Chem 1995; 2:13–22.
148. Curran T. Fos and Jun: oncogenic transcription factors. Tohoku J Exp Med 1992; 168:169–174.
149. Forrest D, Curran T. Crossed signals: oncogenic transcription factors. Curr Opin Genet Dev 1992; 2:19–27.
150. Angel P, Karin M. The role of Jun, Fos and the AP-1 complex in cell proliferation and transformation. Biochim Biophys Acta 1991; 1072:129–157.
151. Angel P, Imagawa M, Chiu R, Stein B, Imbra RJ. Phorbol ester-inducible genes contain a common cis element recognized by a TPA-modulated trans-acting factor. Cell 1987; 49:729–39.
152. Lee W, Mitchell P, Tjian R. Purified transcription factor AP-1 interacts with TPA inducible elements. Cell 1987; 49:741–52.
153. Angel P, Allegretto EA, Okino ST, Hattori K, Boyle WJ. Oncogene jun encodes a sequence-specific trans-activator similar to AP-1. Nature 1988; 332:166–71.
154. Hollander MC, Fornace AJJ. Induction of fos RNA by DNA-damaging agents. Cancer Res 1989; 49:1687–1692.
155. Amstad PA, Krupitza G, Cerutti PA. Mechanism of c-fos induction by active oxygen. Cancer Res 1992; 52:3952–3960.
156. Cerutti P, Shah G, Peskin A, Amstad P. Oxidant carcinogenesis and antioxidant defense. Ann NY Acad Sci 1992; 663:158–166.
157. Janssen YMW, Van Houten B, Borm PJA, Mossman BT. Cell and tissue responses to oxidative damage. Lab Invest 1993; 69:261–274.
158. Djavaheri-Mergny M, Mergny JL, Bertr F, Santus R, Mazière C. Ultraviolet-A includes activation of AP-1 in cultured human keratinocytes. FEBS Lett 1996; 384:92–96.
159. Heintz NH, Janssen YMW, Mossman BT. Persistent induction of c-fos and c-jun protooncogene expression by asbestos. Proc Natl Acad Sci USA 1993; 90:3299–3303.
160. Puga A. Dioxin induces expression of c-fos and c-jun pro-oncogenes and a large increase in transcription factor AP-1. DNA Cell Biol 1992; 11:269–281.
161. Hoffer A, Chang C-Y, Puga A. Dioxin induces fos and jun gene expression by Ah receptor-dependent and -independent pathways. Toxicol Appl Pharmacol 1996; 141:238–247.
162. Zawaski K. Evidence for enhanced expression of c-fos, c-jun, and the Ca2+-activated neutral protease in rat liver following carbon tetrachloride administration. Biochem Biophys Res Commun 1993; 197:585–590.
163. Pinkus R. Phenobarbital induction of AP-1 binding activity mediate activation of glutathione S-transferase and quinone reductase gene expression. Biochem J 1993; 290:637–640.
164. Datta R, Hallahan DE, Kharbanda SM, Rubin E, Sherman ML. Involvement of reactive oxy-

gen intermediates in the induction of c-jun gene transcription by ionizing radiation. Biochemistry 1992; 31:8300–8306.
165. Stevenson MA, Pollock SS, Coleman CN, Calderwood SK. X-irradiation, phorbol esters, and H2O2 stimulate mitogen-activated protein kinase activity in NIH-3T3 cells through the formation of reactive oxygen intermediates. Cancer Res 1994; 54:12–15.
166. Schreiber M, Baumann B, Cotten M, Angel P, EF Wagner. Fos is an essential component of the mammalian UV response. EMBO J 1995; 14:5338–5349.
167. Keyse SM, Tyrrell RM. Heme oxygenase is the major 32-kDa stress protein induced in human skin fibroblasts by UVA radiation, hydrogen peroxide, and sodium arsenite. Proc Natl Acad Sci USA 1989; 86:99–103.
168. Saran M, Bors W. Oxygen radicals acting as chemical messengers: A hypothesis. Free Rad Res Commun 1989; 7:213–220.
169. Cerutti P, Larsson R, Krupitza G. Mechanisms of oxidant carcinogenesis. In: Haws CC, Liotta LA, eds. Genetic Mechanisms in Carcinogenesis and Tumor Promotion. New York: Wiley-Liss, 1990:69–82.
170. Pahl HL, Baeuerle PA. Oxygen and the control of gene expression. BioEssays 1994; 16: 497–502.
171. Cowan DB, Weisel RD, Williams WG, Mickle DG. Identification of oxygen responsive elements in the 5′-flanking region of the human glutathione peroxidase gene. J Biol Chem 1993; 268:26904–26910.
172. Suzuki YJ, Forman HJ, Sevanian A. Oxidants as stimulators of signal transduction. Free Rad Biol Med 1997; 22:269–285.
173. Loewenstein WR. The cell-to-cell channel of gap junctions. Cell 1987; 48:725–726.
174. Pitts JD, Finbow ME. The gap junction. J Cell Sci 1986; 4:239–266.
175. Revel JP, Karnovsky MJ. Hexagonal array of subunits in intercellular junctions of the mouse heart and liver. J Cell Biol 1967; 33:C7–C12.
176. Tsien RW, Weingart R. Ionotropic effects of cyclic AMP I calf ventricular muscle studied by a cut end method. J Physiol 1976; 260:117–141.
177. Pitts JS, Sims JW. Permeability of junctions between animal cells. Intercellular transfer of nucleotides but not of macromolecules. Exp Cell Res 1977; 104:153–163.
178. Cornell-Bell AH, Finkbeiner SM, Cooper MS, Smith SJ. Glutamate induces calcium waves in cultured astrocytes: Long-range glial signaling. Science 1990; 247:470–473.
179. Trosko JE, Ruch R. Cell-cell communication and carcinogenesis. Front Biosci 1998; 3:208–236.
180. Yamasaki H. Gap junctional intercellular communication and carcinogenesis. Carcinogenesis 1990; 11:1051–1058.
181. Krutovskikh V, Yamasaki H. The role of gap junctional intercellular communication (GJIC) disorders in experimental and human carcinogenesis. Histol Histopath 1997; 12:761–768.
182. Mesnil M, Yamasaki H. Cell-cell communication and growth control of normal and cancer cells: evidence and hypothesis. Carcinogenesis 1993; 7:14–17.
183. Trosko JE, Chang CC. In: Nongenotoxic mechanisms in carcinogenesis: Role in inhibited intercellular communication. Bangurg Report 31. Carcinogen Risk Assessment: New Directions in the Qualitative and Quantitative Aspects. New York: Cold Spring Harbor Laboratory, 1988:139–170.
184. Klaunig JE, Hartnett JA, Ruch RJ, Weghorst CM, Hampton JA, Schafer LD. Gap junctional intercellular communication in heptic carcinogenesis. Progr Clin Biol Res 1990; 340D:165–174.
185. Klaunig JE, Ruch RJ. Biology of disease: Role of inhibition of intercellular communication in carcinogenesis. Lab Invest 1990; 62:135–146.
186. Williams GM. Liver carcinogenesis: The role for some chemicals of an epigenetic mechanism of liver-tumor promotion involving modification of the cell membrane. Food Cosmet Toxicol 1981; 19:577–581.

187. Budunova IV, Williams GM. Cell culture assays for chemicals with tumor-promoting or tumor inhibiting activity based on the modulation of intercelllular communication. Cell Biol Toxicol 1984; 10:71–116.
188. Klaunig JE, Siglin JC, Schafer LD, Hartnett JA, Weghorst CM, Olson MJ, Hampton JA. Correlation between species and tissue sensitivity to chemical carcinogenesis in rodents and the induction of DNA synthesis. Prog Clin Biol Res 1991; 369:185–194.
189. Yamasaki H. Aberrant expression and function of gap junctions during carcinogenesis. Environ Health Persp 1991; 93:191–197.
190. Klaunig JE, Ruch RJ. Strain and species effects on the inhibition of hepatocyte intercellular communication by liver tumor promoters. Cancer Lett 1987; 36:161–168.
191. Trosko JE, Chang CC, Medcalf A. Mechanisms of tumor promotion: Potential role of intercellular communication. Cancer Invest 1983; 1:511–526.
192. Diwan BA, Rice JM, Ohshima M, Ward JM, Dove LF. Comparative tumor-promoting activities of phenobarbital, amobarbital, barbital sodium, and barbituric acid on livers and other organs of male F344/NCr rats following initiation with N-nitrosodiethylamine. J Natl Cancer Inst 1985; 74:509–516.
193. Ruch RJ, Klaunig JE. Kinetics of phenobarbital inhibition of intercellular communication in mouse hepatocytes. Cancer Res 1988; 48:2519–2523.
194. Cerutti P, Ghosh R, Oya Y, Amstad P. The role of the cellular antioxidant defense in oxidant carcinogenesis. Environ Health Perspect 1994; 102:123–129.
195. Upham BL, Sun Kang K, Cho H-Y, Trosko JE. Hydrogen peroxide inhibits gap junctional intercellular communication in glutathione sufficient but not glutathione deficient cells. Carcinogenesis 1997; 18:37–42.
196. Robertson FM, Beavis AJ, Obersyzyn T, O'Connell SM, Dokidos A, Laskin DL, Laskin JD, Reiners JJ. Production of hydrogen peroxide by murine epidermal keratinocytes following treatment with the tumor promoter 12-0-tetradeconoylphorbol-13-acetate. Cancer Res 1990; 50:6062–6067.
197. Ruch RJ, Klaunig JE. Antioxidant prevention of tumor promoter induced inhibition of mouse hepatocyte intercellular communication. Cancer Lett 1988; 33:137–150.
198. Sáez JC, Bennett MVL, Spray DC. Carbon tetrachloride at hepatotoxic levels blocks reversibly gap junctions between rat hepatocytes. Science 1987; 236:967–969.
199. Trosko JE. Challenge to the simple paradigm that 'carcinogens' are 'mutagens' and to the in vitro and in vivo assays used to test the paradigm. Mutat Res 1997; 373:245–249.
200. Neveu MJ, Babcock KL, Hertzberg EL, Paul D, Nicholson BJ, Pitot HC. Colocalized alterations in connexin 32 and cytochrome P4350IIB1/2 by phenobarbital and related liver tumor promoters. Cancer Res 1994; 54:3145–3152.
201. Scholz W, Schutze K, Kunz W, Schwarz M. Phenobarbital enhances the reactive oxygen in neoplastic rat liver nodules. Cancer Res 1990; 50:7015–7022.

# 5

# Proposed Mechanisms of Arsenic Toxicity Carcinogenesis

**MICHAEL I. LUSTER and PETIA P. SIMEONOVA**

*National Institute for Occupational Safety and Health, Morgantown, West Virginia*

## I. INTRODUCTION

Trivalent and pentavalent forms of inorganic arsenic are ubiquitous elements found in nature that unfortunately result in significant human exposure. Oral exposure to arsenic occurs primarily from contamination of drinking water and food constituents, and is particularly high in certain regions of the world including areas of the southwestern United States, eastern Europe, India, China, Taiwan, and Mexico (1,2). Humans can also be exposed to arsenic through inhalation. This occurs primarily in occupations involved in mining/smelting operations, agriculture, or microelectronics (3,4). Epidemiological studies have demonstrated that exposure to inorganic arsenic is associated with increased risk of cancers of the skin and internal organs, including the urinary bladder, respiratory tract, liver, and kidney in populations from Finland, Taiwan, China, Bangladesh, Mexico, southwestern United States, and Central and South America (3–8). Arsenic-induced skin cancers usually develop 20 to 30 years after exposure, and occur in sun-exposed as well as nonexposed areas. The types of skin tumors found include either Bowen's disease, squamous cell carcinomas, basal cell carcinomas, or combined lesions (9–11). The key to identifying patients with arsenic-induced skin tumors is that they normally occur at multiple sites and unusual locations. Internal tumors are also common and are most frequently associated with the bladder. The association between arsenic exposure and urinary bladder cancers, typically transitional cell carcinomas, has been observed in the same endemic areas of the world where skin cancer populations have been identified. Lung tumors from arsenic are often associated with occupational exposure, such as smelters or agriculture workers, and occur from inhalation (12). In addition to neoplasia, additional pathological manifestations of chronic arsenic exposure include skin hyperpigmentation and hyperkeratosis (9,13), as well as vascular disease (14,15). Hyperpigmentation is the most common effect in individuals and can occur at any body site, and already pigmented areas are more accentuated. Arsenic-induced hyperpigmentation occurs almost exclusively in individuals of Oriental descent, although the genetic basis for this is not understood (16). Hyperkeratosis, which can appear within 4 years of exposure to arsenic, is manifested primarily in the form of hyperkeratotic papules or plaques and are most commonly found on the palms and soles (17). There are reports of cellular atypia at the base of these papules and on occasion their transformation into basal or squamous cell carcinoma. In contrast to carcinogenicity, little is known regarding the vascular effects of arsenic and most of the reports originate from individuals living in inner Mongolia, Xinjiiang, Toroku, and Nakajo (14,15,18). Circulatory manifestations of arseniasis include increased prevalence of ischemic heart disease and peripheral vascular disease. The latter is commonly known as blackfoot disease in southwestern Taiwan. In addition, dose-response relationships between arsenic exposure and hypertension prevalence have been reported in Southwestern Taiwan (19) and Bangladesh (20).

On the basis of numerous epidemiological studies, arsenic has been classified as a potent human carcinogen, and population cancer risk due to arsenic has been suggested to be comparable to environmental tobacco smoke and radon in homes with risk estimates of around 1 per 1000 (11). It has been estimated that over 350,000 people in the United States consume drinking water containing over 50 μg/L of arsenic, the current EPA standard, and more than 2.5 million people use water containing more than 25 μg/L of arsenic (21). Subsequently, there is significant regulatory pressure to lower the acceptable levels. However, epidemiological studies, where exposure levels have been collected, suggest that the current EPA cancer slope factor (CSF) for arsenic may actually overpredict cancer cases at relatively low exposure levels (22). This may be due to the fact that the CSF was calculated assuming a standard linear dose-response relationship while a nonlinear or sublinear dose response may be more appropriate. Human epidemiological data are available providing empirical evidence supporting both a linear (23) and nonlinear (24) association between excess cancer and arsenic exposure. As will be discussed in the following sections, although the precise carcinogenic mechanism for arsenic has not been established, molecular, cellular, and metabolic studies suggest that a nonlinear relation may be most appropriate.

## II. ISSUES IN ARSENIC TOXICITY

Although arsenic is classified as a human carcinogen, until recently animal models were considered either negative or equivocal, thus limiting mechanistic studies. We recently established an unconventional animal model in which increased numbers of papillomas develop on the skin of Tg.AC transgenic mice, which contain the human v-Ha ras structural gene, following exposure to arsenite in the drinking water (25). Although this would suggest that arsenic acts as a tumor promoter, it is also necessary to administer low doses of phorbol ester, a classical promoter, to obtain increased papilloma load. This suggests that arsenic acts as a copromoter and is consistent with in vitro studies, implying that arsenic may enhance cell growth rather than activating genotoxic mechanisms. This is also consistent with studies demonstrating that urinary bladder cancers can develop in arsenic-treated rats following exposure to *N*-butyl-*N*-(4-hydroxybutyl) nitrosamine (26), a potent chemical tumor initiator. However, it has been recently reported that bladder tumors develop in rats when administered greater than 50 ppm of dimethylarsenic acid (DMA) in the feed for 2 years (27). This is somewhat inconsistent with the current thinking regarding arsenic toxicity, as it suggests that DMA not only acts as a complete carcinogen (i.e., tumor initiator and promoter) but also is more potent than inorganic arsenic. With reference to the latter, the majority of evidence indicates that the inorganic forms of arsenic, particularly $iAs^{3+}$ (arsenite), are more reactive, more toxic, and less readily excreted in the urine than the methylated forms (28). Arsenite and arsenate, in contrast to the methylated forms, have a predilection to react with tissue rich in vicinal dithiols, such as keratins in the skin (29). In mammals, arsenic metabolism first involves the reduction of arsenate ($As^{5+}$) to arsenite ($As^{3+}$) by reactions involving glutathione (GSH). Arsenite is then enzymatically methylated, primarily in the liver to monomethylarsonic acid (MMA) and then to DMA, presumably resulting in detoxification and urinary excretion (30). It has been argued that interhuman variability, as well as animal species differences, in arsenic toxicity are due to differences in metabolism. There is even circumstantial evidence in humans that suggests polymorphisms in the methyltransferase enzymes exist which affect efficacy of the oxidative addition of methyl groups to arsenic (31). In addition to the role that methylation plays in arsenic toxicities, another major question, which will be discussed in more detail in the following sections, is the role that reactive oxygen species (ROS) play. It is known that arsenite binds avidly to dithiols, and, when added to cell cultures, has a particular affinity to glutathione resulting in its reduction (32). In this respect, arsenic has been employed historically to elicit heat shock responses, induce oxidant-sensitive enzymes such as heme oxygenase, and stimulate oxidant-sensitive MAP kinase pathways (33–37). The ability to diminish many arsenic-associated cellular responses by addition of *N*-acetylcysteine (NAC) or enhance toxicity by addition of GSH inhibitors, such as L-buthionine sulfoximine, further support these observations. Once GSH is depleted, arsenic either becomes available to interact more freely with so-called ''arsenic targets'' or there is an increased availability of cell-derived ROS to induce oxidative damage. One current hypothesis that has been actively pursued is that arsenic increases cellular oxidants, which subsequently activate oxidant-sensitive transcription factors, such as AP-1 and NF-κB and affect gene transcription (38–42). It has even been argued that differences observed in cancer rates between populations highly exposed to arsenic may be related to the oxidant levels in their diets (43,44).

## III. PROPOSED CLASSICAL GENETIC MECHANISMS OF CARCINOGENICITY

Although there is ample evidence that certain forms of heavy metals, such as cadmium oxide and chromium VI, are carcinogenic because the act as classical genotoxic agents (45,46), this is not supported in studies with arsenic. Arsenic fails to interact directly with DNA to induce mutations, nor is it a DNA-reactive electrophile (47,48). Arsenic causes chromatid abnormalities, such as sister chromatid exchanges, but only at high concentrations that suggest cytotoxicity (45,49). Arsenic induces amplification of the dihydrofolate reductase gene in mouse 3T6 cells and, although gene amplification has been suggested as a possible mechanism of arsenic carcinogenicity (50), it has not been substantiated in other models. DNA repair enzymes are inhibited by arsenic, resulting in a comutagenic response with x-rays, ultraviolet radiation, or alkylating agents (51–53), but epidemiological studies do not support that these other agents are necessary. Since the concentrations of arsenic required to inhibit DNA ligase activity in vitro are higher than those needed to inhibit repair within cells, it has been argued that arsenic may modulate the control of cellular DNA repair processes (51).

## IV. PROPOSED NONCLASSICAL GENETIC MECHANISMS OF CARCINOGENICITY

Considerable studies have focused on the relationship between arsenic metabolism and its ability to alter DNA methylation patterns (54,55). DNA methylation contributes to the control and expression of a number of genes, including protooncogenes such as *c-myc*. Methyltransferase enzymes, which are responsible for the methylation of both DNA as well as arsenic, require cofactors, such as *S*-adenosyl-methionine (SAM) (56). It has been shown that as the level of arsenic exposure in humans increases, the urinary excretion of methylated forms decreases while inorganic arsenic increases correspondingly (57). This is believed to be due either to saturation of methyltransferase, to depletetion of cofactors such as SAM, or to depletion of intracellular GSH. Independent of the cause, the net result would be increased availability of inorganic arsenic that can then react with target tissues. Direct evidence for the hypomethylation hypothesis has been provided in vitro using rat epithelial cells where arsenic-induced cell transformation was found to parallel global DNA hypomethylation in the presence of reduced cellular SAM levels. This was associated with overexpression of the metallothionein gene, a gene controlled by DNA hypomethylation and *c-myc* expression, a marker of cell proliferation (54).

The transcription of some genes is sensitive to hypermethylation, and in apparent contrast to the hypomethylation hypothesis, other studies have suggested that hypermethylation is involved in arsenic carcinogenicity (55). This was first suggested in studies with A549 cells, a type II lung epithelial cell line. When cultured in the presence of arsenic, a dose-responsive cytosine methylation occurs in a portion of the *p53* tumor suppressor gene where transcription is methylation-sensitive. Hypermethylation can be observed within a 341-base pair fragment of the promoter of the cell cycle regulator molecule. Based upon the previously described events involved in arsenic metabolism, the hypermethylation hypothesis cannot be easily explained, but may be due to the existence of methyltransferases with varying sensitivities to arsenic.

## V. PROPOSED EPIGENETIC MECHANISMS OF CARCINOGENICITY

As was eluded to earlier, increasing evidence supports the hypothesis that arsenic shares many properties of tumor promoters by affecting specific cell signal transduction pathways involved in cell proliferation. Similar to classic tumor promoters, such as phorbol esters, okadaic acid, and UV radiation, arsenic activates transcription factors, such as AP-1 and AP-2, and induces immediate early genes including *c-fos*, *c-jun*, and *c-myc* (38,58,59), whose products stimulate cell proliferation. Consistent with these observations, arsenic induces a moderate, albeit persistent increase in keratinocyte cell proliferation in vitro as evidenced by increases in thymidine incorporation (25,60), cell cycling (61), labeling of Ki-67, a proliferating cell marker (61), and ornithine decarboxylase activity (62). Recently, it has been demonstrated that fibroblasts (63,64) and human urinary bladder epithelial cells (65) also respond to arsenic in vitro by moderate enhanced cell growth. Electromobility shift assays (EMSAs) and proliferating cell nuclear antigen (PCNA) immunostaining have established that AP-1 activation and hyperplasia can occur concurrently in urinary bladder epithelial cells and epidermis of mice and rats within 8 weeks following exposure to arsenite (25,65,66). The ability of arsenic to activate AP-1 in vivo has recently been confirmed in transgenic mice which contain an AP-1 luciferase reporter construct (65). Characterization of arsenic-induced AP-1 DNA binding complex has demonstrated that the complex consists of fos/jun heterodimers (58,65), a common heterodimer responsible for regulating cell mitogenesis (67). Of particular relevance to these studies is evidence that *c-jun* expression occurs simultaneously with urinary bladder transitional carcinoma (68,69).

The mechanisms by which arsenic activates these transcription factors may be explained by some of its physicochemical properties. Arsenite accumulates in tissues rich in sulfhydryl-containing molecules such as keratin (70,71), which may explain the accumulation of arsenic in epithelial cells from the skin and bladder and development of carcinogenicity in these tissues. As mentioned earlier, arsenic also reduces intracellular levels of reduced GSH (72). The altered redox state of the cells results in oxidative stress, which can activate oxidant-sensitive transcription factors, such as AP-1 and NF-κB (73). Thus cellular effects of arsenic can be exacerbated by agents, like buthionine-sulfoximine (BSO), which reduce intracellular GSH levels and attenuate in the presence of NAC, a precursor of GSH (74,75). GSH, in addition to serving as an antioxidant, can detoxify and methylate arsenic by direct binding (76). Arsenic-induced GADD 153 expression, a gene whose product is associated with growth arrest and cell damage, although inhibited by NAC, is not affected by ROS scavengers such as *o*-phenanthroline, a metal iron chelator, or mannitol, a hydroxyl radical scavenger, which inhibits $H_2O_2$-induced GADD153. This suggests that direct arsenic–GSH interactions, such as GSH reduction, are more likely involved in gene expression than direct induction by ROS. Recent studies reported a high frequency of 8-hydroxy-2′-deoxyguanosine (8-OHdG), a sensitive marker of oxidative DNA damage, in arsenic-related skin cancers (77). While this can be due to the generation of ROS, 8-OHdG can also be formed by direct electron transfer without the participation of ROS. Additional studies that measure the ability of arsenic to directly generate ROS will be necessary to clarify the involvement of oxidative stress in arsenic toxicity.

Several studies have suggested that arsenic activates gene expression by modulation of intracellular phosphorylation events and mitogen-activated protein kinases (MAPK) (58). It has been demonstrated that arsenite activates *c-jun* N-terminal kinase (JNK) and p38 kinase in HeLa cells in parallel with AP-1 activation and *c-jun*/*c-fos* gene expression. It was suggested that arsenite may interact with sulfhydryl groups on cysteine at the catalytic site of JNK phosphatase to inhibit its activity and prolong JNK and p38 activation. Studies conducted with PC12 cells, used commonly to examine MAPK activation, demonstrated that arsenite is a potent activator of both JNK and p38 kinase, but only moderately activates ERK (40). The activation of all three kinases by arsenic could be prevented by addition of NAC, suggesting a role of GSH and/or oxidative stress in this response. There is evidence that arsenite also activates ERK in PC-12 cells by binding to cysteine-rich domains of the epidermal growth factor receptor (EGFR), which subsequently activates the Ras-dependent pathway (40,78). Previous studies established that arsenic serves as a ligand for receptors that have vicinal thiols in their binding sites, such as glucocorticoid receptors (79), and EGFR would also fall into this category. Recently, it was shown that arsenite activates MAPK in JB6 mouse epidermal cell line, as evidenced by ERK phosphorylation, at doses as low as 0.8 μM while only high doses (>50 μM) activates JNK (41). Furthermore, overexpression of a dominant negative ERK blocked arsenite-induced cell transformation in this cell line, indicating a direct role of ERK in arsenic-induced cell transformation. The variability of the specific kinase responses detected in these studies may depend on the specific cell type and concentration of arsenic employed. Activation of different members of MAPK has been related to specific stimuli. For example, ERK is activated strongly by mitogenic stimuli, but only moderately by stress, while JNK and p38 are strongly activated by stressors and moderately by growth factors (80,81).

Arsenic has been shown to influence not only early-immediate gene expression, such as *c-fos, c-jun*, and *c-myc*, whose products are directly involved in cell cycle progression, but also genes that regulate late mitogenic signals including growth factors and certain cytokines. For example, we have demonstrated that transforming growth factor (TGF)-α and granulocyte-macrophage colony-stimulating factor (GM-CSF) expression increase in keratinocytes cultured with arsenic as well as in the skin of rodents and humans exposed to arsenic in drinking water (25,82). Immunohistochemical staining has localized TGF-α overexpression to the hair follicles, a site where arsenic tends to concentrate. Overex-

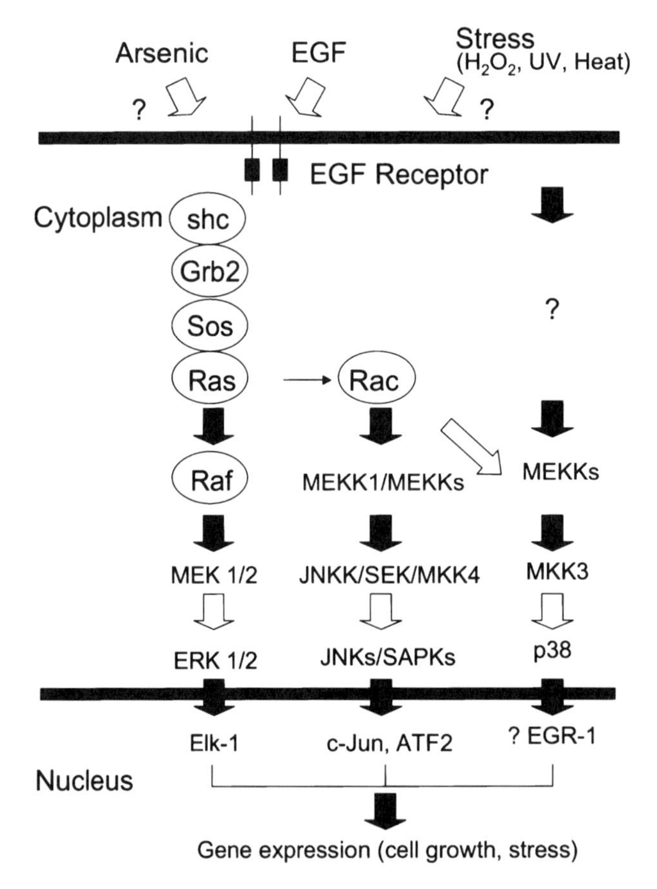


**Figure 1** Overview of transduction by MAPKS, ERK1/2, JNKs, SAPKs, and p38. Extracellular signals received by EGF and other stimuli, including possibly arsenic, $H_2O_2$, UV, or heat, are transduced into the nucleus via the EGF receptor and, subsequently, three cascades that lead to gene expression. Question marks represent unknowns or uncertainties. *Abbreviations*: EGF, epidermal growth factor; ERK extracellular signal-regulated kinase; JNK, Jun *N*-terminal kinase; MAPK, mitogen-activated protein kinase; MEK, MAPK/ERK kinase; MEKK, MEK kinase; SAPK, stress-activated protein kinase; UV, ultraviolet.

pression of TGF-α, and to a lesser extent GM-CSF, has been associated with neoplastic transformation in the skin (83) and TGF-α transgenic mice exhibit hyperkeratoses and increased spontaneous skin and internal tumors (84), suggesting that TGF-α overexpression, like arsenic, has the unique ability to complement both tumor initiation and promotion.

Recently, cDNA microarray technology was employed to establish the profile of gene expression induced by arsenite in UROsta cells, a human uro-epithelial cell line (65). These DNA microarrays demonstrated that arsenic consistently activated 16 genes at a concentration of 50 μM sodium arsenite, 7 of which were also induced by a concentration of 10 μM arsenite. In addition to previously reported early-immediate genes modulated by arsenic, such as AP-1 and *c-myc*, the DNA microarray demonstrated a strong induction of early growth response gene-1 (EGR-1). This gene, which encodes for zinc finger DNA binding transcription factors, has been related to the cell-proliferative effects of mitogenic factors such as epidermal growth factor (EGF), nerve growth factor (NGF), or serum (85). Recently, overexpression of EGR-1 has been associated with human prostate cancer, with its expression correlating with the pathomorphological stage of malignancy. Functional EGR-1 binding sites are found in the promoter domains of a large number of genes involved in cell growth, including TGF-α, insulin growth factor II (IGF-II), *c-myc*, thymidine kinase, and cyclin D (86). Arsenite also activated genes implicated in cellular stress and growth arrest responses, such as GADD153 and GADD45. Activation of these genes is an integral part of endoplasmic reticulum and is associated with activation of C/EBP and modulation of pathways leading to cell death and regeneration (75). Arsenic was also found to alter genes that encode antiapoptotic proteins (BCL-2 binding protein and BAG-1), repair associated protein, and proteins involved in cytoskeleton reorganization. Taken together, these data suggest that arsenic initiates cell signaling pathways that lead to transcription factors, such as AP-1, by binding to EGFR and activating the RAS-dependent pathway or by altering the redox state of the cells, which directly activates nuclear transcription factors that regulate genes involved in stress and mitogenic response (Fig. 1). It should be cautioned, however, that the precise role of these genes in arsenic-induced malignancies is yet to be defined.

## VI. ROLE IN CANCER THERAPY

In stark contrast to the carcinogenic effects of chronic arsenic exposure, arsenic trioxide ($As_2O_3$) has been used therapeutically to successfully treat patients with acute promyelocytic leukemia (APL) without causing severe toxicity (87). In vitro studies have shown that $As_2O_3$ induces apoptosis at low concentrations in APL cells in contrast to other leukemic cells, which requires considerably higher concentrations (88). The mechanisms responsible for this cell-specific response appear to be associated with the low constitutive levels of GSH peroxidase and higher content of $H_2O_2$ in APL cells (89). In studies with normal human keratinocytes, it was found that the mitogenic effects of arsenic occur at concentrations only slightly below that which produced cytotoxicity (82). Thus, it would appear that the efficacy of arsenic as a cancer therapy may depend upon the pharmacokinetics of treatment as well as the oxidant status of the target cell.

## VII. CONCLUDING REMARKS

Overwhelming epidemiological evidence indicates that arsenic is involved in the development of cancers of the skin and internal organs, as well as certain noncancer toxicities including dermatotoxicity and vascular disease. Although the mechanisms responsible for its toxicity have not been fully defined, alterations in epigenetic events involving cell growth control appear to be intimately involved. It will be important to identify the precise pathways through which arsenic affects these processes in order to provide potential targets for therapeutic intervention or prevention. The potential need to use nonlinear models in risk assessment dictated either because of unique epigenetic mechanisms or dose-dependent changes in methylation capabilities will provide challenges in conducting accurate low-dose risk assessment. The tissue and species diversity in arsenic metabolism, the role of noncancer pathologies, such as vascular disease, as well as the role of nutrition, will also need to be considered in determining safe levels.

## REFERENCES

1. Welch AH, Helsel DR, Focazio MJ, Watkins SA. Arsenic in ground water supplies of the United States. In: Chappell WR, Abernathy CO, Calderon RI, eds. Arsenic Exposure and Health Effects. Oxford, UK: Elsevier Science Ltd, 1999:9–17.
2. Thornton I. Arsenic in the global environment: Looking towards the millennium. In: Chappell WR, Abernathy CO, Calderon RL, eds. Arsenic Exposure and Health Effects. Oxford, UK: Elsevier Science, 1999:1–7.
3. Tsuda T, Babazono A, Yamamoto E, Kurumatini N, Mino Y, Ogawa T, Kishi Y, Aoyama H. Ingested arsenic and internal cancer: a historical cohort study followed for 33 years. Am J Epidemiol 1995; 141:198–209.
4. Nriagu JO. Human health and ecosystem effects. In: Nriagu JO, ed. Arsenic in the Environment. New York: Wiley, 1994.
5. Smith AH, Hopenhayn-Rich C, Bates MN, Goeden HM, Hertz-Picciotto P, Duggan HM, Wood R, Kornett MJ, Smith MT. Cancer risks from arsenic in drinking water. Environ Health Perspect 1992; 97:259–267.
6. Chiou HY, Hsueh YM, Liaw KF, Horng SF, Chiang MH, Pu YS, Lin JS, Huang CH, Chen CJ. Incidence of internal cancers and ingested inorganic arsenic: A 7 year follow-up study in Taiwan. Cancer Res 1995; 55:1296–1300.
7. Cebrian ME, Albores A, Garcia-Vargas G, Razo LMD, Ostrosky-Wegman P. Chronic arsenic poisoning in humans: the case of Mexico. In: Nriagu JO, ed. Arsenic in the Environment, Part II: Human Health and Ecosystem Effects. New York: Wiley, 1994:93–107.
8. Mazumder DNG, Gupta JD, Santra A, Pal A, Ghose A, Sarkar S. Chronic arsenic toxicity in West Begal—the worst calamity in the world. J Ind Med Assoc 1997; 96:4–8.
9. Maloney ME. Arsenic in dermatology. Dermatol Surg 1996; 22:301–304.
10. Chai C-Y, Yu H-S, Yen H-T, Tsai K-B, Chen S-S, Yu C-L. The inhibitory effect of UVB irradiation on the expression of p53 and Ki-67 proteins in arsenic-induced Bowen's disease. J Cutan Pathol 1997; 24:8–13.
11. IARC. Arsenic and arsenic compounds. IARC monograph on the evaluation ocf carcinogenic risks to humans: Overal evaluations of carcinogenicity. International Agency for Research on Cancer (IARC), 1987:100–106.
12. Viren J, Silvers A. Nonlinearity in the lung cancer dose-response for airborne arsenic: Apparent confounding by year of hire in evaluating lung cancer risks from arsenic exposure in Tacoma smelter workers. Regal Toxicol Pharmacol 1999; 30:117–129.
13. Schwartz RA. Arsenic and the skin. Int J Dermatol 1997; 36:241–250.
14. Engel RR, Hopenhayn-Rich C, Receveur O, Smith AH. Vascular effects of chronic arsneic exposure: a review. Epidemiol Rev 1994; 16:184–209.
15. Tseng CH, Chong CK, Chen CJ, Tai TY. Dose-response relationship between peripheral vascular disease and ingested inorganic arsenic among residents in blackfoot disease endemic villages in Taiwan. Atherosclerosis 1996; 120:125–133.

16. Shannon RL, Strayer DS. Arsenic-induced skin toxicity. Human Toxicol 1989; 8:99–104.
17. Yeh S, How SW, Lin CS. Arsenical cancer of skin: Histologic study with special reference to Bowen's disease. Cancer 1968; 21:312–339.
18. Hsueh Y-M, Wu W-L, Huang Y-L, Chiou H-Y, Tseng C-H, Chen C-J. Low serum carotene level and increased risk of ischemic heart disease related to long-term arsenic exposure. Atherosclerosis 1998; 141:249–257.
19. Chen CJ, Hsueh YM, Lai MS, Shyu MP, Chen SY, Wu MM, Kuo TL, Tai TY. Increased prevalence of hypertension and long-term arsenic exposure. Hypertension 1995; 25:53–60.
20. Rhaman M, Axelson O. Hypertension and arsenic exposure in Bangladesh. Hypertension 1999; 33:74–78.
21. Karagas MR, Tosteson TD, Blum J, Morris JS, Baron JA, Klaue B. Design of an epidemiologic study of drinking water arsenic exposure and skin and bladder cancer risk in a U.S. population. Environ Health Perspect 1998; 106(4):1047–1050.
22. Valberg PA, Beck BD, Boardman PD, Chen JT. Liklihood ratio analysis of skin cancer prevalence associated with arsenic in drinking water in the U.S. Environ Geochem Health 1998; 20:61–66.
23. US EPA. The intergrated information system (IRIS): Inorganic arsneic. Cincinnati: Office of Health and Environmental Assessment, 1992.
24. NHW. Arsenic and its compounds: Canadian Environmental Protection Act, Priority Substances List. Washington, D.C.: National Health and Welfare, Department of the Environment, 1993.
25. Germolec DR, Spalding J, Yu HS, Chen GS, Simeonova PP, Humble MC, Bruccoleri A, Boorman GA, Foley JF, Yoshida T, Luster MI. Arsenic enhancement of skin neoplasia by chronic stimulation of growth factors. Am J Pathol 1998; 153:1775–1785.
26. Yamamoto S, Konishi Y, Matsuda T, Murial T, Shibata MA, Matsui-Yuasa I, Otani S, Kuroda K, Endo G, Fukushima S. Cancer induction by an organic arsenic compound, dimethylarsenic acid (Cacodylic acid), in F344/DuCrj rats after pretreatment with five carcinogens. Cancer Res 1995; 55:1271–1276.
27. Wei M, Wanibuchi H, Yamamoto S, Li W, Fukushima S. Urinary bladder carcinogenicity of dimethylarsenic acid in male F344 rats. Carcinogenesis 1999; 20:1873–1876.
28. Vahter M. Variation in human metabolism of arsenic. In: Chappell W, Abernathy C, Calderon R, eds. Arsenic Exposure and Health Effects. Oxford, UK: Elsevier Science, 1999:267–280.
29. Scott N, Hatlelid KM, MacKenzie NE, Carter DE. Reactions of arsenic(III) and arsenic(V) species with glutathione. Chem Res Toxicol 1993; 6:102–106.
30. Styblo M, Razo LMD, LeCluyse EL, Hamilton GA, Wang C, Cullen WR, Thomas DJ. Metabolism of arsenic in primary cultures of human and rat hepatocytes. Chem Res Toxicol 1999; 12:560–565.
31. Vahter M, Concha G, Nermell B, Nilsson R, Dulout F, Natarajan AT. A unique metabolism of inorganic arsenic in native Andean women. Eur J Pharmacol 1995; 293:455–462.
32. Chang W-C, Chen S-H, Wu H-L, Shi G-Y, Murota S-I, Morita I. Cytoprotective effect of reduced glutathione in arsenical-induced endothelial cell injury. Toxicology 1991; 69:101–110.
33. Lee K-J, Hahn GM. Abnormal proteins as the trigger for induction of stress responses: Heat, diaminde, and sodium arsenite. J Cell Physiol 1988; 136:411–420.
34. Hei TK, Liu SX, Waldren C. Mutagenicity of arsenic in mammalian cells: role of reactive oxygen species. Proc Natl Acad Sci 1998; 95:8103–8107.
35. Ishikawa T, Igarashi T, Hata K, Fujita T. c-fos induction by heat, arsenite, and cadmium is mediated by a heat shock element in its promoter. Biochem Biophys Res Commun 1999; 254: 566–571.
36. Honda K-I, Hatayama T, Takahashi K-I, Yukioka M. Heat shock proteins in human and mouse embryonic cells after exposure to heat shock or teratogenic agents. Teratogen Carcinogene Mutagen 1992; 11:235–244.

37. Guadalupe Aguilar Gonzalez M, Hernandez H, Lourdes Lopez M, Mendoza-Figueroa T, Albores A. Arsenite alters heme synthesis in long-term cultures of adult rat hepatocytes. Toxicol Sci 1999; 49:281–289.
38. Burleson FG, Simeonova PP, Germolec DR, Luster MI. Dermatotoxic chemicals stimulate c-jun and c-fos transcription and AP-1 DNA binding in human keratinocytes. Res Comm Molec Pathol Pharmacol 1996; 93:131–148.
39. Elbirt KK, Whitmarsh AJ, Davis RJ, Bonkovsky HL. Mechanism of sodium arsenite-mediated induction of heme oxygenase-1 in hepatoma cells. J Biol Chem 1998; 273:8922–8931.
40. Liu Y, Guyton KZ, Gorospe M, Xu Q, Lee JC, Holbrook NJ. Differential activation of ERK, JNK/SAPK and P38/CSBP/RK map kinase family members during the cellular response to arsenite. Free Rad Biol Med 1996; 21:771–781.
41. Huang C, Ma WY, Li J, Goranson A, Dong Z. Requirement of Erk, but not JNK, for arsenite-induced cell transformation. J Biol Chem 1999; 274:14595–14601.
42. Kaltreider RC, Pesce CA, Ihnat MA, Lariviere JP, Hamilton JW. Differential effects of arsenic (III) and chromium (VI) on nuclear transcription factor binding. Mol Carcinogen 1999; 25: 219–229.
43. Hsueh Y-M, Cheng G-S, Wu M-M, Yu H-S, Kuo T-L, Chen C-J. Multiple risk factors associated with arsenic-induced skin cancer: Effects of chronic liver disease and malnutritional status. Br J Cancer 1995; 71:109–114.
44. Valentine JL, Cebrian ME, Garcia-Vargas GG, Faraji B, Kuo J, Gibb HJ, Lachenbruch PA. Daily selenium intake estimates for residents of arsenic-endemic areas. Environ Res 1994; 64:1–9.
45. Larramendy ML, Popescu NC, DiPaolo JA. Induction by inorganic metal salts of sister chromatid exchanges and chromosome abberations in human and Syrian hamster cell strains. Environ Mutagen 1981; 3:597–606.
46. Hamilton JW, Kaltreider RC, Bajenova OV, Ihnat MA, McCaffrey J, Turpie BW, Rowell EE, Oh J, Nemeth MJ, Pesce CA, Lariviere JP. Molecular basis for effects of carcinogenic heavy metals on iducible gene expression. Environ Health Perspect 1998; 106:1005–1015.
47. Rossman TG, Stone D, Moling M, Troll W. Absence of arsenite mutagenicity in *E. coli* and Chinese hamster cells. Environ Mutagen 1980; 2:371–379.
48. Jacobson-Kram D, Montalbano D. The reproductive effects assessment group's report on mutagenicity of inorganic arsenic. Environ Mutagen 1985; 7:787–804.
49. Hanston P, Verellen-Dumoulin C, Libouton JM, Leonard A, Leonard ED, Mahieu P. Sister chromatid exchanges in human peripheral blood lymphocytes after ingestion of high doses or arsenicals. Int Arch Occup Environ Health 1996; 68:342–344.
50. Lee TC, Tanaka N, Lamb PW, Gilmer TM, Barrett JC. Induction of gene amplification by arsenic. Science 1988; 241:79–81.
51. Li JH, Rossman TG. Inhibition of DNA ligase activity by arsenite: A possible mechanism of its comutagenesis. Mol Toxicol 1989; 2:1–9.
52. Jha AN, Noditi M, Nilsson R, Natarajan AT. Genotixic effects of sodium arsenite on human cells. Mut Res 1992; 284:215–221.
53. Lee T-C, Kao S-L, Yih L-H. Suppression of sodium arsenite-potentiated cytotoxicity of ultraviolet light by cycloheximide in Chinese hamster ovary cells. Arch Toxicol 1991; 65:640–645.
54. Zhao CQ, Young MR, Diwan BA, Coogan TP, Waalkes MP. Association of arsenic-induced malignant transformation with DNA hypomethylation and aberrant gene expression. Proc Natl Acad Sci USA 1997; 94:10907–10912.
55. Mass MJ, Wang L. Arsenic alters cytosine methylation patterns of the promoter of the tumor suppressor gene p53 in human lung cells: a model for a mechanism of carcinogenesis. Mut Res 1997; 386:263–277.
56. Styblo M, Delnomdedieu M, Thomas DJ. Biological mechanisms and toxicological consequences of the methylation of arsenic. In: Goyer RA, Cherian MG, eds. Handbook of Experi-

mental Pharmacology: toxicology of Metals, Biochemical Effects. New York: Springer-Verlag, 1995:407–434.
57. Buchet JP, Lauwreys R, Roels H. Urinary excretion of inorganic arsenic and its metabolites after repeated ingestion of sodium metaarsenite by volunteers. Int Arch Occup Environ Health 1981; 48:111–118.
58. Cavigelli M, Li WW, Lin A, Su B, Yoshioka K, Karin M. The tumor promoter arsenite stimulates AP-1 activity by inhibiting a JNK phosphatase. EMBO J 1996; 15:6269–6279.
59. Kachinskas DJ, Qin Q, Phillips MA, Rice RH. Arsenate suppression of human keratinocyte programming. Mut Res 1997; 386:253–261.
60. Corsini E, Asti L, Viviani B, Marinovich M, Galli CL. Sodium arsenate induces overproduction of interleukin-1α in murine keratinocytes: role of mitochondria. J Invest Dermatol 1999; 113:760–765.
61. Klimecki WT, Borchers AH, Egbert RE, Nagle RB, Carter DE, Bowden GT. Effects of acute and chronic arsenic exposure of human-derived keratinocytes in an in vitro human skin equivalent system: A novel model of human arsenicism. Toxicol in Vitro 1997; 11:89–98.
62. Brown JL, Kitchin KT. Arsenite, but not cadmium, induces ornithine decarboxylase and heme oxygenase activity in rat liver: relevance to arsenic carcinogenesis. Cancer Lett 1996; 98:227–231.
63. Barchowsky A, Roussel RR, Klei LR, James PE, Ganju N, Smith KR, Dudek EJ. Low levels of arsenic trioxide stimulate proliferative signals in primary vascular cells without activating stress effector pathways. Toxicol Appl Pharmacol 1999; 159:65–75.
64. Trouba KJ, Glanzer JG, Vorce RL. Wild-type and Ras-transformed fibroblasts display differential mitogenic responses to transient sodium arsenite exposure. Toxicol Sci 1999; 50:72–81.
65. Simeonova PP, Wang S, Toriuma W, Kommineni V, Matheson J, Unimye N, Kayama F, Harki D, Ding M, Vallyathan V, Luster MI. Arsenic mediates cell proliferation and gene expression in the bladder epithelium: association with AP-1 transactivation. Cancer Res 2000; 60:3445–3453.
66. Arnold LL, Cano M, St John M, Eldan M, van Gemert M, Cohen SM. Effects of dietary dimethylarsinic acid on the urine and urothelium of rats. Carcinogenesis 1999; 20:2171–2179.
67. Angel P, Karin M. The role of Jun, Fos and the AP-1 complex in cell-proliferation and transformation. Biochim Biophys Acta 1991; 1072:129–157.
68. Tiniakos DG, Mellon K, Anderson JJ, Robinson MC, Neal DE, Horne CH. c-jun oncogene expression in transitional cell carcinoma of the urinary bladder. Br J Urol 1994; 74:757–761.
69. Skopelitou A, Hadjiyannakis M, Dimopoulos D, Kamina S, Krikoni O, Alexopoulou V, Rigas C, Agnantis NJ. p53 and c-jun expression in urinary bladder transitional cell carcinoma: correlation with proliferating cell nuclear antigen (PCNA) histological grade and clinical stage. Eur Urol 1997; 31:464–471.
70. Lindgren A, Vahter M, Dencker L. Autoradiographic studies on the distribution of arsenic in mice and hamsters administered 74As-arsenite or -arsenate. Acta Pharmacol Toxicol (Copenh) 1982; 51:253–265.
71. Yamauchi H, Yamamura Y. Concentration and chemical species of arsenic in human tissue. Bull Environ Contam Toxicol 1983; 31:267–270.
72. Snow ET. Metal carcinogenesis: mechanistic implications. Pharmacol Ther 1992; 53:31–65.
73. Flohe L, Brigelius-Flohe R, Saliou C, Traber MG, Packer L. Redox regulation of NF-kappa B activation. Free Radic Biol Med 1997; 22:1115–1126.
74. Shimizu M, Hochadel JF, Fulmer BA, Waalkes MP. Effect of glutathione depletion and metallothionein gene expression on arsenic-induced cytotoxicity and c-myc expression in vitro. Toxicol Sci 1998; 45:204–211.
75. Guyton KZ, Xu Q, Holbrook NJ. Induction of the mammalian stress response gene GADD153 by oxidative stress: role of AP-1 element. Biochem J 1996; 314:547–554.
76. Delnomdedieu M, Basti MM, Otvos JD, Thomas DJ. Transfer of arsenite from glutathione to dithiols: a model of interaction. Chem Res Toxicol 1993; 6:598–602.

77. Matsui M, Nishigori C, Toyokuni S, Takada J, Akaboshi M, Ishikawa M, Imamura S, Miyachi Y. The role of oxidative DNA damage in human arsenic carcinogenesis: detection of 8-hydroxy-2′-deoxyguanosine in arsenic-related Bowen's disease. J Invest Dermatol 1999; 113: 26–31.
78. Chen W, Martindale JL, Holbrook NJ, Liu Y. Tumor promoter arsenite activates extracellular signal-regulated kinase through a signaling pathway mediated by epidermal growth factor receptor and Shc. Mol Cell Biol 1998; 18:5178–5188.
79. Lopez S, Miyashita Y, Simons SS, Jr. Structurally based, selective interaction of arsenite with steroid receptors. J Biol Chem 1990; 265:16039–16042.
80. Su B, Karin M. Mitogen-activated protein kinase cascades and regulation of gene expression. Curr Opin Immunol 1996; 8:402–411.
81. Waskiewicz AJ, Cooper JA. Mitogen and stress response pathways: MAP kinase cascades and phosphatase regulation in mammals and yeast. Curr Opin Cell Biol 1995; 7:798–805.
82. Germolec DR, Yoshida T, Gaido K, Wilmer JL, Simeonova PP, Kayama F, Burleson F, Dong W, Lange RW, Luster MI. Arsenic induces overexpression of growth factors in human keratinocytes. Toxicol Appl Pharmacol 1996; 141:308–318.
83. Vasunia KB, Miller ML, Puga A, Baxter CS. Granulocyte-macrophage colony-stimulating factor (GM-CSF) is expressed in mouse skin in response to tumor-promoting agents and modulates dermal inflammation and epidermal dark cell numbers. Carcinogenesis 1994; 15:653–660.
84. Sandgren EP, Luetteke NC, Palmiter RD, Brinster RL, Lee DC. Overexpression of TGF alpha in transgenic mice: induction of epithelial hyperplasia, pancreatic metaplasia, and carcinoma of the breast. Cell 1990; 61:1121–1135.
85. Hofer G, Grimmer C, Sukhatme VP, Sterzel RB, Rupprecht HD. Transcription factor Egr-1 regulates glomerular mesangial cell proliferation. J Biol Chem 1996; 271:28306–28310.
86. Eid MA, Kumar MV, Iczkowski KA, Bostwick DG, Tindall DJ. Expression of early growth response genes in human prostate cancer. Cancer Res 1998; 58:2461–2468.
87. Soignet SL, Maslak P, Wang ZG, Jhanwar S, Calleja E, Dardashti LJ, Corso D, DeBlasio A, Gabrilove J, Scheinberg DA, Pandolfi PP, Warrell RPJJ. Complete remission after treatment of acute promyelocytic leukemia with arsenic trioxide. N Engl J Med 1998; 339:1341.
88. Chen GQ, Zhu J, Shi XG, Ni JH, Zhong HJ, Si GT, Jin XL, Tang W, Li XS, Xong SM, Shen ZX, Sun GL, Ma J, Zhang P, Zhang TD, Gazin C, Naoe T, Chen SJ, Wang ZY, Chen Z. In vitro studies on cellular and molecular mechanisms of arsenic trioxide (As203) in the treatment of acute promyelocytic leukemia: $As_2O_3$ induces NB4 cell apoptosis with downregulation of Bcl-2 expression and modulation of PML-RAR alpha/PML proteins. Blood 1996; 88:1052–1061.
89. Jing By, Dai J, Chalmer-Redman RME, Tatton WG, Waxman S. Arsenic trioxide selectively induces acute promyelocytic leukemia cell apoptosis via a hydrogen peroxide-dependent pathway. Blood 1999; 94:2101–2111.

# 6

# Redox Regulation of Gene Expression and Transcription Factors in Response to Environmental Oxidants

**HIROSHI MASUTANI, AKIRA NISHIYAMA, YONG-WON KWON, YONG-CHUL KIM, HAJIME NAKAMURA, and JUNJI YODOI**

*Institute for Virus Research, Kyoto University, Kyoto, Japan*

## I. INTRODUCTION: ENVIRONMENTAL STRESSORS AND HOST DEFENSE

Environmental factors (external agents such as chemicals, radiation, and viruses) are major causes in the majority of human tumors. A wide variety of chemicals can induce neoplasia by interacting with macromolecules (1). One group of such chemicals, polycyclic aromatic hydrocarbons (PAHs) including 2,3,7,8-tetrachlorodibenzo-*p*-dioxin (TCDD) are ubiquitous environmental pollutants that have many untoward effects on humans and wildlife, such as immune suppression, thymic involution, endocrine disruption, wasting syndrome, birth defects, and carcinogenesis (2). To take measures against these noxious effects, the mechanism of the effects of environmental stressors should be elucidated.

Environmental stressors seem to evoke distinct sets of responses, although they overlap. X-ray or ultraviolet (UV) light irradiation or chemicals inducing DNA strand breaks activate p53 (3), which subsequently transactivates genes coding for regulatory proteins of the cell cycle, apoptosis, and repair system. The NF-κB system is activated by reactive oxygen species (ROS) and has been recognized as a central regulator of inflammatory genes (4). Hydrogen peroxide treatment or UV irradiation activate the MAP kinase system to transmit signals to transcription factors such as Ets family of proteins, followed by the induction of immediate early genes such as jun and fos gene families (5). It is well known that UV activates Jun kinase to phosphorylate *c-Jun* (6). The heat shock response is largely mediated by heat shock factors, resulting in upregulation of heat shock proteins (7). Aryl hydrocarbon receptors are responsible for transmitting signals induced by xenobiotic substances to enhance expression of phase I detoxifying enzymes such as cytochrome p450 (2). Electrophile targeting chemicals activate cap'n'collar transcription factors such as Nrf2 to induce phase II detoxifying enzymes (8) (Fig. 1).

Although the mechanisms of effects of environmental stressors appear heterogeneous, some of them are supposed to be related to oxidative stress. ROS is produced in response to environmental stressors (9). Inflammatory cytokines cause a chain of signal transduction events, resulting in the production of intracellular ROS. Mitochondria, NADP (H) oxygenase, xanthine oxygenase, and NO synthetase are major intracellular sources of ROS (10,11). Oxidative stress also directly activates the signal transduction pathway and provokes modification of macromolecules, inducing cellular responses such as activation of p53. To protect cells from damage, various defense mechanisms have been developed against oxidants. Apoptosis is a finely regulated mechanism that maintains genome stability by eliminating cells with severe DNA damage, whereas induction of cell cycle arrest is a mechanism that affords cells time to repair damaged DNA. In addition to well-known antioxidant systems, such as manganese-dependent superoxide dismutase (Mn-SOD) and catalases, thiol reduction is one of the important mechanisms against excessive oxidative stress. Reduction of cysteine residues of various nuclear factors is important for the DNA–protein interaction. Redox factor-1 (Ref-1) (12), glutathione (GSH)/ glutaredoxin, and the thioredoxin (TRX) system (13) maintain an intranuclear reducing environment to favor the function of transcription factors (14) (Fig. 2). Since the oxidative stress response and redox regulation are two sides of the same coin, this paper reviews the current findings concerning the role of TRX and the redox regulation in the environmental stress-evoked cellular signaling and response.

ENVIRONMENTAL STRESSORS

CHEMICALS INDUCING DNA DOUBLE STRAND BREAK UV X-RAY
VIRAL INFECTION HYDROGEN PEROXIDE UV REACTIVE OXYGEN SPECIES
UV ELECTROPHILES METALS HYDROGEN PEROXIDE
HEAT SHOCK METALS
XENOBIOTICS
ELECTROPHILES METALS

P53
NFκB
ETS
JUN/FOS
HEAT SHOCK FACTOR
ARYL HYDROCARBON RECEPTOR
NRF2

CELL CYCLE REGULATING PROTEINS
REPAIR ENZYMES
IMMUNE REGULATORY PROTEINS
APOPTOSIS REGULATING PROTEINS
HEAT SHOCK PROTEINS
DETOXIFYING PROTEINS
ANTIOXIDANTS AND REDOX REGULATING PROTEINS

HOST DEFENSE

**Figure 1** Environmental stressors and activation of transcription factors.

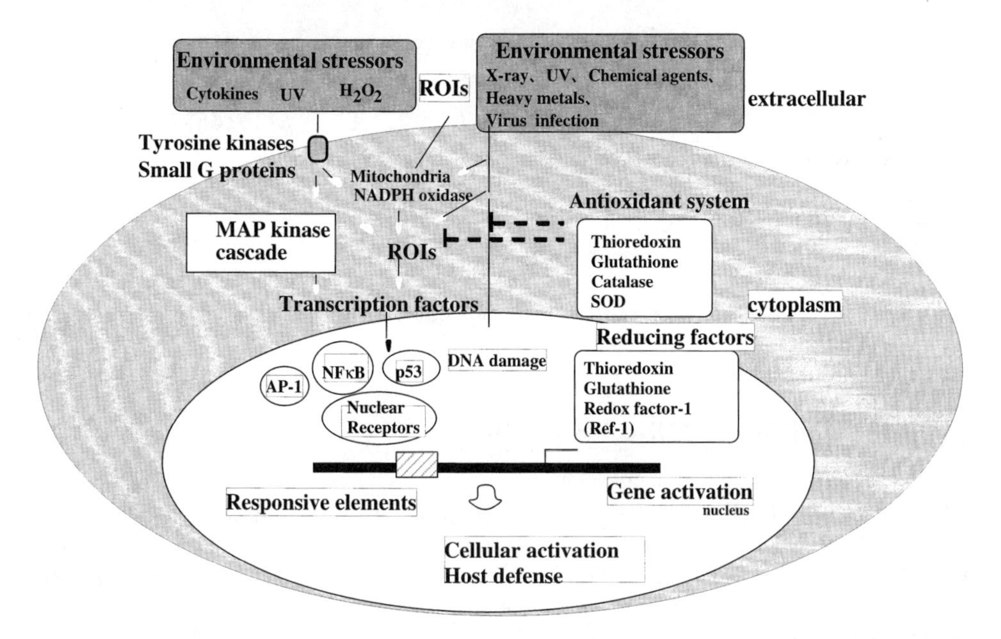


**Figure 2** Redox regulation of oxidative stress response and signaling. Excessive oxidative stress causes damage to biomolecules, whereas mild oxidative stress elicits cellular responses beneficial to cells and individuals. Extracellular or intracellular ROS evoke a response and signaling, including activation of antioxidative enzymes.

## II. TRX AND RELATED MOLECULES

The TRX system in concert with the GSH and glutaredoxin system constitutes a cellular reducing environment. TRX is a small protein with two redox-active cysteine residues in an active center (Cys-Gly-Pro-Cys-) and operates together with NADPH and TRX reductase as an efficient reducing system of exposed protein disulfides. TRX is present in many different prokaryotes and eukaryotes and appears to be present in all living cells (13). We identified and cloned an adult T-cell leukemia (ATL)-derived (ADF) factor from the HTLV-I positive cell line ATL-2, which was found to be a human homolog of TRX (15,16). Several cytokinelike factors such as 3B6-IL-1, eosinophil cytotoxicity enhancing factor (ECEF), T-cell hybridoma MP6-derived B-cell growth factor, and early pregnancy factor are identical or related to TRX. TRX has also been reported to have chemokinelike activity (17). These studies indicate that TRX plays a multifunctional role (18). Mammalian thioredoxin 2 (TRX2) has high homology with TRX and has an active site C-G-P-C with thiol-reducing activity and is specifically localized in mitochondria (19). TRX2 may be related to the protection of oxidative stress in mitochondria. Thioredoxin reductase has a selenium-containing active center in the C-terminus and there exist several isoforms of thioredoxin reductase (20). The recently identified thioredoxin reductase shows 54% identity to that of previously identified TRXR (TRXR1) and is localized in mitochondria. Thioredoxin-dependent peroxidases (peroxiredoxin) are considered to be members of a family scavenging hydrogen peroxide as well as glutathione peroxidase. So far, six isoforms of the peroxiredoxin family have been identified. The mitochondrial-specific thioredoxin-dependent peroxidase (peroxiredoxin III) has also been reported (21–23). Thus, the

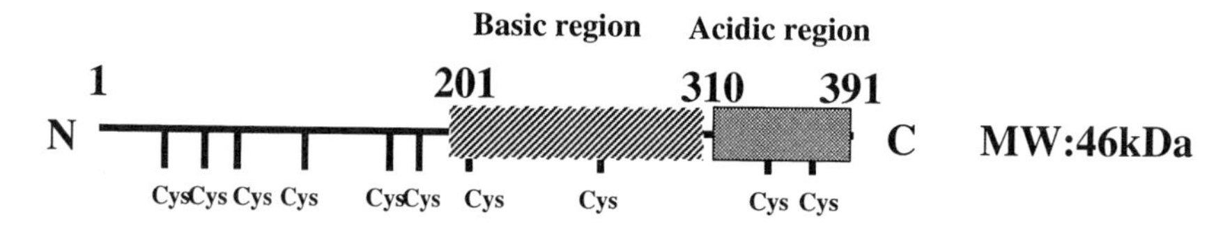


**Figure 3** Structure of thioredoxin binding protein-2 (TBP-2).

TRX system is composed of several related molecules forming a network of recognition and interaction with its active site cysteine residues.

## III. THIOREDOXIN BINDING PROTEINS

In our search for the target molecules interacting with TRX, we isolated vitamin $D_3$ upregulated protein 1 (VDUP1) as a TRX binding protein (TBP-2), using a yeast two-hybrid system (24). VDUP1 was originally reported as an upregulated gene in HL-60 cells stimulated by 1α, 25-dihydroxyvitamin $D_3$ (25). The association of TRX with TBP-2/VDUP1 was observed in vitro and in vivo. TRX treated with oxidative reagents was incapable of binding with TBP-2/VDUP1. In addition, TRX (C32S/C35S), a mutant of TRX, was not able to interact with TBP-2/VDUP1, strongly suggesting that the TRX-TBP-2/VDUP1 interaction required cysteine residues of the TRX active site. Moreover, the insulin-reducing activity of TRX was inhibited by the addition of recombinant TBP-2/VDUP1 protein. Additionally, transient transfection with TBP-2/VDUP1 expression vector suppressed the reducing activity and protein expression of TRX. These results suggested that TBP-2/VDUP1 serves as a negative regulator of the function and expression of TRX. In 1-α, 25-dihydroxyvitamin $D_3$-treated HL-60 cells, TBP-2/VDUP1 expression was enhanced, whereas TRX expression and the reducing activity were downregulated. 1-α, 25-dihydroxyvitamin $D_3$ is important for regulation of calcium homeostasis and hormone secretion (26) and is a potent inducer of myeloid cell differentiation. TRX-TBP-2/VDUP1 interaction may play an important role in the redox regulation of growth and differentiation of the cells sensitive to a variety of inducers, including 1-α, 25-dihydroxyvitamin $D_3$ responses (Fig. 3).

Physical interactions of TRX with other proteins have been reported. TRX is directly associated with Ref-1 (27), glucocorticoid receptor (28), and NF-κB (29,30). TRX was also isolated as an ASK1-binding protein by a yeast two-hybrid system (31). Interestingly, the interaction of glutathione S-transferase p (GSTp) with Jun kinase has been reported recently (32). Therefore, direct interaction between redox regulating proteins and its binding proteins may be a basic mechanism of the redox regulation of cellular processes.

## IV. REDOX REGULATION OF TRANSCRIPTION FACTORS BY TRX

Transcription factors are important sensing and signaling components of oxidative signaling. Redox regulation appears to be involved in various steps of activation of transcription factors (Fig. 4). In bacteria, the OxyR transcription factor is activated through the formation of a disulfide bond and is deactivated by enzymatic reduction with glutaredoxin, the expression of which is regulated by OxyR (33,34). Thus, a transcription factor stably bound to DNA serves as a redox sensor of oxidative stress. In higher organisms, oxidative

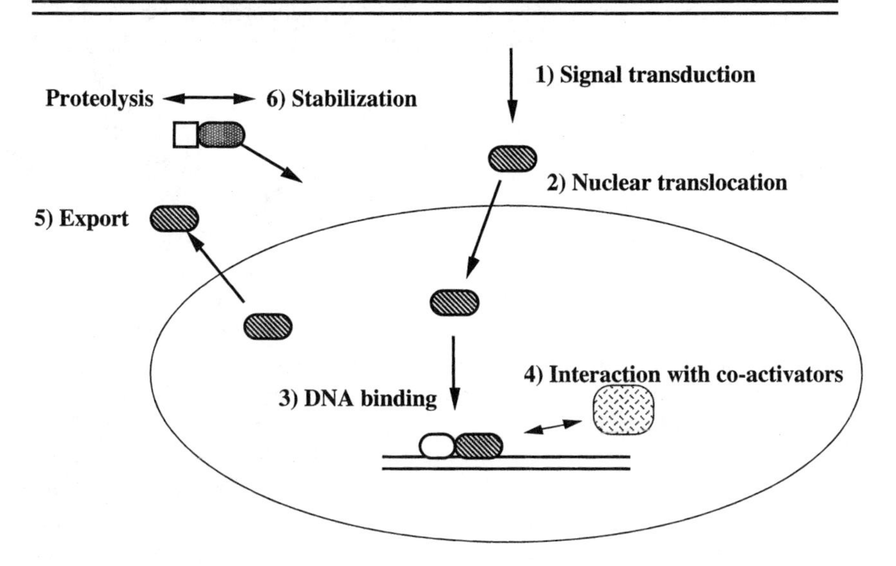


**Figure 4** Redox regulation of transcription factors.

stress signaling is more complex. In yeast, Yap1, an AP-1-like transcription factor, and Skn7 cooperate on the yeast TRX2 promoter to induce transcription in response to oxidative stress (35,36). Yap1 is translocated to the nucleus upon oxidative stress, and its cysteine residues are essential for the translocation, indicating that the redox state of the cysteine residues is important for the Yap1 activity (37). In the TRX-deficient mutant, Yap1p was constitutively concentrated in the nucleus and the level of expression of the Yap1 target genes was high (38). Cysteine residues in the nuclear export signal (NES)-like sequence of Yap1p are essential for the regulated nuclear localization, suggesting that the Yap1p regulatory domain can confer the oxidative stress sensor function (39).

In the mammalian system, one of the earliest studies in the redox regulation of transcription factors appeared in the study of the glucocorticoid receptor. Rees and Bell showed that sulfhydryl groups are required for glucocorticoid receptors to bind steroid (40). The endogenous heat-stable glucocorticoid receptor-activating factor was reported to be TRX (41). The sulfhydryl moiety on the glucocorticoid receptors is absolutely required for the receptor to bind DNA (42). Accumulating evidence shows that interaction of RNA- or DNA-binding proteins with their recognition sequences is regulated by redox regulation. Free sulfhydryl groups of iron-responsive element (IRE)-binding protein (IRE-BP) are required for the specific interaction between IRE-BP and IRE. Treatment of lysates with reducing agents increases the binding activity, whereas agents that block sulfhydryls inhibit binding (43). The importance of conserved cysteine residues of transcription factors was revealed by a study showing that DNA binding of the Fos-Jun heterodimer was modulated by reduction-oxidation (redox) of a single conserved cysteine residue in the DNA-binding domains of the two proteins (44). Ref-1 was identified as a factor to facilitate AP-1 DNA binding activity (12). Interestingly, Ref-1 is identical to formally described apurinic/ apyrimidinic (AP) endonuclease (45,46), although the redox and DNA-repair activities of Ref-1 are encoded by nonoverlapping domains (47). DNA binding of NF-κB to the κB site was also proved to be regulated by the redox status (48–50). However, previous studies showed that transient overexpression of TRX activated AP-1 but not NF-κB (51,52). The discrepancy of the effect of TRX on NF-κB may be explained by a dual and opposing role of TRX in the regulation of NF-κB. A previous study showed that ROS serve as messengers mediating the release of I-κB from NF-κB (53). In the cytoplasm, TRX interferes with the signals to I-κB kinase and blocks the degradation of I-κB, whereas in the nucleus TRX enhances NF-κB transcriptional activities by enhancing its ability to bind DNA (27). Several other transcription factors, including c-Myb and Ets have been shown to be modulated by the cellular redox state (54,55). The interaction between dioxin receptor and xenobiotic responsive element (XRE) is also modulated by TRX (56). Related bHLH transcription factors may be a target for redox regulation. Indeed, activation of the hypoxia-inducible transcription factor depends upon redox-sensitive stabilization of its alpha subunit (57). Nuclear receptors like glucocorticoid receptors or estrogen receptors are also activated by TRX (58,59). We previously showed that the in vitro DNA binding activity of PEBP2, which belongs to a family (RUNX family) of heterodimeric transcription factors, is enhanced by the addition of either TRX or Ref-1 (60). TRX and/ or Ref-1 regulate the DNA binding activity of p53 in vitro and in vivo (61). Interaction between HIF and coactivators may also be a target of redox regulation. The C-terminal activation domain of hypoxia-inducible factor 1$\alpha$ and its related factor have a specific cysteine. Expression of TRX and Ref-1 enhanced the interaction of these factors with a coactivator, CREB-binding protein (CBP)/p300 (62).

TRX translocation from cytosol to nucleus was induced by a wide variety of oxidative stresses including UV irradiation (30), hydrogen peroxide (63), hypoxia (62), and treatment with CDDP (61). Therefore, it is assumed that TRX is translocated to the nuclear compartment upon oxidative stress to interact with Ref-1. TRX translocation may also be involved in the activation of glucocorticoid receptors (28). Nuclear localization of TRX is often observed in pathological tissue. In cervical tissue, TRX expression is observed in human papilloma virus–infected cells and TRX is localized in the nucleus (64). In the renal proximal tubules, TRX is induced and translocated to nuclei by oxidative damage mediated by Fe-nitrilotriacetate (65). Although the significance and mechanisms of these observations are unclear, TRX translocation may be related to the cyotoprotection and pathogenesis of oxidative stress-related disorders.

## V. REDOX REGULATION OF p53

The tumor suppressor protein p53 is rapidly induced by various kinds of oxidative stress including ultraviolet, ionic irradiation, or treatment with genotoxic chemical compounds such as *cis*-diamminedichloroplatinum (II) (CDDP). Its overexpression arrests cell cycle progression in the G1 phase and suppresses cell proliferation through p21 induction. p53 exerts its tumor suppressor effect by controlling the expression of cell-cycle-related genes after DNA damage. Therefore, p53 is thought to be a gatekeeper against DNA damage (66). Enhanced phosphorylation of p53 by ataxia–telangiectasia gene product (ATM) or DNA-dependent protein kinase appears an important regulatory mechanism for the response to DNA damage (3). Involvement of redox regulation in the p53 gene has also been reported. The site-specific DNA binding activity of p53 is dependent upon its highly conserved central DNA binding domain that contains a zinc ion and cysteine residues. Mutagenesis of cysteine residues, especially around the metal center, abolished its sequence-specific DNA binding activity (67). Agents such as dithiothreitol (DTT) and metal chelators modulate the DNA binding capacity of p53 (68). Thus, the DNA binding activity of p53 can be regulated by the redox state. Ref-1, which was reported to be a redox

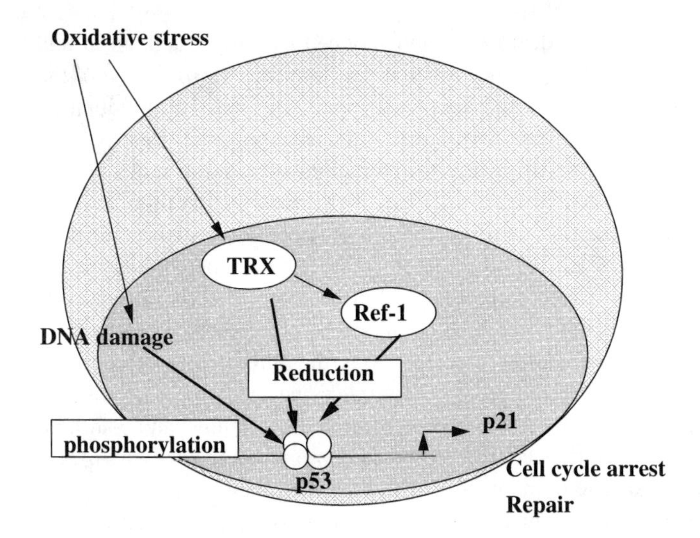


**Figure 5** Redox regulation of p53 by TRX and Ref-1.

regulator of AP-1 (12,45,46,69), augmented p53 function (70). Based on these previous studies, we have investigated the role of TRX in the regulation of p53 activity. Recombinant TRX or Ref-1 augmented the sequence-specific DNA binding of p53. The effect of TRX and Ref-1 was seen at almost the same concentration in an electrophoretic mobility shift assay (EMSA). In a luciferase assay, TRX enhanced Ref-1-mediated p53 activation, further indicating a functional coupling between TRX and Ref-1 in the p53 activation system. Furthermore, Western blot analysis revealed that p53-dependent induction of p21 protein was also facilitated by transfection with TRX. Overexpression of transdominant negative mutant TRX (mTRX) suppressed the effects of TRX or Ref-1, suggesting that at least part of the Ref-1 mediation of p53 activation is modulated by the redox activity of TRX. CDDP induced p53 activation and p21 transactivation. The p53-dependent p21 transactivation induced by CDDP was inhibited by mTRX overexpression, suggesting that TRX-dependent redox regulation is physiologically involved in p53 regulation. CDDP also stimulated translocation of TRX from the cytosol into the nucleus. TRX-dependent redox regulation of p53 activity indicates coupling of the oxidative stress response and p53-dependent repair mechanism (Fig. 5). In a yeast system, the importance of the TRX system in p53 activity has also been reported where deletion of the thioredoxin reductase gene inhibited p53-dependent reporter gene expression (71). Ref-1 overexpression enhances the ability of p53 to stimulate a number of p53 target promoters and increase the ability of p53 to stimulate endogenous p21 and cyclin G expression. Downregulation of Ref-1 causes a diminished ability of p53 to transactivate the p21 and Bax promoters. Ref-1 levels are correlated with the extent of apoptosis induced by p53 (72). These studies collectively suggest that the redox regulation is an important regulatory mechanism of p53.

## VI. REDOX REGULATION OF APOPTOSIS

Apoptosis can be considered as a finely regulated mechanism to maintain genome stability by eliminating cells with severe DNA damage caused by oxidative stress. It is well known that excessive oxidation causes cellular death (73). Extracellular treatment with the thiol-oxidizing agent, diamide, evoked cell death (74) and induced caspase-3 activation and apoptosis (75–77). Cytochrome c release from mitochondria into the cytosol is one of the central mechanisms of caspase activation and apoptosis induction (78). The modulation of thiol content largely influences the oxidative stress–induced apoptosis process, suggesting that reducing molecules are involved in the interaction with some molecules in the mitochondria to protect against oxidative stress (75). TRX2 localized in mitochondria may modulate apoptosis-inducing signal via mitochondria in cooperation with other thiol molecules. Diamide-induced cell death is either apoptotic or necrotic, depending on the concentration of diamide (74,75). When cells were cultured in a high concentration of diamide, the processing of caspase-3 was not detected in spite of the release of mitochondrial cytochrome c, indicating that diamide also decreases the activity of caspase-9 or interferes with the cytochrome c/ Apaf-1/ caspase 9 complex, the likely activator of caspase-3 (75). Apoptosis was induced by a low concentration of xanthine and xanthine oxidase (XO), whereas necrosis was induced by a high concentration of XO (79). Caspases have cysteine residues in the catalytic domain (80). Although the precise mode of regulation of the apoptosis/ necrosis switch is uncertain, the disruption of enzymatic activity of caspases induced by excessive oxidation may cause a shift from apoptosis to necrosis (81).

Other than cytochrome c, mitochondrial intermembrane spaces contain several other proteins that participate in the degradation phase of apoptosis. Among them, apoptosis-inducing factor (AIF) has been cloned. Curiously, AIF is a flavoprotein and has high homology with benzene 1,2-dioxygenase system ferredoxin NADH reductase from *Pseudomonas putida*, although the reducing activity of AIF has not yet been reported. Recombinant AIF causes characteristic chromatin formation and fragmentation of DNA (82).

Jun kinase/ SAP kinase and p38 kinase play an important role in oxidative stress-

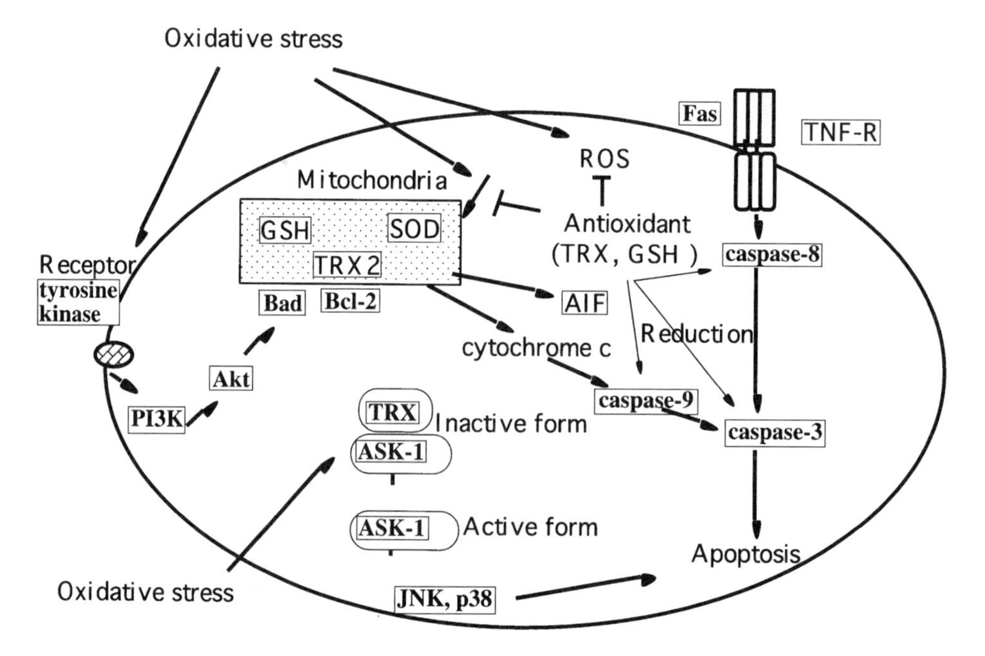


**Figure 6** Redox regulation of apoptosis.

induced apoptosis (83). TRX has been found to bind to ASK1, a MAPKKK, and inhibits the apoptosis process (31). When TRX is oxidized by oxidative stress, ASK1 is dissociated from oxidized TRX and activated to induce an apoptosis signal. Indeed, we previously showed that TRX prevents apoptosis induced by TNF (84) or L-cystine and glutathione depletion (85). In addition, overexpression of TRX negatively regulates p38 Map kinase activation and p38 Map kinase-mediated IL-6 production by TNF-α-stimulated cells, indicating that TRX is critical for p38 Map kinase activation (86). Thioredoxin peroxidase (peroxiredoxin II) is reported to be an inhibitor of apoptosis with a mechanism distinct from that of Bcl-2 (87). Taken together, redox regulation is deeply involved in the regulation of cell death (Fig. 6).

## VII. DYSREGULATION OF THE REDOX SYSTEM IN VIRAL INFECTION

TRX and glutaredoxin play a role in the host–virus relationship and life cycle of M13, T7, T4 *E. coli* phages. A principal hydrogen donor for the herpes simplex virus type 1-encoded ribonucleotide reductase is a cellular TRX (88). Human TRX has been reported as a growth-promoting factor derived from adult T-cell leukemia and is involved in lymphocyte immortalization by HTLV-I and EBV (89,90). In the EBV system, TRX inhibits PMA-induced EBV replication and suppresses proliferation that is lethal to the host (91). A previous study also showed the coexpression of TRX and human papilloma virus DNA in neoplastic cervical squamous epithelium (64). Hence, it can be supposed that TRX is an important factor for cohabitation between host and virus.

Accumulating evidence has shown downregulation of the antioxidant system in HIV infection. Systemic depletion of cystine and cysteine was partly responsible for deterioration of the immune system in HIV-infected individuals (92–94). The intracellular GSH level in T-cell subsets decreased in HIV-infected individuals (95). *N*-acetylcysteine (NAC), a precursor of GSH, counteracts the effects of ROS, which activate NF-κB and expression and replication of HIV-1 (53,96). In addition, cells highly producing TRX decreased in HIV-infected lymph nodes (97). HIV Tat represses expression of manganese superoxide dismutase (Mn-SOD) in HeLa cells (98). Decreased Mn-SOD expression has been associated with decreased levels of GSH and a lower ratio of reduced:oxidized GSH (99). Downregulation of Bcl-2 expression was demonstrated in T-lymphocytes from HIV-infected individuals and might ultimately be a possible cause of their increased rate of apoptosis (100). A balanced state composed of compensated expression of Bcl-2 and TRX, cell survival, and low viral production seems to be important for the establishment of HIV chronic persistent infection (101,102). The plasma TRX level was found to be elevated in HIV-infected patients, indicating that the elevated level may reflect an involvement of oxidative stress in advanced disease conditions (103). Taken together, dysregulation of the redox and redox-regulating mechanisms and the Bcl-2 system may contribute to the establishment of chronic persistent HIV infection.

## VIII. CYTOPROTECTIVE ACTION OF TRX

TRX plays a cytoprotective role against various oxidative stresses in a variety of systems (104). TRX has radical scavenging activity (105). TRX can protect cells from TNF or anti-Fas antibody (84), hydrogen peroxide, activated neutrophils (63), and ischemic reperfusion injury (106,107). Homozygous mutants carrying a targeted disruption of TRX gene were lethal shortly after implantation, suggesting that TRX expression is essential for early differentiation and morphogenesis of the mouse embryo (108). TRX is overexpressed in decidua and trophoblast cells in the tissues of pregnant uterus (109) and has been reported as an early pregnancy factor (110). TRX as well as superoxide dismutase restore mouse two-cell stage developmental block by restoring defective cdc2 kinase (111,112). Collectively, TRX may be beneficial in protecting the fertilized egg and placental trophoblasts from the cytotoxic effects of ROS.

TRX can effectively reduce lens-soluble protein disulfide bonds generated by $H_2O_2$ and act as a free radical quencher based on studies with a stable free radical system generated by ascorbic acid and 2, 6-dimethoxy-*p*-benzoquinone (113). TRX expression is induced in the retinal pigment epithelium upon oxidative stress (114,115). Prostaglandins may be involved in this activation (116). The role of thioredoxin reductase in the reduction of free radicals at the surface of the epidermis has been reported (117). Recombinant TRX promotes survival of primary cultured neurons in vitro (118). The focal cerebral ischemia was attenuated in TRX transgenic mice (119). TRX has also been shown to have a protective effect against chemotherapeutic agents such as *cis*-diamminedichloroplatinum (II) CDDP-induced cytotoxicity (120–123). Decreased incidence of experimentally induced diabetes was observed in nonobese diabetic transgenic mice that overexpress TRX in their pancreatic β-cells, presumably because of reduced production of ROS (124). These data collectively indicate that TRX physiologically protects cells against oxidative stress.

## IX. TRX INDUCTION BY OXIDATIVE STRESS

TRX expression is induced by a variety of stresses, including virus infection, mitogens, phorbol myristate acetate (PMA), X-ray and UV irradiation (125), hydrogen peroxide, and ischemic reperfusion (126). We have analyzed the 5′-upstream region of the TRX gene and identified an oxidative responsive element that was inducible by various oxidative agents (127). The isolation of the binding proteins to this sequence as well as the further elucidation of the mechanism of the transduction of signals to this element is underway. In an erythroleukemic cell line K562, TRX is induced transcriptionally by hemin (128). HSF-2 was suggested to mediate hemin-induced activation of the TRX genes by analyses using stable transfectants of HSF-2 (128). However, we could not identify any consensus sequence for HSFs in the TRX gene. In the 5′-upstream sequence of the human TRX gene, there exist putative binding sites such as AP-1, CRE, AP-4, ARE, XRE, and SP-1. The major stimulatory effect on the TRX gene by hemin seemed to be mediated by the ARE (128a). TRX is also induced by polycyclic aromatic hydrocarbons (PAHs) (128b). Elucidation of the regulation of the TRX promoter upon various oxidative stresses seems important for understanding how cells integrate and differentiate oxidative signals (Fig. 7).

Groups of redox regulating enzymes such as γ-glutamylcysteine synthetase, NAD (P) H: quinone oxidoreductase, glutathione S-transferase Ya genes are known to contain the antioxidant responsive element (ARE)/ EpRE (129,130) for response to electrophile targeting xenobiotics. The ARE/ EpRE has marked sequence homology with the NF-E2 binding site (131,132) and Maf recognition element (MARE) (8,133). The cap'n'collar transcription factors including NF-E2p45 (132), Nrt1 (134,135), and Nrf2 (136–138) form heterodimers with small Maf proteins such as MafK (139), MafF (139), and MafG (133), binding to the NF-E2 recognition element/ ARE/ EpRE/ MARE (8). The importance of Nrf2 in the oxidative-stress-induced gene activation has been shown by several studies.

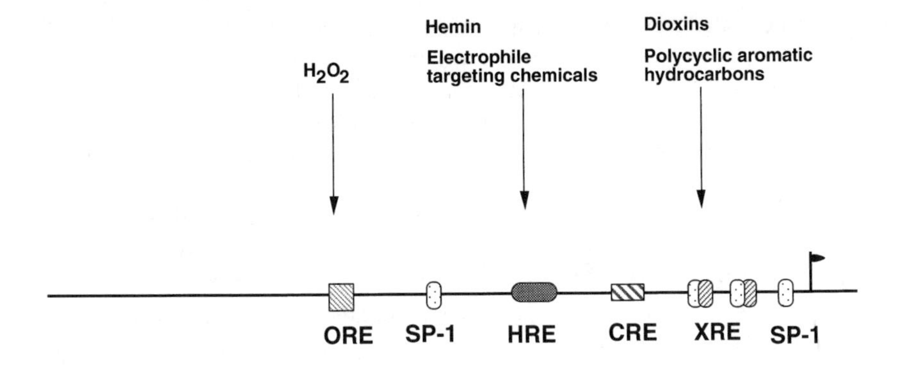


**Figure 7** Activation of the TRX gene by oxidative stress.

In nrf2 −/− mice, the induction of phase II detoxifying enzymes by butylated hydroxyanisole (BHA) was decreased (140). The hemin-mediated induction of heme oxygenase-1 gene was inhibited by overexpression of Nrf2 dominant negative expression vector (141). The heme oxygenase-1 gene and γ-glutamylcysteine synthetase subunit genes are regulated by Nrf2 (141,142). Therefore, TRX, an important redox regulator, and these redox enzymes all have common regulatory mechanisms and may play coordinated roles against a subset of oxidative stress to which the ARE-mediated activation pathway is involved in the response. Heme release from hemoglobin (143) and iron release from the heme moiety of cytochrome p-450 (144) are implicated in the pathogenesis of reperfusion injury. Heme oxygenase 1 plays a role against endothelial cell damage caused by reperfusion injury. Previously, we showed that administration of TRX protein can protect cells in ischemic reperfusion injury (106,107), and that TRX expression is induced in ischemic reperfusion (104,144a). Considering the similarity of the regulation mechanism by environmental stressors among TRX, heme oxygenase 1, and phase II detoxifying enzymes, TRX may have protective roles against pathological conditions such as reperfusion injury and chemical carcinogenesis (128a).

## X. ENVIRONMENTAL POLLUTANTS OR CHEMICALS AND SIGNAL

Polycyclic aromatic hydrocarbons (PAHs) including 2,3,7,8-tetrachlorodibenzo-*p*-dioxin (TCDD) are ubiquitous environmental pollutants that have many untoward effects in humans and wildlife, such as immune suppression, thymic involution, endocrine disruption, wasting syndrome, birth defects, and carcinogenesis (2). PAHs are very strong promutagens and procarcinogens, since the cytochrome p-450 1A1 (CYP1A1) enzyme first oxidizes these compounds into reactive intermediates during metabolism, which then can cause oxidative DNA damage and form adducts with DNA, starting the mutagenic chain of events responsible for tumor initiation (2). Most of the biological effects of PAHs are mediated by aryl hydrocarbon receptor (AhR) (2,145). AhR is a ligand-activated basic helix–loop–helix transcription factor (bHLH), which displays high-affinity binding to compounds including PAHs. The ligand–AhR complex is translocated to nuclei, where it switches its partner from heat shock protein (hsp) 90 to AhR nuclear translocator (Arnt) and binds to the cognate enhancer sequence, xenobiotic-responsive element (XRE). XRE is found in the promoter regions of several genes including members of the cytochrome P-450 family and aldo-keto reductases (AKRs), which are known to oxidize PAHs to form genotoxic (DNA damaging) metabolites (146). NADP (H):quinone oxidoreductase, cytosolic aldehyde dehydrogenase 3, UDP glucuronyltransferase, and glutathione transferase contain the XRE in the promoter region. In addition, the TRX gene has a sequence similar to the XRE (128b). Therefore, TRX and other redox regulating enzymes may also play roles against the xenobiotic-induced stress response.

## XI. CONCLUSION AND FUTURE PERSPECTIVE

Environmental stressors cause damage to macromolecules, often accompanied by production of ROS. Responses to environmental stressors are related to the oxidative stress response. We hypothesize that oxidative stress causes several subsets of responses, although they are mutually interconnected. TRX appears to be a regulator/modulator involved in cellular signaling against oxidative stress. The TRX system can also be considered as a redox sensoring system against oxidative stress at various steps: (1) interaction between TRX and TRX-binding proteins can be changed in the redox status; (2) the selenocysteine residues of TRX reductase may directly sense the redox status; (3) the TRX gene is activated by a variety of oxidative stresses; (4) TRX enhances oxidative stress-induced p53 DNA binding activity and may serve as a redox sensoring system of DNA damage; and (5) TRX2 may regulate the mitochondrial apoptotic signal induced by oxidative stress. Further analysis of the role of TRX in the oxidative stress response and the mechanism of the regulation of the TRX gene by oxidative stress should help to elucidate how cells link the oxidative stress response to gene regulation. Environmental stressors including environmental pollutants and chemicals are a serious threat to human health. The elucidation of the mechanisms of the stress response induced by these stressors is required to take measures against the untoward effects such as carcinogenesis.

## ACKNOWLEDGMENTS

We thank the members of our laboratory for helpful discussion as well as Ms. Y. Kanekiyo for secretarial help. This work was supported by a grant-in-aid for Scientific Research from the Ministry of Education, Science and Culture of Japan and a grant-in-aid of Research for the Future from the Japan Society for the Promotion of Science, Japan.

## REFERENCES

1. Miller EC, Miller JA. Mechanisms of chemical carcinogenesis: nature of proximate carcinogens and interactions with macromolecules. Pharmacol Rev 1966; 18:805–838.
2. Nebert DW, Puga A, Vasiliou V. Role of the Ah receptor and the dioxin-inducible [Ah] gene battery in toxicity, cancer, and signal transduction. Ann NY Acad Sci 1993; 685:624–640.
3. Matlashewski G. p53: Twenty years on, meeting review. Oncogene 1999; 18:7618–7620.
4. Baeuerle PA, Baltimore D. NF-kappa B: ten years after. Cell 1996; 87:13–20.

5. Treisman R. Journey to the surface of the cell: Fos regulation and the SRE. EMBO J 1995; 14:4905–4913.
6. Hibi M, Lin A, Smeal T, Minden A, Karin M. Identification of an oncoprotein- and UV-responsive protein kinase that binds and potentiates the c-Jun activation domain. Genes Dev 1993; 7:2135–2148.
7. Wu C. Heat shock transcription factors: structure and regulation. Ann Rev Cell Dev Biol 1995; 11:441–469.
8. Motohashi H, Shavit JA, Igarashi K, Yamamoto M, Engel JD. The world according to Maf. Nucl Acids Res 1997; 25:2953–2959.
9. Shertzer HG, Nebert DW, Puga A, Ary M, Sonntag D, Dixon K, Robinson LJ, Cianciolo E, Dalton TP. Dioxin causes a sustained oxidative stress response in the mouse. Biochem Biophys Res Commun 1998; 253:44–48.
10. Finkel T. Signal transduction by reactive oxygen species in non-phagocytic cells. J Leukoc Biol 1999; 65:337–340.
11. Halliwell B, Gutteridge JM. Role of free radicals and catalytic metal ions in human disease: an overview. Methods Enzymol 1990; 186:1–85.
12. Xanthoudakis S, Curran T. Identification and characterization of Ref-1, a nuclear protein that facilitates AP-1 DNA-binding activity. EMBO J 1992; 11:653–665.
13. Holmgren A, Björnstedt M. Thioredoxin and thioredoxin reductase. Methods Enzymol 1995; 252:199–208.
14. Packer L, Yodoi J. Redox Regulation of Cell Signaling and Its Clinical Application. New York: Marcel Dekker, 1999.
15. Tagaya Y, Maeda Y, Mitsui A, Kondo N, Matsui H, et al. ATL-derived factor (ADF), an IL-2 receptor/Tac inducer homologous to thioredoxin; possible involvement of dithiol-reduction in the IL-2 receptor induction. EMBO J 1989; 8:757–764.
16. Wollman EE, d'Auriol L, Rimsky L, Shaw A, Jacquot JP, et al. Cloning and expression of a cDNA for human thioredoxin. J Biol Chem 1988; 263:15506–15512.
17. Bertini R, Howard OM, Dong HF, Oppenheim JJ, Bizzarri C, Sergi R, Caselli G, Pagliei S, Romines B, Wilshire JA, Mengozzi M, Nakamura H, Yodoi J, Pekkari K, Gurunath R, Holmgren A, Herzenberg LA, Ghezzi P. Thioredoxin, a redox enzyme released in infection and inflammation, is a unique chemoattractant for neutrophils, monocytes, and T cells. J Exp Med 1999; 189:1783–1789.
18. Yodoi J, Uchiyama T. Diseases associated with HTLV-I: virus, IL-2 receptor dysregulation and redox regulation. Immunol Today 1992; 13:405–411.
19. Spyrou G, Enmark E, Miranda VA, Gustafsson J. Cloning and expression of a novel mammalian thioredoxin. J Biol Chem 1997; 272:2936–2941.
20. Liu SY, Stadtman TC. Heparin-binding properties of selenium-containing thioredoxin reductase from HeLa cells and human lung adenocarcinoma cells. Proc Natl Acad Sci USA 1997; 94:6138–6141.
21. Yamamoto T, Matsui Y, Natori S, Obinata M. Cloning of a housekeeping-type gene (MER5) preferentially expressed in murine erythroleukemia cells. Gene 1989; 80:337–343.
22. Watabe S, Hiroi T, Yamamoto Y, Fujioka Y, Hasegawa H, Yago N, Takahashi SY. SP-22 is a thioredoxin-dependent peroxide reductase in mitochondria. Eur J Biochem 1997; 249: 52–60.
23. Miranda Vizuete A, Damdimopoulos AE, Pedrajas JR, Gustafsson JA, Spyrou G. Human mitochondrial thioredoxin reductase cDNA cloning, expression and genomic organization. Eur J Biochem 1999; 261:405–412.
24. Nishiyama A, Matsui M, Iwata S, Hirota K, Masutani H, Nakamura H, Takagi Y, Sono H, Gon Y, Yodoi J. Identification of thioredoxin-binding protein-2/vitamin D(3) up-regulated protein 1 as a negative regulator of thioredoxin function and expression. J Biol Chem 1999; 274:21645–21650.
25. Chen KS, DeLuca HF. Cloning of the human 1 alpha,25-dihydroxyvitamin D-3 24-hydroxy-

lase gene promoter and identification of two vitamin D-responsive elements. Biochim Biophys Acta 1995; 1263:1–9.
26. Suda T, Shinki T, Takahashi N. The role of vitamin D in bone and intestinal cell differentiation. Annu Rev Nutr 1990; 10:195–211.
27. Hirota K, Matsui M, Iwata S, Nishiyama A, Mori K, Yodoi J. AP-1 transcriptional activity is regulated by a direct association between thioredoxin and Ref-1. Proc Natl Acad Sci USA 1997; 94:3633–3638.
28. Makino Y, Yoshikawa N, Okamoto K, Hirota K, Yodoi J, Makino I, Tanaka H. Direct association with thioredoxin allows redox regulation of glucocorticoid receptor function. J Biol Chem 1999; 274:3182–3188.
29. Qin J, Clore GM, Kennedy WP, Kuszewski J, Gronenborn AM. The solution structure of human thioredoxin complexed with its target from Ref-1 reveals peptide chain reversal. Structure 1996; 4:613–620.
30. Hirota K, Murata M, Sachi Y, Nakamura H, Takeuchi J, Mori K, Yodoi J. Distinct roles of thioredoxin in the cytoplasm and in the nucleus. A two-step mechanism of redox regulation of transcription factor NF-κB. J Biol Chem 1999; 274:27891–27897.
31. Saito M, Nishitoh H, Fujii M, Takeda K, Tobiume K, Sawada Y, Kawabata M, Miyazono K, Ichijo H. Mammalian thioredoxin is a direct inhibitor of apoptosis signal-regulating kinase (ASK) 1. EMBO J 1998; 17:2596–2606.
32. Adler V, Yin Z, Fuchs SY, Benezra M, Rosario L, Tew KD, Pincus MR, Sardana M, Henderson CJ, Wolf CR, Davis RJ, Ronai Z. Regulation of JNK signaling by GSTp. EMBO J 1999; 18:1321–1334.
33. Storz G, Tartaglia LA, Ames BN. Transcriptional regulator of oxidative stress-inducible genes: direct activation by oxidation. Science 1990; 248:189–194.
34. Zheng M, Storz G. Activation of the OxyR transcription factor by reversible disulfide bond formation. Science 1998; 279:1718–1721.
35. Kuge S, Jones N. YAP1 dependent activation of TRX2 is essential for the response of Saccharomyces cerevisiae to oxidative stress by hydroperoxides. EMBO J 1994; 13:655–664.
36. Morgan BA, Banks GR, Toone WM, Raitt D, Kuge S, Johnston LH. The Skn7 response regulator controls gene expression in the oxidative stress response of the budding yeast Saccharomyces cerevisiae. EMBO J 1997; 16:1035–1044.
37. Kuge S, Jones N, Nomoto A. Regulation of yAP-1 nuclear localization in response to oxidative stress. EMBO J 1997; 16:1710–1720.
38. Izawa S, Maeda K, Sugiyama K, Mano J, Inoue Y, Kimura A. Thioredoxin deficiency causes the constitutive activation of Yap1, an AP-1-like transcription factor in *Saccharomyces cerevisiae*. J Biol Chem 1999; 274:28459–28465.
39. Kuge S, Toda T, Iizuka N, Nomoto A. Crm1 (XpoI) dependent nuclear export of the budding yeast transcription factor yAP-1 is sensitive to oxidative stress. Genes Cells 1998; 3:521–532.
40. Rees AM, Bell PA. The involvement of receptor sulphydryl groups in the binding of steroids to the cytoplasmic glucocorticoid receptor from rat thymus. Biochim Biophys Acta 1975; 411:121–132.
41. Grippo JF, Tienrungroj W, Dahmer MK, Housley PR, Pratt WB. Evidence that the endogenous heat-stable glucocorticoid receptor-activating factor is thioredoxin. J Biol Chem 1983; 258:13658–13664.
42. Bodwell JE, Holbrook NJ, Munck A. Sulfhydryl-modifying reagents reversibly inhibit binding of glucocorticoid-receptor complexes to DNA-cellulose. Biochemistry 1984; 23:1392–1398.
43. Hentze MW, Rouault TA, Harford JB, Klausner RD. Oxidation-reduction and the molecular mechanism of a regulatory RNA-protein interaction. Science 1989; 244:357–359.
44. Abate C, Patel L, Rauscher FD, Curran T. Redox regulation of fos and jun DNA-binding activity in vitro. Science 1990; 249:1157–1161.

45. Seki S, Akiyama K, Watanabe S, Hatsushika M, Ikeda S, Tsutsui K. cDNA and deduced amino acid sequence of a mouse DNA repair enzyme (APEX nuclease) with significant homology to *Escherichia coli* exonuclease III. J Biol Chem 1991; 266:20797–20802.
46. Demple B, Herman T, Chen DS. Cloning and expression of APE, the cDNA encoding the major human apurinic endonuclease: definition of a family of DNA repair enzymes. Proc Natl Acad Sci USA 1991; 88:11450–11454.
47. Xanthoudakis S, Miao GG, Curran T. The redox and DNA-repair activities of Ref-1 are encoded by nonoverlapping domains. Proc Natl Acad Sci USA 1994; 91:23–27.
48. Toledano MB, Leonard WJ. Modulation of transcription factor NF-kappa B binding activity by oxidation-reduction in vitro. Proc Natl Acad Sci USA 1991; 88:4328–4332.
49. Okamoto T, Ogiwara H, Hayashi T, Mitsui A, Kawabe T, Yodoi J. Human thioredoxin/adult T cell leukemia-derived factor activates the enhancer binding protein of human immunodeficiency virus type 1 by thiol redox control mechanism. Int J Immunol 1992; 4:811–819.
50. Hayashi T, Ueno Y, Okamoto T. Oxidoreductive regulation of nuclear factor kappa B. Involvement of a cellular reducing catalyst thioredoxin. J Biol Chem 1993; 268:11380–11388.
51. Meyer M, Schreck R, Baeuerle PA. $H_2O_2$ and antioxidants have opposite effects on activation of NF-kappa B and AP-1 in intact cells: Ap-1 as secondary antioxidant-responsive factor. EMBO J 1993; 12:2005–2015.
52. Schenk H, Klein M, Erdbrugger W, Dröge W, Schulze Osthoff K. Distinct effects of thioredoxin and antioxidants on the activation of transcription factors NF-kappa B and AP-1. Proc Natl Acad Sci USA 1994; 91:1672–1676.
53. Schreck R, Rieber P, Baeuerle PA. Reactive oxygen intermediates as apparently widely used messengers in the activation of the NF-κB transcription factor and HIV-1. EMBO J 1991; 10:2247–2258.
54. Myrset AH, Bostad A, Jamin N, Lirsac PN, Toma F, Gabrielsen OS. DNA and redox state induced conformational changes in the DNA-binding domain of the Myb oncoprotein. EMBO J 1993; 12:4625–4633.
55. Wasylyk B, Hagman J, Gutierrez Hartmann A. Ets transcription factors: nuclear effectors of the Ras-MAP-kinase signaling pathway. Trends Biochem Sci 1998; 23:213–216.
56. Ireland RC, Li SY, Dougherty JJ. The DNA binding of purified Ah receptor heterodimer is regulated by redox conditions. Arch Biochem Biophys 1995; 319:470–480.
57. Huang LE, Arany Z, Livingston DM, Bunn HF. Activation of hypoxia-inducible transcription factor depends primarily upon redox-sensitive stabilization of its alpha subunit. J Biol Chem 1996; 271:32253–32259.
58. Makino Y, Okamoto K, Yoshikawa N, Aoshima M, Hirota K, Yodoi J, Umesono K, Makino I, Tanaka H. Thioredoxin: a redox-regulating cellular cofactor for glucocorticoid hormone action. Cross talk between endocrine control of stress response and cellular antioxidant defense system. J Clin Invest 1996; 98:2469–2477.
59. Hayashi S, Hajiro NK, Makino Y, Eguchi H, Yodoi J, Tanaka H. Functional modulation of estrogen receptor by redox state with reference to thioredoxin as a mediator. Nucl Acids Res 1997; 25:4035–4040.
60. Akamatsu Y, Ohno T, Hirota K, Kagoshima H, Yodoi J, Shigesada K. Redox regulation of the DNA binding activity in transcription factor PEBP2. The roles of two conserved cysteine residues. J Biol Chem 1997; 272:14497–14500.
61. Ueno M, Masutani H, Arai RJ, Yamauchi A, Hirota K, Sakai T, Inamoto T, Yamaoka Y, Yodoi J, Nikaido T. Thioredoxin-dependent redox regulation of p53-mediated p21 activation. J Biol Chem 1999; 274:35809–35815.
62. Ema HK, Mimura J, Abe H, Yodoi J, Sogawa K, Poellinger L. Molecular mechanisms of transcription activation by HLF and HIF1 alpha in response to hypoxia: their stabilization and redox signal-induced interaction with CBP/p300. EMBO J 1999; 18:1905–1914.
63. Nakamura H, Matsuda M, Furuke K, Kitaoka Y, Iwata S, Toda K, Inamoto T, Yamaoka Y, Ozawa K, Yodoi J. Adult T cell leukemia-derived factor/human thioredoxin protects endo-

thelial F-2 cell injury caused by activated neutrophils or hydrogen peroxide. Immunol Lett 1994; 42:75–80.
64. Fujii S, Nanbu Y, Nonogaki H, Konishi I, Mori T, Masutani H, Yodoi J. Coexpression of adult T-cell leukemia-derived factor, a human thioredoxin homologue, and human papillomavirus DNA in neoplastic cervical squamous epithelium. Cancer 1991; 68:1583–1591.
65. Tanaka T, Nishiyama Y, Okada K, Hirota K, Matsui M, Yodoi J, Hiai H, Toyokuni S. Induction and nuclear translocation of thioredoxin by oxidative damage in the mouse kidney: independence of tubular necrosis and sulfhydryl depletion. Lab Invest 1997; 77:145–155.
66. Levine AJ. p53, the cellular gatekeeper for growth and division. Cell 1997; 88:323–331.
67. Rainwater R, Parks D, Anderson ME, Tegtmeyer P, Mann K. Role of cysteine residues in regulation of p53 function. Mol Cell Biol 1995; 15:3892–3903.
68. Hainaut P, Rolley N, Davies M, Milner J. Modulation by copper of p53 conformation and sequence-specific DNA binding: role for Cu(II)/Cu(I) redox mechanism. Oncogene 1995; 10:27–32.
69. Xanthoudakis S, Miao G, Wang F, Pan YC, Curran T. Redox activation of Fos-Jun DNA binding activity is mediated by a DNA repair enzyme. EMBO J 1992; 11:3323–3335.
70. Jayaraman L, Murthy KG, Zhu C, Curran T, Xanthoudakis S, Prives C. Identification of redox/repair protein Ref-1 as a potent activator of p53. Genes Dev 1997; 11:558–570.
71. Pearson GD, Merill GF. Detection of the *Saccharomyces cerevisiae* TRR1 gene encoding thioredoxin reductase inhibits p53-dependent reporter gene expression. J Biol Chem 1998; 273:5431–5434.
72. Gaiddon C, Moorthy NC, Prives C. Ref-1 regulates the transactivation and pro-apoptotic functions of p53 in vivo. EMBO J 1999; 18:5609–5621.
73. Buttke TM, Sandstrom PA. Oxidative stress as a mediator of apoptosis. Immunol Today 1994; 15:7–10.
74. Sato N, Iwata S, Nakamura K, Hori T, Mori K, Yodoi J. Thiol-mediated redox regulation of apoptosis. Possible roles of cellular thiols other than glutathione in T cell apoptosis. J Immunol 1995; 154:3194–3203.
75. Ueda S, Nakamura H, Masutani H, Sasada T, Yonehara S, Takabayashi A, Yamaoka Y, Yodoi J. Redox regulation of caspase-3 (-like) protease activity: regulatory roles of thioredoxin and cytochrome c. J Immunol 1998; 161:6689–6695.
76. Hampton MB, Orrenius S. Dual regulation of caspase activity by hydrogen peroxide: implications for apoptosis. FEBS Lett 1997; 414:552–556.
77. Marchetti P, Decaudin D, Macho A, Zamzami N, Hirsch T, Susin SA, Kroemer G. Redox regulation of apoptosis: impact of thiol oxidation status on mitochondrial function. Eur J Immunol 1997; 27:289–296.
78. Reed JC. Cytochrome c: can't live with it—can't live without it. Cell 1997; 91:559–562.
79. Higuchi M, Honda T, Proske RJ, Yeh ET. Regulation of reactive oxygen species-induced apoptosis and necrosis by caspase 3-like proteases. Oncogene 1998; 17:2753–2760.
80. Wilson KP, Black JA, Thompson JA, Kim EE, Griffith JP, Navia MA, Murcko MA, Chambers SP, Aldape RA, Raybuck SA et al. Structure and mechanism of interleukin-1 beta converting enzyme. Nature 1994; 370:270–275.
81. Hirsh T, Marchetti P, Susin SS, Dallaporta B, Zamzami N, Marzo I, Geuskins M, Kroemer G. The apoptosis-necrosis paradox. Apoptogenic proteases activated after mitochondrial permeability transition determine the mode of cell death. Oncogene 1997; 15:1573–1581.
82. Susin SA, Lorenzo HK, Zamzami N, Marzo I, Snow BE, Brothers GM, Mangion J, Jacotot E, Costantini P, Loeffler M, Larochette N, Goodlett DR, Aebersold R, Siderovski DP, Penninger JM, Kroemer G. Molecular characterization of mitochondrial apoptosis-inducing factor. Nature 1999; 397:441–446.
83. Ichijo H, Nishida E, Irie K, Ten DP, Saitoh M, Moriguchi T, Takagi M, Matsumoto K, Miyazono K, Gotoh Y. Induction of apoptosis by ASK1, a mammalian MAPKKK that activates SAPK/JNK and p38 signaling pathways. Science 1997; 275:90–94.

84. Matsuda M, Masutani H, Nakamura H, Miyajima S, Yamauchi A, Yonehara S, Uchida A, Irimajiri K, Horiuchi A, Yodoi J. Protective activity of adult T cell leukemia-derived factor (ADF) against tumor necrosis factor-dependent cytotoxicity on U937 cells. J Immunol 1991; 147:3837–3841.
85. Iwata S, Hori T, Sato N, Ueda TY, Yamabe T, Nakamura H, Masutani H, Yodoi J. Thiol-mediated redox regulation of lymphocyte proliferation. Possible involvement of adult T cell leukemia-derived factor and glutathione in transferrin receptor expression. J Immunol 1994; 152:5633–5642.
86. Hashimoto S, Matsumoto K, Gon Y, Furuichi S, Maruoka S, Takeshita I, Hirota K, Yodoi J, Horie T. Thioredoxin negatively regulates p38 MAP kinase activation and IL-6 production by tumor necrosis factor-alpha. Biochem Biophys Res Commun 1999; 258:443–447.
87. Zhang P, Liu B, Kang SW, Seo MS, Rhee SG, Obeid LM. Thioredoxin peroxidase is a novel inhibitor of apoptosis with a mechanism distinct from that of Bcl-2. J Biol Chem 1997; 272: 30615–30618.
88. Holmgren A. Thioredoxin and glutaredoxin systems. J Biol Chem 1989; 264:13963–13966.
89. Makino S, Masutani H, Maekawa N, Konishi I, Fujii S, Yamamoto R, Yodoi J. Adult T-cell leukaemia-derived factor/thioredoxin expression on the HTLV-I transformed T-cell lines: heterogeneous expression in ALT-2 cells. Immunology 1992; 76:578–583.
90. Yodoi J, Tursz T. ADF, a growth-promoting factor derived from adult T cell leukemia and homologous to thioredoxin: involvement in lymphocyte immortalization by HTLV-I and EBV. Adv Cancer Res 1991; 57:381–411.
91. Sono H, Teshigawara K, Sasada T, Takagi Y, Nishiyama A, Ohkubo Y, Maeda Y, Tatsumi E, Kanamaru A, Yodoi J. Redox control of Epstein-Barr virus replication by human thioredoxin/ATL-derived factor: Differential regulation of lytic and latent infection. Antioxidants Redox Signal 1999; 1:155–165.
92. Dröge W, Eck HP, Mihm S. HIV-induced cysteine deficiency and T-cell dysfunction—a rationale for treatment with *N*-acetylcysteine. Immunol Today 1992; 13:211–214.
93. Eck HP, Gmunder H, Hartmann M, Petzoldt D, Daniel V, Dröge W. Low concentrations of acid-soluble thiol (cysteine) in the blood plasma of HIV-1-infected patients. Biol Chem Hoppe Seyler 1989; 370:101–108.
94. Buhl R, Jaffe HA, Holroyd KJ, Wells FB, Mastrangeli A, Saltini C, Cantin AM, Crystal RG. Systemic glutathione deficiency in symptom-free HIV-seropositive individuals. Lancet 1989; 2:1294–1298.
95. Roederer M, Staal FJ, Osada H, Herzenberg LA. CD4 and CD8 T cells with high intracellular glutathione levels are selectively lost as the HIV infection progresses. Int Immunol 1991; 3:933–937.
96. Roederer M, Staal FJ, Raju PA, Ela SW, Herzenberg LA. Cytokine-stimulated human immunodeficiency virus replication is inhibited by N-acetyl-L-cysteine. Proc Natl Acad Sci USA 1990; 87:4884–4888.
97. Masutani H, Nakamura H, Ueda Y, Kitaoka Y, Kawabe T, Iwata S, Mitsui A, Yodoi J. ADF (adult T cell leukemia-derived factor)/human thioredoxin and viral infection: possible new therapeutic approach. Adv Exp Med Biol 1992; 319:265–274.
98. SC Flores, JC Marecki, KP Harper, SK Bose, SK Nelson, JM McCord. Tat protein of human immunodeficiency virus type 1 represses expression of manganese superoxide dismutase in HeLa cells. Proc Natl Acad Sci USA 1993; 90:7632–7636.
99. Westendorp MO, Shatrov VA, Schulze Osthoff K, Frank R, Kraft M, Los M, Krammer PH, Dröge W, Lehmann V. HIV-1 Tat potentiates TNF-induced NF-kappa B activation and cytotoxicity by altering the cellular redox state. EMBO J 1995; 14:546–554.
100. Boudet F, Lecoeur H, Gougeon ML. Apoptosis associated with ex vivo down-regulation of Bcl-2 and up-regulation of Fas in potential cytotoxic CD8+ T lymphocytes during HIV infection. J Immunol 1996; 156:2282–2293.
101. Aillet F, Masutani H, Elbim C, Raoul H, Chene L, Nugeyre MT, Paya C, Barre Sinoussi F,

Gougerot Pocidalo MA, Israel N. Human immunodeficiency virus induces a dual regulation of Bcl-2, resulting in persistent infection of CD4(+) T- or monocytic cell lines. J Virol 1998; 72:9698–9705.
102. Elbim PS, Prevost MH, Preira A, Girard PM, Rogine N, Masutani H, Hakim IN, Gougerot-Pocidalo MA. Redox and activation status of monocytes from human immunodeficiency virus-infected patients: relationship with viral load. J Virol 1999; 73:4561–4566.
103. Nakamura H, De Rosa S, Roederer M, Anderson MT, Dubs JG, Yodoi J, Holmgren A, Herzenberg LA. Elevation of plasma thioredoxin levels in HIV-infected individuals. Int Immunol 1996; 8:603–611.
104. Masutani H, Ueno M, Ueda S, Yodoi J. Role of thioredoxin and redox regulation in oxidative stress response and signaling. In: Sen CK, Sies H, Baeuerle PA, eds. Antioxidant and Redox Regulation of Genes. San Diego: Academic Press, 1999:297–311.
105. Mitsui A, Hirakawa T, Yodoi J. Reactive oxygen-reducing and protein-refolding activities of adult T cell leukemia-derived factor/human thioredoxin. Biochem Biophys Res Commun 1992; 186:1220–1226.
106. Wada H, Muro K, Hirata T, Yodoi J, Hitomi S. Rejection and expression of thioredoxin in transplanted canine lung. Chest 1995; 108:810–814.
107. Aota M, Matsuda K, Isowa N, Wada H, Yodoi J, Ban T. Protection against reperfusion-induced arrhythmias by human thioredoxin. J Cardiovasc Pharmacol 1996; 27:727–732.
108. Matsui M, Oshima M, Oshima H, Takaku K, Maruyama T, Yodoi J, Taketo MM. Early embryonic lethality caused by targeted disruption of the mouse thioredoxin gene. Dev Biol 1996; 178:179–185.
109. Kobayashi F, Sagawa N, Nanbu Y, Kitaoka Y, Mori T, Fujii S, Nakamura H, Masutani H, Yodoi J. Biochemical and topological analysis of adult T-cell leukaemia-derived factor, homologous to thioredoxin, in the pregnant human uterus. Hum Reprod 1995; 10:1603–1608.
110. Tonissen KF, Wells JR. Isolation and characterization of human thioredoxin-encoding genes. Gene 1991; 102:221–228.
111. Natsuyama S, Noda Y, Narimoto K, Umaoka Y, Mori T. Release of two-cell block by reduction of protein disulfide with thioredoxin from Escherichia coli in mice. J Reprod Fertil 1992; 95:649–656.
112. Natsuyama S, Noda Y, Yamashita M, Nagahama Y, Mori T. Superoxide dismutase and thioredoxin restore defective p34cdc2 kinase activation in mouse two-cell block. Biochim Biophys Acta 1993; 1176:90–94.
113. Spector A, Yan GZ, Huang RR, McDermott MJ, Gascoyne PR, Pigiet V. The effect of $H_2O_2$ upon thioredoxin-enriched lens epithelial cells. J Biol Chem 1988; 263:4984–4990.
114. Ohira A, Honda O, Gauntt CD, Yamamoto M, Hori K, Masutani H, Yodoi J, Honda Y. Oxidative stress induces adult T cell leukemia derived factor/thioredoxin in the rat retina. Lab Invest 1994; 70:279–285.
115. Gauntt CD, Ohira A, Honda O, Kigasawa K, Fujimoto T, Masutani H, Yodoi J, Honda Y. Mitochondrial induction of adult T cell leukemia derived factor (ADF/hTx) after oxidative stresses in retinal pigment epithelial cells. Invest Ophthalmol Vis Sci 1994; 35:2916–2923.
116. Yamamoto M, Sato N, Tajima H, Furuke K, Ohira A, Honda Y, Yodoi J. Induction of human thioredoxin in cultured human retinal pigment epithelial cells through cyclic AMP-dependent pathway; involvement in the cytoprotective activity of prostaglandin E1. Exp Eye Res 1997; 65:645–652.
117. Schallreuter KU, Wood JM. The role of thioredoxin reductase in the reduction of free radicals at the surface of the epidermis. Biochem Biophys Res Commun 1986; 136:630–637.
118. Hori K, Katayama M, Sato N, Ishii K, Waga S, Yodoi J. Neuroprotection by glial cells through adult T cell leukemia-derived factor/human thioredoxin (ADF/TRX). Brain Res 1994; 652:304–310.

119. Takagi Y, Mitsui A, Nishiyama A, Nozaki K, Sono H, Gon Y, Hashimoto N, Yodoi J. Overexpression of thioredoxin in transgenic mice attenuates focal ischemic brain damage. Proc Natl Acad Sci USA 1999; 96:4131–4136.
120. Sasada T, Iwata S, Sato N, Kitaoka Y, Hirota K, Nakamura K, Nishiyama A, Taniguchi Y, Takabayashi A, Yodoi J. Redox control of resistance to cis-diamminedichloroplatinum (II) (CDDP): protective effect of human thioredoxin against CDDP-induced cytotoxicity. J Clin Invest 1996; 97:2268–2276.
121. Wang J, Kobayashi M, Sakurada K, Imamura M, Moriuchi T, Hosokawa M. Possible roles of an adult T-cell leukemia (ATL)-derived factor/thioredoxin in the drug resistance of ATL to adriamycin. Blood 1997; 89:2480–2487.
122. Yokomizo A, Ono M, Nanri H, Makino Y, Ohga T, Wada M, Okamoto T, Yodoi J, Kuwano M, Kohno K. Cellular levels of thioredoxin associated with drug sensitivity to cisplatin, mitomycin C, doxorubicin, and etoposide. Cancer Res 1995; 55:4293–4296.
123. Kawahara N, Tanaka T, Yokomizo A, Nanri H, Ono M, Wada M, Kohno K, Takenaka K, Sugimachi K, Kuwano M. Enhanced coexpression of thioredoxin and high mobility group protein 1 genes in human hepatocellular carcinoma and the possible association with decreased sensitivity to cisplatin. Cancer Res 1996; 56:5330–5333.
124. Hotta M, Tashiro F, Ikegami H, Niwa H, Ogihara T, Yodoi J, Miyazaki J. Pancreatic beta cell-specific expression of thioredoxin, an antioxidative and antiapoptotic protein, prevents autoimmune and streptozotocin-induced diabetes. J Exp Med 1998; 188:1445–1451.
125. Wakita H, Yodoi J, Masutani H, Toda K, Takigawa M. Immunohistochemical distribution of adult T-cell leukemia-derived factor/thioredoxin in epithelial components of normal and pathologic human skin conditions. J Invest Dermatol 1992; 99:101–107.
126. Sachi Y, Hirota K, Masutani H, Toda K, Okamoto T, Takigawa M, Yodoi J. Induction of ADF/TRX by oxidative stress in keratinocytes and lymphoid cells. Immunol Lett 1995; 44: 189–193.
127. Taniguchi Y, Taniguchi UY, Mori K, Yodoi J. A novel promoter sequence is involved in the oxidative stress-induced expression of the adult T-cell leukemia-derived factor (ADF)/human thioredoxin (Trx) gene. Nucl Acids Res 1996; 24:2746–2752.
128. Leppa S, Pirkkala L, Chow SC, Eriksson JE, Sistonen L. Thioredoxin is transcriptionally induced upon activation of heat shock factor 2. J Biol Chem 1997; 272:30400–30404.
128a. Kim YC, Masutani H, Yamaguchi Y, Itoh K, Yamamoto M, Yodoi J. Hemin-induced activation of the thioredoxin gene by Nrf2: A differential regulation of the antioxidant responsive element (ARE) by switch of its binding factors. Submitted.
128b. Kwon YW, Masutani H, Yamaguchi Y, Yodoi J. Activation of the thioredoxin gene by polycyclic aromatic hydrocarbons (PAHs). In preparation.
129. Rushmore TH, Morton MR, Pickett CB. The antioxidant responsive element. Activation by oxidative stress and identification of the DNA consensus sequence required for functional activity. J Biol Chem 1991; 266:11632–11639.
130. Friling RS, Bensimon A, Tichauer Y, Daniel V. Xenobiotic-inducible expression of murine glutathione S-transferase Ya subunit gene is controlled by an electrophile-responsive element. Proc Natl Acad Sci USA 1990; 87:6258–6262.
131. Mignotte V, Eleouet JF, Raich N, Romeo PH. Cis- and trans-acting elements involved in the regulation of the erythroid promoter of the human porphobilinogen deaminase gene. Proc Natl Acad Sci USA 1989; 86:6548–6552.
132. Andrews NC, Erdjument Bromage H, Davidson MB, Tempst P, Orkin SH. Erythroid transcription factor NF-E2 is a haematopoietic-specific basic-leucine zipper protein. Nature 1993; 362:722–728.
133. Kataoka K, Igarashi K, Itoh K, Fujiwara KT, Noda M, Yamamoto M, Nishizawa M. Small Maf proteins heterodimerize with Fos and may act as competitive repressors of the NF-E2 transcription factor. Mol Cell Biol 1995; 15:2180–2190.
134. Caterina JJ, Donze D, Sun CW, Ciavatta DJ, Townes TM. Cloning and functional character-

ization of LCR-F1: a bZIP transcription factor that activates erythroid-specific, human globin gene expression. Nucl Acids Res 1994; 22:2383–2391.
135. Chan JY, Han XL, Kan YW. Cloning of Nrf1, an NF-E2-related transcription factor, by genetic selection in yeast. Proc Natl Acad Sci USA 1993; 90:11371–11375.
136. Chui DH, Tang W, Orkin SH. cDNA cloning of murine Nrf2 gene, coding for a p45 NF-E2 related transcription factor. Biochem Biophys Res Commun 1995; 209:40–46.
137. Itoh K, Igarashi K, Hayashi N, Nishizawa M, Yamamoto M. Cloning and characterization of a novel erythroid cell-derived CNC family transcription factor heterodimerizing with the small Maf family proteins. Mol Cell Biol 1995; 15:4184–4193.
138. Moi P, Chan K, Asunis I, Cao A, Kan YW. Isolation of NF-E2-related factor 2 (Nrf2), a NF-E2-like basic leucine zipper transcriptional activator that binds to the tandem NF-E2/AP1 repeat of the beta-globin locus control region. Proc Natl Acad Sci USA 1994; 91:9926–9930.
139. Fujiwara KT, Kataoka K, Nishizawa M. Two new members of the maf oncogene family, mafK and mafF, encode nuclear b-Zip proteins lacking putative trans-activator domain. Oncogene 1993; 8:2371–2380.
140. Itoh K, Chiba T, Takahashi S, Ishii T, Igarashi K, Katoh Y, Oyake T, Hayashi N, Satoh K, Hatayama I, Yamamoto M, Nabeshima Y. An Nrf2/small Maf heterodimer mediates the induction of phase II detoxifying enzyme genes through antioxidant response elements. Biochem Biophys Res Commun 1997; 236:313–322.
141. Alam J, Stewart D, Touchard C, Boinapally S, Choi AM, Cook JL. Nrf2, a cap'n'collar transcription factor, regulates induction of the heme oxygenase-1 gene. J Biol Chem 1999; 274:26071–26078.
142. Wild AC, Moinova HR, Mulcahy RT. Regulation of γ-glutamyl cysteine synthetase subunit gene expression by the transcription factor Nrf2. J Biol Chem 1999; 274:33627–33636.
143. Balla J, Jacob HS, Balla G, Nath K, Eaton JW, Vercellotti GM. Endothelial-cell heme uptake from heme proteins: induction of sensitization and desensitization to oxidant damage. Proc Natl Acad Sci USA 1993; 90:9285–9289.
144. Paller MS, Jacob HS. Cytochrome P-450 mediates tissue-damaging hydroxyl radical formation during reoxygenation of the kidney. Proc Natl Acad Sci USA 1994; 91:7002–7006.
144a. Kasuno K, Muso E, Ono T, Masutani H, Nakamura H, Mitsui A, Sasayama S, Yodoi J. Thioredoxin protects ischemic renal tubular injury by rapid hypoxia-induced mRNA and protein stabilization mechanism. Submitted.
145. Hankinson O. The aryl hydrocarbon receptor complex. Annu Rev Pharmacol Toxicol 1995; 35:307–340.
146. Safe SH. Modulation of gene expression and endocrine response pathways by 2,3,7,8-tetrachlorodibenzo-*p*-dioxin and related compounds. Pharmacol Ther 1995; 67:247–281.

# 7

# Cigarette Smoking and Pulmonary Oxidative Damage

**CHING K. CHOW**

*University of Kentucky, Lexington, Kentucky*

## I. INTRODUCTION

Cigarette smoking has been implicated as a significant contributing factor in the etiology of respiratory, cardiovascular, and other disorders (1–4). Lung cancer and chronic emphysema are the most serious respiratory diseases attributable to cigarette smoking. Smokers have a much higher mortality rate from lung cancer, bronchitis, and emphysema than nonsmokers (2,3). Also, coronary thrombosis and myocardial infarction are significantly more prevalent in smokers than in nonsmokers, and the death rate from coronary heart disease is higher in smokers than in nonsmokers (3). While cigarette smoking has long been recognized as playing an important role in the pathogenesis of these disorders, the underlying mechanisms by which cigarette smoking is involved remain poorly understood.

Oxidative stress has been implicated in the pathogenesis of numerous diseases of the lung, including cystic fibrosis, chronic obstructive airway disease, and asthma. All these conditions are characterized by an increased formation of reactive oxygen species and decreased antioxidant potential. In the lung, reactive oxygen species can arise from both endogenous sources, such as mitochondria and inflammatory cells, and exogenous sources, such as oxidant air pollutants and cigarette smoke. The lung, like other organs, has a wide variety of antioxidant defense systems—present in the intracellular, vascular, and extracellular respiratory tract ling fluid compartments—that help to maintain a balanced redox status. Oxidative damage, however, can result from increased oxidative stress and/or decreased antioxidant potential (5,6).

Among the many hypotheses of smoking-associated disorders, the possible association between free-radical-induced oxidative damage and the development of lung cancer, emphysema, and other smoking-related disorders has received considerable attention. The possible role of cigarette smoking in causing/initiating oxidative damage to pulmonary tissue is the focus of this chapter.

## II. CIGARETTE SMOKING AND OXIDATIVE STRESS

Smokers are subjected to two distinct types of oxidative stress. The first type of oxidative stress is due to oxidants and free radicals present in cigarette smoke; the second type results from reactive oxygen species generated by the recruited inflammatory cells onto the pulmonary tissue.

### A. Oxidants and Free Radicals Present in Cigarette Smoke

Cigarette smoke is a composite of numerous pollutants in rather high concentrations. Thousands of smoke constituents, including many oxidants, prooxidants, free radicals, as well as reducing agents, have been identified (7,8). Cigarette smoke can be divided into gas phase and particulate matter (tar), and two different populations of free radicals, one in the tar and one in the gas phase, have been identified in cigarette smoke (9,10). The principal radical in the tar phase, a quinone/hydroquinone complex, is capable of reducing molecular oxygen to superoxide radicals. The gas phase of cigarette smoke contains small oxygen-, nitrogen-, and carbon-centered radicals that are much more reactive than are the tar-phase radicals. Nitrogen dioxide, one of the major oxidant air pollutants present in photochemical smog, for example, is found in cigarette smoke at levels as high as 250 ppm (2). Panda et al. (10a) suggested that the gas-phase cigarette smoke contains unstable reactive oxygen species such as superoxide and hydrogen peroxide that can cause substantial oxidation of pure proteinlike albumin but is unable to produce significant oxidative damage of microsomal protein; in addition, stable oxidants in the cigarette smoke are not present in tobacco and are produced by the interaction of superoxide, hydrogen peroxide, and hydroxyl radicals of the gas phase with some components of the tar phase during/following the burning of tobacco. Thus, cigarette smoke contains many oxidants, free radicals, and metastable products derived from radical reactions that are capable of reacting with or inactivating essential cellular constituents.

### B. Reactive Oxygen Species Indirectly Resulting from Cigarette Smoke Exposure

In smokers, the cumulative smoking history is highly correlated with both leukocytosis and elevation of acute-phase reactants (11–14). This reflects a smoking-induced inflammatory response with increasing accumulation of alveolar macrophages and neutrophils in the lungs (12). In addition, cigarette smoking stimulates an activation of oxidative metabolism of macrophages and neutrophils (13–16). The increased oxidative metabolism of phagocytes is accompanied by increased generation of reactive oxygen species, such as hydrogen peroxide, oxides of nitrogen, hydroxyl radicals, and superoxide radicals. Also, smokers have higher neutrophil myeloperoxidase activity than nonsmokers (17). Morrison et al. (18) found higher neutrophil numbers in bronchoalveolar lavage fluid and superoxide radical release from mixed bronchoalveolar lavage fluid leukocytes than nonsmokers. The findings suggest an increase in permeability and the number of neutrophils in the air spaces of cigarette smokers concomitant with evidence of increased oxidative stress.

The effect of cigarette smoking on the status of the important free radical mediator, nitric oxide (NO), has also been investigated. Deliconstantios et al. (19) showed that rat alveolar macrophages challenged with cigarette smoke released both superoxide radicals and NO, as well as their reaction product peroxynitrite. The results suggest that exposure of alveolar macrophages to cigarette smoke evokes formation of nitric oxide, oxides of nitrogen, peroxynitrite, and hydrogen peroxide in the lung, and that these substances stimulate alveolar macrophages and perhaps endothelium of the alveolar vessels to produce more oxidant, resulting in lung damage. Similarly, Chambers et al. (20) found increased lower respiratory tract NO concentration following cigarette smoking and suggested a novel mechanism for the handling of NO in the human lung. However, Kharitonov et al. (21) found that smokers exhaled lower levels of NO than nonsmokers and suggested that cigarette smoking may inhibit the activity of NO synthase. Since endogenous NO is important in defending the respiratory tract against infection, counteracting bronchoconstriction and vasoconstriction, and inhibiting platelet aggregation, this effect may contribute to the increased risks of chronic respiratory and cadiovascular diseases in cigarette smokers.

Chronic bronchitis in cigarette smokers shares many clinical and histological features with environmental lung diseases attributable to bacteria endotoxin inhalation. Bioactive endotoxin was detected in both the smoke portion and filter tips of cigarettes, as well as in both mainstream and sidestream smoke (22). Bacterial endotoxin plays an important role in the pathogenesis of infection, and can stimulate a septic state that leads to multiple organ failure. The toxic effect of endotoxin is partly attributable to its ability to stimulate generation of reactive oxygen species by endogenous mediators of inflammation (23,24). Thus, the presence of bioactive endotoxin in cigarette smoke may also contribute to increased oxidative stress that leads to the pathogenesis of chronic bronchitis in susceptible cigarette smokers.

## III. CIGARETTE SMOKING AND OXIDATIVE DAMAGE

### A. In Vitro and Cell Culture Studies

As cigarette smoke contains a number of oxidants and oxidizing agents, it is not surprising that smoke or smoke components are capable of oxidizing cells or cellular constituents. Results obtained from in vitro and cell culture studies generally support the idea that cigarette smoke can cause or promote oxidative damage. For example, incubation of sonicated rabbit alveolar macrophages and pulmonary protective factor with an aqueous extract of cigarette smoke results in increased formation of lipid peroxidation products (25). An unidentified factor in cigarette smoke is capable of oxidizing thiols (26). The levels of lipid peroxidation products, thiobarbituric acid reactants (TBAR), mainly malondialdehyde, are found to be higher in the oxidized low-density lipoproteins (LDL) of smokers than in nonsmokers, possibly reflecting lower levels of vitamin E in LDL of smokers (27). Also, following activation with the synthetic chemotactic tripeptide *N*-formylmethionine-leucyl-phenylalanine, potentiated by cytochalasin B, blood neutrophils from smokers inflicted increased damage to the DNA of cocultured mononuclear leukocytes, and the damage to DNA was preventable by the inclusion of superoxide dismutase and catalase, either individually or in combination (28).

Cigarette smoke has also been shown to promote oxidative reaction of other agents. Kamp et al. (29) showed that DNA damage of cultured alveolar epithelial cells by asbestos was augmented by cigarette smoke, supporting the view that asbestos and cigarette smoke are genotoxic to relevant target cells in the lung and that iron-induced free radicals may in part cause these effects. Similarly, Yoshi and Ohishima (30) found that incubation of pBR322 plasmid DNA with aqueous extract of cigarette tar and a nitric-oxide-releasing compound, diethyamine NONOate, caused synergistic induction of DNA single-strand breakage, whereas either cigarette tar alone or NO alone induced much less breakage, and the synergistic effect was prevented by superoxide dismutase, nitric oxide trapping agent, carboxy-PTIO, or *N*-acetylcysteine. The findings also suggest a possible formation of potent free radical peroxynitrite between cigarette tar and nitric oxide.

Nicotine, the major component of cigarette smoke, has been implicated as playing an important role in the development of cardiovascular diseases. While nicotine has been shown to promote low-density-lipoprotein susceptibility to oxidative modification by copper sulfate and to increase the formation of secondary oxidation products, it had no effect on the lipoprotein oxidative susceptibility in smokers and nonsmokers (31). Also, the production of oxygen free radicals by neutrophils was decreased by smoke components nicotine and cotinine (32).

## B. In Vivo Studies

While in vitro and cell culture studies generally support the idea that oxidative damage is a consequence of cigarette smoking, the results obtained from in vivo studies are not as clear. Assessments of cigarette-smoking-induced oxidative damage can roughly be divided into three types: (1) alteration of antioxidant status; (2) formation of oxidation products; and (3) pulmonary and related dysfunctions that are attributable to an oxidative damage mechanism.

### *1. Alteration of Antioxidant Status*

Smokers are being subjected to oxidative stress resulting from oxidants and free radicals present in smoke, as well as reactive oxygen species generated by increased, activated phagocytes. Therefore, the antioxidant status of a cigarette smoker is likely to be adversely affected if increased oxidative stress is a consequence of cigarette smoking.

*Vitamin C.* It has long been established that human smokers have lower blood vitamin C levels and decreased urinary excretion of vitamin C (33–37). Decreased plasma/serum and leukocyte concentrations of vitamin C in smokers are associated with increased numbers and activity of neutrophils (38–41). Similar to the human studies, the plasma levels of ascorbic acid were lower in cigarette-smoked rats than in the sham group, and the degree of decline was relatively greater in animals fed a vitamin-E-deficient diet than in the supplemented group (42,43). These findings suggests that increased consumption of vitamin C is associated with increased oxidants and free radicals generated (40,43a,43b), although decreased intake/bioavailability of vitamin C in smokers may also be responsible for their lower status.

Relatively little information is available concerning the pulmonary effects of cigarette smoking on vitamin C. Contrary to the expectation, available data obtained from experimental studies showed that the levels of ascorbic acid were either increased or unaltered in the lungs of smoked animals (44–47). As the rodent is capable of synthesizing ascorbic acid, increased levels of ascorbic acid may be due to increased synthesis to meet an increased need. A stimulation of hepatic ascorbic acid biosynthesis in rats and mice has been reported following exposure to various xenobiotics (48,49). Guinea pigs, like humans, cannot synthesize ascorbic acid. However, when guinea pigs were used as experimental animals, the levels of vitamin C in plasma, liver, kidneys, and lungs were not different significantly among cigarette-smoked, sham-smoked, or room control groups

(45). Since the guinea pigs employed for this experiment were fed a nutritionally adequate diet, dietary ascorbic acid might be sufficient to replenish the vitamin consumed. Further studies using guinea pigs as experimental animals and using defined amounts of ascorbic acid in the diet are needed to better understand the impact of cigarette smoking on vitamin C status in the lung and other organs.

*Vitamin E.* Vitamin E is the most important fat-soluble antioxidant and free radical scavenger (5,6). Efforts to correlate the nutritional status of vitamin E with the incidence of smoking-related respiratory diseases, however, have resulted largely in negative findings. There is no significant correlation between plasma vitamin E levels and indices of smoking effects, such as pulmonary function abnormalities and cytogenetic changes (11,38,50). Also, there is no significant difference in plasma vitamin E levels between smokers and nonsmokers (11,33). In addition, cigarette smoking did not significantly alter the plasma levels of vitamin E or potentiate vitamin E deficiency in either vitamin-E-deficient or supplemented rats (42,44). Besides, vitamin E and selenium deficiencies do not enhance lung inflammation from cigarette smoke in the hamster. While Liu et al. (37) found smokers had lower plasma vitamin E levels than nonsmokers above 45 years old, their plasma levels of vitamin E were not correlated with the measurement of reactive oxidants from circulating phagocytes.

Since the respiratory system is the primary target of cigarette smoke, the vitamin E content of pulmonary tissue is likely to be adversely affected. However, similar to vitamin C, the influence of cigarette smoking on vitamin E status in the lung and extrapulmonary organs remains controversial. The levels of vitamin E in the lungs of chronically smoked animals are higher, rather than lower, than in the controls. Relative to the sham group, the vitamin E concentration was increased over threefold in the lungs of guinea pigs exposed to mainstream smoke, and over twofold in the group exposed to sidestream smoke for 17 or 20 weeks (45). The levels of vitamin E in the plasma, liver, and kidney were not significantly different between the treatment groups. Similarly, the levels of vitamin E were significantly increased in the lungs, but not in the plasma, liver, or kidney of rats exposed to mainstream smoke for 8, 16, 24, or 32 weeks (46). Compared with the sham-exposed group, the levels of vitamin E were not significantly altered in the lung, plasma, or bronchoalveolar lavage fluid of 10-week-old female Sprague–Dawley rats exposed to cigarette smoke daily for 65 weeks (47). Since those experimental animals were fed a diet containing an adequate amount of vitamin E, it is possible that unaltered or higher levels of vitamin E found in the lungs of smoked animals may result from increased tissue uptake, mobilization of body stores, or the combination of both.

Ascorbic acid, reduced glutathione (GSH), lipoic acid, and coenzyme Q have been shown to be involved in the regeneration of vitamin E (6). Therefore, the status of these compounds may impact the regeneration and subsequently the status of vitamin E. Rabbits exposed to sidestream smoke, 30 min daily for 3 weeks, resulted in decreased levels of coenzyme Q10 and Q9 in the heart, along with impaired oxidative phosphorylation, diminished cytochrome oxidase activity, and increased mitochondrial F1-ATPase protein concentration (51).

*GSH and Antioxidant Enzymes.* In addition to being the source of reducing equivalents for GSH peroxidase, GSH plays an important role in detoxification of xenobiotics and may be involved in the regeneration of vitamin E (5,6). Cotgreave et al. (52) have shown that acute cigarette smoke inhalation for 1 h caused significant depletion of GSH in the lungs, bronchoalveolar lavage cells, and lavage fluid of rats. They also showed that acute cigarette smoke inhalation had no effect on the blood or hepatic GSH redox balance, and that the levels of total cysteine were unaffected in the lungs, increased in the liver, and decreased in plasma. These results suggest that acute cigarette smoke exposure causes a transient depletion of GSH and that the lungs may respond to the stress by increasing cysteine uptake. Banerjee et al. (36) also found smokers had lowered levels of GSH in their blood than nonsmokers. On the other hand, the levels of GSH were found to be either unchanged or increased in the lungs of rats chronically exposed to cigarette smoke (42,44,47). Similarly, Zappacosta et al. (52a) found higher levels of GSH in the saliva of healthy smokers than nonsmokers, although smoking a single cigarette induced a reduction of GSH concentration.

GSH peroxidase, which utilizes the reducing equivalent of GSH, is an important system responsible for the reduction of lipid hydroperoxides that may be formed. The levels of GSH and activities of GSH peroxidase and metabolically related enzymes, GSSG reductase and glucose 6-phosphate dehydrogenase, are either higher or unchanged in lungs of cigarette smoked rats, depending upon the exposure conditions (42,53). Also, compared with the sham-exposed group, the activities of GSH peroxidase and catalase were not significantly altered in the lung, plasma, or bronchoalveolar lavage fluid of rats exposed to cigarette smoke daily for 65 weeks (47).

McCuster et al. (54) have shown that the activities of superoxide dismutase and catalase from alveolar macrophages of smokers and smoke-exposed hamsters were twice that found in control subjects, but there was no change in the activity of GSH peroxidase. They also found that smoked hamsters had prolonged survival in normobaric hyperoxia (>95% O). The activity of superoxide dismutase was also found to be significantly higher in the lung of rats exposed to cigarette smoke for 24 or 32 weeks (55). Similarly, Hulea et al. (56) showed that smokers aged 46 to 80 years had significantly altered plasma lipid peroxidation products, leukocyte activation, thiol concentration, total antioxidant capacity, and red cell activity of superoxide dismutase and GSH peroxidase as compared to that of nonsmokers. Additionally, after 6 weeks of supplementation with vitamins E and C and beta-carotene, the serum levels of these micronutrients were similarly increased in smokers and nonsmokers, and vitamin E and beta-carotene levels were increased in the bronchoalveolar cells following supplementation (57). However, there was no significant downregulation in the activities of superoxide dismutase, catalase, or GSH peroxidase in the lung lavage cells. Thus it appears that a differential regulation of these defense systems and that the augmented antioxidant enzyme activity may serve as a mechanism to limit oxidant-mediated damage to alveolar structure.

### 2. *Formation of Oxidation Products*

If an oxidative damage mechanism plays a key role in the pathogenesis of smoking-induced disorders, an increased formation of oxidation products is expected to result following cigarette smoke exposure. While in vitro and cell culture studies generally support the fact that cigarette smoke induces/enhances formation of oxidation products, experimental evidence for increased formation of oxidation products in smokers is not conclusive. Most available studies have focused their assessments on the oxidation products of lipids and DNA.

*Lipid Peroxidation Products.* A number of studies have shown that increased formation of lipid peroxidation products is a consequence of cigarette smoking. Hoshino et al. (58), for example, reported that smokers had significantly higher outputs of breath pentane than nonsmokers, and that supplementation with vitamin E (800 mg/day for 2 weeks) decreased exhaled pentane in smokers. Habib et al. (58a) also found higher ethane in the exhaled breath of smokers than nonsmokers; however, vitamin E supplement did not reduce ethane significantly. Richards et al. (16) measured the levels of chemiluminescence as an indicator for the release of reactive oxidants and found that cigarette smoking was associated with elevated intracellular extracellular chemiluminescence responses. Also, the levels of conjugated dienes in plasma are significantly higher in smokers than nonsmokers (59), and smokers have higher levels of TBAR, mainly malondialdehyde, in plasma than nonsmokers (18,36). Similarly, Liu et al. (37) found that smokers had higher plasma malondialdehyde levels than nonsmokers over 45 years of age. These studies provide evidence for increased formation of oxidation products following cigarette smoke exposure.

On the other hand, some reports have shown that cigarette smoking either decreased or had no significant effect on the formation of lipid peroxidation products. For example, Surmen-Gur et al. (60) reported that cigarette smoking and/or vitamin E supplementation (400 IU daily for 28 days) did not influence postexercise plasma TBAR or antioxidant status at rest; Kawakami et al. (61) showed that smokers had lower levels of lipid hydroperoxides in bronchoalveolar lavage fluid than nonsmokers; and Harats et al. (27) found that the levels of TBAR in freshly isoalted LDL are no different in smokers and nonsmokers. Similarly, Duthie et al. (59) found that plasma levels of malondialdehyde were not different between smokers and nonsmokers, and Niewoehner et al. (52) reported that exposure of vitamin-E- and selenium-deficient Syrian hamsters to cigarette smoke for 8 weeks did not enhance inflammation. Compared to the supplemented animals, no alterations were found in the histological appearance of smoke-induced inflammatory lesions, in the number of phagocytes recruited, or in the oxidative metabolism of these phagocytes (62). Also, the levels of malondialdehyde are not significantly altered by cigarette smoking in the plasma or bronchoalveolar lavage fluid of rats exposed to cigarette smoke daily for 65 weeks (47).

Whether cigarette smoking enhances peroxidative damage in pulmonary tissue has been examined in experimental animals. The levels of TBAR, however, were either decreased or unchanged, rather than increased, in the lungs of cigarette-smoked rats as compared with those of the room-control or sham-smoke-exposed animals (42,44,47). Similar results were observed whether rats were fed a vitamin-E-deficient or supplemented diet (42), although the magnitude of the differences was relatively greater in the vitamin-E-deficient group. In another study, Wistar rats exposed to cigarette smoke, twice daily for 20 min, for 27 days had a significant decrease in liver weight and increase in lung weight; however, the smoke-exposed group showed higher levels of lipid peroxidation products in the liver but not in the lung (63). On the other hand, compared with the sham-exposed group, the levels of conjugated dienes and alpha-tocopherylquinone, oxidation products of alpha-tocopherol, were significantly higher in the lung, but not in plasma or bronchoalveolar lavage fluid, of 10-week-old female Sprague–Dawley rats exposed to cigarette smoke daily for 65 weeks (47). Thus, despite the presence of many free radicals and oxidants, increased formation of lipid peroxidation products in the lungs of smoked animals does not seem to be clearly evident.

*DNA Oxidation Products*. Results obtained from a number of recent studies suggest that increased oxidative damage to DNA is associated with cigarette smoking. Asami et al. (64) evaluated the oxidative stress of cigarette smoking by measuring the levels of 8-hydroxydeoxyguanosine in the central sites of lungs from current smokers, ex-smokers, and nonsmokers. They found that the levels of 8-hydroxydeoxyguanosine in the lung tissues of smokers were 1.43-fold higher than that of the nonsmokers, and that a positive correlation exists between the levels of 8-hydroxydeoxyguanosine, one of the oxidation products of DNA damage, in normal tissues and the number of cigarettes smoked per day. Also, increased formation of 8-hydroxydeoxyguanosine is found in human peripheral leukocytes (65). These results suggest that oxidative DNA damage is induced in lung DNA by cigarette smoking.

Izzotti et al. (66) studied the tissue-selective formation and persistence of DNA adducts in Sprague–Dawley rats exposed to a mixture of mainstream and sidestream smoke 6 h per day, 5 days per week. DNA adducts were measured by P-32 postlabeling in rat organs (lung, heart, liver, bladder, and testis), tracheal epithelia, and cells isolated from bronchoalveolar lavage. The top levels of P-32 postlabeled DNA modifications were found after 4 to 5 weeks of exposure. The increases of smoke-induced DNA adducts averaged 10.2-fold in the tracheal epithelium, 9.4-fold in bronchoalveolar lavage cells, 6.3-fold in the heart, 5.3-fold in the lung, 4.1-fold in the bladder, 0.9-fold in the testis, and 0.1-fold in the liver. Significant increases of 8-hydroxy-2′-deoxyguanosine level and of aryl hydrocarbon hydroxylase activity, and decreases in GSH level were also observed in the lung. One week after cessation of smoke exposure, DNA adduct levels were significantly decreased in the lung, tracheal epithelia, and bladder, but not in the bronchoalveolar lavage cells and heart. Similarly, rats exposed to cigarette smoke showed similar qualitative patterns of DNA adducts by P-32 postlabeling assay in various respiratory (lung, trachea, larynx) and nonrespiratory (heart and bladder) tissues, and the order of DNA adducts found is heart > lung > trachea > larynx > bladder (67). The level of lung DNA adducts increased with the duration of exposure up to 23 weeks and reverted to control levels 19 weeks after the cessation of exposure. Whole-body exposure of rats to sidestream smoke also enhanced the preexisting DNA adducts by severalfold in different tissues. These data suggest that cigarette smoke poses a potential risk of developing mutation-related diseases.

Wang et al. (68) examined the relationship between plasma concentrations of beta-carotene and alpha-tocopherol and lifestyle factors and DNA adducts in lymphocytes. They found that DNA adduct levels of low beta-carotene or alpha-tocopherol groups were not significantly different from DNA adduct levels in high-plasma beta-carotene and alpha-tocopherol groups among current smokers and nonsmokers. Lee et al. (68a), on the other hand, found that supplementing smokers with vitamin E, vitamin C, and beta-carotene for 4 weeks resulted in a time-dependent decrease of 8-hydroxydeoxyguanosine and carbonyl content in plasma.

While 8-hydroxydeoxyguanosine has been widely used as a biomarker of oxidative DNA damage in both animal and human studies, controversial data exist on the relationship between 8-hydroxydeoxyguanosine formation and age, sex, and cigarette smoking in humans. van Zeeland et al. (69) examined the level of 8-hydroxydeoxyguanosine in DNA from peripherial leukocytes and found current smokers had lower 8-hydroxydeoxyguanosine values than subjects who never smoked, and an inverse relationship was found between 8-hydroxydeoxyguanosine and lifetime smoking. It thus appears that 8-hydroxydeoxyguanosine levels in leukocytes may not serve as a sensitive marker of smoke exposure.

The mitochondria electron transport system consumes over 85% of oxygen used by the cells, and up to 5% of the oxygen consumed by mitochondria is converted to reactive oxygen species. Proximal to a large flux of reactive oxygen species, mitochondrial DNA is particularly susceptible to oxidative damage and mutation because it lacks protective histones and effective repair systems (70–72). In agreement with this view, Ballinger et al. (73) found smokers had 5.6 times the level of mitochondrial DNA damage, 2.6 times the damage at a nuclear locus (beta-globin gene cluster), and almost 7 times the levels of 4.9-kb mitochondrial DNA deletion compared to nonsmokers, although the latter increase was not significant. They suggest that the mitochondrial DNA is a sensitive biomarker for smoke-induced genetic damage and mutation. Fahn et al. (74) investigated the effect of cigarette smoking on mitochondrial DNA mutation and lipid peroxidation in lung resection samples, and found that the frequency of occurrence and the proportion of the 4839-bp mitochondrial DNA deletion in the lung increased significantly with the smoking index in terms of pack-years. Also, the incidence and proportion of the 4839-bp mitochondrial DNA deletion in the lung tissues of current smokers were higher than in those of nonsmokers, and the content of lipid peroxidation products in the lungs of smokers was higher than in nonsmokers. Multiple regression analysis showed that the smoking index, lipid peroxidation product content, and FEV1/FVC ratio were correlated with the proportion of the 4839-bp mitochondrial DNA deletion in the lung. These results suggest that cigarette smoking increases mitochondrial DNA mutation and lipid peroxidation in the lung tissue. Similarly, Liu (75) reported a higher incidence of the 4977-bp mitochondrial DNA deletion of hair follicles is not associated with aging, but correlated with the amount of cigarette smoking.

*Protein Oxidation Products*. Experimental evidence of oxidative damage to protein and DNA caused by cigarette smoke was obtained by Park et al. (76). Rats exposed to cigarette smoke for 30 days resulted in a significant decrease of GSH, and increase in GSSG protein S-thiolation and 8-oxo-2′-deoxyguanosine in the lung. They also found that the pulmonary oxidative effects of cigarette smoke were greatly potentiated in buthionine sulfoximine-treated rats. The study suggests that the lung is a primary target of cigarette-smoke-induced oxidative damage, and that GSH plays a crucial role in protecting protein and DNA from oxidation caused by cigarette smoking. Compared with the sham-exposed group, the levels of protein carbonyls, however, were not significantly altered in the lung, plasma, or bronchoalveolar lavage fluid of 10-week-old female Sprague–Dawley rats exposed to cigarette smoke daily for 65 weeks (47).

### *3. Pulmonary and Related Dysfunctions*

The assessment of pathophysiological effects produced by environmental tobacco smoke in humans is complicated and is often controversial in epidemiological studies (66).

*Pulmonary Functions*. The respiratory system is the primary target of cigarette smoke exposure. Cigarette smoking is the main risk factor for the development of chronic obstructive pulmonary disease (77,78) and its incidence continues to increase as the population of developed countries ages. Cigarette smoking causes a decline in forced expiratory volume, airway hyperreactivity, eosinophilia and atopy, and smoking cessation is essential for reducing or preventing chronic obstructive pulmonary disease. In addition to chronic obstructive pulmonary disease, regular smoking increases the risk of coronary heart disease and cancers of the upper aerodigestive system and the lung (79). Pulmonary dystrophy in smokers is highly correlated with the extent of generation of oxygen species by activated neutrophils (16).

Cigarette smoking has been implicated as the most frequent factor responsible for the development of chronic obstructive pulmonary disease by leading to oxidant overload in the lower airways. Nowak et al. (80,81) found higher concentrations of TBAR and hydrogen peroxide in the expired breath condensate of patients with chronic obstructive pulmonary disease than in health controls. However, they also observed that current smokers with chronic obstructive pulmonary disease did not exhale significantly more TBAR and hydrogen peroxide than ex-smokers with chronic obstructive pulmonary disease or those who had never smoked. Also, no correlation was found between TBAR levels in the whole chronic obstructive pulmonary disease group. Thus, current cigarette smoking does not distinguish chronic obstructive pulmonary disease subjects with respect to TBAR and hydrogen peroxide exhalation.

In a population-based study, Ness et al. (82) examined the cross-section relationship between respiratory function and plasma vitamin C, and found that vitamin C was positively correlated, after adjustment for age and height, with both $FEV_1$ and FVC in men, but not in women, and suggested that vitamin C may be protective for lung function. Similarly, Schwartz and Weiss (83) assessed the relationship between dietary vitamin C intake and the level of pulmonary function in 2526 adults seen as part of the first National Health and Nutrition Examination Survey, and found that vitamin C intake was positively and significantly associated with the level of $FEV_1$, but interaction terms for vitamin C intake and smoking and vitamin C intake and respiratory disease were not significant. Also, Scherer et al. (84) found that supplementation with vitamin E orally (1000 IU/day) for 3 weeks did not alter pulmonary clearance nor its correlation with expired carbon monoxide, carboxyhemoglobin, urinary cotinine, and lung function parameters of smokers. However, the strong correlations between pulmonary clearance and indices of acute cigarette smoking, which reflect the amount of inhaled smoke and the resultant oxidant stress, do not allow exclusion of the involvement of oxidative stress in the pulmonary clearance increase observed in smokers.

Grievink et al. (84a) found that subjects with higher plasma beta-carotene levels tended to have a higher $FEV_1$ and FVC; however, there was no relation between plasma concentration of alpha-tocopherol and lung function. Theron et al. (85) also showed that cigarette smoking, but not mineral dust exposure, was associated with increased numbers and prooxidative activity of circulating neutrophils and monocytes, decreased concentrations of vitamin C, and pulmonary dysfunction. Liu et al. (37) reported that smokers had lower plasma vitamins C and E levels than nonsmokers older than 45; however, plasma levels of vitamins C and E are not correlated with the measurement of reactive oxidants from circulating phagocyte or spirometric abnormalities in cigarette smokers. Neunteufl et al. (85a) found that oral supplementation of vitamin E (600 IU/day) can attenuate transient impairment of endothelial function after heavy smoking due to an improvement of oxidative status, but cannot restore chronic endothelial dysfunction within 4 weeks in health male smokers.

*Alpha-1-Protease Inhibitor and Chronic Obstructive Pulmonary Disease*. In the early 1960s, the concept of the elastase/antielastase hypothesis was introduced. Neutrophil elastase and alpha-1-antiprotease balance was found to be critical, predisposing patients deficient in alpha-1-antiprotease to panacinar emphysema. However, the causative elastases in common cigarette-smoke-induced pulmonary emphysema remain unclear. The possible association of oxidative inactivation of alpha-1-protease inhibitor (also known as antiprotease, antielastase, or antitrypsin) and the incidence of emphysema in smokers has been extensively studied. Alveolar macrophages in bronchoalveolar lavages obtained from smokers produce more elastase than nonsmokers (86). Lung lavage collected from cigarette-smoked rats has lower elastase-inhibitory capacity and the loss could be reversed by a reducing agent (87). Decreased alpha-1-antiprotease activity by oxidative mechanisms has been suggested as playing a key role in the pathogenesis of obstructive lung disease in smokers (88,89). Cox and Billingsley (90), however, did not find a difference in the plasma-inhibitory activity of alpha-1-antiprotease between smokers and nonsmokers, and suggested that sufficient antioxidant in the plasma may prevent the detection of oxidized inactivated alpha-1-antiprotease.

Ofulue and Ko (91) examined the putative roles of neutrophils and macrophages in the pathogenesis of cigarette-smoking-induced emphysema on the basis of effects of antineutrophil antibody or antimonocyte/macrophage antibody on the development of emphysema in rats exposed to cigarette smoke daily for 2 months. They found that cigarette-smoke exposure induced lung elastin breakdown and emphysema in the lung was not prevented in the lungs of antineutrophil-antibody-treated smoke-exposed rats but was clearly prevented in lungs of the antimonocyte/macrophage-antibody-treated smoke-exposed rats. These findings implicate macrophages, rather than neutrophils, as the critical pathogenic factor in cigarette-smoke-induced emphysema. Ofulue et al. (92) compared the time course of neutrophil and macrophage elastinolytic potentials in the lungs of rats exposed to cigarette smoke daily for up to 6 months. They found that the number of lung neutrophils was increased 1 month after exposure, but was reduced to control levels at months 2 to 6, while increased numbers of lung macrophages were evident in the smoked rats at month 2 and continued to month 6. The elastinolytic activity of lung neutrophils was not altered by cigarette smoking, but that of lung macrophages was increased from month 2 to month 6. Lung elastin breakdown, judged by increased levels of elastin-derived peptides and desmosine in lavage fluid, was increased in the smoked rats at months 2 to 6. These data indicate that increased macrophage-directed elastinolytic activity in the lung, not that of neutrophils, is more closely associated with the evolution of smoking-induced emphysema.

Terashima et al. (93) showed that rabbits exposed to cigarette smoke for 11 days did not increase circulating polymorphonuclear leukocyte counts but had a higher percent of band cells and bromo-deoxyuridine-labeled polymorphonuclear leukocytes in the circulation, and more polymorphonuclear leukocytes were sequestered in the lung than the controls. The results suggested that younger polymorphonuclear leukocytes released from bone marrow by cigarette smoking are preferentially sequestered in pulmonary microvessels, and that these polymorphonuclear leukocytes may contribute to the alveolar wall damage associated with smoke-induced lung emphysema.

Dhami et al. (94) recently created transgenic mice with expression of the human alpha-1-antitrypsin gene directed by a human surfactant protein C promoter fragment or a rat Clara cell 10-kDa protein promoter. Transgene-mediated expression of alpha-1-antitrypsin in pulmonary epithelial cells resulted in diffuse expression of the transgene in the alveolar parenchyma and reproducibly led to transfer of protein to the interstitium. The animal model, when further characterized, should aid in studying the efficacy of alpha-1-antitrypsin in preventing cigarette-smoke-induced lung damage, and the therapeutic potential of increased of alpha-1-antitrypsin expression against chronic obstructive pulmonary disease.

*Interleukins and Nitric Oxide Synthase Expression*. Interleukin-8 is a proinflammatory peptide and a potent chemotactic factor for neutrophils, and is produced by both immune and nonimmune cells, including monocytes and alveolar macrophages. Ohta et al. (95) investigated the effect of cigarette smoking on the secretion of interleukin-8 and found that the interleukin-8 concentration in bronchoalveolar lavage fluid was higher in smokers than in nonsmokers. However, spontaneous interleukin-8 secretion by cultural alveolar macrophages was lower in smokers than in nonsmokers. These observations suggest that decreased interleukin-8 secretion by alveolar macrophages of smokers may modify or decrease the inflammatory response in the lung. Also, Tappia et al. (96) reported that cigarette smokers had higher blood tumor necrosis factor production and plasma interleukin-6 concentration, and lower levels of plasma vitamin C, but not vitamins A and E, than nonsmokers. Smoking-induced elevation of tumor necrosis factor production and plasma interleukin-6 concentration, and compromised antioxidant status, may play a role in the biological mechanisms underlying the pathology associated with smoking.

Nishikawa et al. (97) examined the hypothesis that superoxide radicals mediate infiltration of neutrophils to the airways through nuclear factor-kappa B and interleukin-8 after acute cigarette-smoke exposure. Hartley guinea pigs exposed to cigarette smoke for 20 puffs caused neutrophil accumulation to the airways and parenchyma, increased DNA-binding activity of nuclear factor-kappa B and increased interleukin-8 mRNA expression in the lung. Pretreatment of guinea pigs with recombinant human superoxide dismutase aerosol reduced the cigarette-smoke-induced neutrophil accumulation to the airways, and inhibited the activation of nuclear factor-kappa B and increased interleukin-8 mRNA expression. The results obtained suggest that cigarette-smoke exposure initiates a superoxide-dependent mechanism that, through nuclear factor-kappa B and increased interleukin-8 mRNA expression, produce infiltration of neutrophils to the airways, and that the alveolar macrophage is one of the potential source of nuclear factor-kappa B and increased interleukin-8 mRNA expression following cigarette smoking.

Morimoto et al. (98) investigated the combined effect of cigarette smoke and mineral fiber on the gene expression of cytokine mRNA from alveolar macrophage and lungs. Wistar rats exposed to sidestream smoke 5 days per week for 4 weeks resulted in higher levels of interleukin-1 αmRNA in alveolar macrophages only. Rats instilled intratracheally with chrysotile stimulated the gene expression of interleukin-1α mRNA, interleukin-6, and tumor necrosis factor-alpha in alveolar macrophages, and the expression of basic fibroblast growth factor in lungs. Rats instilled intratracheally with refractory ceramic fiber stimulated the expression of interleukin-1α mRNA and tumor necrosis factor-alpha in alveolar macrophages. Smoked rats treated with chrysotile increased expression of inducible nitric oxide synthase in alveolar macrophages and expression of basic fibroblast growth factor and interleukin-6 in lungs. On the other hand, smoked rats treated with refractory ceramic fiber increased expression of interleukin-1 αmRNA in alveolar macrophages, and basic fibroblast growth factor in lungs. While the mechanism of the stimulating effect of smoking and mineral fiber is not yet clear, altered redox state may play a role.

Rats exposed to cigarette smoke for up to 28 days did not alter nitric oxide synthase-1 gene expression or protein levels, while the levels of nitric oxide synthase-2 expression, but not protein, were more than double at day 1 and below the control after 28 days (99). On the other hand, nitric oxide synthase-3 expression was increased after 2 days of smoke exposure and remained increased to 28 days, and the protein levels were elevated from day 7. The results suggest that cigarette smoke can rapidly affect the expression of nitric oxide synthase, and thus potentially affect the function of pulmonary vasoculature.

*Tumorigenesis.* A/J mice exposed to a mixture of mainstream and sidestream smoke for 5 months and allowed to recover for 4 months in filtered room air had a higher average number of lung tumors than the sham-exposed animals (100). The incidence and multiplicity of papillomas in the upper respiratory track were lower in Syrian hamsters fed the beta-carotene-supplemented diet treated with diethylnitrosamine followed by cigarette-smoke exposure than the unsupplemented group (101).

Studies of relationship between vitamin E status and risk of lung cancer among smokers have yielded conflicting results. To clarify this association, Woodson et al. (101a) examine the prospective relationship between collected serum alpha-tocopherol and lung cancer in the Alpha-Tocopherol Beta-Carotene Cancer Prevention Study cohort and found that higher serum alpha-tocopherol status was associated with lower lung cancer risk, and that the relationship was stronger among younger persons and among those with less cumulative smoke exposure. The authors suggest that higher levels of alpha-tocopherol, if present during the early critical stages of tumorigenesis, may inhibit lung cancer development.

Whether beta-carotene is protective or harmful for smokers is controversial. Intervention trials designed to test the hypothesis of the protective nature of beta-carotene resulted in data suggestive of an increased rather than a decreased risk of cancer with beta-carotene supplementation (102,103). Based on these findings, it is hypothesized that beta-carotene can promote lung carcinogenesis by acting as a prooxidant in the smoke-exposed lung. Baker et al. (104) recently examined the interaction of cigarette smoke and beta-carotene in a model system and found that both whole smoke and gas-phase smoke oxidized beta-carotene and formed several products. A major product of the reaction was identified as 4-nitro-beta-carotene, which was formed by nitrogen oxides in smoke. Wang et al. (105) studied the possible mechanism by which high beta-carotene and cigarette smoking increase the risk of lung cancer among smokers. Ferrets maintained in a beta-carotene-supplemented diet had a strong proliferative response in lung tissue and squamous metaplasia, and this response was enhanced by exposure to cigarette smoke. These animals had three- to four-fold elevated expression of the *c-jun* and *c-fos* genes. The results obtained suggest that diminished retinoid signaling, resulting from the suppression of retinoic acid receptor beta gene expression and overexpression of activator protein-1, could be a mechanism to enhance tumorigenesis after high-dose beta-carotene supplementation and smoke exposure.

Dilsiz et al. (106) showed that cigarette smoking increased the risk of cataract formation as assessed by the alteration of intracellular ionic environment, and supplementation of vitamin C, selenium, or selenium plus vitamin E in the diet prevented the adverse effect.

## IV. FACTORS AFFECTING CIGARETTE-SMOKE-INDUCED OXIDATIVE DAMAGE

As stated previously, cigarette smoke contains many oxidants and free radicals (9,10). Also, increased reactive oxygen species are generated by activated phagocytes in the respiratory system following cigarette smoking (11–16). Results obtained from in vitro studies (25–27) are generally supportive of the idea that increased oxidative damage to cellular components is a consequence of cigarette smoking. However, data obtained from in vivo studies are not conclusive. Nutritional status of experimental animals, types, age of experimental subjects, as well as smoke exposure conditions (Table 1) employed can all contribute significantly to the differential results obtained.

The nutritional status of experimental subjects plays a role in modulating the action and metabolism of chemicals, drugs, and environmental agents. Administration of vitamin E, for example, has been shown to lessen the toxicity of a large variety of environmental agents (5,6). If the oxidative damage mechanism is involved in the toxicity of cigarette smoking, the nutritional status of vitamins C and E should mediate the development of smoking-related disorders, and increased utilization of vitamins C and E is expected to associate with cigarette smoking. Pacht et al. (107) indeed have shown that the levels of vitamin E were significantly lower in alveolar fluid of smokers than in that of nonsmokers. They suggested that smokers had a faster rate of vitamin E utilization and that smoking

**Table 1** Factors Affecting Cellular Susceptibilities to Cigarette Smoking

| |
|---|
| Type of cigarette used |
| Numerous smoke components |
| Highly variable |
| Experimental subjects |
| Species |
| Age |
| Sex |
| Nutritional and health status |
| Exposure conditions |
| Type of filter, if used |
| Smoke frequency (puffs/session; sessions/day; days of exposure/week) |
| Smoke duration (continuous or intermittent) |
| Smoke concentration |
| Mode of smoke (active or passive) |
| Animal restraint device used, if any |
| Type of smoke machine for animals |
| Sham or control exposure employed |
| Samples examined |
| Lung |
| Bronchoalveolar lavage fluid |
| Plasma/serum |
| Blood cells |
| Other organs |

might predispose them to enhanced oxidant attack on their lung parenchymal cells. However, contrary to this expectation, the levels of vitamins C and E in the lungs of chronically smoked animals are either unchanged or higher, rather than lower, than in the controls (45–47). The higher or unaltered levels of vitamins C and E observed in the lungs of chronically smoked guinea pigs and rats suggest an adaptive response that may protect pulmonary tissues against oxidative damage. Since those experimental animals were fed a diet containing an adequate amount of vitamin E, it is possible that higher levels of vitamin E found in the lungs of smoked animals may result from increased tissue uptake, mobilization of body stores, or the combination of both. The view that the uptake of vitamin E may be increased in the lungs is supported by the findings that while decreased levels of vitamin E were not found in the lungs of chronically smoked rats, increased formation of alpha-tocopherylquinone was detected (47). Increased levels of vitamin E have also been found in the blood of human subjects during exercise (108) and in the lungs of ozone-exposed animals (109), and in tissues of polychlorinated biphenyls, chlorobutanol or phenobarbital-treated rats (110,111).

While increased formation of oxidation products in pulmonary tissues of cigarette-smoked animals is not clearly evident (45–47), failure to detect increased formation of oxidation products in the pulmonary tissue of smoked subjects does not indicate that oxidative damage did not occur. A metabolic adaptation of pulmonary tissue following cigarette smoking may explain this seemingly paradoxical view. Higher levels of vitamin E in the lungs of chronically cigarette smoked animals (45,46) suggest that increased vitamin E may result from increased retention from the dietary source or mobilized from other tissues to the lung. Also, the activities of superoxide dismutase, GSH peroxidase, and metabolically related enzymes are either higher or unaltered in the lungs of smoked rats (42,44,47). Similarly, the activities of superoxide dismutase and catalase in the alveolar macrophages of smokers or smoked hamsters are higher than in control subjects (55). Furthermore, smokers have higher pulmonary activity of dithio-diaphorase, which catalyzes the two-electron reduction of quinones to less harmful hydroquinones, than nonsmokers (112). Thus, sustained or enhanced antioxidant defense potential observed in the pulmonary tissue following smoke exposure may enable experimental subjects to metabolite reactive oxygen species generated, to repair the damage occurred, and to resist further damage or limit oxidant-mediated damage to alveolar structures.

Sex and age differences of experimental subjects also result in differential findings. In a population-based study, Ness et al. (82) examined the cross-section relationship between respiratory function and plasma vitamin C, and found that vitamin C was positively correlated, after adjustment for age and height, with both $FEV_1$ and FVC in men, but not in women. Also, Liu et al. (37) showed that smokers had lower plasma vitamin C and E levels and higher malondialdehyde levels than nonsmokers older than age 45, but not younger than age of 35. The results suggest that the antioxidant potential may be overwhelmed by oxidative stress elicited by cigarette smoking in elderly subjects. Similarly, Hulea et al. (57) found that in smokers aged 18 to 45 years, the changes of the plasma lipid peroxidation products, leukocyte activation, thiol concentration, and total antioxidant capacity were not significantly different from those of the controls, whereas in the 46 to 80 age group they were, and that the activities of superoxide dismutase and GSH peroxidase were increased in the red cells of smokers aged 18 to 45, and decreased in the 46 to 80 age group. The findings suggest an adaptive process takes place in younger smokers, but not in older smokers.

## V. SUMMARY AND CONCLUSIONS

Cigarette smoke contains a large variety of compounds, including many oxidants and free radicals that are capable of initiating or promoting oxidative damage. Also, oxidative stress may result from reactive oxygen species generated by the increased and activated phagocytes following cigarette smoking. In vitro studies are generally supportive of the hypothesis that cigarette smoke can initiate or promote oxidative damage. However, information obtained from in vivo studies is inconclusive. This is partly due to the fact that cigarette smoke exerts complex cellular effects, and that factors such as species, sex, age, and nutritional status of experimental subjects, type of cigarette, exposure conditions employed, as well as specific tissues examined can all significantly affect the results obtained. The levels of oxidation products were found to be increased, decreased, or unchanged in the lungs of chronically smoked animals. Metabolic adaptation, such as increased uptake of vitamin E in the lung, and increased activities of antioxidant enzymes in alveolar macrophages and pulmonary tissues of chronically smoked animals may enable smokers to counteract oxidative stress and to resist further damage to smoke exposure. However, it is also possible that the metabolic adaptation may be secondary to the inflammatory response and the injury repair process following smoking exposure. More studies are needed to better understand the role of oxidative damage in the etiology of smoking-related disorders.

## REFERENCES


1. U.S. Department of Public Health Service. Smoking and health: the report of the Surgeon General. U.S. DHEW Publication no. (PHS) 79-50066. Washington, DC: Government Printing Office, 1979.
2. U.S. Department of Health and Human Service. The health consequences of smoking. Chronic obstructive lung disease. A report of the Surgeon General. DHHS publication no. 84–50205. Washington, DC: Government Printing Office, 1984.
3. Hammond EC. Smoking in relation to the death rates of I million men and women. Monograph 19, 127–204. Washington, DC: National Cancer Institute, 1966.
4. Doyle JT, Dawler TR, Kannel WB, Kinch SH, Kahn HA. The relationship of cigarette smoking to coronary heart disease. J Am Med Assoc 1964; 190:886–890.
5. Chow CK. Nutritional influence on cellular antioxidant systems. Am J Clin Nutr 1979; 32: 1061–1083.
6. Chow CK. Vitamin E and oxidative stress. Free Radical Biol Med 1991; 11:215–232.
7. Schumacher JM, Green CR, Best FW, Newell MP. Smoke composition: an extensive investigation of the water-soluble portion of cigarette smoke. J Agric Food Chem 1977; 25:310–320.
8. Sakuma H, Ohsumi T, Sugawara S. Particulate phase of cellulose cigarette smoke. Agric Biol Chem 1980; 44:555–561.
9. Church DF, Pryor WA. Free radical chemistry of cigarette smoke and its toxicological implications. Environ Health Perspect 1985; 64:111–126.
10. Pryor WA, Hales BJ, Premovic PI, Church DF. The radicals in cigarette tar: their nature and suggested physiological implications. Science 1983; 220:425–427.

10a. Panda K, Chattopadhyay R, Ghosh MK, Chattopadhyay DJ, Chatterjee IB. Vitamin C prevents cigarette smoke-induced oxidative damage of proteins and increased proteolysis. Free Rad Biol Med 1999; 27:1064–1079.

11. Bridges RB, Chow CK, Rehm SR. Micronutrients and immune function in smokers. Ann NY Acad Sci 1990; 587:218–231.
12. Hunninghake CW, Crystal RG. Cigarette smoking and lung destruction: accumulation of neutrophils in the lungs of cigarette smokers. Am Rev Respir Dis 1982; 128:833–838.
13. Hoidal JR, Fox RB, LeMarbe PA, Perri R, Repine JE. Altered oxidative metabolic responses in vitro of alveolar macrophages from asymptomatic cigarette smokers. Am Rev Respir Dis 1981; 123:85–89.
14. Hoidal JR, Niewoehner DE. Cigarette-smoke-induced phagocyte recruitment and metabolic alterations in humans and hamsters. Am Rev Respir Dis 1982; 126:548–552.
15. Ludwig PW, Hoidal JR. Alterations in leukocyte oxidative metabolism in cigarette smokers. Am Rev Respir Dis 1982; 126:977–980.
16. Richards GA, Theron AJ, Van der Merwe CA, Anderson R. Spirometric abnormalities in young smokers correlate with increased chemiluminescence responses of activated blood phagocytes. Am Rev Respir Dis 1989; 139:181–187.
17. Bridges RB, Fu MC, Rehm SR. Increased neutrophil myeloperoxidase activity associated with cigarette smoking. Eur J Respir Dis 1985; 67:84–93.
18. Morrison D, Rahman I, Lannan S, MacNec W. Epithelial permeability, inflammation and oxidative stress in the air spaces of smokers. Am J Respir Crit Care Med 1999; 159:473–479.
19. Deliconstantinos G, Villiotou V, Stavrides JC. Scavenging effects of hemoglobin and related heme containing compounds on nitric oxide, reactive oxidants and carcinogenic voltile nitrosocompounds of cigarette smoke. A new method for protection against the dangerous cigarette constituents. Anticancer Res 1994; 14:2717–2726.
20. Chambers DC, Tunnicliffe WS, Ayres JG. Acute inhalation of cigarette smoke increases lower respiratory tract nitric oxide concentrations. Thorax 1998; 53:677–679.

21. Kharitonov SA, Robbins RA, Yates D, Keatings V, Barnes PJ. Acute and chronic effects of cigarette smoking on exhaled nitric oxide. Am J Respir Crit Care Med 1995; 152:609–612.
22. Hasday JD, Bascom R, Costa JJ, Fitzgerald T, Dubin W. Bacterial endotoxin is an active component of cigarette smoke. Chest 1999; 115:829–835.
23. Minamiya Y, Abo S, Kitamura M, Izumi K, Kimura Y, Tozawa K, Saito S. Endotoxin induced hydrogen peroxide production in intact pulmonary circulation of rat. Am J Respir Crit Care Med 1995; 152:348–354.
24. Supinski G, Nethery D, DoMarco A. Effect of free radical scavengers on endotoxin-induced respiratory muscle dysfunction. Am Rev Respir Dis 1993; 148:1318–1324.
25. Lentz PE, Di Luzio NRD. Peroxidation of lipids in alveolar macrophages: production by aqueous extracts of cigarette smoke. Arch Environ Health 1974; 28:279–282.
26. Fenner ML, Braven J. The mechanism of carcinogenesis by tobacco smoke: further experimental evidence and a prediction from the thio-defense hypothesis. Br J Cancer 1968; 22: 474–479.
27. Harats D, Naim M, Dabach Y, Hollander G, Stein O, Stein Y. Cigarette smoking renders LDL susceptible to peroxidative modification and enhanced metabolism by macrophages. Atherosclerosis 1989; 79:245–252.
28. Schwalb G, Anderson R. Increased frequency of oxidant-mediated DNA strand breaks in mononuclear leukocytes exposed to activated neutrophils from cigarette smokers. Nutr Res 1989; 225:95–99.
29. Kamp DW, Greenberger MJ, Sbalchierro JS, Preusen SE, Weitzman SA. Cigarette smoke augments asbestos-induced alveolar epithelial cell injury: role of free radicals. Free Rad Biol Med 1998; 25:728–739.
30. Yoshie Y, Ohshima H. Synergistic induction of DNA strand breakage by cigarette tar and nitric oxide. Carcinogenesis 1997; 18:1359–1363.
31. Gouaze V, Dousset N, Dousset JC, Valdiguie P. Effect of nicotine on the susceptibility to in vitro oxidation of LDL in healthy non-smokers and smokers. Clin Chim Acta 1998; 277: 25–37.
32. Srivastava ED, Hallett MB, Rhodes J. Effect of nicotine and cotinine on the production of oxygen free radicals by neutrophils in smokers and non-smokers. Human Toxicol 1989; 8: 461–463.
33. Chow CK, Thacker R, Bridges RB, Rehm SR, Humble J, Turbek J. Lower levels of vitamin C and carotenes in plasma of cigarette smokers. J Am Coll Nutr 1986; 5:305–312.
34. Pelletier O. Vitamin C status of cigarette smokers and nonsmokers. Am J Clin Nutr 1970; 23:520–528.
35. Pelletier O. Smoking and vitamin C levels in humans. Am J Clin Nutr 1969; 21:1259–1267.
36. Banerjee KK, Marimuthu P, Sarkar A, Chauduri RN. Influence of cigarette smoking on vitamin C, glutathione and lipid peroxidation status. Ind J Publ Health 1998; 42:20–23.
37. Liu CS, Chen HW, Lii CK, Chen SC, Wei YH. Alterations of small molecular weight antioxidants in the blood of smokers. Chem Biol Interact 1998; 116:143–154.
38. Theron AJ, Richards GA, Van Rensburg AJ, Van der Merwe CA, Anderson R. Investigation of the role of phagocytes and antioxidant nutrients in oxidant stress mediated by cigarette smoke. Int J Vitam Nutr Res 1990; 60:261–266.
39. Barton GM, Roath OS. Leukocyte ascorbic acid in abnormal leukocyte states. Int J Vitam Nutr Res 1976; 46:271–274.
40. Hemila H, Roberts P, Wikstrom M. Activated polymorphonuclear leukocytes consume vitamin C. FEBS Lett 1984; 178:25–30.
41. Wayner DD, Burton GW, Ingold KW, Barclay LRC, Locke SJ. The relative contributions of vitamin E, urate, ascorbate and protein to the total peroxy radical-trapping antioxidant activity of human blood plasma. Biochim Biophys Acta 1987; 924:408–419.
42. Chow CK, Chen LH, Thacker RR, Griffith RB. Dietary vitamin E and pulmonary biochemical responses of rats to cigarette smoking. Environ Res 1984; 34:8–17.

43. Chen LH, Chow CK. Effect of cigarette smoking and dietary vitamin E on plasma levels of vitamin C in rats. Nutr Rep Int 1980; 22:301–309.
43a. Dallongeville J, Marceaux N, Fruchart JC, Amouyel P. Cigarette smoking is associated with unhealthy patterns of nutrient intake: meta analysis. J Nutr 1998; 128:1450–1457.
43b. Lykkesfeldt J, Christen S, Wallock LM, Chang HH, Jacob RA, Ames BN. Ascorbate is depleted by smoking and repleted by moderate supplementation: a study in male smokers and nonsmokers with matched dietary antioxidant intakes. Am J Clin Nutr 2000; 71:530–536.
44. Chow CK. Dietary vitamin E and cellular susceptibility to cigarette smoking. Ann NY Acad Sci 1982; 393:426–436.
45. Airriess G, Changchit C, Chen LC, Chow CK. Increased levels of vitamin E in the lungs of guinea pigs exposed to mainstream or sidestream smoke. Nutr Res 1988; 8:653–661.
46. Chow CK, Airriess GR, Changchit C. Increased vitamin E content in the lungs of chronic cigarette-smoked rats. Ann NY Acad Sci 1989; 570:425–427.
47. Wuzel H, Yeh CC, Gairola C, Chow CK. Oxidative damage and antioxidant status in the lungs and bronchoalveolar lavage fluid of rats exposed chronically to cigarette smoke. J Biochem Toxicol 1995; 10:1–7.
48. Boyland E, Grove PL. Stimulation of ascorbic acid synthesis and excretion by carcinogenic and other foreign compounds. Biochem J 1961; 81:163–168.
49. Conney AH, Burns JJ. Stimulatory effects of foreign compounds on ascorbic acid biosynthesis and on drug-metabolizing enzymes. Nature 1959; 184:363–364.
50. Van Rensburg CEJ, Theron AJ, Richards GA, Van der Merwe CA, Anderson R. Investigation of the relationships between plasma levels of ascorbate, vitamin E and beta-carotene and the frequency of sister-chromatid exchanges and release of reactive oxidants by blood leukocytes from cigarette smokers. Mutat Res 1989; 215:167–172.
51. Gvozdjakova A, Simko F, Kucharska J, Braunova Z, Psenck P, Kyselovic J. Captopri increased mitochondrial coenzyme Q10 level, improved respiratory respiratory chain function and energy production in the left ventricle in rabbits with smoke mitochondrial cardiomyopathy Biofactor 1999; 10:61–65.
52. Cotgreave IA, Johansson U, Moldeus P, Brattsand R. Effect of acute cigarette smoke inhalation on pulmonary and systemic cysteine and glutathione redox states in the rat. Toxicology 1987; 45:203–212.
52a. Zappacosta B, Perschilli S, De Sole P, Mordente A, Giardina B. Effect of smoking one cigarette on antioxidant metabolites in the saliva of healthy smokers. Arch Oral Biol 1999; 44:485–488.
53. York GK, Pierce TH, Schwartz LS, Cross CE. Stimulation by cigarette smoke of glutathione peroxidase system enzyme activities in rat lung. Arch Environ Health 1976; 31:286–290.
54. McCuster K, Hoidal J. Selective increase of antioxidant enzyme activity in the alveolar macrophages from cigarette smokers and smoke-exposed hamsters. Am Rev Respir Dis 1990; 141:678–682.
55. Changchit C, Chow CK. Unpublished results.
56. Hulea SA, Olinescu R, Nita S, Croenan D, Kummerow FA. Cigarette smoking causes biochemical changes in blood that are suggestive of oxidative stress: a case-control study. J Environ Pathol Toxicol Oncol 1995; 14:173–180.
57. Hilbert J, Mohsenin V. Adaption of lung antioxidant to cigarette smoking in humans. Chest 1996; 110:916–920.
58. Hoshino E, Shariff R, Van Gossum A, Allard JP, Pichard C, Kurian R, Jeejeebhoy KN. Vitamin E suppresses increased lipid peroxidation in cigarette smokers. J Parenter Enteral Nutr 1990; 14:300–305.
58a. Habib MP, Tank LJ, Lane LC, Garewal HS. Effect of vitamin E on exhaled ethane in cigarette smokers. Chest 1999; 115:684–690.
59. Duthie GG, Arthur IR, James WP, Vint HM. Antioxidant status of smokers and nonsmokers. Effects of vitamin E supplementation. Ann NY Acad Sci 1989; 570:435–438.

60. Surmen-Gur E, Ozturk E, Gur H, Punduk Z, Tuncel P. Effect of vitamin E supplementation on post-exercise plasma lipid peroxidation and blood antioxidant status in smokers: with special reference to haemoconcentration effect. Eur J Appl Physiol 1999; 79:472–478.
61. Kawakami M, Kameyama S, Takizawa T. Lipid peroxidation in bronchoalveolar lavage fluid in interestitial lung diseases in relation to other components and smoking. Nippon Kyobu Shikkan Gakkai Zasshi 1989; 27:422–427.
62. Niewoehner DE, Peterson FI, Hoidal JR. Selenium and vitamin E deficiencies do not enhance lung inflammation from cigarette smoke in the hamster. Am Rev Respir Dis 1983; 127:227–230.
63. Watanabe K, Eto K, Furuno K, Mori T, Kawasaki H, Gomita Y. Effect of cigarette smoke on lipid peroxidation and liver function tests in rats. Acta Med Okayama 1995; 49:271–274.
64. Asami S, Manabe H, Miyake J, Tsurudome Y, Hirano T, Yamaguchi R, Itoh H, Kasai H. Cigarette smoking induces an increase in oxidative DNA damage, 8-hydroxydeoxyguanosine, in a central site of the human lung. Carcinogenesis 1997; 18:1763–1766.
65. Kiyosawa H, Suka M, Okudaira H, Murata K, Miyamoto T, Chung HH, Kasai H, Nishimura S. Cigarette smoking induces formation of 8-hydroxydeoxyguanosine, one of the oxidative products of DNA damage in human peripheral leukocytes. Free Rad Res Commun 1990; 11:1–3.
66. Izzotti A, Bagnasco M, D'Agostini F, Cartiglia C, Lubet RA, Kelloff GL, De Flora S. Formation and persistence of nucleotide alterations in rats exposed whole body to environmental cigarette smoke. Cracinogenesis 1999; 20:1499–1505.
67. Gupta RC, Arif M, Gairola CG. Enhancement of pre-existing DNA adducts in rodents exposed to cigarette smoke. Mutat Res 1999; 424:195–205.
68. Wang Y, Ichiba M, Oishi H, Iyadomi M, Shono N, Tomokuni K. Relationship between plasma concentrations of beta-carotene and alpha-tocopherol and life-style factors and levels of DNA adducts in lymphocytes. Nutr Cancer 1997; 27:69–73.
68a. Lee BM, Lee SK, Kim HS. Inhibition of oxidative DNA damage, 8OHdG, and carbonyl contents in smokers treated with antioxidants (vitamin E, vitamin C, beta-carotene and red ginseng). Cancer Lett 1998; 132:219–227.
69. van Zeeland AA, de Groot AJ, Hall J, Donato F. 8-Hydroxydeoxyguanosine in DNA from leukocytes of health adults: relationship with cigarette smoking environmental tobacco smoke, alcohol and coffee consumption. Mutat Res 1999; 439:249–257.
70. Cullinane C, Bohr VA. DNA inter-strand cross-link induced by psoralen are not repaired in mammalian mitochondria. Cancer Res 1998; 58:1400–1404.
71. Miquel J. An update on the oxygen stress-mitochondrial mutation theory of aging: genetic and evolutionary implication. Exp Gerontol 1998; 33:113–126.
72. Richter C, Park J-W, Ames BN. Normal oxidative damage to mitochondrial and nuclear DNA is extensive. Proc Natl Acad Sci USA 1988; 85:6465–6467.
73. Ballinger SW, Bouder TG, Davis GS, Judice SA, Nicklas JA, Albertini RJ. Mitochondrial genome damage associated with cigarette smoking. Cancer Res 1996; 56:5692–5697.
74. Fahn HJ, Wang LS, Kao SH, Chang SC, Huang MH, Wei YH. Smoking-associated mitochondrial DNA mutations and lipid peroxidation in human lung tissues. Am J Respir Cell Mol Biol 1998; 19:901–909.
75. Liu CS, Kao SH, Wei YH. Smoking-associated mitochondrial DNA mutation in human hair follicles. Environ Mol Mutagen 1997; 30:47–55.
76. Park EM, Park YM, Gwak YS. Oxidative damage in tissues of rats exposed to cigarette smoke. Free Rad Biol Med 1998; 25:70–86.
77. George RB. Course and prognosis of chronic obstructive pulmonary disease. Am J Med Sci 1999; 318:103–106.
78. Markewitz BA, Owens MW, Payne DK. The pathogenesis of chronic obstructive pulmonary disease. Am J Med Sci 1999; 318:74–78.
79. Iribarren C, Takawa IS, Sidney S, Friedman GD. Effect of cigar smoking on the risk of cardiovascular disease, chronic obstructive pulmonary disease, and cancer in men. N Engl J Med 1999; 340:1773–1780.

80. Nowak D, Kasielski M, Antezak A, Pietras T, Bialasiewiez P. Increased content of thiobarbituric acid reactive substances and hydrogen peroxide in the expired breath condensate of patients with stable chronic obstructive pulmonary disease: no significant effect of cigarette smoking. Respir Med 1999; 93:389–396.
81. Nowak D, Kasielski M, Pietras T, Bialasiewiez P, Antezak A. Cigarette smoking does not increase hydrogen peroxide levels in expired breath condensate of patients with COPD. Monaldi Arch Chest Dis 1998; 53:268–273.
82. Ness AR, Khaw KT, Bingham NE Day. Vitamin C status and resoiratory function. Eur J Clin Nutr 1996; 50:573–579.
83. Schwartz J, Weiss ST. Relationship between dietary vitamin C intake and pulmonary function in the First National Health and Nutrition Examination Survey (NHANES I). Am J Clin Nutr 1994; 59:110–114.
84. Scherrer-Crosbie M, Paul M, Meignan M, Dahan E, Lagrue G, Atlan G, Lorino AM. Pulmonary clearance and lung function: influence of acute tobacco intoxification and vitamin E. J Appl Physiol 1996; 81:1071–1077.
84a. Grievink L, Smit HA, Veer P, Brunekreef B, Kromhout D. Plasma concentrations of the antioxidants beta-carotene and alpha-tocopherol in relation to lung function. Eur J Clin Nutr 1999; 53:813–817.
85. Theron AJ, Richards GA, Myer MS, van Antwerpen VL, Sluis-Cremer GK, Wolmarans L, van der Merwe CA, Anderson R. Investigation of the relative contributions of cigarette smoking and mineral exposure to activation of circulating phagocytes, alterations in plasma concentrations of vitamin C, vitamin E, and beta-carotene, and pulmonary dysfunction in South African gold miners. Occup Environ Med 1994; 51:564–567.
85a. Neunteufl T, Priglinger U, Heher S, Zehetgruber M, Soregi G, Lehr S, Huber K, Maurer G, Weidinger F, Kostner K. Effects of vitamin E on chronic and acute endothelial dysfunction in smokers. J Am Coll Cardiol 2000; 35:277–283.
86. Rodriquez RJ, White RR, Senior RM, Levine EA. Elastase release from human alveolar macrophages: comparison between smokers and nonsmokers. Science 1977; 198:313–314.
87. Janoff A, Carp H, Lee DK. Cigarette smoking induces functional antiprotease deficiency in the lower respiratory tract of humans. Science 1979; 206:1313–1314.
88. Carp H, Janoff A. Inactivation of bronchial mucous protease inhibitor by cigarette smoke and phagocyte-derived oxidants. Exp Lung Res 1980; 1:225–237.
89. Carp H, Janoff A. Possible mechanisms of emphysema in smokers: in vitro suppression of serum elastase inhibitory capacity by fresh cigarette smoke and its prevention by antioxidants. Am Rev Respir Dis 1978; 118:617–621.
90. Cox AW, Billingsley GD. Oxidation of plasma alpha-1-antitrypsin in smokers and nonsmokers and by an oxidizing agent. Am Rev Respir Dis 1984; 594–599.
91. Ofulue AF, Ko M. Effects of depletion of neutrophils or macrophages on development of cigarette smoke-induced emphysema. Am J Physiol 1999; 277:L97–L105.
92. Ofulue AF, Ko M, Abboud RT. Time course of neutrophil and macrophage elastinolytic activities in cigarette smoke-induced emphysema. Am J Physiol 1999; 275:L1134–L1144.
93. Terashima T, Klut ME, English D, Hards J, Hogg JC, van Ecden SF. Cigarette smoking causes sequestration of polymorphonuclear leukocytes released from the bone marrow in lung microvessels. Am J Respir Cell Mol Biol 1999; 20:171–177.
94. Dhami R, Zay K, Gilks B, Porter S, Wright JL, Churg A. Pulmonary epithelial expression of human alphal-antitrypsin in transgenic mice results in delivery of alphal-antitrypsin protein to the interstitium. J Mol Med 1999; 77:377–385.
95. Ohta T, Yamashita N, Maruyama M, Sugiyama E, Kobayashi M. Cigarette smoking decreases interlukin-8 secretion by human alveolar macrophages. Respir Med 1998; 92:922–927.
96. Tappia PS, Troughton KL, Langeley-Evans SC, Grimble RF. Cigarette smoking influences cytokine production and antioxidant defense. Clin Sci Colch 1995; 88:485–489.

97. Nishiwawa M, Kakemizu N, Ito T, Kudo M, Kaneko T, Suzuki M, Udaka N, Ikeda H, Okubo T. Superoxide mediates cigarette smoke-induced infiltration of neutrophils into the airway through nuclear factor-kappaB activation and IL-8 mRNA expression in guinea pigs in vivo. Am J Respir Cell Mol Biol 1999; 20:189–198.
98. Morimoto Y, Tsuda T, Hori H, Yamato H, Ohgami A, Higashi T, Nagata N, Kido M, Tanaka T. Combined effect of cigarette smoke and mineral fibers on the gene expression of cytokine mRNA. Environ Health Perspect 1999; 107:495–500.
99. Wright JL, Dai J, Zay K, Price K, Gilks CB, Churg A. Effects of cigarette smoking on nitric oxide synthase expression in the rat lung. Lab Invest 1999; 79:975–983.
100. Witschi H, Espiriyu I, Uyeminami D. Chemoprevention of tobacco smoke-induced lung tumors in A/J strain mice with dietary myo-inositol and dexamethasone. Carcinogenesis 1999; 20:1375–1378, 1999.
101. Furukawa F, Nishikawa A, Kasahara K, Lee IS, Wakabayashi K, Takahashi M, Hirose M. Inhibition by beta-carotene of upper respiratory tumorigenesis in hamsters receiving diethylnitrosamine followed by cigarette smoke exposure. Jpn J Cancer Res 1999; 90:154–161.
101a. Woodson K, Tangrea JA, Barrett MJ, Virtamo J, Taylor PR, Albanes D. Serum alpha-tocopherol and subsequent risk of lung cancer among male smokers. J Natl Cancer Inst 1999; 91: 1738–1743.
102. Alpha Tocopherol Beta Carotene Cancer Prevention Study Group. The effect of vitamin E and beta carotene on the incidence of lung cancer and other cancers in male smokers. N Engl J Med 1994; 330:1029–1035.
103. Omenn GS, Goodman GE, Thornquist MD, Cullen MR, Glass A, Sammar S. Effects of a combination of beta carotene and vitamin A on lung cancer and cardiovascular disease. N Engl J Med 1966; 334:1150–1155.
104. Barker DL, Krol ES, Jacobsen N, Liebler DC. Reactions of beta-carotene with cigarette smoke oxidants. Identification of carotenoid oxidation products and evaluation of the prooxidant/antioxidant effect. Chem Res Toxicol 1999; 12:535–543.
105. Wang XD, Liu C, Bronson RT, Smith DE, Krinsky NI, Russell M. Retinoid signaling and activator protein-1 expression in ferrets given beta-carotene supplements and exposed to tobacco smoke. J Natl Cancer Inst 1999; 91:60–66.
106. Dilsiz N, Olecucu A, Cay M, Naziroglu M, Cobanoglu D. Protective effects of selenium, vitamin C and vitamin E against oxidative stress of cigarette smoke in rats. Cell Biochem Funct 1999; 17:1–7.
107. Pacht ER, Kaseki H, Mohammed JR, Cornwell DG, Davis WR. Deficiency of vitamin E in the alveolar fluid of cigarette smokers. Influence on alveolar macrophage cytotoxicity. J Clin Invest 1988; 77:789–796.
108. Pincemail J, Deby C, Camus G, Pirnay F, Bouchez R, Massaux L, Goutier R. Tocopherol mobilization during intensive exercise. Eur J Appl Physiol 1988; 57:189–191.
109. Elsayed NM, Mustafa MG, Mead JF. Increased vitamin E content in the lung after ozone exposure: a possible mobilization in response to oxidative stress. Arch Biochem Biophys 1990; 282:263–269.
110. Katayama T, Momota Y, Watanabe Y, Kato N. Elevated concentrations of α-tocopherol, ascorbic acid, and serum lipids in rats fed polychlorinated biphenyls, chlorobutanol, or phenobarbital. J Nutr Biochem 1991; 2:92–96.
111. Kato N, Momota Y, Kusuhara T. Changes in distributions of α-tocopherol and cholesterol in serum lipoproteins and tissues of rats by dietary PCB and dietary level of protein. J Nutr Sci Vitaminol 1989; 35:655–660.
112. Schlager JJ, Powis G. Cytosolic NAD(P)H:(quinone-acceptor)oxidreductase in human normal and tumor tissue: effects of cigarette smoking and alcohol. Int J Cancer 1990;45:403–409.

# 8

# Health Effects of Diesel Exhaust and Diesel Exhaust Particles

**MASARU SAGAI**

*Aomori University for Health and Welfare, Aomori City, Japan*

**TAKAMICHI ICHINOSE**

*Ohita University of Nursing and Health Sciences, Ohita-Notsuhara, Japan*

## I. INTRODUCTION

In recent years, there has been a progressive increase in urban air pollution, characterized by high concentrations of atmospheric suspended particulate matter (SPM). This is primarily a result of a steady increase in the number of automobiles worldwide, particularly diesel-engine-powered cars. Although the mechanisms of underling respiratory morbidity due to SPM are not clear, it is thought that the fine particles (PM 2.5) in SPM are of greatest concern to health, since they are breathed into the lungs most deeply and contain large amounts of toxic compounds. On the other hand, coarse particles cannot be breathed deeply into the lungs and do not contain so many toxic compounds because the particles are composed mainly of inert soil dust.

Furthermore, many epidemiological studies have reported that there was a clear association between episodes of PM 2.5 and impaired lung function, coughing, infections of the lower respiratory tract, and respiratory symptoms in asthmatics (1–3). Recent studies have also reported very high associations between the atmospheric concentrations of PM 2.5 and daily mortality rates (4–6). Important parts of PM 2.5 in big cities in Japan consist of diesel exhaust particles (DEP).

It is well known that DEP causes lung cancers and allergic rhinitis in experimental animals (7), but it is not known whether it can cause asthma. Some epidemiological studies, however, report a clear association between air pollution caused by diesel exhaust and hospital admissions for asthma, an increase in allergic respiratory diseases, and chronic respiratory symptoms (8–11).

Recently, we have found out that DEP produces large amounts of reactive oxygen in the lungs (12–14), causes lung damage (15), and that DEP or diesel exhaust (DE) induces lung cancer (16–18) and asthmalike features in experimental animals via the production of reactive oxygen (14,19–21). Others have also reported that reactive oxygen is involved in asthmalike symptoms by using experimental animals. In this chapter, we will introduce their results and a possible mechanism of DEP toxicity, lung cancer, and asthmalike symptoms by DEP or DE via oxidative stress.

## II. CHEMICAL COMPOSITIONS IN DE AND DEP

Diesel engine emissions are highly complex mixtures, consisting mainly of gaseous and particulate phases. The gaseous phase contains irritant and nonirritant gases such as NOx, SOx, and COx, as well as hydrocarbons such as alcohol, aldehyde, and ketone-type derivatives. The particulate phase is diesel exhaust particles, which are very small (90% are less than 1 μm by mass), and respirable. DEP are composed of elemental carbon and organic compounds such as polyaromatic hydrocarbons (PAHs), nitroarene, and dioxine derivatives (7,22–25). DEP contains $SO_4^-$, $NO_3^-$, and their ammonium salts (7). Large amounts of light metals and trace amounts of heavy metals are also contained in DEP. Concentrations of the light metals in our DEP were 2710, 944, 374, 228, 204, and 173 ppm for Na, Ca, K, Zn, Mg, and Al, respectively. Those of the heavy metals were 126, 12, and 6 ppm for iron, copper, and chromium, respectively.

We found that DEP produced reactive oxygen species by redox reaction without any biological stimulating factors in vitro (12). ESR signals obtained in this system suggest the existence of superoxide ($O_2^-$) and hydroxyl radical (·OH). The trace amount of transient heavy metals seems to be sufficient to catalyze the formation of ·OH by the Fenton reaction. We tried to separate organic compounds from DEP by stepwise extraction with *n*-hexane, benzene, and methanol. The composition ratios (w/w) were 28%, 12%, 25%, and 31% for *n*-hexane, benzene, methanol, and residual soot, respectively. On the other hand, superoxide production ratios were 9%, 20%, 58%, and 14% for *n*-hexane, benzene, methanol, and residual soot, respectively. Methanol and benzene fractions showed high activity of superoxide production. This production might be induced by redox cycling reaction of quinonelike compounds (26–28). It is considered that *n*-hexane layer contains long-chain aliphatic hydrocarbons with carbon numbers from 6 to 26 anthraquinonelike compounds. Benzene fraction contains polyaromatic hydrocarbons (PAHs) and nitro-PAHs, which are very well-known mutagens and carcinogens. Methanol fraction contains hydrophobic compounds such as alcohol, aldehyde, ketone, and quinone-type derivatives (13,22–25).

## III. TOXICITY OF DIESEL EXHAUST PARTICLES (DEP) VIA REACTIVE OXYGEN SPECIES

### A. Mortality of Mice by DEP and Suppression by SOD

We examined the mortality of mice to estimate the toxicity of DEP. DEP was instilled intratracheally. The death rate of ICR-strain mice increased with the concentration of DEP (12), as shown in Figure 1; 0.9 mg of DEP caused death to all the mice. $LC_{50}$ was 0.6 mg/mouse.

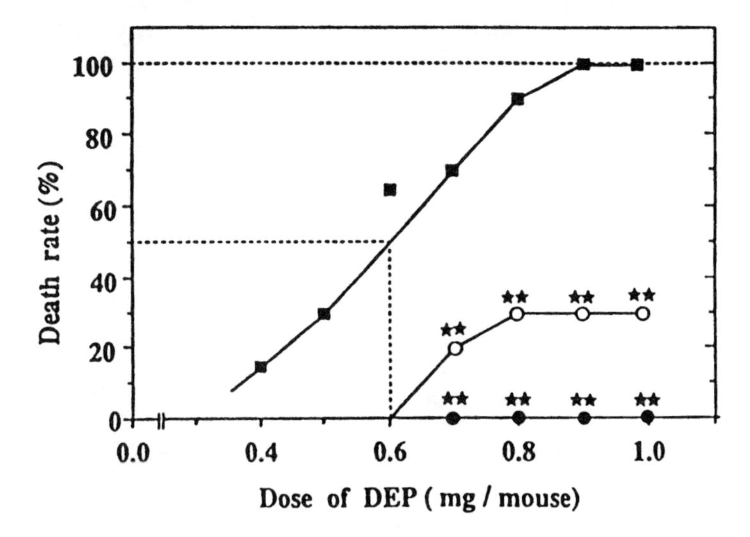


**Figure 1** Death rate of ICR mice intratracheally instilled with DEP and the depression of death rate by PEG-SOD (12). ■, DEP; ●, methanol-washed DEP; ○, DEP + PEG-SOD (2 mg per a mouse from tail vein). Each mouse was used in each dose group of DEP. The death rate of mice pretreated with PEG-SOD showed statistically significant decreases from the rate of mice instilled with DEP. ** $p < 0.001$.

The cause of death was lung edema due to pulmonary vascular endothelial cell damage (15). Endothelial cells are generally sensitive to reactive oxygen species (29). Furthermore, an intravenous injection of PEG-SOD into a tail vein of the mice 1 h before DEP instillation depressed the death rate from 100% to 30% (Fig. 1). Furthermore, intratracheal instillation of PEG-SOD did not cause the death of any mice. This suggests that DEP produces a great deal of superoxide in the lungs, and that PEG-SOD inhibits $O_2^-$ toxicity in pulmonary endothelial or epithelial cells. Residual soot extracted by methanol also did not cause any deaths (Fig. 1). BHT injected intraperitoneally on the third and second days prior to DEP instillation significantly suppressed the death rate and intraperitoneal injection of desferal or EDTA, which are metal chelating agents, as also effective. This evidence suggests that the organic compounds and metals extracted by methanol may be sources of $O_2^-$ and ·OH productions and the reactive oxygen species produced from the organic compounds may be toxic to the endothelial and epithelial cells in lungs to cause lung edema.

The ICR mouse was the most resistant strain to the toxicity of DEP. The order of resistancy to DEP was as follows: ICR > ddy > BDF1 > CDF1 > BALB/c > DBA/2 > C57BL/6 > C3H/He. $LC_{50}$ of the C3H/He mouse was 0.15 mg, and 100% lethal concentration was 0.2 mg (Fig. 2). We had previously reported the concentration of glutathione (GSH) and the activities of glucose-6-phosphate dehydrogenase (G6PD), 6-phosphogluconate dehydrogenase (6PGD), glutathione reductase (GR), glutathione peroxidase (GPx), glutathione transferase, CuZn-SOD, and sulfide reductase in lungs of ICR, BALB/c, ddy, and C57BL/6 strain mice (30). The order of the GSH concentration and the enzyme activities was C57BL/6 > BALB/c > ddy > ICR. This order was correlated inversely with the order of mortality by the exposure to a lethal dose of $NO_2$. The results suggest that antioxidative ability may be important to protect edematous damage by $NO_2$. The

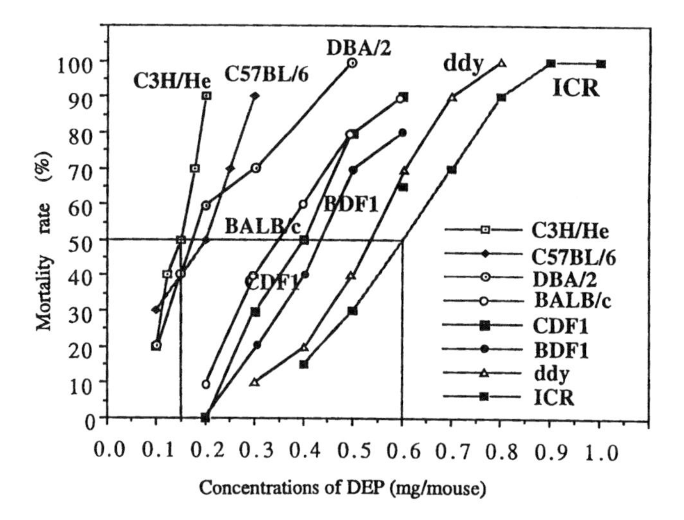


**Figure 2** Comparisons of mortality rate and $LC_{50}$ of 8 strain mice instilled once intratracheally with DEP.

order of mortality by DEP, however, correlated positively with the antioxidative protective ability. This appears to be contradictory because we mentioned earlier that the toxicity of DEP was due to reactive oxygen species.

The activity of antioxidative protective enzymes in the lungs of mice instilled once intratracheally with DEP was examined (12). The activities of reactive oxygen scavenging enzymes such as SOD and GPx were depressed significantly. On the other hand, the activities of GSH recycling and NADPH-producing enzymes such as GR, G6PD, and 6-PGD did not change or increased slightly as G6PD. This result shows that DEP instillation decreases oxidant scavenging ability in the lungs. Therefore, the protection ability of DEP toxicity may not be due to the latent activity of antioxidative ability, but due to remaining activity of the ability. The correct mechanism remains to be made clear.

## B. Production of Reactive Oxygen Species from DEP under the Presence of Lung Enzymes

Until now, we have assumed that the production of reactive oxygen species from DEP is due to chemical reaction alone without any biochemical activation (12). This assumption has emanated indirectly from evidence on the suppression of the mortality rate of mice through PEG-SOD pretreatment.

In the present study, the involvement of lung enzymes in reactive oxygen production by DEP was shown by enzymatic reaction and ESR. Under the presence of NADPH, cytochrome P-450 reductase (CYP-450 reductase), and methanol extracts of DEP, $O_2^-$ production increased 260 times higher than in the nonenzymatic reaction system (13). The ESR signal obtained in this system clearly showed the production of $O_2^-$ and ·OH, and SOD suppressed completely the $O_2^-$ signal, but not catalase (Fig. 3). In the assay of enzyme activity in lungs, enzyme activity in microsome (Ms), mitochondria (Mt), cytosol, and whole homogenate fractions was examined. Specific activity and total activity were the highest in Ms and cytosol fractions, respectively.

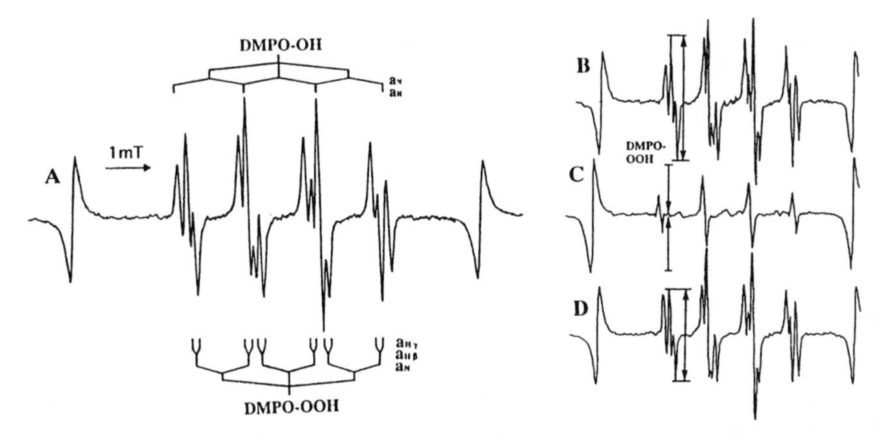


**Figure 3** ESR spectra of superoxide (DMPO-OOH) and hydroxyl radical (DMPO-OH) (A, B), and the effects of superoxide dismutase (C) and catalase (D) on DMPO-OOH spectra (13).

Kappus (26) has suggested that CYP-450 reductase in microsome and nuclear membrane, NADH dehydrogenase and NADPH nitro-reductase in mitochondria, and xanthine oxidase, GSH reductase (GR), and ferredoxin reductase in cytosol fraction could produce $O_2^-$ during metabolism of the substrates such as anthraquinone anticancer drugs (27,28), aromatic-nitro compounds, paraquat, and aldehyde. $O_2^-$ production from DEP may be due to the enzymes in Ms, Mt, nuclear, and cytosol fractions in lungs.

## C. Examination of the Superoxide-Producing Fraction of DEP

Next, we examined which fractions in DEP are responsive to producing $O_2^-$ under the presence of lung enzymes. DEP was extracted stepwise with *n*-hexane, benzene, dichloromethane, ethyl ether, and methanol. Then the ability of $O_2^-$ production in each fraction was examined in the presence of NADPH and CYP-450 reductase as a representative enzyme well known to produce $O_2^-$ using xenobiotics as substrates. The percentage of $O_2^-$ production per unit weight of DEP was 22%, 23%, 10%, 13%, and 32% for *n*-hexane, benzene, dichloromethane, ethylether, and methanol layers, respectively (13). $O_2^-$ was widely produced from several fractions, although methanol fraction was the major. This suggests that quinonelike compounds and other hydrophilic organic compounds in DEP may be substrate in the presence of biological activation systems such as NADPH, CYP-450 reductase, and others. A possible mechanism of $O_2^-$ production is proposed in Figure 4.

## D. DEP Toxicity in the Human Pulmonary Endothelial Cell Line

DEP has been proved to induce pulmonary edema mediated by vascular endothelial cell damage. The cytotoxic mechanism of DEP on human pulmonary artery endothelial cells was investigated by focusing on the role of reactive oxygen species (31).

DEP extracts damaged endothelial cells under both subconfluent and confluent conditions. The extract-induced cytotoxicity was markedly reduced by treatment with SOD, catalase, MPG [N-(2-mercaptopropionyl)-glycine, a hydroxyl radical scavenger], and ebselen (a selenium-containing compound with glutathione peroxidaselike activity). Thus at least, $O_2^-$, $H_2O_2$, and $\cdot OH$ are likely to be implicated in DEP extract-induced endothelial

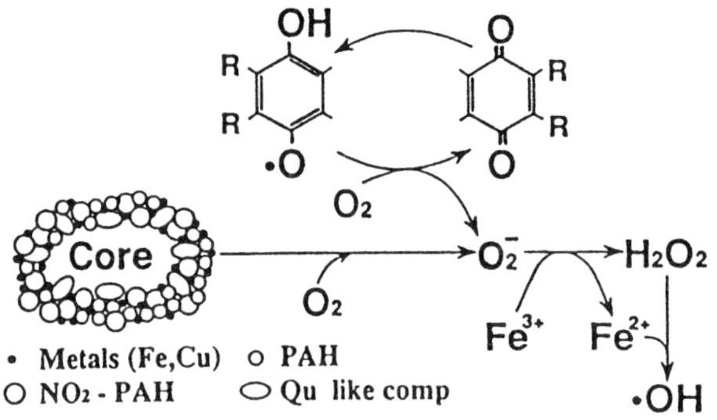

**Figure 4** Possible mechanism on the production of active oxygens such as superoxide ($O_2^-$) and hydroxyl radical (·OH) from diesel exhaust particles (DEP).

cell damage. It was made clear that benzene fraction in DEP extracts was the most toxic in this assay system. Moreover, L-NAME and L-NMA, inhibitors of NO synthase, also attenuated DEP extract-induced cytotoxicity, while sepiaputerin, the precursor of tetrahydrobiopterin (BH4, a NO synthase cofactor), interestingly enhanced DEP extract-induced cell damage. These findings suggest that NO is also involved in DEP extract-mediated cytotoxicity. These reactive oxygen species, including peroxynitrite ($ONOO^-$), are a clue to explaining the mechanism of endothelial cell damage upon DEP exposure, as shown in Figure 5.

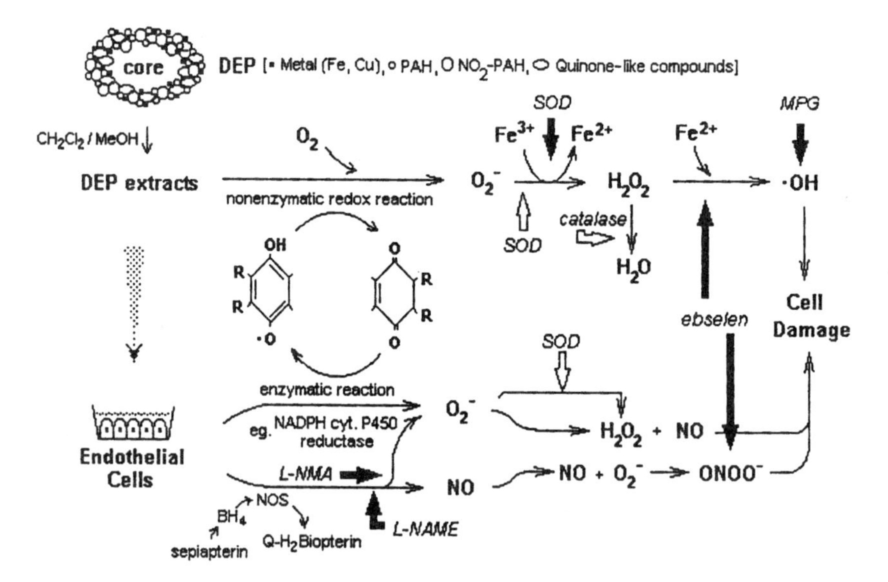


**Figure 5** A schematic view of the mechanism of DEP-induced cytotoxicity to human pulmonary vascular endothelial cells (HPVEC) and the effects of antioxidative reagents (solid arrow, inhibition; open arrow, stimulation) used in this research (31).

## IV. MANIFESTATION OF ASTHMALIKE FEATURES BY DEP OR DIESEL EXHAUST (DE)

### A. Nonallergic Mechanism via Involvement of Reactive Oxygen Species

We described that DEP produces large amounts of $O_2^-$ and ·OH via xenobiotic metabolizing reactions of quinonelike compounds, polyaromatic hydrocarbons, polyaromatic nitrocompounds, and others in DEP (13). Furthermore, it is well known that $O_2^-$ is also produced during phagocytic reaction of DEP by alveolar macrophages (32). These reactive oxygen species damage endothelial and epithelial cells in lungs (29,31,33), and this damage may then lead to characteristic features of asthma. The essential features of asthma are defined as follows: (1) chronic airway inflammation with marked infiltration of eosinophils (this is the most important feature of asthma); (2) mucus hypersecretion with hyperplasia of goblet cells; and (3) airway hyperresponsiveness with smooth muscle constriction. It is interesting to see whether or not the characteristic features will relate to reactive oxygen species.

#### 1. *Chronic Airway Inflammation*

We examined whether or not the asthmatic features could be induced by DEP. Mice were instilled with 0.1 or 0.2 mg of DEP once a week for 16 weeks, and then the essential features of asthma were observed (19).

To confirm the evidence of chronic airway inflammation, we examined by light microscopy the number of eosinophils and neutrophils under the mucosal layer of airway in the lungs stained with Diff-Quick solution to distinguish them (Fig. 6). The number of neutrophils reached the maximum level at the eighth week after DEP instillation, and then dropped back to near the control level. On the other hand, the number of eosinophils

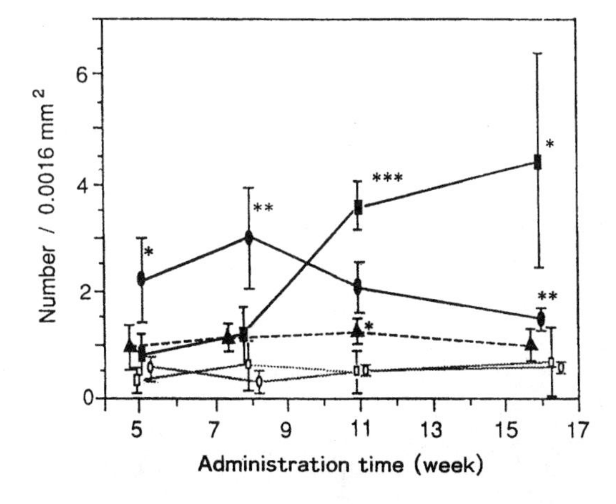


**Figure 6** Number of neutrophils in the submucosa of medium and small bronchioles of mice repeatedly instilled with DEP or PEG-SOD plus DEP. Vales are expressed as mean ± SD. Significance was examined against each control value at each time point. Elongated circles (○, ●) show the number of neutrophils and squares (□, ■) that of eosinophils. ○, □, numbers in the control group; ●, ■, numbers in the 0.2-mg DEP group; and ▲, numbers in l-mg PEG-SOD pretreatment + 0.2-mg DEP group. $*p < 0.05$; $**\ p < 0.01$; $***p < 0.001$.

reached maximum levels at the eleventh week, where they remained. In this experiment, some of the mice were pretreated with 1 mg of PEG-SOD 1 day before DEP instillation, and in this case the number of eosinophils did not increase (19). This suggests that $O_2^-$ may play an important role in the infiltration of eosinophils into the submucosal layer of the bronchiole. Inflammatory cells such as neutrophils, lymphocytes, and plasma cells, as well as eosinophils, were also observed by electron microscopy under the submucosal layer of the airway (19).

### 2. *Mucus Hypersecretion and Airway Constriction*

Figure 7 shows light microscopic photographs of airway constriction in a mouse instilled with vehicle or 0.1 mg of DEP for 16 weeks (19). At left is the vehicle group, and at right are those in the DEP group. The bronchiole in the vehicle group is covered smoothly with a single epithelial cell layer, and the alveolar area is also normal. In contrast, marked bronchoconstriction in the DEP group was observed 24 h after the final DEP instillation. Airway epithelial cells were replaced with goblet cells. These cells contained sizeable amounts of mucus which were stained red by PAS staining. The muscle layer around the airway thickened markedly, and a lymphocyte cluster could also be seen near the site of vessels.

### 3. *Airway Hyperresponsiveness*

We examined the changes of respiratory resistance (Rrs) as a marker of airway hyperresponsiveness to acetylcholine (ACh) stimulation (Fig. 8) and found that it increased with DEP dose. We also carried out experiments with a group of mice pretreated weekly with PEG-SOD prior to 0.2 mg DEP instillation (19). The PEG-SOD pretreatment decreased Rrs. These data also suggest that $O_2^-$ may play an important role in airway hyperresponsiveness as an asthmalike feature.

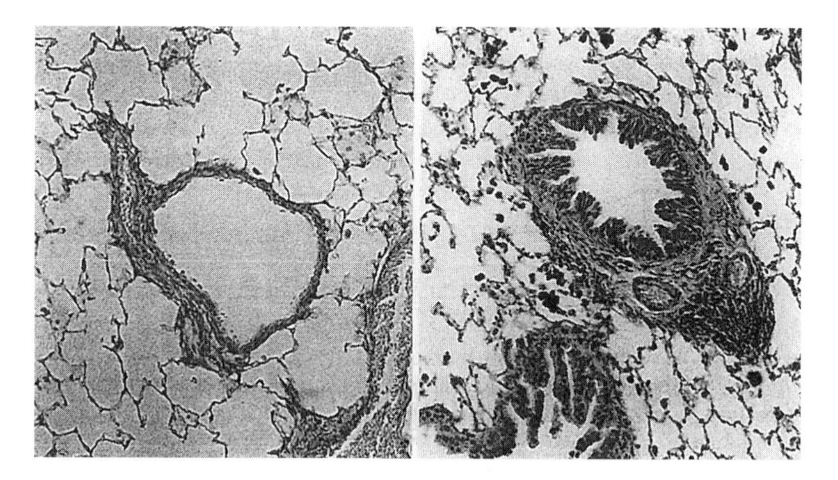

**Figure 7** Airway constrictions of mice instilled with vehicle or DEP. Left: airway in a vehicle-treated mouse once a week for 16 weeks. Right: airway in a mouse after 16 weekly treatments of 0.1-mg DEP. Muscle constriction, hyperplasia of goblet cells, infiltration of inflammatory cells, and clusters of lymphocytes are observed in the DEP group.

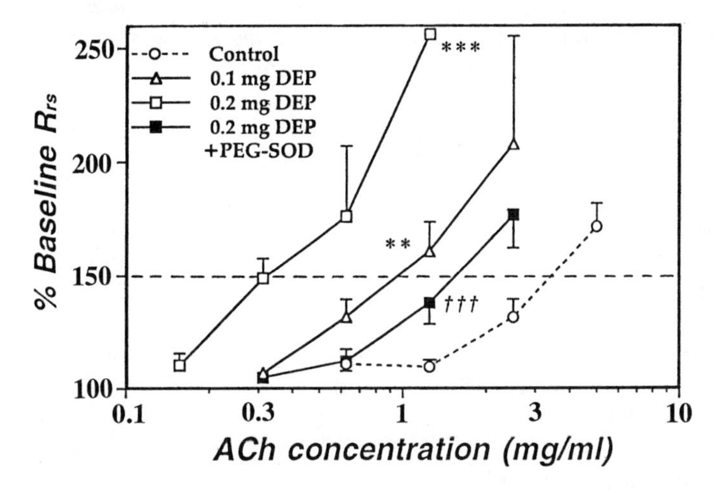


**Figure 8** Concentration-response curves of respiratory resistance (Rrs) to aerosolized acetylcholine (19). ○, vehicle alone ($n$ = 15); △, 0.1 mg DEP ($n$ = 15); □, 0.2 mg DEP ($n$ = 12); ■, PEG-SOD (1 mg) + 0.2 mg DEP ($n$ = 12). Each curve represents data collected from 15 or 12 different animals. Rrs is expressed as mean ± SE of the percentage of the response to aerosolized 0.9% NaCl solution (for 2 min) given prior to administration of acetylcholine.

### 4. *Activity Changes of CYP-450 Reductase in Epithelial Cells in the Airway*

We examined whether or not enough of the enzyme exists to produce $O_2^-$ in the lungs, focusing on CYP-450 reductase as a representative of the enzymes, as mentioned in Section III.C Table 1 compares CYP-450 reductase activity and CYP-450 contents in the lungs and liver of ICR mice. The CYP-450 content in liver is six times higher than that

**Table 1** CYP-450 Contents and NADPH CYP-450 Reductase Activity in the Liver and Lungs of ICR Mice

| Tissue | CYP-450 (nmol/mg prot.) | CYP-450 reductase (nmol/mg prot./min) |
|---|---|---|
| Liver[a] | 0.82 ± 0.08 | 132.8 ± 19.6[c] |
| | | 50.6 ± 2.8[d] |
| Lungs[a] | 0.14 ± 0.02 | 197.9 ± 16.9[c] |
| | | 60.1 ± 10.4[d] |
| Lungs[b] | Vehicle group | 45.3 ± 1.6[d] |
| | DEP group | 55.8 ± 1.8[d] |

All values are expressed as mean ± SD.
[a] Ms fraction was separated from three nontreated mice among a total of 18 mice ($n$ = 6) at 6 weeks of age.
[b] Ms fraction from mice instilled intratracheally with vehicle or 0.1 mg DEP for 11 weeks was obtained from the same number.
[c] The values were obtained using cytochrome c as a substrate.
[d] Substrate was acetylated cytochrome c.
*Source*: Ref. 14.

in the lungs. However, CYP-450 reductase activity in the lungs is higher than that in the liver (14). This suggests that the lungs are able to produce a great deal of $O_2^-$ in the presence of DEP.

Furthermore, from the viewpoint of pathogenesis of asthma, distribution of CYP-450 reductase in mouse lungs was thus examined using the immunohistochemical method to determine whether CYP-450 reductase actually exists within the airway tracts (14). Positive and negative stain cells exist in the epithelium in the airway; the positive are ciliated cells and the negative are goblet cells, which produce mucus in the proximal bronchus and medium bronchiole. Clara cells in the airway of mice instilled with DEP exhibit more intense staining than the cells in the mice instilled with vehicle.

In this experiment, it is characteristic that the goblet cells were hyperplasia. This increased stain-intensity of ciliated cells by DEP instillation corresponded to the enzyme activity in Table 2. The DEP instillations increased CYP-450 reductase activity in the lungs (14). This suggests that the enzyme in the epithelial cells and Clara cells can produce a large amount of $O_2^-$ in the presence of DEP.

### 5. *Decrease of Enzyme Activities of CuZn-SOD and Mn-SOD and Their Distribution in Lungs*

Table 2 shows a summary of the enzyme activities of the two types of SOD in lungs of mice instilled with DEP for 11 weeks (14). The activities of CuZn-SOD and Mn-SOD were reduced by repeated DEP instillation, with the reduction of Mn-SOD being greater. The changes of immunohistochemical staining corresponded to the activity changes. CuZn-SOD stained immunohistochemically in a proximal bronchus and bronchiole. The decrease of CuZn-SOD stain-intensity in the DEP group was observed, and the decrease of Mn-SOD was severe. These results mean that the $O_2^-$ scavenging ability in the DEP group was decreased, and that increased activity of the $O_2^-$-producing enzyme, CYP-450 reductase, also may enhance the oxidative stress.

### 6. *Distribution of Nitric Oxide Synthase (NOS) and NO Exhalation*

The change of stain intensity of the two types NOS in lung tissue was also examined, because NO may play an important role in inducing asthma (34,35). The stain intensity of cNOS in the DEP group was higher than in the control group, and the stain intensity of iNOS in macrophages also increased markedly (14). These increased intensities of the two types of NOS may cause the increase in exhaled NO. Indeed, the concentration of exhaled NO in the DEP group was two times higher than that in the control group. How does the increased NO impact the pathogenesis of asthma? Is it a good effect or a bad

**Table 2** Two Types of SOD Activities in the Lungs of Mice Instilled Intratracheally with DEP

| | Vehicle group | DEP group |
|---|---|---|
| Cu, Zn-SOD (nmol/mg prot./min) | 21.1 ± 0.3 (100%) | 18.1 ± 1.4 (86%) |
| Mn-SOD (nmol/mg prot./min) | 6.4 ± 0.5 (100%) | 4.3 ± 0.4 (67%) |

All values are expressed as mean ± SE. The lungs of five mice in each group among a total of 30 mice ($n = 6$) were pooled. Mice were instilled weekly with vehicle or 0.1 mg of DEP for 11 weeks.
*Source*: Ref. 14.

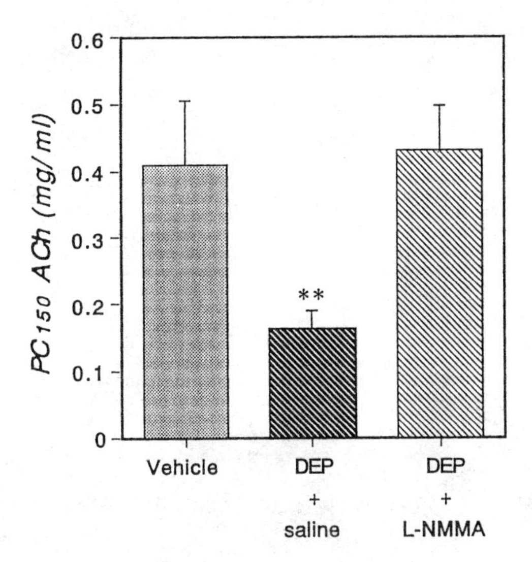


**Figure 9** Effect of L-NMMA on airway responsiveness to ACh in ICR mice. Airway responsiveness was determined from changes in respiratory resistance (Rrs) (14). ($PC_{150}$ = the provocative concentration of acetylcholine causing a 50% increase in Rrs.) ** $p < 0.01$ vs. vehicle control.

effect? (36). To examine the meaning of the increased NO, we examined the effect of DEP and NOS inhibitor on airway hyperresponsiveness. Figure 9 (14) shows the changes of $PC_{150}$ as an indicator of respiratory resistance (Rrs) to acetylcholine (ACh). $PC_{150}$ is ACh concentration that causes an increase of 50% in Rrs. $PC_{150}$ decreased by repeated DEP instillation, and $PC_{150}$ in the DEP group was 2.5 times lower than that in the control group. The $PC_{150}$ was increased by the administration of NOS inhibitor (L-NMMA), and the response returned to the control level (14). This result suggests that the increased NO may be no good, and it is also deeply involved in the pathogenesis of asthma.

### 7. *Possible Mechanism of the Involvement of $O_2^-$ and NO in the Manifestation of Asthmalike Features by DEP*

Figure 10 shows the possible mechanism of the involvement of $O_2^-$ and NO on the manifestation of asthmalike features in mice instilled intratracheally with DEP.

For example, quinonelike and other compounds in DEP are metabolized to semiquinone or other radicals by CYP-450 reductase and other enzymes in Ms, Mt, and cytosol fractions (14) in epithelial cells. We detected an ESR signal of a semiquinone radical with a $g = 2.0047$ in vitro system containing DEP, NADPH, and CYP-450 reductase. Therefore, large amounts of $O_2^-$ may be produced during this process. The increase of NOS activity in epithelial cells and the DEP phagocytosis by macrophages also produce $O_2^-$ and NO.

$O_2^-$ easily reacts with NO to produce peroxynitrite ($ONOO^-$) (36,37), which may be toxic against epithelial cells (14), and which may change to ·OH. $O_2^-$, NO, $ONOO^-$, and ·OH may damage epithelial cells, and this damage in turn may lead to the induction of asthmalike features such as chronic airway inflammation, mucus hypersecretion, and airway hyperresponsiveness (12,14). We assume that this may be a possible mechanism of the manifestation of nonallergic asthmalike features induced by reactive oxygen species including NO.

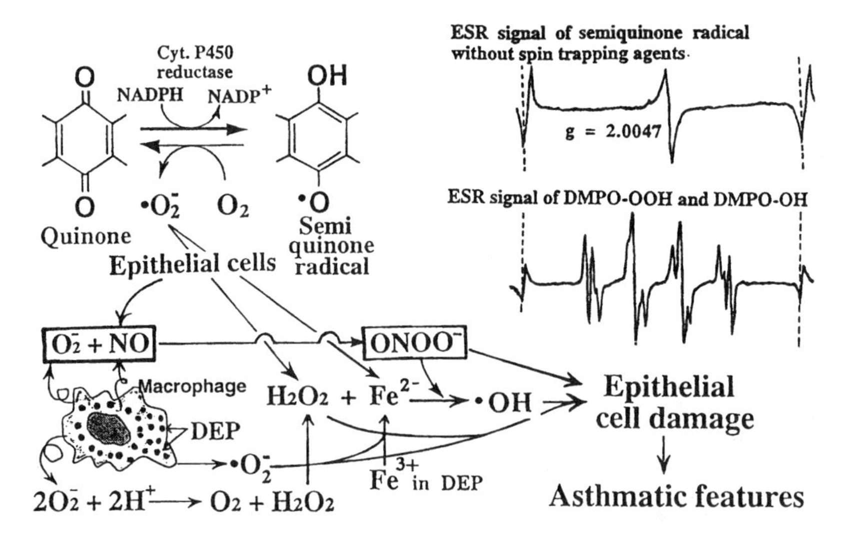


**Figure 10** A schematic view of the putative mechanisms underlying the involvement of $O_2^-$ and NO in the manifestation of asthmalike features, by repeated exposure to DEP (14).

Recently, it was reported that the $ONOO^-$ scavenging reagent, ebselen, was more effective in preventing asthmalike features than steroid hormones (38,39). This evidence strongly supports that our mechanism may be correct. Furtheremore, these reactive oxygen species may induce the infiltration of eosinophils. The eosinophils and reactive oxygen species may release toxic granule proteins such as major basic protein (MBP), eosinophil cationic protein (ECP), eosinophil peroxidase (EPO), and neurotoxin (NT). For example, Horie et al. (21) reported that there is a positive relationship between the release of toxic granule protein (NT) and the release of superoxide from eosinophils stimulated by cytokines, chemokines, PAF, and immunoglobulins (Fig. 11). It is well known that these toxic

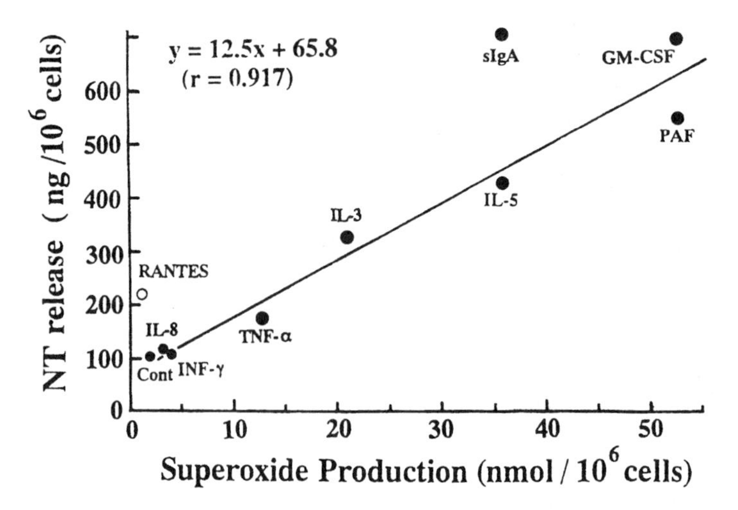


**Figure 11** A relationship between degranulation of neurotoxin (NT) from eosinophils and superoxide production via stimulation by cytokines, chemokines, and immunoglobulin (21).

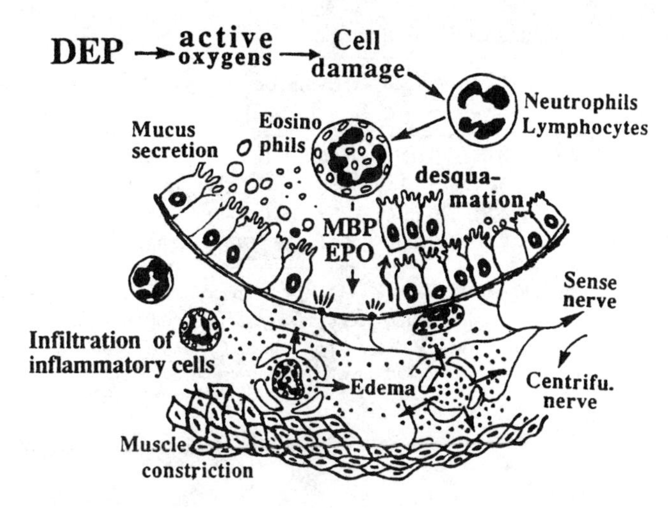


**Figure 12** A schematic view of the putative mechanism on development of asthmalike symptoms by DEP administration via the formation of reactive oxygen species.

granule proteins cause desquamation of epithelial cells from the basement of the airway (20,21). Numerous nerves lying under the epithelium may therefore be sensitive, and the smooth muscle would constrict markedly. This may be a possible mechanism inducing the asthmalike features mediated by reactive oxygen species derived from DEP (Fig. 12).

There is further evidence that reactive oxygen species cause pathogenic features of asthma (40). Mucus hypersecretion by $O_2^-$ has been reported by Alder et al. (41). Bronchial hyperresponsiveness to $O_2^-$ from cigarette smoke has been shown by Dusser et al. (42). Ambrosio et al. (43) have reported that reactive oxygen species inactivated PAF-acetylhydrolase to cause an increase in PAF, which may cause inflammation and airway hyperresponsiveness (44). Katsumata et al. (45) have reported that inhalation of xanthine and xanthine oxidase resulted in airway hyperresponsiveness to ACh in cats. Reactive oxygen species may also cause the release of chemical mediators and deterioration of receptors released to asthma symptoms. Massini et al. (46,47), Manaioni et al. (48), and Weiss et al. (49,50) have shown that chemical mediators such as histamine, prostaglandins, and leukotrienes were increased by $O_2^-$. Engels et al. (51) found that $O_2^-$ generated from pulmonary macrophages induced deterioration of β-adrenoceptor function in the guinea pig trachea. This evidence also shows that reactive oxygen species are deeply involved in the pathogenesis of asthma.

## B. Allergic Mechanism via Involvement of IgG-1 Antibody, IL-5, and Eosinophils

We investigated the effects of DEP instilled intratrachially on antigen-induced airway infiltration, local expression of cytokines, and antigen-specific immunoglobulin (Ig) production in mice, although it remains unclear whether reactive oxygen species are involved in the manifestation of allergic asthma. Mice were received vehicle, ovalbumin (OA, 1 μg, at 3-week intervals), DEP (0.1 mg weekly), and OA + DEP for 6 weeks, respectively (52). DEP enhanced OA-induced airway inflammation characterized by infiltration of eosinophils and lymphocytes and an increase of goblet cells in the bronchial layer (Table

**Table 3** Numbers of Inflammatory Cells and Goblet Cells in Lung Tissue[a]

| Group | Animals ($n$) | Eosinophils ($n$/mm) | Neutrophils ($n$/mm) | Lymphocytes ($n$/mm) | Goblet cells ($n$/mm) |
|---|---|---|---|---|---|
| Vehicle | 7 | 0.016 ± 0.010 | 0.038 ± 0.010 | 0.060 ± 0.013 | 0.309 ± 0.117 |
| OA | 9 | 0.744 ± 0.569 | 0.348 ± 0.172 | 1.30 ± 0.529 | 0.957 ± 0.377 |
| DEP | 9 | 0.150 ± 0.087 | 0.428 ± 0.117 | 1.32 ± 0.209 | 3.92 ± 1.10 |
| OA + DEP | 9 | 5.24 ± 1.74[b,d,g] | 1.85 ± 0.706[b] | 8.11 ± 1.76[c,e,h] | 13.0 ± 2.23[c,f,h] |

[a] Animals received intratracheal instillation of vehicle, OA, DEP, or OA + DEP for 6 weeks. Lungs were removed and fixed 24 h after the last intratracheal administration. Sections were stained with Diff-Quik for measurement of inflammatory cells around the airways or with PAS for goblet cells in the bronchial epithelium. Results are expressed as number of cells per length of basement of airways. Values are mean ± SE.
[b] $p < 0.05$ versus vehicle.
[c] $p < 0.0001$ versus vehicle.
[d] $p < 0.05$ versus OA.
[e] $p < 0.001$ versus OA.
[f] $p < 0.0001$ versus OA.
[g] $p < 0.01$ versus DEP.
[h] $p < 0.001$ versus DEP.
*Source*: Ref. 52.

3). DEP with OA markedly increased interleukin-5 (IL-5) in lung tissue compared with either antigen alone or DEP alone, and the combination of DEP and OA induced significant increases in local expression of IL-4 and GM-CSF (Table 4). DEP exhibited adjuvant activity for the antigen-specific production of IgG-1, but not IgE (Table 5). It is reported that IgG-1 can cause degranulation from eosinophils via the binding of FcγRII receptor on eosinophils (53). Although IgE is important in many allergic reactions, antigen-specific

**Table 4** Protein Levels of Th2 Cytokines in Lung Tissue and BAL Supernatants[a]

| | | Lung tissue supernatants | | BAL supernatants | |
|---|---|---|---|---|---|
| Group | Animals ($n$) | IL-5 (pg/total lung tissue supernatants) | IL-4 (pg/total lung tissue supernatants) | IL-5 (pg/total BAL supernatants) | IL-4 (pg/total BAL supernatants) |
| Vehicle | 12 | 11.3 ± 5.46 | 123.1 ± 20.5 | 21.2 ± 7.62 | 79.4 ± 17.2 |
| OA | 12 | 12.4 ± 4.52 | 391.3 ± 88.0[c,g] | 48.7 ± 15.9 | 76.7 ± 18.0 |
| DEP | 12 | 11.1 ± 3.71 | 112.1 ± 15.3 | 23.8 ± 5.92 | 95.3 ± 17.2 |
| OA + DEP | 12 | 90.7 ± 33.8[c,e,g] | 308.5 ± 63.5[b,f] | 86.4 ± 18.8[d,g] | 78.9 ± 14.9 |

[a] Four groups of mice were intratracheally inoculated with vehicle, OA, DEP, or the combination of OA and DEP for 6 weeks. BAL was conducted 24 h after the last intratracheal instillation. In other animals, lungs were removed and frozen 24 h after the last intratracheal administration. Protein levels in BAL and lung tissue supernatants were analyzed using ELISA. Results are shown as mean ± SE.
[b] $p < 0.05$ versus vehicle.
[c] $p < 0.01$ versus vehicle.
[d] $p < 0.001$ versus vehicle.
[e] $p < 0.01$ versus OA.
[f] $p < 0.05$ versus DEP.
[g] $p < 0.01$ versus DEP.
*Source*: Ref. 52.

**Table 5** OA-Specific Immunoglobulin Titers[a]

| Group | Animals ($n$) | $IgG_{2a}$ (antibody titers) | $IgG_1$ (antibody titers) | IgE (antibody titers) |
|---|---|---|---|---|
| Vehicle | 36 | 100.4 ± 0.42 | 111.3 ± 6.0 | 16.0 ± 0 |
| OA | 36 | 161.3 ± 61.3 | 175.7 ± 24.7 | 16.3 ± 0.17 |
| DEP | 36 | 100.0 ± 0 | 131.7 ± 20.7 | 16.0 ± 0 |
| OA + DEP | 48 | 100.0 ± 0 | 1483 ± 454[a,b,c] | 16.0 ± 0 |

[a] Four groups of mice were administered intratracheally vehicle OA, DEP, or the combination of OA and DEP for 6 weeks. Plasma samples were retrieved 24 h after the last intratracheal instillation. OA-specific immunoglobulins were analyzed using ELISA. Results are expressed as mean ± SE.
[a] $p < 0.01$ versus vehicle.
[b] $p < 0.01$ versus OA.
[c] $p < 0.01$ versus DEP.
*Source*: Ref. 52.

IgG-1 antibody, but not IgE antibody, in patients' sera may contribute to late asthmatic response (54) via antigen-specific eosinophil degradation via FcγRII on the eosinophil surface. IgG-1 with antigen is a strong agonist for eosinophil degradation in vitro (53).

Furthermore, we found that there are marine strain differences in allergic airway inflammation by a combination of DEP + OA, and there is a high relationship between the inflammation and the ability of IgG-1 production (55,56). From this evidence, we would like to propose that DEP can enhance the manifestation of allergic asthmalike features by a mechanism via IgG-1, IL-5, and GM-CSF productions and increased infiltration of Th2 lymphocytes and eosinophils (Fig. 13).

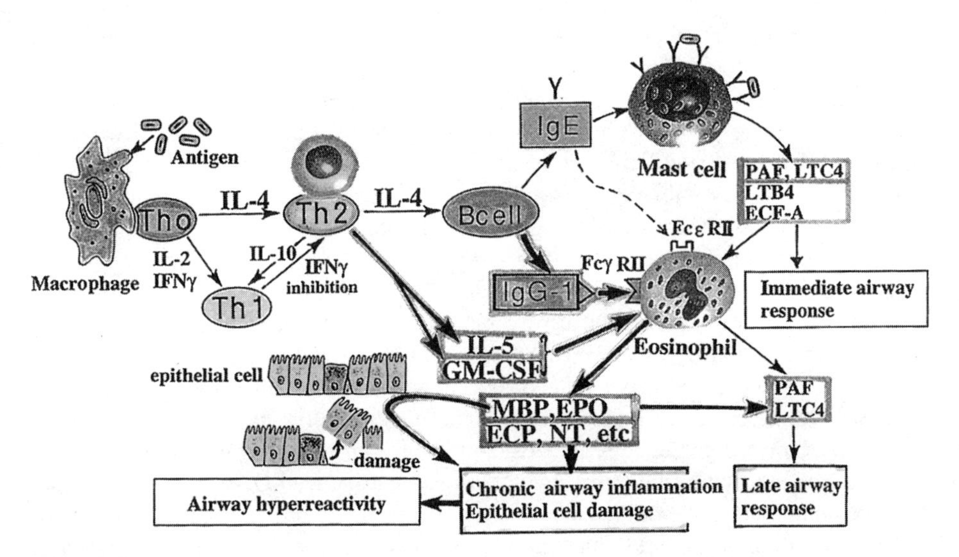


**Figure 13** Proposed mechanism of pathogenesis of asthma via increases of IgG-1 and IL-5 concentrations by repeated intratracheal instillation of DEP and allergen.

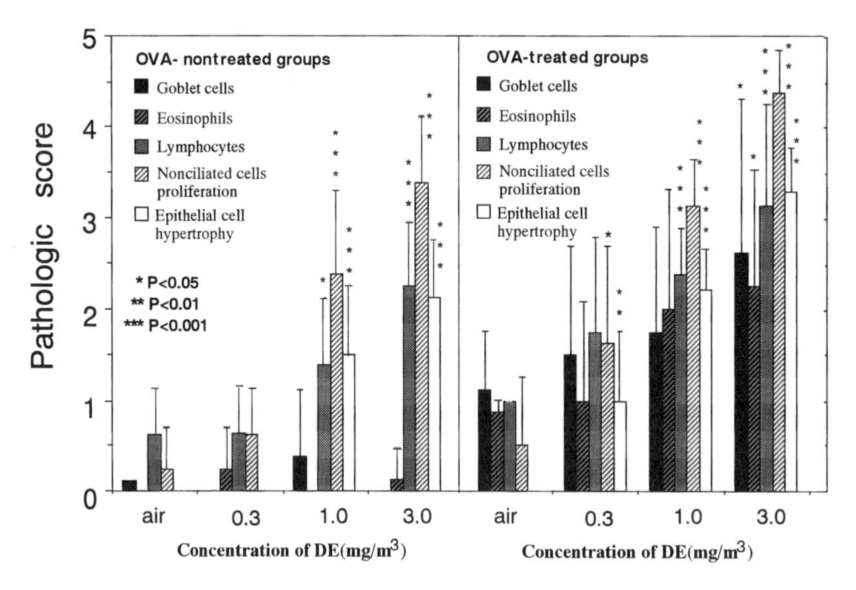


**Figure 14** The changes of histopathological score in the airway of mice exposed to air, diesel exhaust (DE), OA (ovalbumin) + air, or OA + DE (57). Pathological examinations were conducted by eosinophilic infiltration in the degree of airway inflammation characterized by eosinophilic infiltration in the proximal airway to distal airway and periodic acid Schiff (PAS) staining to evaluate the degree of proliferation of goblet cells in the bronchial epithelium.

## C. Diesel Exhaust (DE) Can Induce Allergic Asthmalike Features

The intratracheal instillation of DEP enhances allergic asthmalike features as mentioned above. It is not known, however, whether the effects of such instillation are similar to those of the daily inhalation of DE. Therefore, we examined whether or not the inhalation of DE could cause allergic asthmalike reactions.

ICR mice were exposed to clean air or DE at a soot concentration of 0.3 mg/m$^3$, 1.0 mg/m$^3$, and 3 mg/m$^3$ for 12 h daily for up to 8 months. Four months after exposure to DE, mice were sensitized intraperitoneally with 10 μg of OA, and then challenged by an aerosol of 1% OA six times at 3-week intervals during the last 4 months of the DE exposure. One day after the final challenge of OA aerosol, mice were sacrificed and examined on the allergic changes (57,58).

DE exposure alone caused a dose-dependent increase of nonciliated cell proliferation and epithelial cell hypertrophy in the airway, but showed no effects of goblet cell proliferation in the bronchial epithelium and eosinophil recruitment in the submucosa of the airway. OA treatment alone induced very slight changes in goblet cell proliferation and eosinophil recruitment. On the other hand, the combination of OA and DE exposure produced dose-dependent increases of goblet cells and eosinophils, in addition to further increases of the typical changes induced by DE (Fig. 14). OA treatment induced OA-specific IgG-1 and IgE production whereas the adjuvant effects of DE exposure on Ig production were not observed. Inhalation of DE led to increased levels of IL-5 in the lung at a soot concentration of 1.0 and 3.0 mg/m$^3$ with OA, although these increases did not reach statistical significance. We conclude that the combination of antigen challenge and chronic exposure to DE produces increased eosinophilic inflammation, and cell damage on the airway epithelium may depend on the degree of eosinophilic inflammation in the airway.

From these results, we believe DE exposure combined with antigen produces asthmalike features via a similar mechanism obtained with the instillation of DEP + OA (52).

## V. CARCINOGENESIS BY ACTIVE OXYGEN SPECIES DERIVED FROM DIESEL EXHAUST PARTICLES (DEP) OR DIESEL EXHAUST (DE)

### A. Increase of Death Rate of Lung Cancer and Involvement of Active Oxygen Species

The number of people dying of lung cancer is constantly increasing throughout the world. Smoking and an inadequate diet are without question the leading causes of developing lung cancer, far outweighing other factors (59). The increased type of lung cancers, however, is not squamous cell carcinoma caused mainly by smoking but adenocarcinoma in which the risk of smoking is the lowest. Air pollutants, therefore, are considered to be another important factor, because the number of diesel engine cars has rapidly increased in recent years, and DEP contains a vast number of carcinogens. In fact, there are many studies to show that diesel exhaust causes lung cancer in animal experiments (60–65) and human exposure (66–73).

The carcinogenicity of DE has been demonstrated in experiments performed in six institutes in the United States, Germany, Switzerland, and Japan, in which F-344 Fisher rats were exposed to DE for 2 years (7,60–65). Figure 15 summarizes the results of the DE-exposure experiments conducted in the six institutes (7). Although the data are scattered to some extent, there is an overall correlation between the amount and the incidence, clearly

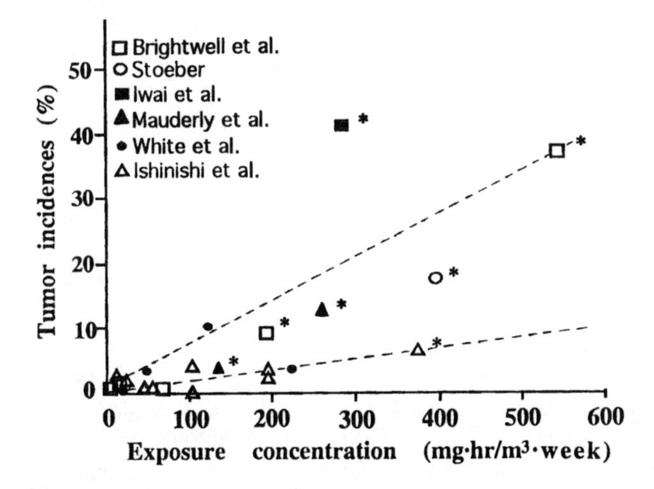


**Figure 15** Lung tumor incidence in Fisher F-344 rats inhaled diesel exhausts for 2 years in six institutes (7). Values with asterisk (*) are different from the control value by statistical test. The abscissa indicates the total exposure amount per week calculated by multiplying the DEP concentration and exposure time, and the ordinate shows the incidence of lung tumors.

indicating the carcinogenicity of DE. Heinrich et al. (61), Brightwell et al. (62), and Iwai et al. (63) found that the gaseous components in DE eliminated DEP had no carcinogenicity, and Kawabata et al. (72) and Ichinose et al. (16) reported that DEP instilled intratracheally caused lung tumors in rats and mice, respectively.

Epidemiological studies have demonstrated a slightly higher risk in the population of those who were occupationally exposed to DEP, and some of the data have demonstrated a statistical significance (66–73). Among several cohort studies, the large-scale and long followed study by Garshick (66) showed a significantly higher relative risk of 1.45 for all subjects and 1.72 for the population exposed for 7 to 17 years. Recently, the Environmental Protection Agency (EPA) in California recognized that DE is carcinogenic to humans.

On the other hand, other environmental factors, without smoking or air pollutants, may also be involved in the increase of lung cancer. Epidemiological studies indicate that a dietary change to a high-fat or a high-calorie diet may play an important part (59,73–75) in carcinogenesis. This is probably because high-fat or high-calorie diets increase reactive oxygen species and lipid peroxides (76,77). Many epidemiological studies have shown that the incidence of lung cancer is low in populations where large amounts of antioxidative substances such as β-carotene, vitamins A, E, and C, and selenium are taken and most of these studies emphasize the effectiveness of β-carotene (78–87). It is well known that anthraquinone carcinostatics produce a large amount of $O_2^-$ in the metabolic process by the drug-metabolizing enzyme, cytochrome P-450 reductase (27,28) and the derivative of $O_2^-$, ·OH, does damage cancer cells. This is one of the well-known mechanisms of carcinostatics and these results suggest that active oxygen species are involved in carcinogenesis.

## B. DNA Damage in Lungs Caused by DEP

Production of DNA adducts in the respiratory tract of rats exposed to diesel exhaust is reported by using $^{32}P$ postlabeling assay (88–91). There is evidence that both localization of diesel-exhaust-induced tumor and the distribution of DNA-adducts are located exclusively in the peripheral lung (92,93). It has therefore been considered that DEP is involved in carcinogenesis by forming DNA adducts with carcinogens such as PAHs and nitro-PAHs in DEP and that the formed DNA adducts may cause errors in genetic information and produce tumors (88–91).

As an alternative mechanism of carcinogenesis by DEP, we are able to suppose an involvement of reactive oxygen species, in addition to the formation of DNA adducts, because we previously mentioned the generation of reactive oxygen species such as $O_2^-$ and ·OH from DEP in the presence of lung enzymes (13–18) and during phagocytosis of DEP by macrophages (17,32,92,93).

Does the active oxygen species from DEP actually cause DNA damage in lungs? We found that the reactive oxygen species generated by DEP in the presence of NADPH and CYP-450 reductase in vitro caused DNA strand breaks. The breaks then decreased in the presence of SOD, catalase, or ·OH scavenging agents (13). 8-Hydroxydeoxyguanosine (8-OHdG) is one of the most critical DNA lesions generated from deoxyguanosine (dG) by ·OH (94,95). 8-OHdG formation in the lungs of mice instilled intratracheally with DEP was also examined (16–18,96). The mice were fed four kinds of diets: basal fat (BF) diet (4%), high-fat (HF) diets (16%), BF + 0.2% β-carotene, and HF + 0.2% β-carotene, respectively. As shown in Figure 16, after a single instillation of DEP, a significant increase of 8-OHdG in mouse lung DNA was observed (96). High dietary fat enhanced the

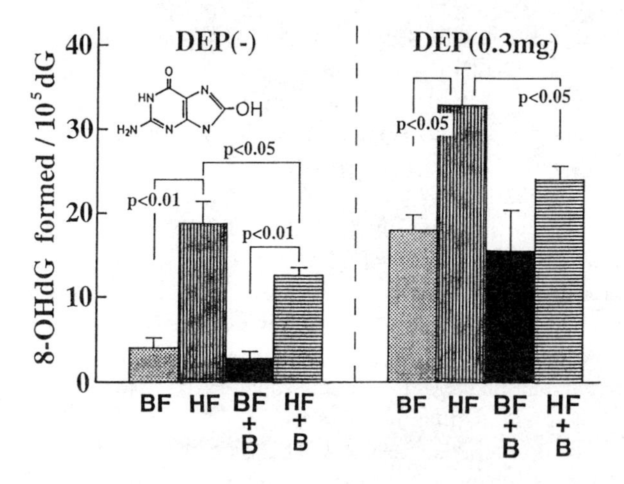


**Figure 16** 8-Hydroxydeoxyguanosine (8-OHdG) formation by single intratracheal instillation of DEP in lungs of mice fed four kinds of diets (96). BF, basal fat diet group (4%); HF, high fat diet group (16%); BF + B, BF + β-carotene diet group (0.02%); HF + B, HF + β-carotene diet group (0.02%).

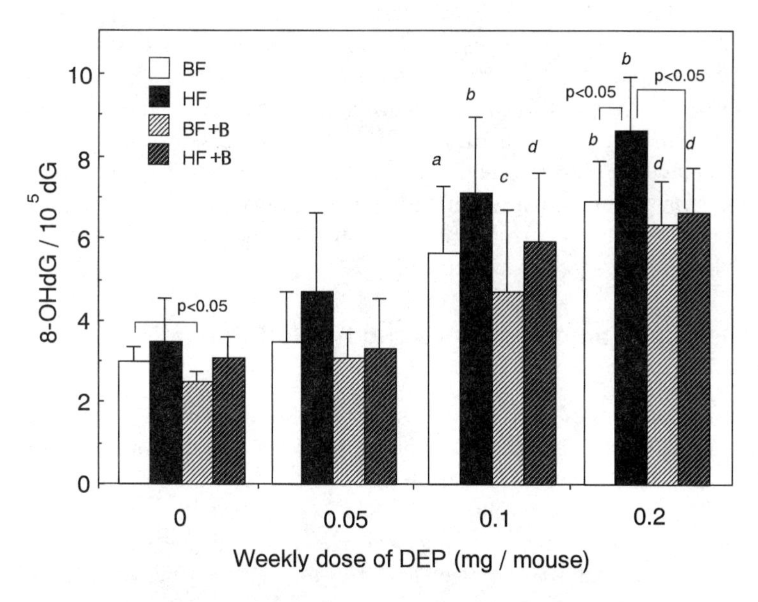


**Figure 17** Formation of 8-OHdG in mouse lung DNA after 10 weekly intratracheal instillations of DEP at three doses, and the effect of high dietary fat and β-carotene (B) on its formation (16). Values are shown as mean ± SD. The mean values in the BF and HF groups treated with 0 mg of DEP were 3.00 ± 0.42 and 3.49 ± 1.21, respectively. The mean values in the BF + B and HF + B groups treated with 0 mg were 2.52 ± 0.25 and 3.07 ± 0.621, respectively. Significant difference from the BF + B + 0 mg DEP group: (a) $p < 0.01$; (b) $p < 0.001$. Significant difference from the BF + B + 0 mg DEP group: (c) $p < 0.05$; (d) $p < 0.001$.

formation of 8-OHdG and an intake of β-carotene suppressed the formation of 8-OHdG in lung DNA. This result suggests that 8-OHdG formation in lung DNA was induced by ·OH produced by DEP and the suppressive effect of β-carotene may be due to the scavenging ability against reactive oxygen species such as singlet oxygen ($^1O_2$), ·OH, and lipid peroxides formed via $O_2^-$ (79).

In an experiment with chronic instillation of DEP, significant increases of 8-OHdG in mouse lung DNA were observed. Figure 17 shows the formation of 8-OHdG in mouse lung DNA after 10 weekly intratracheal instillations of DEP at four doses of 0 mg, 0.05 mg, 0.1 mg, and 0.2 mg/mouse/week (16–18). The rodents were fed four kinds of foods for 10 weeks, as mentioned above, and then sacrificed the following week. The 8-OHdG formation in lung DNA showed a dose-dependent increase. High dietary fat enhanced the formation of 8-OHdG in lung DNA and an intake of β-carotene suppressed the formation, the same as in an experiment with a single instillation (16).

Thus, it seems that 8-OHdG is a promutagenic lesion in DEP-induced lung tumorigenesis in mice and that high dietary fat enhances this process through the generation of 8-OHdG in mouse lung DNA (16). 8-OHdG is considered to cause G:C to T:A transversion in vitro and in vivo (95). The resultant damage to DNA may induce genetic changes in all stages of carcinogenesis, that is, initiation, promotion, and progression.

### C. Tumor Formation by DEP and Dietary Effects Concerned with the Formation of Active Oxygen Species

It is commonly known that DEP causes carcinogenesis by both inhalation and intratracheal instillation, as mentioned above (16–18,60–65,96). DEP were instilled intratracheally to mice once a week for 10 weeks with four doses and the mice were fed four kinds of diets, as mentioned above. The rodents were sacrificed after 1 year and a tumor examination was conducted. As shown in Figure 18 (16), the lung tumor incidence in mice treated with 0.05 mg and 0.1 mg in both groups of BF and HF diet increased significantly from the control, but was decreased at 0.2-mg group. High dietary fat enhanced the incidence of both benign and malignant tumors. On the other hand, β-carotene partially prevented the tumor development, especially malignant tumors. During this time, inflammatory reaction was observed in the respiratory tract and alveoli (57,58).

### D. Relationship Between Tumor Incidence and 8-OHdG Formation by DEP

In order to clarify the involvement of reactive oxygen in lung carcinogenesis induced by DEP, we examined the relationship between lung tumor response and the formation of 8-OHdG in lungs (Fig. 19). The tumor incidence at doses of 0.05 and 0.1 mg showed a significant correlation with the measured values of 8-OHdG ($p < 0.001$; $p < 0.05$). The tumor incidence at 0.2 mg did not show a significant correlation with the values of 8-OHdG (16). This result suggests that active oxygen may play an important role in carcinogenesis by DEP.

### E. A Possible Mechanism of Lung Tumor by DEP

A mechanism of lung carcinogenesis induced by diesel exhaust is not fully understood. It is thought, however, that the carcinogenic compounds present in DEP contribute to the development of lung cancer, since the carcinogenic compounds such as PAHs and nitro-

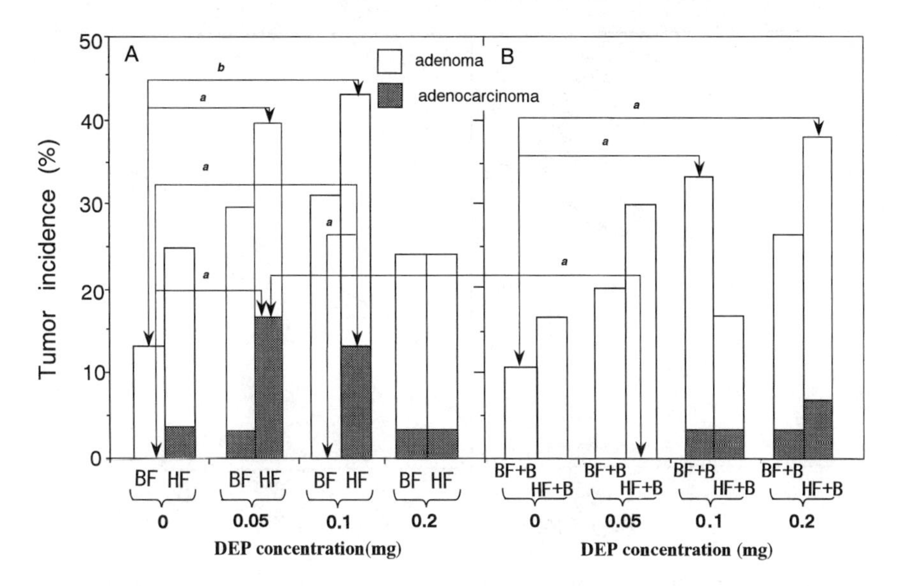


**Figure 18** Lung tumor incidences in the mice by diesel exhaust particles (DEP) and effects of dietary high fat and β-carotene (16). The mice were repeatedly instilled DEP intratracheally 10 times once per week and they were fed the four kinds of diets for 1 year, respectively. Column A shows lung tumor incidences in the mice fed the diet containing 4% fat or 16% fat, respectively. Column B shows lung tumor incidences in the mice fed the basal fat diet containing 4% corn oil + 0.02% β-carotene (BF + B) and 16% corn oil + 0.02% β-carotene (HF + B), respectively. BF means 4% crude fat diet and HF means high-fat diet enriched with 12% corn oil to BF. (a) $p < 0.05$; (b) $p < 0.01$.

PAHs could form DNA adducts which are involved in carcinogenesis (89–91), since these chemicals have been shown to induce lung carcinoma in animals. However, Kawabata et al. (72) found that activated charcoal, which is nontoxic and devoid of genotoxic PAH, caused lung tumors. Furthermore, it was reported that there are no differences in the incidence of lung cancer between rats exposed for 2 years to DE containing DEP and rats exposed for the same period to noncarcinogenic carbon black (CB) or titanium dioxide ($TiO_2$) at the same concentration(97,98), as shown in Table 6 and 7. From these results, it is unclear whether the carcinogenic compounds such as PAHs and nitro-PAHs contained in DEP are actually involved in the pulmonary carcinogenesis induced by diesel exhaust.

Gallagher et al. (98) also examined the amounts of DNA adducts in the lungs of rats exposed for 2 years to a high concentration of DE equivalent to 7.5 mg DEP/$m^3$ by using the $^{32}P$ postlabeling method. However, they have not found a correlation between the amount of DNA adducts and the tumor incidence. Furthermore, a carcinogenic mechanism of particles is controversial, since carbon black and titanium dioxide, both of which are nontoxic and devoid of genotoxic compounds (97,98), have been found to cause lung tumor, as does DEP (see Tables 6 and 7). From these results, we think that particles such as CB and $TiO_2$ are phagocytosed by macrophages and produce large amounts of $O_2^-$ and ·OH by overloading in high-concentration levels, so large amounts of reactive oxygen species may not be produced later by low-level exposure.

We observed that large amounts of $O_2^-$ and ·OH were enzymatically generated from the organic compounds in DEP. The process is shown in Figure 20: soot-associated quinonelike compounds are reduced to the semiquinone radical by CYP-450 reductase and

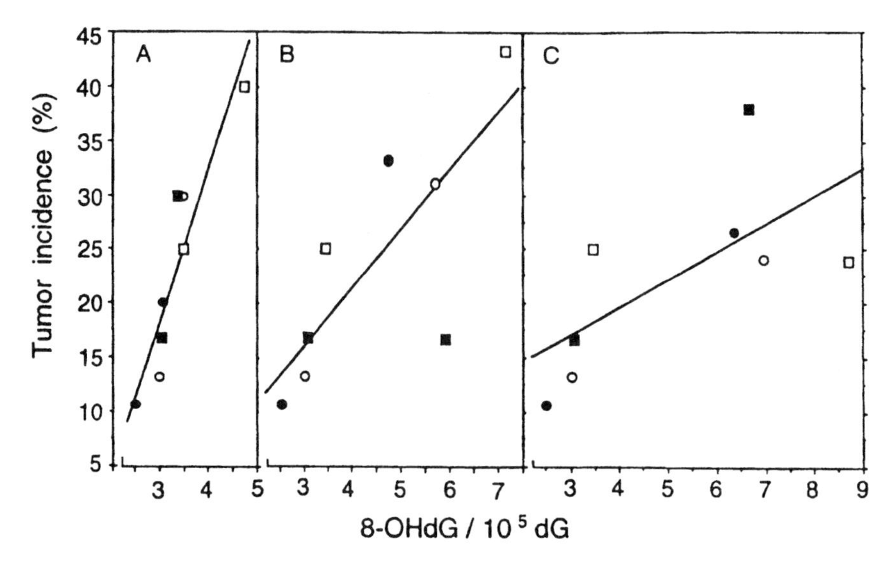


**Figure 19** Correlation between the lung tumor incidence after 12 months and the mean values of 8-OHdG in mouse lung DNA after 10 weekly injections of DEP (16). (A) Correlation in the 0.05-mg DEP group ($r = 0.916$; $p < 0.001$); (B) 0.1-mg DEP group ($r = 0.765$; $p < 0.05$); and (C) 0.2-mg DEP group ($r = 0.677$; $p = 0.065$). The data in the vehicle group are also given in (A), (B), and (C). The tumor incidence at doses of 0.05 mg and 0.1 mg showed a significant correlation with the measured values of 8-OHdG, but not at the 0.2-mg dose. ○, BF; ●BF-β; □, HF; ■, HF-β.

**Table 6** Lung Tumor Formation by Chronic Inhalation Experiments of Diesel Exhaust Particles (DEP) and Carbon Black (CB)[a]

| | Conc. DEP | Tumor incidence | Conc. of B[a]P | Conc. of 1-NP |
|---|---|---|---|---|
| DEP | 6.5 $mg/m^3$ | 17.9% | 3900 ng/g | 19000 ng/g |
| CB | 6.5 $mg/m^3$ | 15.2% | 0.6 ng/g | 0.5 ng/g |

[a] Exposure period and animals: 16 h/day, 5 day/week for 24 months to Fisher F-344 rats (97).

**Table 7** Lung Tumor Formation by Chronic Inhalation Experiments of Diesel Exhaust Particles (DEP), Carbon Black (CB), and Titanium Oxide ($TiO_2$)[a]

| | Exposure conc. | Tumor incidence | Deposition wt. in lung | Conc. of organics | Particle diameter |
|---|---|---|---|---|---|
| DEP | 7.5 $mg/m^3$ | 22% | 63.9 mg/lung | 33.9% | 0.3 μm |
| CB | 11.3 $mg/m^3$ | 39% | 43.9 mg/lung | 0.039% | 0.65 μm |
| $TiO_2$ | 10.4 $mg/m^3$ | 32% | 39.3 mg/lung | ND | 0.82 μm |

[a] Exposure period and animals: 18 h/day, 5 day/week for 24 months to Wistar rats (98).

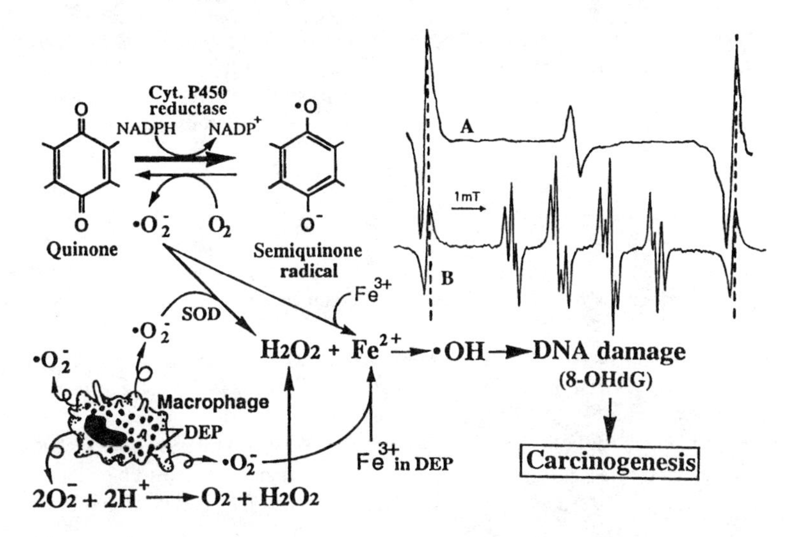


**Figure 20** A proposed mechanism for reactive-oxygen-species-induced lung carcinogenesis in mice by repeated intratracheal instillation of DEP (16).

these semiquinone radicals reduce $O_2$ to $O_2^-$, and the produced $O_2^-$ reduces ferric ions to ferrous ions, which catalyzes the homolitic cleavage of $H_2O_2$ dismutated by SOD or spontaneous reactions, to produce ·OH, which causes oxidative DNA damage such as formation of 8-OHdG in vitro (16,96) and DNA strand breaks (13). Furthermore, it is presumed that the reactive oxygen species released during phagocytosis of DEP in the lungs are also associated with the DNA damage. This idea was, therefore, supported by a clear relationship between DEP dose-dependent increase of 8-OHdG in lungs and the tumor incidences, as shown in Figure 19. The accumulation of 8-OHdG induced by the repeated DEP treatment would be an important factor in enhancing the mutation rate leading to lung cancer, as shown in Figure 20.

## REFERENCES


1. Dockery DW, Brunekreef B. Longitudinal studies of air pollution effects on lung function. Am J Respir Crit Care Med 1996; 154:S 250–256.
2. Xu X, Wang L. Association of indoor and outdoor particulate level with chronic respiratory disease. Am Rev Respir Dis 1993; 148:1516–522.
3. Schwartz J. Particulate air pollution and chronic respiratory disease. Environ Res 1993; 62: 7–13.
4. Dockery DW, Pope CA, Xu P, Spengler JD, Ware JH, Fay ME, Ferris BG, Speixer FE. An association between air pollution and mortality in six U.S. cities. N Engl J Med 1993; 329: 1753–1759.
5. Pope CA. Particulate air pollution as a predictor of mortality in a prospective study of U.S. adults. Am J Respir Crit Care Med 1995; 151:669–674.
6. Xu X, et al. Air pollution and daily mortality in residential areas of Beijing, China. Arch Environ Health 1994; 49:216–222.
7. McClellan RO. Health effects of exposure to diesel exhaust particles. Ann Rev Pharmacol Toxicol 1989; 27:279–300.

8. Peterson B, et al. Global increases in allergic respiratory diseases: the possible role of diesel exhaust particles. Ann Allergy Asthma Immunol 1996; 77:263–268 (quiz 269–270).
9. Edwards J, Walters S, Griffiths PK. Hospital admissions for asthma in pre-school children: Relationship to major road in Birmingham, United Kingdom. Arch Environ Health 1994; 49: 223–227.
10. Brunekreef B, Janssen NAH, de Hartog J, Harssenma H, Knape M, Van Vliet P. Air pollution from truck traffic and lung function in children living near motorways. Epidemiology 1997; 8:298–303.
11. Oosterlee A, Drijver M, Lebret E, Brunkreef B. Chronic respiratory symptoms of children and adults living along streets with high traffic density. Occup Environ Med 1996; 53:241–247.
12. Sagai M, Saito H, Ichinose T, Kodama M, Mori Y. Biological effects of diesel exhaust particals. I. In vitro production of superoxide and in vivo toxicity in mouse. Free Radical Biol Med 1993; 14:37–47.
13. Kumagai Y, Arimoto T, Shinyashiki M, Shimojyo N, Nakai Y, Yoshikawa T, Sagai M. Generation of reactive oxygen species during interaction of diesel exhaust particles components with NADPH-cytochrome P-450 reductase and involvement of the bioactivation in the DNA damage. Free Rad Biol Med 1996; 22:479–487.
14. Lim HB, Ichinose T, Miyabara Y, Takano H, Kumagai Y, Shimojyo N, Devaria JL, Sagai M. Involvement of superoxide and nitric oxide on airway inflammation and hyperresponsiveness induced by diesel exhaust particles in mice. Free Radical Biol Med 1998; 25:635–644.
15. Ichinose T, Furuyama A, Sagai M. Biological effect of diesel exhaust particles (DEP). II. Acute toxicity of DEP introduced into lung by intratracheal instillation. Toxicology 1995; 99: 153–167.
16. Ichinose T, Yajima Y, Takenoshita S, Nagashima M, Nagamachi E, Sagai M. Lung carcinogenesis and formation of 8-hydroxydeoxyguanosine in mice by diesel exhaust particles. Carcinogenesis 1996; 18:101–108.
17. Ichinose T, Yamanushi T, Seto H, Sagai M. Oxygen radicals in lung carcinogenesis accompanying phagocytosis of diesel exhaust particles. Int J Oncol 1997; 11:571–575.
18. Tokiwa H, Sera N, Nakanishi Y, Sagai M. 8-Hydroxydeoxyguanosine formed in human lung tissues and the association with diesel exhaust particles. Free Rad Biol Med 1999; 27:1251–1258.
19. Sagai M, Furuyama A, Ichinose T. Biological effects of diesel exhaust particles (DEP). III. Pathogenesis of asthma like symptoms in mice. Free Rad Biol Med 1996; 21:199–209.
20. Hirata A, Motojima S, Fukuda T, Makino S. Damage to respiratory epithelium by guinea pig eosinophiles stimulated with IgG-coated sepharose beads. Clin Exp Allergy 1996; 26:848–858.
21. Horie S, Glich GJ, Kita H. Cytokine directly induce degranulation and superoxide production from human eosinophils. J Allergy Clin Immunol 1996; 98:371–381.
22. Schuetzle D, Lee FSC, Prater TJ. The identification of polynuclear aromatic hydrocarbons (PAH) derivatives in mutagenic fractions of diesel particulate extracts. Int J Environ Anal Chem 1981; 9:93–144.
23. Schuetzle D. Sampling of vehicle emissions for chemical analysis and biological testing. Environ Health Perspect 1983; 47:65–80.
24. Scheepers PTJ, Bos RP. Combustion of diesel fuel toxicological perspective. I. Origin of incomplete combustion products. Arch Environ Health 1992; 64: 149–161.
25. Oehme M, Larssen S, Bbrevik EM. Emission factors of PCDD and PCDF for road vehicles obtained by tunnel experiment. Chemosphere 1991; 23:1699–1708.
26. Kappus H. Overview of enzyme systems involved in bioreduction of drugs and in redox cycling. Biochem Pharmacol 1986; 35:1–6.
27. Handa K, Sato S. Generation of free radicals of quinone group containing anti-cancer chemicals in NADPH-microsome system as evidence by initiation of sulfite oxidation. Gann 1975; 66:43–47.

28. Berlin V, Haseltine WAR. Reduction of adriamycin to semiquinone free radical by NADPH cytochrome P-450 reductase produce DNA cleavage in a reaction mediated by molecular oxygen. J Biol Chem 1981; 256:4747–4756.
29. Autor AP, Bonham AC, Thies RL. Toxicity of oxygen radicals in cultured pulmonary endothelial cells. J Toxicol Environ Health 1984; 13:387–395.
30. Ichinose T, Suzuki AK, Tsubone H, Sagai M. Biochemical studies on strain difference of mice in the susceptibility to nitrogen dioxide. Life Sci 1982; 31:1963–1972.
31. Bai Y, Suzuki AK, Sagai M. The cytotoxic effects of diesel exhaust particles on human pulmonary artery endothelial cells in vitro: On the role of active oxygen species. Free Radical Biol Med 2001; 30 (in press).
32. Hiura TS, et al. Chemicals in diesel exhaust particles generate reactive oxygen radicals and induce apoptosis in macrophages. J Immunol 1999; 163:5582–5591.
33. Boland S, et al. Diesel exhaust particles are taken up by human airway epithelial cells in vitro and alter cytokine production. Am J Physiol 1999; 276:L604–613.
34. Hamid Q, Springall DR, Riveros-Moreno V, Chanez P, Howarth P, Redington A, Bousquet J, Godard P, Holgate S, Polak JM. Induction of nitric oxide synthase in asthma. Lancet 1993; 342:1510–1513.
35. Kharitonov SA, Yates D, Robbins RA, Logan-Sinclair R, Shinebourne EA, Barnes PJ. Increased nitric oxide in exhaled air of asthmatic patients. Lancet 1994; 343:133–135.
36. Beckman JS, Koppenol WH. Nitric oxide, superoxide, and peroxynitrite, the good, the bad, and the ugly. Am J Physiol 1996; 271 (Cell Physiol. 40):C1424–C1437.
37. Radi R, Beckman JS, Bush KM, Freeman BA. Peroxynitrite-induced membrane lipid peroxidation: The cytotoxic potential of superoxide and nitric oxide. Arch Biochem Biophys 1991; 288:481–487.
38. Sugiura H, et al. Role of peroxynitrite in airway microvascular hyperpermeability during late allergic phase in guinea pigs. Am J Respir Crit Care Med 1999; 160:663–671.
39. Nomura A, Uchida Y, Iijima H, et al. Ebselen suppresses late airway responses and airway inflammation in a guinea pig asthma model (in preparation).
40. Doelman CJA, Bast A. Oxygen radicals in lung pathology. Free Rad Biol Med 1990; 9:381–400.
41. Alder KB, Holden WJ, Repine JE. Oxygen metabolites stimulate release of high molecular weight glycoconjugates by cell organ cultures of rodent respiratory epithelium via an arachidonic acid-dependent mechanism. J Clin Invest 1990; 85:75–85.
42. Dusser DJ, Djokic TD, Borson DB, Nadel JA. Cigarette smoke induces bronchoconstrictor hyperresponsiveness to substance P and inactivates airway neutral endopeptidase in the guinea pig. J Clin Invest 1989; 84:900–906.
43. Ambrosio G, Oriente A, Napoli C, Palumbo G, Chiriello G, Marone G, Condorelli M, Chiariello M, Triggiani M. Oxygen radicals inhibit human plasma acetylhydrolase, the enzyme that catabolizes platelet-activating factor. J Chin Invest 1994; 93:2408–2416.
44. Cuss FM, Dixon CM S, Barnes PJ. Effects of inhaled platelet activating factor on pulmonary function and bronchial responsiveness in man. Lancet 1986; 26:189–192.
45. Katsumata U, Miura M, Ichinose M, Kimura K, Takahashi T, Inoue H, Takishima T. Oxygen radicals produce airway constriction and hyperreactivity in anesthetized cats. Am Rev Respir Dis 1990; 141:1158–1161.
46. Masini E, Giannella E, Palmerani B, Pistelli A, Gambassi F, Mannaioni PF. Prostaglandin H-synthetase activates xenobiotics into free radicals, histamine release and biochemical characteristics. Int Arch Allergy Appl Immunol 1989; 88:134–135.
47. Masini E, Palmerani B, Gaambass F. Histamine release from rat mast cells induced by metabolic activation of polyunsaturated fatty acids into free radicals. Biochem Pharmacol 1990; 39:879–889.
48. Mannaioni PE, Giannella E, Palmerani B. Free radicals as endogenous histamine releasers. Agents Actions 1988; 23:129–142.

49. Weiss EB, Bellino JR. Superoxide anion generation during airway anaphylaxis. Int Arch Allergy App Immunol 1986; 80:211–213.
50. Weiss EB, Bellino JR. Leukotriene-associated oxygen metabolites induces airway hyperreactivity. Chest 1986; 89:709–716.
51. Engles F, Oosting RS, Nijkamp EP. Pulmonary macrophages induce deterioration of guinea pig tracheal beta-adrenergic function through release of oxygen radicals. Eur J Pharmacol 1985; 111:143–144.
52. Takano H, Yoshikawa T, Ichinose T, Miyabara Y, Sagai M. Diesel exhaust particles enhance antigen-induced airway inflammation and hyperresponsiveness via Th2 cytokine expression and specific IgG1 production in mice. Am J Respir Crit Care Med 1997; 15:36–42.
53. Kaneko M, Swanson MC, Gleich GJ, Kita H. Allregen specific IgG1 and IgG3 through Fc γ R II induce eosinophil degranulation. J Clin Invest 1995; 95:2813–2821.
54. Ito K, Kudo K, Okudaira H, et al. IgG1 antibodies to house dust mite (Dermatophagoides farinae) and late asthmatic response. Int Arch Allergy Appl Immunol 1986; 81:69–74.
55. Miyabara Y, Yanagisawa R, Shimojyo N, Takano H, Lim HB, Ichinose T, Sagai M. Murine strain difference in airway inflammation caused by diesel exhaust particles. Eur Resp J 1998; 11:291–298.
56. Ichinose T, Takano H, Miyabara Y, Sagai M. Strain difference of mice in allergic airway inflammation and immunoglobulin production by combination of antigen and diesel exhaust particles. Toxicology 1997; 122:183–192.
57. Ichinose T, Takano H, Miyabara Y, Sagai M. Long-term exposure enhances antigen-induced eosinophilic inflammation and epithelial damage in the airway. Toxicol Sci 1998; 44:70–79.
58. Takano H, Ichinose T, Miyabra Y, Shibuya T, Lim HB, Yoshikawa T, Sagai M. Inhalation of diesel exhaust enhance allergen-induced eosinophil recruitment and airway hyperresponsiveness in mice. Toxicol Appl Pharmacol 1998; 150:328–337.
59. Wyder EL, Gori GB. Contribution of the environment to cancer incidence. An epidemologic exercise. J Natl Cancer Inst 1977; 58:825–832.
60. Mauderly JL, Jones RK, Griffith WG, Henderson RF, McClellan RO. Diesel exhaust is a pulmonary carcinogen in rats exposed chronically by inhalation. Fund Appl Toxicol 1987; 9: 208–291.
61. Heinrich U, Muhle H, Takenaka S, Ernst H, Fuhst R, Mohr U, Port F, Stober W. Chronic effects on the respiratory tract of hamsters, mice and rats after long-term inhalation of high concentrations of filtered and unfiltered diesel engine emissions. J Appl Toxicol 1986; 6:383–395.
62. Brightwell J, Fouillet X, Cassano-Zoppi AL, Gatz R, Duchosal F. Neoplastic and functional changes in rodents after chronic inhalation of engine exhaust emissions. In: Ishinishi N, Koizumi A, McClellan RO, Stoeber W, eds. Carcinogenic and Mutagenic Effects of Diesel Engine Exhaust. Amsterdam: Elsevier, 1986:471–488.
63. Iwai K, Udagawa T, Yamagishi M, Yamada H. Long-term inhalation studies on F-344 SPF rats. In: Ishinishi N, Koizumi A, McClellan RO, Stoeber W, eds. Carcinogenic and Mutagenic Effects of Diesel Engine Exhaust. Amsterdam: Elsevier, 1986:349–360.
64. White H, Vostal JJ, Kaplan HL, MacKenzie WF. A long-term inhalation study evaluates the pulmonary effects of diesel emissions. J Appl Toxicol 1983; 3:332–338.
65. Ishinishi N, Kuwabata N, Nagase S, Suzuki T, Ishikawa S, Kohno T. Long-term inhalation studies on effects of exhaust from heavy and light duty diesel engines on F344 rats. In: Ishinishi N, Koizumi A, McClellan RO, Stoeber W, eds. Carcinogenic and Mutagenic Effects of Diesel Engine Exhaust. Amsterdam: Elsevier, 1986; 329–348.
66. Garshick E, Schenker M, Munoz A, Smith TJ, Woskie SH, Hammond SK, Speizer FE. A retrospective cohort study of lung cancer and diesel exhaust exposure in railroad workers. Am J Respir Dis 1988; 137:820–825.
67. Howe GR, Fraser D, Lindsay J, Presnal B, Yu SZ. Cancer mortality (1965 ∼ 77) in relation

to diesel fume and coal exposure in cohort of retired railway workers. J Natl Cancer Inst 1983; 70:1015–1019.
68. Boffetta P, Stellman SD, Garfinkel L. Diesel exhaust exposure and mortality among males in the American cancer society prospective study. Am J Indust Med 1988; 14:403–415.
69. Garshick E, Schenker MB, Munoz A, Segal M, Smith TJ, Woskie SR, Hammond K, Speizer FE. A case-control study of lung cancer and diesel exposure in railroad workers. Am Rev Respir Dis 1987; 135:1242–1248.
70. Gustafsson L, Plato N, Lidnstrom EB, Hogstedt C. Lung cancer and exposure to diesel exhaust among bus garage workers. Scand J Work Environ Health 1990; 16:348–354.
71. Steenland NK, Siverman DT, Hornung RW. Case-control study of lung cancer and truck driving in the teamsters union. Am J Indust Med 1990; 17:577–591.
72. Kawabata Y, Udagawa T, Higuchi K, Yamada H, Hashimoto H, Iwai K. Lung injury and carcinogenesis following trans-tracheal instillation of diesel soot particule to the lung. J Jpn Soc Air Pollut 1988; 23:32–40.
73. Carrol KK. Experimental evidence of dietary factors and hormone dependent cancers. Cancer Res 1975; 35:3374–3383.
74. Lands WC, Kulmacz RJ, Marshall PJ. Lipid peroxide actions in the regulation of prostaglandin biosynthesis. In: Pryor WA, ed. Free Radical Biology. New York: Academic Press, 1984:39–61.
75. Newberene PM. Dietary fat, immunological respose and cancer in rats. Cancer Res 1981; 41: 3783–3785.
76. Gorog P. Activation of human blood monocytes by oxidized polyunsaturated fatty acids: A possible mechanism for the generation of lipid peroxides in the circulation. Int J Exp Pathol 1991; 72:227–237.
77. German JB. Food processing and lipid oxidation. Adv Exp Med Biol 1999; 459:23–50.
78. Ames BN. Dietary carcinogens and anticarcinogens. Science 1983; 221:1256–1264.
79. Rouseau EJ, Davidon AJ, Dunn B. Protection by β-carotene and related compounds against oxygen-mediated cytotoxicity and genotoxicity: Implications for carcinogenesis and anticarcinogenesis. Free Radical Biol Med 1992; 13:407–433.
80. Totter JR. Spontaneous cancer and its possible relationship to oxygen metabolism. Proc Natl Acad Sci USA 1980; 77:1763–1767.
81. Menkes MS, Comstock GW, Vuilleumer JP, Helsing KJ, Rider AA, Brookmeyer R. Serum beta-carotene, vitamin A and E, selenium, and the risk of lung cancer. N Engl J Med 1986; 315:1250–1254.
82. Nomura AMY, Stemmermann GN, Heilbrun LK, Salkeld RM, Vuileumier JP. Serum vitamin levels and the cancer of ecific sites in men of Japanese ancestry in Hawaii. Cancer Res 1985; 45:2369–2372.
83. LeGardeur BY, Lopez-S A, Johnson WD. A case-control study of serum vitamin A, E and C in lung cancer patients. Nutr Cancer 1990; 14:133–140.
84. Shekelle RB, Lepper M, Liu S, Maliza C, Raynor WJ, Rossof AH. Dietary vitamin A and risk of cancer in the Western Electric Study. Lancet 1981; 28:1185–1190.
85. Stahelin HB, Gey KF, Eichholzer M, Ludin E, Bernasconi F, Thurneysen J, Brubacher G. Plasma antioxidant vitamins and subsequent cancer mortality in the 12-year follow-up of the prospective basel study. Am J Epidemiol 1991; 133:766–775.
86. Wald NJ, Thompson SG, Densem JW, Boreham J, Bailey A. Serum vitamin E and β-carotene and subsequent risk of cancer: Results from the BUPA study. Br J Cancer 1988; 57:428–433.
87. Knekt P, Jarvinen R, Seppanen R, Rissanen A, Aromaa A, Heinonen OP, Albanes D, Heinonen M, Pukkala E, Teppo L. Dietary antioxidants and the risk of lung cancer. Am J Epidemiol 1991; 134:471–479.
88. Wong D, Mitchell CE, Wolff RK, Mauderly JL, Jeffrey AM. Identification of DNA damage as a result of exposure of rats to diesel exhaust. Carcinogenesis 1986; 7:1595–1597.

89. Bond JA, Wolff RK, Harkema JR, Mauderly JL, Henderson RF, Griffith WC, McClellan RO. Distribution of DNA adducts in the respiratory tract of rats exposed to diesel exhaust. Toxicol Appl Pharmacol 1988; 96:336–346.
90. Bond JA, Mauderly JL, Wolff RK. Concentration and time dependent formati ~ on of DNA-adducts in lungs of rats exposed to diesel exhaust. Toxicol 1990; 60:127–135.
91. Gallagher J, Lewtas J. Detection of DNA-adducts and mutagens after exposure to complex environmental mixture. Environ Mol Mutagen 1988; 11:36–37.
92. Bond JA, et al. The role of DNA adducts in diesel exhaust-induced pulmonary carcinogenesis. Prog Clin Biol Res 1990; 340:259–269.
93. Cui XS, et al. Early formation of DNA adducts compared with tumor formation in a long-term tumor study in rats after administration of 2-nitrofluorene. Carcinogenesis 1995; 16: 2135–2141.
94. Kasai H, Nishimura S. Formation of 8-hydroxydeoxyguanosine in DNA by oxygen radicals and its biological significance. In: Sies H, ed. Oxidative Stress, Oxidant and Antioxidants. London: Academic Press; 1991.
95. Cheng KC, Cahill DS, Kasai H, Nishimura S, Loeb L. 8-Hydroxyguanosine, an abundant form of oxidative DNA damage, causes G-T and C-A substitutions. J Biol Chem 1992; 267:166–172.
96. Nagashima M, Kasai H, Yokota J, Nagamachi Y, Ichinose T, Sagai M. Formation of an oxidative DNA damage, 8-hydroxydeoxyguanosine, in mouse lung DNA after intratracheal instillation of diesel exhaust particles and effects of high dietary fat and beta-carotene on this process. Carcinogenesis 1995; 16:1441–1445.
97. Mauderly JL, Snipes MB, Barr EB, Belinsky SA, Bond JA, Brooks AL, Cheng IY, Gillet NA, Griffith WC, Henderson RF, Mitchell CE, Nokula KJ, Thomassen DG. Pulmonary toxicity of inhaled diesel exhaust and carbon black in chronically exposed rats. Part I. Neo-plastic lung lesions. Health Effect Inst Res Rep 1994; 68:1–75.
98. Gallagher J, Heinrich U, George M, Hendee L, Phillips DH, Lewtas J. Formation of DNA adducts in rat lung following chronic inhalation of diesel emissions, carbon black and titanium dioxide particles. Carcinogenesis 1994; 15:1291–1299.

# 9

# The Role of Free Radicals in Asbestos- and Silica-Induced Fibrotic Lung Diseases

**ARTI SHUKLA, CYNTHIA R. TIMBLIN, and BROOKE T. MOSSMAN**

*University of Vermont College of Medicine, Burlington, Vermont*

**ANDREA HUBBARD**

*University of Connecticut School of Pharmacy, Storrs-Mansfield, Connecticut*

## I. INTRODUCTION

Occupational exposures to asbestos fibers and crystalline silica particles are associated with the development of fibrotic lung disease or pulmonary fibrosis (i.e., asbestosis and silicosis) (1,2). Asbestosis and silicosis are distinct lesions histologically in the lung, but are fundamentally similar in their outcome—a buildup of collagen and extracellular matrix material that may progress to ventilatory and restrictive abnormalities with reduced diffusing capacity (3,4). These manifestations may be progressive and lead to disability and death.

Asbestos and silica are different minerals, both chemically and physically. For example, there are the major crystalline forms of silica: quartz, tridymite, and cristobalite, in addition to other silicates that consist of silica bound to various metal cations (5). Silica and silicates generally exist as particles (i.e., having a $\leq 3{:}1$ length-to-diameter ratio). In contrast, asbestos family members are fibers having a $>3{:}1$ length-to-diameter ratio by definition (6) (Fig. 1). There are six chemically, physically, and geometrically dissimilar types of asbestos: chrysotile [$Mg_6Si_4O_{10}(OH)_8$], crocidolite [$Na_2(Fe^{3+})_2(Fe^{2+})_3Si_8O_{22}(OH)_2$], amosite [$(Fe,Mg)_7Si_8O_{22}(OH)_2$], tremolite [$Ca_2Mg_5Si_8O_{22}(OH)_2$], actinolite [$Ca_2(Mg,Fe)_5Si_8O_{22}(OH)_2$], and anthophyllite [$(Mg,Fe)_7Si_8O_{22}(OH)_2$]. The chemical constitution, size, and durability of asbestos fibers and silica particles may govern their biolog-

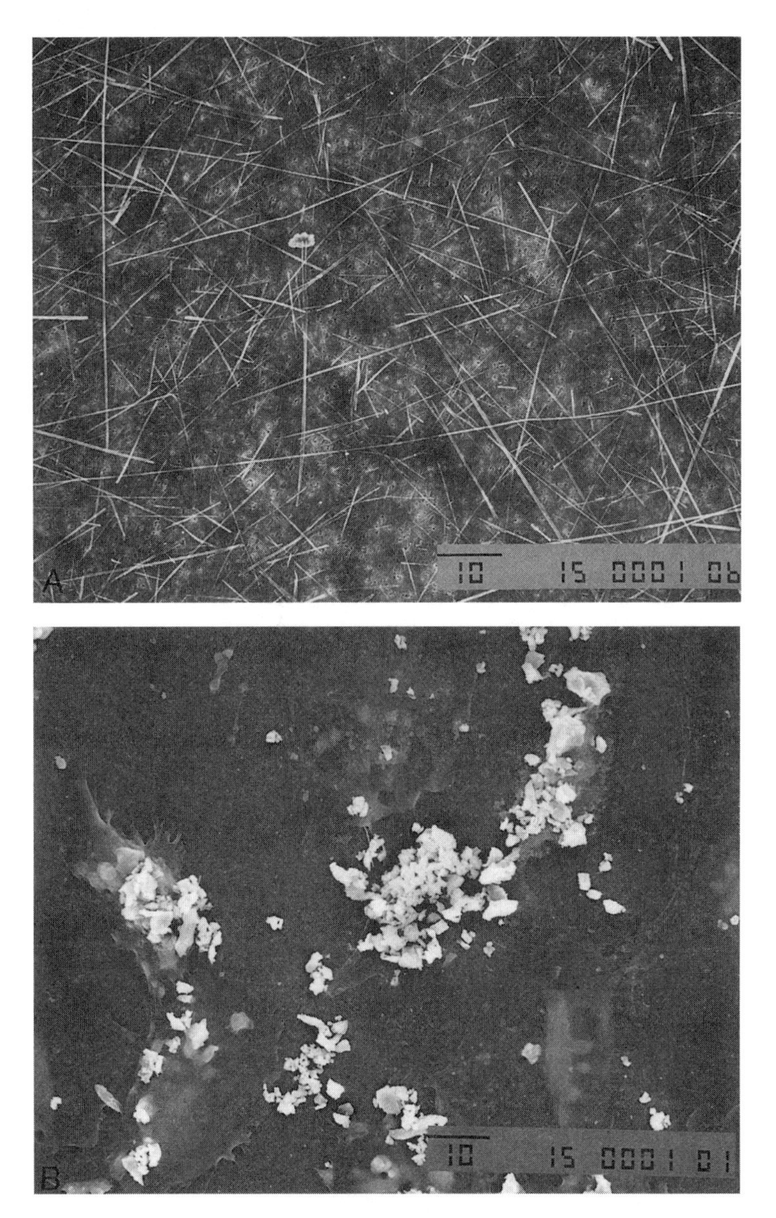


**Figure 1** Scanning electron microscopy showing different morphological features of asbestos fibers and silica particles. (top) Crocidolite asbestos, a rodlike, thin fiber, × 1000. (bottom) α-Quartz silica particles on the surfaces of pulmonary epithelial cells in culture, × 1000.

ical reactivity, propensity to cause pulmonary fibrosis, and ability to generate reactive oxygen species (ROS) (7).

In addition to causing pulmonary fibrosis (i.e., asbestosis), workers exposed to asbestos develop malignant mesotheliomas, nonmalignant pleural diseases, and lung cancer (2,6–9). In addition, crystalline silica has recently been classified by the International Agency for Research on Cancer (IARC) as a human carcinogen in lung (10). For these reasons, mechanisms of asbestos- and silica-induced cell injury, genotoxicity, and DNA damage have been studied extensively. Moreover, the altered gene transcription and expression of a variety of proteins that have an important modulatory role in cell proliferation, cell death, and inflammation have been characterized in a number of models. Work cited in this chapter suggests that many of these biological ramifications of exposures to these minerals may involve generation of ROS and reactive nitrogen species (RNS).

## II. MECHANISMS OF ASBESTOSIS

### A. Asbestos and Oxidative Stress

Many investigators (reviewed in Refs. 7,11–13) have suggested that ROS and RNS catalyzed by asbestos minerals have a role in early cell injury. Alveolar epithelial cell injury is an important early event in the pathogenesis of asbestosis, and both chrysotile and $H_2O_2$ directly affect alveolar epithelial resistance and permeability (14). However, the increase in epithelial permeability may not involve oxidants exclusively, but also tyrosine kinase activation (15).

There could be two possible mechanisms attributed to ROS generation by asbestos. The first mechanism involves the available iron on the fiber, which augments hydroxyl radical ($OH^{\bullet}$) formation from $H_2O_2$ in a Fenton-like reaction as ferrous ions are oxidized to ferric ions by $H_2O_2$ (16,17). This reaction depends upon the availability of $H_2O_2$. A second mechanism is the release of ROS during phagocytosis of long asbestos fibers by alveolar macrophages (AM) and neutrophils (18). During phagocytosis, the sustained generation of superoxide radical ($O_2^{\bullet}$) produces an increased level of $H_2O_2$, which may lead to a chain reaction of hydroxyl ($OH^{\bullet}$) generation catalyzed by iron present in the fibers. The asbestos types, amosite and crocidolite, contain a greater concentration of iron than chrysotile (5). Since iron-containing amphibole fibers are rodlike and more durable than chrysotile fibers in lung, a protracted release of ROS may contribute to their increased pathogenicity in the lung.

Asbestos-induced ROS and RNS can modify macromolecules such as DNA and RNA, leading to cellular dysfunction, cytotoxicity, and possibly malignant transformation (reviewed in 6,7,13). Electron spin resonance (ESR) spin trapping methods reveal that ROS are produced by asbestos in cell free systems (16,17). These studies suggest an important role of iron in asbestos toxicity and pathogenicity by providing evidence in $OH^{\bullet}$ radical generation and a stronger ESR signal from the high iron-containing types of asbestos (i.e., amosite and crocidolite) than chrysotile. The importance of iron in asbestos-generated $OH^{\bullet}$ has also been confirmed using deferoxamine, an iron chelator (19). Furthermore, Aust and associates (13,20) have demonstrated that the rate of mobilization of iron from asbestos and its reactivity in a cell-free system depends upon the pH and the chelator used. Under in vivo conditions, lung lining fluids may coat native fibers and therefore affect fiber surface chemistry. For example, surfactant-coated fibers release more $Fe^{3+}$ than native fibers at both pH 4.5 and 7.2 (21).

Interactions between asbestos fibers and cigarette smoke increase risks of bronchogenic carcinoma and, possibly, pulmonary fibrosis, and may be dependent upon iron-catalyzed free radicals. In addition, asbestos and cigarette tar may act in a cooperative or synergistic way to generate $OH^{\bullet}$ (22,23). Furthermore, amosite asbestos and aqueous whole cigarette smoke extracts induce DNA single-strand breaks (DNA-SB) in cultured alveolar epithelial cells by a mechanism that is, at least in part, due to the production of iron-catalyzed free radicals (24).

Phagocytosis of asbestos activates and enhances the release of significant amounts of $O_2^{\bullet-}$ from AM and neutrophils (18,25). Release of $O_2^{\bullet-}$ from phagocytes occurs spontaneously after in vitro and in vivo exposure to asbestos and continues for several hours. Long fibers of asbestos, which persist in cells, may cause ''frustrated'' phagocytosis and continued $O_2^{\bullet-}$ production. In addition to AMs, neutrophils may play a role in asbestos-induced inflammation and lung pathology. For example, abnormalities of pulmonary gas exchange in asbestos workers directly correlate with the percentage of neutrophils in bronchoalveolar lavage fluids, but not with the number or percentage of AMs (26,27). Whereas AMs from patients with asbestosis release increased levels of ROS, cytokines, and growth factors that may have a harmful effect on lung (reviewed in Ref. 7), AMs may also decrease asbestos-induced alveolar epithelial toxicity as they are better equipped to sequester the fibers from the epithelial cell (28). Both chrysotile asbestos and man-made fibers increase the production of $O_2^{\bullet-}$ and $H_2O_2$, deplete the concentration of glutathione (GSH), and increase the level of free intracellular calcium in alveolar macrophages (29).

Asbestos fibers may also directly activate ROS-producing enzymes such as nicotinamide adenine dehydrogenase (NADPH) oxidase and/or phospholipase C pathways which activate NADPH oxidase (30). In addition, staurosporin, an inhibitor of PKC and NADPH oxidase, attenuates crocidolite-induced $O_2^{\bullet-}$ release from human neutrophils (31). Asbestos can initiate lipid peroxidation in various cell types (32), and in rat liver and lung microsomes (33,34). More importantly, lipid peroxidation occurs in lung after inhalation of asbestos, particularly in alveolar macrophages (35).

In addition to ROS-induced effects, pulmonary toxicity by asbestos may be modulated by $NO^\bullet$. For example, asbestos induces the expression and activity of inducible $NO^\bullet$ synthase (iNOS) in AMs (36), A549 epithelial cells (37), and mesothelial cells (38). AMs from rats inhaling either crocidolite or chrysotile asbestos release increased nitrite and nitrate levels, and this effect is reduced by the iNOS inhibitor, NG-monomethyl-L-arginine (NMMA) (39). Inhalation of asbestos for 1 to 6 weeks augments the formation of RNS and nitrotyrosine in rat lungs and pleura (40). These data suggest that $NO^\bullet$-derived free radicals are produced after activation of iNOS by asbestos in various cell types and are responsible for pulmonary toxicity. Mutagenicity caused by asbestos in Chinese hamster V79 cells involves both iron and $NO^\bullet$, which may lead to the generation of another reactive mutagenic species, such as peroxynitrite ($ONOO^\bullet$), or inhibit a DNA repair enzyme(s) (41).

## B. Cellular and Molecular Changes by Asbestos

A number of cell types including AMs, pulmonary epithelial cells, mesothelial cells, endothelial cells, and fibroblasts are susceptible to the toxic effects of asbestos (7,42). Several studies also have shown that asbestos-induced cellular damage is mediated by oxidant stress in a variety of cell types (43–47). Sometimes adaptive and defense mechanisms also are activated following exposure to asbestos. For example, exposure of hamster tracheal epithelial cells in vitro to crocidolite and chrysotile asbestos increases total endogenous superoxide dismutase (SOD) activities over a time frame of several days (43). Also, overexpression of murine MnSOD cDNA results in increased mRNA levels and MnSOD activity that renders the cells less susceptible to the cytotoxic effects of crocidolite (48). Furthermore, transfecting WI-26 (type-1-like) epithelial cells with catalase also significantly protects these cells against both $H_2O_2$ and asbestos-induced DNA damage (49). Like crocidolite, rockwool, a man-made fiber, increases MnSOD gene expression in human bronchial epithelial cells (BEAS 2B), but this is not accompanied by elevations in MnSOD activity (50).

The molecular targets of ROS and RNS generated by asbestos include critical biological macromolecules such as lipids, DNA, and signal transduction proteins. As already discussed, asbestos-mediated lipid peroxidation is one mechanism that modifies cell membrane structure and function, and thiobarbituric acid reactive substance, a lipid peroxidation product, is present in the plasma of asbestos-exposed workers (51).

Asbestos is also considered genotoxic as it can cause DNA damage in cell-free systems as well as in pulmonary epithelial and pleural mesothelial cells (24,52,53). A murine mesothelial cell line spontaneously acquiring a point mutation in p53 gene shows increased sensitivity to DNA damage induced by crocidolite asbestos fibers (54). Based on this study, p53-deficient mice may show increased sensitivity to the genotoxic effects and accelerated development of malignant mesotheliomas. Though studies by some laboratories indicate that asbestos-induced mutagenic lesions in DNA (53) and/or apoptosis in mesothelial cells (54a) are due to oxidative damage, other studies (55) show that asbestos-induced DNA-SB in human mesothelial cells is not oxidant-dependent. These disparate results may reflect differences between the antioxidant status of cells or medium.

In addition to mutagenesis or transformation, DNA damage may result in necrosis or apoptosis, a major pathway responsible for the resolution of alveolar type II cell hyperplasia in acute lung injury (56). Asbestos-induced ROS generation could be one of the critical stimuli for induction of apoptosis, as ROS are known to induce apoptosis either directly or indirectly (57,58). Asbestos causes apoptosis in mesothelial cells, alveolar macrophages, and alveolar epithelial cells (24,54a,59,60). Yet, mesothelioma cell lines are highly resistant to apoptosis induced by either asbestos or $H_2O_2$ when compared with primary cultured mesothelial cells (61). One possible explanation for increased resistance of mesothelioma cell lines is their increased levels of antioxidant enzymes (i.e., MnSOD and catalase), which render them resistant to oxidative stress–induced effects (62). Human breast cancer cells (MCF-7) and IB-3 lung epithelial cells overexpressing MnSOD also are resistant to tumor necrosis factor-α (TNF-α)-, $H_2O_2$- and irradiation-induced apoptosis, also suggesting a possible role of ROS in the process of apoptosis (63,64).

### C. Cell Signaling by Asbestos

Asbestos can also stimulate signal transduction cascades as first shown by studies in which asbestos caused increased hydrolysis of phosphoinositides and elevations in diacylglycerol in tracheal epithelial cells (65). Increased activation of protein kinase C (66,67) and extracellular signal-regulated kinases (ERKs 1 and 2) (68) have also been demonstrated in rat pleural mesothelial (RPM) cells exposed to asbestos. Further studies demonstrated that asbestos-induced apoptosis in RPM cells is in part due to an activation of ERKs, but not c-Jun–N-terminal kinases (JNK/SAPK), while $H_2O_2$-induced apoptosis was associated with activation of both signaling pathways in this cell type (69). A role for iron-catalyzed ROS in induction of apoptosis is possible as asbestos-associated ERK activity was reduced by addition of catalase, *N*-acetylcysteine (NAC) or deferoxamine.

Nuclear factor kappa-B (NF-κB), a transcription factor associated with oxidative stress, is known to be involved in the activation of various genes including cytokines, growth factors, adhesion molecules, and iNOS. Crocidolite asbestos causes prolonged dose-dependent transcriptional activation of NF-κB-dependent gene expression in hamster tracheal epithelial (HTE) cells (70). Moreover, elimination of asbestos-induced NF-κB DNA binding by pretreatment with NAC suggested a role for ROS. Recently, it has been demonstrated that brief inhalation exposures of rats to crocidolite or chrysotile asbestos result in marked increases in p65 (a subunit of NF-κB) immunofluorescence in bronchiolar and alveolar duct epithelial cells at the sites of initial fiber deposition (71).

Asbestos-induced ROS generation activates NF-κB and IL-6 in A549 and normal human bronchial epithelial cells (72,73). Moreover, long fiber amosite asbestos increases NF-κB and AP-1 transcriptional activity in rat alveolar macrophages via ROS (74). Support for the hypothesis that ROS generated from asbestos causes transcription factor activation also is suggested by studies showing that NF-κB and AP-1 binding or transactivation is reduced by vitamin E, an inhibitor of lipid peroxidation (75). Lipoxygenase metabolites of arachidonic acid, which are produced following lipid peroxidation, may be involved in NF-κB transcription factor activation by asbestos. Thus, data summarized above indicate a role of free radicals in regulating transcription factor activation by asbestos, which in turn may initiate several cellular responses.

### D. Cell Signaling and Proliferation

Asbestos and asbestos-derived free radicals can augment cellular proliferation, increases in ornithine decarboxylase, squamous metaplasia, and inflammation. Thus fibers may act as conventional tumor promoters such as phorbol esters (reviewed in Refs. 6, 77–79). Amosite, crocidolite, and chrysotile asbestos also promote morphological transformation and elevate the expression of an array of genes invariably induced by tumor promoters in C3H/10T1/2 cells (80). These effects are reduced by addition of NAC, suggesting a possible role of ROS.

Asbestos activates the activator-protein-1 (AP-1) family of transcription factors, including homodimeric (Jun/Jun) and heterodimeric (Fos/Jun) proteins derived from the *c-fos* and *c-jun* proto-oncogene families (74,75,81–84). Moreover, asbestos causes proliferation and transformation in HTE cells by an AP-1-dependent mechanism (83).

### E. Oxidative Stress and Cytokines in Asbestosis

A large number of chemokines and cytokines have been implicated in asbestos-induced inflammation and asbestosis (reviewed in Ref. 7). These factors amplify cellular injury and may attract and activate inflammatory cells, stimulate proliferation of epithelial cells and fibroblasts, and promote collagen deposition.

TNF-α, an oxidant stress that increases MnSOD expression in a variety of cell types, is a key cytokine involved in asbestos-induced pulmonary toxicity (85,86). Alveolar macrophages from patients with asbestosis and idiopathic pulmonary fibrosis have been shown to release increased levels of TNF-α (87). Iron-catalyzed free radicals are demonstrated to be involved in asbestos-induced TNF-α release from the AMs (88). Studies using TNF-receptor (TNFR) knockout mice suggest that TNF-α is a key proximal mediator of asbestos-induced pulmonary toxicity (86). In these studies, asbestos causes inflammation, cell proliferation, and fibrosis in the wild-type and single TNFR knockout mice; however, double TNFR knockout showed no asbestos-induced damage. Recently, work from the same laboratory (89) using an in situ hybridization technique showed that the expression of the genes encoding transforming growth factor-α (TGF-α) and platelet-derived growth factor-A (PDGF-A) chain are reduced in the C57BL/6-129 TNF-α knockout mice in comparison to control animals. These findings confirm earlier studies showing that TNF-α is essential for the development of interstitial pulmonary fibrosis and indicate that TNF-α mediates its effects through activation of other growth factors such as PDGF and TGF-α that control cell growth and matrix production. Involvement of TNF-α as a key mediator in asbestos-induced lung injury indicates that reducing TNF-α expression either by antioxidants, iron chelators, or anti-TNF-α antibodies could inhibit fibrosis as has already been shown by others in models of bleomycin- and silica-induced fibrosis in mice (90,91).

A novel group of proinflammatory cytokines known as chemokines, which include IL-8, macrophage inflammatory protein-2 (MIP-2), macrophage inflammatory protein-1α and β (MIP-1α and β), and monocyte chemoattractant proteins 1, 2, and 3 (MCP-1, -2, -3) has been implicated as key contributors to the inflammatory response in bleomycin-induced fibrosis (92) and asbestosis (93,94). The bronchiolar and alveolar epithelial cells, in addition to AMs, synthesize MIP-2, which is upregulated after in vitro addition of TNF-α, asbestos, or silica to rat alveolar type II epithelial cells (95). This shows that pulmonary epithelial cells are mediators as well as targets of inflammation, an additional source of ROS/RNS production in lung, and a feature of early asbestosis (7).

## III. MECHANISMS OF SILICOSIS

### A. Silica-Induced Generation of Free Radicals

The physical and chemical properties of silica particles also impact on the potential of this inorganic crystal to elicit cell injury, lung damage, and resulting fibrosis. Upon interaction of silica particles with pulmonary cells, cytotoxicity may be initiated by membranolysis after adsorption of constituents of the cell membrane onto the silica particles (96). The degree of adsorption is related to the distribution and abundance of silanol (SiOH) groups at the surface of the crystal (97,98) and to silanol groups dissociated in water (99). The role of these silanol groups in cell viability is confirmed by studies in which the silica has been coated with polymers (100), chemically modified (101), or treated with aluminum salts (102). Alternatively, the generation of free radicals by silica may be more important in noncytolytic cell responses including apoptosis (102a).

Silica-derived ROS include $OH^\bullet$, $O_2^{\bullet-}$, and peroxides (103), which contribute to increased oxidative stress in cells and lung tissue. These species play a role in initiating injury, cellular inflammation, and eventual long-term consequences (103,104–106). Freshly ground (fractured) dusts are more inflammatory and fibrogenic than aged crystals (107). This is in part due to the greater generation of silica-derived free radicals in freshly ground material where surface peroxides or hydroperoxides are formed (104,105,108). Upon grinding, the SiO bonds are ruptured by heterolytic cleavage, and $Si^\bullet$ and $SiO^\bullet$ radicals are produced (109). These react with oxygen to form $SiO_2^\bullet$, $SiO_3^\bullet$, and $SiO_2^{\bullet-}$. These free radicals may also react with $H_2O_2$ to produce $O_2^{\bullet-}$ and $OH^\bullet$ (110). The presence of iron in some samples of silica can also enhance the generation of ROS leading to DNA damage, cell transformation, and pulmonary injury (111). These trace amounts of iron can generate free radicals through Fenton chemistry (110,112).

The contribution of particle surface-derived ROS to pulmonary injury is evidenced in studies in which the surface of silica particles is modified by different masking agents. For example, masking active surface sites of quartz with aluminum compounds is thought to modify the redox and acid-base protection of silica and reduce surface radical species that result in a decrease in biological activity (113). In addition, adsorption of the polymer polyvinyl-pyridine-*N*-oxide (PVPNO) to quartz surface appears to scavenge particle-generated OH$^•$ and to decrease macrophage cytotoxicity (114).

## B. Cellular Generation of ROS in Response to Silica

Generation of cellular oxidants by silica-activated cells has resulted in cell and lung injury, activation of MAP kinase pathways, activation of specific transcription factors, and increased expression of inflammatory cytokines.

Crystalline silica is a potent stimulant of the respiratory burst in phagocytic cells and results in increased oxygen consumption and production of ROS. Silica, when added to AMs, stimulates $O_2^{•-}$ production, $H_2O_2$ release, and enhanced chemiluminescence (115). By electron spin resonance (ESR) spin trapping, the enhanced generation of ROS is directly related to the known toxicity of different particles when normalized to equal surface areas. In addition, bronchoalveolar lavage cells from silica-exposed rats show enhanced oxygen consumption, $H_2O_2$ release, and chemiluminescence in response to an in vitro stimulation with unopsonized zymosan particles (116). In rats, silica exposure by intratracheal injection elicits increased OH$^•$ production in lung tissue compared to rats receiving the nontoxic particle, titanium dioxide (117). This was the first study to describe an association between acute inflammation and OH$^•$ generation in the lungs of rats exposed to silica. Exposure to freshly fractured silica also results in enhanced generation of ROS, lung injury, and inflammation (107). In addition, the presence of surface complexed iron appears to potentiate silica-induced cell damage and lung injury. For example, increased amounts of surface iron enhance the ability of silica to generate oxidants in vitro, to stimulate the respiratory burst by rat AMs in vitro, and to elicit acute pulmonary inflammation in rats exposed by intratracheal instillation (118).

Increased oxidative stress elicited by silica is also evidenced by increased expression of antioxidant enzymes. In lungs of rats exposed by inhalation to cristobalite silica, a significant increase in MnSOD protein occurs which is localized primarily to type II epithelial cells (119). This is reflected by a significant increase in steady-state mRNA levels of MnSOD and glutathione peroxidase (120). However, no increases in activities of these antioxidant enzymes occur in the lungs of these animals. More recently, it was noted in isolated rat AMs that silica led to a dose- and time-dependent decrease in intracellular GSH (121). Incubation with the GSH precursor, NAC, decreased silica-induced ROS formation as well as changes in membrane permeability and DNA-SB.

Finally, silica instillation into rat lungs elicits increased mRNA for inducible nitric oxide synthase (iNOS) in alveolar macrophages and an enhanced iNOS-dependent chemiluminescence (122). This increased production of NO$^•$ is implicated in increased lung injury since the reaction of NO$^•$ with $O_2^•$ forms the potent free radical, peroxynitrite, also capable of causing cell damage (123).

## C. Role of Cellular Oxidants in MAP Kinase Activation Following Silica Exposure

Silica-induced oxidative stress can initiate the activation of cell signaling pathways. In the rat fibroblast cell line, Rat2, silica exposure stimulated intracellular ROS production as well as increased ERK phosphorylation (124). This phosphorylation of ERK could be attenuated by catalase, but not by SOD, suggesting a role of silica-induced $H_2O_2$ production. Interestingly, SOD treatment appeared to enhance the phosphorylation of these enzymes. Recent work by our laboratory shows that α-quartz stimulates JNK phosphorylation and activity as well as ERK phosphorylation in a murine alveolar type II epithelial cell line. These changes are associated with transient production of oxidants and precede alterations in expression of AP-1 family members (102a).

## D. Role of Cellular Oxidants in NF-κB and AP-1 Regulation Following Silica Exposure

Silica-induced oxidative stress can also activate specific transcription factors including NF-κB and AP-1. In the mouse macrophage cell line, RAW 264.7, silica exposure causes the activation of NF-κB 2 to 12 h after exposure as detected by gel mobility shift analyses (125). In this model, the presence of the antioxidant NAC did not affect silica-induced NF-κB activation, suggesting a minimal role for ROS in activation of this transcription factor. In subsequent studies, catalase, formate, and deferoxamine inhibited NF-κB activation whereas SOD enhanced activation, suggesting that $OH^{\bullet}$ radicals play a key role in silica-induced NF-κB activation (126). In contrast, the presence of the inhibitor of iNOS, *N*-monomethyl-L-arginine (NMMA), enhanced silica-induced NF-κB activation, suggesting that $NO^{\bullet}$ is important in a negative-feedback regulation of NF-κB (127). In more recent work, bronchoalveolar lavage (BAL) cells from rats instilled with silica demonstrated enhanced NF-κB binding to DNA over an 18-h time course (128). Treatment with the anti-inflammatory agent, dexamethasone, decreased NF-κB activation and concomitantly decreased luminol-dependent chemiluminescence in PMA-stimulated BAL cells.

Few studies have evaluated activation of AP-1 by silica. Using AP-1 luciferase reporter transgenic mice, Ding et al. (129) demonstrated that IT administration of silica increased luciferase activity 3 days postexposure, indicating AP-1 activation in lung tissue of exposed mice. In addition, exposure of a rat epithelial cell line (RLE) stably transfected with an AP-1 luciferase reporter plasmid to silica showed AP-1 activation.

## E. Cytokine Expression Induced by Silica

One result of transcription factor activation by silica might be increased cytokine expression, another event influenced by silica-induced ROS. AMs lavaged from silica-exposed rats show increased zymosan or phorbol ester (PMA)–triggered chemiluminescence as well as increased TNF-α mRNA expression and protein secretion (130). Pretreatment of these animals with a free radical scavenger (*N*-ter-butyl-α-phenylnitrone) reversed lung pathology and decreased ROS and TNF-α production from the AMs. In another study, pretreatment in vitro of mouse macrophages with antioxidants [dimethylsulfoxide (DMSO), GSH, or NAC] prior to exposure to cristobalite silica significantly decreased TNF-α mRNA and protein production. Many of these same antioxidants also decreased mRNA levels for macrophage inflammatory protein (MIP)-2, MIP-1α, MIP-1β, and MCP-1 (131). Comparable observations were made by these investigators in a murine lung epithelial (MLE)-15 cell line exposed to cristobalite silica (132). Again, various antioxidants [extracellular GSH, DMSO, NAC, and buthionine sulfoximine (BSO)] decreased TNF-α-induced and cristobalite-induced mRNA expression of MIP-2 and MCP-1. However, these antioxidants did not reduce silica-induced TNF-α mRNA expression. These results suggest that the role of oxidants in enhancing cytokine expression is more pronounced in a mouse macrophage cell line than in a mouse epithelial cell line.

The role of oxidants in silica-induced cytokine expression has also been investigated in human cells. In human lung epithelial cells (A549) primed with TNF-α, IL-8 production is increased after addition of silica (133). This enhanced cytokine response could be, in part, attributed to ROS, since pretreatment with NAC decreased IL-8 production by approximately 50%.

## IV. SUMMARY

Asbestos fibers and silica particles are chemically and physically dissimilar, yet a common feature of asbestos- and silica-induced cell signaling and injury involves the elaboration of ROS/RNS by acellular and cellular mechanisms. The insolubility of these pathogenic minerals and their retention in lung over time may be a source of chronic release of oxidants. Since lung diseases associated with occupational exposure to asbestos or silica have protracted latency periods and are generally related to the concentrations of these minerals in the workplace and duration of exposures, antioxidant responses important in lung repair from mineral-induced injury may be overwhelmed at high concentrations of minerals. Thus, the balance of antioxidants in relationship to oxidant stress may govern the development of asbestos or silica-associated diseases. In support of this hypothesis, a direct relationship between oxidant stress, inflammation, and asbestosis has been demonstrated in a rodent inhalation model of asbestosis in which these endpoints were inhibited after administration of catalase (134). These studies and others suggest that administration of antioxidants may be a feasible approach clinically for prevention and/or therapy of asbestos- and silica-induced lung diseases (2,6,7).

## REFERENCES


1. Beckett W, Abraham J, Becklake M, Christiani D, Cowie R, Davis G, Jones R, Kreiss K, Parker J, Wagner G. Adverse effects of crystalline silica exposure. Am J Respir Crit Care Med 1997; 155:761–765.
2. Mossman BT, Gee JBL. Medical Progress. Asbestos-related diseases. N Engl J Med 1989; 320:1721–1730.
3. Craighead JE, Abraham JL, Churg A, Green FHY, Kleinerman J, Pratt PC, Seemayer TA, Vallyathan V, Weill H. Asbestos-associated diseases. Arch Pathol Lab Med 1982; 106:544–596.
4. Craighead JE, Kleinerman J, Abraham JL, Gibbs AR, Green FHY, Harley RA, Ruettner JR, Vallyathan NV, Juliano EB. Diseases associated with exposure to silica and nonfibrous silicate minerals. Arch Pathol Lab Med 1988; 112:673–790.
5. Guthrie GD Jr., Mossman BT. Health Effects of Mineral Dusts. Reviews in Mineralogy, Vol. 28 (Series Editor: PH Ribbe). Washington, DC: Mineralogical Society of America, 1993:1–584.
6. Mossman BT, Bignon J, Corn M, Seaton A, Gee JB. Asbestos: scientific developments and implications for public policy. Science 1990; 247:294–301.
7. Mossman BT, Churg A. Mechanisms in the pathogenesis of asbestosis and silicosis. Am J Respir Crit Care Med 1998; 157:1666–1680.
8. Mossman BT, Kamp DW, Weitzman SA. Mechanisms of carcinogenesis and clinical features of asbestos-associated cancers. Cancer Invest 1996; 14:464–478.
9. Kamp DW, Weitzman SA. Asbestosis: clinical spectrum and pathogenic mechanisms. Proc Soc Exp Biol Med 1997; 214:12–26.
10. IARC Monographs on the Evaluation of Carcinogenic Risks to Humans. Silica, Some Silicates, Coal Dust and Para-Aramid Fibrils, Vol. 68, Lyon, France, 1999.
11. Kamp DW, Graceffa P, Pryor WA, Weitzman SA. The role of free radicals in asbestos-induced diseases. Free Rad Biol Med 1992; 12:293–315.

12. Vallyathan V, Shi X. The role of oxygen free radical in occupational and environmental lung diseases. Environ Health Perspect 1997; 105:165–177.
13. Hardy JA, Aust AE. Iron in asbestos chemistry and carcinogenicity. Chem Rev 1995; 95: 97–118.
14. Gardner SY, Brody AR, Mangum JB, Everitt JI. Chrysotile asbestos and $H_2O_2$ increase permeability of alveolar epithelium. Exp Lung Res 1997; 23:1–16.
15. Peterson MW, Kirschbaum J. Asbestos induced lung epithelial permeability: potential role of nonoxidant pathways. Am J Physiol 1998; 275:L262–L268.
16. Weitzman SA, Graceffa P. Asbestos catalyzes hydroxyl and superoxide radical generation from hydrogen peroxide. Arch Biochem Biophys 1984; 228:373–376.
17. Gulumian M, Van Wyk JA. Hydroxyl radical production in the presence of fibers by a Fenton-type reaction. Chem Biol Interact 1987; 62:89–97.
18. Hansen K, Mossman BT. Generation of superoxide ($O_2^{\bullet -}$) from alveolar macrophages exposed to asbestiform and nonfibrous particles. Cancer Res 1987; 47:1681–1686.
19. Weitzman SA, Chester JF, Graceffa P. Binding of deferoxamine to asbestos fibres in vitro and in vivo. Carcinogenesis 1988; 9:1643–1645.
20. Lund LG, Aust AE. Iron mobilization from asbestos by chelators and ascorbic acid. Arch Biochem Biophys 1990; 278:60–64.
21. Fisher CE, Brown DM, Shaw J, Beswick PH, Donaldson K. Respirable fibres: surfactant coated fibres release more $Fe^{3+}$ than native fibers at both pH4.5 and 7.2. Ann Occup Hyg 1998; 42:337–345.
22. Jackson JH, Schraufstatter IU, Hyslop PA, Vosbeck K, Sauerheber R, Weitzman SA, Cochrane CG. Role of oxidants in DNA damage: hydroxyl radical mediates the synergistic DNA damaging effects of asbestos and cigarette smoke. J Clin Invest 1987; 80:1090–1095.
23. Valavanidis A, Balomenou H, Macropoulou I, Zarodimos I. A study of the synergistic interaction of asbestos fibers with cigarette tar extracts for the generation of hydroxyl radicals in aqueous buffer solution. Free Rad Biol Med 1996; 20:853–858.
24. Kamp DW, Greenberger MJ, Sbalchierro JS. Cigarette smoke augment and asbestos induced alveolar epithelial cell injury: role of free radicals. Free Rad Biol Med 1998; 25:728–739.
25. Vallyathan V, Mega JF, Shi X, Dalal NS. Enhanced generation of free radicals from phagocytes induced by mineral dusts. Am J Respir Cell Mol Biol 1992; 6:404–413.
26. Garcia JGN, Griffith DE, Cohen AB, Callahan KS. Alveolar macrophages from patients with asbestos exposure release increased levels of leukotriene B4. Am Rev Respir Dis 1989; 139: 1494–1501.
27. Schwartz DA, Galvin JR, Merchant RK, Dayton CS, Burmeister LF, Merchant JA, Hunninghake GW. Influence of cigarette smoking on bronchoalveolar lavage cellularity in asbestos induced lung disease. Am Rev Respir Dis 1992; 145:400–405.
28. Kamp DW, Dunn MM, Sbalchiero JS, Knap AM, Weitzman SA. Contrasting effects of alveolar macrophages and neutrophils on asbestos induced pulmonary epithelial cell injury. Am J Physiol (Lung Cell Mol Physiol) 1994; 266:L84–L91.
29. Wang QE, Han CH, Wu WD, Wang HB, Liu SJ, Kohyama N. Biological effects of man-made mineral fibers. I. Reactive oxygen species production and calcium homeostasis in alveolar macrophages. Ind Health 1999; 37:62–67.
30. Roney PL, Holian A. Possible mechanisms of chrysotile asbestos-stimulated superoxide anion production in guinea pig alveolar macrophages. Toxicol Appl Pharmacol 1989; 100: 132–144.
31. Ishizaki T, Yano E, Evans PH. Cellular mechanisms of reactive oxygen metabolite generation from human polymorphonuclear leukocytes induced by crocidolite asbestos. Environ Res 1997; 75:135–140.
32. Gabor S, Anca Z. Effects of asbestos on lipid peroxidation in the red cells. Br J Ind Med 1975; 32:39–41.

33. Gulumian M, Sardianos F, Kilroe-Smith T, Ockerse G. Lipid peroxidation in microsomes induced by crocidolite fibres. Chem Biol Interact 1983; 44:111–118.
34. Gulumian M, Kilroe-Smith TA. Crocidolite induced lipid peroxidation in rat lung microsomes. Environ Res 1987; 43:267–273.
35. Petruska JM, Wong SHY, Sunderman Jr. FW, Mossman BT. Detection of lipid peroxidation in lung and in bronchoalveolar lavage cells and fluid. Free Rad Biol Med 1990; 9:51–58.
36. Thomas G, Ando T, Verma K, Kagan E. Asbestos fibres and interferon upregulate nitric oxide production in rat alveolar macrophages. Am J Respir Cell Mol Biol 1994; 11:707–715.
37. Chao CC, Park SH, Aust AE. Participation of nitric oxide and iron in the oxidation of DNA in asbestos-treated human lung epithelial cells. Arch Biochem Biophys 1996; 326: 152–157.
38. Choe N, Tanaka S, Kagan E. Asbestos fibres and interleukin-1 upregulate the formation of reactive nitrogen species in rat pleural mesothelial cell. Am J Respir Cell Mol Biol 1998; 19:226–236.
39. Quinlan TR, BeruBe KA, Hacker MP, Taatjes DJ, Timblin CR, Goldberg J, Kimberley P, O'Shaughnessy X, Hemenway D, Torino J, Jimenez LA, Mossman BT. Mechanisms of asbestos induced nitric oxide production by rat alveolar macrophages in inhalation and in vitro models. Free Rad Biol Med 1998; 24:778–788.
40. Tanaka S, Choe N, Hemenway DR, Zhu S, Matalon S, Kagan E. Asbestos inhalation induces reactive nitrogen species and nitrotyrosine formation in the lungs and pleura of the rat. J Clin Invest 1998; 102:445–454.
41. Park SH, Aust AE. Participation of iron and nitric oxide in the mutagenicity of asbestos in hgprt−, gpt+ Chinese hamster V79 cells. Cancer Res 1998; 58:1144–1148.
42. Churg A. The uptake of mineral particles by pulmonary epithelial cells. Am J Respir Crit Care Med 1996; 154:1124–1140.
43. Mossman BT, Marsh JP, Shatos MA. Alteration of superoxide dismutase activity in tracheal epithelial cells by asbestos and inhibition of cytotoxicity by antioxidants. Lab Invest 1986; 54:204–212.
44. Shatos MA, Doherty JM, Marsh JP, Mossman BT. Prevention of asbestos induced cell death in rat lung fibroblasts and alveolar macrophages by scavengers of active oxygen species. Environ Res 1987; 44:103–116.
45. Garcia JGN, Gray LD, Dodson RF, Callahan KS. Asbestos-induced endothelial cell activation and injury. Am Rev Respir Dis 1988; 138:958–964.
46. Goodglick L, Kane A. Role of reactive oxygen metabolites in crocidolite asbestos toxicity to mouse macrophages. Cancer Res 1986; 46:5558–5566.
47. Goodglick LA, Kane AB. Cytotoxicity of long and short crocidolite asbestos fibers in vitro and in vivo. Cancer Res 1990; 50:5153–5163.
48. Mossman BT, Surinrut P, Brinton BT, Marsh JP, Heintz NH, Lindau-Shepard B, Shaffer JB. Transfection of a manganese-containing superoxide dismutase gene into hamster tracheal epithelial cells ameliorates asbestos-mediated cytotoxicity. Free Rad Biol Med 1996; 21: 125–131.
49. Kamp DW, Pollack N, Yeldandi A, et al. Catalase reduces asbestos-induced DNA damage in pulmonary epithelial cells. Am J Respir Crit Care Med 1997; 155:687A.
50. Marks-Konczalik J, Gillissen G, Jaworska M, Lascke S, Voss B, Eisseler-Eckhoff A, Schmitz I, Schultze-Werninghaus G. Induction of manganese superoxide dismutase gene expression in bronchoepithelial cells after rockwool exposure. Lung 1998; 176:165–180.
51. Kamal AA, Gomaa A, el Khafif M, Hammad AS. Plasma lipid peroxides among workers exposed to silica or asbestos dusts. Environ Res 1989; 49:173–180.
52. Jaurand M-C. Mechanisms of fibre-induced genotoxicity. Environ Health Perspect 1997; 105(suppl 5):1073–1084.
53. Fung H, Kow YW, Van Houten B, Mossman BT. Patterns of 8-hydroxydeoxyguanosine

formation in DNA and indications of oxidative stress in rat and human pleural mesothelial cells after exposure to crocidolite asbestos. Carcinogenesis 1997; 18:825–832.
54. Marsella JM, Liu BL, Vaslet CA, Kane AB. Susceptibility of p53-deficient mice to induction of mesothelioma by crocidolite asbestos fibers. Environ Health Perspect 1997; 105(suppl 5): 1069–1072.
54a. Broaddus VC, Yang L, Scavo LM, Ernst JD, Boylan AM. Crocidolite asbestos induces apoptosis of pleural mesothelial cells: role of reactive oxygen species and poly (ADP-robosyl) polymerase. Environ Health Perspect 1997; 105(suppl 5):1147–1152.
55. Ollikainen T, Linnainmaa K, Kinnula VL. DNA strand breaks induced by asbestos fibers in human pleural mesothelial cells in vitro. Environ Mol Mutagen 1999; 33:153–160.
56. Uhal BD. Cell cycle kinetics in the alveolar epithelium. Am J Physiol (Lung Cell Mol Physiol) 1997; 272:L1031–1045.
57. Hagar H, Ueda N, Shah SV. Role of reactive oxygen metabolites in DNA damage and cell death in chemical hypoxic injury to LLC-KK1 cells. Am J Physiol (Renal Fluid Electrolyte Physiol) 1996; 271:F209–F215.
58. McGowan AJ, Ruiz-Ruiz MC, Gorman AM, Lopez-Rivas A, Cotter TG. Reactive oxygen intermediate(s) (ROI): common mediator(s) of poly CADP-ribose polymerase (PARP) cleavage and apoptosis. FEBS Lett 1996; 392:299–303.
59. BeruBe KA, Quinlan TR, Fung H, Magae J, Vacek P, Taatjes DJ, Mossman BT. Apoptosis is observed in mesothelial cells after exposure to crocidolite asbestos. Am J Respir Cell Mol Biol 1996; 15:141–147.
60. Hamilton RF, Lyer R, Holian A. Asbestos induces apoptosis in human alveolar macrophages. Am J Physiol (Lung Cell Mol Physiol) 1996; 271:L813–L819.
61. Narasimhan SR, Yang L, Gerwin BI, Broaddus C. Resistance of pleural mesothelioma cell lines to apoptosis: relation to expression of Bcl-2 and Bax. Am J Physiol (Lung Cell Mol Physiol) 1998; 275:L165–L171.
62. Kahlos K, Anttila S, Asikainen T, Kinnula K, Raivio KO, Mattson K, Linnainmaa K, Kinnula VL. Manganese superoxide dismutase in healthy human pleural mesothelium and in malignant pleural mesotheliomas. Am J Respir Cell Mol Biol 1998; 18:570–580.
63. Manna SK, Zhang HJ, Yan T, Oberley LW, Aggarwal BB. Overexpression of manganese superoxide dismutase suppresses tumor necrosis factor-induced apoptosis and activation of nuclear transcription factor-κB and activated protein-1. J Biol Chem 1998; 273:13245–13254.
64. Zwacka RM, Dudus L, Epperly MW, Greenberger JS, Engelhardt JF. Redox gene therapy protects human IB-3 lung epithelial cells against ionizing radiation-induced apoptosis. Human Gene Therapy 1998; 9:1381–1386.
65. Sesko A, Cabot M, Mossman BT. Hydrolysis of phosphoinositides precedes cellular proliferation in asbestos-stimulated tracheobronchial epithelial cells. Proc Natl Acad Sci USA 1990; 87:7385–7389.
66. Perderiset M, Marsh JP, Mossman BT. Activation of protein kinase C by asbestos in hamster tracheal epithelial cells. Carcinogenesis 1991; 12:1499–1502.
67. Fung H, Quinlan TR, Janssen YMW, Timblin CR, Marsh JP, Heintz NH, Taatjes DJ, Vacek P, Jaken S, Mossman BT. Inhibition of protein kinase C (PKC) prevents asbestos-induced c-fos and c-jun protooncogene expression in mesothelial cells. Cancer Res 1997; 57:3101–3105.
68. Zanella CL, Posada J, Tritton TR, Mossman BT. Asbestos causes stimulation of the ERK-1 mitogen-activated protein kinase cascade after phosphorylation of the epidermal growth factor receptor. Cancer Res 1996; 56:5334–5338.
69. Jimenez LA, Zanella C, Fung H, Janssen YMW, Vacek P, Charland C, Goldberg J, Mossman BT. Role of extracellular signal-regulated protein kinase in apoptosis by asbestos and $H_2O_2$. Am J Physiol 1997; 273:L1029–L1035.
70. Janssen YMW, Barchowsky A, Treadwell M, Driscoll KE, Mossman BT. Asbestos induces

nuclear factor κB (NF-κB) DNA binding activity and NF-κB dependent gene expression in tracheal epithelial cells. Proc Natl Acad Sci USA 1995; 92:8458–8462.
71. Janssen YMW, Driscoll KE, Howard B, Quinlan TR, Treadwell M, Barchowsky A, Mossman BT. Asbestos causes translocation of p65 protein and increases NF-κB DNA binding activity in rat lung epithelial and pleural mesothelial cells. Am J Pathol 1997; 151:389–401.
72. Simeonova PP, Toriumi W, Kommineni C, Erkan M, Munson AE, Rom WN, Luster MI. Molecular regulation of IL-6 activation by asbestos in lung epithelial cells: role of reactive oxygen species. J Immunol 1997; 159:3921–3928.
73. Luster MI, Simeonova PP. Asbestos induces inflammatory cytokines in the lung through redox sensitive transcription factors. Toxicol Lett 1998; 102/103:271–275.
74. Gilmour PS, Brown DM, Beswick PH, MacNee W, Rahman I, Donaldson K. Free radical activity of industrial fibers: role of iron in oxidative stress and activation of transcription factors. Environ Health Perspect 1997; 105:1313–1317.
75. Faux SP, Howden PJ. Possible role of lipid peroxidation in the induction of NF-kappa B and AP-1 in RLF-6 cells by crocidolite asbestos: evidence following protection by vitamin E. Environ Health Perspect 1997; 105:1127–1130.
76. Woodworth CD, Mossman BT, Craighead JE. Induction of squamous metaplasia in organ cultures of hamster trachea by naturally occurring and synthetic fibers. Cancer Res 1983; 43:4906–4913.
77. Mossman BT, Craighead JE, MacPherson BV. Asbestos-induced epithelial changes in organ cultures of hamster trachea: inhibition by retinyl methyl ether. Science 1980; 207:311–313.
78. Marsh JP, Mossman BT. Mechanisms of induction of ornithine decarboxylase activity in tracheal epithelial cells by asbestiform minerals. Cancer Res 1988; 48:709–714.
79. Marsh JP, Mossman BT. Role of asbestos and active oxygen species in activation and expression of ornithine decarboxylase in hamster tracheal epithelial cells. Cancer Res 1991; 51: 167–173.
80. Parfett CL, Pilon R, Caldeirra AA. Asbestos promotes morphological transformation and elevated expression of a gene family invariably induced by tumor promoters in C3H/10T1/2 cells. Carcinogenesis 1996; 17:2719–2726.
81. Heintz NH, Janssen YMW, Mossman BT. Persistent induction of c-fos and c-jun expression by asbestos. Proc Natl Acad Sci USA 1993; 90:3299–3303.
82. Mossman BT, Faux S, Janssen Y, Jimenez LA, Timblin C, Zanella C, Goldberg J, Walsh E, Barchowsky A, Driscoll K. Cell signalling pathways elicited by asbestos. Environ Health Perspect 1997; 105(suppl 5):1121–1125.
83. Timblin CR, Janssen YMW, Mossman BT. Transcriptional activation of the proto-oncogene c-jun, by asbestos and $H_2O_2$ is directly related to increased proliferation and transformation of tracheal epithelial cells. Cancer Res 1995; 55:2723–2726.
84. Timblin CR, Janssen YM, Goldberg JL, Mossman BT. GRP78, HSP 72/73, and c-Jun stress protein levels in lung epithelial cells exposed to asbestos, cadmium or $H_2O_2$. Free Rad Biol Med 1998; 24:632–642.
85. Driscoll KE, Carter JM, Hassenbein DG, Howard B. Cytokine and particle induced inflammatory cell recruitment. Environ Health Perspect 1997; 105(suppl 5):1159–1164.
86. Brody AR, Liu JY, Brass D, Corti M. Analyzing the genes and peptide growth factors expressed in lung cells in vivo consequent to asbestos exposure and in vitro. Environ Health Perspect 1997; 105(suppl 5):1165–1171.
87. Zhang Y, Lee TC, Guillemin B, Yu MC, Rom WN. Enhanced IL-1 beta and tumor necrosis factor-alpha release and messenger RNA expression in macrophages from idiopathic pulmonary fibrosis or after asbestos exposure. J Immunol 1993; 150:4188–4196.
88. Simeonova PP, Luster MI. Iron and reactive oxygen species in the asbestos induced tumor necrosis factor alpha response from alveolar macrophages. Am J Respir Cell Mol Biol 1995; 12:676–683.
89. Liu JY, Brass DM, Hoyle GW, Brody AR. TNF-α receptor knockout mice are protected

from the fibroproliferative effects of inhaled asbestos fibers. Am J Pathol 1998; 153:1839–1847.
90. Piguet PF, Vesin C. Treatment by human recombinant soluble TNF receptor of pulmonary fibrosis induced by bleomycin or silica in mice. Eur Respir J 1994; 7:515–518.
91. Goldstein RH, Fine A. Potential therapeutic initiatives for fibrogenic lung disease. Chest 1995; 108:848–855.
92. Smith RE, Strieter RM, Phan SH, Kunkel SL. C-C chemokines: novel mediators of the profibrotic inflammatory response to bleomycin challenge. Am J Respir Cell Mol Biol 1996; 15: 693–702.
93. Driscoll KE, Hassenbein DG, Carter J, Poynter J, Asquith TN, Grant RA, Whitten J, Purdon MP, Takigiku R. Macrophage inflammatory proteins 1 and 2: expression by rat alveolar macrophages, fibroblasts, and epithelial cells and in rat lung after mineral dust exposure. Am J Respir Cell Mol Biol 1993; 8:311–318.
94. Rosenthal GJ, Germolec DR, Blazka ME, Corsini E, Simeonova P, Pollock P, Kong LY, Kwon J, Luster MI. Asbestos stimulates IL-8 production from human lung epithelial cells. J Immunol 1994; 153:3237–3244.
95. Driscoll KE, Howard BW, Carter JM, Asquith T, Johnston C, Detilleux P, Kunkel SL, Isfort RJ. Alpha-quartz-induced chemokine expression by rat lung epithelial cells: effects of in vivo and in vitro particle exposure. Am J Pathol 1996; 149:1627–1637.
96. Nash T, Allison AC, Harington JS. Physico-chemical properties of silica in relation to its toxicity. Nature 1966; 210:259–261.
97. Hemenway D, Absher M, Fubini B, Bolis V. What is the relationship between hemolytic potential and fibrogenicity of mineral dusts? Arch Environ Health 1993; 48:343–347.
98. Fubini B. Surface reactivity in the pathogenic response to particulates. Environ Health Perspect 1997; 105:1013–1020.
99. Nolan RP, Langer AM, Harington JS, Oster G, Selikoff IJ. Quartz haemolysis as related to its surface functionalities. Environ Res 1981; 26:503–520.
100. Mao Y, Daniel LN, Knapton AD, Shi X, Saffiotti U. Protective effects of silanol group binding agents on quartz toxicity to rat lung alveolar cells. Appl Occup Environ Hyg 1995; 10:1132–1137.
101. Wiessner JH, Mandel NS, Sohnle PG, Hasegawa A, Mandel GS. The effect of chemical modification of quartz surfaces on particulate-induced pulmonary inflammation and fibrosis in the mouse. Am Rev Respir Dis 1990; 141:111–116.
102. Brown GM, Donaldson K. Modulation of quartz toxicity by aluminum. In: Silica and Silica Induced Lung Diseases. Boca Raton, FL: CRC Press, 1996:299–304.
102a. Shukla A, Timblin CR, Hubbard AK, Bravman J, Mossman BT. Silica-induced activation of c-Jun $NH_2$-terminal amino kinases (JNKS), protracted expression of the activator protein-1 (AP-1) protooncogene, *fra*-1, and S-phase alterations are mediated via oxidative stress. Cancer Res (in press).
103. Vallyathan V, Leonard S, Kuppusamy P, Pack D, Chzhan M, Sanders SP, Zweir JL. Oxidative stress in silicosis: evidence for the enhanced clearance of free radicals from whole lungs. Mol Cell Biochem 1997; 168:125–132.
104. Fubini B, Giamello E, Volante M, Bolis V. Chemical functionalities at the silica surface determining its reactivity when inhaled: formation and reactivity of surface radicals. Toxicol Ind Health 1990; 6:571–594.
105. Giamello E, Fubini B, Volante M, Costa D. Surface oxygen radicals originating via redox reactions during the mechanical activation of crystalline $SiO_2$ in hydrogen peroxide. Colloids Surfaces 1990; 45:155–165.
106. Dalal NS, Shi X, Vallyathan V. Role of the free radicals in the mechanisms of hemolysis and lipid peroxidation by silica: comparative ESR and cytotoxicity studies. J Toxicol Environ Health 1990; 29:307–316.
107. Vallyathan V, Castranova V, Pack D, Leonard S, Shumaker J, Hubbs AF, Shoemaker DA,

Ramsay DM, Pretty JR, McLaurin JL, Khan A, Teass A. Freshly fractured quartz inhalation leads to enhanced lung injury and inflammation in rats. Am J Respir Crit Care Med 1995; 152:1003–1009.
108. Volante M, Giamello E, Merlo E, Mollo L, Fubini B. Enhanced surface reactivity on mechanically activated covalent solids and its relationship with the toxicity of freshly ground dusts. In: Tkachova K, ed. An EPR study. Proceedings of the 1st International Conference on Mechanochemistry. Cambridge, UK: Cambridge Intersci Pub 1994:125–130.
109. Fubini B. Surface chemistry and quartz hazard. Ann Occup Hyg 1998; 42:521–530.
110. Fubini B, Mollo L, Giamello E. Free radical generation at the solid/liquid interface in iron containing minerals. Free Radic Res 1995; 23:593–614.
111. Castranova V, Vallyathan V, Ramsey DM, McLaurin JL, Pack D, Leonard S, Barger MW, Ma JY, Dalal NS, Teass A. Augmentation of pulmonary reaction to quartz inhalation by trace amounts of iron-containing particles. Environ Health Perspect 1997; 105:1319–1324.
112. Gilmour PS, Beswick PH, Brown DM, Donaldson K. Detection of surface free radical activity of respirable industrial fibres using supercoiled X 174RF1 plasmid DNA. Carcinogenesis 1995; 16:2973–2979.
113. Brown GM, Donaldson K, Brown DM. Bronchoalveolar leukocyte response in experimental silicosis: Modulation by a soluble aluminum compound. Toxicol Appl Pharmacol 1989; 101: 95–105.
114. Kaw JL, Beck EC, Bruck J. Studies of quartz cytotoxicity on peritoneal macrophages of guinea pigs pretreated with polyvinyl pyridine N-oxide. Environ Res 1975; 9:313–320.
115. Vallyathan V, Mega JF, Shi X, Dalal NS. Enhanced generation of free radicals from phagocytes induced by mineral dusts. Am J Respir Cell Mol Biol 1992; 6:404–413.
116. Castranova V, Kang JH, Moore MD, Pailes WH, Frazer DG, Schwegler-Berry D. Inhibition of stimulant-induced activation of phagocytic cells with tetrandrine. J Leukoc Biol 1991; 50: 412–422.
117. Schapira RM, Ghio AJ, Effros RM, Morrisey J, Almagro UA, Dawson CA, Hacker AD. Hydroxyl radical production and lung injury in the rat following silica or titanium dioxide instillation in vivo. Am J Respir Cell Mol Biol 1995; 12:220–226.
118. Ghio AJ, Kennedy TP, Whorton AR, Crumbliss AL, Hatch GE, Hoidal JR. Role of surface complexed iron in oxidant generation and lung inflammation induced by silicates. Lung Cell Mol Physiol 1992; 7:L511–L518.
119. Holley JA, Janssen YMW, Mossman BT, Taatjes DJ. Increased manganese superoxide dismutase protein in type II epithelial cells of rat lungs after inhalation of crocidolite asbestos or cristobalite silica. Am J Pathol 1992; 141:475–485.
120. Janssen YMW, Marsh JP, Absher MP, Hemenway D, Vacek PM, Leslie KO, Borm PJA, Mossman BT. Expression of antioxidant enzymes in rat lungs after inhalation of asbestos or silica. J Biol Chem 1992; 267:10625–10630.
121. Zhang Z, Shen H-M, Zhang Q-F, Ong CN. Critical role of GSH in silica-induced oxidative stress, cytotoxicity and genotoxicity in alveolar macrophages. Lung Cell Mol Physiol 1999; 277:L743–L748.
122. Blackford JA, Antonini JM, Castranova V, Dey RD. Intratracheal instillation of silica upregulates inducible nitric oxide synthase gene expression and increases nitric oxide production in alveolar macrophages and neutrophils. Am J Respir Cell Mol Biol 1994; 11:426–431.
123. Blough NV, Safiriou OC. Reaction of superoxide with nitric oxide to form peroxynitrite in alkaline aqueous solution. Inorg Chem 1985; 24:3502–3504.
124. Cho Y-J, Seo M-S, Kim JK, Lim Y, Chae G, Ha K-S, Lee K-H. Silica induced generation of reactive oxygen species in Rat2 fibroblasts: role in activation of mitogen activated protein kinase. Biochem Biophys Res Commun 1999; 262:708–712.
125. Chen F, Sun S-C, Kuhn DC, Haydos LJ, Demers LM. Essential role of NFκB activation in silica induced inflammatory mediator production in macrophages. Biochem Biophys Res Commun 1995; 214:985–992.

126. Shi X, Dong Z, Huang C, Ma W, Kejian L, Ye J, Chen F, Leonard SS, Ding M, Castranova V, Vallyathan V. The role of hydroxyl radical as a messenger in the activation of nuclear transcription factor NF-κB. Mol Cell Biochem 1999; 194:63–70.
127. Chen F, Kuhn DC, Sun S-C, Gaydos LJ, Demers LM. Dependence and reversal of nitric oxide production on NFκB in silica and lipopolysaccharide induced macrophages. Biochem Biophys Res Commun 1995; 214:839–846.
128. Sacks M, Gordon J, Bylander J, Porter D, Shi XL, Castranova V, Kaczmarczyk W, Van Dyke K, Reasor MJ. Silica-induced pulmonary inflammation in rats: activation of NFκB and its suppression by dexamethasone. Biochem Biophys Res Commun 1998; 253:181–184.
129. Ding M, Shi X, Dong Z, Chen F, Lu Y, Castranova V, Vallyathan V. Freshly fractured crystalline silica induces activator protein-1 activation through ERKs and p38 MAPK. J Biol Chem 1999; 274:30611–30616.
130. Gossart S, Cambon C, Orfila C, Seguelas M-H, Lepert J-C, Rami J, Carre P, Pipy B. Reactive oxygen intermediates as regulators of TNFa production in rat lung inflammation induced by silica. J Immunol 1996; 156:1540–1548.
131. Barrett EG, Johnston C, Oberdorster G, Finkelstein JN. Antioxidant treatment attenuates cytokine and chemokine levels in murine macrophages following silica exposure. Toxicol Appl Pharmacol 1999; 158:211–220.
132. Barrett EG, Johnston C, Oberdorster G, Finkelstein JN. Silica induced chemokine expression in alveolar type II cells is mediated by TNFα-induced oxidant stress. Lung Cell Mol Physiol 1999; 20:L979–L988.
133. Stringer B, Kobzik L. Environmental particulate-mediated cytokine production in lung epithelial cells (A549): role of preexisting inflammation and oxidant stress. J Toxicol Environ Health 1998; 55:31–44.
134. Mossman BT, Marsh JP, Sesko A, Hill S, Shatos MA, Doherty J, Adler KB, Hemenway D, Mickey R, Vacek P, Kagan E. Inhibition of lung injury, inflammation, and interstitial pulmonary fibrosis by polyethylene glycol-conjugated catalase in a rapid inhalation model of asbestosis. Am Rev Respir Dis 1990; 141:1266–1271.

# 10

# Reactive Oxygen Species and Silica-Induced Carcinogenesis

**XIANGLIN SHI, MIN DING, FEI CHEN, VAL VALLYATHAN, and VINCE CASTRANOVA**

*National Institute for Occupational Safety and Health, Morgantown, West Virginia*

Epidemiological and pathological studies have established that occupational exposure to crystalline silica leads to the development of pulmonary fibrosis (1,2). Increasing evidence from epidemiological and animal studies has also implicated crystalline silica as a potential carcinogen (3,4) (e.g., inhalation of silica has been shown to be carcinogenic in rats) (5–9). Intrapleural administration of crystalline silica in rats leads to the induction of localized malignant histiocytic lymphomas. Epidemiological studies also show that there appears to be an increased lung cancer risk in many, but not all, human subjects with silicosis (4,9). Based on current evidence obtained from studies on laboratory animals and epidemiological studies on humans, the International Agency for Research on Cancer has classified crystalline silica as a human class 1 carcinogen (9).

Because silica is a newly established carcinogen, there have been few investigations concerning the mechanisms of silica-induced carcinogenesis. Studies have demonstrated that freshly fractured silica particles generate silicon-based free radicals such as $Si^\bullet$, $SiO^\bullet$, and $SiOO^\bullet$ (10). Upon reaction with aqueous medium, these particles generate hydrogen peroxide ($H_2O_2$), superoxide radical ($O_2^{\bullet-}$), singlet oxygen ($^1O_2$), and hydroxyl radical ($^\bullet OH$) (10). Oxygen free radicals, such as $O_2^{\bullet-}$ and $^\bullet OH$, and related oxygen reduction products, such as $H_2O_2$ and $^1O_2$, are collectively called reactive oxygen species (ROS). Because ROS are known to be involved in the carcinogenicity of a variety of substances, we hypothesize that silica-mediated free radical reactions may cause a persistent oxidative stress in the lung and play a key role in the mechanism of silica-induced carcinogenesis (10,11).

Most knowledge of the role of ROS in silica-induced carcinogenesis comes from in vitro studies. The generation of ROS represents one of the main mechanisms by which phagocytes kill invading organisms. ROS production increases in response to phagocyte activation by microorganisms, particulates, and chemicals, resulting in a sudden increase in oxygen consumption called the ''respiratory burst.'' Crystalline silica is a potent stimulant of respiratory burst. Respiratory burst (i.e., an increase in oxygen consumption) is associated with an elevated production of ROS (12–14). Silica particles are able to cause lipid peroxidation in vitro and in exposed workers (15). Through free radical reactions, silica particles are also able to cause DNA strand breaks, dG hydroxylation, and thymine glycol formation (11,16).

In the past one or two decades, chemical and cellular studies have contributed enormously to our understanding of the role of free radical reactions in the mechanism of diseases and cancer induced by silica. Recently, a subdiscipline of molecular toxicology and carcinogenesis has developed. New techniques are now available to understand the mechanism of silica-induced cellular injury and the role of free radical reactions in precise molecular terms. For example, what molecules are the mediators? What are the signal transduction pathways? This chapter will focus on silica-induced activation of activator protein (AP-1) and nuclear transcription factor (NF)-κB and the role of free radical reactions in these activation processes.

AP-1 was chosen due to its important role involved in a diversity of biological processes. This factor is a complex protein composed of homodimers and heterodimers of oncogene proteins of the Jun and Fos families. The genes encoding these proteins, *c-jun* and *c-fos*, can be induced by a variety of extracellular stimuli and function as intermediary transcriptional regulators in signal transduction processes leading to proliferation and transformation. This transcription factor interacts with regulatory DNA sequences known as TPA response elements or AP-1 sites. Such AP-1 binding to DNA results in the induction of mRNAs for a number of growth factors. The activity of AP-1 is modulated by several factors, including the redox state of the cell. The direct involvement of ROS in AP-1 activation was demonstrated by using defined ROS generating systems to challenge cultured cells. Both $H_2O_2$ and $O_2^{\bullet -}$ are capable of inducing the expression of several early-response genes including *c-jun* and *c-fos*. While the detailed mechanism of ROS-mediated AP-1 activation has not been elucidated, it has been suggested that AP-1 activation under oxidative conditions may be in part mediated by phosphorylation of jun proteins (17).

With regard to NF-κB, this transcription factor is found in many different cell types and is involved in the transmission of signals from the cytoplasm to the nucleus. It regulates a variety of genes involved in inflammatory or acute phase responses, such as expression of various cytokines and surface receptors (18–23). In cells that have inducible NF-κB activity, the active form of this factor is composed of two different subunits, p50 and p65 (24). In resting cells, NF-κB is retained in the cytoplasm in an inactive form and its DNA binding activity and nuclear/cytoplasmic distribution are controlled by binding to an inhibitory protein known as inhibitor-α (IκBα) and -β of NF-κB, p105 (precursor of p50), and p100 (precursor of p52). Upon activation with extracellular stimuli, these inhibitory proteins are proteolytically degraded or processed by proteasomes and certain proteases, which allows NF-κB to be released and then translocated into the nucleus in an activated form. This translocation initiates or regulates early response gene transcription by binding to a decameric motif GGGRNNYYCC (κB elements) found in promoter regions of cellular or viral genes.

This article summarizes our recent studies (25,26) on silica-induced activations of AP-1 and NF-κB, the role of ROS, and the implication in silica-induced carcinogenesis.

## I. FRESHLY FRACTURED SILICA CAUSES AP-1 ACTIVATION IN JB6 CELLS AND RLE/AP02 CELLS

To explore the effects of silica on the induction of AP-1 activity, $5 \times 10^4$ JB6/AP/κB cells were exposed to varying doses (10 ~ 300 μg/mL) of freshly fractured silica for 24 h. Freshly fractured silica caused a significant dose-dependent AP-1 activation in JB6 cells (Fig. 1). The AP-1 activation attained significance at a low concentration of 80 μg/mL of silica (≈40 $\mu g/cm^2$) and was maximal at 200 μg/mL (≈100 $\mu g/cm^2$). Based on this result, 200 μg/mL silica was selected as the concentration to be used for time course studies. At intervals from 12 to 72 h, relative AP-1 activity was tested using the luciferase assay. Induction of AP-1 activity was first observed after 12 h of incubation with silica. Thereafter, AP-1 activity increased to a maximal eightfold activation at 24 h (data not shown). Further incubation of cells with silica for 48 and 72 h resulted in a decrease of AP-1 activation.

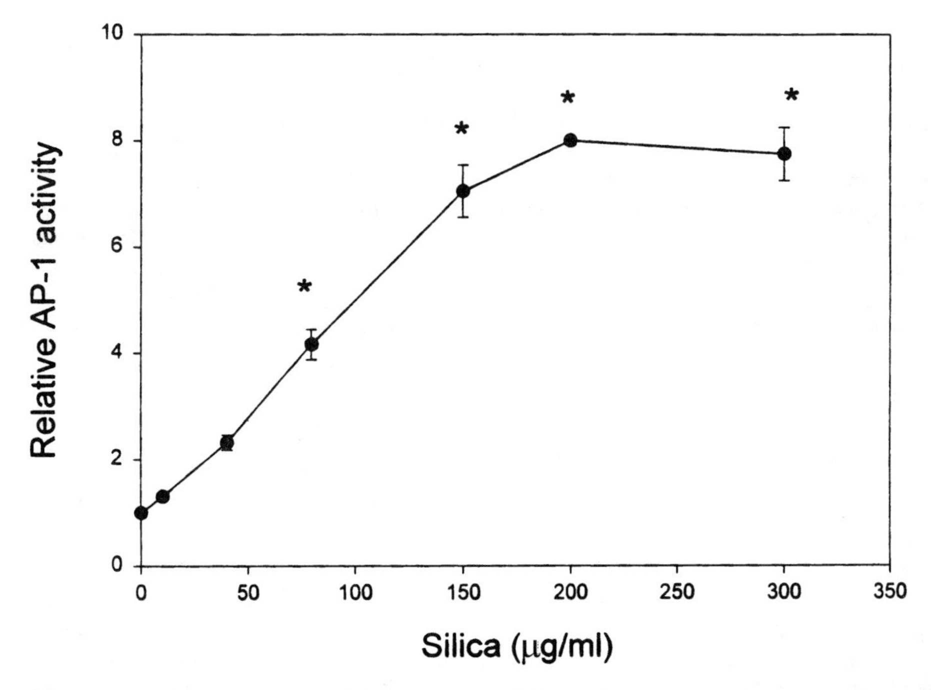


**Figure 1** Freshly fractured silica-induced AP-1 activation in JB6 $P^+$ cells. JB6/AP/κB cells ($5 \times 10^4$ cells in 1 mL of MEM medium with 5% fetal bovine serum) were seeded into each well of a 24-well plate. After overnight culture at 37°C, the cells were cultured in MEM plus 0.5% fetal bovine serum for 12 h. Then the cells were treated for 24 h with various concentrations of silica suspended in the same medium. The AP-1 activity was measured by the luciferase activity assay as described previously (25). Results, presented as relative AP-1 induction compared to the untreated control cells, are means and standard errors of nine assay wells from three independent experiments. *Indicates a significant increase from control ($p \leq 0.05$).

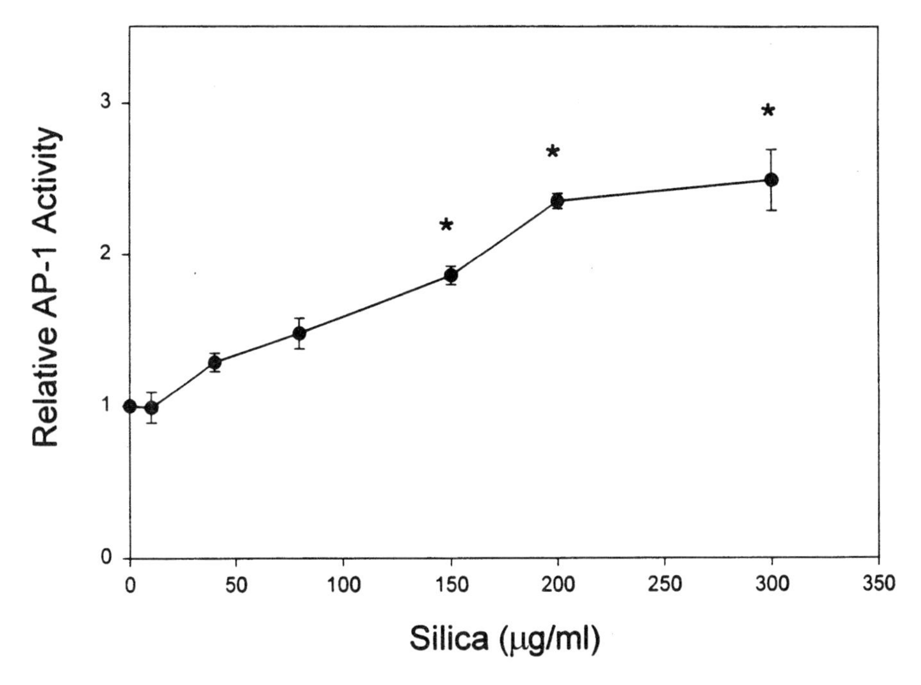


**Figure 2** Induction of AP-1 activity in RLE cells. RLE/AP02 cells ($5 \times 10^4$ cells in 1 mL of F12K medium containing 10% of fetal bovine serum), stably transfected with AP-1 luciferase reporter plasmid, were seeded into each well of a 24-well plate. After overnight culture at 37°C, cells were cultured in F12K medium plus 0.5% fetal bovine serum for 12 h. The cells were then exposed to various concentrations of silica suspended in the same medium for 72 h. Other experimental conditions were the same as those described in the legend to Figure 1. Results, presented as relative AP-1 induction compared to the untreated control cells, are means ± SEM of 12 assay wells from three independent experiments. * Indicates a significant increase from control ($p \leq 0.05$).

Since crystalline silica causes pulmonary epithelial hyperplasia and neoplastic lesions, we next asked whether freshly fractured silica would induce AP-1 activation in rat lung epithelial cells. Freshly fractured silica was incubated for 72 h with $5 \times 10^4$ RLE/AP02 cells stably transfected with AP-1-luciferase reporter plasmid. Freshly fractured silica causes a dose-dependent induction of AP-1 activation in RLE/AP02 cells (Fig. 2). At a silica concentration of 300 μg/mL (≈150 μg/cm$^2$), RLE cells exhibited AP-1 levels 2.5-fold greater than those observed in control cells. Time course studies in which a silica concentration of 200 μg/mL (≈100 μg/cm$^2$) was used indicated that significant induction of AP-1 activity was achieved after 24 h of exposure to silica (data not shown). The AP-1 induction in RLE/AP02 cells was different from that observed in JB6 cells. In RLE cells, the induction of AP-1 activity in response to silica occurred more slowly and persisted for at least 72 h. In addition, the maximal level of AP-1 induced by freshly fractured silica was lower in RLE cells (2.5-fold increase) than that in JB6 cells (eightfold increase). Because JB6 cells respond to a greater degree than RLE cells, JB6 cells were chosen for the further studies.

## II. TRANSACTIVATION OF AP-1 BY FRESHLY FRACTURED SILICA IN AP-1-LUCIFERASE REPORTER TRANSGENIC MICE

To investigate whether similar mechanisms exist in vivo, we used AP-1-luciferase reporter transgenic mice for these studies. The transgenic mice were exposed to freshly fractured silica (5 mg/mouse) by intratracheal aspiration of a silica suspension (70 mg/mL in 0.9% sterile NaCl). At the intervals of 1, 2, 3, and 4 days postexposure, animals were anesthetized with sodium pentobarbital and sacrificed by exsanguination. Lungs were removed and their luciferase activities were measured as described previously (25). Elevated AP-1 transactivation was not detected at 1-day postexposure (data not shown). However, AP-1 activation increased significantly at 2 and 3 days postexposure, and decreased toward control levels at 4 days postexposure (Fig. 3). At day 3 postexposure, the induction of AP-1 activation in lung tissue by freshly fractured silica was 22 times higher than that of the control group.

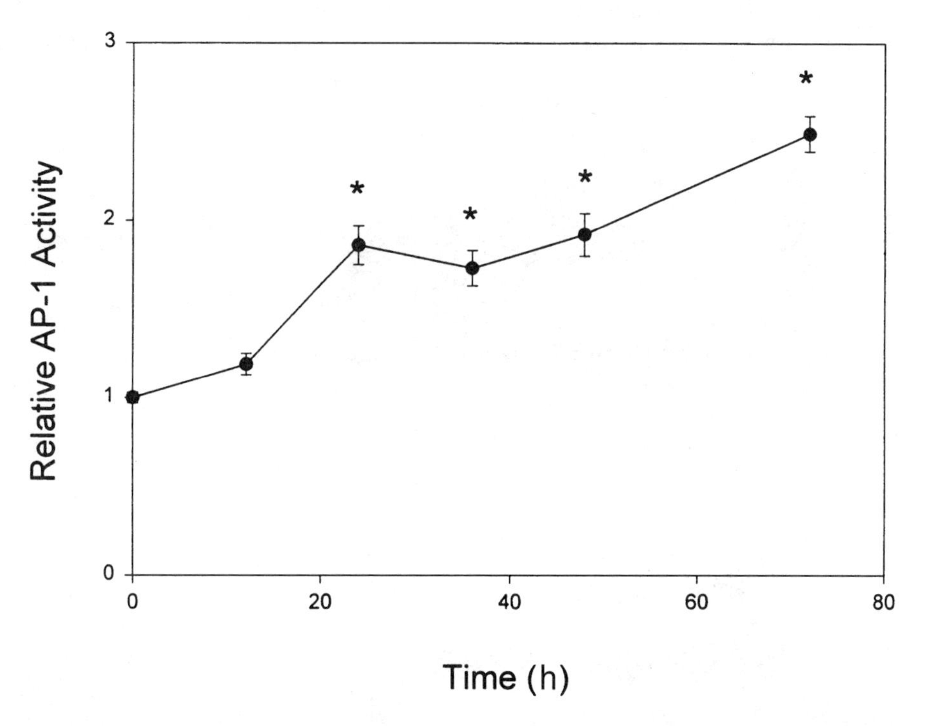


**Figure 3** Freshly fractured silica induces the transactivation of AP-1 in AP-1-luciferase reporter transgenic mice. The AP-1-luciferase transgenic mice were intratracheally instilled with 5 mg freshly fractured silica suspended in 0.07 mL of 0.9% sterile saline. At 2, 3, or 4 days postexposure, the mice were sacrificed and the lung tissue was removed. The luciferase activity of the tissue was measured as described previously (25). The results, presented relative to the level of luciferase activity of control groups, are means ± SEM of eight mice. * Indicates a significant increase from controls ($p \leq 0.05$).

## III. ACTIVATION OF ERKs AND p38 KINASE BY FRESHLY FRACTURED SILICA IN JB6 CELLS

Since mitogen-activated protein kinases (MAP), including p38 kinase, ERKs, as well as JNKs, are the upstream kinases responsible for c-Jun phosphorylation and AP-1 activation (27–30), we tested which classes of MAPK were involved in the AP-1 activation by silica. Using antibodies specific for the above MAP kinase family and phospho-specific for the phosphorylated MAP kinases, we studied ERK1, ERK2, JNKs, and p38 kinase proteins and the protein phosphorylation of ERK1, ERK2, JNKs, and p38 kinase in JB6 $P^+$ cells. Exposure to freshly fractured silica significantly stimulated the phosphorylation of p38 kinase and ERKs. The time course of p38 kinase phosphorylation induced by silica (150 μg/mL) is shown in Figure 4A. Phosphorylation of p38 kinase was first apparent after 15 min of exposure to silica and its maximal activation occurred after 2 h of exposure. To examine the dose dependence of the p38 kinase response in cells exposed to silica, JB6 cells were treated for 2 h with various concentrations of silica. A dose-related increase in p38 kinase phosphorylation was observed in cells treated with increasing concentrations of silica with prominent increases at 100 μg/mL–200 μg/mL silica (Fig. 4B). Freshly fractured silica (150 μg/mL) also causes phosphorylation of ERK1 and ERK2 in a time-dependent manner (Fig. 5). In contrast, silica did not affect the phosphorylation levels of JNKs (Fig. 6). Similar results were obtained by using the RLE cell line (data not shown).

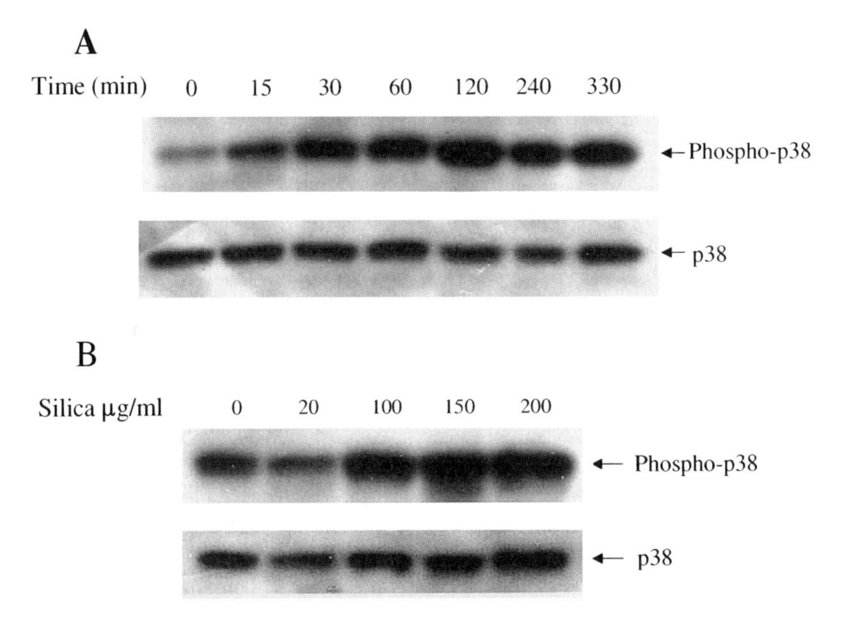


**Figure 4** Freshly fractured silica stimulates the phosphorylation of p38 MAP kinase. JB6 $P^+$ cells were cultured in 5% FBS MEM medium in 6-well (35-mm-diam) plates until 80% confluent and then cultured in 0.5% FBS MEM medium for 24 h. After this time, the cells were exposed to 150 μg/mL (47 μg/cm$^2$) silica suspended in the same medium for different times as indicated (A) or to various concentrations of silica for 2 h (B). The cells were lysed and phosphorylated p38 kinase proteins and nonphosphorylated p38 kinase proteins were assayed using a PhosphoPlus MAPKs kit from New England Biolabs. The phosphorylated proteins and nonphosphorylated proteins were analyzed using the same transferred membrane blot.

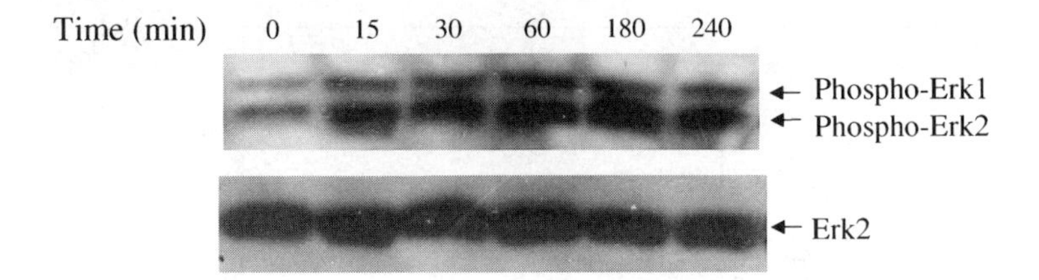


**Figure 5** Freshly fractured silica induces activation of Erk1 and Erk2. JB6 $P^+$ cells were cultured in 5% FBS MEM medium in 6-well (35-mm-diam) plates until 80% confluent. The cells were cultured in 0.5% FBS MEM medium for 24 h and then exposed to 150 μg/mL (47 $\mu g/cm^2$) silica suspended the same medium for different times as indicated. The cells were lysed and phosphorylated Erk1 and Erk2 proteins and nonphosphorylated Erk2 proteins were assayed using a PhosphoPlus MAPKs kit from New England Biolabs. The phosphorylated proteins and nonphosphorylated proteins were analyzed using the same transferred membrane blot.

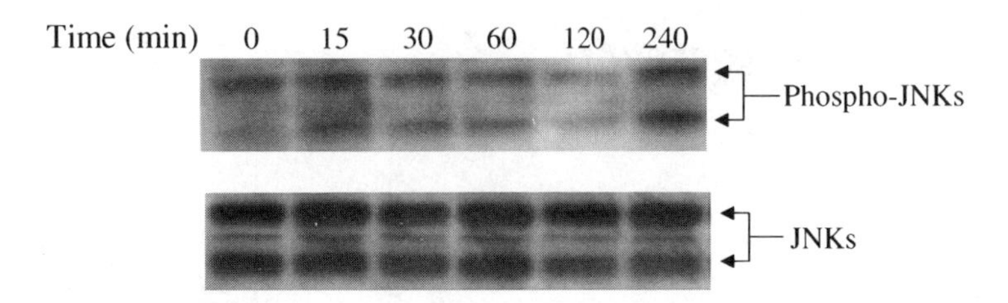


**Figure 6** Effect of freshly fractured silica on JNK activity. JB6 $P^+$ cells were cultured in 5% FBS MEM medium in 6-well (35-mm-diam) plates until 80% confluent. The cells were cultured in 0.5% FBS MEM medium for 24 h and then exposed to 150 μg/mL (47 $\mu g/cm^2$) silica suspended in the same medium for different times as indicated. The cells were lysed and phosphorylated JNK proteins and nonphosphorylated JNK proteins were assayed using a PhosphoPlus MAPKs kit from New England Biolabs. The phosphorylated and nonphosphorylated proteins were analyzed using the same transferred membrane blot.

These results suggested that ERKs and p38 kinase, but not JNK, may be involved in silica-induced AP-1 activation in JB6 cells as well as in RLE cells.

## IV. INHIBITION OF ERKs OR p38 KINASES BY SPECIFIC INHIBITORS BLOCKS FRESHLY FRACTURED SILICA-INDUCED AP-1 ACTIVATION

To further confirm that activation of AP-1 by silica is mediated through p38- and ERKs-dependent signal transduction pathways, we examined the effects of PD 98059 and SB 203580 on silica-induced AP-1 activation. PD 98059 has been shown to act as a highly selective inhibitor of MEK1 activation, while SB 203580 has been shown to be a specific inhibitor of p38 kinase. MEK1 is an upstream activator of ERKs. Silica-induced AP-1 activation was significantly inhibited by 20 to 50 μM PD 98059 or 2 μM SB 203580 (Fig. 7A and 7B).

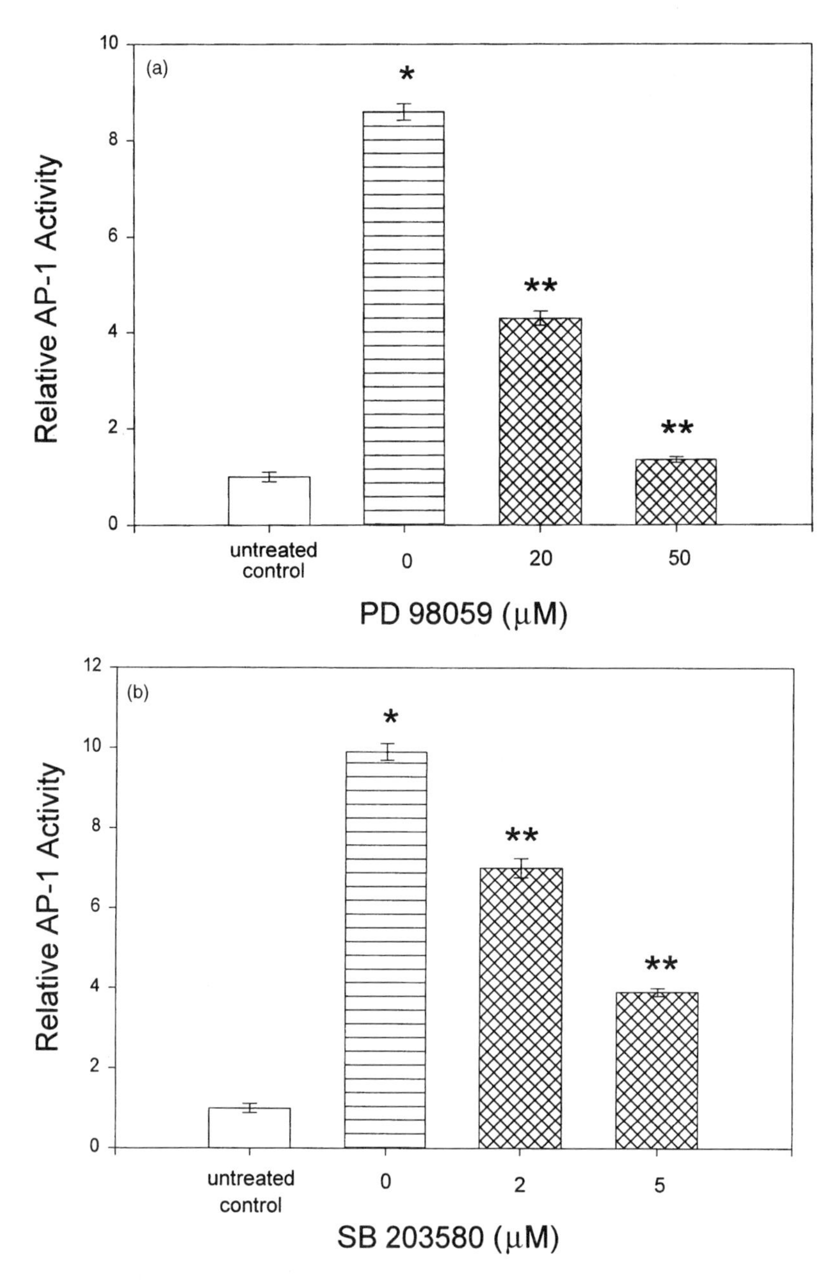


**Figure 7** Inhibition of silica-induced AP-1 activation by PD 98059 or SB 203580. JB6 cells ($5 \times 10^4$) were seeded into each well of a 24-well plate. After overnight culture at 37°C, the cells were cultured in MEM plus 0.5% fetal bovine serum for 12 h. Then the cells were pretreated with various concentrations of ERKs inhibitor, PD 98059 (a), or p38 inhibitor, SB 203580 (b) for 2 h and then exposed to 150 μg/mL silica in the presence of the inhibitors for 24 h. The AP-1 activity was measured by the luciferase activity assay as described in the ''Materials and Methods.'' Results, presented as relative AP-1 induction compared to the control cells, are means and standard errors of twelve assay wells from two independent experiments. * Indicates a significant increase from untreated control and ** indicates significant decrease from silica alone ($p \leq 0.05$).

## V. PD 98059 AND SB 203580 INHIBITED FRESHLY FRACTURED SILICA-INDUCED AP-1 DNA BINDING ACTIVITY

To study the molecular basis of the induction on AP-1 activity by silica and further confirm the above findings, the AP-1 DNA binding activity was analyzed by the gel-shift assay. As shown in Figure 8, silica induces AP-1 DNA binding activity and PD 98059 or SB 203580 inhibited silica-induced AP-1 DNA binding activity. These data provide further support that ERKs and p38 kinase are involved in silica-induced AP-1 activation.

## VI. INDUCTION OF AP-1 ACTIVATION BY FRESHLY FRACTURED SILICA VERSUS AGED SILICA

Early studies have demonstrated that freshly ground silica generates more ROS than aged silica (10). To explore the differential effects of freshly fractured versus aged silica on the induction of AP-1 activity, $5 \times 10^4$ JB6 cells were exposed for 24 h to varying concentrations (10 ~ 300 μg/mL) of freshly fractured silica or fractured silica aged for 12 months before use. The AP-1 activation induced by freshly fractured silica was significantly higher than that of aged silica in JB6 cells (Fig. 9). The maximum AP-1 induction by freshly fractured silica was eightfold compared to controls, whereas the maximum AP-1 induction by aged silica was only twofold. This result indicated that freshly fractured silica exhibited a greater effect on AP-1 induction than aged silica.

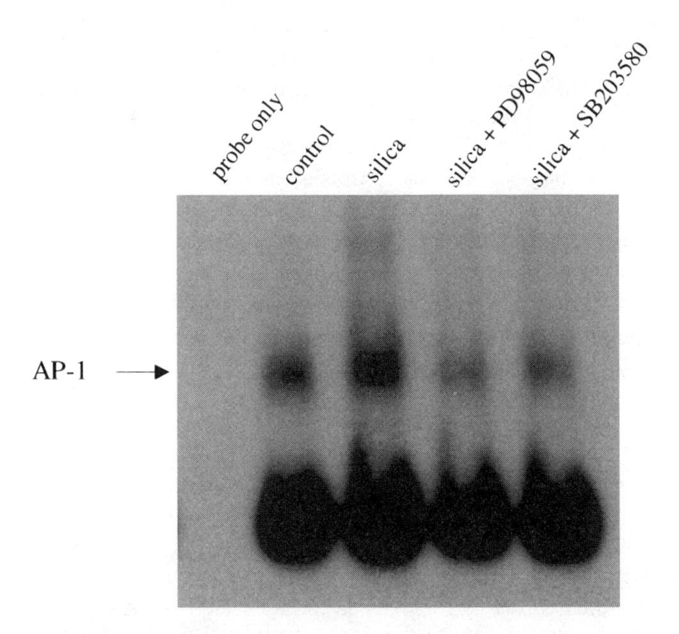


**Figure 8** Electrophoretic mobility shift assay. JB6 cells were seeded into each well of a 6-well plate until 80% confluent. Then the cells were cultured in MEM plus 0.5% fetal bovine serum for 24 h. The cells were pretreated with 20 μM of PD 98059 or 5 μM of SB 203580 for 2 h and then exposed to 150 μg/mL silica in the presence of the inhibitors for another 2 h. The AP-1 DNA binding activity was determined by gel-shift assay.

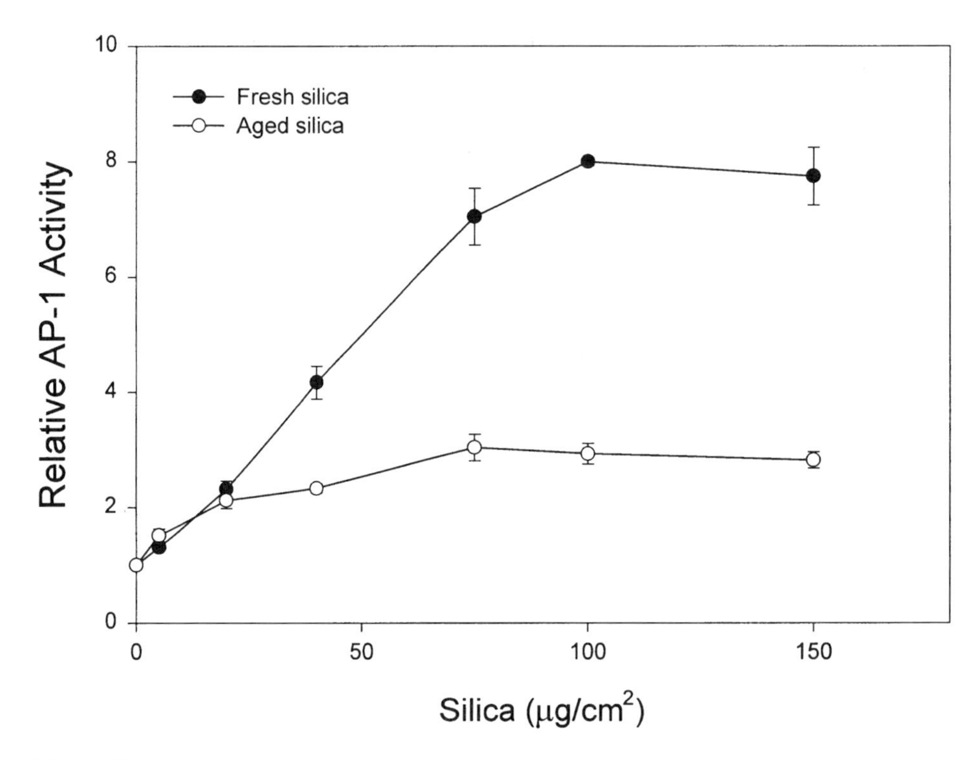


**Figure 9** Dose-dependent induction of AP-1 activation by freshly fractured silica versus aged silica. JB6 cells ($5 \times 10^4$ in 1 mL of MEM medium with 5% fetal bovine serum), stably transfected with AP-1 luciferase reporter plasmid, were seeded into each well of a 24-well plate. After overnight culture at 37°C, the cells were cultured in MEM plus 0.1% fetal bovine serum for 24 h. Then the cells were treated with 75 $\mu g/cm^2$ freshly fractured or 1-year-old silica for 24 h. The AP-1 activity was measured by the luciferase activity assay. Results, presented as relative AP-1 induction compared to the untreated control cells, are means and standard errors of eight assay wells from two independent experiments.

## VII. EFFECTS OF ANTIOXIDANTS AND CHELATING REAGENTS ON SILICA-INDUCED AP-1 ACTIVATION AND MAPKs PHOSPHORYLATION

The effects of antioxidant reagents on silica-induced AP-1 activation and MAPKs phosphorylation were investigated to further elucidate if ROS mediate AP-1 activation by silica. The effects of various antioxidants and chelating agents on silica-induced AP-1 activation are shown in Figure 10. Catalase, an enzyme for $H_2O_2$ decomposition, inhibited silica-induced AP-1 activation by 90%. Sodium formate, a scavenger for $^{\bullet}OH$ radical, had little or no effect. SOD, a $O_2^{\bullet -}$ radical scavenger, enhanced the AP-1 activation by 70%. Deferoxamine, a metal ion chelator that reduces the ability of metal ions to react with $H_2O_2$ and generate $^{\bullet}OH$, had no significant inhibitory effects. Both *N*-acetylcysteine, a nonspecific antioxidant, and PVPNO, a silanol blocker, significantly decreased AP-1 induction.

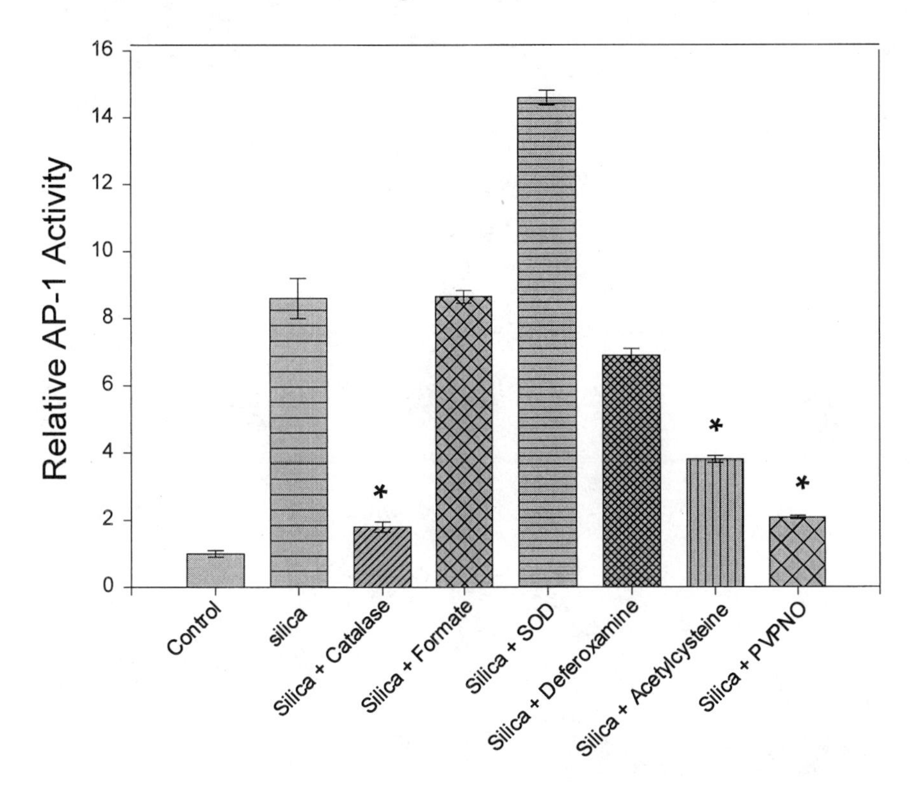


**Figure 10** Effect of antioxidant reagents on silica-induced AP-1 activation. JB6 cells ($5 \times 10^4$ cells in 1 mL of MEM medium containing 5% of fetal bovine serum) were seeded into each well of a 24-well plate. After overnight culture at 37°C, cells were cultured in the same medium plus 0.1% fetal bovine serum for 24 h. The cells were then exposed to 75 $\mu g/cm^2$ of freshly fractured silica and various reagents as indicated for 24 h. The concentrations of the reagent used were: catalase, 10,000 U/mL; sodium formate, 2 mM; SOD, 500 U/mL; deferoxamine, 1 mM; *N*-acetylcysteine, 1 mM; PVPNO, 50 mg/mL. Other experimental conditions were the same as those described in the legend to Figure 1. Results, presented as relative AP-1 induction compared to the untreated control cells, are means ± SEM of eight assay wells from two independent experiments. * Indicates a significant decrease from cells treated with freshly fractured silica ($p < 0.05$).

## VIII. ACTIVATION OF NF-κB

The mouse macrophage cell line, RAW 264.7 cells, was used to detect NF-κB activation by silica and other reagents. The cells were exposed for 6 h and NF-κB was analyzed in the nuclear extracts. As shown in Figure 11a, lane 1, the untreated cells did not exhibit any NF-κB activity. Upon treatment with LPS (5 μg/mL), the cells showed enhanced NF-κB binding activity (Fig. 11a, lane 2). Tetrandrine, a drug reported to decrease silica-stimulated oxidant production and cytokine release (31,32), inhibited LPS-induced NF-κB activation (Fig. 11a, lane 3). Figure 11b, lane 1 again showed the untreated cells as a control. Silica induced significant NF-κB activation (Fig. 11b, lane 2). Tetrandrine inhibited silica-induced NF-κB activation (Fig. 11b, lane 3). Similar experiments were carried out using PMA as a stimulant. As shown in Figure 11c, lanes 1–3, PMA also induced NF-κB activation and tetrandrine exhibited an inhibitory effect.

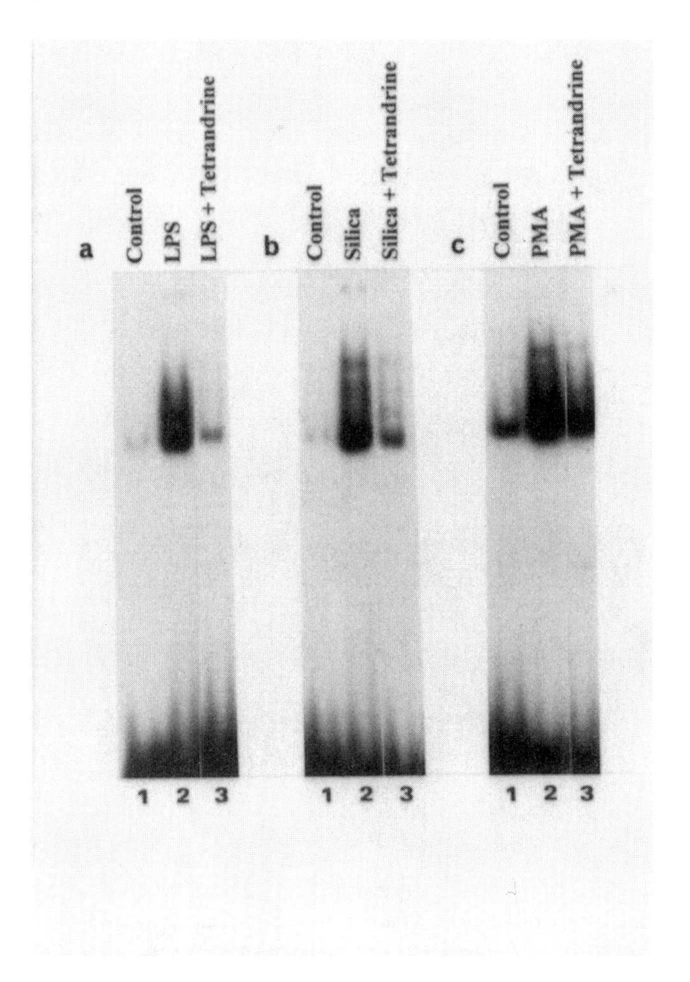


**Figure 11** Induction of DNA binding activity of NF-κB protein by LPS, silica, and/or PMA and the effect of tetrandrine. RAW 264.7 cells were adjusted to a density of $5 \times 10^6$/mL and treated for 6 h with different stimuli, then subjected to extraction of the nuclear proteins as described previously (26). DNA binding activity of the NF-κB protein was detected with a probe of $^{32}$P-labeled double-stranded NF-κB binding oligonucleotide by EMSA. (a) Lane 1, untreated cells; lane 2, cells + 5 μg/mL LPS; lane 3, cells + 5 μg/mL LPS + 150 μM tetrandrine. (b) Lane 1, untreated cells; lanes 2, cells + 100 μg/mL silica; lanes 3, cells + 100 μg/mL silica + 150 μM tetrandrine. (c) Lane 1, untreated cells; lane 2, cells + 10 nM PMA; lane 3, cells + 10 nM PMA + 150 μM tetrandrine.

## IX. EFFECT OF ANTIOXIDANTS ON NF-κB ACTIVATION

The effect of catalase on silica-induced NF-κB activation was evaluated to study the role of $H_2O_2$. As shown in Figure 12a, catalase caused a dose-dependent inhibition of silica-induced NF-κB activation. In contrast, SOD enhanced silica-induced NF-κB activation (Fig. 12b). A stronger enhancement effect was observed at a higher SOD concentration. Inactivated catalase or SOD did not exhibit any observable effects (data not shown). The metal chelator, deferoxamine, also inhibited silica-induced NF-κB activation (Fig. 12c, lanes 1–3). Deferoxamine itself did not cause any NF-κB activation (Fig. 12c, lane 4).

The involvement of iron-catalyzed reactions in the NF-κB activation is further implicated by the ability of Fe(II) but not Fe(III) to induce NF-κB activation (Fig. 12d).

The effects of the antioxidant, ascorbate, and the •OH radical scavenger, formate, were also tested for their effect on NF-κB activation induced by LPS or silica. Both ascorbate and formate significantly inhibited silica-induced NF-κB activation (data not shown). PVPNO, a silanol (SiOH) blocker, was used to test for a possible role of SiOH in the mechanism of silica-induced NF-κB activation. As shown in Figure 13, PVPNO caused a dramatic inhibition of NF-κB activity.

Similar to the results using a peritoneal monocyte cell line outlined above, silica was able to cause NF-κB activation in rat primary alveolar macrophage cells. Formate, deferoxamine, and PVPNO inhibited this activation (data not shown).

Occupational exposure to silica is associated with the development of silicosis and lung cancer (3–9). The molecular mechanisms involved in silica-induced carcinogenesis are unclear. We hypothesize that activation of nuclear transcription factors induced by silica is a primary event in the initiation of signal transduction cascades at the cell membrane level leading to the induction of early response genes that are critical in carcinogenesis. We have studied the effect of silica on the activation of AP-1 and the signal transduction pathways involved in AP-1 activation in cell culture models and in transgenic mice, and the silica-induced NF-κB activation. We have demonstrated that silica stimulates AP-1 DNA binding activity as well as AP-1 transactivation activity. Silica induced an eightfold increase of AP-1 activity in JB6 epidermal cells and a 2.5-fold increase in rat lung epithelial cells. Silica also stimulated AP-1 transactivation in pulmonary tissues of transgenic mice. At 3 days after intratracheal aspiration of silica, the AP-1 activity was elevated 23-fold as compared to the controls. Most importantly, we have found that phosphorylation of ERK1, ERK2, and p38 kinases was induced by freshly fractured silica. Phosphorylation is involved in silica-induced AP-1 activation. These data demonstrate that freshly fractured silica induces AP-1 activation through MAPK signal transduction pathways.

Previous studies using different model systems have suggested an important role of AP-1 activation in preneoplastic-to-neoplastic transformation in cell culture and animal models (33–36). AP-1 is a critical mediator of tumor promotion and is involved in a diversity of processes. This transcription factor is able to alter gene expression in response to a number of stimuli, including the tumor promoter (TPA), EGF, TNF-α, IL-1, and UV irradiation (33). Some of the genes regulated by AP-1 are involved in immune and inflammatory responses, tumor promotion, and tumor progression. These include cytokines, such as IL-1, TNF-α, GM-CSF, collagenase IV, and stromelysin (37–39). Overexpression of *c-jun* in JB6 $P^+$ cells causes neoplastic transformation. Inhibition of AP-1 activity by either pharmaceutical agents, such as fluocinolone acetonide or retinoic acid, or molecular biological inhibitors, such as dominant negative *c-jun* and dominant negative phosphatidylinositol-3 kinase, was found to block tumor promoter–induced neoplastic transformation (34,35,40–43).

AP-1 is a complex protein composed of homodimers and heterodimers of oncogene proteins of the Jun and Fos families. The genes encoding these proteins, *c-jun* and *c-fos*, are inducible in response to a variety of extracellular stimuli and function as intermediary transcriptional regulators in signal transduction processes leading to proliferation and transformation. The activation of AP-1 may trigger downstream signal cascades, such as *jun*, *fos*, and other target genes. The members of the Jun and Fos protein families may couple cell signaling events at the cell surface to changes in gene expression which modu-

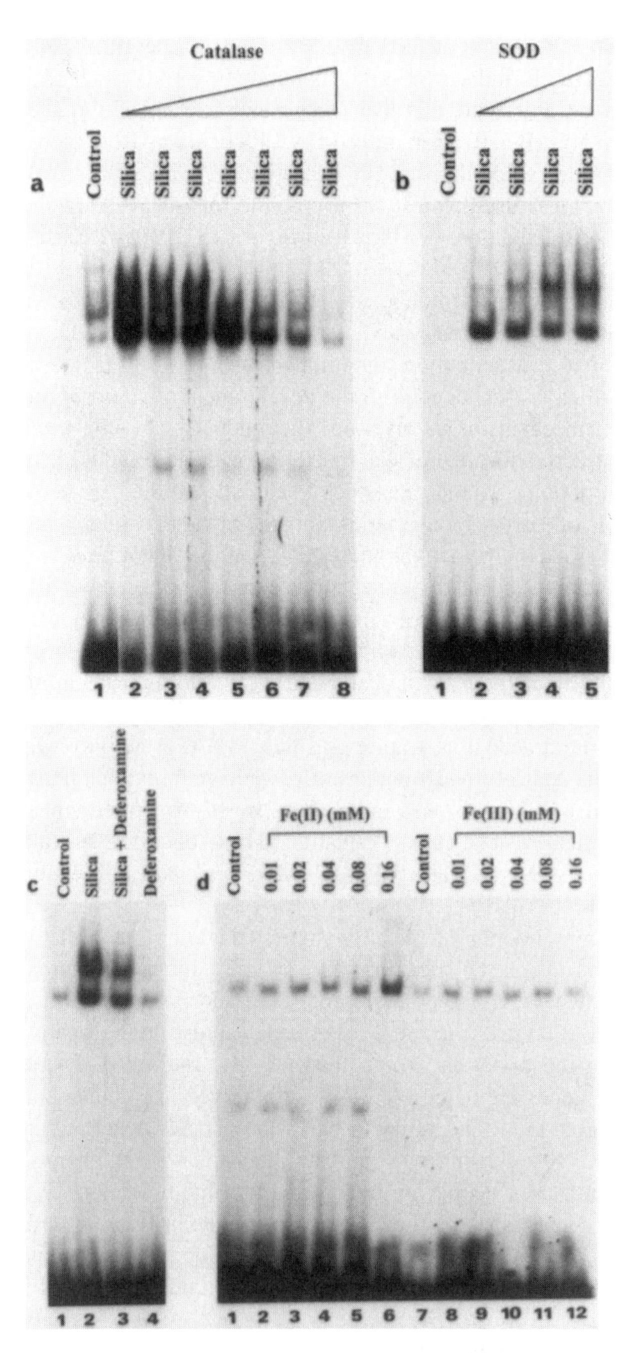


**Figure 12** Effect of antioxidants on silica-induced NF-κB activation in macrophages. (a) Effect of catalase on silica-induced NF-κB activation. Lane 1, untreated cells ($5 \times 10^6$/mL); lane 2, cells + 100 μg/mL silica; lane 3, cells + 100 μg/mL silica + 1250 units/mL catalase; lane 4, cells + 100 μg/mL silica + 2500 units/mL catalase; lane 5, cells + 100 μg/mL silica + 5000 units/mL catalase; lane 6, cells + 100 μg/mL silica + 10,000 units/mL catalase; lane 7, cells + 100 μg/mL silica + 20,000 units/mL catalase; lane 8, cells + 100 μg/mL silica + 40,000 units/mL catalase. (b) Effect

late cell responses, including proliferation and changes in phenotype. The important element of this study is that freshly fractured silica stimulates AP-1 activation, which may be one of the critical mechanisms for silica-induced carcinogenesis. The level of AP-1 induction by silica was lower in rat lung epithelial cells than that in JB6 epidermal cells. This may be due to the following differences between these two cell lines: (1) the rate of luciferase gene expression; (2) the half-life of luciferase; (3) the basal MAPK or AP-1 activities; or (4) the antistress enzyme activity.

The signal transduction pathways leading to transcription factor activation have been extensively studied in the last several years. It is believed that stress-related signals, such as UV light or ROS, induce the activation of MAP kinase pathways (ERKs, JNKs, and p38). AP-1 is a downstream target of these three MAP kinase members (44). We have studied the possible role of the MAPK family, including p38 kinase, ERKs, and JNKs, in silica-induced AP-1 activation. We have found that freshly fractured silica induced phosphorylation of ERKs and p38 kinase, but not JNKs. Pretreatment of cells with p38 and ERK inhibitors, PD 98059 or SB 203580, inhibited AP-1 transactivation as well as AP-1 DNA binding activity induced by silica. Thus, these results suggest that silica-induced AP-1 activation may be mediated through MAPK p38 and ERK pathways.

The development of AP-1 luciferase transgenic mice makes it possible to study the role of AP-1 activation in tumor promotion in vivo (45). We have shown that freshly fractured silica is able to cause AP-1 activation in transgenic mice. Maximal AP-1 activation was increased by 23-fold in pulmonary tissues at 3 days after intratracheal aspiration of silica. However, the cell types involved in this AP-1 activation response have not yet been identified. Additional studies are required to answer this question.

Our studies also show that freshly fractured silica particles are much more potent in inducing AP-1 activation than aged silica. As demonstrated earlier, freshly fractured silica particles generate more ROS upon reaction with cells or aqueous medium than aged silica (46). Due to the enhanced ROS generation, freshly fractured silica is more potent in causing lipid peroxidation, DNA damage, and other cell injuries. It appears that silica-induced AP-1 activation involves ROS-mediated reactions. Among ROS, $H_2O_2$ plays a major role in silica-induced AP-1 activation as supported by the following results: (1) freshly fractured silica is able to generate $H_2O_2$ upon reaction with aqueous medium as reported earlier (14); (2) catalase, whose function is to remove $H_2O_2$, blocked the AP-1 activation; (3) sodium formate, a scavenger of $^{\bullet}OH$ radical, did not exhibit any effect; and (4) SOD, whose function is to remove $O_2^{\bullet -}$ and generate $H_2O_2$, enhanced the AP-1 activation.

---

of SOD on silica-induced NF-κB activation. Lane 1, untreated RAW 264.7 cells ($5 \times 10^6$/mL); lane 2, cells + 100 μg/mL silica; lane 3, cells + 100 μg/mL silica + 160 units/mL SOD; lane 4, cells + 100 μg/mL silica + 640 units/mL SOD; lane 5, cells + 100 μg/mL silica + 1280 units/mL SOD. (c) Effect of deferoxamine on silica-induced NF-κB activation. Lane 1, untreated RAW 264.7 cells ($5 \times 10^6$/mL); lane 2, cells + 100 μg/mL silica; lane 3, cells + 100 μg/mL silica + 1.6 mM deferoxamine; lane 4, cells + 1.6 mM deferoxamine. (d) NF-κB activation by Fe(II) and Fe(III). Lane 1, untreated 264.7 cells ($5 \times 10^6$/mL); lane 2, cells + 0.01 mM Fe(II); lane 3, cells + 0.02 mM Fe(II); lane 4, cells + 0.04 mM Fe(II); lane 5, cells + 0.08 mM Fe(II); lane 6, cells + 0.16 mM Fe(II), lane 7, untreated cells; lane 8, cells + 0.01 mM Fe(III); lane 9, cells + 0.02 mM Fe(III); lane 10, cells + 0.04 mM Fe(III); lane 11, cells + 0.08 mM Fe(III); lane 12, cells + 0.16 mM Fe(III).

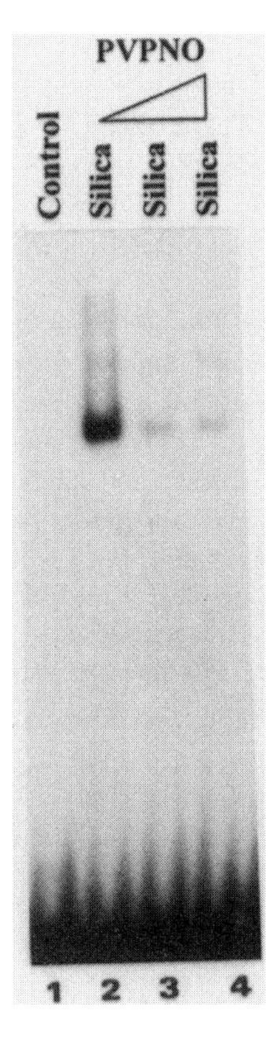


**Figure 13** Effect of PVPNO on silica-induced NF-κB activation. Lanes 1, untreated RAW 264.7 cells ($5 \times 10^6$/mL); lane 2, cells + 100 μg/mL silica; lane 3, cells + 100 μg/mL silica + 20 μg/mL PVPNO; lane 4, cells + 100 μg/mL silica + 50 μg/mL PVPNO.

We have demonstrated that silica particles are also able to activate NF-κB and that $^{\bullet}OH$ radicals may play a key role in silica-induced NF-κB activation. The following experimental observations support this conclusion.

1. Silica is able to generate $^{\bullet}OH$ radical in the presence and absence of $H_2O_2$ as demonstrated by spin trapping measurements in earlier (14,46,47) and present studies.
2. Catalase blocked the NF-κB activation.
3. SOD exhibited an opposite effect. It may be noted that earlier studies have shown that molecular oxygen was consumed in the generation of $O_2^{\bullet -}$ by silica suspensions (12,48). Silica-induced DNA damage was inhibited in an argon atmosphere, indicating that the oxygen radicals responsible for DNA damage were generated from $O_2$ via $O_2^{\bullet -}$ and $H_2O_2$ as intermediates [Eq. (1)] (12,48).

$$O_2 \rightarrow O_2^{\bullet -} \rightarrow H_2O_2 \rightarrow {}^{\bullet}OH. \tag{1}$$

4. Metal ions, Fe(II) but not Fe(III), enhanced the NF-κB activation. It is known that Fe(II) generates $^{\bullet}OH$ from $H_2O_2$ via the Fenton reaction [Eq. (2)].

$$Fe(II) + H_2O_2 \rightarrow Fe(III) + OH^- + {}^{\bullet}OH. \quad (2)$$

Fe(III), on the other hand, is unable to generate $^{\bullet}OH$ radical without being first reduced to Fe(II).

5. Metal chelator, deferoxamine, also reduced the NF-κB activation. Deferoxamine chelates metal ions, such as Fe(II) or Fe(III), to make them less reactive toward $H_2O_2$ and thus attenuated the generation of $^{\bullet}OH$ radicals.
6. The antioxidant, ascorbate, and an $^{\bullet}OH$ radical scavenger, formate, inhibited NF-κB activation.
7. Tetrandrine, which has been reported to inhibit $H_2O_2$ release from macrophages (49), and also function as an $^{\bullet}OH$ radical scavenger (13), inhibited NF-κB activation.

It may be noted that silica is a fibrogenic agent due to its ability to elicit resident macrophages to release inflammatory mediators and cytokines that can promote fibroblast proliferation and collagen deposition. It has been suggested that NF-κB activation is crucial in cytoplasmic/nuclear signaling that occurs when cells are exposed to injury-producing conditions (50). NF-κB serves as a second messenger to induce a series of cellular genes in response to an environmental perturbation. Among cellular genes regulated by NF-κB are several proinflammatory or fibrogenic cytokines, including IL-2, IL-6, and TNF-α (22). NF-κB activates these genes by acting as a transcriptional factor and binding to the NF-κB consensus sequence in their promoters. Reactive oxygen intermediates have been suggested to be mediators of NF-κB activation in response to a variety of initiators such as Cr(VI) and phorbal esters (51–54). Since it is possible that silica particles cause NF-κB activation via free radical reactions, it is proposed that signals for a variety of silica-induced responses are due to a common signaling component, which is regulated by reactive oxygen species. It may be noted that a recent study has shown that *N*-acetylcysteine did not block silica-induced NF-κB activation (18). Although *N*-acetylcysteine is considered an antioxidant, it is not an efficient $^{\bullet}OH$ scavenger and thus may not inhibit silica-induced NF-κB activation via $^{\bullet}OH$ initiated reactions.

The results obtained from our studies have shown that tetrandrine is able to inhibit silica-induced NF-κB activation. While tetrandrine has been reported to retard and reverse the fibrotic lesions of silicosis in humans and in rats, its mechanism of action is unclear. As mentioned earlier, tetrandrine is an effective inhibitor of silica-induced $H_2O_2$ production by macrophages (31). In addition, recent studies have shown that tetrandrine is capable of scavenging $^{\bullet}OH$ radicals and inhibiting silica-induced lipid peroxidation (13). Tetrandrine has been shown to be an effective inhibitor of IL-1 secretion from alveolar macrophages activated by silica or LPS (32). We have demonstrated that tetrandrine inhibited LPS-induced NF-κB activation. It is possible that silica or LPS causes increased secretion of IL-1 or other cytokines via activation of NF-κB. It appears that the mechanism of NF-κB activation induced by LPS, although it can be inhibited by tetrandrine, is different from that for silica. For silica, both ascorbate and formate exhibited inhibitory effects. For LPS, the effect of these antioxidants is weak. While it is likely that one of the steps involves antioxidant activity of tetrandrine toward $^{\bullet}OH$ radical, other mechanisms may also exist. For example, tetrandrine decreases the stimulant-induced increase in intracellular calcium concentration due to its channel blocker effect and inhibits the production of NO in activated macrophages (55). The latter is especially important, since the production of NO has a direct effect on silica-induced NF-κB activation (19). It may be noted that although other transcription factors, such as AP-1, are also important, NF-κB may be particularly relevant in the inflammatory, immune, and acute phase responses after silica exposure.

Our studies have also shown that deferoxamine reduces silica-mediated ·OH radical generation and attenuates silica-induced NF-κB activation. Deferoxamine is widely used for the prevention and treatment of iron overload (56,57). It inhibits ·OH radical generation from $H_2O_2$ by transition metals, including Fe, Cr, and V (58,59). A large dose of deferoxamine (50 mg/kg/day) can be safely injected into humans (57). Thus, further investigation on the use of deferoxamine or other metal chelators may offer a possible preventative strategy against silica-induced fibrosis and carcinogenesis.

Another important result obtained from our recent studies is that PVPNO inhibited silica-induced NF-κB activation. SiOH groups on the silica surface have been considered to be involved in silica-induced cellular damage (4,10,60,61). Chemical modification of the silica surface can be used to reduce toxicity in vitro and fibrosis in vivo. It is known that when silica particles are exposed to water, surface silicon-oxygen bonds (Si–O) are hydrated, resulting in the formation of SiOH groups. PVPNO is able to bind to SiOH groups. It has been reported to inhibit silica-induced toxicity (62), to decrease and delay the development of silicosis in experimental animals and in humans (63,64), and to block the interaction of the silica surface with the phosphate groups of DNA in vitro (16). It has also been reported that PVPNO inhibits silica-induced production of oxygen radicals in cells (65,66). The ESR spin trapping measurements in the present study have demonstrated the inhibitory effect of PVPNO on ·OH radical generation by silica plus $H_2O_2$,

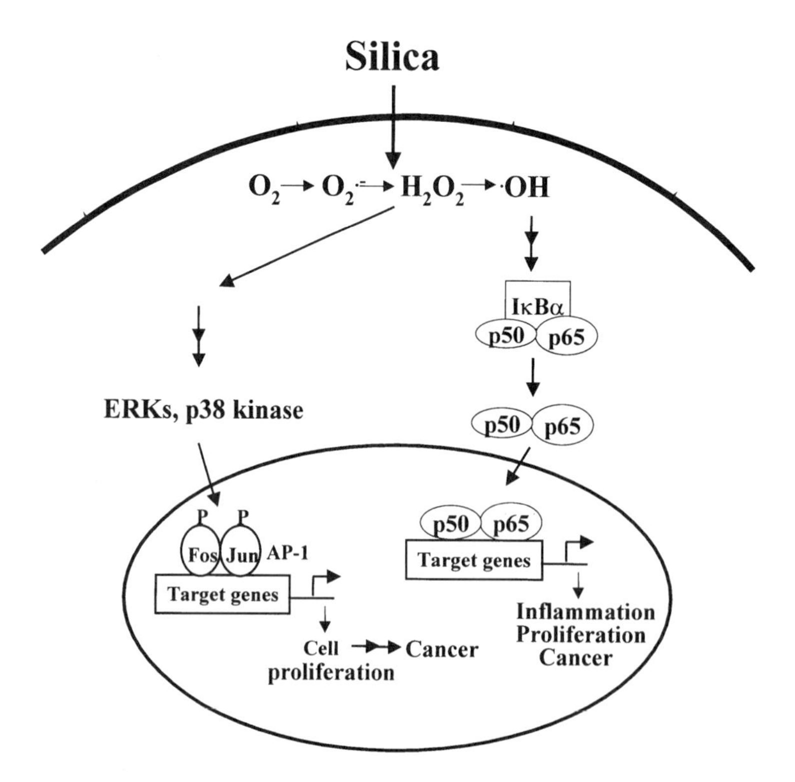


**Figure 14** Schematic representation of silica-induced generation of ROS and mechanistic scheme for silica-induced carcinogenesis.

implying the involvement of SiOH group in the generation of ·OH radical from $H_2O_2$ by silica.

Our studies present a molecular model for the elucidation of events involved in cell proliferation and carcinogenesis by crystalline silica (Fig. 14). By activating transcription factors, such as AP-1 and NF-κB, silica may induce chronic cell proliferation, which subsequently contributes to silicosis and carcinogenesis in the lung. It is possible that activation of AP-1 is a crucial event that initiates cell proliferation and progression through the cell cycle. Biopersistent silica particles may provide prolonged redox signals and growth stimulus during the long latency periods of tumorigenesis and, thereby, contribute to the eventual fixation of genetic changes caused by silica itself or other agents. Furthermore, the induction of AP-1 and NF-κB may affect changes in cell phenotype, which contribute to neoplastic transformation.

## REFERENCES

1. Silicosis and Silicate Disease Committee. Disease associated with exposure to silica and silicosis. Arch Pathol Lab Med 1988; 112:663–720.
2. Reiser KM, Last JA. Silicosis and fibrogenesis: fact and artifact. Toxicology 1979; 13:15–72.
3. IRAC. Silica and some silicates. In: IARC Monographs on the Evaluation of Carcinogenic Risk of Chemicals to Humans. Lyon: International Agency for Research on Cancer, 1987: 39–143.
4. Saffiotti U, Williams AO, Daniel LN, Kaighn ME, Mao Y, Shi X. Carcinogenesis by crystalline silica: animal, cellular, and molecular studies. In: Castranova V, Vallyathan V, Wallace WE, eds. Silica and Silica-Induced Lung Diseases. Boca Raton: CRC Press, 1985:345–381.
5. Wehner AP, Dagle GE, Clark ML, Buschbom RL. Lung changes in rats following inhalation exposure to volcanic ash for two years. Environ Res 1996; 40:499–517.
6. Johnson NF, Smith DM, Sebring R, Holland LM. Silica-induced alveolar cell tumors in rats. Am J Ind Med 1987; 11:93–107.
7. Muhle H, Takenaka S, Mohr U, Dasenbrock C, Mermelstein R. Lung tumor induction upon long-term low-level inhalation of crystalline silica. Am J Ind Med 1989; 15:343–346.
8. Muhle H, Kittel B, Ernst H, Mohr U, Mermelstein R. Neoplastic lung lesions in rat after chronic exposure to crystalline silica. Scand J Work Environ Health 1995; 21(suppl 2):27–29.
9. IRAC. Silica and Some Silicates. In: IARC Monographs on the Evaluation of Carcinogenic Risk of Chemicals to Humans. Lyon: International Agency for Research on Cancer, 1987: 39–143. Silica and Some Silicates, Coal Dust and Para-Aramid Fibrils. 1997; 68:1–475.
10. Shi X, Dalal NS, Hu XN, Vallyathan V. The chemical properties of silica particle surface in relation to silica-cell interaction. J Toxicol Environ Health 1989; 27:435–454.
11. Shi X, Castranova V, Halliwell B, Vallyathan V. Reactive oxygen species and silica-induced carcinogenesis. J Toxicol Environ Health B Crit Rev 1998; 1:181–197.
12. Shi X, Mao Y, Daniel LN, Saffiotti U, Dalal NS, Vallyathan V. Silica-induced DNA damage and lipid peroxidation. Environ Health Perspect 1994; 102(suppl 10):149–154.
13. Shi X, Mao Y, Saffiotti U, Wang L, Rojanasakul Y, Leonard SS, Vallyathan V. Antioxidant activity of tetrandrine and its inhibition of quartz-induced lipid peroxidation. J Toxicol Environ Health 1995; 46:233–248.
14. Shi X, Dalal NS, Vallyathan V. ESR evidence for the hydroxyl formation in aqueous suspension of quartz particles and its possible significance to lipid peroxidation in silicosis. J Toxicol Environ Health 1988; 25:237–245.
15. Kamal AM, Goma A, Khafif ME, Hammad AS. Plasma lipid peroxides among workers exposed to silica or asbestos dusts. Environ Res 1989; 49:173–180.
16. Shi X, Mao Y, Daniel LN, Saffiotti U, Dalal NS, Vallyathan V. Generation of reactive oxygen

species by quartz particles and its implication for cellular damage. Appl Occup Environ Hyg 1995; 10:1138–1144.

17. Sun Y, Oberley LW. Redox regulation of transcriptional activators. Free Rad Biol Med 1996; 21:335–348.
18. Chen F, Sun SC, Kuh DC, Gaydos LJ, Demers LM. Essential role of NF-κB activation in silica-induced inflammatory mediator production in macrophages. Biochem Biophys Res Commun 1995; 214:985–992.
19. Chen F, Kuh DC, Sun SC, Gaydos LJ, Demers LM. Dependence and reversal of nitric oxide production on NF-κB in silica and lipopolysaccharide-induced macrophages. Biochem Biophys Res Commun 1995; 214:839–846.
20. Siebenlist U, Franzoso G, Brown K. Structure, regulation and function of NF-κB. Annu Rev Cell Biol 1994; 10:405–455.
21. Baeuerle PA, Henkel T. Function and activation of NF-κB in the immune system. Annu Rev Immunol 1994; 12:141–179.
22. Schrect R, Meier B, Mannel DM, Droge W, Baeuerle PA. Dithiocarbamates as potent inhibitors of nuclear factor-κB activation in intact cells. J Exp Med 1992; 175:1181–1192.
23. Kessler DJ, Duyao MP, Spicer DB, Gail E, Sonenshein GE. NF-κB-like factors mediated interleukin 1 induction of c-myc gene transcription in fibroblasts. J Exp Med 1992; 176:787–792.
24. Baldwin AS. The NF-κB and IκB proteins: new discoveries and insights. Annu Rev Immunol 1996; 14:649–683.
25. Ding M, Shi X, Dong Z, Castranova V, Vallyathan V. Freshly fractured crystalline silica induces activator protein-1 activation through erks and p53 mitogen-activated kinase. J Biol Chem 1999; 274:30611–30616.
26. Chen F, Lu Y, Rojanasakul Y, Shi X, Vallyathan V, Castranova V. Role of hydroxyl in silica-induced NF-κB activation in macrophages. Ann Clin Lab Sci 1998; 28:1–13.
27. Sturgill TW, Ray LB, Erikson E, Maller JL. Insulin-stimulated MAP-2 kinase phosphorylates and activates ribosomal protein S6 kinase II. Nature 1988; 334:715–718.
28. Cowley S, Paterson H, Kemp P, Marshall CJ. Activation of MAP kinase is necessary and sufficient for PC12 differentiation and for transformation of NIH 3T3 cells. Cell 1994; 77: 841–852.
29. Derijard B, Hibi M, Wu IH, Barrett T, Su B, Deng T, Karin M, Davis RJ. JNK 1: a protein kinase stimulated by UV light and HaRas that binds and phosphorylates the c-Jun activation domain. Cell 1994; 76:1025–1037.
30. Bernstein LR, Ferris DK, Colburn NH, Sobel ME. A family of mitogen-activated protein kinase-related proteins interacts in vivo with activator protein-1 transcription factor. J Biol Chem 1994; 269:9401–9404.
31. Castranova V, Ma JKH, Ma JYC. Bisbenzylisoquinoline alkaloids: animal studies. In: Castranova V, Vallyathan V, Wallace WE, eds. Silica and Silica-Induced Lung Diseases. Boca Raton: CRC Press, 1996:305–321.
32. Kang JH, Lewis DW, Castranova V, Rojanasakul Y, Ma JKH, Ma JYC. Inhibitory action of tetrandrine on macrophage production of interleukin-1 (IL-1)-like activity and thymocyte proliferation. Exp Lung Res 1992; 18:715–729.
33. Angel P, Karin M. The role of Jun, Fos and the AP-1 complex in cell-proliferation and transformation. Biochim Biophys Acta 1991; 1072:129–157.
34. Dong Z, Birrer MJ, Watts RG, Matrisian LM, Colburn NH. Blocking of tumor promoter-induced AP-1 activity inhibits induced transformation in JB6 mouse epidermal cells. Proc Natl Acad Sci USA 1994; 91:609–613.
35. Dong Z, Huang C, Brown RE, Ma WY. Inhibition of activator protein 1 activity and neoplastic transformation by aspirin. J Biol Chem 1997; 272:9962–9970.
36. Huang C, Ma WY, Hanenberger D, Cleary MP, Bowden GT, Dong Z. Inhibition of ultraviolet B-induced activator protein-1 (AP-1) activity by aspirin on AP-1-luciferase transgenic mice. J Biol Chem 1997; 272:26325–26331.

37. Angel P, Baumann I, Stein B, Delius H, Rahmsdorf HJ, Herrlich P. 12-O-tetradecanoyl-phorbol-13-acetate induction of the human collagenase gene is mediated by an inducible enhancer element located in the 5′-flanking region. Mol Cell Biol 1987; 7:2256–2266.
38. Kerr LD, Miller DB, Matrisian LM. TGF-beta 1 inhibition of transin/stromelysin gene expression is mediated through a Fos binding sequence. Cell 1990; 61:267–278.
39. Foletta VC, Segal DH, Cohen DR. Transcriptional regulation in the immune system: all roads lead to AP-1. J Leukoc Biol 1998; 63:139–152.
40. Li JJ, Dong Z, Dawson MI, Colburn NH. Inhibition of tumor promoter-induced transformation by retinoids that transrepress AP-1 without transactivating retinoic acid response element. Cancer Res 1996; 56:483–489.
41. Li JJ, Westergaard C, Ghosh P, Colburn NH. Inhibitors of both nuclear factor-kappa B and activator protein-1 activation block the neoplastic transformation response. Cancer Res 1997; 57:3569–3576.
42. Barthelman M, Chen W, Gensler HL, Huang C, Dong Z, Bowden GT. Inhibitory effects of perullyl alcohol on UVB-induced murine skin cancer and AP-1 transactivation. Cancer Res 1998; 58:711–716.
43. Dong Z, Lavrovsky V, Colburn NH. Transformation reversion induced in JB6 RT101 cells by AP-1 inhibitors. Carcinogenesis 1995; 16:749–756.
44. Whitmarsh AJ, Davis RJ. Transcription factor AP-1 regulation by mitogen-activated protein kinase signal transduction pathways. J Mol Med 1996; 74:589–607.
45. Ding M, Dong Z, Chen F, Pack D, Ma WY, Ye J, Shi X, Castranova V, Vallyathan V. Asbestos induces activator protein-1 transactivation in transgenic mice. Cancer Res 1999; 59:1884–1889.
46. Vallyathan V, Shi XL, Dalal NS, Irr W, Castranova V. Generation of free radicals from freshly fractured silica dust. Potential role in acute silica-induced lung injury. Am Rev Respir Dis 1988; 138:1213–1219.
47. Vallyathan V, Mega JF, Shi X, Dalal NS. Enhanced generation of free radicals from phagocytes induced by mineral dusts. Am J Respir Cell Mol Biol 1992; 6:404–413.
48. Daniel LN, Mao Y, Saffiotti U. Oxidative DNA damage by crystalline silica. Free Radical Biol Med 1993; 14:463–472.
49. Castranova V, Kang JH, Moore MD, Pails WH, Frazer DG, Schuegler-Berry D. Inhibition of stimulant-induced activation of phagocytic cells with tetrandrine. J Leukocyte Biol 1991; 50: 412–422.
50. Baeuerle PA. The inducible activator NF-κB: regulation of distinct protein subunits. Biochim Biophys Acta 1991; 1072:63–80.
51. Schmidt KN, Amstad P, Cerutti P, Baeuerle PA. The role of hydrogen peroxide and superoxide as messengers in the activation of transcription factor NF-κB. Chem Biol 1995; 2:13–22.
52. Myer M, Pahl HL, Baeuerle PA. Regulation of the transcription factors NF-κB and AP-1 by redox changes. Chem Biol Interactions 1994; 91:91–100.
53. Ye J, Zhang X, Young HA, Mao Y, Shi X. Chromium(VI)-induced nuclear factor-κB activation in intact cells via free radical reactions. Carcinogenesis 1995; 16:2401–2405.
54. Schreck R, Albermann K, Baeuerle PA. Nuclear factor-κB: an oxidative stress-responsive transcription factor of eukaryotic cells (a review). Free Rad Res Commun 1992; 17:221–237.
55. Kondo Y, Takano F, Hojo H. Inhibitory effect of bisbenzylisoquinoline alkaloids on nitric oxide production in activated macrophages. Biochem Pharmacol 1993; 46:1887–1892.
56. Halliwell B. Protection against tissue damage in vivo by deferoxamine: what is its mechanism of action? Free Rad Biol Med 1989; 7:645–651.
57. McLaren GD, Muir WA, Kellermeyer RW. Iron overload disorders: natural history, pathogenesis, diagnosis, and therapy. CRC Crit Rev Clin Lab Sci 1983; 19:205–265.
58. Keller RJ, Rush JD, Grover TA. Spectrophotometric and ESR evidence for vanadium(IV) deferoxamine complexes. J Inorg Biochem 1991; 41:269–276.

59. Shi X, Sun X, Gannett PM, Dalal NS. Deferoxamine inhibition of Cr(V)-mediated radical generation and deoxyguanine hydroxylation: ESR and HPLC evidence. Arch Biochem Biophys 1992; 293:281–286.
60. Nash T, Allison AC, Harington JS. Physio-chemical properties of silica in relation to its toxicity. Nature 1981; 210:259–261.
61. Castranova V. Generation of oxygen radicals and mechanisms of injury prevention. Environ Health Perspect 1994; 102(suppl 10):65–68.
62. Mao Y, Daniel LN, Knapton AD, Shi X, Saffiotti U. Protective effects of silanol group binding agents on quartz toxicity to rat lung alveolar cells. Appl Occup Environ Hyg 1995; 10:1132–1137.
63. Holt PE. Poly(vinylpyridine oxides) in pneumoconiosis research. Br J Ind Med 1971; 28:72–77.
64. Zhao JD, Liu JD, Li GZ. Long-term follow-up observations of the therapeutic effect of PVPNO on human silicosis. Zentralbl Bakteriol Mikrobiol Hyg [B] 1983; 178:259–261.
65. Klochars M, Hedenborg M, Vanhala E. Effect of two particle surface-modifying agents, polyvinylpyridine-*N*-oxide and carboxymethylcellulose, on the quartz and asbestos mineral fiber-induced production of reactive oxygen metabolites by human polymorphonuclear leukocytes. Arch Environ Health 1990; 42:8–14.
66. Nyberg P. Polyvinlpyridine-*N*-oxide and carboxymethylcellulose inhibit dust-induced production of reactive oxygen species by human macrophages. Environ Res 1991; 55:157–164.

# 11

# Atmospheric Oxidants and Respiratory Tract Surfaces

**CARROLL E. CROSS, SAMUEL LOUIE, SUNYE KWACK, PAT WONG, SHARANYA REDDY, and ALBERT VAN DER VLIET**

*University of California, Davis, California*

## I. INTRODUCTION

Atmospheric pollutants, largely arising as primary and secondary products of combustion, represent an important source of oxidative and nitrosative stress to Earth's biosystem. Since the various biosurfaces are directly exposed to these pollutant stresses, it is not surprising that living organisms have developed complex integrated extracellular and intracellular defense systems against stresses related to atmospheric reactive oxygen species (ROS) and reactive nitrogen species (RNS), including ozone ($O_3$) and nitrogen dioxide ($NO_2$). Animal epithelial surfaces, including the delicate respiratory tract surfaces, contain an antioxidant network that would be expected to provide defense against environmental stress caused by ambient ROS and RNS, thus ameliorating their injurious effects on underlying cellular constituents.

The reactive substances at these biosurfaces not only represent an important protective system against oxidizing environments, but products of their reactions with ROS/RNS may also serve as second messengers that induce injury to underlying cells or cause cell activation resulting in production of proinflammatory substances including chemokines and cytokines. Additionally, such products could represent useful biomarkers of environmental oxidative stress. In this chapter, we discuss antioxidant defense systems against oxidative environmental toxins in respiratory tract lining fluids (RTLF). A somewhat more complete description of these defenses, emphasizing the parallels that exist among plant and cutaneous tissue biosurfaces and RT biosurfaces, is available elsewhere (1).

Widespread attempts have been made to evaluate mammalian responses to environmental oxidants, of which hyperbaric $O_2$ and $O_3$, the latter one of a number of potential atmospheric pollutant oxidants, have perhaps been the most characterized. Many studies have focused on documenting the effects of atmospheric pollutant oxidants on human health (2). Pathobiological studies have focused on the involvement of ROS (3) and the biological antioxidant protective mechanisms induced by oxidative stress, especially with regard to antioxidant enzyme induction in the context of adaptive tolerance (e.g., glutathione peroxidase, superoxide dismutases and catalases, and heme oxygenases (4–7). However, it is the outermost portions of the human RT, the RTLFs, that initially are directly exposed to gaseous environmental toxicants, including environmental oxidants. Their extracellular and largely nonenzymatic antioxidant defenses will be the first to encounter insult and may, in fact, be solely responsible for antioxidant defenses in situations where environmental oxidative stress fails to reach levels that penetrate these defenses to reach underlying cell membrane surfaces, thereby eliciting adaptive or injurious responses in RT epithelia. Thus, in a sense, it is failure of this outermost defense system that brings oxidants to the surface of underlying plasma membranes, causing injurious or adaptive responses to RT cells.

Although there are numerous other environmental oxidants that contribute to RT surface oxidative stress (e.g., oxides of nitrogen, cigarette smoke, and urban particulates), we focus here on $O_3$ as a representative environmental oxidant and on the antioxidants with which it can be expected to react at RT surfaces. Although in most circumstances the oxidant can be expected to be neutralized, products of reactions with constituents of the extracellular fluids, such as cytotoxic aldehydes and bioactive lipids produced by reactions of $O_3$ with lipids, may be largely responsible for RT cell or tissue injury (8–15). The potential of lipids, including oxidized lipids (other than prostanoids and leukotrienes), as pluripotent effectors of signaling cascades and as mediators of communication among lung cells, remains to be more fully explored (16).

## II. ANATOMICAL CONSIDERATIONS

The interfaces between the environment and the RT are simplistically depicted in Figure 1. The uptake of environmental toxins can be expected to occur mainly via the RT surface

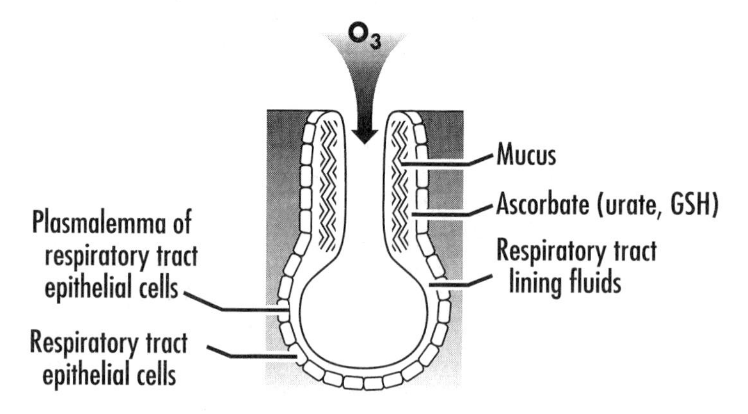


**Figure 1** Environmental oxidants such as $O_3$ first react with components of the respiratory tract lining fluids (RTLF) and their contained antioxidants including ascorbic acid.

**Table 1** Human Respiratory Tract Lining Fluids

| | Thickness (μm) | Surface area ($cm^2$) | Volume (mL) | Turnover |
|---|---|---|---|---|
| Nasal | 5–10 | 180 | 0.15 | ? > 1000×/day |
| Airways | 1–10 | 4500 | 3.5 | ? > 20×/day |
| Alveoli | 0.5–0.2 | 885,000 | 9 | ??? |

Based largely on morphological data (17).

fluids. This will be especially true for reactive oxidant pollutants such as $O_3$ and $NO_2$ (8,11). Numerous studies have documented the effects of oxidants on lungs and, to some extent, have directed attention to the importance of the RT lining fluids (1,17–19).

The RTLFs vary in different levels of the RT. The thickness of these layers has only been estimated, not rigorously quantified, and much remains uncertain concerning the RT surface fluid volume, composition, and turnover (20–22). For example, as shown in Table 1, RTLF depth (both sol and gel layers) in the upper RT has been estimated to range from 1 to 10 μm, whereas in the distal bronchoalveolar regions RTLF depth is only 0.2 to 0.5 μm (17,23). The composition of RTLFs also varies widely, with mucins being present in upper RTLFs and surfactant proteins and lipids being present in lower RTLFs. Modeling studies have suggested that upper airway mucus does not flow evenly but preferentially concentrates along troughs or grooves (24). The quantitative description of antioxidants present at any given level of the respiratory tract is compromised by the complex gel–sol nature of the upper RTLFs and the probability that the mucin-containing gel layer is covered by a lipid layer that may be derived in part from the lower bronchoalveolar regions (24–27).

Thus RTLFs represent an important site of environmental–biosystem interaction, being the site of continuous reactive absorption of oxidant pollutants. The external extracellular compartment can be thought of as an important sink for atmospheric oxidants, screening out to some extent the effects of inhaled oxidants on the underlying RT epithelial cells.

## III. RESPIRATORY TRACT LINING FLUIDS (RTLFs) AND THEIR ANTIOXIDANTS

Because antioxidants in RTLFs may form the first line of defense against inhaled oxidants, it is important to understand the nature, chemistry, and fate of these antioxidants. Interest in the role of RTLF antioxidants in protecting against environmental oxidant pollutants has been stimulated by several observations. For example, uric acid, an important extracellular antioxidant, is thought to be secreted by the same upper RT epithelial cells that secrete mucin (28,29), and lower RT epithelial cells secrete alpha-tocopherol together with surfactant (30), the latter being itself a target for oxidative injury (31–34). Bronchoalveolar RTLFs contain glutathione (GSH) in considerable excess of plasma levels and may, in some species at least, contain more ascorbic acid than plasma (17,35). Mucin itself (36–38), together with proteoglycans such as heparin sulfate and hyaluronic acid (39,40), has significant antioxidant properties, and in fact many toxicant irritants, including oxidants, can stimulate mucin secretion (41). In the case of mucins, both the abundant sugar moieties and the cysteine-rich domains would be expected to provide a rich antioxidant pool (36).

It is regrettable that we know so little about the specific and quantitative nature of the antioxidant functions of the RT mucins.

Interest in the antioxidant functions of RTLFs has been further aroused by the observation that the GSH content of the lower RTLF appears to be subnormal in AIDS (42), idiopathic pulmonary fibrosis (43,44), farmer's lung (45), cystic fibrosis (46), adult respiratory distress syndrome (ARDS) (47), and lung allograft patients (48), but appears to be supranormal in cigarette smokers (49), asthmatics (50,51), asymptomatic farmers with lung exposure similar to those experiencing farmer's lung (45), and 1 day following a single day 4 h/day or a 4 day 4 h/day exposure to 0.2 ppm $O_3$ under exercise conditions (52). While it seems possible that the high GSH levels observed in smokers can be ascribed to activation of a defense mechanism increasing GSH synthesis in cells of the lower airways (53), the precise mechanisms of apparent GSH dysregulation in the other conditions remains enigmatic, although almost certainly related to the duration and magnitude of the underlying inflammatory–immune and fibroproliferative processes occurring in the various RT disorders. Nonetheless, the recognition that RT GSH levels are reduced in many conditions, including idiopathic pulmonary fibrosis and cystic fibrosis, has given impetus to therapeutic strategies designed to increase RT GSH levels (53), and in patients with early septic shock (a frequent prodrome of ARDS), administration of high doses of GSH, along with *N*-acetyl-L-cysteine, seems to provide some protection against the peroxidative stress of septic shock (54). Although aerosolized GSH is known to induce significant bronchospasm in some individuals (55), and theoretical constructs of just how it may be beneficial are incomplete, it is apparent that further studies of strategies that would increase airway surface "antioxidants" are clearly warranted. Of relevance, it has been speculated that variations in the composition of antioxidants contained in the RTLFs could contribute to the interspecies and human variability in sensitivity to RT injury by inhaled oxidants (1,17,56,57).

As in other extracellular fluids (58,59), the antioxidants in RTLFs are variable in nature and include low molecular mass antioxidants, metal-binding proteins, certain antioxidant enzymes, and several proteins and unsaturated lipids that may act as sacrificial antioxidants. The oxidant dose reaching the underlying RT epithelial cells is almost certainly influenced by the RTLF volume (including surface area and thickness) and composition, and presumably also by the turnover rate of the various antioxidants present in the RTLS (1,17,56).

Several investigators have attempted to determine antioxidant concentrations in human RTLFs, which are commonly collected by nasal or bronchoalveolar lavage techniques. As summarized in Table 2, both nasal and bronchoalveolar lavage RTLFs are found to contain ascorbic acid, uric acid, and GSH as major water-soluble antioxidants. Moreover, it is readily apparent that lavaged nasal RTLFs contain relatively high levels of uric acid, while uric acid and GSH appear to be the most predominant low-molecular-mass antioxidants in lavaged distal RTLFs. Calculation of RTLF dilution by these lavage procedures has been attempted with the use of various dilution markers, such as urea concentration. Measurements of urea are based on the hypothesis that extracellular urea concentrations are similar in all compartments, so that dilution of RTLF urea can be calculated by simultaneous measurement of plasma urea. However, this procedure is hindered by the fact that instillation of urea-free saline causes a high concentration gradient, resulting in rapid diffusion of intracellular, interstitial, or intravascular urea into the lavage fluid. This compromises quantitation of RTLF dilution by sequential lavage procedures (63–66), which have been commonly used in studies of RTLF antioxidants. There is also

**Table 2** Approximate Values for the Concentrations of Nonenzymatic Antioxidants in Human Plasma Compared to Lavages of Distal Respiratory Tract (RTLF) and Nasal Lining Fluids (NLF)

| Antioxidant | Plasma (μM)[a] | Lavaged Distal RTLF (μM)[b] | Lavaged NLF (μM)[c] |
|---|---|---|---|
| Ascorbic acid | 30–150 | 0.5–1.8 | 0.1–2.5 |
| Glutathione | 0.5–2 | 1.2–6.7 | 0.0–1.3 |
| Uric acid | 160–450 | 0.3–3.0 | 1.5–17.0 |
| Bilirubin | 5–20 | — | — |
| α-Tocopherol | 15–25 | 0.02–0.05 | — |
| β-Carotene | 0.3–0.6 | — | — |
| Ubiquinol-10 | 0.4–1.0 | — | — |
| Albumin-SH | 500 | 0.7 | — |

[a] Plasma values derived from Refs. 58 and 59.
[b] Distal RTLF lavage values largely derived from Refs. 17 and 60.
[c] NLF data derived from Refs. 17, 29, 47, 60, 61, and 62.

the confounding problem that blood-to-RT permeability to urea may vary among different subject populations (67). RTLF dilution by commonly performed bronchoalveolar lavage procedures has been estimated to be about 100-fold, and lower RTLF concentrations of GSH have been reported to be in the range of 100 to 500 μM (35,43,60,68). Extrapolation of this dilution to reported ascorbic and uric acid concentrations would yield lower RTLF ascorbic acid levels of 50 to 180 μM and uric acid levels of 30 to 300 μM (i.e., similar to or lower than in plasma) (Table 2).

Measurements of urea have also been used to estimate RTLF dilution by nasal lavage procedures (69). Collection of nasal RTLFs by instillation of 5 mL saline into one nostril for 10 s has thus been estimated to result in approximately 10- to 15-fold dilution of nasal RTLFs. This implies that local antioxidant concentrations, as estimated from reported nasal lavage levels listed in Table 2, may range from 2 to 30 μM ascorbic acid, 20 to 300 μM uric acid, and 0 to 15 μM GSH. This suggests that antioxidant levels in nasal RTLFs are lower compared to bronchoalveolar RTLFs (especially true for GSH, and to a lesser extent ascorbic acid), but, of course, they contain considerable quantities of high-molecular-weight mucins, themselves potent antioxidants (36).

Lower RTLFs are also known to contain several antioxidant enzymes and other proteins (70,71), as summarized in Table 3. Recently, it has become clear that lung cells secrete antioxidant enzymes such as extracellular glutathione peroxidase (eGP$_x$) (72) and extracellular CuZn SOD (SOD-3) (73), which may contribute to the antioxidant defense in RTLFs. Recently, intracellular CuZn SOD (SOD-1) has been found to be detectable in BALF, in addition to very small amounts of SOD-3 (74). Although some investigators have considered catalase (70,71,75) to be physiologically important in the RTLFs, there remains uncertainty as to whether it plays a significant role in the antioxidant defenses of RTLFs. It is possible that cytolysis, occurring as a natural process during cell turnover or occurring subsequent to lavage procedures themselves, may contribute to the detection of ''extracellular'' enzymatic antioxidants demonstrated to be found in RTLFs (this holds for GSH and ascorbic acid as well, as intracellular concentrations of these low-molecular-weight antioxidants are generally much higher than those found extracellularly). Because of its significant concentration in upper RTLFs, it is likely that secreted lactoferrin plays

**Table 3** Antioxidant Proteins Known to be Present in RTLFs[a]

| Metal-binding proteins | Other antioxidant proteins |
|---|---|
| Lactoferrin | Catalase |
| Ceruloplasmin | SOD-1 |
| | SOD-3 |
| Transferrin | Glutathione reductase |
| Albumin | Ceruloplasmin (ferroxidase activity) |
| | Albumin (e.g., SH activity) |

[a] These proteins either bind metal ions in safe forms (unable to catalyze damaging free radical reactions) or possess other significant antioxidant properties (58,70,71). Note that albumin can both sequestrate metals and contribute a free thiol RSH group to antioxidant pools.

a significant antioxidant function by binding iron and inhibiting iron-dependent free radical reactions. Interestingly, although one would consider transferrin to be protective, the hypotransferrinemic mouse seems to have diminished injury after exposure to metal-rich particulates (76) associated with an adaptive increase in other metal transport and storage proteins.

## IV. RTLFs AND POLLUTANTS

There is much theoretical and some experimental evidence that reactive gases, such as inhaled $O_3$ and $NO_2^{\bullet}$, react with RTLF components and may never reach the underlying RT epithelial cells, at least in areas where they are covered by RTLFs (1,8,18,77–79). Therefore, toxic actions of $O_3$ or $NO_2^{\bullet}$ toward RT epithelial cells are most likely mediated, at least in part, by products of the reaction of these inhaled toxins with RTLF constituents. For the case of $O_3$, such toxins may include not only products from reactions of $O_3$ with water, but also products from reactions of $O_3$ with lipids and proteins. These products would include lipid hydroperoxides, cholesterol ozonization products, ozonides, aldehydes, and other oxidation products of antioxidants or proteins, as illustrated in Figure 2 (8–15,79,80).

As $O_3$ and $NO_2^{\bullet}$ are relatively insoluble, interactions of these inhaled gases with RTLFs are primarily governed by reactive absorption (i.e., the more oxidizable substrate that is present in RTLFs, the more $O_3$ will be absorbed by the RTLFs) (81,82). Therefore, inhaled $O_3$ or $NO_2^{\bullet}$ may be effectively removed by antioxidants present in the more abundant, proximal RTLFs, to protect more distal RTLFs and RT epithelial cells. However, when these toxic gases are not removed, reaction of $O_3$ or $NO_2^{\bullet}$ may injure RT epithelial cells in the upper RT, result in cell activations designed to augment defense systems (e.g., increase vascular permeability, increase mucin secretion), or initiate inflammatory–immune processes (83).

Studies with solutions of various RTLF antioxidants, or with human blood plasma as a model of extracellular fluid at least somewhat resembling RTLFs, have demonstrated that $O_3$ reacts readily with ascorbic acid, uric acid, and thiols (82,84–87). Exposure of plasma to $O_3$ resulted in rapid depletion of its major water-soluble antioxidants—ascorbic acid and uric acid—at similar rates (85,86). As uric acid appears to be the most predominant low-molecular-mass antioxidant in proximal RTLFs (e.g., nasal) (28,29), it has been

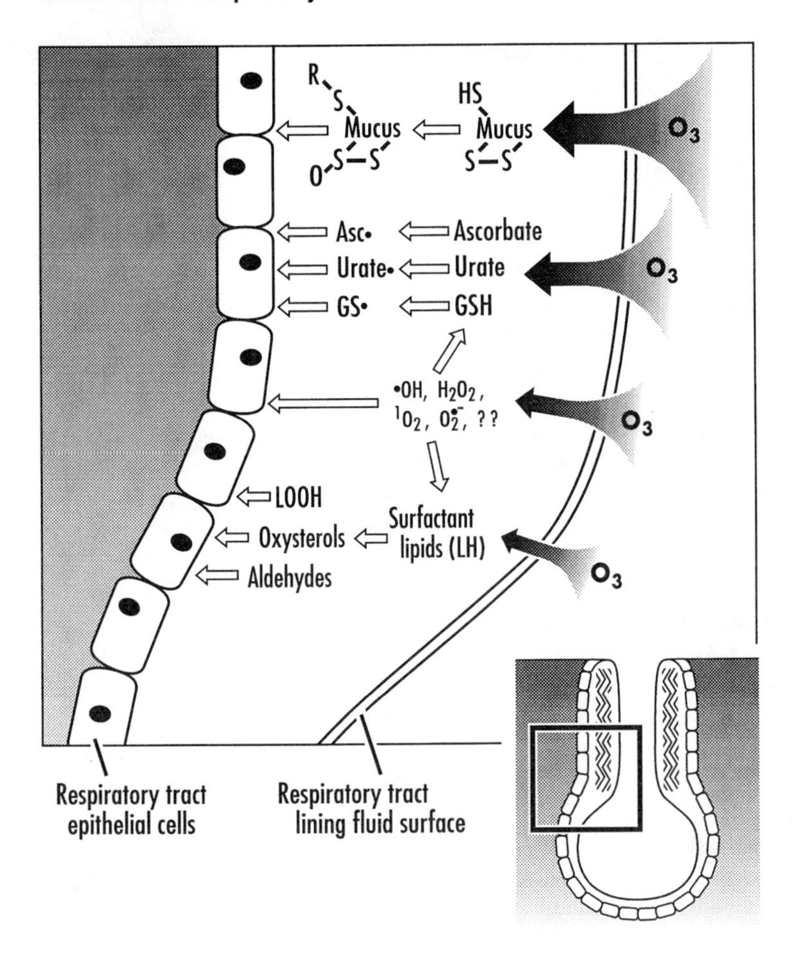


**Figure 2** $O_3$ reactions with constituents of RTLFs. Note that as inspired $O_3$ is inhaled, its concentration is decreased secondary to its chemical reactions with the more proximal RTLFs including reactions with mucus glycoproteins and antioxidants. Dissolved $O_3$ may form ROS itself or react with the surface lipids to generate toxic and/or bioactive lipids and aldehydes.

hypothesized that uric acid may provide the most significant antioxidant defense against $O_3$ in at least the nasal RTLFs.

Studies with blood plasma (as a model extracellular fluid) have indicated that exposure of plasma to $O_3$ or $NO_2$ results in rapid depletion of ascorbic and uric acid, suggesting that these antioxidants provide the major defense system against both of these oxidant gases (88,89). A recent study with a composite mixture of RTLF antioxidants has demonstrated that exposure to ambient $O_3$ concentrations results in depletion primarily of uric and ascorbic acid, and, to a lesser extent, GSH. Moreover, these antioxidants were found to prevent oxidative modification of proteins present in the mixture (87). It follows that uric and ascorbic acid provide a protective screen against inhaled $O_3$, and that GSH may be relatively less effective, at least in these model systems.

A study with human volunteers has demonstrated that exposure to 2 ppm $NO_2$ for 9 h results in transient depletion of RTLF ascorbic and uric acid (90). RTLF GSH levels were not depleted but rather increased, again suggesting that ascorbic and uric acid are major reactants with inhaled oxidants, forming the major first line of antioxidant defense in the RT.

## V. FUTURE CONSIDERATIONS

Mechanistic understanding of RT responses to environmental oxidant pollutants have blossomed over the past decade and continue to grow, receiving strong impetus from emerging academic disciplines of molecular environmental toxicology and driven by such resources as the NIEHS Environmental Genome Project. Existing animal models (78,81,87), and others to be generated, should provide important systems for studying the role of surface antioxidant species in modulating environmental effects on RT cell-signaling systems and gene expression pathways. The use of cell culture models for studying environmental oxidant effects must recognize both the role of antioxidant species external to cellular surfaces in modifying potential endpoints in vivo, and also the role of extracellular antioxidants in maintaining intracellular levels of such important antioxidants as ascorbic acid and tocopherols. When extrapolations are made from data generated in such ''clean'' systems as in vitro primary, secondary, and immortalized cell culture systems exposed to environmental oxidants, investigators should take into consideration the contributions of extracellular agents at the environment–biosystem boundary under in vivo conditions. Much remains to be characterized regarding the precise nature of the RT surface liquids and solutes, not only with regard to volume, composition, and turnover, but also with regard to RT trans-plasma membrane transport systems and oxidoreductase electron transport systems that may be active in the regulation of RT lining fluid antioxidants such as ascorbic acid and GSH (91). Finally, there is no doubt that studies of the effects of environmental oxidants on animal transgenics (92), in which components of antioxidant defense systems have been over- or underexpressed, will contribute to further understanding of mechanisms of oxidative stress at animal RT biosurfaces.

## ACKNOWLEDGMENTS

Supported, in part, by grants from the National Institutes of Health and the University of California Tobacco Related Disease Prevention Program.

## REFERENCES

1. Cross CE, van der Vliet A, Louie S, Thiele JJ, Halliwell B. Oxidative stress and antioxidants at biosurfaces: plants, skin, and respiratory tract surfaces. Environ Health Perspectives 1998; 106:1241–1251.
2. Committee of the Environmental and Occupational Health Assembly of the American Thoracic Society. Health effects of outdoor air pollution. Am J Respir Crit Care Med 1996; 153:3–50.
3. Elstner EF, Obwald W. Air pollution: involvement of oxygen radicals (a mini review). Free Rad Res Commun 1991; 12–13:795–807.
4. Freeman BA, Crapo JD. Free radicals and tissue injury. Lab Invest 1982; 47:412–426.
5. Wiese AG, Pacifici RE, Davies KJA. Transient adaptation to oxidative stress in mammalian cells. Arch Biochem Biophys 1995; 318:231–240.
6. Remacle J, Raes M, Toussaint O, Renard P, Rao G. Low levels of reactive oxygen species as modulators of cell function. Mutat Res 1995; 316:103–122.
7. Ryter S, Tyrrell R. Heme synthesis and degradation pathways: role in oxidant sensitivity.

Heme oxygenase has both pro- and antioxidant properties. Free Rad Biol Med 2000; 28:289–308.
8. Pryor WA. Mechanisms of radical formation from reactions of ozone with target molecules in the lung. Free Rad Biol Med 1994; 17:451–465.
9. Hamilton RF, Jr, Li L, Eschenbacher WL, Szweda L, Holian A. Potential involvement of 4-hydroxynonenal in the response of human lung cells to ozone. Am J Physiol 1998; 274:L8–L16.
10. Kafoury R, Pryor WA, Squadrito GL, Salgo MG, Zou X, Friedman M. Lipid ozonation products (LOP) initiate signal transduction by activating phospholipases A2, C and D. Toxicol Appl Pharmacol 1998; 150:338–349.
11. Postlethwait EM, Cueto R, Velsor LW, Pryor WA. Ozone-induced formation of bioactive lipids: estimated surface concentrations and lining layer effects. Am J Physiol 1998; 274: L1006–L1016.
12. Pryor WA, Cueto R, Squadrito GL. Pulmonary biomarkers for ozone exposure (HEI Investigators' Revised Final Report). Cambridge, MA: Health Effects Institute, 1999.
13. Frampton MW, Pryor WA, Cueto R, Cox C, Morrow PE, Utell MJ. Ozone exposure increases aldehydes in human lung epithelial lining fluid. Am J Respir Crit Care Med 1999; 159:1134–1137.
14. Frampton MW, Pryor WA, Cueto R, Cox C, Morrow PE, Utell MJ. Aldehydes (nonanal and hexanal) in rat and human bronchoalveolar lavage fluid after ozone exposure. (HEI Investigators' Research Report: Research Report #90). MA: Health Effects Institute, 1999.
15. Kafoury RM, Pryor WA, Squadrito GL, Salgo MG, Zou X, Friedman M. Induction of inflammatory mediators in human airway epithelial cells by lipid ozonation products. Am J Respir Crit Care Med 1999; 160:1934–1942.
16. Peters-Golden M. Playing ''telephone'': bioactive lipids as mediators of intercompartmental communication in the alveolus. J Clin Invest 2000; 105:857–858.
17. Hatch GE. Comparative biochemistry of the airway lining fluid. In: Parent RA, ed. Treatise on Pulmonary Toxicology, vol 1. Comparative Biology of the Normal Lung. Boca Raton, FL: CRC Press, 1992:617–632.
18. Pryor WA. How far does ozone penetrate into the pulmonary air/tissue boundary before it reacts? Free Rad Biol Med 1992; 12:83–88.
19. Cross CE, van der Vliet A, Eiserich JP and Wong J. Oxidative stress and antioxidants in respiratory tract lining fluids. In: Clerch LB, Massaro DJ, eds. Oxygen, Gene Expression, and Cellular Function. New York: Marcel Dekker, 1997:367–398.
20. Widdicombe J. Airway and alveolar permeability and surface liquid thickness: theory. J Appl Physiol 1997; 82:3–12.
21. Boucher RC. Topical review: Molecular insights into the physiology of the 'thin film' of airway surface liquid. J Phys 1999; 516:631–638.
22. Hannrahan JW. Airway plumbing. J Clin Invest 2000; 105:1343–1344.
23. Bastacky J, Goerke J, Lee CY, Yager D, Kenaga L, Koushafar H, Hayes TL, Chen Y, Clements JA. Alveolar lining liquid layer is thin and continuous: low-temperature scanning electron microscopy of normal rat lung. Am Rev Respir Dis 1993; 147:A148.
24. Sims DE, Horne MM. Heterogeneity of the composition and thickness of tracheal mucus in rats. Am J Physiol 1997; 273:L1036–L1041.
25. Girod S, Zahm J-M, Plotkowski C, Beck G, Puchelle E. Role of the physicochemical properties of mucus in the protection of the respiratory epithelium. Eur Respir J 1992; 5:477–487.
26. Kim KC, Singh BN. Hydrophobicity of mucin-like glycoproteins secreted by cultured tracheal epithelial cells: association with lipids. Exp Lung Res 1990; 16:279–292.
27. Rose MC. Mucins: structure, function, and role in pulmonary diseases. Am J Physiol 1992; 263:L413–L429.
28. Kaliner MA. Human nasal respiratory secretions and host defense. Am Rev Respir Dis 1991; 144:S52–S56.

29. Peden DB, Swiersz M, Ohkubo K, Hahn B. Nasal secretion of the ozone scavenger uric acid. Am Rev Respir Dis 1993; 148:455–461.
30. Rustow B, Haupt R, Stevens PA, Kunze D. Type II pneumocytes secrete vitamin E together with surfactant lipids. Am J Physiol 1993; 265:L133–L139.
31. Merritt TA, Amirkhanian JD, Helbock H, Halliwell B, Cross CE. Reduction of the surface-tension-lowering ability of surfactant after exposure to hypochlorous acid. Biochem J 1993; 295:19–22.
32. Cifuentes J, Ruiz-Oronoz J, Myles C, Nieves B, Carlo WA, Matalon S. Interaction of surfactant mixtures with reactive oxygen and nitrogen species. J Appl Physiol 1995; 78:1800–1805.
33. Putman E, Liese W, Voorhout WF, van Bree L, van Golde LMG, Haagsman HP. Short-term ozone exposure affects the surface activity of pulmonary surfactant. Toxicol Appl Pharm 1997; 142:288–296.
34. Putman E, van Golde LMG, Haagsman HP. Toxic oxidant species and their impact on the pulmonary surfactant system. Lung 1997; 175:75–103.
35. Cantin AM, North SL, Hubbard RC, Crystal RG. Normal alveolar epithelial lining fluid contains high levels of glutathione. J Appl Physiol 1987; 63:152–157.
36. Cross CE, Halliwell B, Allen A. Antioxidant protection: a function of tracheobronchial and gastrointestinal mucus. Lancet 1984; 1:1328–1330.
37. Grisham MB, Von Ritter C, Smith BF, LaMont JT, Granger DN. Interaction between oxygen radicals and gastric mucin. Am J Physiol 1987; 253:G93–G96.
38. Hiraishi H, Terano A, Ota S, Mutoh H, Sugimoto T, Harada T, Razandi M, Ivey KJ. Role for mucous glycoprotein in protecting cultured rat gastric mucosal cells against toxic oxygen metabolites. J Lab Clin Med 1993; 121:570–578.
39. Saari H, Konttinen YT, Friman C, Sorza T. Differential effects of reactive oxygen species on native synovial fluid and purified human umbilical cord hyaluronate. Inflammation 1993; 17: 403–415.
40. Raats CJ, Bakker MA, van den Born J, Berden JH. Hydroxyl radicals depolymerize glomerular heparan sulfate in vitro and in experimental nephrotic syndrome. J Biol Chem 1997; 272: 26734–26741.
41. Adler KB, Holden-Stauffer WJ, Repine JE. Oxygen metabolites stimulate release of high-molecular-weight glycoconjugates by cell and organ cultures of rodent respiratory epithelium via an arachidonic acid-dependent mechanism. J Clin Invest 1990; 85:75–85.
42. Buhl R, Jaffe HA, Holroyd KJ, Wells FB, Mastrangeli A, Saltini C, Cantin AM, Crystal RG. Systemic glutathione deficiency in symptom-free HIV-seropositive individuals. Lancet 1989; 2:1294–1298.
43. Cantin AM, Hubbard RC, Crystal RG. Glutathione deficiency in the epithelial lining fluid of the lower respiratory tract in idiopathic pulmonary fibrosis. Am Rev Respir Dis 1989; 139: 370–373.
44. Behr J, Degenkolb B, Maier K, Braun B, Beinert T, Krombach F, Vogelmeier C, Fruhmann G. Increased oxidation of extracellular glutathione by bronchoalveolar inflammatory cells in diffuse fibrosing alveolitis. Eur Respir J 1995; 8:123–138.
45. Behr J, Degenkolb B, Beinart T, Krombach F, Voglemeier C. Pulmonary glutathione levels in acute episodes of farmer's lung. Am J Respir Crit Care Med 2000; 161:1968–1971.
46. Roum JH, Buhl R, McElvaney NG, Borok Z, Crystal RG. Systemic deficiency of glutathione in cystic fibrosis. J Appl Physiol 1993; 38:491–513.
47. Pacht ER, Timerman AP, Lykens MG, Merola AJ. Deficiency of alveolar fluid glutathione in patients with sepsis and the adult respiratory distress syndrome. Chest 1991; 100:1397–1403.
48. Baz MA, Tapson VF, Roggli VL, Van Trigt P, Piantadosi CA. Glutathione depletion in epithelial lining fluid of lung allograft patients. Am J Respir Crit Care Med 1996; 153:742–746.
49. Linden M, Hakansson L, Ohlsson K, Sjodin K, Tegner H, Tunek A, Venge P. Glutathione in bronchoalveolar lavage fluid from smokers is related to humoral markers of inflammatory cell activity. Inflammation 1989; 13:651–658.

50. Smith LJ, Houston M, Anderson J. Increased levels of glutathione in bronchoalveolar lavage fluid from patients with asthma. Am Rev Respir Dis 1993; 147:1461–1464.
51. Kelly FJ, Mudway I, Blomberg A, Frew A, Sandstrom T. Altered lung antioxidant status in patients with mild asthma. Lancet 1999; 354:482–483.
52. Jorres RA, Holz O, Zachgo W, Timm P, Koschyk S, Muller B, Grimminger F, Seeger W, Kelly FJ, Dunster C, Frischer T, Lubec G, Waschewski M, Niedorf A, Magnussen H. The effect of repeated ozone exposures on inflammatory markers in bronchoalveolar lavage fluid and mucosal biopsies. Am J Respir Crit Care Med 2000; 161:1855–1861.
53. Rahman O, Abidi P, Afaq F, Schiffmann D, Mossman BT, Kamp DW, Athar M. Glutathione redox system in oxidative lung injury. Crit Rev 1999; 29:543–568.
54. Ortolani O, Conti A, De Gaudio AR, Moraldi E, Cantini Q, Novelli G. The effect of glutathione and N-acetylcysteine on lipoperoxidative damage in patients with early septic shock. Am J Respir Crit Care Med 2000; 161:1907–1911.
55. Marrades RM, Roca J, Barbera JA, de Jover L, MacNee W, Rodriguez-Roisin R. Nebulized glutathione induces bronchoconstriction in patients with mild asthma. Am J Respir Crit Care Med 1997; 156:425–430.
56. Slade R, Crissman K, Norwood J, Hatch G. Comparison of antioxidant substances in bronchoalveolar lavage cells and fluid from humans, guinea pigs, and rats. Exp Lung Res 1993; 19:469–484.
57. Kelly FJ, Mudway IS. Sensitivity to ozone: could it be related to an individual's complement of antioxidants in lung epithelium lining fluid? Redox Report 1997; 3:199–206.
58. Halliwell B, Gutteridge JMC. The antioxidants of human extracellular fluids. Arch Biochem Biophys 1990; 280:1–8.
59. Stocker R, Frei B. Endogenous antioxidant defences in human blood plasma. In: Sies H, ed. Oxidative Stress: Oxidants and Antioxidants. London: Academic Press, 1991:213–243.
60. van der Vliet A, O'Neill CA, Cross CE, Halliwell B, Louie S. Determination of low-molecular-mass antioxidant concentrations in human respiratory tract lining fluids. Am J Physiol 1999; 276:L289–L296.
61. Housley DG, Mudway I, Kelly FJ, Eccles R, Richards RJ. Depletion of urate in human nasal lavage following in vitro ozone exposure. Int J Biochem Cell Biol 1995; 27:1153–1159.
62. Testa B, Mesolella M, Testa D, Giulliano A, Costa G, Maione F, Iaccarino F. Glutathione in the upper respiratory tract. Ann Otol Rhinol Laryngol 1995; 104:117–119.
63. Marcy TW, Merrill WW, Rankin JA, Reynolds HY. Limitations of using urea to quantify epithelial lining fluid recovered by bronchoalveolar lavage. Am Rev Respir Dis 1987; 135: 1276–1280.
64. Rennard SI, Basset G, Lecossier D, O'Donnell KM, Pinkston P, Martin PG, Crystal RG. Estimation of volume of epithelial lining fluid recovered by lavage using urea as marker of dilution. J Appl Physiol 1986; 60:532–538.
65. Dargaville PA, South M, Vervaart P, McDougall PN. Validity of markers of dilution in small volume lung lavage. Am J Respir Crit Care Med 1999; 160:778–784.
66. Peterson BT, Griffith DE, Tate RW, Clancy SJ. Single-cycle bronchoalveolar lavage to determine solute concentrations in epithelial lining fluid. Am Rev Respir Dis 1993; 147:1216–1222.
67. Ward C, Thien F, Secombe J, Gollant S, Walters EH. Bronchoalveolar lavage fluid urea as a measure of pulmonary permeability in healthy smokers. Eur Respir J 2000; 15:285–290.
68. Behr J, Degenkolb B, Maier K, Braun B, Beinert T, Krombach F, Bogelmeier C, Fruhmann G. Increased oxidation of extracellular glutathione by bronchoalveolar inflammatory cells in diffuse fibrosing alveolitis. Eur Respir J 1995; 8:1286–1292.
69. Kaulbach HC, White MV, Igarashi Y, Hahn BK, Kaliner MA. Estimation of nasal epithelial lining fluid using urea as a marker. J Allergy Clin Immunol 1993; 92:457–465.
70. Cantin AM, Fells GA, Hubbard RC, Crystal RG. Antioxidant macromolecules in the epithelial lining fluid of the normal human lower respiratory tract. J Clin Invest 1990; 86:962–971.

71. Davis WB, Pacht ER. Extracellular antioxidant defenses. In: Crystal RG, West JB, Barnes PJ, Cherrniack NS, Weibel ER, eds. The Lung: Scientific Foundations. New York: Raven Press, 1991:1821–1826.
72. Avissar N, Finkelstein JN, Horowitz S, Willey JC, Coy E, Frampton MW, Watkins RH, Khullar P, Xu Y, Cohen HJ. Extracellular glutathione peroxidase in human lung epithelial lining fluid and in lung cells. Am J Physiol 1996; 270:L173–L182.
73. Oury TD, Day BJ, Crapo JD. Extracellular superoxide dismutase: a regulator of nitric oxide bioavailability. Lab Invest 1996; 75:617–636.
74. DiSilvestro RA, Pacht E, Davis B, Jarjour N, Joung H, Trela-Fulop K. BAL fluid contains detectable superoxide dismutase 1 activity. Chest 1998; 113:401–404.
75. Matalon S, Holm BA, Baker RR, Whitfield MK, Freeman BA. Characterization of antioxidant activities of pulmonary surfactant mixtures. Biochim Biophys Acta 1990; 1035:121–127.
76. Ghio A, Carter J, Richards J, Crissman K, Bobb H, Yang F. Diminished injury in hypotransferrinemic mice after exposure to a metal-rich particle. Am J Physiol 2000; 278:L1051–L1061.
77. Kanofsky JR, Sima PD. Reactive absorption of ozone by aqueous biomolecule solutions: implications for the role of sulfhydryl compounds as targets for ozone. Arch Biochem Biophys 1995; 316:52–62.
78. Postlethwait EM, Bidani A. Mechanisms of pulmonary $NO_2$ absorption. Toxicology 1994; 89: 217–237.
79. Kanofski JR, Sima P. Singlet oxygen production from the reactions of ozone with biological molecules. J Biol Chem 1991; 266:9039–9042.
80. Velsor LW, Postlethwait EM. $NO_2$-induced generation of extracellular reactive oxygen is mediated by epithelial lining layer antioxidants. Am J Physiol 1997; 273:L1265–L1275.
81. Langford SD, Bidani A, Postlethwait EM. Ozone-reactive absorption by pulmonary epithelial lining fluid constituents. Toxicol Appl Pharmacol 1995; 132:122–130.
82. van der Vliet A, O'Neill CA, Eiserich JP, Cross CE. Oxidative damage to extracellular fluids by ozone and possible protective effects of thiols. Arch Biochem Biophys 1995; 321:43–53.
83. Devlin RB, Folinsbee LJ, Biscardi F, Hatch G, Becker S, Madden MC, Robbins M, Koren HS. Inflammation and cell damage induced by repeated exposure of humans to ozone. Inhal Toxicol 1997; 9:211–235.
84. Giamalva DH, Church DF, Pryor WA. A comparison of the rates of ozonation of biological antioxidants and oleate and linoleate esters. Biochem Biophys Res Commun 1985; 133:773–779.
85. Cross CE, Motchnik PA, Bruener BA, Jones DA, Kaur H, Ames BN, Halliwell B. Oxidative damage to plasma constituents by ozone. FEBS Lett 1992; 298:269–272.
86. O'Neill CA, van der Vliet A, Hu M-L, Kaur H, Cross CE, Louis S, Halliwell B. Oxidation of biological molecules by ozone: the effect of pH. J Lab Clin Med 1993; 122:491–505.
87. Mudway IS, Kelly FJ. Modeling the interactions of ozone with pulmonary epithelial lining fluid antioxidants. Toxicol Appl Pharmacol 1998; 148:91–100.
88. Cross CE, Motchnik PA, Bruener BA, Jones DA, Kaur H, Ames BN, Halliwell B. Oxidative damage to plasma constituents by ozone. FEBS Lett 1992; 298:269–272.
89. Halliwell B, Hu M, Louie S, Duval TR, Tarkington BK, Motchnik P, Cross CE. Interaction of nitrogen dioxide with human plasma: antioxidant depletion and oxidative damage. FEBS Lett 1992; 313:62–66.
90. Kelly FJ, Blomberg A, Frew A, Holgate ST, Sandstrom T. Antioxidant kinetics in lung lavage fluid following exposure of humans to nitrogen dioxide. Am J Respir Crit Care Med 1996; 154:1700–1705.
91. Kelly FJ, Buhl R, Sandstorm T. Measurement of antioxidants and oxidation products in bronchoalveolar lavage fluid. Eur Respir Rev 1999; 9:66–98.
92. Ho Y-S, Magnenat J-L, Gargano M, Cao J. The nature of antioxidant defence mechanism: a lesson from transgenic studies. Environ Health Perspect 1998; 106:1219–1228.

# 12

# Redox Modulation and Oxidative Stress in Dermatotoxicology

**JÜRGEN FUCHS, MAURIZIO PODDA, and THOMAS ZOLLNER**

*J.W. Goethe University, Frankfurt, Germany*

## I. INTRODUCTION

Human exposure to foreign chemical compounds is continuously increasing due to growing industrial and agricultural activities, as well as increased use of body and home products. Environmental and occupational exposure can lead to various diseases and significant morbidity (1) and compelling evidence has emerged for oxidative stress-related pathologies from numerous foreign substances (2–5). Oxidative stress is defined as an imbalance between prooxidants and antioxidants in favor of the former, resulting in an overall increase of reactive oxygen species (ROS) (6). The intracellular redox state is a tightly regulated parameter that provides the cell with an optimal ability to counteract the highly oxidizing extracellular environment. Intracellular redox homeostasis is regulated by antioxidants, particularly by thiol-containing molecules such as glutathione and thioredoxin. Key contributors to altered redox state are ROS that are formed by inflammatory cytokines and toxic xenobiotics, for example. Small variations in the basal level of ROS have been shown to modulate cell metabolism, gene expression, as well as post-translational modification of proteins. The skin is one of the largest body organs, it serves as a major portal of entry for many xenobiotics, and is a target organ for several environmental and occupational toxins (7–9). Parallel to the growing burden of chemical exposure, the incidence of skin disease suspected to be triggered by xenobiotics seems to be increasing. Common clinical forms of environmental and occupational skin diseases include allergic and irritant contact dermatitis as well as photosensitivity reactions. Chemical-induced leukoderma, comedogenesis, scleroderma, and skin cancer are rare cases of environmental and occupational skin diseases. The purpose of this review is to provide the reader with current information on the pathophysiology of selected environmental and occupational skin diseases caused by prooxidant chemicals from the viewpoint of redox-mediated toxicology.

## II. CONTACT DERMATITIS

Both irritant contact dermatitis (ICD) and allergic contact dermatitis (ACD) are very common and important conditions in clinical and occupational dermatology, with ICD accounting for 50 to 80% of all cases. Oxidative stress and inflammation have recently been linked to cutaneous damage in contact dermatitis (10–13). Free radicals are known to play a major role in inflammation and earlier reports have concentrated on the direct tissue-destructive effects of these species. If generated in large amounts and near critical cellular targets, reactive species (RS), which comprise oxidizing intermediates such as free radicals, ROS, reactive nitrogen species (RNS), and reactive halogen species (RHS), have the potential to damage these structures directly, thus triggering cytotoxicity and genotoxicity. Within the last 10 years, experimental evidence has accumulated suggesting that ROS play an important role as second messengers in signal transduction processes and in gene regulation. The modulation of redox-sensitive kinases and transcription factors by ROS has been suggested to be a central and early event in the induction of inflammatory reactions (14).

### A. Pathophysiology of Contact Dermatitis

Contact dermatitis can be classified by their etiology into ICD and ACD. ICD is a nonimmunological, local inflammatory skin reaction in response to chemical exposure, while ACD is a cell-mediated immune type IV hypersensitivity reaction. ACD requires preexisting genetic susceptibility, an immunocompetent individual, and a chemical antigen capable of transepidermal absorption (15). Although the definitions of ICD and ACD seem to be distinct, both the histological, molecular, and clinical features of ICD and ACD are strikingly similar. One of the reasons ACD and ICD are so similar is likely related to the limited repertoire of the pathological responses to skin injury (15). There are several thousand chemicals that can cause ICD (e.g., acids, alkalis, and oxidants) (16), but allergens are less abundant. Chemical irritants exert different degrees and types of irritation and different forms of ICD can be distinguished clinically. Acute, delayed-acute, and cumulative irritation are the most frequently observed types of ICD. Acute ICD develops when the skin is exposed to a potent irritant [e.g., chromium (VI) salt, acids], the inflammatory reaction reaches its maximum quickly and then starts to heal completely, or with a defect. A corrosion, or chemical burn, is an acute, severe irritant reaction in which necrosis usually develops. Delayed-acute ICD usually occurs 8 to 24 h after a single exposure. Examples of chemicals causing delayed-acute ICD are alkylating agents such as sulfur mustard, ethylene oxide, epichlorhydrin, 1,3-propane sultone, and other compounds such as hydrofluoric acid, tributyltin oxide, and the mycotoxin trichothecene. Cumulative ICD is a consequence of multiple subthreshold damages to the skin if the frequency of exposure is too high in relation to the skin recovery time. Cumulative ICD is linked to exposure of weak irritants. Nonerythematous and subjective irritation can be observed as reactions to cosmetic and topical pharmaceutical products. Nonerythematous ICD is characterized only by changes in the stratum corneum barrier function (e.g., increase in transepidermal water loss) without a clinical correlate. Subjective irritation is also characterized by the lack of clinical signs, but individuals report a stinging or burning feeling after contact with certain chemicals. Different mechanisms are presumably involved in the induction of these distinct forms of ICD (17–19). The potency of skin irritants can be characterized by their $ID_{50}$ value, which is the concentration in percent of a chemical, which causes an irritant skin reaction after one application in half the persons or animals tested. Unfortunately, a comprehensive database of $ID_{50}$ values of skin irritants (oxidizing vs. nonoxidizing chemicals) is not yet available.

Skin irritation is a complex phenomenon that involves resident epidermal cells, dermal fibroblasts, and endothelial cells as well as invading leukocytes interacting with each other under the control of a network of cytokines, neuropeptides, and eicosanoids. Keratinocytes presumably play a central role in the pathophysiology of ICD as well as ACD via generation of signals leading to attraction of leukocytes (20–22). They respond to irritants or sensitizers by synthesis and release of proinflammatory cytokines such as IL-1 and TNF-α, which induce endothelial cells to express adhesion molecules necessary for accumulation of inflammatory cells. As little as 2 h after contact with irritants or allergens, human skin activates cellular mechanisms to increase T-lymphocyte infiltration. These mechanisms are not dependent upon specific immune sensitivity and reflect a capacity of skin cells to respond to chemical provocation for the presumed purpose of immune surveillance (23). Although ICD and ACD have different pathogenetic mechanisms, both types of reaction are characterized by an almost identical dermal infiltrate after 72 h, consisting mainly of CD4+ T-lymphocytes, many of which are activated, and macrophages (24). In particular, CD4+ T-helper cells have a decisive role in metal-specific immune responses regardless of the ultimate type of effector mechanisms (25). Most recently, however, evidence for an effector role of IFN-γ producing CD8+ T-cells and a regulator role of Th2 lymphocytes in contact hypersensitivity in response to haptens such as 2,4-dinitro-fluorobenzene, oxazolone, or nickel (26–30) has accumulated. In ACD, the proliferation of antigen-specific lymphocytes allows secondary immune response upon reexposure to the antigen leading to a long-lasting state of boosted defense against this specific antigen. Furthermore, the antigen-specific T-cells trigger a B-cell-mediated reaction as well as a cytotoxic response.

In the development of ACD the *sensitization* phase is distinguished from the *elicitation* phase (31,32). For sensitization, a low-molecular-weight contact allergen (hapten) has to penetrate the skin and, in most instances, combine with a protein carrier. Haptens can be classified by their chemical reactivity. They can bind to proteins by nucleophilic, electrophilic, or free radical reactions. Metal ions intercalate with proteins, thus forming noncovalent complexes. Free radical reactions are important for hapten–protein binding involving hydroperoxides (33–36), alkyl phenols and alkyl resorcinols (37–39), quinones (40), dinitrohalo(chloro, bromo,iodo) benzenes (41), azo compounds such as 2,2-azobis (2-aminopropane) (AAPH) (42), and primary aromatic amines such as *p*-phenylenediamine (43–45). In addition, free radical reactions are particularly significant for photoallergy reactions. For instance, chlorpromazine and related phenothiazines can cause severe cutaneous and ocular phototoxic and photoallergic reactions in animals and humans treated with these drugs. Upon irradiation with ultraviolet light, chlorpromazine yields a variety of free radicals. The neutral promazinyl radical was suggested to be the reactive haptene, covalently conjugating to macromolecules and thereby forming a complete antigen (46–48). The determinants of antigenicity are multifactorial, including the type of binding that the hapten undergoes with the carrier and the final three-dimensional configuration of the conjugate (49). The hapten–protein complex will be taken up and processed by Langerhans cells, which leads to presentation of hapten-peptide complexes via MHC class II molecules. In the presence of keratinocyte-derived granulocyte macrophage colony-stimulating factor (GM-CSF), epidermal Langerhans cells migrate via the afferent lymph into the paracortical region of local lymph nodes where they meet naive lymphocytes with hapten-specific T-cell receptors. The T-cell receptor (TCR) responsible for T-lymphocyte recognition of foreign antigen in association with molecules by the MHC is a disulfide-linked 90-kD heterodimer consisting of two polypeptide chains, a and b, expressed on the surface of the majority of T-cells (TCR a/b+ T-cells). Like the immunoglobulin genes, the TCR genes consist of noncontinuous variable (V), joining (J), diversity (D), and constant regions. The TCR a/b chains are assembled by somatic recombination of discontinuous germline gene segments during T-cell development.

Polymerization of template-independent N region and template-dependent P-nucleotide insertions at the junctions, D b usage, the precise joining of the germline gene segments (selected from a large pool), and the combinatorial association of TCR a/b polypeptides help to generate an extensive diversity in the association of TCR a/b repertoire allowing recognition and response to any (foreign) antigen (50). The hapten-specific cells are subsequently activated, proliferate, and acquire the predilection for the recruitment into specific tissues that is determined by the expression of a specific set of homing receptors.

The *elicitation* phase is characterized by an inflammatory reaction mainly mediated by skin infiltration with antigen-specific T-lymphocytes. Application of contact allergens such as nickel initiates keratinocyte cytokine production such as TNF-α (20). These cytokines induce endothelial expression of adhesion molecules such as E-selectin, intercellular adhesion molecule-1 (ICAM-1), and vascular cell adhesion molecule-1 (VCAM-1), resulting in the recruitment of skin-specific memory T-lymphocytes. Memory T-lymphocytes that infiltrate the skin express a unique skin-homing receptor called cutaneous lymphocyte-associated antigen (CLA), a carbohydrate epitope that facilitates the targeting of T-lymphocytes to inflamed skin (51,52). CLA binds specifically to E-selectin and CLA is present on most T-lymphocytes at sites of cutaneous immune response. Antioxidants such as *N*-acetylcysteine, α-tocopherol, and ascorbate blocked the expression and functional activity of the CLA in isolated human T-lymphocytes (53). The redox sensitivity of CLA expression on human T-cells may allow attenuation of T-lymphocyte-mediated skin inflammation by antioxidants and may open up a new avenue in therapy of T-lymphocyte-mediated skin diseases, such as contact dermatitis, atopic dermatitis, and psoriasis.

#### *1. Stress-Sensitive Protein Kinases*

Exposure of keratinocytes to chemical irritants and allergens triggers activation of several stress-sensitive protein kinases, involving ROS as mediators, and leading to enhanced synthesis of cytokines and to expression of antigens. There is evidence that stress-activated signal cascades are affected by altered redox potential and key contributors to altered redox potential are ROS (54). ROS can directly alter kinases, phosphatases, and transcription factors, or modulate cysteine-rich redox-sensitive proteins (55,56). ROS have been shown to activate protein kinase signal pathways in human keratinocytes. ROS enhanced epidermal growth factor (EGF), receptor phosphorylation and activated extracellular regulated kinase (ERK), and *c-jun N*-terminal kinase (JNK) activities (57,58). MAP kinases are important mediators of the cellular stress response and can also be activated by IL-1 and TNF-α. Different forms of cellular stress triggered distinct signaling cascades involving either oxidative stress or GTPase-coupled pathways in MAP kinase activation (59), suggesting that antioxidants may only influence MAP kinase activation triggered by oxidative stress. Protein kinase C (PKC) regulates a variety of signal transduction events implicated in differentiation, proliferation, and pathogenesis of inflammation, including the biosynthesis of inflammatory cytokines, superoxide anion radical, and activation of phospholipase A2 (60). In mouse epidermal keratinocytes, six different isoforms of PKC are expressed and it was demonstrated that overexpression of PKC-α increases the expression of specific proinflammatory mediators such as cyclooxygenase (COX), COX-2, the neutrophil chemotactic factor macrophage inflammatory protein, and TNF-α (61). In human keratinocytes, PKC regulates production of the neutrophil chemotactic cytokine IL-8 (62). The activity of PKC is regulated by the intracellular redox status (63), and isolated PKC is inhibited by α-tocopherol in vitro (64). It remains to be seen whether PKC modulation contributes to the anti-inflammatory activity of α-tocopherol (65).

### *2. Transcription Factor NF-κB*

NF-κB transcription factor complex is tightly regulated through its cytoplasmic retention inhibitory proteins known as IκB, which includes a group of related proteins, IκB-α being the best characterized example. Associated with its inhibitor, NF-κB resides as an inactive form in the cytoplasm of most cells. The activity of NF-κB has been shown to be regulated by the intracellular redox state (56,66–69). Upon stimulation by various agents, many of which are probably mediated by various ROS (e.g., singlet oxygen, peroxides) and proinflammatory cytokines (e.g., IL-1, TNF-α), IκB-α undergoes phosphorylation and is finally proteolyzed. Removal of IκB-α allows the complex to translocate into the nucleus, where it activates its target genes (70). The transduction pathways that lead to IκB-α inactivation remain poorly understood (71). The target genes for NF-κB comprise a list of genes intrinsically linked to a coordinate inflammatory response, including genes encoding TNF-α, IL-1, IL-6, IL-8, inducible nitric oxide synthase, MCH class I antigens, E-selectin, and VCAM-1 (72–75). Some of the cytokines whose genes are switched on by NF-κB, such as TNF-α and IL-1, are themselves activators of NF-κB, giving them the potential for a positive feedback cycle. The activation of NF-κB by ROS has been suggested to be an early event in the induction of inflammatory reactions, because NF-κB induces a range of important inflammatory genes (76). Even if the initial cellular stimulus for NF-κB activation is not ROS, the apparent merging of all the pathways of signal transduction involving NF-κB on a ROS-dependent step suggests a possible therapeutic target (12). Recently, a ROS-independent pathway of NF-κB activation has been discovered (77,78), and it is believed that activation of NF-κB can occur through redox-sensitive as well as through redox-insensitive pathways (79), similar to what has been described for the MAP-kinase pathways. Furthermore, the effects of oxidants and reductants on NF-κB activation can be tissue- or cell-type-specific. The activation of NF-κB by oxidative stress provides numerous examples of cell and tissue specificity (56). Several studies have demonstrated NF-κB activation in skin induced by common irritants and allergens. NF-κB is activated in human keratinocytes and fibroblasts by oxidizing agents such as ozone (80), hydrogen peroxide (81), UVA (82), and UVB (83,84). The skin irritant anthralin induces NF-κB in murine keratinocytes (85) and in BALB-c mice (86) via a mechanism involving ROS. The corrosive irritant and potent allergen Cr(VI) and its lower oxidation state Cr(IV) activated NF-κB via free radical reactions. In Jurkat cells, Cr(VI) and Cr(IV) induced the activation of NF-κB and enhanced NF-κB-DNA binding (87,88). Ni(II), which is a skin irritant and allergen, activated NF-κB and subsequently triggered the production of proinflammatory cytokines such as I1-1, TNF-α, IL-6, and the expression of adhesion molecules in human endothelial cells and keratinocytes. Ni(II) caused a strong increase of NF-κB–DNA binding and subsequently induced gene transcription of ICAM-1, VCAM-1, and E-selectin as well as mRNA production and protein secretion of IL-6 in endothelial cells in a redox-dependent mechanism (89). In cultured human vascular endothelial cells (HUVECS), Ni(II) activated the translocation of NF-κB into the nucleus and enhanced NF-κB–DNA binding (90).

### 3. *Cytokines*

Almost every kind of cytokine known so far can be detected in skin under certain physiological or pathological conditions (91). While resting epidermal keratinocytes produce some cytokines constitutively, a variety of chemicals can induce epidermal keratinocytes to synthesize and release proinflammatory cytokines. This is accomplished by activation of several stress-sensitive protein kinases involving ROS as mediators (92). In human and animal skin, proinflammatory cytokines induced by irritants comprise TNF-α, IL-1, IL-2, and IL-6, chemotactic cytokines such as IL-8, growth-promoting cytokines such as IL-6, IL-7, and IL-15, GM-CSF, TGF-α, and cytokines regulating humoral versus cellular immunity such as IL-10, IL-12, IL-18 (93–97). IL-1 and TNF-α have been considered as primary cytokines, whereas others such as IL-6 and IL-8 are secondary, because they are insufficient to induce an inflammatory response in absence of other stimuli or primary cytokines. In addition to TNF-α (98), IL-1 is considered an extraordinarily important cytokine, which links innate and acquired immunity in human skin (99). IL-1a is predominantly produced and stored in keratinocytes, while IL-1b is primarily synthesized by macrophages and monocytes. IL-1 and TNF-α are known to induce cellular production of ROS, thereby perpetuating their own formation and action (100–103). In particular, the expression of IL-8, which is a polymorphonuclear leukocyte chemoattractant, is upregulated by oxidative stress (104,105). Many allergens and irritants have been shown to induce keratinocyte activation with cytokine synthesis and secretion. In the mouse, TNF-α is induced in epidermal keratinocytes and in the dermal infiltrate after induction of contact hypersensitivity reactions and irritant reaction (106). Seemingly, TNF-α plays a key role in these two types of reactions, and administration of antibodies to TNF-α or soluble TNF receptors blocked irritant reactions and contact hypersensitivity in the mouse (106). It is controversial whether the cytokine profile induced by irritants differs from that induced by allergens. It was not possible to identify specific cytokine profiles for different classes of dermatotoxic agents such as sensitizers, irritants, corrosives, and carcinogens, but nonsensitizing contact irritants produce relative increases in the synthesis and secretion of the cytokines IL-1, IL-8, and TNF-α, compared to the other chemical agents (107). Allergens such as Ni(II) and 2.4-dinitrofluorobenzene (108,109) as well as irritants such as sodium dodecyl sulfate (108) induced TNF-α expression in the epidermis of sensitized mice or in isolated keratinocytes, respectively. The irritant anthralin induced expression of IL-6, IL-8, and TNF-α in skin of BALB-c mice (110) and in human keratinocytes (86) via a mechanism involving ROS. The irritant sulfur mustard induced the synthesis and release of IL-1, IL-6, IL-8, and TNF-α in keratinocytes (111–113). Noncytotoxic concentrations of phenol, which induces oxidative stress in human keratinocytes via a redox cycling mechanism (114), promoted the expression of TNF-α, IL-8, and IL-1a in cultured human keratinocytes (115). Ni(II) and Cr(VI), both potent generators of ROS in noncellular and cellular systems, induced in noncytotoxic concentrations TNF-α production in cultured human keratinocytes (116). Topical application of the thiol antioxidant *N*-acetylcysteine inhibited Ni(II), induced ACD, and reduced the cutaneous expression of mRNA for TNF-α in mice (117). However, in human skin, topical *N*-acetylcysteine failed to inhibit ICD induced by either sodium lauryl sulfate or dimethylsulfoxide (118). In humans, it needs to be determined whether TNF-α is induced in the epidermis after application of sodium lauryl sulfate or dimethylsulfoxide, whether TNF-α plays a major role in the development of the inflammatory reaction in human skin, and whether the *N*-acetylcysteine molecule can sufficiently penetrate human epidermis.

### 4. Adhesion Molecules

VCAM-1 and ICAM-1 are important for the recruitment of leukocytes into inflamed tissues by promoting their strong adhesion. Induction of E-selectin by cytokines or ROS is vital to the initial tethering and rolling of T-cells on endothelial cells, which is followed by strong adhesion-mediated (e.g., by ICAM-1 and VCAM-1) expression on endothelial cells. Therefore, upregulation of these adhesion molecules by IL-1, TNF-α, and ROS plays an important role in cutaneous inflammation (119,120). ICAM, VCAM, and E-selectin are expressed on endothelial cells of inflamed skin in both ICD and ACD (121). Allergens and irritants can induce these adhesion molecules either by direct action or through induction of proinflammatory keratinocyte cytokines (122,123). The expression of endothelial leukocyte adhesion molecules has been postulated to be regulated by redox-sensitive events. Expression of ICAM-1, E-selectin, and VCAM-1 appear in large part to be controlled by the actions of cytokines, such as TNF-α and IL-1b, and ROS via activation of NF-κB (75,124,125). Chemical irritants (anthralin) as well as allergens [Ni(II)] induced the expression of E-selectin, VCAM-1, and ICAM-1 in human skin and increased the number of CLA-positive T-cells (126). *N*-acetylcysteine inhibited IL-1-induced mRNA and cell surface expression of both E-selectin and VCAM-1 in endothelial cells, and reduced NF-κB binding to the NF-κB binding site of the VCAM-1 gene, but not of the E-selectin gene (127). Thus NF-κB binding to its consensus sequences in the VCAM-1 and E-selectin gene exhibits marked differences in redox sensitivity (127). TNF-α and IL-1-induced E-selectin expression in human vascular endothelial cells is regulated by a mechanism involving vicinal thiols. These thiols must be in the reduced state; oxidation attenuates the cell's response to the cytokines (128). Keratinocytes can express ICAM-1 (129), and IFN-γ, a Th1-lymphocyte-derived cytokine, expressed ICAM-1 on the keratinocyte cell surface (122). In addition, ICAM-1 expression has been shown to be induced by oxidative signals such as singlet oxygen in isolated human keratinocytes (130). As already outlined above, the regulated expression of ICAM-1 is of key importance to the initiation and evolution of localized inflammatory processes in the skin as it facilitates recruitment of T-cells at sites of inflammation and interaction of keratinocytes with T-cells (131). However, studies on the involvement of oxidative stress in the induction of ICAM-1 expression in transformed human keratinocytes (HaCaT) were inconclusive, showing that oxidizing agents such as ferric chloride and hydrogen peroxide upregulated ICAM-1 expression, but there was no clear relationship between the ability of the agents to induce ICAM-1 expression and their ability to alter the levels of reduced glutathione (132).

### 5. Cox-2

Chemicals can directly or indirectly trigger release of arachidonic acid (AA) from the plasma membrane by the action of phospholipase A2 (PLA2), which translocates from the cytosol to the membrane upon activation by $Ca^{2+}$ and phosphorylation by the MAP kinase cascade (133). Subsequently, AA undergoes oxygenation by different enzymatic pathways [e.g., via lipoxygenase and cyclooxygenase (COX)], and is converted into biologically active eicosanoids. These enzymatic transformations of AA are catalyzed by free radical processes, which can be minimized therapeutically by the presence of antioxidants. The arachidonic acid cascade has been shown to be modulated by antioxidants such as α-tocopherol (134), but the tocopherol dosage seems to be critical for an anti-inflammatory effect. ROS have been demonstrated to activate cytosolic and membrane-bound PLA2 activities (135). Activation of PLA2 also stimulates extracellular regulated kinase (ERK) and *c-jun N*-terminal kinase (JNK) activities, leading to the activation of transcriptional factors and the ultimate stimulation of the transcription of several mitogen-stress-responsive genes (135). COX exists as two distinct but similar isoenzymes, COX-1 and COX-2. COX-1 is mainly expressed constitutively, while COX-2 is predominantly the inducible form. Prostaglandins (PGs) formed by the enzymatic activity of COX-1 are primarily involved in the regulation of homeostatic functions throughout the body, whereas PGs formed by COX-2 principally mediate inflammation (136,137). However, there is strong evidence that COX-1 also contributes to inflammation, but presumably to a minor extent (138). COX-2 expression is mediated by the activation of transcriptional factors, especially NF-κB (139). Proinflammatory cytokines, such as IL-1 and TNF-α rapidly induce COX-2 activity in many cell types including skin-derived cells (140,141). Leukotriene LTB4, which is a powerful chemotactic agent for polymorphonuclear leukocytes, has been established as a clinically important mediator of skin inflammation (142). Recently, a link was established between peroxynitrite, nitric oxide, and prostaglandin biosynthesis (144,145) and the clinical relevance of this connection awaits characterization. In contrast to earlier reports, proinflammatory prostaglandins do not seem to play a central role in the cause of inflammatory skin diseases, because nonsteroidal anti-inflammatory drugs, which inhibit both COX-1 and COX-2 activity, clinically are not very effective in the treatment of inflammatory skin diseases, including contact dermatitis, atopic dermatitis, and psoriasis (143). It remains to be seen whether selective COX-2 inhibitors are more potent anti-inflammatory agents in skin than the conventional nonsteroidal anti-inflammatory agents.

### 6. Reactive Oxygen, Nitrogen, and Halogen Species

ROS and RNS are constitutively produced in most cell types, including epidermal keratinocytes and dermal fibroblasts (146,147). In addition to stimulated ROS/RNS production by resident epidermal and dermal cells, these species as well as reactive halogen species (RHS) can be produced and released into skin by invading macrophages as well as polymorphonuclear and eosinophilic leukocytes. Several chemical agents can induce ROS generation in skin, either directly or mediated by proinflammatory cytokines. ROS, which are an ubiquitous component of inflammation, may exacerbate inflammation through generation of cytokines, thus initiating a positive feedback cycle. These ROS potentially regulate levels and activity of phosphorylated proteins and protein kinases within the keratinocytes (148) and could be modulated by enzymatic and low-molecular-weight antioxidants. Nitric oxide (NO) is a critical mediator of various biological functions. It plays a role in both the vasodilatory component of the inflammatory response and in the modulation of immune responses in the skin. In high concentrations, peroxynitrite ($ONOO^-$), the reaction product of nitric oxide and superoxide anion radical, has the potential to cause cytotoxic effects in skin. In a human keratinocyte cell line (HaCaT), the copper/zinc superoxide dismutase gene was found to be upregulated by NO (149), implicating an autoprotective regulatory mechanism. NO is generated from L-arginine by nitric oxide synthase (NOS), which has three isoforms: endothelial-type NOS (eNOS) and brain-type

NOS (bNOS) are constitutive enzymes; inducible-type NOS (iNOS) is expressed after stimulation. Epidermal keratinocytes in normal human skin contain both eNOS and iNOS (150). It was recently demonstrated that iNOS expression in rat astrocytes and macrophages is mediated by the activation of transcriptional factors, especially NF-κB, but an NF-κB-independent activation step was also discovered in rat glial cells and primary astrocytes (151). IL-2 is a potent inducer of NO synthesis in mice and humans (152). A combination of IL-8 and IFN-γ induced the expression of iNOS-specific mRNA and of the functional enzyme in cultured human keratinocytes (153). In human skin, a significant increase in iNOS was found immunohistochemically in both irritant and allergic contact dermatitis (154). In guinea pig skin, the inhibition of the production of endogenous NO inhibits both leukocyte accumulation and edema formation induced by different mediators of inflammation (155), indicating that NO plays a role as neutrophil chemoattractant. NO specific trapping agents such as iron chelates may be useful as topical modulators of the inflammatory response in skin.

### 7. Polymorphonuclear Leukocytes

The inflammatory infiltrate of ICD is characterized predominantly by CD4+ T-lymphocytes, some CD8+ T-lymphocytes, polymorphonuclear leukocytes (PMN), and monocytes. There is accumulating evidence indicating that PMNs have the capacity to recruit T-lymphocytes and to modulate T-lymphocyte-mediated delayed hypersensitivity responses. PMNs have the potential to damage surrounding tissue by releasing ROS produced via NADPH oxidase and RHS via myeloperoxidase. In addition, they can release large amounts of proteolytic enzymes before self-necrosis. PMN necrosis further exacerbates inflammation and promotes chemotaxis and PMN recruitment. ROS are known to inactivate protease inhibitors, thus enhancing the destructive effects of PMNs. Furthermore, PMN recruitment is stimulated by oxidative stress because ROS increases adhesion of monocytes and PMN to the endothelium (156,157). The movement of PMN from the mainstream of blood to the extravascular space is also triggered by several proinflammatory molecules, including histamine, LTB4, platelet activating factor, and complement products (158). In particular, proinflammatory cytokines in concert with other mediators augment PMN cytotoxicity. This is illustrated by two examples. IL-6 at pathophysiologically relevant concentrations enhanced both basal and formyl-Met-Leu-Phe-stimulated elastase release by human PMNs (159). Incubation of rat PMN with TNF-α, or a mixture of TNF-α and IFN-γ resulted in a dose-dependent enhancement of ROS production in response to phorbol myristate acetate. The incubation with both IFN-γ and TNF-α caused a significantly higher increase of ROS production than with TNF-α alone (160). Nitric oxide regulates PMN function (161), but whether or not human PMNs express NOS activity is controversial (162–164).

### 8. Matrix Metalloproteinases

Matrix metalloproteinases (MMPs) and specific tissue inhibitors of matrix metalloproteinase (TIMPs) play an important role in physiological as well as in inflammatory processes, particularly in prolonged skin inflammation or photoaging (165,166). These enzymes are produced by fibroblasts, keratinocytes, mast cells, endothelial cells, and leukocytes. MMPs are not constitutively expressed in skin but are induced in response to cytokines and growth factors, for example. The genes that code for MMP and TIMP are modulated by the redox-sensitive transcription factor AP-1. TIMPs should be susceptible to redox modulation because their complex tertiary structure is dependent upon six disulfide bonds. It was recently demonstrated that the polymorphous leukocyte-derived oxidant HOCl inactivated TIMP-1 at concentrations achieved at sites of inflammation (167). ROS are known to activate MMP in human dermal fibroblasts (168–170). ROS-activated MMPs can degrade and inactivate alpha-1-protease inhibitor (alpha-1-PI) by proteolysis. Thus, the activation of MMPs, accompanied by the inactivation of alpha-1-PI, will bring about enhanced proteolytic damage to tissues infiltrated with polymorphous leukocytes by both MMPs and elastase (171). In MMP-3(stromelysin)-deficient mice the contact hypersensitivity response to the allergen 2.4-dinitrofluorobenzene was significantly impaired, while the irritant reaction to phenol remained unaffected (172). This may indicate that MMP-3 serves an important function in ACD.

### 9. T-Lymphocytes

The cellular infiltrate of ACD is characterized predominantly by CD4+ T-lymphocytes, some CD8+ T-lymphocytes, a few PMNs, and some eosinophilic leukocytes. Naive T-lymphocytes migrate through the B-cell areas of the lymph nodes where they may encounter antigen-presenting dendritic cells. In case these naive T-cells carry an antigen receptor (T-cell receptor) specific for the presented antigen, the T-cell becomes activated, proliferates, and is induced to express a specific set of adhesion molecules. Although the molecular nature of the signals required for the induction of recirculation (homing) molecules is not yet known in detail, several factors seem to be involved: antigen, certain cytokines such as TGF-β and IL-12, and possibly ROS/RNS. Sources of ROS/RNS are keratinocytes and Langerhans cells, which have to be drained to the lymph nodes if produced by keratinocytes, and can be directly released in close anatomical context with the T-cell if produced by dendritic cells. Following homing to skin, activated T-lymphocytes induce pathology in the dermis and epidermis. In ACD, the proportion of T-lymphocytes specific for the provoking antigen is less than 1:1000 (173), indicating that T-lymphocytes are extremely potent and must provoke significant amplification (32). This is further demonstrated since the transfer of one single (174) or a few antigen-specific (175) T-cells into a naive recipient animal is sufficient to provoke contact dermatitis. In ACD CD4+ and CD8+ T-lymphocytes are found in skin, but usually CD4+ T-lymphocytes predominate in the dermal infiltrate. CD4+ T-lymphocytes have many potential mechanisms for inducing inflammation, including oxidative burts, cell-mediated cytotoxicity, recruitment of other inflammatory cells, and direct effects of cytokines such as IL-2 and IFN-γ.

### 10. Eosinophilic Leukocytes

Eosinophils are attracted by several humoral and cellular factors such as products of the complement cascade and AA metabolism, as well as by factors derived from lymphocytes, mast cells, and polymorphonuclear leukocytes (176). The chemokines RANTES and eotaxin were identified as the most important eosinophil-attracting chemokines (177), RANTES representing a potent eosinophil-specific activator of oxidative metabolism (176). At sites of inflammation, eosinophils release toxic proteins, such as eosinophilic cation protein, major basic protein, eosinophil-derived neurotoxin, ROS, and RHS leading to tissue damage of the host. They are able to produce two to three times as much superoxide anion radicals as polymorphonuclear leukocytes (176). Eosinophils play a critical role in the late-phase reaction of allergic inflammatory responses. Eosinophilic infiltration in susceptible individuals was identified to be a specific property of the allergen (178).

### *11. Macrophages*

Among leukocytes, macrophages are considered the most important active inflammatory cells in secretion of degrading enzymes and production of ROS (179,180). The production of ROS in macrophages depends significantly on the specific stimulus, site of macrophage isolation, and the extent of differentiation of the cells. Monocytes contain a myeloperoxidase and produce oxidizing halogens and oxidizing radicals. During conversion of monocytes to macrophages they lose their myeloperoxidase, and microbial killing by macrophages may therefore be accomplished only through oxidizing oxygen species. The ability of cultured monocytes to produce ROS decreases with their differentiation into macrophages, histiocytes, epitheloid cells, and giant cells. The order of declining potency is: monocyte > histiocyte (tissue macrophage) > epitheloid cell > giant cell (181).

## B. Oxidative Stress in Contact Dermatitis

Recently a direct correlation between antioxidant levels and skin reactivity to irritants was found, which suggested that oxidative stress plays a general role in the pathophysiology of acute ICD (14). However, it seems too early to draw this conclusion based on these findings alone. The majority of skin irritants are redox inactive compounds. Their metabolism does not involve generation of RS, these chemicals do not stimulate cellular production of ROS, nor inhibit antioxidant defense systems. Consequently, their mechanism of toxicity does not involve oxidative stress as a primary and initiating event. However, secondary reactions may trigger oxidative stress and contribute to pathology. For example, most organic solvents are believed to trigger skin inflammation by defatting the epidermis. Disturbed epidermal lipid composition can result in increased production and release of prooxidant cytokines from keratinocytes, such as TNF-$\alpha$. Thus formation of ROS is a late event in the inflammatory cascade triggered by organic solvents. The irritating surfactant sodium lauryl sulfate (SLS), which does not generate ROS directly, is believed to cause skin irritation through its action as surfactant and membrane pertubator, thereby disrupting the epidermal barrier. Recently it was shown that SLS inhibits human epidermal Cu,Zn SOD activity (182), which is thought to be a consequence of oxidative stress caused by increased production and release of proinflammatory epidermal cytokines. A few skin irritants (e.g., hydroperoxides/peroxides, metal salts such as iron (II), chromium (VI), and nickel (II), phenols/quinones, azo-compounds) and primary amines can generate free radicals/ROS directly through metabolic activation, redox cycling, or other mechanisms and presumably cause ICD primarily through action of these reactive species. Some skin irritants, such as the alkylating agents sulfur mustard, 1,3-propan sultone, ethylene oxide, and epichlorhydrin, which can deplete cellular levels of reduced glutathione, have the potential to shift the thiol equilibrium thus modulating redox-sensitive signal transduction processes involved in the inflammatory response.

Several studies have shown depletion of low-molecular-weight antioxidants and increased production of ROS in the inflammatory phase of ACD (13,183,184). Antioxidants inhibited ACD of mice to 2,4,5-trinitrochlorobenzene (117) and urushiol (37), as well as ICD to anthralin (86,185,186), and to sulfur mustard (187,188). In unsensitized mice and in mice sensitized to 2,4-dinitrochlorobenzene (DNCB), topical exposure with DNCB (0.5%) caused either ICD (unsensitized group) or ACD (sensitized group). In both groups cutaneous and hepatic levels of reduced glutathione decreased immediately after challenge and the total reduction of hepatocutaneous GSH was twofold higher than the dose of DNCB in ACD, while the molar ratio is 1:1 in ICD. This implies that in ICD direct conjugation of DNCB occurred with GSH, while in ACD both conjugation and enhancement of extracellular excretion of GSH occurred, indicating oxidative stress (13). In contrast, in mice sensitized to 2,4-dinitrofluorobenzene and 3,3,4,5-tetrachlorosalicylanilide free radical reactions seemed neither to play a significant role in the sensitization phase nor in the elicitation phase of the contact hypersensitivity response (189).

In patients with contact allergy to $Ni^{2+}$ and $Co^{2+}$, the immune response in peripheral blood mononuclear cells (measured parameters: increased [methyl-3H]thymidine uptake and IFN-$\gamma$ production) upon addition of $Ni^{2+}$ and $Co^{2+}$ to the cell cultures was not influenced by ascorbate (190). The contact hypersensitivity response to nickel in nickel-sensitized patients was not significantly inhibited by systemic $\alpha$-tocopherol/ascorbate in human skin (191). High concentrations of topical $\alpha$-tocopherol/ascorbate only slightly inhibited (10% inhibition) the contact hypersensitivity reaction caused by nickel sulfate in nickel-sensitive subjects, while chelating agents such as clioquinol and EDTA caused 100% or 40% inhibition, respectively (192). These findings argue against a dominant role of free radicals and ROS in the elicitation phase of nickel-induced contact hypersensitivity response.

## III. ATOPIC DERMATITIS

Atopic dermatitis is a genetically determined T-cell-mediated inflammatory skin disease. T-cells are believed to play a pivotal role in many inflammatory and certain malignant skin diseases, including psoriasis, atopic dermatitis, contact dermatitis, and cutaneous T-cell lymphoma. Particularly in atopic dermatitis activated T-cells with skin-homing properties play a specific and decisive role in the pathogenesis and exacerbation of the disease (193). The expression of E-selectin and other adhesion molecules such as VCAM-1 and ICAM-1 are constitutively upregulated in healthy skin of patients with atopic dermatitis, which is presumably mediated by the release of IL-4 from cells that reside in atopic skin (194). The exact role of environmental pollution in the manifestation of atopic dermatitis is not clear (195), but some studies point to a role of environmental allergens and irritants (196–198). Acute exposure to a major chemical burden was identified as a risk factor contributing to the exacerbation of latent and to the development of new atopic dermatitis (199). In particular, air pollution was recognized as a significant risk factor for the incidence of atopic dermatitis in children and young adults (200–202). Parental smoking increased the prevalence and severity of the atopic state in children (203). Maternal smoking during pregnancy or lactation or both seem to play a role in the development of atopic dermatitis (204). Recently, an epidemiological study showed that traffic-related air pollution, especially exposure to oxides of nitrogen, leads to increased prevalence of atopic sensitizations, allergic symptoms, and diseases in children (205).

## IV. CHEMICAL SCLERODERMA

Scleroderma is a chronic disease with autoimmune responses, which occurs as a progressive and often fatal systemic mesenchymal disease [progressive systemic sclerosis (PSS)], or as a localized form [localized scleroderma (LS)]. The early pathological events include dysfunction in the immune system, as well as vascular alterations, while dysregulation of connective tissue metabolism is a late event leading to massive organ sclerosis. LS- and PSS-like symptoms can be induced by a large number of chemically unrelated agents (206–209), including bleomycin, paraquat, quartz dust, halogenated aliphatic hydrocarbons, and aliphatic and aromatic hydrocarbons. Most of these sclerodermic agents have the potential to induce oxidative stress in connective tissue through their metabolism or activation of cellular ROS production and it was speculated that these chemicals may trigger chemical scleroderma through a redox-sensitive mechanism (146,210–212). In particular, the revelation of immunocryptic epitopes in self-antigens may initiate the autoimmune response in scleroderma. Several of the autoantigens targeted in scleroderma are uniquely susceptible to cleavage by ROS in a metal-dependent manner, indicating a pathophysiological role of oxidative stress in the development of scleroderma (213). Recently, a mouse model for localized scleroderma was established by repeated intradermal injections of bleomycin (214). Bleomycin-induced skin fibrosis in rats is an obligate toxic reaction and resembles skin changes in humans with PSS, without immunological alterations (215). Bleomycin is known to generate ROS in vitro (216–218). Cytokines and growth factors can promote fibroblast proliferation and collagen deposition (219–222). In particular, transforming growth factor-β (TGF-β) was suggested to be a crucial determinant in the pathogenesis of PSS and TGF-β was shown to play a crucial role in bleomycin-induced skin fibrosis in C3H mice (223). Systemic applications of lecithinized SOD protected from bleomycin-induced scleroderma and decreased the number of infiltrating mast cells and eosinophils in skin, indicating an involvement of ROS in the cellular infiltrate and in the development of sclerotic skin lesions (224). Specific genes and signal transduction pathways have been shown to be regulated or affected by oxidants (56,225). For instance, changes in the cellular redox state can modulate transcriptional activation of collagen (226) and collagenase (227). Collagen synthesis is stimulated by elevated steady-state concentrations of ROS (e.g., in mouse hyperoxic periodontal tissue culture, rat hyperoxic lung organ culture, and rat lung by ozone (146). It must be pointed out that genetic susceptibility is a significant determinant in the development of chemical scleroderma, because not everybody exposed to these chemicals will acquire the sclerosis. In particular, vinylchloride and toxic oil-induced scleroderma are associated with such a genetic susceptibility. In conclusion, experimental evidence points to a role of ROS in the pathogenesis of bleomycin-induced scleroderma, but their role remains speculative for most other sclerodermic agents.

## V. CHEMICAL LEUKODERMA

Several industrial chemicals have depigmenting properties (228,229) and chemical leukoderma resembles idiopathic vitiligo. Not everybody exposed to the chemicals will develop leukoderma, suggesting involvement of an individual genetic susceptibility. Primarily phenols and alkylphenols have been identified as depigmenting chemicals in mouse and human skin (230–240). Other potent skin depigmenting agents include the glutathione reductase inhibitor carmustin (1,3-bis-(2-chlorethyl)-1-nitrosourea) (241), bleomycin (242), mercaptoamines, thiols (232), and mercurials. Phenols and alkylphenols are frequently used as industrial antioxidants, and they are metabolized into reactive free radical products. Carmustin, bleomycin, thiols, and mercurials unequivocally have the potential to cause oxidative stress in tissues. The specific cellular target of the depigmenting chemicals is the melanocyte and it was assumed that oxidative reactions are involved in melanocyte toxicity of these compounds. Tyrosinase is the critical enzyme that ultimately regulates melanogenesis (243,244), and the selective metabolism of phenols/alkylphenols by tyrosinase and formation of cytotoxic redox cycling semiquinone radicals may partially explain the pigment cell specificity of these compounds (245–247). It was demonstrated that the stability and reactivity of the semiquinone radical determines the extent of the depigmenting effect. The most potent phenols are those containing an alkyl substitution in the 4 position (232). The antioxidant capacity of melanocytes seems to be less effective than the antioxidant defense of other skin cells such as fibroblasts and keratinocytes (248). This could contribute to the pigment cell–specific toxicity of phenols/alkylphenols. However, it is clinically well known that non-redox-active chemicals, which induce ICD or ACD as well as physical trauma, can cause leukoderma. This points to a distinct mechanism in melanocyte toxicity, one of them being redox/free radical–mediated. Antioxidant intervention will presumably be effective only during the initial phase of toxic leukoderma caused by prooxidant xenobiotics that are bioactivated by tyrosinase to redox cycling free radicals. However, this approach is presumably of limited clinical value, because most patients seek medical help with already established leukoderma.

## VI. CHEMICAL COMEDOGENESIS

Occupational and environmental agents can cause or exacerbate acne and folliculitis. Although acne vulgaris and chemically induced acne share some common clinical features, significant differences exist, permitting separation of the two entities (249). Chloracne can be a severe and disfiguring skin disease, which in some cases persists for decades after exposure has ceased. Common comedogenic agents are oils, grease, coal tar, and halogenated hydrocarbons, particularly polyhalogenated aromatic hydrocarbons (PHAH). Chloracne is believed to be the most sensitive indicator of toxicity resulting from exposure to PHAH such as polychloronaphthalenes, polychlorinated (brominated) biphenyls, polychloro(bromo)dibenzofurans, and tetrachloroazo(azoxy) benzene (250). Chloracne was the most frequent cutaneous manifestation in 2,3,7,8-tetrachlorodibenzo-p-dioxin (TCDD) toxicity in humans (251), occurring 30 to 60 days after exposure. TCDD is comedogenic by topical contact or systemic administration in humans (252,253). Little is known on the molecular mechanism of chloracne. Most biological effects of PHAH-like compounds are thought to be mediated by binding of the substance to the cytoplasmic arylhydrocarbon receptor (AHR). Particularly the toxicity of TCDD and its bioisosters involves binding to a TCDD-specific AHR (254). The interaction of this complex with chromatin and the production of a pleiotropic response (255,256) leading to abnormal differentiation of follicular epithelium and disturbed lipid metabolism could be the molecular basis of the observed comedogenesis. TCDD is not converted to free radicals, but it is known to stimulate oxidative stress pathways and synergize the toxicity of other environmental contaminants that produce free radicals (257–268). These prooxidant effects are presumably a secondary effect triggered by TCDD and not the direct cause of comedogenesis.

## VII. CHEMICAL CARCINOGENESIS

Epidemiological evidence indicates that smoking, alcohol intake, and intense sun exposure are the major environmental factors that contribute to human carcinogenesis. It was pointed out that environmental pollutants appear to account for less than 1% of human cancer, and that there is no convincing evidence that synthetic chemical pollutants are quantitatively relevant as a cause of human cancer (269,270). Nevertheless, the skin is a potential target organ for chemical carcinogens that exist in our environment. From a qualitative point of view, it is clinically established that arsenic, PAH, and sulfur mustard cause nonmelanoma skin cancer in humans (271), while other potential human chemical carcinogens [e.g., chromium (VI), nickel (II), and benzene] are clinically not recognized as skin carcinogens. In comparison to photocarcinogenesis, chemical-induced nonmelanoma skin cancer is in general not a troublesome clinical issue that urgently requires large-scale intervention strategies. In rapidly dividing epithelium, such as the epidermis, nuclear damage triggered by some xenobiotics may not be so important because of the constant introduction of new healthy cells, whereas a DNA mutation has a much higher probability to become fixed to a transformed phenotype in tissues (e.g., liver) with slow cell turnover. This may explain at least in part why the absolute number of clinically well-recognized human skin carcinogens is so small. It is widely believed that cancer development in humans and laboratory animals is caused by sequential mutations and clonal outgrowth of somatic cells. Increasing evidence has implicated a role for free radicals and oxidative stress in all three stages of the carcinogenic process. Radicals may be involved in the initiation step, either in the oxidative activation of a procarcinogen to its carcinogenic form or in the binding of the carcinogenic species to DNA, or both (272–274), thus making oxidative stress an important cofactor for carcinogen activation. The fraction of initiation events that involve radicals, as opposed to two-electron steps, is not known, but radicals probably are involved in a substantial number, although probably not a majority, of cancer initiation reactions (274). Promotion always involves radicals, at least to some extent (275–285), while their role in progression is controversial (274). Although in vitro and animal studies implicate free radicals as playing a significant role in carcinogenesis, clinical intervention trials have not shown any effectiveness of antioxidant supplements on prevention of nonmelanoma skin cancer. A randomized, placebo-controlled clinical trial on the efficacy of oral β-carotene (50 mg/day over 5 years) in prevention of skin cancer in patients with recent nonmelanoma skin cancer showed no significant effect of β-carotene on either number or time of occurrence of new nonmelanoma skin cancer (286). In a separate trial among healthy men, 12 years of supplementation with β-carotene (50 mg on alternate days) produced no reduction of the incidence of malignant neoplasms, including nonmelanoma skin cancer (287). It must be pointed out that these intervention trials were conducted with patients whose skin cancer was primarily UV-induced and it remains to be seen whether antioxidants are clinically effective in prevention of cutaneous chemocarcinogenesis.

## A. Cytochrome P450

Most chemical carcinogens require metabolic activation to exert their carcinogenic effects. Carcinogen metabolism is carried out by large groups of xenobiotic-metabolizing enzymes that include the phase I cytochromes P450 and phase II enzymes that include various transferases. The cytochrome P450-dependent mono-oxygenase system has evolved as one of the primary defenses against toxic chemicals present in our environment. The cytochrome enzyme family is the most isoenzyme-rich family known (288); in particular, cytochrome P450 I–IV play an important role in xenobiotic metabolism. This multienzyme system functions as an adaptive response to environmental challenge in that exposure to specific toxic agents induces the expression of cytochrome P450 isoenzymes active in their metabolism (289). Polycyclic aromatic hydrocarbons, polyhalogenated aromatic hydrocarbons such as TCDD, and several pesticides including pyrethroids are among the many compounds capable of inducing cytochrome P450 isoenzymes (254,290,291). Rats living in a chemically polluted environment have higher monooxygenase activities than control animals in both liver and lung (292). To our knowledge, human data on monooxygenase activities in target organs, such as the lung or the skin, are not available for subjects living in chemically contaminated environments or workplaces. In most cases, increased monooxygenase activities lead to an increased rate of chemical detoxification, but in certain cases they can also lead to an increased rate of chemical activation toward toxic products and concomitant formation of ROS. By this mechanism, cytochrome P450 is believed to be involved in intolerance reactions and in chemocarcinogenesis (293–296). The high correlation between the induction of cytochrome isoenzymes and the superoxide anion levels supports the previously reported observation that the tumor-promoting ability of cytochrome inducers is mainly mediated by superoxide anions (296). The total activity of cytochrome P450 in skin is assumed to be about 1 to 5% of that in liver. Thus the quantitative turnover of xenobiotics in skin is expected to be low, unless the enzymes are induced. In skin, multiple cytochrome P450 isoenzymes exist, which are expressed quantitatively and qualitatively differently compared to the liver. Benzo[a]pyrene is preferentially metabolized in skin by cytochrome P450I, while in the liver other cytochrome P450-dependent metabolic pathways exist that do not produce carcinogenic metabolites. Consequently, benzo[a]pyrene can be detoxified more effectively in the liver than in the skin, making skin a target organ of benzo[a]pyrene-induced carcinogenicity (297). This is in accordance with a recent publication, indicating that extrahepatic cytochrome P450 CYP1B1 mediates the carcinogenicity of 7,12-dimethylbenz(a)anthracene in mice (294). It should be pointed out that there is a remarkable species specificity with regard to metabolic pathways, different isoenzymes, and substrate specificities in cutaneous xenobiotic metabolism. This makes it extremely difficult to draw conclusions from animal data to the human situation (297). The activation of chemical carcinogens is a complex balance between metabolic activation by cytochrome P450 monooxygenases and detoxification by enzymes such as glutathione-S-transferase. Increased skin tumorigenesis induced by PAH is observed in mice lacking glutathione-S-transferase activity (298). There may be an association between expression of cytochrome P450 CYP1B1 and cancer risk in humans (299). Particularly in the case of cutaneous basal cell carcinomas, the expression of specific cytochrome P450 isoenzymes and glutathione-S-transferase influences skin cancer risk in humans (300,301). It is clear that polymorphisms in detoxifying enzyme genes are important in determining susceptibility to skin cancer (302).

### B. Oncogenes and Tumor Suppressor Genes

In carcinogenesis, three different types of genes have been identified to play a critical role: oncogenes that stimulate cell proliferation, tumor suppressor genes that act as inhibitors, and genes that contribute to tumor progression. Oncogenes act as positive growth regulators and overexpression of the normal protein or the production of a more potent mutant form often leads to increased growth of transformed cells. Tumor suppressor genes act as negative growth regulators. Upregulation of tumor suppressor genes results in growth arrest, followed by DNA repair, or apoptosis. It is widely believed that combinations of activation of proto-oncogenes to oncogenes and inactivation of tumor suppressor genes leads to carcinogenesis and the generation of the first tumor cell (303). Angiogenesis is considered a key step in tumor growth, invasion, and metastasis. Regulation of angiogenesis is dependent upon a balance of several proangiogenic and antiangiogenic factors. Vascular endothelial growth factor (VEGF) and basic fibroblast growth factor (bFGF) are one of the most potent angiogenic growth factors identified to date. VEGF is a p53-

regulated endothelial cell-specific mitogen that plays an essential role both in normal and pathological angiogenesis (304). IL-8, which is known to be upregulated by oxidative stress (104,105), has been shown to be a potent angiogenic factor. Genes that specify cell adhesion molecules, such as cadherins, and those for proteases that specialize in the degeneration of extracellular matrices, such as MMPs, are likely to be involved in tumor metastasis. MMPs and TIMPs play a complex role in regulating angiogenesis (305,306). The oncogene c-H-ras and the tumor suppressor gene p53 are genes whose involvement in the steps of epithelial skin cancers are duly established (307). The development of nonmelanoma skin cancers in humans often involves inactivation of the p53 tumor suppressor gene (located on chromosome 17) through various mechanisms (308,309). The p53 protein is a transcription factor that plays a critical role in regulating progression of cells through the cell cycle, as well as of cells into the apoptosis pathways. Loss of p53 relieves a normal block to cell growth, giving mutant cells an advantage. Loss of p53 function also allows cells to escape apoptosis. The physiological role of p53 in limiting mutagenic damage makes it likely that mutation of p53 will allow an increased rate of accumulation of genetic damage in the cell. Mutations of the p53 gene play a role in chemically induced skin carcinogenesis in mice (310). ROS may cause genetic changes in proto-oncogenes and tumor suppressor genes. ROS have been implicated to play several distinct roles in the p53 pathway. First, they are important activators of p53 through their capacity to induce DNA strand breaks. Second, they regulate the DNA binding activity of p53 by modulating the redox status of a critical set of cysteines in the DNA binding domain. Third, they play a role in the signaling pathways regulated by p53, as several genes encoding redox effectors are transcriptionally controlled by p53 (311). p53 DNA binding and transcriptional activities are controlled by the thiol redox state (312–314) and thiol oxidation abolishes its interaction with its specific DNA target frequency (315). Furthermore, p53 mRNA translation is also regulated by the thiol redox status (316).

## VIII. SYNOPSIS

The scientific data on redox-modulated molecular and cellular events in skin function and cutaneous diseases have increased dramatically during the last decade. It becomes more and more evident that the redox status of the cell is importantly involved in several basic cellular processes, such as signal transduction, gene expression, inflammation, and apoptosis. Similarly, according to the number of publications, the scientific concern for redox-sensitive processes and their role in toxicology is strongly increasing. According to the huge and exponentially increasing interest in both fields, redox modulation of cell function in cutaneous biology and toxicology, there is a need to build bridges between these different but connected research areas, thus creating a new perspective. This approach clearly shows that modulation of the cellular redox state and oxidative stress, which are interrelated closely, is often involved in the pathology following chemical exposure to skin, but the clinical relevance of these molecular events is not yet fully understood in most, if not all, cases. Gene–environment interactions are thought to be critical for the manifestation of several multifactorial skin disorders induced or aggravated by xenobiotics. The development and application of microarrays and DNA chip technology will substantially help to identify individuals with increased susceptibility to environmental pollutants (317). Recent advances in the understanding of redox-sensitive mechanisms in cutaneous toxicity of environmental and occupational chemicals have been translated into novel interventional therapies, which are mostly still experimental. As shown for selected examples of contact dermatitis, which is a very common and important condition in clinical and occupational dermatology, oxidative stress and redox modulation can make a significant contribution to the pathology at various time points and at distinct molecular levels. If oxidative stress is involved early and at several different key steps of the inflammatory pathway, antioxidants may have a reasonable chance to effectively modulate these reaction cascades. In the patent literature, antioxidants such as ascorbic acid and α-tocopherol have been considered for topical therapy of ICD and ACD (318–321). Inhibition of NF-κB (322,323), direct blocking of ICAM-1 expression on endothelial cells (324), interference with the T-lymphocyte and PMN response, and interaction with the hapten are among the proposed mechanisms. While in vitro and animal studies generally support an attenuation of inflammatory skin reactions by antioxidants, the results obtained from clinical studies are less clear and do not in general support the claims made in the patent literature. In particular, the effectiveness of tocopherol, ascorbate, and other small molecular antioxidants for the treatment of contact dermatitis has not been confirmed in prospective double-blind clinical trials. It remains to be seen whether antioxidants or antioxidant combinations will prove beneficial against distinct aspects of contact dermatitis in humans. The skin as the most readily accessible human organ invites for clinical studies challenging the concept of oxidative stress and redox-modulated toxicity. It seems logical to extend the search for antioxidant intervention strategies to other, less frequently occurring clinical forms of chemical-induced skin disorders, but the pathological scenario is even less clear in these cases. In chemical scleroderma, the role of redox-sensitive events and oxidative stress remains to be defined, while in chemical comedogenesis it seems to represent an epiphenomenon. In chemical leukoderma, oxidative injury is an early and presumably irreversible event and is thus probably not responsive to antioxidant intervention. In cutaneous chemical carcinogenesis (e.g., arsenic, sulfur mustard, and PAH-induced nonmelanoma skin cancer), experimental evidence points to some role of redox modulation and oxidative injury in the molecular scenario, but it will be difficult to conduct clinical studies proving that oxidative events are biologically significant in the pathology.

## REFERENCES

1. LaDou J. Occupational and Environmental Medicine. Stamford: Appleton & Lange, 2nd ed., 1997.
2. Chignell CF. Structure activity relationships in the free radical metabolism of xenobiotics. Environ Health Perspec 1985; 61:133–137.
3. Stohs SJ. The role of free radicals in toxicity and disease. J Basic Clin Physiol Pharmacol 1995; 6:205–228.
4. Ahmad S. Oxidative stress from environmental pollutants. Arch Insect Biochem Physiol 1995; 29:135–157.
5. Toraason M. 8-Hydroxy-deoxyguanosine as a biomarker of workplace exposure. Biomarkers 1999; 4:3–26.
6. Sies H. Oxidative Stress: Introductory remarks. In: Sies H, ed. Oxidative Stress. London: Academic Press, 1985:1–8.
7. Shupack JL. The skin as a target organ for systemic agents. In: Drill VA, Lazar P, eds. Cutaneous Toxicity. New York: Academic Press, 1976:43–52.
8. Taylor JS, Parrish JA, Blank IH. Environmental reactions to chemical, physical and biologic agents. J Am Acad Dermatol 1984; 11:1007–1021.
9. Adams RM. Occupational Skin Disease, 3rd ed. Philadelphia: Saunders, 1999.
10. Hirai A, Minamiyama Y, Hamada T, Ishii M, Inoue M. Glutathione metabolism in mice is

enhanced more with haptene induced allergic contact dermatitis than with irritant contact dermatitis. J Invest Dermatol 1997; 109:314–318.

11. Camera E, Jensen C, Stab LS, Scala G, Baadsgaard O, Picardo M. Correlation between antioxidant levels and skin reactivity to irritants. J Dermatol Sci 1998; 16(suppl 1):S193.
12. Kimura J, Hayakari M, Kumano T, Nakano H, Satoh K, Tsuchida S. Altered glutathione transferase levels in rat skin inflamed due to contact hypersensitivity: induction of the alpha-class subunit 1. Biochem J 1998; 335:605–610.
13. Sarnstrand B, Jansson AH, Matuseviciene G, Scheynius A, Pierrou S, Bergstrand H. N,N′-Diacetyl-L-cystine—the disulfide dimer of N-acetylcysteine—is a potent modulator of contact sensitivity/delayed type hypersensitivity reactions in rodents. J Pharmacol Exp Ther 1999; 288:1174–1184.
14. Winyard PG, Blake DR. Antioxidants, redox-regulated transcription factors, and inflammation. In: Sies H, ed. Advances in Pharmacology: Antioxidants in Disease Mechanism and Therapy. San Diego: Academic Press, 1997:403–421.
15. Wakem P, Gaspari AA. Mechanisms of allergic and irritant contact dermatitis. In: Kydonieus AF, Wille JJ, eds. Biochemical Modulations of Skin Reactions. Boca Raton: CRC Press, 2000:83–106.
16. Bagley DM, Gardner JR, Holland G, Lewis RW, Regnier JF, Stringer DA, Walker AP. Skin irritation: Reference chemicals data bank. Toxicol In Vitro 1996; 10:1–6.
17. Berardesca E, Distante F. Mechanism of skin irritation. In: Elsner P, Maibach H, eds. Current Problems in Dermatology—Irritant Dermatitis. New Clinical and Experimental Aspects. Basel: Karger, 1995:1–8.
18. Weltfriend S, Bason M, Lammintausta K, Maibach HI. Irritant dermatitis (irritation). In: Marzulli FN, Maibach HI, eds. Dermatotoxicology. Washington, DC: Taylor & Francis, 1996:87–118.
19. Wigger-Alberti W, Iliev D, Elsner P. Contact dermatitis due to irritation. In: Adams RM, ed. Occupational Skin Disease, 3rd ed. Philadelphia: Saunders, 1999:1–9.
20. Barker JN, Mitra RS, Griffiths CE, Dixit VM, Nickoloff BJ. Keratinocytes as initiators of inflammation. Lancet 1991; 337:211–214.
21. Enk AH, Katz SI. Early molecular events in the induction phase of contact sensitivity. Proc Natl Acad Sci USA 1992; 89:1398–1402.
22. Marks F, Fürstenberger G, Müller-Decker K. Arachidonic acid metabolism as a reporter of skin irritancy and target of cancer chemoprevention. Toxicol Lett 1998; 96–97:111–118.
23. Friedmann PS, Strickland I, Memon AA, Johnson PM. Early time course of recruitment of immune surveillance in human skin after chemical provocation. Clin Exp Immunol 1993; 91:351–356.
24. Brasch J, Burgard J, Sterry W. Common pathogenic pathways in allergic and irritant contact dermatitis. J Invest Dermatol 1992; 98:166–170.
25. Griem P, Gleichmann E. Metal ion induced autoimmunity. Current Opin Immunol 1995; 7: 831–838.
26. Bour H, Pyron E, Gaucherand M, Garrigue JL, Desvignes C, Kaiserlian D, Revvilard JP, Nicolas JF. Major histo-compatibility complex class I-restricted CD8+ T cells and class II restricted CD4+ T cells, respectively, mediate and regulate contact sensitivity to dinitrofluorobenzene. Eur J Immunol 1995; 25:3006–3010.
27. Xu H, Dilulio NA, Fairchild RL. T cell populations primed by hapten sensitization in contact sensitivity are distinguished by polarized patterns of cytokine production: interferon gamma-producing (Tc1) effector CD8+ T cells and interleukin (Il)4/Il-10-producing (Th2) negative regulatory CD4+ T cells. J Exp Med 1996; 183:1001–1012.
28. Cavani A, Mei D, Guerra E, Corinti S, Giani M, Pirotta L, Puddu P, Girolomoni G. Patients with allergic contact dermatitis to nickel and nonallergic individuals display different nickel-specific T-cell responses. Evidence for the presence of effector CD8+ and regulatory CD4+ T cells. J Invest Dermatol 1998; 111:621–628.

29. Moulon C, Wild D, Dormoy A, Weltzien HU. MHC-dependent and -independent activation of human nickel-specific CD8+ cytotoxic T cells from allergic donors. J Invest Dermatol 1998; 111:360–366.
30. Kalish RS, Askenase PW. Molecular mechanisms of CD8+ T-cell mediated delayed hypersensitivity: implications for allergies, asthma and autoimmunity. J Allergy Clin Immunol 1999; 103:192–199.
31. Krasteva M, Kehren J, Ducluzeau MT, Sayag M, Cacciapuoti M, Akiba H, Descotes J, Nicolas JF. Contact dermatitis I. Pathophysiology of contact sensitivity. Eur J Dermatol 1999; 9:65–77.
32. Kalish RS. Pathogenesis of allergic contact dermatitis: Role of T cells. In: Kydonieus AF, Wille JJ, eds. Biochemical Modulations of Skin Reactions. Boca Raton: CRC Press, 2000: 145–156.
33. Gäfvert E, Shao LP, Karlberg AT, Nilson U, Nilson JLG. Contact allergy to resin acid hydroperoxides. Hapten binding via free radicals and epoxides. Chem Res Toxicol 1994; 7:260–266.
34. Lepoittevin JP, Karlberg AT. Interaction of allergic hydroperoxides with proteins. A radical mechanism. Chem Res Toxicol 1994; 7:130–133.
35. Payne MP, Wals PT. Structure activity relationships for skin sensitization potential: development of structural alerts for use in knowledge-based toxicity prediction systems. J Chem Inf Com Sci 1994; 34:154–161.
36. Bezard M, Karlberg AT, Montelius J, Lepoittevin JP. Skin sensitization to linayl hydroperoxide: support for radical intermediates. Chem Res Toxicol 1997; 10:987–993.
37. Schmidt RJ, Khan L, Chung LY. Are free radicals and not quinones the haptenic species derived from urushiols and other contact allergenic mono- and dihydric alkylbenzenes? The significance of NADH, glutathione, and redox cycling in the skin. Arch Dermatol Res 1990; 282:56–64.
38. Barratt MD, Basketter DA. Possible origin of the skin sensitization potential of isoeugenol and related compounds. (I). Preliminary studies of potential reaction mechanisms. Contact Dermatitis 1992; 27:98–104.
39. Scholes EW, Pendlington RU, Sharma RK, Basketter DA. Skin metabolism of contact allergens. Toxicol In Vitro 1994; 8:551–553.
40. Barratt MD, Basketter DA. Structure activity relationships for skin sensitization: an expert system. In: Rougier A, Goldberg AM, Maibach HI, eds. In Vitro Toxicology: Irritation, Phototoxicity, Sensitization. New York: Ann Liebert, 1994:293–301.
41. Schmidt RJ, Chung LY. Biochemical responses of skin to allergenic and non-allergenic nitrohalobenzenes. Evidence that an NADPH dependent reductase in skin may act as a prohapten activating enzyme. Arch Dermatol Res 1992; 284:400–408.
42. Takiwaki H, Arase S, Nakayama H. Contact dermatitis due to 2,2′-azobis(2-amidinopropane) dihydrochloride: an outbreak in production workers. Contact Dermatitis 1998; 39:4–7.
43. Rudzki E. Pattern of hypersensitivity to aromatic amines. Contact Dermatitis 1975; 1:248–249.
44. Santucci B, Cristaudo A, Cannistraci C, Amantea A, Picardo M. Hypertrophic allergic contact dermatitis from hair dye. Contact Dermatitis 1994; 31:169–171.
45. Picardo M, Zompetta C, Marchese C, De Luca C, Faggioni A, Schmidt RJ, Santucci B. Paraphenylenediamine, a contact allergen, induces oxidative stress and ICAM-1 expression in human keratinocyte. Br J Dermatol 1996; 126:450–455.
46. Chignell CF, Motten AG, Buettner GR. Photoinduced free radicals from chlorpromazine and related phenothiazines: relationship to phenothiazine-induced photosensitization. Environ Health Perspec 1985; 64:103–110.
47. Motten AG, Buettner GR, Chignell CF. Spectroscopic studies of cutaneous photosensitizing agents—VIII. A spin trapping study of light induced free radicals from chlorpromazine and promazine. Photochem Photobiol 1985; 42:9–15.

48. Schoonderwoerd SA, Beijersbergen GMJ, van Henegouwen X, Van Belkum S. In-vivo photodegradation of chlorpromazin. Photochem Photobiol 1989; 50:659–664.
49. Lepoittevin JP, Berl V. Molecular basis of allergic contact dermatitis. In: Marzulli FN, Maibach HI, eds. Dermatotoxicology. Washington DC: Taylor & Francis, 1996:147–160.
50. Davis MM, Bjorkman PJ. T-Cell antigen receptor genes and T-Cell recognition. Nature 1988; 334:395–402.
51. Santamaria Babi LF, Perez Soler MT, Hauser C, Blaser K. Skin-homing T cells in human cutaneous allergic inflammation. Immunol Res 1995; 14:317–324.
52. Fuhlbrigge RC, Kieffer JD, Armerding D, Kupper TS. Cutaneous lymphocyte antigen is a specialized form of PSGL-1 expressed on skin-homing T cells. Nature 1997; 389:978–981.
53. Zollner TM, Diehl S, Kaufmann R, Podda M. Antioxidants block the expression and functional activity of the skin-selective homing receptor cutaneous lymphocyte-associated antigen. J Invest Dermatol 1998; 110:476.
54. Monteiro HP, Stern A. A redox modulation of tyrosine phosphorylation-dependent signal transduction pathways. Free Rad Biol Med 1996; 21:323–333.
55. Adler V, Yin Z, Tew TD, Ronai Z. Role of redox potential and reactive oxygen species in stress signaling. Oncogene 1999; 18:6104–6111.
56. Allen RG, Tresini M. Oxidative stress and gene regulation. Free Radic Biol Med 2000; 28: 463–499.
57. Peus D, Vasa RA, Meves A, Pott M, Beyerle A, Squillace K, Pittelkow MR. $H_2O_2$ is an important mediator of UVB induced EGF receptor phosphorylation in cultured keratinocytes. J Invest Dermatol 1998; 110:966–971.
58. Peus D, Vasa RA, Beyerle A, Meves A, Krautmacher C, Pittelkow MR. UVB activates ERK1/2 and p38 signalling pathways via reactive oxygen species in cultured human keratinocytes. J Invest Dermatol 1999; 112:751–756.
59. Wesselborg S, Bauer MKA, Vogt M, Schmitz ML, Schulze-Osthoff K. Activation of transcription factor NF-kappa-B and p38 mitogen-activated protein kinase is mediated by distinct and separate stress effector pathways. J Biol Chem 1997; 272:12422–12429.
60. Jacobson PB, Kuchera SL, Metz A, Schachtele C, Imre K, Schrier DJ. Anti-inflammatory properties of Go 6850: a selective inhibitor of protein kinase C. J Pharmacol Exp Ther 1995; 275:995–1002.
61. Wang HQ, Smart RC. Overexpression of protein kinase C-alpha in the epidermis of transgenic mice results in striking alterations in phorbol ester-induced inflammation and COX-2, MIP-2 and TNF-alpha expression but not tumor promotion. J Cell Sci 1999; 112: 3497–3506.
62. Chabot-Fletcher M, Breton J, Lee J, Young P, Griswold DE. Interleukin-8 production is regulated by protein kinase C in human keratinocytes. J Invest Dermatol 1994; 103:509–515.
63. Orrenius S, Burkitt MJ, Kass GE, Dypbukt JM, Nicotera P. Calcium ions and oxidative cell injury. Ann Neurol 1992; 32(suppl):S33–S42.
64. Azzi AM, Bartoli G, Boscoboinik D, Hensey C, Szewczyk A. α-Tocopherol and protein kinase C regulation of intracellular signalling. In: Fuchs J, Packer L, eds. Vitamin E in Health and Disease. New York: Marcel Dekker, Inc, 1993:371–383.
65. Blankenhorn G, Clewing S. Human studies on vitamin E in rheumatic inflammatory disease. In: Fuchs J, Packer L, eds. Vitamin E in Health and Disease. New York: Marcel Dekker, Inc, 1993:563–576.
66. Meyer M, Pahl HL, Baeuerle PA. Regulation of the transcription factors NF-kappa-B and AP-1 by redox changes. Chem Biol Interact 1994; 91:91–100.
67. Sen CK, Packer L. Antioxidant and redox regulation of gene transcription. FASEB J 1996; 10:709–720.
68. Flohe L, Brigelius-Flohe R, Saliou C, Traber MG, Packer L. Redox regulation of NF-kappa B activation. Free Rad Biol Med 1997; 22:1115–1126.

69. Gius D, Botero A, Shah S, Curry HA. Intracellular oxidation/reduction status in the regulation of transcription factors NF-kappaB and AP-1. Toxicol Lett 1999; 106:93–106.
70. Piette J, Piret B, Bonizzi G, Schoonbroodt S, Merville MP, Legrand-Poels S, Bours V. Multiple redox regulation in NF-kappaB transcription factor activation. Biol Chem 1997; 378: 1237–1245.
71. Courtois G, Whiteside ST, Sibley CH, Israel A. Characterization of a mutant cell line that does not activate NF-kappaB in response to multiple stimuli. Mol Cell Biol 1997; 17:1441–1449.
72. Schreck R, Baeuerle PA. NF-κB as inducible transcriptional activator of the granulocyte macrophage colony stimulating factor gene. Mol Cell Biol 1996; 10:1281–1286.
73. Schreck R, Rieber P, Baeuerle PA. Reactive oxygen intermediates as apparently widely used second messengers in the activation of the NFκB transcription factor and HIV-1. Eur Mol Biol Org J 1991; 10:2247–2258.
74. Schreck R, Albermann K, Baeuerle PA. Nuclear transcription factor κB: an oxidative stress responsive transcription factor of eukaryotic cells (a review). Free Rad Res Commun 1992; 17:221–237.
75. Baeuerle PA, Henkel T. Function and activation of NFκB in the immune system. Annu Rev Immunol 1994; 12:141–179.
76. Kaltschmidt C, Kaltschmidt B, Lannes-Vieira J, Kreuzberg GW, Wekerle H, Baeuerle PA, Gehrmann J. Transcription factor NFκB is activated in microglia during experimental autoimmune encephalomyelitis. J Neuroimmunol 1994; 55:99–106.
77. Bonizzi G, Dejardin E, Piret B, Piette J, Merville MP, Bours V. Interleukin-1β induces nuclear factor κB in epithelial cells independently of the production of reactive oxygen intermediates. Eur J Biochem 1996; 242:544–549.
78. Brennan P, O'Neill LAJ. Effects of oxidants and antioxidants on nuclear factor κB activation in three different cell lines: evidence against an universal hypothesis involving oxygen radicals. Biochem Biophys Acta 1995; 1260:167–175.
79. Legrand-Poels S, Zecchinon L, Piret B, Schoonbroodt S, Piette J. Involvement of different transduction pathways in NF-κB activation by several inducers. Free Rad Res 1997; 27:301–309.
80. Podda M, Koh B, Thiele J, Milbradt R, Packer L. Ozone activates the transcription factor NFκB in keratinocytes via reactive oxygen species. Austral J Dermatol 1997; 38(suppl 2): 185.
81. Ikeda M, Miyoshi K, Chikazawa M, Kodama H. Activation of nuclear factor κB by hydrogen peroxide in epidermal keratinocytes and its inhibition by antioxidants. Austral J Dermatol 1997; 38(suppl 2):190.
82. Vile GF, Tyrrell RM. UVA radiation induced oxidative damage to lipids and proteins in vitro and in human skin fibroblasts is dependent on iron and singlet oxygen. Free Radic Biol Med 1995; 18:721–730.
83. Devary Y, Gottlieb RA, Lau LF, Karin M. Rapid and preferential activation of the c-jun gene during the mammalian UV response. Mol Cell Biol 1991; 11:2804–2811.
84. Radler-Pohl A, Sachsenmaier C, Gebel S, Auer H, Bruder JT, Rapp U, Angel P, Rahmsdorf HJ, Herrlich P. UV induced activation of AP-1 involves obligatory extranuclear steps including RAF-1 kinase. EMBO J 1993; 12:1005–1012.
85. Schmidt KN, Podda M, Packer L, Baeuerle PA. Anti-psoriatic drug anthralin activates transcription factor NF-kappa-B in murine keratinocytes. J Immunol 1996; 156:4514–4519.
86. Lange RW, Germolec DR, Foley JF, Luster MI. Antioxidants attenuate anthralin-induced skin inflammation in BALB-c mice: Role of specific proinflammatory cytokines. J Leukocyte Biol 1998; 64:170–176.
87. Ye J, Zhang X, Young HA, Mao Y, Shi X. Chromium(VI)-induced nuclear factor-kappaB activation in intact cells via free radical reactions. Carcinogenesis 1995; 16:2401–2405.

88. Shi X, Ding M, Ye J, Wang S, Leonard SS, Zang L, Castranova V, Vallyathan V, Chiu A, Dalal N. Cr(IV) causes activation of nuclear transcription factor-kappa-B, DNA strand breaks and dG hydroxylation via free radical reactions. J Inorgan Biochem 1999; 75:37–44.
89. Goebeler M, Roth J, Broecker EB, Sorg C, Osthoff KS. Activation of nuclear factor-kappa-B and gene expression in human endothelial cells by the common haptens nickel and cobalt. J Immunol 1995; 155:2459–2467.
90. Wagner MK, Klein CL, Van Kooten GT, Kirkpatrick J. Mechanisms of cell activation by heavy metal ions. J Biomed Mat Res 1998; 42:443–452.
91. Luger TA, Lotti T. Neuropeptides: role in inflammatory skin diseases. J Eur Acad Dermatol Venereol 1998; 10:207–211.
92. Koy A. Initiation of acute phase response and synthesis of cytokines. Biochim Biophys Acta 1996; 1317:84–94.
93. Lewis RW, McCall JC, Botham PA, Kimber I. Investigation of TNF-alpha release as a measure of skin irritancy. Toxicol In Vitro 1993; 7:393–395.
94. Corsini E, Galli CL. Cytokines and irritant contact dermatitis. Toxicol Lett 1998; 102–103: 277–282.
95. Larsen CG, Ternowitz T, Larsen FG, Zachariae G, Thestrup-Pedersen K. ETAF/interleukin-1 and epidermal lymphocyte chemotactic factor in epidermis overlying an irritant patch test. Contact Dermat 1989; 20:335–340.
96. Hunziker T, Brand CU, Kapp A, Waelti ER, Braathen LR. Increased levels of inflammatory cytokines in human skin lymph derived from sodium lauryl sulphate induced contact dermatitis. Br J Dermatol 1992; 127:254–257.
97. Kimber I. Epidermal cytokines in contact hypersensitivity: Immunological roles and practical applications. Toxicol In Vitro 1993; 7:295–298.
98. Luster MI, Simeonova PP, Gallucci R, Matheson J. Tumor necrosis factor alpha and toxicology. Crit Rev Toxicol 1999; 29:491–511.
99. Murphy JE, Robert C, Kupper TS. Interleukin-1 and cutaneous inflammation: A crucial link between innate and acquired immunity. J Invest Dermatol 2000; 114:602–608.
100. Adamson GM, Billings RE. Tumor necrosis factor induced oxidative stress in isolated mouse hepatocytes. Arch Biochem Biophys 1992; 294:223–229.
101. Goucherot-Pocidalo MA, Revillard JP. Oxidative stress, cytokines and lymphocyte activation. In: Fuchs J, Packer L, eds. Oxidative Stress in Dermatology. New York: Marcel Dekker, Inc, 1993:187–205.
102. Lo YYC, Conquer JA, Grinstein S, Cruz TF. Interleukin-1-beta induction of c-fos and collagenase expression in articular chondrocytes: Involvement of reactive oxygen species. J Cell Biochem 1998; 69:19–29.
103. Hennet T, Richter C, Peterhans E. Tumor necrosis factor-alpha induces superoxide anion generation in mitochondria of L929 cells. Biochem J 1993; 289:587–592.
104. Deforge LE, Preston AM, Takeuchi E, Kenny J, Boxter LA, Remick DG. Regulation of interleukin 8 gene expression by oxidant stress. J Biol Chem 1993; 268:25568–25576.
105. Shimada T, Watanabe N, Hiraishi H, Terano A. Redox regulation of interleukin-8 expression in MKN28 cells. Dig Dis Sci 1999; 44:266–273.
106. Piguet PF, Grau GE, Hauser C, Vasalli P. Tumor necrosis factor is a critical mediator in hapten induced irritant and contact hypersensitivity reactions. J Exp Med 1991; 173:673–679.
107. Luster MI, Wilmer JL, Germolec DR, Spalding J, Yoshida T, Gaido K, Simeonova PP, Burleson FG, Bruccoleri A. Role of keratinocyte-derived cytokines in chemical toxicity. Toxicol Lett 1995; 82–83:471–476.
108. Lisby S, Muller KM, Jongeneel CV, Saurat JH, Hauser C. Nickel and skin irritants up-regulate tumor necrosis factor-alpha mRNA in keratinocytes by different but potentially synergistic mechanisms. Int Immunol 1995; 7:343–349.

109. Little MC, Gawkrodger DJ, MacNiel S. Chromium- and nickel-induced cytotoxicity in normal and transformed human keratinocytes: an investigation of pharmacological approaches to the prevention of Cr(VI) induced cytotoxicity. Br J Dermatol 1996; 134:199–205.
110. Lange RW, Hayden PJ, Chignell CF, Luster MI. Anthralin stimulates keratinocyte-derived proinflammatory cytokines via generation of reactive oxygen species. Inflamm Res 1998; 47:174–181.
111. Kurt EM, Schafer RJ, Arroyo CM. Effects of sulfur mustard on cytokines released from cultured human epidermal keratinocytes. Int J Toxicol 1998; 17:223–229.
112. Tsuruta J, Sugisaki K, Dannenberg AM, Yoshimura T, Abe Y, Mounts P. The cytokines NAP-1 (IL-8), MCP-1 IL-1 beta, and GRO in rabbit inflammatory skin lesions produced by the chemical irritant sulfur mustard. Inflammation 1996; 20:293–318.
113. Arroyo CM, Schafer RJ, Kurt EM, Broomfield CA, Carmichael AJ. Response of normal human keratinocytes to sulfur mustard (HD): Cytokine release using a non-enzymatic detachment procedure. Hum Exp Toxicol 1999; 18:1–11.
114. Shedova AA, Kommineni C, Jeffries BA, Castranova V, Tyurina YY, Tyurin VA, Serbonova E, Fabisiak JB, Kagan V. Redox cycling of phenol induces oxidative stress in human epidermal keratinocytes. J Invest Dermatol 2000; 114:354–364.
115. Wilmer JL, Burleson FG, Kayama F, Kanno J, Luster MI. Cytokine induction in human epidermal keratinocytes exposed to contact irritants and its relation to chemical induced inflammation in mouse skin. J Invest Dermatol 1994; 102:915–922.
116. Gueniche A, Viac J, Lizard G, Charveron M, Schmitt D. Effect of various metals on intercellular adhesion molecule-1 expression and tumour necrosis factor alpha production by normal human keratinocytes. Arch Dermatol Res 1994; 286:466–470.
117. Senaldi G, Pointaire P, Piguet PF, Grau GE. Protective effect of N-acetylcysteine in hapteneinduced irritant and contact hypersensitivity reactions. J Invest Dermatol 1994; 102:934–937.
118. Pasche-Koo F, Arechalde A, Arrighi JF, Hauser C. Effect of N-acetylcysteine, an inhibitor of tumor necrosis factor, on irritant contact dermatitis in the human. In: Elsner P, Maibach HI, eds. Irritant Dermatitis. Freiburg: Karger, 1995:198–206.
119. Manning AM, Chen CC, Rosenbloom CL, McNab AR, Cruz R, Anderson DC. Redox and proteolytic signal transduction events in the activation of E-selectin, VCAM-1 and ICAM-1 gene expression. J Leuk Biol 1994; 10(suppl):38.
120. Marui N, Offermann MK, Swerlick R, Kunsch C, Rosen CA, Ahmad M, Alexander RW, Medford RM. Vascular cell adhesion molecule-1 (VCAM-1) gene transcription and expression are regulated through an antioxidant-sensitive mechanism in human vascular endothelial cells. J Clin Invest 1993; 92:1866–1874.
121. Das PK, de Boer OJ, Visser A, Verhagen CE, Bos JD, Pals ST. Differential expression of ICAM-1, E-selectin and VCAM-1 by endothelial cells in psoriasis and contact dermatitis. Acta Dermatol Venereol 1994; 186(suppl):21–22.
122. Griffiths CEM, Nickoloff BJ. Keratinocyte intracellular adhesion molecule-1 (ICAM-1) expression preceeds dermal T lymphocyte infiltration in allergic contact dermatitis (Rhus dermatitis). Am J Pathol 1989; 135:1045–1053.
123. Wildner O, Lipkow T, Knop J. Increased expression of ICAM-1, E-selectin, and VCAM-1 by cultured human endothelial cells upon exposure to haptens. Exp Dermatol 1992; 1:191–198.
124. Pober JS, Cotran RS. Cytokine and endothelial cell biology. Phys Rev 1990; 70:427–451.
125. Ghosh S. NF-kappa-B and Rel proteins: Evolutionarily conserved mediators of immune responses. Annu Rev Immunol 1998; 16:225–260.
126. Jung K, Imhof BA, Linse R, Wollina Y, Neumann C. Adhesion molecules in atopic dermatitis: Upregulation of alpha-6 integrin expression in spontaneous lesional skin as well as in atopen, antigen and irritative induced patch test reactions. Int Arch All Immunol 1997; 113: 495–504.

127. Farugi RM, Poptic EJ, Farugi TR, De La Motte C, Dicorleto PE. Distinct mechanisms for N-acetylcysteine inhibition of cytokine induced E-selectin and VCAM-1 expression. Am J Physiol 1997; 273:H817–H826.
128. Friedrichs B, Mueller C, Brigelius-Flohe R. Inhibition of tumor necrosis factor-alpha and interleukin-1 induced endothelial E-selectin expression by thiol-modifying agents. Arterio Thromb Vasc Biol 1998; 18:1829–1837.
129. Willis CM, Stephens CJM, Wilkinson JD. Selective expression of immune associated surface antigen by keratinocytes in irritant contact dermatitis. J Invest Dermatol 1991; 96:505–511.
130. Grether-Beck S, Olaizola-Horn S, Schmitt H, Grewe M, Jahnke A, Johnson JP, Briviba K, Sies H, Krutmann J. Activation of transcription factor AP-2 mediates UVA radiation- and singlet oxygen-induced expression of the human intercellular adhesion molecule 1 gene. Proc Natl Acad Sci USA 1996; 93:14586–14591.
131. Caughman SW, Li LJ, Degitz K. Human intercellular adhesion molecule-1 gene and its expression in the skin. J Invest Dermatol 1992; 98:61S–65S.
132. Little MC, Metcalfe RA, Haycock JW, Healy J, Gawkrodger DJ, Mac Neil S. The participation of proliferative keratinocytes in the preimmune response to sensitizing agents. Br J Dermatol 1998; 138:45–56.
133. Leslie CC. Properties and regulation of cytosolic phospholipase A2. J Biol Chem 1997; 272: 16709–16714.
134. Cornwell DG, Panganamala RV. Vitamin E action in modulating the arachidonic acid cascade. In: Fuchs J, Packer L, eds. Oxidative Stress in Dermatology. New York: Marcel Dekker, Inc, 1993:385–416.
135. Chakraborti S, Chakraborti T. Oxidant-mediated activation of mitogen-activated protein kinases and nuclear transcription factors in the cardiovascular system: a brief overview. Cell Signal 1998; 10:675–683.
136. Simon LS. Role and regulation of cyclooxygenase-2 during inflammation. Am J Med 1999; 106:37S–42S.
137. Adelizzi RA. COX-1 and COX-2 in health and disease. J Am Osteopath Assoc 1999; 99(suppl 11):S7–S12.
138. Wallace JL. Distribution and expression of cyclooxygenase (COX) isoenzymes, their physiological roles, and the categorization of nonsteroidal antiinflammatory drugs (NSAIDs). Am J Med 1999; 107(6A):11S–17S.
139. Sahnoun Z, Jamoussi K, Zeghal KM. Radicaux libres et anti-oxydants: physiologie, pathologie humaine et aspects therapeutiques (ileme partie). Therapie 1998; 53:315–339.
140. Vane JR, Bakhle YS, Botting RM. Cyclooxygenases 1 and 2. Annu Rev Pharmacol Toxicol 1998; 38:97–120.
141. Crofford LJ. COX-1 and COX-2 tissue expression: implications and predictions. J Rheumatol 1997; 24(suppl 49):15–19.
142. Camp RDR, Russel JR, Brain D, Woollard PM. Production of intradermal microabscesses by topical application of leukotriene B. J Invest Dermatol 1984; 82:202–207.
143. Ikai K. Proinflammatory mediators of the arachidonic acid cascade. In: Kydonieus AF, Wille JJ, eds. Biochemical Modulations of Skin Reactions. Boca Raton: CRC Press, 2000:189–201.
144. Landino LM, Crews BC, Timmons MD, Morrow JD, Marnett JL. Peroxynitrite, the coupling product of nitric oxide and superoxide, activates prostaglandin biosynthesis. Proc Natl Acad Sci USA 1996; 93:15069–15074.
145. Goodwin DC, Landino LM, Marnett LJ. Effects of nitric oxide and nitric oxide-derived species on prostaglandin endoperoxide synthase and prostaglandin biosynthesis. FASEB J 1999; 13:1121–1136.
146. Fuchs J, ed. Oxidative Injury in Dermatopathology. Heidelberg: Springer Verlag, 1992.
147. Darr D, Fridovich I. Free radicals in cutaneous biology. J Invest Dermatol 1994; 102:671–675.

148. Coquette A, Berna N, Poumay Y, Pittelkow MR. The keratinocyte in cutaneous irritation and sensitization. In: Kydonieus AF, Wille JJ, eds. Biochemical Modulations of Skin Reactions. Boca Raton: CRC Press, 2000:125–143.
149. Frank S, Kampfer H, Podda M, Kaufmann R, Pfeilschifter J. Identification of copper/zinc superoxide dismutase as a nitric oxide-regulated gene in human (HaCaT) keratinocytes: implications for keratinocyte proliferation. Biochem J 2000; 346:719–728.
150. Shimizu Y, Sakai M, Umemura Y, Ueda H. Immunohistochemical localization of nitric oxide synthase in normal human skin: expression of endothelial-type and inducible-type nitric oxide synthase in keratinocytes. J Dermatol 1997; 24:80–87.
151. Pahan K, Raymond JR, Singh I. Inhibition of phosphatidyl-inositol 3-kinase induces nitric-oxide synthase in lipopolysaccharide- or cytokine-stimulated C6 glial cells. J Biol Chem 1999; 274:7528–7536.
152. Yim CY, McGregor JR, Kwon OD, Bastian NR, Rees M, Mori M, Hibbs JB, Samlowski WE. Nitric oxide synthesis contributes to IL-2-induced antitumor responses against intraperitoneal Meth A tumor. J Immunol 1995; 155:4382–4390.
153. Bruch-Gerharz D, Fehsel K, Suschek C, Michel G, Ruzicka T, Kolb-Bachofen VA. A proinflammatory activity of interleukin 8 in human skin: Expression of the inducible nitric oxide synthase in psoriatic lesions and cultured keratinocytes. J Exp Med 1996; 184:2007–2012.
154. Ormerod AD, Dwyer CM, Reid A, Copeland P, Thompson WD. Inducible nitric oxide synthase demonstrated in allergic and irritant contact dermatitis. Acta Dermatol Venereol 1997; 77:436–440.
155. Teixeira MM, Williams TJ, Hellewell PG. Role of prostaglandins and nitric oxide in acute inflammatory reactions in guinea-pig skin. Br J Pharmacol 1993; 110:1515–1521.
156. Rattan V, Sultana C, Shen Y, Kalra VK. Oxidant stress-induced transendothelial migration of monocytes is linked to phosphorylation of PECAM-1. Am J Physiol 1997; 273:E453–E461.
157. Lo SK, Janakidevi K, Lai L, Malik AB. Hydrogen peroxide-induced increase in endothelial adhesiveness is dependent on ICAM-1 activation. Am J Physiol 1993; 264:L406–L412.
158. Kubes P. Polymorphonuclear leukocyte-endothelium interactions: A role for pro-inflammatory and anti-inflammatory molecules. Can J Physiol Pharmacol 1993; 71:88–97.
159. Johnson JL, Moore EE, Tamura DY, Zallen G, Biffl WL, Silliman CC. Interleukin-6 augments neutrophil cytotoxic potential via selective enhancement of elastase release. J Surg Res 1998; 76:91–94.
160. Vondracek J. Effects of recombinant rat tumor necrosis factor-alpha and interferon-gamma on the respiratory burst of rat polymorphonuclear leukocytes in whole blood. Folia Biologica 1997; 43:115–121.
161. Belenky SN, Robbins RA, Rennard SI, Gossman GL, Nelson KJ, Rubinstein I. Inhibitors of nitric oxide synthase attenuate human neutrophil chemotaxis in vitro. J Lab Clin Med 1993; 122:388–394.
162. Su Z, Ishida H, Fukuyama N, Todorov R, Genka C, Nakazawa H. Peroxynitrite is not a major mediator of endothelial cell injury by activated neutrophils in vitro. Cardiovasc Res 1998; 39:485–491.
163. Goode HF, Webster NR, Howdle PR, Walker BE. Nitric oxide production by human peripheral blood polymorphonuclear leucocytes. Clin Sci 1994; 86:411–415.
164. Yan L, Vandivier RW, Suffredini AF, Danner RL. Human polymorphonuclear leukocytes lack detectable nitric oxide synthase activity. J Immunol 1994; 153:1825–1834.
165. Maillard JL, Favreau C, Reboud-Ravaux M. Role of monocyte/macrophage derived matrix-metalloproteinases (gelatinases) in prolonged skin inflammation. Clin Chim Acta 1995; 233: 61–74.
166. Kahari VM, Saarialho-Kere U. Matrix metalloproteinases in skin. Exp Dermatol 1997; 6: 199–213.

167. Shabani F, McNeil J, Tippett L. The oxidative inactivation of tissue inhibitor of metalloproteinase-1 (TIMP-1) by hypochlorous acid (HOCI) is suppressed by anti-rheumatic drugs. Free Rad Res 1998; 28:115–123.
168. Zaw KK, Yokoyama Y, Ishikawa O, Miyachi Y. The reactive oxygen species (ROS) regulate the gene expressions involved in the biosynthesis and biodegradation of collagens in three-dimensional culture of normal human dermal fibroblasts. J Invest Dermatol 1999; 112:609.
169. Brenneisen P, Briviba K, Wlaschek M, Wenk J, Scharffetter-Kochanek K. Hydrogen peroxide ($H_2O_2$) increases the steady-state mRNA levels of collagenase/MMP-1 in human dermal fibroblasts. Free Rad Biol Med 1997; 22:515–524.
170. Kawaguchi Y, Tanaka H, Okada T, Konishi H, Takahashi M, Ito M, Asai J. The effects of ultraviolet A and reactive oxygen species on the mRNA expression of 72-kDa type IV collagenase and its tissue inhibitor in cultured human dermal fibroblasts. Arch Dermatol Res 1996; 288:39–44.
171. Maeda H, Okamoto T, Akaike T. Human matrix metalloprotease activation by insults of bacterial infection involving proteases and free radicals. Biol Chem 1998; 379:193–200.
172. Wang M, Qin X, Mudgett JS, Ferguson TA, Senior RM, Welgus HG. Matrix metalloproteinase deficiencies affect contact hypersensitivity: stromelysin-1 deficiency prevents the response and gelatinase B deficiency prolongs the response. Proc Natl Acad Sci USA 1999; 96:6885–6889.
173. Kalish RS, Johnson KL. Enrichment and function of urushiol (poison ivy) specific T-lymphocytes in lesions of allergic contact dermatitis to urushiol. J Immunol 1990; 145:3706–3713.
174. Marchal G, Seman M, Milon G, Truffa-Bachi P, Zilberfarb V. Local adoptive transfer of skin delayed-type hypersensitivity initiated by a single T lymphocyte. J Immunol 1982; 129: 954–958.
175. Bianchi AT, Hooijkaas H, Benner R, Tees R, Nordin AA, Schreier MH. Clones of helper T cells mediate antigen-specific, $H_2$-restricted DTH. Nature 1981; 290:62–63.
176. Elsner J, Kapp A. Activation and modulation of eosinophils by chemokines. Allergologie 2000; 23:59–72.
177. Schroder JM, Mochizuki M. The role of chemokines in cutaneous allergic inflammation. Biol Chem 1999; 380:889–896.
178. Litchfield TM, Smith CH, Atkinson BA, Norris PG, Elliott P, Haskard DO, Lee TH. Eosinophil infiltration into human skin is antigen-dependent in the late-phase reaction. Br J Dermatol 1996; 134:997–1004.
179. Nathan CF, Murray HW, Cohn ZA. The macrophage as an effector cell. N Engl J Med 1980; 303:622–626.
180. Gemsa D, Leser HG, Seitz M, Debatin M, Bärlin E, Deimann W, Kramer W. Cells in inflammation. The role of macrophages in inflammation. Agents Action 1982; 11(suppl):93–114.
181. Nakagawara A, Nathan CF, Cohn ZA. Hydrogen peroxide metabolism in human monocytes during differentiation in-vitro. J Clin Invest 1981; 68:1243–1252.
182. Willis CM, Reiche L, Wilkinson JD. Immunocytochemical demonstration of reduced Cu,Zn-superoxide dismutase levels following topical application of dithranol and sodium lauryl sulphate: an indication of the role of oxidative stress in acute irritant contact dermatitis. Eur J Dermatol 1998; 8:8–12.
183. Miyachi Y, Uchida K, Komura J, Asada Y, Niwa Y. Auto oxidative damage in cement dermatitis. Arch Dermatol Res 1985; 277:288–292.
184. Sharkey P, Eedy DJ, Burrows D, McCaigue MD, Bell AL. A possible role for superoxide production in the pathogenesis of contact dermatitis. Acta Dermato-Venereol 1991; 71:156–159.
185. Finnen MJ, Lawrence CM, Shuster S. Inhibition of dithranol inflammation by free radical scavengers. Lancet 1984; 1129–1130.

186. Fuchs J, Milbradt R. Antioxidant inhibition of skin inflammation induced by reactive oxidants: Evaluation of the redox couple dihydrolipoate/lipoate. Skin Pharmacol 1994; 7:278–284.
187. Colardyn F, De Keyser H, Ringoir S, De Bersaques J. Clinical observation and therapy of injuries with vesicants. J Toxicol Clin Exp 1986; 6:237–246.
188. Somani SM, Babu SR. Toxicodynamics of sulfur mustard. Int J Clin Pharmacol Ther Toxicol 1989; 27:419–435.
189. Miyachi Y, Imamura S, Niwa Y, Tokura Y, Takigawa M. Mechanisms of contact photosensitivity in mice. VI. Oxygen intermediates are involved in contact photosensitization but not in ordinary contact sensitization. J Invest Dermatol 1986; 86:26–28.
190. Van Den Broeke LT, Graslund A, Larsson PH, Nilsson JL, Wahlberg JE, Scheynius A, Karlberg AT. Free radicals as potential mediators of metal allergy: effect of ascorbic acid on lymphocyte proliferation and IFN-γ production in contact allergy to $Ni^{2+}$ and $Co^{2+}$. Acta Dermatol Venereol 1998; 78:95–98.
191. Fuchs J, Packer L. Antioxidant protection from solar-simulated radiation induced suppression of contact hypersensitivity to the recall antigen nickel sulfate in human skin. Free Radic Biol Med 1999; 27:422–427.
192. Memon AA, Molokhia MM, Friedmann PS. The inhibitory effects of topical chelating agents and antioxidants on nickel-induced hypersensitivity reactions. J Am Acad Dermatol 1994; 30:560–565.
193. Mastrandrea E, Cadario G, Bedello PG, Nicotra MR, Natali PG. Expression of T-lineage early developmental markers by cells establishing atopic dermatitis skin infiltrates. J Invest Allergol Clin Immunol 1998; 8:359–364.
194. Jung K, Linse F, Heller R, Moths C, Goebel R, Neumann C. Adhesion molecules in atopic dermatitis: VCAM-1 and ICAM-1 expression is increased in healthy-appearing skin. Allergy 1996; 51:452–460.
195. Ring J. Haut und Umwelt. Hautarzt 1993; 44:625–635.
196. Niwa Y, Iizawa O. Abnormalities in serum lipids and leucocyte superoxide dismutase and associated cataract formation in patients with atopic dermatitis. Arch Dermatol 1994; 130: 1387–1392.
197. Werfel T, Kapp A. Atopic dermatitis: Asthma of the skin? Atemwegs Lungenkrankheiten 1997; 23:592–599.
198. Wuethrich B. Why does the number of atopic patients increase? Ann Dermatol Venereol 1997; 124:831–837.
199. Traupe H, Menge G, Kandt I, Karmaus W. Higher frequency of atopic dermatitis and decrease in viral warts among children exposed to chemicals liberated in a chemical accident in Frankfurt, Germany. Dermatology (Basel) 1997; 195:112–118.
200. Schultz-Larsen F. Atopic dermatitis: a genetic-epidemiologic study in a population based twin sample. J Am Acad Dermatol 1993; 24:719–723.
201. Dotterud LK, Kvammen B, Bolle R, Falk ES. A survey of atopic diseases among school children in Sor-Varanger community: Possible effects of subarctic climate and industrial pollution from Russia. Acta Dermatol Venereol 1994; 74:124–128.
202. Shamov BA, Malanicheva TG, Zakiev RZ. Air pollution in residential districts and incidence of allergodermatoses in adolescents. Vest Dermatol Venereol 1997; 1:10–12.
203. Ronchetti R, Bonci E, Cutrera R, De Castro G, Indinnimeo L, Midulla F, Tancredi G, Martinez FD. Enhanced allergic sensitization related to parental smoking. Arch Dis Child 1992; 67:496–500.
204. Schaefer T, Dirschedl P, Kunz B, Ring J, Ueberla K. Maternal smoking during pregnancy and lactation increases the risk for atopic eczema in the offspring. J Am Acad Dermatol 1997; 36:550–556.
205. Kramer U, Koch T, Ranft U, Ring J, Behrendt H. Traffic-related air pollution is associated with atopy in children living in urban areas. Epidemiology 2000; 11:64–70.

206. Haustein UF, Ziegler V. Environmentally induced systemic sclerosis like disorders. Int J Dermatol 1985; 24:147–151.
207. Czirjak L, Danko K, Schlammadinger J, Suranyi P, Tamasi L, Szegedi GY. Progressive systemic sclerosis occurring in patients exposed to chemicals. Int J Dermatol 1987; 26:374–378.
208. Flindt-Hansen H, Isager H. Scleroderma after occupational exposure to trichlorethylene and trichlorethane. Acta Derm Venereol 1987; 67:263–264.
209. Espriella DJ, Cricks B. Scleroderma and environment. I. Scleroderma and sclerodermiform states induced by silica and chemical agents or drugs. Ann Dermatol Venereol 1991; 118: 948–953.
210. Murrell DF. A radical proposal for the pathogenesis of scleroderma. J Am Acad Dermatol 1993; 28:78–85.
211. Zangerle R, Wachter H, Fuchs D. Activated macrophages and radicals in scleroderma. J Am Acad Dermatol 1994; 30:1045–1047.
212. Parke DV, Sapta A. Chemical toxicity and reactive oxygen species. Int J Occupat Med Environ Health 1996; 9:331–340.
213. Casciola-Rosen L, Wigley F, Rosen A. Scleroderma autoantigens are uniquely fragmented by metal-catalyzed oxidation reactions: implications for pathogenesis. J Exp Med 1997; 185: 71–79.
214. Yamamoto T, Takagawa S, Katayama I, Hamazaki Y, Shinkai H, Nishioka K. Animal model of sclerotic skin. I: Local injections of bleomycin induce sclerotic skin mimicking scleroderma. J Invest Dermatol 1999; 112:456–462.
215. Mountz JD, Downs-Minor MB, Turner R, Tohams MB, Richaros F, Pisko E. Bleomycin induced cutaneous toxicity in the rat: analysis for histopathology and ultrastructure compared with progressive systemic sclerosis (scleroderma). Br J Dermatol 1983; 108:679–686.
216. Oberley LW, Buettner GR. The production of hydroxyl radicals by bleomycin and iron (II). FEBS Lett 1979; 97:47–49.
217. Ekimoto H, Kuramochi H, Takahashi K, Matsuda A, Umezawa H. Kinetics of the reaction of bleomycine Fe(II)-02 complex with DNA. J Antibio 1980; 33:426–434.
218. Gutteridge JMC, Ziai-Chang F. Protection of iron catalyzed free radical damage to DNA and lipids by copper (II) bleomycin. Biochem Biophys Res Commun 1981; 99:1354–1360.
219. Leroy EC. Pathogenesis of scleroderma (systemic sclerosis). J Invest Dermatol 1982; 79(suppl 1):87–89.
220. Leroy EC, Kahaleh MB, Mercuric S. A fibroblast mitogen present in scleroderma but not control sera. Rheumatol Int 1983; 3:35–38.
221. Leroy EC, Smith EA, Kahaleh B, Trojanowska M, Silver RM. A strategy for determining the pathogenesis of systemic sclerosis. Is transforming growth factor $\beta$ the answer? Arthritis Rheum 1989; 32:817–825.
222. Krieg T, Aumailley M. Connective tissue diseases in the skin. J Dermatol 1990; 17:67–84.
223. Yamamoto T, Takagawa S, Katayama I, Nishioka K. Anti-sclerotic effect of transforming growth factor-beta antibody in a mouse model of bleomycin-induced scleroderma. Clin Immunol 1999; 92:6–13.
224. Yamamoto T, Takagawa S, Katayama I, Mizushima Y, Nishioka K. Effect of superoxide dismutase on bleomycin induced dermal sclerosis: implications for the treatment of systemic sclerosis. J Invest Dermatol 1999; 113:843–847.
225. Allen RG. Oxidative stress and superoxide dismutase in development, aging and gene regulation. Age 1998; 21:47–76.
226. Chojkier M, Houglum K, Solis-Herruzo J, Brenner DA. Stimulation of collagen gene expression by ascorbic acid in cultured human fibroblasts. J Biol Chem 1989; 264:16957–16962.
227. Brenneisen P, Briviba K, Wlaschek M, Wenk J, Scharfetter-Kochanek K. Hydrogen peroxide ($H_2O_2$) increases the steady state mRNA level of collagenase/MMP-1 in human dermal fibroblasts. Free Radic Biol Med 1997; 22:515–524.

228. Malten KE, Seutter E, Hara I, Nakajima T. Occupational vitiligo due to para tertiary butylphenol and homologues. Trans St Johns Hosp Dermatol Soc 1971; 57:115–134.
229. Stevenson CL. Occupational vitiligo: Clinical and epidemiological aspects. Br J Dermatol 1981; 105(suppl 21):51–56.
230. Spencer MC. Hydroquinone bleaching. Arch Dermatol 1961; 84:131–134.
231. Snell RS. Monobenzylether of hydroquinone. Arch Dermatol 1964; 90:63–70.
232. Bleehen SS, Pathak MA, Hori Y, Fitzpatrick TB. Depigmentation of skin with 4-isopropylcatechol, mercaptoamines and other compounds. J Invest Dermatol 1968; 50:103–117.
233. Riley PA. Hydroxyanisole depigmentation: in-vitro studies. J Pathol 1969; 97:193–206.
234. Riley PA. Hydroxyanisole depigmentation: in-vivo studies. J Pathol 1969; 97:185–191.
235. Kahn G. Depigmentation caused by phenolic detergent germicides. Arch Dermatol 1970; 102:177–187.
236. Gellin GA, Possick PA, Peronne VB. Depigmentation from 4-tertiary butyl catechol—an experimental study. J Invest Dermatol 1970; 55:190–197.
237. Vollum DI. Hypomelanosis from antioxidant in polyethylene film. Arch Dermatol 1971; 104: 70–72.
238. Bentley-Phillips B, Bayles MAH. Cutaneous reactions to topical application of hydroquinone. S Afr Med J 1975; 1391–1395.
239. Shelley WB. P-cresol: cause of ink-induced hair depigmentation in mice. Br J Dermatol 1974; 90:169–174.
240. Penney KB, Smith CJ, Allen JC. Depigmenting action of hydroquinone depends on disruption of fundamental cell processes. J Invest Dermatol 1984; 82:308–310.
241. Hilger R, Fukuyama K, Zackheim HS, Epstein JH. Increased melanocytes in hairless mice following topical treatment with carmustine (BCNU) and nitrogen mustard (NM). Clin Res 1974; 22:159A.
242. Bork K. Medikamentöse Zahnverfärbungen. In: Bork K, ed. Kutane Arzneimittelnebenwirkungen. Stuttgart: Schattauer Verlag, 1985:297–302.
243. Hearing VJ, Jimenez M. Mammalian tyrosinase—The critical regulatory control point in melanocyte pigmentation. Int J Biochem 1987; 19:1141–1147.
244. Riley PA. Melanin. Int J Biochem Cell Biol 1997; 29:1235–1239.
245. Yamazaki I, Mason HS, Piette L. Identification, by electron paramagnetic resonance spectroscopy, of free radicals generated from substrates by peroxidase. J Biol Chem 1960; 235:2444–2449.
246. Riley PA. Acquired hypomelanosis. Br J Dermatol 1971; 84:290–293.
247. Bolognia LJ, Pawelek JM. Biology of hypopigmentation. J Am Acad Dermatol 1988; 19: 217–255.
248. Yohn JJ, Norris DA, Yrastorza DG, Buno IJ, Leff JA, Hake SS. Disparate antioxidant enzyme activities in cultured human cutaneous fibroblasts, keratinocytes and melanocytes. J Invest Dermatol 1991; 97:405–409.
249. Fowler JF. Acne, folliculitis, and chloracne. In: Adams RM, ed. Occupational Skin Disease, 3rd ed. Philadelphia: Saunders, 1999:135–141.
250. Suskind RR. The ''hallmark'' of dioxin intoxication. Scand J Work Environ Health 1985; 11:165–168.
251. Caputo R, Monti M, Ermacora E, Carminati G, Gelmetti C, Gianotti R, Gianni E, Puccinelli V. Cutaneous manifestations of tetrachlorodibenzo-p-dioxin in children and adolescents. J Am Acad Dermatol 1988; 19:812–819.
252. Neuberger M, Kundi M, Jager R. Chloracne and morbidity after dioxin exposure (preliminary results). Toxicol Lett 1998; xx:96–97.
253. Mukerjee D. Health impact of polychlorinated dibenzo-p-dioxins: a critical review. J Air Waste Manag Assoc 1998; 48:157–165.
254. Okey AB, Riddick DS, Harper PA. The Ah receptor: Mediator of the toxicity of 2,3,7,8-tetrachlorodibenzo-p-dioxin (TCDD) and related compounds. Toxicol Lett 1994; 70:1–22.

255. Berkers JA, Hassing I, Spenkelink B, Brouwer A, Blauboer BJ. Interactive effects of 2,3,7,8-tetrachlorodibenzo-p-dioxin and retinoids on proliferation and differentiation in cultured human keratinocytes: quantification of cross-linked envelope formation. Arch Toxicol 1995; 69:368–378.
256. Panteleyev AA, Thiel R, Wanner R, Zhang J, Roumak VS, Paus R, Neubert D, Henz BM, Rosenbach T. 2,3,7,8-tetrachlorodibenzo-p-dioxin (TCDD) affects keratin 1 and keratin 17 gene expression and differentially induces keratinization in hairless mouse skin. J Invest Dermatol 1997; 108:330–335.
257. Sweeney GD, Jones KG, Cole FM, Basford D, Krestynski F. Iron deficiency prevents liver toxicity of 2,3,7,8-tetrachlorodibenzo-p-dioxin. Science 1979; 204:332–335.
258. Poland A, Knutson JC. 2,3,7,8-tetrachlorodibenzo-p-dioxin and related halogenated aromatic hydrocarbons: examination of the mechanism of toxicity. Ann Rev Pharmacol Toxicol 1982; 22:517–554.
259. Hassan MQ, Mohammadpour H, Hermansky SJ, Murray WJ, Stohs SJ. Comparative effects of BHA and ascorbic acid on the toxicity of 2,3,7,8-tetrachlorodibenzo-p-dioxin (TCDD) in rats. Gen Pharmacol 1987; 18:547–550.
260. Nohl H, Silva DD, Summer KH. 2,3,7,8-tetrachloro-dibenzo-p-dioxin induces oxygen activation associated with cell respiration. Free Rad Biol Med 1989; 6:369–374.
261. Stohs SJ. Oxidative stress induced by 2,3,7,8-tetrachloro-dibenzo-p-dioxin (TCDD). Free Rad Biol Med 1990; 9:79–90.
262. Al-Bayati ZAF, Stohs SJ. The possible role of phospholipase A2 in hepatic microsomal lipid peroxidation induced by 2,3,7,8-tetrachlorodibenzo-p-dioxin in rats. Arch Environ Contam Toxicol 1991; 20:361–365.
263. Shara MA, Dickson PH, Bagchi D, Stohs SJ. Excretion of formaldehyde, malondialdehyde, acetaldehyde and acetone in the urine of rats in response to 2,3,7,8-tetrachlorodibenzo-p-dioxin, paraquat, endrin and carbon tetrachloride. J Chromatogr 1992; 576:221–233.
264. Bagchi M, Stohs SJ. In vitro induction of reactive oxygen species by 2,3,7,8-tetrachlorodibenzo-p-dioxin, endrin, and lindane in rat peritoneal macrophages, and hepatic mitochondria and microsomes. Free Radic Biol Med 1993; 14:11–18.
265. Alsharif NZ, Lawson T, Stohs SJ. Oxidative stress induced by 2,3,7,8-tetrachlorodibenzo-p-dioxin is mediated by the aryl hydrocarbon (Ah) receptor complex. Toxicology 1994; 92: 39–51.
266. Albro PW, Corbett JT, Harris M, Lawson LD. Effects of 2,3,7,8-tetrachlorodibenzo-p-dioxin on lipid profiles in tissues of the Fischer rat. Chem Biol Interact 1978; 23:315–320.
267. Hassoun EA, Stohs SJ. TCDD, endrin and lindane induced oxidative stress in fetal and placental tissues of C57BL/6J and DBA/2J mice. Comp Biochem Physiol C Pharmacol Toxicol Endocrinol 1996; 11:11–18.
268. Hassoun EA, Wilt SC, Devito MJ, Van Birgelen A, Alsharif NZ, Birnbaum LS, Stohs SJ. Induction of oxidative stress in brain tissues of mice after subchronic exposure to 2,3,7,8-tetrachlorodibenzo-p-dioxin. Toxicol Sci 1998; 42:23–27.
269. Ames BN, Gold LS. Environmental pollution, pesticides, and the prevention of cancer: Misconceptions. FASEB J 1997; 11:1041–1052.
270. Ames BN, Gold LS. The causes and prevention of cancer: The role of environment. Biotherapy 1998; 11:205–220.
271. Rugo HS, Fischman ML. Occupational cancer. In: LaDou J, ed. Occupational and Environmental Medicine. Stamford: Appleton & Lange, 1997:235–271.
272. Guyton KZ, Kensler TW. Oxidative mechanisms in carcinogenesis. Br Med Bull 1993; 49: 523–544.
273. Trush MA, Kensler TW. An overview of the relationship between oxidative stress and chemical carcinogenesis. Free Rad Biol Med 1991; 10:201–209.
274. Pryor WA. Cigarette smoke radicals and the role of free radicals in chemical carcinogenicity. Environ Health Perspect 1997; 105(suppl 4):875–882.

275. Cerutti PA. Prooxidant states and promotion. Science 1985; 227:375–381.
276. Troll W, Wiesner R. The role of oxygen radicals as a possible mechanism of tumor promotion. Ann Rev Pharmacol Toxicol 1985; 25:509–528.
277. Marks F, Fürstenberger G. Tumor promotion in skin: Are active oxygen species involved? In: Sies H, ed. Oxidative Stress. London: Academic Press, Inc, 1985:437–475.
278. Kensler TW, Taffe BG. Free radicals and tumor promotion. Adv Free Rad Biol Med 1986; 2:347–387.
279. Crawford D, Zbinden I, Amstad P, Cerutti P. Oxidant stress induces the proto-oncogenes c-fos and c-myc in mouse epidermal cells. Oncogene 1988; 3:27–32.
280. Sun Y. Free radicals, antioxidant enzymes and carcinogenesis. Free Rad Biol Med 1990; 8: 583–597.
281. Cheng KC, Cahill DS, Kasa H, Nishimura S, Loeb LA. 8-Hydroxyguanosine, an abundant form of oxidative DNA damage, causes G-T and A-C substitutions. J Biol Chem 1992; 267: 166–172.
282. Agarwal R, Mukhtar H. Oxidative stress in skin chemical carcinogenesis. In: Fuchs J, Packer L, eds. Oxidative Stress in Dermatology. New York: Marcel Dekker, Inc, 1993:207–241.
283. Feig DI, Reid TM, Loeb LA. Reactive oxygen species in tumorigenesis. Cancer Res 1994; 54(suppl 7):18902–18904.
284. Kensler T, Guyton K, Egner P, McCarthy T, Lesko S, Akman S. Role of reactive intermediates in tumor promotion and progression. Prog Clin Biol Res 1995; 391:103–116.
285. Slaga TJ. Tumor promotion and-or enhancement models. Int J Toxicol 1998; 17(suppl 3): 109–127.
286. Greenberg ER, Baron JA, Stukel TA, Stevens NM, Mandel JS, Spencer SK, Elias PM, Lowe N, Nierenberg DW, Bayrd G, Vance JC, Freeman DH, Clendenning WE, Kwan T, and the alpha-tocopherol, beta-carotene cancer prevention study group. A clinical trial of beta-carotene to prevent basal-cell and squamous-cell cancer on the skin. N Engl J Med 1990; 323:789–795.
287. Hennekens CH, Buring JE, Manson JE, Stampfer M, Rosner B, Cook NR, Belanger C, LaMotte F, Gaziano JM, Ridker PM. Lack of effect of long-term supplementation with beta carotene on the incidence of malignant neoplasms and cardiovascular disease. N Engl J Med 1996; 334:1145–1149.
288. Nelson DR, Kamataki T, Waxman DJ, Guengerich FP, Estabrook RW, Feyereisen R, Gonzales FJ, Coon MJ, Gunsalus IC, Gotoh O, Okuda K, Nebert DW. The P450 superfamily: Update on new sequences, gene mapping, accession numbers, early trivial names of enzymes, and nomenclature. DNA Cell Biol 1993; 12:1–51.
289. Simpson AECM. The cytochrome P450 4 (CYP4) family. Gen Pharmacol 1997; 28:351–359.
290. Schrenk D. Impact of dioxin-type induction of drug-metabolizing enzymes on the metabolism of endo- and xenobiotics. Biochem Pharmacol 1998; 55:1155–1162.
291. Dogra SC, Whitelaw ML, May BK. Transcriptional activation of cytochrome P450 genes by different classes of chemical inducers. Clin Exp Pharmacol Physiol 1998; 25:1–9.
292. Fouchecourt MD, Riviere JL. Activities of cytochrome P450-dependent monooxygenases and antioxidant enzymes in different organs of Norway rats (Rattus norvegicus) inhabiting reference and contaminated sites. Chemosphere 1995; 31:4375–4386.
293. Bast A. Is formation of reactive oxygen by cytochrome P450 perilous and predictable. Trends Pharmacol Sci 1986; 7:266–270.
294. Buters JTM, Sakai S, Richter T, Pineau T, Alexander DL, Savas U, Doehmer J, Ward JM, Jefcoate CR, Gonzalez FJ. Cytochrome P450 CYP1B1 determines susceptibility to 7,12-dimethylbenz(a)anthracene-induced lymphomas. Proc Natl Acad Sci USA 1999; 96:1977–1982.
295. Doehmer J, Goeptar AR, Vermeulen NE. Cytochromes P450 and drug resistance. Cytotechnology 1993; 12:357–366.

296. Paolini M, Pozetti L, Pedulli GF, Cipollne M, Mesirca R, Cantelli-Forti G. Paramagnetic resonance in detecting carcinogenic risk from cytochrome P450 overexpression. J Invest Med 1996; 44:470–473.
297. Merk HF, Jugert FK, Frankenberg S. Biotransformations in the skin. In: Marzulli FN, Maibach HI, eds. Dermatotoxicology. Washington, DC: Taylor & Francis, 1996:61–73.
298. Henderson CJ, Smith AG, Ure J, Brown K, Bacon EJ, Wolf CR. Increased skin tumorigenesis in mice lacking pi class glutathione S-transferases. Proc Natl Acad Sci USA 1998; 95:5275–5280.
299. Murray GI, Taylor MC, McFadyen MC, McKay JA, Greenlee WF, Burke MD, Melvin WT. Tumor-specific expression of cytochrome P450 CYP1B1. Cancer Res 1997; 57:3026–3031.
300. Lear JT, Heagerty AH, Smith A, Bowers B, Payne CR, Smith CA, Jones PW, Gilford J, Yengi L, Alldersea J. Multiple cutaneous basal cell carcinomas: glutathione S-transferase (GSTM1, GSTT1) and cytochrome P450 (CYP2D6, CYP1A1) polymorphisms influence tumour numbers and accrual. Carcinogenesis 1996; 17:1891–1896.
301. Lear JT, Smith AG, Bowers B, Heagearty AH, Jones PW, Gilford J, Alldersea J, Strange RC, Fryer AA. Truncal tumor site is associated with high risk of multiple basal cell carcinoma and is influenced by glutathione S-transferase, GSTT1, and cytochrome P450, CYP1A1 genotypes, and their interaction. J Invest Dermatol 1997; 108:519–522.
302. Lear JT, Smith AG, Strange RC, Fryer AA. Detoxifying enzyme genotypes and susceptibility to cutaneous malignancy. Br J Dermatol 2000; 142:8–15.
303. Bishop MJ. The molecular genetics of cancer. Science 1987; 235:305–311.
304. Zetter BR. Angiogenesis and tumor metastasis. Ann Rev Med 1998; 49:407–424.
305. Sang QX. Complex role of matrix metalloproteinases in angiogenesis. Cell Res 1998; 8:171–177.
306. Coussens LM, Werb Z. Matrix metalloproteinases and the development of cancer. Chem Biol 1996; 3:895–904.
307. Basset-Seguin N, Dereure O, Guillot B. Genetic bases of cutaneous tumors. Ann Dermatol Venereol 1995; 122:217–225.
308. Daya-Grosjean L, Sumaz N, Sarasin A. The specificity of p53 mutation spectra in sunlight induces human cancers. J Photochem Photobiol B Biol 1995; 28:115–124.
309. Harris CC. p53 tumor suppressor gene: at the crossroads of molecular carcinogenesis, molecular epidemiology, and cancer risk assessment. Environ Health Perspect 1996; 104(suppl 3): 435–439.
310. Wang XJ, Greenhalgh DA, Jiang A, He D, Zhong L, Medina D, Brinkley BR, Roop DR. Expression of a p53 mutant in the epidermis of transgenic mice accelerates chemical carcinogenesis. Oncogene 1998; 17:35–45.
311. Mepla C, Richard MJ, Hainut P. Redox signalling and transition metals in the control of the p53 pathway. Biochem Pharmacol 2000; 59:25–33.
312. Hainaut P, Milner J. Redox modulation of p53 conformation and sequence specific DNA binding in vitro. Cancer Res 1993; 53:4469–4473.
313. Rainwater R, Parks D, Anderson ME, Tegtmeyer P, Mann K. Role of cysteine residues in regulation of p53 function. Mol Cell Biol 1995; 15:3892–3903.
314. Verhaegh G, Richard MJ, Hainaut P. Regulation of p53 by metal ions and antioxidants. Dithiocarbamate down regulates p53 DNA binding activity by increasing the intracellular level of copper. Mol Cell Biol 1997; 17:5699–5706.
315. Parks D, Bolinger R, Mann K. Redox state regulates binding of p53 to sequence specific specific DNA, but not to non-specific or mismatched DNA: Nucl Acid Res 1997; 25:1289–1295.
316. Liu M, Pelling JC, Ju J, Chu E, Brash DE. Antioxidant action of p53 mediated apoptosis. Cancer Res 1998; 58:1723–1729.
317. Haugen A. Progress and potential of genetic susceptibility to environmental toxicants. Scand J Work Environ Health 1999; 25:537–540.

318. Roshdy I, German Patent DE 3,522,572, February 1987.
319. Nagy CF, Quick TW, Shaapiro SS. US Patent 5,252,604, October 1993.
320. Duffy J. US Patent 5,516,793, May 1996.
321. Hall BJ, Baur JA, Deckner GE. US Patent 5,545,407, August 1996.
322. Roederer M, Staal FTJ, Raju PA, Ela SW, Herzenberg LA. Cytokine stimulated human immunodeficiency virus replication is inhibited by *N*-acetyl-L-ysteine. Proc Natl Acad Sci USA 1990; 87:4884–4888.
323. Staal FJT, Roederer M, Herzenberg LA, Herzenberg LA. Intracellular thiols regulate activation of nuclear factor κB and transcription of human immunodeficiency virus. Proc Natl Acad Sci USA 1990; 87:9943–9947.
324. Ikeda M, Schroeder KK, Mosher LB, Woods CW, Akeson AL. Suppressive effect of antioxidants on intracellular adhesion molecule-1 (ICAM-1) expression in human epidermal keratinocytes. J Invest Dermatol 1994; 103:791–796.

# 13

# Dermatotoxicology of Environmental and Occupational Chemical Hazards: Agents and Action

**JÜRGEN FUCHS**

*J.W. Goethe University, Frankfurt, Germany*

## I. INTRODUCTION

It is becoming increasingly clear that free radicals and reactive oxygen species (ROS) are formed and play a role in the toxic manifestations of many xenobiotics and pollutants (1–8), among which are several industrial chemicals (9). Skin as the outermost barrier of the body must constantly withstand direct exposure to environmental pollutants and occupational hazards. Prooxidant dermatotoxic chemicals comprise metal salts, phenols, quinones, primary amines, azo-compounds, hydroperoxides, peroxides, alkylating agents, polyhalogenated and polycyclic aromatic hydrocarbons, some pesticides, and organic solvents. There is cumulative experimental evidence that suggests, but does not prove, that redox-sensitive events and oxidative stress are a biologically significant factor in the toxicity of some environmental and occupational chemical hazards. However, the clinical relevance of these studies is often difficult to substantiate conclusively. In some conditions outlined in this chapter, oxidative stress induced by environmental and occupational xenobiotics may represent a primary mechanism of skin injury, while in other cases it represents an epiphenomenon, or is simply a consequence of skin damage. This article focuses on the molecular actions of selected dermatotoxic xenobiotics and should be seen and understood in conjunction with Chapter 12. No attempt was made to be exhaustive in this very large body of literature.

## II. METALS AND METALLOIDS

Due to industrial activity, huge amounts of metals and metalloids are released into the atmosphere or redistributed into different environmental compartments. For instance, thousands of tons of oxidized transition metals are released in the fly ash generated during fossil fuel burning (10). Vanadium and nickel are the main metallic elements found in natural oil and they account for about three-quarters of the fly ash mass (11). Coal contains significant amounts of iron, copper, and manganese in addition to vanadium. The toxicities produced by metal ions mainly involve neurotoxicity, hepatotoxicity, nephrotoxicity, and dermatotoxicity. The mechanisms of metal ion–induced dermatotoxicity are not well understood, but experimental evidence points to some role of oxidative stress. Redox cycling is a characteristic of transition metals (12) and Fenton-like production of ROS appear to be involved for iron-, copper-, chromium-, and vanadium-mediated tissue damage. Inorganic salts of mercury, nickel, lead, arsenic, cadmium, and gold deplete cellular thiols and thus have the ability to affect redox-sensitive signal transduction pathways and gene expression. Excellent reviews and books have presented a comprehensive overview on the toxic effects of metals on the skin (13,14). Skin manifestations of metal toxicity mainly comprise irritant contact dermatitis (ICD) and allergic contact dermatitis (ACD). Metal ions are important causes of human allergies and the majority of metals induce allergic reactions in the skin. Through their chelating abilities, mercury (II), palladium (II), platinum (II), and gold (III) may directly form protein adducts and thus alter self-proteins (15). In particular, nickel (II), cobalt (II), chromium (III,VI), mercury (I,II), gold (III), platinum (II), and palladium (II) cause ACD in human skin. In rare cases, immediate contact urticaria (ICU), systemic allergic reactions (SAR), or granulomatous reactions to metals (e.g., aluminium, beryllium) have been observed. At least nickel, cadmium, lead, chromium, beryllium, and arsenic are proven or putative human carcinogens and a possible role of oxidative damage in metal-induced carcinogenesis is emerging (12,16–18).

### A. Chromium (Cr)

Chromium is an essential trace element ubiquitous in the environment. It is a widely used industrial chemical and is found in cement, steel, alloy, paints, metal finishes, leather tanning, pigment production, and wood treatment. Chromium occurs in the workplace primarily in the valence form Cr(VI) and Cr(III), while Cr(V), Cr(IV), and Cr(II) are rather unstable and transient states. In its hexavalent state, chromium is cytotoxic, genotoxic, and carcinogenic in animals. The reduction of Cr(VI) to its lower oxidation states and related free radical reactions was suggested to play an important role in chromium toxicity (19). Cr(VI) (20,21), Cr(IV) (22,23), and Cr(III) (24,25) can promote formation of free radical species in noncellular systems. Cr(VI) has been shown to generate ROS in cultured Jurkat cells (26) and in rodents in vivo (27). Cr(VI) is reduced by cellular reductants to the reactive Cr(V) species, which readily reacts in a Fenton-like reaction with hydrogen peroxide to generate hydroxyl radicals (28,29). An in vivo EPR study showed that topical application of Cr(VI) on the skin of rats generated the transient species Cr(V) (30). Cr(III), which was thought to be relatively nontoxic, can be reduced to Cr(II) by biological reductants, and Cr(II) may then act as a Fenton reagent (31). Cr(VI) and Cr(V) produced formation of 8-hydroxy-2-deoxyguanosine in isolated DNA (23,32) and Cr(III) triggered DNA strand breakage in the presence of hydrogen peroxide (33). Cr(VI) caused concentration and time-dependent oxidative damage and apoptosis in cultured J774A.1 macrophages (34), and induced NF-κB activation in Jurkat cells via free radical reactions (26). In Jurkat cells, Cr(IV) also caused NF-κB activation, produced DNA strand breaks and deoxyguanosine hydroxylation through hydroxyl-radical-initiated reactions (35). In rodents, chronic low-dose Cr(VI) induced oxidative stress, as determined by increased hepatic lipid peroxidation, hepatic nuclear DNA damage, and excretion of oxidized urinary lipids (36). Chromium-induced oxidative DNA damage, DNA–DNA interstrand cross-links, mutation of the tumor suppressor gene p53, and formation of ROS were suggested to be among the major factors playing a significant role in chromium-induced carcinogenesis (17,35,37). Increased rates of lung cancer have been observed in industry workers chronically exposed to Cr(VI), but skin cancer incidence is not increased in these subjects. Many Cr(VI) compounds are corrosive irritants and potent sensitizers (38,39). Cr(VI) causes ICU as well as ACD (39–41), and SAR to Cr(VI) have also been observed. The allergenicity of chromium salts differs significantly between oxidation states. It is believed that Cr(III) is the true hapten responsible for ACD, although Cr(VI) is more cytotoxic than Cr(III) in human keratinocytes due to its greater membrane permeability and oxidizing potential (42). In cultured human fibroblasts, the cytotoxicity of Cr(VI) was two to three orders of magnitude higher than that shown by Cr(III) (43). Chromium adheres to the skin and traces of the metal are found bound to the skin surface following even the most casual contact with common everyday objects (44). Both Cr(III) and Cr(VI) penetrate the skin, but Cr(VI) is a much better penetrant (45). Cr(VI) does not complex with organic substances, whereas Cr(III) shows strong affinity for epithelial and dermal tissues forming stable complexes, presumably via binding to protein sulfhydryl groups (46). Cr(VI) applied topically is readily converted into Cr(III) during passage through the skin. When the reducing capacity of the skin is exhausted, Cr(VI) passes unchanged through the skin (47). In summary, both Cr(VI) and Cr(III) are biologically the most active oxidation states of chromium, Cr(VI) being the more toxic valent form that produces greater oxidative stress (48), and Cr(III) being the true hapten responsible for ACD (42).

### B. Cobalt (Co)

Cobalt is an essential trace element used extensively in the steel industry, in pottery, glass, and porcelain manufacture, as well as in hair dyes, cosmetics, and paints. Co(II) produces ROS in vitro (49,50), induces formation of 8-hydroxy-2-deoxyguanosine in isolated DNA (51,52), and generates ascorbyl radicals in vivo as demonstrated by EPR spectroscopy (53). Upon skin contact with cobalt metal, cobalt salts are readily formed in human skin and Co(II) slowly penetrates into the epidermis. Co(II) can cause significant skin toxicity due to its irritant and allergenic potential (54). Co(II) is classified as a moderate sensitizer (55), causing ICU (56,57), ACD (54), and photocontact dermatitis (58,59).

### C. Copper (Cu)

Copper is an essential trace element, present in many alloys, water pipelines, jewelry, coins, and food. Ionic copper is tightly bound to ceruloplasmin in plasma and to metallothionein intracellularly to prevent its toxic effects, which are presumably mediated by catalyzing free radical reactions involving ROS (60). In humans, acute copper poisoning causes hepatic necrosis. Wilson's disease is an autosomal recessive form of chronic copper intoxication, caused by the lack of the copper carrier protein ceruloplasmin. Liver, kidney, brain, and eye damage are observed in Wilson's disease, but the skin is not significantly affected. Mobilization of copper from various copper-containing utensils in synthetic sweat has been demonstrated (61). Metallic copper worn as jewelry is measurably oxidized by sweat forming organometallic salts. These can penetrate skin and exert anti-inflammatory activity, presumably acting as superoxide dismutase mimetics. Lipophilic copper compounds penetrate human skin in significant amounts (62), and percutaneous absorption of copper from copper chloride and copper sulfate has also been demonstrated in human skin (63). Copper is rated a rare sensitizer in the guinea pig maximization test (64). Allergic reactions such as ACD and ICU to copper from nonoccupational sources are infrequently observed. However, occupational exposure to copper through electroplating or use of copper-containing fungicides, fertilizers, and mordants in fur dying have shown a high incidence of ICD and ACD (65–67).

### D. Iron (Fe)

Iron is one of the most important essential metals and highly abundant in the earth's crust. In daily life, humans are continuously exposed topically to iron-containing utensils as well as systemically through the normal diet. Iron exists in various bound forms in biological systems, including heme iron, iron attached to ferritin, hemosiderin and transferrin, and ''low-molecular-weight iron,'' such as iron bound to ATP, citrate, deoxyribose, and membrane lipids. When iron is not bound to transport or storage proteins, it becomes acutely toxic. It is widely believed that iron causes complications in several diseases by catalyzing free radical reactions involving ROS as well as ferryl and perferryl species that produce cell and tissue injuries (60,68,69). Hydroxyl radicals were detected by the spin trapping technique in an acute iron-poisoning animal model (70). Hemochromatosis is a classic example of chronic iron toxicity. The massive iron overload causes dysfunction of the liver, heart, and pancreas, finally leading to hepatic cirrhosis, liver cancer, diabetes, and cardiac insufficiency. Skin symptoms are diffuse hyperpigmentation mainly due to increased melanin formation, which is most apparent in light-exposed skin sites. There are an abundance of experimental data (71,72), and some evidence from epidemiological reports (73,74) that indicate that cancer risk is correlated with elevated body iron levels. It was recently shown that iron overload augmented 12-O-tetradecanoylphorbol-13-acetate-initiated and 7,12-dimethylbenz(a)anthracene-promoted skin tumorigenesis in Swiss mice (75,76). It was concluded that oxidative stress generated by iron overload may be responsible for the augmentation of cutaneous carcinogenesis (76). However, an increased incidence of skin cancer or inflammatory skin reactions is not observed in hemochromatosis patients. But some in vitro and animal studies indicate that liberated tissue iron can contribute significantly to inflammation, particularly in UV-induced dermatitis (77–80). In rodents loaded with iron dextran intravenously, which was associated with an increased iron content in the skin, a cutaneous inflammatory response could be induced by intradermal injection of glucose oxidase attached to polyethylene glycol. In the same model, skin inflammation without iron loading also increased skin iron levels (80). A clinical study suggests that skin iron levels may contribute to the pathogenesis of skin inflammation and ulceration caused by chronic venous insufficiency (81). Thus dermatopathological conditions may exist, in which iron contributes significantly to the disease process. However, it is evident that environmental iron does not constitute a cutaneous health risk. In humans, iron is eliminated through the integument (82), and relatively high iron levels may be suspected to be present physiologically at the epidermal surface. Percutaneous absorption of iron has been reported only for chelated iron in mice (83). Iron readily binds with proteins forming a complete antigen; however, ACD due to iron compounds is extremely rare (84,85). This naturally occurring tolerance to iron is presumably caused by genetic adaptation because of its ubiquitous presence in cells since the origin of life.

### E. Nickel (Ni)

Nickel is an essential trace element present in most alloys used in the manufacture of common utensils, including coins, and casual skin contact is virtually unavoidable. Ni(II) can be a promotor of free radical species in noncellular and cellular systems (86–95), and in animals in vivo (53). In the presence of hydrogen peroxide, Ni(II) induced oxidation of 2-deoxyguanosine to 8-hydroxy-2-deoxyguanosine (96). Ni(II) acitvated NF-κB and subsequently triggered the production of proinflammatory cytokines IL-1, IL-6, TNF-α and the expression of adhesion molecules in human endothelial cells and keratinocytes in a redox-dependent mechanism (97). In cultured human vascular endothelial cells (HUVECS), Ni(II) activated the translocation of NF-κB into the nucleus and enhanced NF-κB–DNA binding (98). In noncytotoxic concentrations, Ni(II) induced TNF-α production in cultured human keratinocytes and induced ICAM-1 expression (99). In endothelial cells, Ni(II) either promoted or suppressed the expression of ICAM-1 depending on its concentration (subtoxic concentrations inhibit and toxic concentrations stimulate ICAM-1 expression) (100). Ni(II) is released in significant amounts from various metal objects including metal fashion accessories, orthodontic appliances, and cooking utensils. Nickel intake is almost unavoidable through the normal human diet, as nickel occurs in most plants and is especially concentrated in nuts and cocoa beans. Upon skin contact, the metal is easily dissolved by sweat (101,102) and binds reversibly to the epidermis (103). Nickel has a particular affinity for keratin (104) and accumulates in the epidermis. The strong binding to keratin reduced the amount available for percutaneous penetration (105). Nickel likely causes malignancies in nickel refinery workers in the respiratory and gastrointestinal tract as well as in the kidneys (106,107). It was recently pointed out that there are few epidemiological data to indicate that exposure to metallic nickel increases the risk of cancer outside the lung and the nasal cavity (108). Nickel salts are potential skin irritants and the irritation potential depends on the counter ion (109), the most irritating nickel salts showing the highest skin penetration rate (105). Nickel compounds frequently induce ACD, while ICU and SAR are less frequently reported. The steadily increasing incidence of allergic reactions to nickel is becoming a significant health problem. The incidence has more than doubled in most industrialized countries over the last 10 years (110) and is now around 30% of the total female population in some countries. Both Ni(III) and Ni(IV) can be generated from Ni(II) by ROS, and Ni(III) as well as Ni(IV), but not Ni(II), were able to sensitize naive T-cells in mice (111). This may explain why in humans nickel-induced ACD develops more readily in irritated than in normal skin.

### F. Platinum (Pt) and Palladium (Pd)

Platinum and palladium are released into the environment by the widespread use and disposal of automobile catalytic converters (112), and are considered as important environmental toxins (113,114). Pd(II) and Pt(II) augmented the production of hydroxyl radicals in the presence of a Fenton system, but did not substitute for iron in such a system. Pd(II) and Pt(II) caused strand breakage of supercoiled DNA in the presence of Fe(II) and hydrogen peroxide (115) and induced oxidative damage in human blood cells (116). The increased use of palladium in the industry as a catalyst and in daily life as dental prostheses or jewelry has resulted in an increased incidence of palladium sensitization (117). Pt(II) has a strong binding capacity for amino acids in proteins and has the ability to form chelates. Pt(II) is highly immunogenic, being capable of causing type I respiratory hypersensitivity, ICU, and ACD in mice and humans (118–120). Not all platinum compounds are equally potent immunogens; compounds containing strongly bound ligands are immunologically inactive. Although Pd(II) and Pt(II) can cause generation of ROS in noncellular systems and induce oxidative damage in isolated human cells, there are presently no reports pointing to a significant role of oxidative stress in Pd/Pt(II)-induced skin toxicity.

## G. Vanadium (V)

Occupational exposure to the essential trace element vanadium occurs in the mining and steel industry, as well as in petroleum refining, while environmental contact is usually through fly ash. V(V) competes with phosphate, inhibiting the action of phosphates and the Na/K-ATPase activity. As V(IV) it competes with other transition metal ions for binding sites in metalloproteins. V(IV), but not V (V), acts in a Fenton-type reaction generating hydroxyl radicals (121). V(IV) triggers lipid peroxidation in micelles, in the presence of hydrogen peroxide via formation of the hydroxyl radical (122). In the presence of hydrogen peroxide, V(IV) oxidizes deoxyguanosine to 8-hydroxy-2-deoxyguanosine and induces single strand breaks as well as formation of 8-hydroxy-2-deoxyguanosine in isolated DNA (123,124). V(IV) inhibits SOD activity in vitro (125) and triggers programmed cell death in human liver cells as a consequence of oxidative stress (126). V(IV) transactivates AP-1 in mouse epidermal JB6P+ cells, and this activation is dependent on the generation of superoxide anion radical and hydrogen peroxide (127). V(V) activates both NF-κB and c-Jun N terminal kinase (JNK) in macrophages, and both reactions are inhibited by pretreatment of cells with *N*-acetylcysteine indicating that the cellular redox state plays a central role in vanadate-induced activation of NF-κB and JNK (128). Chronic intratracheal exposure of rats to vanadium pentoxide causes lung injury presumably via bioreduction of V(V) and reoxidation of V(IV). This one-electron redox cycling initiates lipid peroxidation in lung biomembranes (129). Ascorbate readily reduces V(V) to V(IV) and this mechanism was suggested to be an important V(V) reduction pathway in vivo (130). A study suggested that a strong correlation exists between V(IV)-induced hepatotoxicity and the induction of oxidative stress and lipid peroxidation in this particular organ (131). Carbon-centered free radical species were detected by in vivo spin trapping in rat lung exposed to V(IV) (10). Vanadium compounds penetrate human skin to a small extent and can irritate skin through occupational exposure (132). Vanadium is categorized as a strong allergen in guinea pigs, but in humans ACD is observed very rarely. Prooxidant effects of vanadium compounds in skin have not yet been reported.

## H. Aluminum (Al)

Aluminum is not an essential element. It is abundant in the earth's crust and industrial activities have led to a large mobilization of Al. It is widely used in various metal alloys, leather tanning, medicines, and cosmetics. Aluminum is a neurotoxic metal and evidence is accumulating that its mechanism of toxicity may involve formation of ROS and peroxidation of membrane lipids (133–141). There are no reports on prooxidant activities of Al(III) in keratinocytes or intact skin, but recently it was shown that Al(III) enhanced melanin-initiated lipid peroxidation in part through an interaction with superoxide anion radicals generated from the autoxidation of melanin (142). Al(III) is a weak antigen and can be a skin irritant, depending on the associated counterion. Aluminum nitrate and chloride salts can cause profound pathological changes in pig, rabbit, and mouse skin. The sulfate, hydroxide, acetate, and chlorohydrate are not irritating at comparable concentrations. Following intradermal injection, Al(III) can cause granulomatous reactions at the injection site. The different skin penetration abilities of the various Al(III) salts determines the degree of skin damage, where Al(III) exerts mainly protein-precipitating properties (143,144). Oxidative events seem not to play a primary role in Al(III)-induced ICD.

### I. Arsenic (As)

Chronic environmental arsenic poisoning is a serious problem in developing countries and caused by drinking water naturally contaminated with arsenic. Heavy contamination of arsenic in drinking water has also been caused by human carelessness (e.g., by the mining industry). In the agricultural setting, arsenicals have been widely used in the past as wood preservatives, insecticides, and herbicides. Exposure to pesticides containing arsenic compounds, working in the semiconductor industry or in electroplating may also result in chronic arsenic poisoning (145). In industrialized countries, the most significant exposure of humans to arsenic results from nutritional intake (e.g., seafood containing organic arsenicals or through drinking water naturally contaminated with arsenic). Methylation of inorganic arsenicals is a detoxifying process; however, inorganic arsenicals can be regenerated from methylated arsenicals in aquatic biosystems (146). Inorganic arsenic is more toxic than organic arsenic compounds, and As(III) is more toxic than As(V). Arsenic-induced generation of free radicals and ROS have been implicated in arsenic carcinogenesis (147–153); however, it seems questionable whether these reactions occur at pathophysiologically relevant concentrations of arsenic, and presumably do not represent a biologically relevant event in arsenic carcinogenesis (154). As already summarized previously, arsenic-mediated chronic stimulation of keratinocyte-derived growth factors may play a significant role in arsenic-induced skin carcinogenicity (155,156). Arsenic is a human carcinogen only in its inorganic form. It exhibits a clear predeliction for the skin and its appendages (157,158), causing nonmelanoma skin cancer (basalioma, squamous cell carcinoma, Bowen carcinoma), but it also produces bladder, liver, lung, and kidney carcinoma (159). In contrast to lipophilic organoarsenicals, inorganic arsenicals have a poor skin-penetrating ability. They are skin irritants and can also act in rare cases as contact allergens (160,161). Allergic reactions to arsenic are mainly observed following occupational exposure. Arsenic accumulation in skin is reported to increase the skin sensitivity to UVR, causing an exaggerated sunburn effect (162). Other cutaneous manifestations of chronic arsenic poisoning include hypo- and hyperpigmentation and palmoplantar hyperkeratosis (163,164).

### J. Cadmium (Cd)

Cadmium is an ubiquitous nonessential element and is released during the smelting and industrial refining of zinc and lead ores. Dietary intake as well as smoking of cigarettes is the primary source of human exposure to cadmium. In industry, cadmium is widely used in electroplating and galvanizing in color pigment paints and in batteries. Cadmium did not generate free radicals in cultured mammalian cells directly (165), but indirectly induced oxidative stress in isolated human phagocytes and J774.1 macrophages (34,166, 167), and triggered oxidative damage to lipids and DNA in rats in vivo (36,168–170). Its primary mechanism of action presumably involves depletion of cellular thiols. Cadmium is a toxic metal with extremely long biological half-time of 15 to 20 years in humans. Cadmium exposure is known to cause a variety of adverse health effects, among which kidney dysfunction, lung diseases, disturbed calcium metabolism, and bone effects are most prominent. Following long-term exposure, the kidney is the critical organ, but the skin is not a target organ of cadmium toxicity. Cadmium is classified as a human carcinogen; it has been associated with increased incidence of lung cancer in chronically exposed workers (171), presumably in association with exposure to arsenic trioxide (172). Cadmium-induced carcinogenicity may be related to oxidative DNA damage (173). Topically applied cadmium binds tightly to epidermal keratin and does not penetrate in significant amounts through human skin (174). The incidence of ICD due to cadmium compounds seems to be low. In the guinea pig, cadmium did not induce a hypersensitivity response (175) and in humans cadmium chloride has a low allergenicity (176).

### K. Lead (Pb)

Lead is perhaps the longest used and best recognized toxic environmental chemical. It is a nonessential metal and an ubiquitous environmental contaminant. Lead exposure may result from contact with lead-based fuel additives, lead-based paints, and lead-containing lubricants. Other sources of lead exposure are contaminated drinking water, folk remedies, cosmetics, and food supplements. Lead exerts its toxic action through inhibition of sulfhydryl-dependent enzymes and functionally displaces or substitutes for calcium ions in receptor proteins (177). Free radicals have been proposed to be involved in lead poisoning (178). Lead can cause oxidative stress in tissues presumably via depleting intracellular thiols (179), and it potentiates iron-induced formation of ROS (180). Lead affects the reproductive, nervous, gastrointestinal, immune, renal, cardiovascular, skeletal, muscular, and hematopoietic systems. Although the skin is a rich sulfhydryl group organ, it is not a target organ of lead toxicity. Irritant and allergic reactions to lead or its salts are very rare. In contrast to inorganic lead salts, lipid-soluble organo-lead compounds easily penetrate through human skin (181) and may cause systemic toxicity.

### L. Mercury (Hg)

Mercury is a nonessential element ubiquitous in the environment in its elemental state, as salt or methyl mercury. Methyl mercury is a mercury metabolite and accumulates in aquatic biosystems. Organic mercury compounds have been used in agriculture as weed killers and fungicides. All three forms are interconverted easily in the environment. The major sources of environmental mercury are its natural occurrence and abundance as well as industrial activity. Mercury exposure occurs through environmental contamination as the result of mining, smelting, and industrial discharge. The major route of mercury exposure is dietary intake (e.g., through consumption of seafood). In humans all forms of mercury are readily converted into Hg(II), which binds to sulfhydryl groups. Hg(II) depleted reduced glutathione content and inhibited the activities of the antioxidant enzymes SOD, catalase, and glutathione peroxidase in rat kidney, a target organ of Hg(II) toxicity (182). Hg(II) does not generate free radicals directly, but is believed to induce oxidative stress via depletion of sulfhydryl groups (172,183,184) or stimulation of the oxygen burst in human polymorphonuclear leukocytes (185). Elemental, inorganic, and organic mercury may cause immunotoxicity, neurotoxicity, nephrotoxicity, and gastrointestinal toxicity. Inorganic mercury compounds are dermatotoxic, causing ICD and ACD. Exposure to mercury metal or mercury salts induced autoimmune diseases in susceptible animals (186–188) and in humans (189–191). The most common form of cutaneous mercury toxicity is ACD and ICD. In humans, Hg(II) chloride is a very potent sensitizer on the Magnusson Kligman scale, while mercury metal is a moderate topical sensitizer (55). Mercury can also cause SAR and in a few patients ICU caused by Hg(II) has been observed.

## III. ORGANIC SOLVENTS

Organic solvents are commonly used for cleaning, degreasing, and extraction and are ubiquitous environmental pollutants. Solvents can be divided into families according to the chemical structure and the attached functional groups. The basic structures are aliphatic, alicyclic, and aromatic. The functional groups include halogens, alcohols, ketons, glycols, esters, ethers, carboxylic acids, amines, and amides. Solvents that are soluble in both lipid and water phase are readily absorbed through the skin. A few solvents are known to cause ACD (e.g., aldehydes, amines, glycolethers), and in rare cases chemical scleroderma in susceptible patients. However, the majority of organic solvents are primary skin irritants as a result of skin defatting, which is related to their lipophilicity, or act through denaturation of proteins or perturbation of membranes. Defatting the epidermis results in disruption of the skin barrier function and is associated with an increased transepidermal water loss, finally leading to ICD. In particular, alcohols, esters, ethers, chlorinated solvents, alicyclic and aromatic hydrocarbons including toluene and xylene (192,193), glycol ethers (194), chlorofluorohydrocarbons (195), and chloromethylethers (196) are potent skin irritants. It was pointed out that the mechanism by which topical organic solvents disrupt the skin barrier involves more than pure lipid removal (197). Extraction of epidermal lipids has been shown to induce the release and synthesis of proinflammatory cytokines in skin, including IL-1$\alpha$ and TNF-$\alpha$ (198–200). In addition, neuropeptides, which function as mediators of cell proliferation, cytokine and growth factor production, cell surface receptor expression (201), were shown to be involved in solvent-induced ICD in rat skin (202). Topical exposure of Swiss mice to leaded gasoline (3,6,12 mL/kg body weight) for seven consecutive days caused lipid peroxidation, a decrease in reduced glutathione, and inhibition of glutathione-S-transferase in skin and extracutaneous tissues (liver, brain) (203). Whether these effects were mediated by the lead additive or the hydrocarbon component of gasoline was not analyzed in this study. The general health consequences of ethanol consumption with particular emphasis on its effects on the skin and skin function were recently reviewed (204). It is clearly established that the mechanism of ethanol toxicity involves oxidative stress (205), but the skin is not a target organ of systemic ethanol. High concentrations of ethanol ($>$70%) are dermatotoxic only when applied topically under occlusion. Open application of concentrated ethanol causes whitening of the skin, while prolonged skin contact under occlusion induces ICD with blisters (206). Although reactive species are unequivocally involved in ethanol cytotoxicity, their participation in ethanol-induced skin irritation is presumably biologically insignificant. At high concentrations, ethanol will mainly cause protein precipitation and thereby damage the skin.

## IV. HALOGENATED ALIPHATIC HYDROCARBONS

Halogenated aliphatic hydrocarbons (HAH) are ubiquitous environmental pollutants and predominantly used as solvents. They are primarily hepato-, neuro-, and cardiotoxic, but may cause ICD as a result of defatting as described above. Localized scleroderma (LS) and progressive systemic sclerosis (PSS)-like conditions have been observed in subjects exposed to trichloroethylene, perchloroethylene, and carbon tetrachloride (207–212). The toxicity, in particular hepatotoxicity, of most HAH is presumably mediated by reactive metabolites formed during dehalogenation. These substances are subject to single-electron reduction by the mixed function oxidase (MFO) system to generate reactive alkyl- and alkylperoxyl radicals (213,214), but other products such as reactive electrophils may be formed as well. Although a role of ROS has been suggested in the pathogenesis of PSS (215), it should be cautioned that the precise role of reactive species, in particular ROS, in the pathogenesis of HAH-induced scleroderma is yet to be defined.

## V. POLYHALOGENATED AROMATIC HYDROCARBONS

Polychlorinated dibenzo-p-dioxins, dibenzofurans, azoxybenzenes, naphthalenes and biphenyls, collectively termed polyhalogenated aromatic hydrocarbons (PHAH) are important toxic environmental pollutants. These groups of chemicals are usually considered together because their chemical structures are similar, they produce a similar and characteristic pattern of toxic response, and they are believed to act via a common mechanism (216). PHAH are not primary industrial products, they are true contaminants. The combustion of chlorinated aromatics and diverse types of chemical, industrial, and municipal waste containing chlorinated chemical products results in the formation and release of these toxic chemicals into the environment. Polychlorinated dibenzo-p-dioxins are also formed as a contaminant in the synthesis of chlorinated aromatics, such as 2,4,5-trichlorophenoxy acetic acid or pentachlorphenol (PCP). These compounds were in wide use as insecticides, herbicides, fungicides, mold inhibitors, and disinfectants. Polychlorinated biphenyls (PCB) and naphthalenes have been manufactured and used worldwide for a variety of commercial purposes such as cooling fluids in capacitators and transformers, lubricants, heat transfer fluids, plasticizers, and adhesives. Most of the biological effects of PHAH, and in particular 2,3,7,8-tetrachlorodibenzo-p-dioxin (TCDD), are believed to be mediated through binding to the aryl-hydrocarbon receptor (AHR) and modulating specific genes (217–220). TCDD is not metabolized into free radicals, but it causes multiple prooxidant effects in cellular systems (216,221–232), presumably as a consequence of its pleiotropic action. During metabolism of PCB, reactive electrophils are formed which yield adducts with nitrogen and sulfur nucleophils including DNA (233,234). These studies also demonstrated formation of superoxide anion radicals during bioactivation of mono-/dichlorinated biphenyls. Free radical—mediated reactions were suggested as a possible molecular mechanism of the toxicity of halogenated biphenyls (235–239). PCBs are readily absorbed through human skin and the cutaneous sebaceous system is one of the main excretory systems of PCB and polychlorinated quaterphenyls (240). Exposure to PCBs has been linked to significant skin toxicity, such as ICD and acneiform eruptions. Acneiform reactions were the most frequent manifestation in TCDD toxicity in humans (241). A possible association between melanoma risk and occupational exposure to PCB has been examined and positive associations were found (242). However, it was pointed out that this study generally did not control well for other confounding factors (243).

## VI. AROMATIC HYDROCARBONS

Benzene is a major industrial chemical and gasoline additive and has myelotoxic and leukemogenic effects in humans. Background benzene concentrations in air are between 1 $\mu g/m^3$ (rural areas) and 10 $\mu g/m^3$ (cities). Smoking 20 cigarettes releases approximately 500 $\mu g$ benzene into the air and about 800 $\mu g$ benzene can be inhaled when filling the gasoline tank of a car. The finding that benzene is a multiple site carcinogen in rodents following long-term oral or inhalation exposure (244,245) raises the possibility that other tissues could be susceptible to benzene-induced carcinogenicity. The metabolic activation of benzene is extensive and yields a number of reactive intermediates, including the semiquinone and quinone forms of hydroquinone, 1,2,4-benzenetriol, phenol, catechol, and muconaldehyde (246–248). ROS are involved in both the metabolism of benzene and its toxicity. Formation of 8-hydroxy-2-deoxyguanosine has been demonstrated to occur during benzene metabolism, and this kind of oxidative DNA damage was suggested to play a role in benzene-induced carcinogenicity (249–253). Benzene metabolites demonstrate significant dermatotoxicity in human and rodent skin, including ACD, ICD, and chemical leukoderma. Topically applied benzene induced papillomas in transgenic mice that harbor the v-Ha-ras oncogene (254). Mortality data from a large epidemiological study revealed a small but significant excess of melanoma and nonmelanoma skin cancers in workers chronically exposed to benzene (255). In this study, skin cancers were possibly underreported, since nonmelanoma skin cancer rarely causes death owing to early detection and treatment. It was recently pointed that the finding that benzene is a multiple-site carcinogen in rodents raises the possibility that other tissues could be susceptible to benzene-induced carcinogenicity, especially since a significant excess of squamous cell carcinomas and papillomas arise from epidermal and oral keratinocytes in benzene-exposed rats (256). Benzene and many of its phase I metabolites induced inflammatory cytokines and growth factors in cultured human epidermal keratinocytes and this occurs through direct covalent binding or the generation of ROS by autoxidation and reduction. It was suggested that the elaboration of proinflammatory cytokines and growth factors by keratinocytes in response to benzene and its principal metabolites may participate in benzene-induced skin carcinogenesis (256).

Benzene, toluene, xylene, and naphthalene have been associated with localized scleroderma (LS)-like skin changes (257–260), which were limited to areas of direct contact. Furthermore, toluenes, xylenes, and xylidines have been reported to cause progressive systemic sclerosis (PSS)-like symptoms in chronically exposed subjects (261,262). The metabolism of aromatic hydrocarbons through the mixed function oxidase (MFO) system generates free radicals and ROS, and it has been speculated that these reactive species may be involved in chemical-induced scleroderma in susceptible patients.

Naphthalene is widely used in various commercial and industrial applications including lavatory scent disks, soil fumigants, moth repellents, and tanning agents. Naphthalene is a pulmonary, renal, and occular toxicant, and it also causes hemolytic anemia in humans. The skin is not a primary target organ of naphthalene toxicity, but skin irritation can occur following topical exposure to naphthalene. Naphthalene induces concentration-dependent production of ROS in cultured J774A.1 macrophages, resulting in lipid peroxidation and oxidative DNA damage (263). Naphthalene is bioactivated by cytochromes P450 to an electrophilic epoxide intermediate, which subsequently is metabolized to naphthoquinones (NQ) and possibly to a free radical intermediate. These reactive intermediates were suggested to be involved in naphthalene cytotoxicity (264–267).

## VII. POLYCYCLIC AROMATIC HYDROCARBONS

Mainly due to combustion of fossil fuels, polycyclic aromatic hydrocarbons (PAH) such as acenaphthen, benzo[a]pyrene, benzo[b]fluoranthene, benzo[i]fluoranthene, benzo[k]fluoranthene, benzo[ghi]perylen, chrysene, indeno[1,2,3-cd]pyrene, fluoranthene, and pyrene are entering the environment in a quantity that may constitute a danger to human health. In 1969, PAH levels in the air were up to 90 $ng/m^3$ in several heavily industrialized cities in Germany. Twenty years later levels up to 7 $ng/m^3$ and 1 $ng/m^3$ in the air were reported in industrial and rural areas, respectively (268). In special workplaces (e.g., road construction) very high benzo[a]pyrene levels (45 $\mu g/m^3$) have been found in the past while tar was still being used. In the late 1960s, tar was replaced by bitumen for road construction, and this resulted in a significant reduction of benzo[a]pyrene exposure of road workers. Tar, which is a pyrrolysis product of coal, contains 5 to 10 g benzo[a]pyrene/kg. Bitumen, which is the distillation residue of crude oil, contains about 2 mg benzo[a]pyrene/kg. However, skin exposure to bitumen can result in significant absorption of PAH (269). In healthy controls, which were not exposed occupationally to PAH-containing substances, no benzo[a]pyrene was detected in skin, while occupationally PAH-exposed workers had benzo[a]pyrene concentrations in skin with mean values of 1.2 $ng/cm^2$. The mean value of urinary 1-hydroxypyrene concentrations was 6.0 nmol/mmol creatinine for the exposed workers and 0.5 nmol/mmol creatinine for the controls, indicating some background level of PAH exposure (270). The major issue of the PAH risk assessment is the potential carcinogenicity of these compounds. In humans, a limited number of PAH are carcinogenic when applied to the skin or inhaled over a long period. Benzo[a]pyrene is the best characterized PAH compound and is a human skin and lung carcinogen. A correlation between scrotal skin cancer in humans and exposure to chimney soot was already established in 1775, and the first experimental cancer in mouse skin was produced by topical application of coal tar condensate in 1915. Lung cancer risk in humans correlated with cumulative PAH dose (271). ROS are involved in both the metabolism of PAH and their toxicity. Most of the biological effects of PAH are mediated via the aryl hydrocarbon receptor (AHR) (272). Benzo[a]pyrene-7,8-dihydrodiol-9,10-epoxide is thought to be the ultimate carcinogenic metabolite (273), and dihydro-diolepoxides also appear to mediate the genotoxic effects of several other polycyclic hydrocarbons. For most PAH, multiple mechanisms of activation exist, depending on the nature of the activating enzymes present in the various target organs. Metabolic activation of PAH can be understood in terms of two main pathways: (a) monooxygenation to produce dihydrodiol-epoxides and (b) one-electron oxidation to yield reactive intermediate radical cations. The formation of the dihydro-diolepoxide requires sequential epoxidation, hydration, and epoxidation steps. This metabolite is produced by several reactions: cytochrome P-450-dependent MFO, cooxidation with prostaglandin synthetase, or lipoxygenase, and possibly direct catalytic epoxidation with several inorganic and organic peroxyl radicals (274,275). Pollutants, such as $NO_2$, $SO_2$, lipid peroxidation products, and asbestos contribute to the toxicity of PAH by catalyzing oxidation of the aromatic hydrocarbons to the ultimate carcinogens (276,277). Radical cations of PAH (278–281) and redox cycling PAH quinones (282–286) are also believed to play a significant role in PAH carcinogenesis.

Nitrated polycyclic aromatic hydrocarbons (NPAH) are widely distributed in the environment mainly as the result of incomplete combustion processes (e.g., automobile and diesel exhaust, coal power plants, and cigarette smoke). In the atmosphere, they are also produced by photochemical reaction of the oxides of nitrogen with PAH (287). They have potentially irritating and genotoxic properties. After inhalation, potent mutagenic metabolites are formed (288). NPAH are among the most potent Salmonella mutagens (289), although they were found to be less potent mutagens than the parent PAH in human B-lymphoblastoid cells (290). NPAH were carcinogenic in rodents, even without enzymic activation (287). NPAH are powerful inducers of cutaneous monooxygenase activities

(291), and may thereby stimulate production of ROS in skin. In the isolated perfused rat liver, acute cytotoxicity of NPAHs seems not to be mediated by oxidative stress generated by redox cycling metabolites (292).

## VIII. PESTICIDES

Pesticides are farm, forest, and household chemicals used against several kinds of natural parasites. They are an important source of increasing productivity in farms and horticulture. Pesticides comprise compounds that act as acaricides, algicides, bactericides, fumigants, fungicides, herbicides, insecticides, molluscicides, and rhodenticides. There are over 1000 chemical compounds used as pesticides in over 30,000 formulations under different brand names. Many of the pesticides are sold as mixed preparations of biozide chemicals of technical grade with several additives (stabilizers, fillers, surfactants). These factors complicate the analysis of toxic effects attributed to the use of commercial pesticides significantly. Various chemical categories are found in commercial pesticides. Table 1 gives an overview of the main substances used.

Organophosphates, carbamates, dithiocarbamates, and chlorinated hydrocarbons cause mainly neurotoxic reactions or exert their toxic effects on the hematological system. Several chlorinated hydrocarbons are no longer in use as insecticides, mainly because of their accumulation in biomaterials and their neurotoxic properties. In industrialized countries, many people often suppose that their skin complaints are caused by pesticides, even though they are not at risk because their skin is not directly exposed. Despite their extensive use, reports on skin diseases from pesticides are not very frequent and mostly of single cases. However, in the occupational setting, cutaneous reactions to pesticides may cause significant morbidity (293). Approximately one-third of all pesticide-related diseases are dermatological. Persons mainly at risk include manufacturers and formulators, agricultural workers, pest control workers, and swimming pool service personnel (294). Several pesticides are potent skin irritants such as bis-pyridinium and organo-tin compounds. The carbamates/dithiocarbamates (benomyl, ferbam, vapam, maneb, zineb, thiram, ziram) and phthalimides (captan, captafol, folpet) are common causes of both ICD and ACD. Tributyltin oxide and benomyl are known photosensitizers (294–297). Some pesticides, especially paraquat, increased histamine release in rat mast cells (298) and may thus cause ICU. In some instances, cutaneous reactions are due to degradation products or contaminants. For

**Table 1** Pesticides: Chemical Classification and Application

| Algicides | Fumigants | Fungicides | Herbicides | Insecticides |
|---|---|---|---|---|
| Organo-tin compounds | Ethylene oxide | Carbamates | Phenoxy-carbonic acids | Organo phosphates |
| Copper chelates | Methyl bromide | Dithiocarbamates | bis-Pyridinium compounds | Carbamates |
| | Metam sodium | Phthalimides | Triazins | Pyrethroides |
| | 1,2-Dichlorpropane | Phthalonitrils | Triazols | Chlorinated hydrocarbons |
| | 1,3-Dichloropropene | | | |
| | Epichlorhydrin | | | |
| | | | Organic arsenicals | |

example, the fumigant metam sodium (sodium N-methyl-dithiocarbamate) is a very strong sensitizer and toxic vesicant, which has caused numerous cases of contact dermatitis due to the degradation product methyl-isothiocyanate, which was the actual toxic agent (296,299). Similarly, dazomet (3,5-dimethyl-tetrahydro-1,3,5,-thiadiazin-2-thion) can cause severe ACD, the actual contact allergen being the degradation product methyl-isothiocyanate (296). Malathion was classified as a potent contact and photocontact sensitizer (55); however, the sensitizing agent was later identified to be the contaminant diethylfumarate (300).

Several in vitro and animal studies suggest that oxidative stress may be a cofactor in the toxicity of insecticides, including organophosphates (301–304), carbamates, and chlorinated hydrocarbons (228–231,302,305–310), and bi-pyridinium compounds (228, 311–318). However, other pesticides such as dithiocarbamates are well known for their antioxidant properties (319). Several pesticides function as substrates, inhibitors, and inducers of drug-metabolizing enzymes, such as cytochrome P-450 isoenzymes, with the same compound frequently acting in more than one of these roles (320). This complicates the analysis of the health effects of pesticides, particularly in the case of pesticide mixtures (301).

### A. Organo-Tin Compounds

Organo-tin compounds have biocidal properties and their widespread application as marine antifouling paint, as wood preservatives, and as agricultural biocide have led to accumulation of tin in water and soil. Inorganic tin salts are relatively nontoxic, and tin and its inorganic salts do not significantly penetrate through skin. In a few cases tin or inorganic tin salts may cause ACD (321). In contrast, organo-tin compounds readily penetrate through the skin and may cause systemic toxicity (322). Organo-tin compounds are strong skin irritants causing delayed-acute irritation, and may also cause phototoxic reactions (323–326). Tributyltin oxide induces IL-1$\alpha$ expression in mouse keratinocytes via a mechanism involving ROS (327). Several organo-tin compounds can induce ACD in the guinea pig (328), but tributyltin oxide is reported not to be allergenic in human skin (329).

### B. Bis-Pyridinium Compounds

Paraquat (1,1′dimethyl-4,4′bipyridyl) and deiquat (1,1′ethylene-2,2′bipyridyl) are imminium-type contact herbicides. They are almost exclusively used as a dichloride or dibromide salt, respectively. Both the herbicidal and toxicological properties of paraquat and deiquat are dependent on the ability of the parent cation to undergo a single electron addition to form a free radical that reacts with molecular oxygen to reform the cation and concomitantly produce a superoxide anion radical. Furthermore, paraquat generates hydroxyl radicals in the presence of copper ions in rats in vivo (330). Paraquat and deiquat can cause skin and lung fibrosis, both being corrosive skin irritants (331,332). Even at high concentrations and prolonged cutaneous exposure, paraquat causes skin irritation without significant internal organ pathology due to absence of significant percutaneous absorption (333). Some paraquat manufacturing workers exposed to bis-pyridinium compounds developed skin premalignacies as well as skin cancer in sun-exposed areas (334, 335), and a synergistic role of bipyridine in causing these lesions was suggested (334). However, an analysis of 112 paraquat production workers showed no statistically significant increase in skin premalignancies in paraquat-exposed workers (336).

## C. Pentachlorphenol

Pentachlorphenol (PCP) was used for wood preservation and as a fungicide, herbicide, and disinfectant. It is highly toxic because of its interference with mitochondrial oxidative phosphorylation and PCP poisoning affects multiple organ systems. Several fatalities from PCP have occurred among chemical production workers, herbicide sprayers, and wood manufacturers. PCP interferes with the electron transport chain in the endoplasmatic reticulum and in mitochondria, and uncouples oxidative phosphorylation. Uncoupling of oxidative phosphorylation may initially cause inhibition of mitochondrial ROS production (337,338), but as soon as high-energy phosphates are depleted ROS generation may be greatly increased (339). PCP is in part metabolized by microsomal enzymes to tetrachloro-p-hydroquinone (TCHQ), which easily oxidizes to its redox cycling semiquinone radical. ROS were suggested to be involved in some of the toxic effects of PCP (340–343). Skin contamination with PCP is an important route of exposure. PCP can cause skin irritation after single exposures to material containing more than 10% PCP, or after prolonged or repeated contact with a 1% PCP solution. Chloracne in PCP-exposed subjects is probably caused by dioxin contaminants, because in animal studies pure PCP did not cause chloracne.

## D. Halogenated Phenoxycarbonic Acids

Halogenated phenoxycarbonic acids such as 2,4-dichloro-phenoxyacetic acid, 2-methyl-4-chloro-phenoxyacetic acid, and 2,4,5-trichloro-phenoxyacetic acid are selective weedkillers that work by uncoupling oxidative phosphorylation and skin contact can cause ICD. In particular, 2,4-dichlorophenoxyacetic acid was in widespread use as a component of ''agent orange'' during the Vietnam war. As already outlined above, uncouplers of oxidative phosphorylation can induce increased formation of mitochondrial-derived ROS. A second mechanism may contribute to increased ROS formation triggered by halogenated phenoxycarbonic acids; however, its clinical relevance is purely speculative. 2,4-dichlorphenoxyacetic acid belongs to a diverse group of chemicals causing peroxisome proliferation. These peroxisome proliferators (PP) comprise drugs (e.g., hypolipidemic agents), plasticizers (e.g., phthalate esters), herbicides (e.g., 2,4-dichlorphenoxyacetic acid), and solvents (e.g., trichloroethylene). PP induce a dramatic increase in peroxisomes in rodent hepatocytes, followed by hepatocarcinomas. Marked species differences are apparent in response to PP. Rats and mice are extremely sensitive and hamsters show an intermediate response, while guinea pigs, monkeys, and humans appear to be relatively insensitive or nonresponsive at dose levels that produce a marked response in rodents (344–347). At present, it is unknown whether peroxisomal proliferation plays a role in the toxicity of 2,4-dichlorphenoxyacetic acid in human skin (ICD). PP are nongenotoxic carcinogens; they act via a receptor mediated mechanism. PP bind to nuclear receptors belonging to the superfamily of steroid receptors, the peroxisome-proliferator-activated receptor (PPAR) (348). PP also activate genes encoding cytochrome P-450 in the liver and other tissues via interaction with the PPAR receptor (349–351). The increased production of ROS due to unbalanced production of peroxisomal enzymes has been proposed to cause oxidative DNA damage with subsequent tumor formation in the liver (352,353). However, studies demonstrated a lack of correlation between hepatic 8-hydroxy-2-deoxyguanosine levels and the carcinogenicity of PP (354,355). This may indicate that 8-hydroxy-2-deoxyguanosine does not accurately reflect the potential peroxisomal hydrogen-peroxide-dependent DNA damage, or that oxidative damage does not play a role in PP-induced carcinogenesis. PPAR activators accelerate formation of the epidermal permeability barrier (356), and PPAR activity may directly contribute to the hyperkeratinizing response seen with some topically encountered xenobiotics (357). Hyperkeratotic hand eczema is a common occupational problem and it remains to be seen whether peroxisome proliferation plays a significant role in this condition.

## IX. AIRBORNE OXIDANTS

The primary oxidizing pollutants present in the atmosphere include ozone and the oxides of nitrogen. In less important amounts, peroxyacyl nitrates, hydrogen peroxide, alkyl peroxides, and diesel exhaust particles exist. Typical indoor air pollutants are the oxides of nitrogen, ozone, tobacco smoke, volatile organic peroxides, and terpenes, as well as mineral fibers such as asbestos. Reactions among indoor pollutants can produce free radical species and reactive aldehydes, ketones, and peroxyacyl nitrates, some of which are known to be mucocutaneous irritants (358). Of all the gaseous oxidants, the toxicity of ozone and the oxides of nitrogen have been studied in detail because of their abundance and earlier identification as constituents of photochemical smog and as having enormous potential effects on humans and ecosystems.

### A. Ozone

Ozone at ground level (trophosphere) is a noxious, highly reactive oxidant air pollutant. Ozone background levels of up to 0.04 ppm (78 $\mu g/m^3$) are usually found in rural areas, but in large urban areas with automobile-dependent transportation and/or petroleum-dependent energy production average annual levels are between 0.1 to 0.2 ppm (196–392 $\mu g/m^3$), and in summer the levels may be two- to three-fold higher (359). In some workplaces (e.g., water purification plants, power plants, chemical industry, and hazardous waste management utilities), even higher ozone concentrations may be encountered. The biological effects of ozone are attributed to its ability to cause ozonization, oxidation, and peroxidation of biomolecules, both directly and via secondary toxic reactions (360–362). Ozone causes mortality in animals if breathing air contains ozone at 4 to 10 ppm (8–20 $mg/m^3$) due to pulmonary toxicity (359). There were only fragmentary reports on possible ozone-induced oxidative damage in skin (363,364), but a potential role of trophospheric ozone as an environmental stress factor to skin is emerging (365–370). In animal experiments, the stratum corneum as the outermost part of the skin is the primary target of ozone toxicity, resulting in disturbances of the barrier function at high ozone concentrations (1–10 ppm). Human studies are required to analyze the clinical significance of ozone skin toxicity at clinically relevant concentrations alone and in combination with other environmental stressors such as UV light, metals ions, and other oxidizing pollutants.

### B. Oxides of Nitrogen

Nitrogen dioxide ($NO_2$) is one of the important indoor oxidizing gases and the oxides of nitrogen, generally a mixture of nitric oxide (NO) and $NO_2$, are produced by both natural and human-made processes. They are generated during combustion of fossil fuels such as oil, coal, and gasoline. In urban areas, average annual concentrations of $NO_2$ vary from 0.2 to 0.8 ppm (150–376 $\mu g/m^3$), with peaks reaching 1 ppm (1880 $\mu g/m^3$) or up to 2.9

ppm (5500 μg/$m^3$) (Los Angeles, 1962) (359). People living in rural areas are exposed to background levels of 0.05 ppm (100 μg/$m^3$) $NO_2$. High levels of exposure can occur in the workplace (e.g., welding, silos, chemical plants). Indoor concentrations of $NO_2$ can be found up to peak values of 0.1 to 0.5 ppm (e.g., in kitchens with gas stoves or kerosene gas heaters), but average values in homes are 0.025 to 0.075 ppm. $NO_2$ causes mortality in animals if breathing air contains $NO_2$ at 50 to 150 ppm (94–282 mg/$m^3$) due to pulmonary toxicity (359). NO is a weak oxidizing free radical and has a very low reactivity (comparable with molecular oxygen) with most biologically important molecules with the exception of metal chelates (371). The direct toxicity of NO is modest but is enhanced by reaction with superoxide anion radical to form peroxynitrite (372), which reacts relatively slowly with most biological molecules, making it a selective oxidant (373). Short-term exposure (4 h) of healthy human skin to low concentrations of $NO_2$ (0.023–0.030 ppm) caused disturbances of epidermal barrier function with increased transepidermal water loss (374).

## C. Peroxyacyl Nitrates

Peroxyacyl nitrates (PAN), such as peroxyacetyl nitrate and peroxypropionyl nitrate, are formed by the action of sunlight on volatile organic compounds and the oxides of nitrogen. Thus their concentration is highest in photochemical smog (375). Maximum ambient concentrations of PAN usually experienced in North America are 0.003 to 0.078 mg/$m^3$. The acute pulmonary toxicity of PAN is less than that of ozone, similar to $NO_2$ and higher than $SO_2$. In humans, the lowest level causing eye irritation was 0.64 mg/$m^3$ for 2 h (376). Presently, no data are available on the effects of PAN on human skin. Presumably PAN toxicity on human skin may be an insignificant health problem, but PAN may exert additive or synergistic effects on the toxicity of other oxidizing airborne compounds.

## D. Sulfur Dioxide

Sulfur dioxide ($SO_2$) is produced in the atmosphere by natural sources as well as by human activities. In rural areas, average annual $SO_2$ levels of up to 0.0038 ppm (10 μg/$m^3$) are found. In urban areas, average concentrations may be up to 0.018 ppm (50 μg/$m^3$), with peak concentrations ranging from 0.11 ppm (300 μg/$m^3$) (industrial areas in East Germany, 1980) and up to 0.7 ppm (1900 μg/$m^3$) (London, 1952). However, due to effective prevention strategies, average $SO_2$ concentrations are presently of no significant health concern in western Europe. Although $SO_2$ is mainly a reductant in chemical systems, sulfite, its hydrolysis product, is converted in biological systems into oxidizing species such as the sulfite radical and other species (377–381). Following sequential I.V. injections of sodium sulfite into mice, the formation of sulfur trioxide anion radical ($SO_3^{\cdot-}$) was observed by low-frequency (1.2 GHz) EPR spin trapping in vivo (382). Transition metals such as Cr(VI), V(V), Fe(II;III), Mn(II), and Ni(II) enhanced sulfur trioxide anion radical generation from sulfite in the presence of nitrite in vitro (383). In the presence of hydrogen peroxide, $SO_2$ generated 8-hydroxy-2-deoxyguanosine in vitro (384). $SO_2$ triggered oxidative stress in rats exposed to 5 ppm for 5 h a day between 7 and 28 days by depleting reduced glutathione (385). Sulfite can cause allergic reactions, the most common of which is bronchoconstriction in asthmatics (type 1 reaction) (386). In rare cases, sulfites may also cause ACD (387). According to our knowledge, there are presently no reports in the literature on prooxidant effects of $SO_2$ in skin.

## E. Mineral Fibers

Traces of airborne mineral fibers can be present as indoor pollutants and may pose a significant health risk, mainly in the respiratory tract. As discussed previously, there is overwhelming evidence that oxidative stress is involved in the pathogenesis of asbestos- and silica-mediated lung disease, such as lung fibrosis and carcinoma. The larger non-respirable man-made mineral fibers (MMMF), such as glass and carbon fibers, as well as stone wools, are potent skin irritants (388–392). MMMF with the largest diameter are stronger irritants than fibers with a small diameter, although other factors seem to influence the irritation potency as well (389). In vitro cytotoxicity assays using a human mesothelial cell line have demonstrated increased toxicity of thin compared to coarse fibers, with glass wool being more toxic than rock wool (393). The inflammatory skin reaction to these fibers was believed to be entirely mechanical, but surface free radical chemistry, release of transition metal ions such as iron ions, and phagocytosis events may lead to oxidative stress. Glasswool and rockwool fibers dose dependently increased the production of ROS/RNS in human polymorphous leukocytes (394). In comparison to respirable industrial fibers such as amosite and crocidolite asbestos, man-made vitreous fibers (MMVF) released very small amounts of Fe(II) in vitro and produced modest free radical damage to supercoiled plasmid DNA (395–397). However, release of iron from the different fibers was generally not a good correlate of ability to cause free radical injury to the plasmid DNA. MMVF caused some degree of oxidative stress in alveolar macrophages, as revealed by depletion of intracellular GSH. Amosite asbestos upregulated nuclear binding of activator protein 1 transcription factor to a greater level than MMVF and long-fiber amosite was the only fiber to enhance activation of NF-κB. It was suggested that the intrinsic free radical activity is the major determinant of transcription factor activation by the industrial fibers (398). In summary, these findings suggest a possible involvement of ROS in the pathogenesis of fiber-induced ICD.

## F. Quartz Dust

As discussed and summarized earlier in this book, occupational exposure to quartz dust has been linked with progressive systemic sclerosis (PSS)-like symptoms in workers employed in the mining industry (399–404) and with PSS or LS-like lesions in patients with silicon implants (405–409). Quartz dust does not penetrate skin, but enters the respiratory system and is ingested by pulmonary macrophages. The development of skin lesions resulting from inhaled quartz dust requires the systemic circulation of factors that act locally. Silicone, which is composed of polydimethylsiloxane and silica, is known to leak from the implants into the tissue. Although quartz dust and silicon implants are different materials, they are chemically similar and both can catalyze formation of ROS in noncellular and cellular systems. The radical reactive surface properties of silica are directly demonstrable by electron paramagnetic resonance (EPR) studies as described in Chapter 9. It was speculated that activation of the immune system may occur as part of this process. While in vitro studies generally support the idea that free radicals and ROS are involved in the pathogenesis of silica-induced scleroderma, no convincing clinical data are yet available to support the view of a ''radical pathology'' of silica-induced scleroderma.

## G. Tobacco Smoke

Tobacco smoke is a complex mixture of thousands of toxic chemical substances, including several prooxidant compounds. Exposure to tobacco smoke is strongly linked to cardiovascular and lung diseases, and tobacco-smoke-induced oxidative stress may significantly be involved in smoking-induced diseases (410–415). Recently, the cutaneous manifestations and consequences of smoking have attracted some attention. Maternal smoking during pregnancy and lactation was identified a risk factor for development of atopic dermatitis in the offspring (416), and smoking was reported to cause disturbed wound healing and premature skin aging (417). However, it was pointed out that smoking plays only a minor role in skin aging when compared to other compounding factors (418). Oral cancer has been overwhelmingly linked to smoking and smokers are at increased risk for development of squamous cell carcinoma of the skin (417). In contrast to the lung, almost no information is available on the effects of smoking on cutaneous molecular and cell biology. As described in Chapter 7, it may be speculated that oxidative stress contributes via circulating factors to smoking-induced skin pathology.

## X. MISCELLANEOUS ENVIRONMENTAL AND OCCUPATIONAL HAZARDS

Several chemically unrelated compounds are believed to be dermatotoxic, causing ICD and less often ACD, primarily through their prooxidant actions. Examples are hydroperoxides/peroxides, phenols/quinones, and primary amines.

### A. Hydroperoxides and Peroxides

Human skin is frequently exposed to various peroxy and hydroperoxy compounds that are in use in the cosmetic, pharmaceutical, and polymer industries and also generated as a result of the peroxidative metabolic conversion of endogenous skin lipids. In the occupational setting, organic peroxides are widely used as initiators for free radical polymerizations of monomers to thermoplastic polymers, for curing, thermoset polyster resins, and for cross-linking elastomers and polyethylene. Their usefulness is based on their thermal instability. The major structural classes of organic peroxides are diacyl peroxides, peroxyacids, ketone peroxides, dialkyl peroxides, peroxy esters, hydroxy peroxides, and peroxydicarbonates. The toxicity of peroxides is ascribed to their ability to generate free radicals and trigger peroxidation of lipids and other biomolecules (419). In isolated human keratinocytes, the cytotoxic effects of organic hydroperoxides and peroxides such as tert-butylhydroperoxide, tert-butyl-hydroxytoluene hydroperoxide, cumene hydroperoxide, and dicumyl peroxide appear to be mediated by their metabolic activation into alkoxyl, alkyl, or aryl radicals (420–427). But other nonradical products of peroxide metabolism, such as electrophilic quinone methides, are toxicologically important reactive intermediates of some hydroperoxides (428). The main toxic effect of hydroperoxides, peroxides, and their degradation products is irritation of the skin, mucous membranes, upper respiratory tract, and eyes (429–435). Tert-butyl-hydroperoxide activated NF-κB in dermal fibroblast (436), indicating that peroxides may trigger cutaneous inflammation through NF-κB-mediated signal transduction cascades. In some cases, ACD due to hydroperoxides and peroxides were observed (437–441). Oxidation products of linoleic acid released histamine from mast cells by a cytotoxic mechanism (442) and may thus initiate ICU. Some hydroperoxides and peroxides such as benzoylperoxide, decanoylperoxide, and cumenehydroperoxide can cause initiation and promotion of cancer in mouse epidermis purportedly due to free radical generation (443–446). Tumor-promoting efficacy generally showed an inverse association with thermal stability for the compounds tested (benzoylperoxide, di-t-butyl peroxide, dicumyl peroxide, di-m-chlorobenzoyl peroxide), suggesting that besides other factors the rate of free radical formation is a key component contributing to tumor promotion by organic peroxides (447). Although benzoylperoxide can induce promutagenic changes in isolated DNA, cultured murine, and human keratinocytes (448–450), and is a tumor promotor in mouse skin (446), it has been used in the treatment of acne for over 30 years, with no clinical reports of adverse effects that could be related to skin carcinogenesis. Based upon lack of activity from 23 animal carcinogenicity studies, the more than 30 years of safe clinical use, and negative findings from epidemiological studies, there is no evidence to support an association between benzoylperoxide and the development of skin cancer in humans (451).

## B. Phenols, Quinones, and Anthrones

A variety of phenolic compounds are used for industrial production of synthetic polymers, paints, adhesives, lacquers, and cosmetics. Phenolic compounds are potent skin irritants (452), but may also cause ACD and chemical leukoderma. Phenol in particular is corrosive when applied to human skin. Enzymic one-electron oxidation of phenolic compounds leads to formation of reactive phenol radicals (252,453–456) and redox cycling of phenol was demonstrated in human epidermal keratinocytes (457). At noncytotoxic concentrations, phenol induces the expression of proinflammatory cytokines such as IL-1, TNF-$\alpha$, and IL-8 in cultured human keratinocytes (458).

Quinones and catechols are widespread in the environment, especially as constituents of edible plants, but also as industrial and pharmaceutical products. Quinone metabolites of various compounds such as phenols, aromatic hydrocarbons, and PAH were proposed to be the reactive metabolites responsible for toxicity. Because of their molecular structure, quinones may act as electrophiles and/or redox cyclers. Redox cycling of quinones is known to be a potent source of ROS (459–461). NAD(P)H: quinone reductase is an important secondary antioxidant system for achieving protection against toxicity of electrophiles and ROS from quinones. Some quinones are potent skin irritants and sensitizers (452,462). For instance, skin necrosis following inadvertent extravasation of the anticancer drug adriamycin is a serious complication in chemotherapy. Adriamycine generates ROS in two ways, via redox cycling and via a Haber–Weiss-type reaction due to iron–adriamycin complexes. Under anaerobic conditions, it generates a reactive quinone methide. The corrosive skin reactions caused by adriamycin were inhibited by antioxidants (463–465).

Anthrones are distributed widely in nature and are useful as bird repellents and for several other agricultural uses. 9-Anthrone derivatives such as anthralin (1,8-dihydroxy-9-anthrone) are helpful in treating psoriasis and are also known to be tumor promoters in mouse skin. Their therapeutic use is accompanied by side effects of severe skin irritation. Anthralin is a strong reductant that is readily oxidized by light, trace concentrations of metal ions, and oxygen. ROS such as singlet oxygen, superoxide anion, and the hydroxyl radical are formed as reaction intermediates during oxidation of anthralin in noncellular and cellular systems (466–470). In addition, anthrone- and anthrone dimer radicals are formed in isolated keratinocytes, in murine skin in vitro (471–473) and in vivo (474). ROS generated from anthralin are presumably responsible for the induction of NF-$\kappa$B and proinflammatory cytokines in murine keratinocytes (475,476) and Balb-c mice (477), and lipid peroxidation was identified as the mediating event of anthralin-induced c-jun-N-terminal-kinase (JNK) activation in human keratinocytes and mononuclear cells (478).

Topical antioxidants inhibited anthralin-induced ICD in mouse (477,479,480) and human skin (481). NO inhibitors suppressed anthralin-induced erythema in human skin (482). Thus anthralin-induced ICD is believed to be mediated by ROS as well as RNS (470,477).

## C. Primary Aromatic Amines

The metabolism of primary aromatic amines may lead to formation of ROS and amine-derived free radical species (4,278,279,483,484) and para-phenylenediamine-induced oxidative stress in normal human keratinocytes in culture (485). Primary aromatic amines are potent contact allergens (486–488) and irritants in human skin (196).

## D. Sulfur Mustard

Sulfur mustard (bis-(2-chloroethyl)sulfide) (SM) has been used in chemical warfare since World War I. It is still a potential threat to both battlefield and civilian populations. SM causes an acute delayed irritant skin reaction with blisters, developing 1 to 24 h after exposure depending on the amount applied (490). One liquid droplet of about 10 μg will cause vesication in human skin, the estimated $LD_{50}$ of sulfur mustard on human skin is about 100 mg/kg (491). SM acts as a mutagen in various mammalian cell lines and is a carcinogen in several animal species (492). A carcinogenic effect was demonstrated in workers previously exposed occupationally to SM and followed for 20 years (493). In particular, the incidence of nonmelanoma skin cancer (494), as well as bronchial carcinoma, bladder carcinoma, and leukemia was significantly increased in these patients (493,495). Most of the toxic and particularly genotoxic effects of SM are related to alkylation of nucleic acids and depletion of thiols. Sulfur mustard rapidly reacts with intracellular reduced glutathione (496) and diminishes cellular reducing equivalents (497). However, the cellular response to SM is far more complex and involves increased production of cytokines, proteases, ROS, and RNS. Cutaneous exposure to SM results in significant neutrophil infiltration (498), and increased formation of ROS/RNS as well as mitochondrial damage may play a role in SM toxicity (499–504). SM induced release of both nitric oxide and presumably nitrosyl chloride, a known alkylating agent, from human epidermal keratinocytes (505). SM triggered formation of proinflammatory cytokines such as IL-1, TNF-α, IL-6, and IL-8 in human epidermal keratinocytes and monocytes (506–508). Enhanced proteolytic activity has been postulated as the cause of SM-induced dermal-epidermal separation (498,509,510) being involved in blister formation, and ROS are known to activate proteases and inactivate protease inhibitors. SM inhibited DNA binding of transcription factor AP-2 in vitro; however, the interference with binding was a result of alkylation of DNA and not damage to the transcription factor (511). Antioxidants such as ascorbic acid, *N*-acetylcysteine (512), and α-tocopherol (500) were recommended for treatment of SM-induced skin injuries. Intraperitoneal or intralesional administration of CuZn-SOD or Mn-SOD before sulfur-mustard-induced wound infliction dramatically reduced burn lesion areas in guinea pigs. No protective effect was found when SOD was administered after burn infliction (499). However, it must be pointed out that presently no specific antidote exists for treatment of SM lesions in patients (504,513,514).

## E. Lewisit

The arsenic-containing chemical warfare agents Adamsite (10-chloro-9,10-dihydrophenarsazine), Clark 1 (diphenylarsine chlorid), Clark 2 (diphenylarsine cyanide), and Lewisite (a mixture of cis/trans 2-chlorovinylarsindichlorid, di-(2-chlorvinyl)chlorarsin, tris-(2-chlorvinyl)arsin, and arsintrichlorid) are potent mucocutaneous irritants and blistering agents. They produce similar-type skin lesions as sulfur mustard. The mechanism of action of the arsenic blistering agents involves reaction with the sulfhydryl groups of proteins through their arsenic group (515). Depletion of reduced glutathione and shifting the intracellular SH/SS ratio toward the oxidized site will cause oxidative stress as a late event. 2,3-Dimercaptopropanol and water-soluble derivatives are effective and are specific antidotes against Lewisite intoxication.

## F. Ethylene Oxide and Epichlorhydrin

Ethylene oxide is one of the most widely used industrial chemical compounds in the western world. It is utilized as a fumigant in the food industry, for medical sterilization, and in the synthesis of ethylene glycol, from which automobile antifreeze and polyester clothing fibers are manufactured. Epichlorhydrin is a reactive epoxide used in polymer synthesis. Severe cutaneous irritant reactions (delayed-acute type of ICD) to ethylene oxide and epichlorhydrin have been reported (516–519). The mechanism of toxicity presumably involves primarily nucleophilic addition reactions to alcohol, amine, and thiol groups. In analogy to SM, ethylene oxide and epichlorhydrin may deplete cellular thiols, thereby causing oxidative stress as a secondary event. Similarly, the potent skin irritant 1,3-propane sultone, which is used as an alkylating agent in organic synthesis, is a potential thiol-depleting agent and causes a delayed acute type of ICD (516).

## G. Hydrofluoric Acid

Hydrofluoric acid is an important industrial chemical and is frequently used in wood manufacturing and as a household chemical for cleansing purposes. Hydrofluoric acid readily penetrates deep into the skin and subcutaneous tissues causing in a dose-dependent manner severe skin irritation and corrosion (520–523). Fluoride toxicity is generally believed to be mediated through calcium ion binding, thereby inactivating cellular metabolism. Fluoride ions modulate the cellular production of ROS and the function of antioxidant enzymes (524–527), but these reactions presumably play no or only a secondary role in acute hydrofluoric acid skin burns. As already outlined in Chapter 1, even if a cell dies as a result of mechanisms other than oxidative stress, necrotic cell death can impose oxidative stress on the surrounding tissue. Beneficial effects of ascorbic acid on reversal of fluoride toxicity in male rats was reported (528). However, in the clinical setting, treatment of choice for hydrofluoric acid burns is still intralesional administration of calcium ions and/or surgical intervention.

## H. Phosphorus

White phosphorus is used in many types of military munitions, fireworks, industrial and agricultural products, while red phosphorus is mainly utilized in matches. P(III/V) compounds are skin irritants (529). White phosphorus burns causing deep thermal injuries in skin and underlying tissue and may result following spontaneous combustion when contaminated skin surfaces dry. It may also cause multiorgan failure because of its toxic effects on erythrocytes, liver, kidneys, and heart. A clinical study showed that cutaneous thermal injury is followed by oxidative damage to the skin (530). In analogy to hydrofluoric acid skin burns, oxidative skin injury is rather a secondary and late event in acute phosphorus burns. In an animal study, SOD application reduced skin and hepatic damage following white phosphorus burns; however, irrigation with water was much more effective (531).

## REFERENCES


1. Plaa GL, Witschi H. Chemicals, drugs and lipid peroxidation. Ann Rev Pharmacol Toxicol 1976; 16:125–141.
2. Kappus H, Sies H. Toxic drug effects associated with oxygen metabolism: redox cycling and lipid peroxidation. Experientia 1981; 37:1233–1241.
3. Mason RP, Chignell CF. Free radicals in pharmacology and toxicology—selected topics. Pharmacol Rev 1982; 33:189–211.
4. Chignell CF. Structure activity relationships in the free radical metabolism of xenobiotics. Environ Health Perspect 1985; 61:133–137.
5. Roberfroid MB, Viehe HG, Remacle J. Free radicals in drug research. Adv Drug Res 1987; 16:1–84.
6. Comporti M. Three models of free radical induced cell injury. Chem Biol Interact 1989; 72: 1–56.
7. Mason RP. Redox cycling of radical anion metabolites of toxic chemicals and drugs and the Marcus theory of electron transfer. Environ Health Perspect 1990; 87:237–243.
8. Parke DV, Sapta A. Chemical toxicity and reactive oxygen species. Int J Occupat Med Environ Health 1996; 9:331–340.
9. Nriagu JO, Pacyna JM. Quantitative assessment of world wide contamination of air, water and soil by trace metals. Nature 1988; 333:134–139.
10. Kadiiska MB, Mason RP, Dreher KL, Costa DL, Ghio AJ. In vivo evidence of free radical formation in the rat lung after exposure to an emission source air pollution particle. Chem Res Toxicol 1997; 10:1104–1108.
11. Semple KM, Cyr N, Fedorak PM, Westlake DWS. Characterization of asphaltenes from Cold Lake heavy oil: Variations in chemical structure and composition with molecular size. Can J Chem 1990; 68:1092–1099.
12. Klein CB, Frenkel K, Costa M. The role of oxidative processes in metal carcinogenesis. Chem Res Toxicol 1991; 4:592–604.
13. Lansdown ABG. Physiological and toxicological changes in the skin resulting from the action and interaction of metal ions. Crit Rev Toxicol 1995; 25:397–462.
14. Guy RH, Hostynek JJ, Hinz RS, Lorence CR. Metals and the Skin. New York: Marcel Dekker, Inc, 1999.
15. Schuppe HC, Rönnau AC, von Schmiedeberg S, Ruzicka T, Gleichmann E, Griem P. Immunomodulation by heavy metal compounds. Clin Dermatol 1998; 16:149–157.
16. Kasprzak KS. The role of oxidative damage in metal carcinogenicity. Chem Res Toxicol 1991; 4:604–615.
17. Standeven AM, Wetterhahn KE. Is there a role for reactive oxygen species in the mechanism of chromium (VI) carcinogenesis? Chem Res Toxicol 1991; 4:616–625.
18. Kasprzak KS. Possible role of oxidative damage in metal-induced carcinogenesis. Cancer Invest 1995; 13:411–430.
19. Shi X, Chiu A, Chen CT, Halliwell B, Castranova V, Vallyathan V. Reduction of chromium (VI) and its relationship to carcinogenesis. J Toxicol Environ Health B Crit Rev 1999; 2: 87–104.
20. Sugiyama M. Role of physiological antioxidants in chromium (VI)-induced cellular injury. Free Radic Biol Med 1992; 12:397–407.
21. Shi X, Dalal NS. The role of superoxide radical in chromium (VI)-generated hydroxyl radical: The Cr(VI) Haber-Weiss cycle. Arch Biochem Biophys 1992; 292:323–327.

22. Mao Y, Zang L, Shi X. Generation of free radicals by Cr(IV) from lipid hydroperoxides and its inhibition by chelators. Biochem Mol Biol Int 1995; 36:327–337.
23. Luo H, Lu Y, Mao Y, Shi X, Dalal NS. Role of chromium(IV) in the chromium(VI)-related free radical formation, dG hydroxylation, and DNA damage. J Inorgan Biochem 1996; 64: 25–35.
24. Shi X, Dalal NS, Kasprzak KS. Generation of free radicals from hydrogen peroxide and lipid hydroperoxides in the presence of Cr(III). Arch Biochem Biophys 1993; 302:294–299.
25. Tsou TC, Yang JL. Formation of reactive oxygen species and DNA strand breakage during interaction of chromium(III) and hydrogen peroxide in vitro: evidence for a chromium(III)-mediated Fenton-like reaction. Chem Biol Interact 1996; 102:133–153.
26. Ye J, Zhang X, Young HA, Mao Y, Shi X. Chromium(VI)-induced nuclear factor-kappaB activation in intact cells via free radical reactions. Carcinogenesis 1995; 16:2401–2405.
27. Kadiiska MB, Xiang Q-H, Mason RP. In vivo free radical generation by chromium(VI): An electron spin resonance spin-trapping investigation. Chem Res Toxicol 1994; 7:800–805.
28. Shi X, Dalal NS. On the hydroxyl radical formation in the reaction between hydrogen peroxide and biologically generated chromium (V) species. Arch Biochem Biophys 1990; 277: 342–350.
29. Jones P, Kortenkamp A, O'Brien P, Wang G, Yang G. Evidence for the generation of hydroxyl radicals from a chromium (V) intermediate isolated from the reaction of chromate with glutathione. Arch Biochem Biophys 1991; 286:652–655.
30. Liu KJ, Mader K, Shi X, Swartz HM. Reduction of carcinogenic chromium (VI) on the skin of living rats. Magn Reson Med 1997; 38:5245–526.
31. Ozawa T, Hanaki A. Spin trapping studies on the reactions of Cr(III) with hydrogen peroxide in the presence of biological reductants; is Cr(III) nontoxic? Biochem Int 1990; 22:343–352.
32. Faux SP, Gao M, Chipman JK, Levy LS. Production of 8-hydroxydeoxyguanosine in isolated DNA by chromium(VI) and chromium(V). Carcinogenesis 1992; 13:1667–1669.
33. Tsou TC, Chen CL, Liu TY, Yang JL. Induction of 8-hydroxydeoxyguanosine in DNA by chromium(III) plus hydrogen peroxide and its prevention by scavengers. Carcinogenesis 1996; 17:103–108.
34. Bagchi D, Tran MX, Newton S, Bagchi M, Ray SD, Kuszynski CA, Stohs SJ. Chromium and cadmium induced oxidative stress and apoptosis in cultured J774A.1 macrophage cells. In Vitro Mol Toxicol 1998; 11:171–181.
35. Shi X, Ding M, Ye J, Wang S, Leonard SS, Zang L, Castranova V, Vallyathan V, Chiu A, Dalal N. Cr(IV) causes activation of nuclear transcription factor-kappaB, DNA strand breaks and dG hydroxylation via free radical reactions. J Inorgan Biochem 1999; 75:37–44.
36. Bagchi D, Vuchetich PJ, Bagchi M, Hassoun EA, Tran MX, Tang L, Stohs SJ. Induction of oxidative stress by chronic administration of sodium dichromate (chromium VI) and cadmium chloride (cadmium II) to rats. Free Radic Biol Med 1997; 22:471–478.
37. Singh J, Carlisle DL, Pritchard DE, Patierno SR. Chromium-induced genotoxicity and apoptosis: Relationship to chromium carcinogenesis (Review). Oncol Rep 1998; 5:1307–1318.
38. Magnusson B, Kligman AM. Allergic contact dermatitis in the guine pig. Identification of contact allergens. Springfield, IL: Charles C Thomas, 1970.
39. Cavelier C, Foussereau J. Kontaktallergie gegen Metalle und deren Salze. Teil I. Chrom und Chromate. Dermatosen/Occup Environ 1995; 43:100–112.
40. Epstein S. Contact dermatitis to nickel and chromate. Arch Dermatol 1956; 73:236–239.
41. Kanerva L, Aitio A. Dermatotoxicological aspects of metallic chromium. Eur J Dermatol 1997; 7:79–84.
42. Little MC, Gawkrodger DJ, MacNeil S. Chromium- and nickel induced cytotoxicity in normal and transformed human keratinocytes: an investigation of pharmacological approaches to the prevention of Cr (VI) induced cytotoxicity. Br J Dermatol 1996; 134:199–207.

43. Katz SA, Salem M. The toxicology of chromium with respect to its chemical specifications: a review. J Appl Toxicol 1993; 13:217–245.
44. Burrows D. Prognosis in industrial dermatitis. Br J Dermatol 1972; 87:145–148.
45. Samitz MH, Shrager J. Patch test reactions to hexavalent and trivalent chromium compounds. Arch Dermatol 1966; 94:304–306.
46. Samitz MH, Katz SA, Scheiner DM, Gross PR. Chromium–protein interactions. Acta Dermatol Venereol 1969; 49:142–146.
47. Gammelgaard B, Fullerton A, Avnstorp C, Menne T. Permeation of chromium salts through human skin in vitro. Contact Derm 1992; 27:302–310.
48. Stohs SJ, Bagagchi D. Oxidative mechanisms in the toxicity of metal ions. Free Rad Biol Med 1995;18:321–336.
49. Hanna PM, Kadiiska MB, Mason RP. Oxygen-derived free radical and active oxygen complex formation from cobalt(II) chelates in vitro. Chem Res Toxicol 1992; 5:109–115.
50. Leonard S, Gannett PM, Rojanasakul Y, Schwegler-Berry D, Casranova V, Vallyathan V, Shi X. Cobalt-mediated generation of reactive oxygen species and its possible mechanism. J Inorg Biochem 1998; 70:239–244.
51. Kawanishi S, Inoue S, Yamamoto K. Active oxygen species in DNA damage induced by carcinogenic metal compounds. Environ Health Perspec 1994; 102(suppl 3):17–20.
52. Mao Y, Liu KJ, Jiang JJ, Shi X. Generation of reactive oxygen species by Co(II) from $H_2O_2$ in the presence of chelators in relation to DNA damage and 2′-deoxyguanosine hydroxylation. J Toxicol Environ Health 1996; 47:61–75.
53. Wang X, Yokoi I, Liu J, Mori A. Cobalt(II) and nickel(II) ions as promoters of free radicals in vivo: Detected directly using electron spin resonance spectrometry in circulating blood in rats. Arch Biochem Biophys 1993; 306:402–406.
54. Cavelier C, Foussereau J. Kontaktallergie gegen Metalle und deren Salze. Teil II. Nickel, Kobalt, Quecksilber und Palladium. Dermatosen/Occup Environ 1995; 43:152–162.
55. Kligman AM. The identification of contact allergens by human assay. III The maximation test: a procedure for screening and rating contact sensitizers. J Invest Dermatol 1966; 47: 393–409.
56. Nürnberger F, Arnold W. Quincke's edema as a manifestation of cobalt allergy due to a sensitization of cobatous shell splinters. Berufsdermatosen 1969; 17:21–25.
57. Pisati G, Zedda S. Outcome of occupational asthma due to cobalt hypersensitivity. Sci Total Environ 1994; 150:167–171.
58. Camarasa JG, Alomar A. Photosensitization to cobalt in a bricklayer. Contact Derm 1981; 7:154–156.
59. Romaguera C, Lech M, Grimault F. Photocontact dermatitis to cobalt salts. Contact Derm 1982; 8:383–385.
60. Gutteridge JMC. Tissue damage by oxy radicals: The possible involvement of iron and copper complexes. Med Biol 1984; 62:101–104.
61. Boman A, Karlberg AT, Einarsson O, Wahlberg JE. Dissolving copper by synthetic sweat. Contact Dermatitis 1983; 9:159–160.
62. Beveridge SJ, Whitehouse MW, Walker WR. Lipophilic copper(II) formulations: some correlations between their composition and anti-inflammatory/antiarthritic activity when applied to the skin of rats. Agents Actions 1982; 12:225–231.
63. Pirot F, Panisset F, Agache P, Humbert P. Simultaneous absorption of copper and zinc through human skin in vitro: influence of counter ion and vehicle. Skin Pharmacol 1996; 9: 43–52.
64. Karlberg AT, Boman A, Wahlberg JE. Copper—a rare sensitizer. Contact Derm 1983; 9: 134–139.
65. Frykholm KO, Frithiof L, Fernstrom AIB, Moberger G, Blohm SG, Bjorn E. Allergy to copper derived from dental alloys as a possible cause of oral lesions of lichen planus. Acta Dermatol Venereol 1969; 49:268–273.

66. Saltzer EI, Wilson JW. Allergic contact dermatitis due to copper. Arch Dermatol 1968; 98: 375–376.
67. Sucvi L, Prodnan L, Lazar V. Research on copper poisoning. Med Law 1981; 3:190–212.
68. Halliwell B, Gutteridge JMC. Oxygen toxicity, oxygen radicals, transition metals and disease. Biochem J 1984; 219:1–14.
69. Halliwell B, Gutteridge JMC. Role of free radicals and catalytic metal ions in human disease: an overview. Meth Enzymol 1990; 186:1–85.
70. Mason RP, Hanna PM, Burkitt MJ, Kadiiska MB. Detection of oxygen derived radicals in biological systems using electron spin resonance. Environ Health Perspect 1994; 102(suppl 10):33–36.
71. Toyokuni S. Iron-induced carcinogenesis: The role of redox regulation. Free Radic Biol Med 1996; 20:553–566.
72. Okada S. Iron and carcinogenesis in laboratory animals and humans: A mechanistic consideration and a review of literature. Int J Clin Oncol 1998; 3:191–203.
73. Selby JV, Friedman GD. Epidemiologic evidence of an association between body iron stores and risk of cancer. Int J Cancer 1988; 41:677–682.
74. Stevens RG, Graubard BI, Micozzi MS, Neriishi K, Blumberg BS. Moderate elevation of body iron level and increased risk of cancer occurrence and death. Int J Cancer 1994; 56: 364–369.
75. Rezazadeh H, Athar M. Evidence that iron overload promotes 7,12-dimethylbenz(a)anthracene induced skin tumorigenesis in mice. Redox Rep 1997; 3:303–309.
76. Rezazadeh H, Julka PK, Athar M. Iron overload augments 7,12-dimethyl-benz(a)anthracene-initiated and 12-O-tetradecanoylphorbol-13-acetate-promoted skin tumorigenesis. Skin Pharmacol Appl Skin Physiol 1998; 11:98–103.
77. Akimov VT. Contribution of iron to the pathogenesis of free radical processes induced by ultraviolet irradiation of the skin of guinea pigs. Vestnik Dermatol Venerol 1987; 2:7–13.
78. Applegate LA, Frenk E. Oxidative defense in cultured human skin fibroblasts and keratinocytes from sun-exposed and non-exposed skin. Photodermatol Photoimmunol Photomed 1995; 11:95–101.
79. Morliere P, Salmon S, Aubailly M, Risler A, Santus R. Sensitization of skin fibroblasts to UVA by excess iron. Biochim Biophys Acta 1997; 1334:283–290.
80. Trenam CW, Dabbagh AJD, Morris CJ. The role of iron in an acute model of skin inflammation induced by reactive oxygen species (ROS). Br J Dermatol 1992; 126:250–256.
81. Ackermann Z, Seidenbaum M, Loewenthal E, Rubinow A. Overload of iron in the skin of patients with varicose ulcers. Arch Dermatol 1988; 124:1376–1378.
82. Weintraub LR, Demis DJ, Conrad ME, Crosby WH. Iron excretion by the skin. Selective localization of iron 59 in epithelial cells. Am J Pathol 1965; 46:121–127.
83. Minato A, Fukuzawa H, Hirose S, Matsunaga Y. Radioisotopic studies on percutaneous absorption. I. Absorption of water soluble substances from hydrophilic cremes and absorption from ointments through mouse skin. Chem Pharm Bull 1967; 15:1470–1477.
84. Nater JP. Epidermal hypersensitivity to iron. Hautarzt 1960; 11:223–225.
85. Baer R. Allergic contact sensitization to iron. J Allergy Clin Immunol 1973; 51:35–37.
86. Athar M, Hasan SK, Srivastav RC. Evidence for the involvement of hydroxyl radicals in nickel mediated enhancement of lipid peroxidation: implications for nickel carcinogenesis. Biochem Biophys Res Commun 1987; 147:1276–1281.
87. Misra M, Rodriguez RE, Kasprzak K. Nickel induced lipid peroxidation in the rat: correlation with nickel effect on antioxidant defense systems. Toxicology 1990; 64:1–17.
88. Misra M, Rodriguez RE, North SL, Kasprzak K. Nickel-induced renal lipid peroxidation in different strains of mice: concurrence with nickel effect on antioxidant defense systems. Toxicol Lett 1991; 58:121–133.
89. Huang X, Frenkel K, Klein CB, Costa M. Nickel induces increased oxidants in intact cultured

mammalian cells as detected by dichlorofluorescein fluorescence. Toxicol Appl Pharmacol 1993; 120:29–36.
90. Shi X, Dalal NS, Kasprzak KS. Generation of free radicals in reactions of nickel-thiol complexes with molecular oxygen and model lipid hydroperoxides. J Inorgan Biochem 1993; 50:211–225.
91. Tkeshelashvili LK, Reid TM, McBride TJ, Loeb LA. Nickel induces a signature mutation for oxygen free radical damage. Cancer Res 1993; 53:4172–4174.
92. Huang X, Zhuang Z, Frenkel K, Klein CB, Costa M. The role of nickel and nickel-mediated reactive oxygen species in the mechanism of nickel carcinogenesis. Environ Health Perspect 1994; 102:281–284.
93. Novelli EL, Rodrigues NL, Ribas BO. Superoxide radical and toxicity of environmental nickel exposure. Hum Exp Toxicol 1995; 14:248–251.
94. Lynn S, Yew FH, Chen KS, Jan KY. Reactive oxygen species are involved in nickel inhibition of DNA repair. Environ Mol Mutagen 1997; 29:208–216.
95. Chen CY, Huang YL, Lin TH. Association between oxidative stress and cytokine production in nickel-treated rats. Arch Biochem Biophys 1998; 356:127–132.
96. Bal W, Lukszo J, Kasprazak KS. Interactions of nickel (II) with histones: enhancement of 2-deoxyguanosine oxidation by nickel (II) complexes with CH3CO-CYS-ALA-ILE-HIS-NH2, a putative metal binding sequence of histone H3. Chem Res Toxicol 1996; 9:5356–540.
97. Goebeler M, Roth J, Broecker EB, Sorg C, Schulze-Osthoff K. Activation of nuclear factor-kappa-B and gene expression in human endothelial cells by the common haptens nickel and cobalt. J Immunol 1995; 155:2459–2467.
98. Wagner MK, Klein CL, Van Kooten GT, Kirkpatrick J. Mechanisms of cell activation by heavy metal ions. J Biomed Mat Res 1998; 42:443–452.
99. Gueniche A, Viac J, Lizard G, Charveron M, Schmitt D. Effect of various metals on intercellular adhesion molecule-1 expression and tumour necrosis factor alpha production by normal human keratinocytes. Arch Dermatol Res 1994; 286:466–470.
100. Wataha JC, Sun ZL, Hanks CT, Fang DN. Effect of Ni ions on expression of intercellular adhesion molecule 1 by endothelial cells. J Biomed Mat Res 1997; 36:145–151.
101. Buckley WR, Lewis CE. The ''ruster'' in industry. J Occupat Med 1960; 2:23–31.
102. Hemingway JD, Molokhia MM. The dissolution of metallic nickel in artificial sweat. Contact Derm 1987; 16:99–105.
103. Samitz MH, Katz SA. Nickel-epidermal interactions: diffusion and binding. Environ Res 1976; 11:34–39.
104. Fullerton A, Hoelgaard A. Binding of nickel to human epidermis in vitro. Br J Dermatol 1988; 119:675–679.
105. Fullerton A, Andersen JR, Hoelgaard A, Menne T. Permeation of nickel salts through human skin in vitro. Contact Derm 1986; 15:173–177.
106. Lewis R. Metals. In: LaDou J, ed. Occupational and Environmental Medicine, 2nd ed. Stamford: Appleton & Lange, 1997:405–439.
107. Barceloux DG. Nickel. Toxicol Clin Toxicol 1999; 37:239–250.
108. Lewis R. Metals. In: LaDou J, ed. Occupational and Environmental Medicine, 2nd ed. Stamford: Appleton & Lange, 1997:405–439.
109. Wahlberg JE. Nickel: the search for alternative, optimal and non-ittitant patch test preparations. Assessment based on laser Doppler flowmetry. Skin Res Technol 1996; 2:136–141.
110. Gollhausen R, Enders F, Przybilla B, Burg G, Ring J. Trends in allergic contact sensitization. Contact 1988; 18:147–54.
111. Artik S, von Vultee C, Gleichmann E, Schwartz T, Griem P. Nickel allergy in mice: enhanced sensitization capacity of nickel at higher oxidation states. J Immunol 1999; 163:1143–1152.
112. NewKirk HW, Goluba RW. Trace metals in the environment—platinum from catalytic con-

verters on a passenger vehicle. University of California, Lawrence Livermore Laboratory, UCID 16321, 1973.

113. Schroeder HA. Metallic micronutrients and intermediary metabolism. US Gov Res Drug Rep 1967; 70:57–58.
114. Liu TZ, Lee SS, Bhatnagar RS. Toxicity of palladium. Toxicol Lett 1979; 4:469–473.
115. Liu TZ, Lin TF, Chiu DTY, Tsai KJ, Stern A. Palladium or platinum exacerbates hydroxyl radical mediated DNA damage. Free Rad Biol Med 1997; 23:155–161.
116. Chiu DTY, Liu TZ. Free radical and oxidative damage in human blood cells. J Biomed Sci 1997; 4:256–259.
117. Camarasa JG, Setta-Baldrich E. Palladium contact dermatitis. Am J Contact Derm 1990; 1: 114–115.
118. Baker DB, Gann PH, Brooks SM, Gallagher J, Bernstein IL. Cross sectional study of platinum salts sensitization among precious metals refinery workers. Am J Ind Med 1990; 18:653–659.
119. Brooks SM, Baker DB, Gann PH, Jarabek AM, Hertzberg V, Gallagher J, Biagini RE, Bernstein IL. Cold air challenge and platinum skin reactivity in platinum refinery workers: bronchial reactivity precedes skin prick response. Chest 1990; 97:101–106.
120. Schuppe HC, Kulig J, Lerchenmueller C, Becker D, Gleichmann E, Kind P. Contact hypersensitivity to disodium hexachloroplatinate in mice. Toxicol Lett 1997; 93:125–133.
121. Keller RJ, Coulombe RA, Sharma RP, Grove TA, Piette JH. Importance of hydroxyl radical in the vanadium-stimulated oxidation of NADH. Free Radic Biol Med 1989; 6:15–22.
122. Keller RJ, Sharma RP, Grover TA, Piette LH. Vanadium and lipid peroxidation: evidence for involvement of vanadyl and hydroxyl radical. Arch Biochem Biophys 1988; 265:524–533.
123. Shi X, Wang P, Jiang H, Mao Y, Ahmed N, Dalal N. Vanadium(IV) causes 2′-deoxyguanosine hydroxylation and deoxyribonucleic acid damage via free radical reactions. Ann Clin Lab Sci 1996; 26:39–49.
124. Shi X, Jiang H, Mao Y, Ye J, Saffiotti U. Vanadium(IV)-mediated free radical generation and related 2′-deoxyguanosine hydroxylation and DNA damage. Toxicology 1996; 106:27–38.
125. Shainkin-Kestenbaum R, Caruso C, Berlyne GM. Vanadium and oxygen free radicals: Inhibition of superoxide dismutase and enhancement of hydroxydopamine oxidation. Trace Elements Med 1991; 8:6–10.
126. Bay BH, Sit KH, Paramanantham R, Chan YG. Hydroxyl free radicals generated by vanadyl[IV] induce cell blebbing in mitotic human Chang liver cells. Biometals 1997; 10:119–122.
127. Ding M, Li JJ, Leonard SS, Ye JP, Shi X, Colburn NH, Castranova V, Vallyathan V. Vanadate induced activation of activator protein-1: a role of reactive oxygen species. Carcinogenesis 1999; 20:663–668.
128. Chen F, Demers LM, Vallyathan V, Ding M, Lu Y, Castranova V, Shi X. Vanadate induction of NF-kappaB involves IkappaB kinase beta and SAPK/ERK kinase 1 in macrophages. J Biol Chem 1999; 274:20307–20312.
129. Zychlinski L, Byczkowski JZ, Kulkarni AP. Toxic effects of long-term intratracheal administration of vanadium pentoxide in rats. Arch Environ Contam Toxicol 1991; 20:295–298.
130. Ding M, Gannett PM, Rojanasakul Y, Liu K, Shi X. One-electron reduction of vanadate by ascorbate and related free radical generation at physiological pH. J Inorg Biochem 1994; 55: 101–112.
131. Younes M, Strubelt O. Vanadate induced toxicity towards isolated perfused rat livers. The role of lipid peroxidation. Toxicology 1991; 66:63–74.
132. Thomas DLG, Stiebris K. Vanadium poisoning in industry. Med J Aust 1956; 1:607–609.
133. Halliwell B. Oxidants and the central nervous system: some fundamental questions. Is oxidant

damage relevant to Parkinson's disease, Alzheimer's disease, traumatic injury or stroke? Acta Neurol Scand 1989; 126(suppl):23–33.

134. Erasmus RT, Savory J, Wills MR, Herman MM. Aluminum neurotoxicity in experimental animals. Ther Drug Mon 1993; 15:588–592.
135. Stankovic A, Mitrovic DR. Aluminum salts stimulate luminol-enhanced chemiluminescence production by human neutrophils. Free Radic Res Commun 1991; 14:47–55.
136. Abreo K, Glass J. Cellular, biochemical, and molecular mechanisms of aluminum toxicity. Nephrol Dial Transplant 1993; 8(suppl 1):5–11.
137. Strong MJ, Garruto RM, Joshi JG, Mundy WR, Shafer TJ. Can the mechanisms of aluminum neurotoxicity be integrated into a unified scheme? J Toxicol Environ Health 1996; 48:599–613.
138. Mundy WR, Freudenrich TM, Kodavanti PR. Aluminum potentiates glutamate-induced calcium accumulation and iron-induced oxygen free radical formation in primary neuronal cultures. Mol Chem Neuropathol 1997; 32:41–57.
139. Van Rensburg SJ, Daniels WMU, Potocnik FCV, Van Zyl JM, Taljaard JJF, Emsley RA. A new model for the pathophysiology of Alzheimer's disease: Aluminum toxicity is exacerbated by hydrogen peroxide and attenuated by an amyloid protein fragment and melatonin. SAMJ 1997; 87:1111–1115.
140. Abou-Seif MA. Oxidative stress of vanadium-mediated oxygen free radical generation stimulated by aluminum on human erythrocytes. Ann Clin Biochem 1998; 35:254–260.
141. Campbell A, Prasad KN, Bondy SC. Aluminum induced oxidative events in cell lines: glioma are more responsive than neuroblastoma. Free Rad Biol Med 1999; 26:1166–1171.
142. Meglio L, Oteiza PI. Aluminum enhances melanin-induced lipid peroxidation. Neurochem Res 1999; 24:1001–1008.
143. Lyon I, Klotz IM. The interaction of epidermal protein with aluminum salts. J Am Pharmacol Assoc 1958; 47:509–514.
144. Lansdown ABG. Production of epidermal damage in mammalian skin by some simple aluminum compounds. Br J Dermatol 1973; 89:67–76.
145. Maloney ME. Arsenic in dermatology. Derm Surg 1996; 22:301–304.
146. Andreae MO. Arsenic speciation in seawater and interstitial waters: the influence of biological-chemical interactions on the chemistry of a trace element. Limnol Oceanogr 1979; 24: 440–452.
147. Yamanaka K, Hasegawa A, Sawaruma R, Okada S. Dimethylated arsenics induce DNA strand breaks in lung via the production of active oxygen in mice. Biochem Biophys Res Commun 1989; 165:43–50.
148. Nordenson I, Beckman L. Is the genotoxic effect of arsenic mediated by oxygen free radicals? Hum Hered 1991; 41:71–73.
149. Yamanaka K, Okada S. Induction of lung specific DNA damage by metabolically methylated arsenics via the production of free radicals. Environ Health Perspect 1994; 102(suppl 3):37–40.
150. Wang TS, Huang H. Active oxygen species are involved in the induction of micronuclei by arsenite in XRS-5 cells. Mutagenesis 1994; 9:253–257.
151. Lee TC, Ho IC. Expression of heme oxygenase in arsenic-resistant human lung adenocarcinoma cells. Cancer Res 1994; 54:1660–1664.
152. Rin K, Kawaguchi K, Yamanaka K, Tezuka M, Oku N, Okada S. DNA-strand breaks induced by dimethylarsinic acid, a metabolite of inorganic arsenics, are strongly enhanced by superoxide anion radicals. Biol Pharmacol Bull 1995; 18:45–48.
153. Wanibuchi H, Hori T, Meenakshi V. Promotion of rat hepatogenesis by dimethylarsenic acid: association with elevated ornithine decarboxylase activity and formation of 8-hydroxydeoxyguanosine in the liver. Jpn J Cancer Res 1997; 88:1149–1154.
154. Landolph JR. The role of free radicals in chemical carcinogenesis. In: Rhodes CJ, ed. Toxicology of the Human Environment. London: Taylor and Francis, 2000:339–362.

155. Luster MI, Wilmer JL, Germolec DR, Spalding J, Yoshida T, Gaido K, Simeonova PP, Burleson FG, Bruccoleri A. Role of keratinocyte-derived cytokines in chemical toxicity. Toxicol Lett 1995; 82–83:471–476.
156. Germolec DR, Spalding J, Boorman GA, Wilmer JL, Yoshida T, Simeonova PP, Bruccoleri A, Kayama F, Gaido K, Tennant R. Arsenic can mediate skin neoplasia by chronic stimulation of keratinocyte-derived growth factors. Mutat Res 1997; 386:209–218.
157. Lingren A, Vater M, Dencker L. Autoradiographic studies on the distribution of arsenic in mice and hamsters administered 74As-arsenite or arsenate. Acta Pharmacol Toxicol 1982; 51:253–260.
158. Hindmarsh JT, McCurdy RF. Clinical and environmental aspects of arsenic toxicity. Crit Rev Clin Lab Sci 1986; 23:315–347.
159. Smith AH, Hopenhayn-Rich C, Bates MN, Goeden HM, Hertz-Picciotto I, Duggan HM, Wood R, Kosnett MJ, Smith MT. Cancer risks from arsenic in drinking water. Environ Health Perspect 1992; 97:259–267.
160. Holinquist R. Occupational arsenical dermatitis. Acta Dermatol Venereol 1951; 31:26–31.
161. Birmingham DJ, Key MM, Holaday DA, Perone VB. An outbreak of arsenical dermatitis in a mining community. Arch Dermatol 1965; 91:457–452.
162. Lüchtrath H. The consequences of chronic arsenic poisoning among Moselle wine growers. Pathoanatomical investigations of post-mortem examinations performed between 1960 and 1977. J Cancer Res Clin Oncol 1983; 105:173–181.
163. Schwartz RA. Arsenic and the skin. Int J Dermatol 1997; 36:241–250.
164. Piamphongsant T. Chronic environmental arsenic poisoning. Int J Dermatol 1999; 38:401–410.
165. Ochi T, Takahashi K, Ohsawa M. Indirect evidence for the induction of a prooxidant state by cadmium chloride in cultured mammamlian cells and a possible mechanism for the induction. Mutat Res 1987; 180:257–266.
166. Amoruso MA, Witz G, Goldstein BD. Enhancement of rat and human phagocyte superoxide anion radical production by cadmium in vitro. Toxicol Lett 1982; 10:133–138.
167. Hassoun EA, Stohs SJ. Cadmium-induced production of superoxide anion and nitric oxide, DNA single strand breaks and lactate dehydrogenase leakage in J774A.1 cell cultures. Toxicology 1996; 112:219–226.
168. Hussain T, Shukla GS, Chandra SV. Effects of cadmium on superoxide dismutase and lipid peroxidation in liver and kidney of growing rats: in vivo and in vitro studies. Pharmacol Toxicol 1987; 60:355–358.
169. Bagchi D, Bagchi M, Hassoun EA, Stohs SJ. Cadmium-induced excretion of urinary lipid metabolites, DNA damage, glutathione depletion, and hepatic lipid peroxidation in Sprague-Dawley rats. Biol Trace Elem Res 1996; 52:143–154.
170. Yang CF, Shen HM, Shen Y, Zhuang ZX, Ong CN. Cadmium-induced oxidative cellular damage in human fetal lung fibroblasts (MRC-5 cells). Environ Health Perspect 1997; 105:712–716.
171. Hayes RB. The carcinogenicity of metals in humans. Cancer Causes Control 1997; 8:371–385.
172. Sorahan T, Lancashire RJ. Lung cancer mortality in a cohort of workers employed at a cadmium recovery plant in the United States: an analysis with detailed job histories. Occupat Environ Med 1997; 54:194–201.
173. Müller T, Schuckelt R, Jaenicke L. Cadmium/zink metallothionein induces DNA strand breaks in vitro. Arch Toxicol 1991; 65:20–26.
174. Wester RC, Maibach HI, Sedik J, Melendres J, DiZio S, Wade M. In vitro percutaneous absorption of cadmium from water and soil into human skin. Fund Appl Toxicol 1992; 19:1–5.
175. Wahlberg JE, Boman A. Guinea pig maximation test method—cadmium chloride. Contact Derm 1979; 5:405.

176. Wahlberg JE. Routine patch testing with cadmium chloride. Contact Derm 1977; 3:293–296.
177. Richardt G, Federolf G, Habermann E. Affinity of heavy metal ions to intracellular $Ca^{2+}$ binding proteins. Biochem Pharmacol 1986; 35:1331–1335.
178. Hermes-Lima M, Pereira B, Bechara EJ. Are free radicals involved in lead poisoning? Xenobiotica 1991; 21:1085–1090.
179. Donaldson WE, Knowles SO. Is lead toxicosis a reflection of altered fatty acid composition of membranes? Comp Biochem Physiol 1993; 104C:377–379.
180. Bondy SC, Guo SX. Lead potentiates iron-induced formation of reactive oxygen species. Toxicol Lett 1996; 87:109–112.
181. Bress WC, Bidanset JH. Percutaneous in vivo and in vitro absorption of lead. Vet Hum Toxicol 1991; 33:212–217.
182. Gstraunthaler G, Pfaller W, Kotanko P. Glutathione depletion and in vitro lipid peroxidation in mercury or maleate induced acute renal failure. Biochem Pharmacol 1983; 32:2969–2972.
183. Yonaha M, Ohbayashi Y, Ichinose T, Sagai M. Lipid peroxidation stimulated by mercuric chloride and its relation to toxicity. Chem Pharm Bull 1982; 30:1437–1442.
184. Shainkin-Kestenbaum R, Caruso C, Berlyne GM. Effect of mercury on oxygen free radical metabolism; inhibition of superoxide dismutase activity and enhancement of hydroxydopamine oxidation. Trace Elements Med 1992;9:9–13.
185. Jansson G, Harms-Ringdahl M. Stimulating effects of mercuric- and silver ions on the superoxide anion production in human polymorphonuclear leukocytes. Free Radic Res Commun 1993; 18:87–98.
186. Pelletier L, Pasquier R, Rossert J, Vial MC, Mandet C, Druet P. Autoreactive T cells in mercury-induced autoimmunity. Ability to induce the autoimmune disease. J Immunol 1988; 140:750–754.
187. Stiller-Winkler R, Zhang S, Idel H, Brockhaus A, Jansen-Rosseck P, Vohr HW, Pietsch P, Olberding P, Gleichmann D. The immune system is a preferential target of $HgCl_2$ in susceptible mouse strains. Immunobiology 1988; 178:148.
188. Bagenstose LM, Salgame P, Monestier M. Murine mercury-induced autoimmunity: A model of chemically related autoimmunity in humans. Immunol Res 1999; 20:67–78.
189. Röger J, Zillikens D, Burg G. Systemic autoimmune disease in a patient with long standing exposure to mercury. Eur J Dermatol 1992; 2:168–170.
190. Schrallhammer-Benkler K, Ring J, Przybilla B. Acute mercury intoxication with lichenoid drug eruption followed by mercury contact allergy and development of antinuclear antibodies. Acta Dermatol Venereol 1992; 72:294–296.
191. Zelikoff JT, Smialowicz R, Bigazzi PE, Goyer RA, Lawrence DA, Maibach HI, Gardner D. Immunomodulation by metals. Fund Appl Toxicol 1994; 22:1–7.
192. Shibata K, Yoshita Y, Matsumoto H. Extensive chemical burns from toluene. Am J Emerg Med 1994; 12:353–355.
193. Langman JM. Xylene: Its toxicity, measurement of exposure levels, absorption, metabolism and clearance. Pathology 1994; 26:301–309.
194. Zisu D. Experimental study of cutaneous tolerance to glycol ethers. Contact Dermat 1995; 32:74–77.
195. Dekant W. Toxicology of chlorofluorocarbon replacements. Environ Health Perspec 1996; 104(suppl 1):75–83.
196. Bagley DM, Gardner JR, Holland G, Lewis RW, Regnier JF, Stringer DA, Walker AP. Skin irritation: Reference chemicals data bank. Toxicol In Vitro 1996; 10:1–6.
197. Abrams K, Harvell JD, Shriner D, Wertz P, Maibach HI, Rehfeld SJ. Effect of organic solvents on in-vitro human skin water barrier function. J Invest Dermatol 1993; 101:609–613.
198. Wood JC, Jackson SM, Elias PM, Grunfeld C, Feingold KR. Cutaneous barrier pertubation stimulates cytokine production in the epidermis of mice. J Clin Invest 1992; 90:482–487.

199. Nickoloff BJ, Naidu Y. Perturbation of epidermal barrier function correlates with initiation of cytokine cascade in human skin. J Am Acad Dermatol 1994; 30:535–546.
200. Wood LC, Elias P, Calhoun C, Tsai JC, Grunfeld C, Feingold KR. Barrier disruption stimulates interleukin-1a expression and release from a pre-formed pool in murine epidermis. J Invest Dermatol 1996; 106:397–403.
201. Luger TA, Lotti T. Neuropeptides: role in inflammatory skin diseases. J Eur Acad Dermatol Venerol 1998; 10:207–211.
202. Iyadomi M, Higak Y, Ichiba M, Morimoto M, Tomokuni K. Evaluation of organic solvent-induced inflammation modulated by neuropeptides in the abdominal skin of hairless rats. Indust Health 1998; 36:40–51.
203. Raza H, Qureshi MM, Montague W. Alteration of glutathione, glutathione S transferase and lipid peroxidation in mouse skin and extracutaneous tissues after topical application of gasoline. Int J Biochem Cell Biol 1995; 27:271–277.
204. Wolf R. Alcohol and the skin. Clin Dermatol 1999; 17:351–489.
205. Albano A. Free radical mechanism of ethanol toxicity. In: Rhodes CJ, ed. Toxicology of the Human Environment. London: Taylor and Francis, 2000:235–263.
206. Bruze M, Fergert S. Chemical skin burns. In: Menne T, Maibach HI, eds. Hand Eczema. Boca Raton: CRC Press, 1994:21–30.
207. Sparrow GP. A connective tissue disorder similar to vinyl chloride disease in a patient exposed to perchlorethylene. Clin Exp Dermatol 1977; 2:17–22.
208. Schirren SM. Skin lesions caused by trichlorethylen in a metalworking plant. Berufsdermatosen 1971; 19:240–254.
209. Czirjak L, Danko K, Schlammadinger J, Suranyi P, Tamasi L, Szegedi GY. Progressive systemic sclerosis occuring in patients exposed to chemicals. Int J Dermatol 1987; 26:374–378.
210. Flindt-Hansen H, Isager H. Scleroderma after occupational exposure to trichlorethylene and trichlorethane. Acta Dermatol Venereol (Stockholm) 1987; 67:263–264.
211. Lockey JE, Kelly CR, Cannon GW. Progressive systemic sclerosis associated with exposure to trichloroethylene. J Occup Med 1987; 29:493–496.
212. Saihan EM, Burton JL, Heaton KW. A new syndrom with pigmentation, scleroderma, gynaecomastia, Raynaud's phenomenon and peripheral neuropathy. Br J Dermatol 1978; 99:437–440.
213. Janzen EG, Stronks HJ, Dubose CM, Poyer JL, McCay PB. Chemistry and biology of spin trapping radicals associated with halocarbon metabolism in vitro and in vivo. Environ Health Perspec 1985; 64:151–170.
214. Knecht KT, Mason RP. The detection of halocarbon-derived radical adducts in bile and liver of rats. Drug Metab Dispos 1991; 19:325–331.
215. Casciola-Rosen L, Wigley F, Rosen A. Scleroderma autoantigens are uniquely fragmented by metal-catalyzed oxidation reactions: implications for pathogenesis. J Exp Med 1997; 185: 71–79.
216. Poland A, Knutson JC. 2,3,7,8-Tetrachlorodibenzo-p-dioxin and related halogenated aromatic hydrocarbons: examination of the mechanism of tocicity. Ann Rev Pharmacol Toxicol 1982; 22:517–554.
217. Alsharif NZ, Lawson T, Stohs SJ. Oxidative stress induced by 2,3,7,8-tetrachlorodibenzo-p-dioxin is mediated by the aryl hydrocarbon (Ah) receptor complex. Toxicology 1994; 92: 39–51.
218. Nebert DW, Puga A, Vasiliou V. Role of the Ah receptor and the dioxin-inducible (Ah) gene battery in toxicity, cancer and signal transduction. Ann NY Acad Sci 1993; 685:624–640.
219. Berkers JA, Hassing I, Spenkelink B, Brouwer A, Blauboer BJ. Interactive effects of 2,3,7,8-tetrachlorodibenzo-p-dioxin and retinoids on proliferation and differentiation in cultured human keratinocytes: quantification of cross-linked envelope formation. Arch Toxicol 1995; 69:368–378.

220. Panteleyev AA, Thiel R, Wanner R, Zhang J, Roumak VS, Paus R, Neubert D, Henz BM, Rosenbach T. 2,3,7,8-Tetrachlorodibenzo-p-dioxin (TCDD) affects keratin 1 and keratin 17 gene expression and differentially induces keratinization in hairless mouse skin. J Invest Dermatol 1997; 108:330–335.
221. Albro PW, Corbett JT, Harris M, Lawson LD. Effects of 2,3,7,8-tetrachlorodibenzo-p-dioxin on lipid profiles in tissues of the Fischer rat. Chem Biol Interact 1978; 23:315–320.
222. Sweeney GD, Jones KG, Cole FM, Basford D, Krestynski F. Iron deficiency prevents liver toxicity of 2,3,7,8-tetrachlorodibenzo-p-dioxin. Science 1979; 204:332–335.
223. Safe SH. Comparative toxicology and mechanism of action of polychlorinated dibenzo-p-dioxins and dibenzofurans. Ann Rev Pharmacol Toxicol 1986; 26:371–399.
224. Hassan MQ, Mohammadpour H, Hermansky SJ, Murray WJ, Stohs SJ. Comparative effects of BHA and ascorbic acid on the toxicity of 2,3,7,8-tetrachlorodibenzo-p-dioxin (TCDD) in rats. Gen Pharmacol 1987; 18:547–550.
225. Nohl H, Silva DD, Summer KH. 2,3,7,8-Tetrachlorodibenzo-p-dioxin induces oxygen activation associated with cell respiration. Free Rad Biol Med 1989; 6:369–374.
226. Stohs SJ. Oxidative stress induced by 2,3,7,8-tetrachlorodibenzo-p-dioxin (TCDD). Free Radic Biol Med 1990; 9:79–90.
227. Al-Bayati ZAF, Stohs SJ. The possible role of phospholipase A2 in hepatic microsomal lipid peroxidation induced by 2,3,7,8- tetrachlorodibenzo-p-dioxin in rats. Arch Environ Contam Toxicol 1991; 20:361–365.
228. Shara MA, Dickson PH, Bagchi D, Stohs SJ. Excretion of formaldehyde, malondialdehyde, acetaldehyde and acetone in the urine of rats in response to 2,3,7,8-tetrachlorodibenzo-p-dioxin, paraquat, endrin and carbon tetrachloride. J Chromatogr 1992; 576:221–233.
229. Bagchi M, Stohs SJ. In vitro induction of reactive oxygen species by 2,3,7,8-tetrachlorodibenzo-p-dioxin, endrin, and lindane in rat peritoneal macrophages, and hepatic mitochondria and microsomes. Free Radic Biol Med 1993; 14:11–18.
230. Hassoun EA, Stohs SJ. Comparative studies on oxidative stress as a mechanism for the fetotoxicity of TCDD, endrin and lindane in C57BL/6J and DBA/2J mice. Teratology 1995; 51: 186.
231. Hassoun EA, Stohs SJ. TCDD, endrin and lindane induced oxidative stress in fetal and placental tissues of C57BL/6J and DBA/2J mice. Comp Biochem Physiol C Pharmacol Toxicol Endocrinol 1996; 11:11–18.
232. Hassoun EA, Wilt SC, Devito MJ, Van Birgelen A, Alsharif NZ, Birnbaum LS, Stohs SJ. Induction of oxidative stress in brain tissues of mice after subchronic exposure to 2,3,7,8-tetrachlorodibenzo-p-dioxin. Toxicol Sci 1998; 42:23–27.
233. Amaro AR, Oakley GG, Bauer U, Spielmann HP, Robertson LW. Metabolic activation of PCBs to quinones: Reactivity toward nitrogen and sulfur nucleophiles and influence of superoxide dismutase. Chem Res Toxicol 1996; 9:623–629.
234. Oakley GG, Robertson LW, Gupta RC. Analysis of polychlorinated biphenyl-DNA adducts by 32P-postlabeling. Carcinogenesis 1996; 17:109–114.
235. Silberhorn EM, Glauert HP, Robertson LW. Carcinogenicity of polyhalogenated biphenyls: PCBs and PBBs. CRC Crit Rev Toxicol 1990; 20:440–496.
236. Peltola V, Mantyla E, Huhtaniemi I, Ahotupa M. Lipid peroxidation and antioxidant enzyme activities in the rat testis after cigarette smoke inhalation or administration of polychlorinated biphenyls or polychlorinated naphthalenes. J Androl 1994; 15:353–361.
237. Mantyla E, Ahotupa M. Polychlorinated biphenyls and naphthalenes: Long-lasting induction of oxidative stress in the rat. Chemosphere 1993; 27:383–390.
238. Smith AG, Carthew P, Clothier B, Costantin D, Francis JE, Madra S. Synergy of iron in the toxicity and carcinogenicity of polychlorinated biphenyls (PCBs) and related chemicals. Toxicol Lett (Shannon) 1995; 82–83:945–950.
239. Hori M, Kondo H, Akari N, Yamada H, Hiratsuka A, Watabe T, Oguri K. Changes in the hepatic glutathione peroxidase redox system produced by coplanar polychlorinated biphenyls

in Ah-responsive and less-responsive strains of mice: Mechanism and implications for toxicity. Environ Toxicol Pharmacol 1997; 3:267–275.
240. Tanaka K, Tsukazaki N, Yoshida H, Irifune H, Watanabe M, Tanimura T. Polychlorinated biphenyls (PCB) and polychlorinated quaterphenyls (PCQs) concentration in skin surface lipids and blood of patients with Yusho. Fukuoka Acta Medica 1995; 86:202–206.
241. Caputo R, Monti M, Ermacora E, Carminati G, Gelmetti C, Gianotti R, Gianni E, Puccinelli V. Cutaneous manifestations of tetrachlorodibenzo-p-dioxin in children and adolescents. J Am Acad Dermatol 1988; 19:812–819.
242. Sinks T, Steele G, Smith AB. Mortality among workers exposed to polychlorinated biphenyls. Am J Epidemiol 1992; 136:389–345.
243. Longstreth JD. Melanoma genesis. Putative causes and possible mechanisms. In: Balch CM, Houghton AN, Sober AJ, Soong S, eds. Cutaneous Melanoma, 3rd ed. St. Louis: Quality Medical Publishing, 1998:535–550.
244. Huff JE, Haseman JK, Demarini DM, Eustis S, Maronpot PR, Peteres AC, Persing RL, Chrisp CE, Jacobs AC. Multiple site carcinogenicity in Fischer 344 rats and B6C3F1 mice. Environ Health Perspect 1989; 82:125–163.
245. Maltoni C, Ciliberti A, Cotti C, Belpoggi F. Benzene, an experimental carcinogen: Results from the long term bioassays performed at the Bologna Institute of Oncology. Environ Health Perspec 1989; 82:109–124.
246. Witz G, Latriano L, Goldstein BD. Metabolism and toxicity of trans, trans-muconaldehyde, an open ring microsomal metabolite of benzene. Environ Health Perspect 1989; 82:19–22.
247. Yardley-Jones A, Anderson D, Parke DV. The toxicity of benzene and its metabolism and molecular pathology in human risk assessment. Br J Ind Med 1991; 48:437–444.
248. Snyder R, Kalf GF. A perspective on benzene leukaemogenesis. CRC Crit Rev Toxicol 1994; 24:177–209.
249. Khan K, Rishnamurthy R, Pandya KP. Generation of hydroxyl radicals during benzene toxicity. Biochem Pharmacol 1990; 39:1393–1395.
250. Subrahmanyam VV, Kolachana P, Smith MT. Hydroxylation of phenol to hydroquinone catalyzed by a human myeloperoxidase-superoxide complex: possible implications in benzene-induced myelotoxicity. Free Radic Res Commun 1991; 15:285–296.
251. Kolachana P, Subrahmanyam VV, Meyer KB, Zhang L, Smith MT. Benzene and its phenolic metabolites produce oxidative DNA damage in HL60 cells in vitro and in the bone marrow in vivo. Cancer Res 1993; 53:1023–1026.
252. Stoyanovsky DA, Goldman R, Jonnalagadda SS, Day BW, Claycamp HG, Kagan VE. Detection and characterization of the electron paramagnetic resonance-silent glutathionyl-5,5-dimethyl-1-pyrroline N-oxide adduct derived from redox cycling of phenoxyl radicals in model systems. Arch Biochem Biophys 1996; 330:3–11.
253. Shen Y, Shen HM, Shi CY, Ong CN. Benzene metabolites enhance reactive oxygen species generation in HL60 human leukemia cells. Hum Exp Toxicol 1996; 15:422–427.
254. French JE, Libbus BL, Hansen L, Spalding J, Tice RR, Mahler J, Tennant RW. Cytogenetic analysis of malignant skin tumors induced in chemically treated TG.AC transgenic mice. Mol Carcinogen 1994; 11:215–226.
255. Bond GG, McLaren EA, Baldwin CL, Cook RR. An update for mortality among chemical workers exposed to benzene. Br J Ind Med 1986; 43:685–691.
256. Wilmer JL, Simeonova PP, Germolec DR, Luster MI. Benzene and its principal metabolites modulate proinflammatory cytokines and growth factors in human epidermal keratinocyte cultures. In Vitro Toxicol 1997; 10:429–436.
257. Walder BK. Solvents and scleroderma. Lancet 1965; II:436.
258. Walder BK. Do solvents cause scleroderma? Int J Dermatol 1983; 22:157–158.
259. Yamakage A, Ishikawa H. Generalized morphea like scleroderma occurring in people exposed to organic solvents. Dermatologica 1982; 165:186–193.

260. Haustein UF, Ziegler V. Environmentally induced systemic sclerosis like disorders. Int J Dermatol 1985; 24:147–151.
261. Owens GR, Medsger TA. Systemic sclerosis secondary to occupational exposure. Am J Med 1988; 85:114–116.
262. Bottomlay WW, Sheehan-Dare RA, Hughes P. A sclerodermatous syndrome with unusual features following prolonged occupational exposure to organic solvents. Br J Dermatol 1993; 128:203–206.
263. Bagchi M, Balmoori J, Ye X, Stohs SJ. Naphthalene induced oxidative stress and DNA damage in cultured macrophage J774A.1 cells. Free Radic Biol Med 1998; 25:137–143.
264. Wells PG, Wilson B, Lubek BM. In vivo murine studies on the biochemical mechanism of naphthalene cataractogenesis. Toxicol Appl Pharmacol 1989; 99:466–473.
265. Wilson AS, Davis CD, Williams DP, Buckpitt AR, Pirmohamed M, Park BK. Characterisation of the toxic metabolite(s) of naphthalene. Toxicology 1996; 114:233–242.
266. Vuchetich PJ, Bagchi D, Bagchi M, Hassoun EA, Tang L, Stohs SJ. Naphthalene induced oxidative stress in rats and the protective effects of vitamin E succinate. Free Radic Biol Med 1996; 21:577–590.
267. Bagchi D, Bagchi M, Balmoori J, Vuchetich PJ, Stohs SJ. Induction of oxidative stress and DNA damage by chronic administration of naphthalene to rats. Res Commun Mol Pathol Pharmacol 1998; 101:249–257.
268. Pott F, Heinrich U. Staub und Staubinhaltsstoffe / Polycyclische aromatische Kohlenwasserstoffe. In: Wichmann L, Schlipköter M, Füllgraf H, eds. Handbuch Umweltmedizin. Landsberg: Ecomed, 1994:1–23.
269. Riala R, Heikkila P, Kanerva L. A questionnaire study of road pavers' and roofers' work-related skin symptoms and bitumen exposure. Int J Dermatol 1998; 37:27–30.
270. Kuljukka T, Vaaranrinta R, Veidebaum T, Sorsa M, Peltonen K. Exposure to PAH compounds among cokery workers in the oil shale industry. Environ Health Perspect 1996; 104 (suppl 3):539–541.
271. Rühl R. Bedeutung der Leitsubstanz B(a)P für die Gesundheitsbeurteilung bei PAH Exposition. In: Hauptverband der Gewerblichen Berufsgenossenschaften, ed. Arbeitsmedizinisches Kolloqium. Meckenheim: DCM Druck Center 1999:47–62.
272. Hankinson O. The aryl hydrocarbon receptor complex. Ann Rev Pharmacol Toxicol 1995; 35:307–340.
273. Kapitulnik J, Wislocki PG, Levin W, Yagi H, Jerina D, Conney AH. Tumorigenicity studies with diol-epoxides of benzo[a]pyrene which indicate that trans 7,8-dihydroxyl-9-epoxy-7,8,9,10-tetrahydrobenzo[a]pyrene is an ultimate carcinogen in newborn mice. Cancer Res 1978; 38:354–358.
274. Marnett LJ. Prostaglandin synthase-mediated metabolism of carcinogens and a potential role for peroxyl radicals as reactive intermediates. Environ Health Perspect 1990; 88:5–12.
275. Ji C, Marnett LJ. Oxygen radical-dependent epoxidation of (7S,8S)-dihydroxy-7,8-dihydrobenzo[a]pyrene in mouse skin in vivo: Stimulation by phorbol esters and inhibition by anti-inflammatory steroids. J Biol Chem 1992; 267:17842–17848.
276. Dix TA, Marnett LJ. Metabolism of polycyclic aromatic hydrocarbon derivatives to ultimate carcinogens during lipid peroxidation. Science 1983; 221:77–79.
277. Constantin D, Mehrotra K, Wallin A, Moldeus P, Jernstrom B. Studies on the effect of sulfite on benzo(a)pyrene-7,8-dihydrodiol activation to reactive intermediates in human polymorphonuclear leukocytes. Chem Biol Interact 1995; 94:73–82.
278. Cavalieri EL, Rogan E. Role of radical cations in aromatic hydrocarbon carcinogenesis. Environ Health Perspect 1985; 64:69–84.
279. Cavalieri EL, Rogan EG. Radical cations in aromatic hydrocarbon carcinogenesis. Free Radic Res Commun 1990; 11:77–87.
280. Cavalieri EL, Rogan EG. The approach to understanding aromatic hydrocarbon carcinogene-

sis. The central role of radical cations in metabolic activation. Pharmacol Ther 1992; 55: 183–199.
281. Cavalieri EL, Rogan EG. Central role of radical cations in metabolic activation of polycyclic aromatic hydrocarbons. Xenobiotica 1995; 25:677–688.
282. Flowers-Geary L, Bleczinki W, Harvey RG, Penning TM. Cytotoxicity and mutagenicity of polycyclic aromatic hydrocarbon ortho-quinones produced by dihydrodiol dehydrogenase. Chem Biol Interact 1996; 99:55–72.
283. Penning TM, Ohnishi ST, Ohnishi T, Harvey RG. Generation of reactive oxygen species during the enzymatic oxidation of polycyclic aromatic hydrocarbon trans-dihydrodiols catalyzed by dihydrodiol dehydrogenase. Chem Res Toxicol 1996; 9:84–92.
284. Flowers L, Ohnishi ST, Penning TM. DNA strand scission by polycyclic aromatic hydrocarbon o-quinones: role of reactive oxygen species, Cu(II)/Cu(I) redox cycling, and o-semiquinone anion radicals. Biochemistry 1997; 36:8640–8648.
285. Jarabak R, Harvey RG, Jarabak J. Redox cycling of polycyclic aromatic hydrocarbon o-quinones: reversal of superoxide dismutase inhibition by ascorbate. Arch Biochem Biophys 1997; 339:92–98.
286. Joseph P, Klein-Szanto AJ, Jaiswal AK. Hydroquinones cause specific mutations and lead to cellular transformation and in vivo tumorigenesis. Br J Cancer 1998; 78:312–320.
287. Cecinato A, Zagari M. Nitroarenes of photochemical origin: A possible source of risk to human health. J Environ Pathol Toxicol Oncol 1997; 16:93–99.
288. Moller L. In vivo metabolism and genotoxic effects of nitrated polycyclic aromatic hydrocarbons. Environ Health Perspect 1994; 102(suppl 4):139–146.
289. Zeiger E, Ashby J, Bakale G, Enslein K, Klopman G, Rosenkranz HS. Prediction of Salmonella mutagenicity. Mutagenesis 1996; 11:471–484.
290. Durant JL, Bushby WF, Lafleur AL, Penman BW, Crespi CL. Human cell mutagenicity of oxygenated, nitrated and unsubstituted polycyclic aromatic hydrocarbons associated with urban aerosols. Mutat Res 1996; 371:123–157.
291. Asokan P, Das M, Rosenkranz HS, Bickers DR, Mukhtar H. Topically applied nitropyrenes are potent inducers of cutaneous and hepatic monooxygenases. Biochem Biophys Res Commun 1985; 129:134–140.
292. Hillesheim W, Jaeschke H, Neumann HG. Cytotoxicity of aromatic amines in rat liver and oxidative stress. Chem Biol Interact 1995; 98:85–95.
293. Li WM. The role of pesticides in skin disease. Int J Dermatol 1986; 25:295–297.
294. Schubert HJ. Pesticides. In: Rycroft RJG, Menne T, Frosch PJ, eds. Textbook of Contact Dermatitis. Berlin: Springer, 1995:527–538.
295. Bajnova A, Kaloyanova F. Study on allergenic and irritating effect of synthetic pyrethroids on skin. Khigiena Zdraveopazvane 1985; 28:19–25.
296. Jung HD, Rothe A, Heise H. Zur Epikutantestung mit Pflanzenschutz-und Schädlingsbekämpfungsmitteln (Pestiziden). Dermatosen 1987; 35:43–51.
297. Hogan DJ, Grafton LH. Pesticides and other agricultural chemicals. In: Adams RM, ed. Occupational Skin Disease, 3rd ed. Philadelphia: W.B. Saunders, 1999:597–622.
298. Sato T, Taguchi M, Nagase H, Kito H, Niikawa M. Augmentation of allergic reactions by several pesticides. Toxicology 1998; 126:41–53.
299. Schubert H. Contact dermatitis to sodium N-dimethyldithiocarbamate. Contact Derm 1978; 4:370–371.
300. Hjorth N, Wilkinson DS. Contact dermatitis. Sensitization to pesticides. Br J Dermatol 1968; 80:272–274.
301. Lodovici M, Aiolli S, Monserrat C, Dolara P, Medica A, Di Simplicio P. Effect of a mixture of 15 commonly used pestizides on DNA levels of 8-hydroxy-2-deoxyguanosine and xenobiotic metabolizing enzymes in rat liver. J Environ Pathol Toxicol Oncol 1994; 13:163–168.
302. Bagchi D, Bagchi M, Hassoun EA, Stohs SJ. In vitro and in vivo generation of reactive

oxygen species, DNA damage and lactate dehydrogenase leakage by selected pesticides. Toxicology 1995; 104:129–140.

303. Hai DO, Varga SI, Matkovics B. Organophosphate effects on antioxidant system of carp (Cyprinus carpio) and catfish (Ictalurus nebulosus). Comp Biochem Physiol C Pharmacol Toxicol Endocrinol 1997; 117:83–88.
304. Sarin S, Gill KD. Potential biomarkers of dichlorous induced neural injury in rats. Biomarkers 1998; 3:169–176.
305. Numan IT, Hassan MQ, Stohs SJ. Protective effects of antioxidants against endrin-induced lipid peroxidation, glutathione depletion, and lethality in rats. Arch Environ Contam Toxicol 1990; 19:302–306.
306. Junqueira VBC, Bainy ACD, Arisi ACM, Azzalis LA, Simizu K, Pimentel R, Barros SBM, Videla LA. Acute lindane intoxication: A study on lindane tissue concentration and oxidative stress-related parameters in liver and erythrocytes. J Biochem Toxicol 1994; 9:9–15.
307. Junqueira VBC, Koch OR, Fuzaro AP, Azzalis LA, Barros SBM, Cravero A, Farre S, Videla LA. Regression of morphological alterations and oxidative stress-related parameters after acute lindane-induced hepatotoxicity in rats. Toxicology 1997; 117:199–205.
308. Samanta L, Chainy GBN. Age-related differences of hexachlorocyclohexane effect on hepatic oxidative stress parameters of chicks. Ind J Exp Biol 1997; 35:457–461.
309. Samanta L, Chainy GBN. Comparison of hexachlorocyclohexane-induced oxidative stress in the testis of immature and adult rats. Comp Biochem Physiol C Pharmacol Toxicol Endocrinol 1997; 118:319–327.
310. Sahoo A, Chainy GBN. Acute hexachlorocyclohexane-induced oxidative stress in rat cerebral hemisphere. Neurochem Res 1998; 23:1079–1084.
311. Howard JK. A clinical study of paraquat formulation workers. Br J Ind Med 1979; 36:220–223.
312. Suntres ZE, Sheek PN. Intratracheally administered liposomal alpha-tocopherol protects the lung against long-term toxic effects of paraquat. Biomed Environ Sci 1995; 8:289–300.
313. Kim BS, Eun HC, Lee HG, Chung JH. A study of a selection of antidotes for paraquat-induced skin damage. Ann Dermatol 1998; 10:13–19.
314. Fabisiak JP, Kagan VE, Tyurina YY, Tyurin VA, Lazo JS. Paraquat-induced phosphatidylserine oxidation and apoptosis are independent of activation of PLA2. Am J Physiol 1998; 274:L793–L802.
315. Nakagawa I, Suzuki M, Imura N, Naganuma A. Enhancement of paraquat toxicity by glutathione depletion in mice in vivo and in vitro. J Toxicol Sci 1995; 20:557–564.
316. Berisha HI, Pakbaz H, Absood A, Said SI. Nitric oxide as a mediator of oxidant lung injury due to paraquat. Proc Natl Acad Sci USA 1994; 91:7445–7449.
317. Ali S, Jain SK, Abdulla M, Athar M. Paraquat induced DNA damage by reactive oxygen species. Biochem Mol Biol Int 1996; 39:63–67.
318. Xi S, Chen LH. Dose effects of paraquat on antioxidant defense and detoxifying enzymes in rats. Biochem Arch 1997; 13:143–150.
319. Moellering D, McAndrew J, Jo H, Darley-Usmar VM. Effects of pyrrolidine dithiocarbamate on endothelial cells: Protection against oxidative stress. Free Rad Biol Med 1999; 26:1138–1145.
320. Hodgson E, Rose RL, Ryu DY, Falls G, Blake BL, Levi PE. Pesticide-metabolizing enzymes. Toxicol Lett 1995; 82–83:73–81.
321. Olivarius DFF, Balslev E, Menne T. Skin reactivity to tin chloride and metallic tin. Contact Derm 1993; 29:110–111.
322. Colosio C, Tomasini M, Cairoli S, Foa V, Minoia C, Marinovich M, Galli CL. Occupational triphenyltin acetate poisoning: a case report. Br J Ind Med 1991; 48:136–139.
323. Barnes JM, Stoner HB. Toxic properties of some dialkyl and trialkyl tin salts. Br J Ind Med 1958; 15:15–22.

324. Ascher KRS, Nissim S. Organotin compounds and their potential use in insect control. World Rev Pest Control 1964; 3:188–211.
325. Stoner HB. Toxicity of triphenyltin. Br J Ind Med 1996; 23:222–229.
326. Cavelier C, Foussereau J. Kontaktallergie gegen Metalle und deren Salze. Teil III. Weitere Metalle. Dermatosen/Occupat Environ 1995; 43:202–209.
327. Corsini E, Viviani B, Marinovich M, Galli CL. Role of mitochondria and calcium ions in tributyltin-induced gene regulatory pathways. Toxicol Appl Pharmacol 1997; 145:74–81.
328. Menne T, Andersen KE, Kaaber K, Osmundsen PE, Andersen JR, Yding F, Valeur G. Tin: an overlooked contact sensitizer. Contact Derm 1987; 16:9–13.
329. Gammeltoft M. Tributyltinoxide is not allergenic. Contact Derm 1978; 4:238.
330. Kadiiska MB, Hanna PM, Mason RP. In vivo ESR spin trapping evidence for hydroxyl radical mediated toxicity of paraquat and copper in rats. Toxicol Appl Pharmacol 1993; 123: 187–192.
331. Smith LL. Paraquat toxicity. Phil Trans R Soc Lond B 1985; 311:647–657.
332. Ronnen M, Klin B, Suster S. Mixed diquat/paraquat induced burns. Int J Dermatol 1995; 34:23–25.
333. Srikrishna V, Riviere JE, Monteiro-Riviere NA. Cutaneous toxicity and absorption of paraquat in porcine skin. Toxicol Appl Pharmacol 1992; 115:89–97.
334. Jee CC, Shiou-Hwa K, Hsien-Wen DWP, Chang CH, Sun CC, Wang J. Photodamage and skin cancer among paraquat workers. Int J Dermatol 1995; 34:466–469.
335. Jee SH, Tsai TF, Chen CC, Sun CC, Chang CH, Chang CC. Bipyridyl dihydrochloride inhibits tumor necrosis factor-alpha secretion by human keratinocytes on ultraviolet irradiation. J Formosan Med Assoc 1996; 95:706–708.
336. Cooper SP, Downs T, Burau K, Buffler PA, Tucker S, Whitehead L, Wood S, Delclos G, Huang CC, Davidson T. A survey of actinic keratoses among paraquat production workers and a nonexposed friend reference group. Am J Indust Med 1994; 25:335–347.
337. Korshunov SS, Korkina OV, Ruuge EK, Skulachev VP, Starkov AA. Fatty acids as natural uncouplers preventing generation of $O_2$ and $H_2O_2$ by mitochondria in the resting state. FEBS Lett 1998; 435:215–218.
338. Demin OV, Kholodenko BN, Skulachev VP. A model of $O_2^{\bullet-}$ generation in the complex III of the electron transport chain. Mol Cell Biochem 1998; 184:21–33.
339. Lemasters JJ, Nieminen AL. Mitochondrial oxygen radical formation during reductive and oxidative stress to intact hepatocytes. Biosci Rep 1997; 17:281–291.
340. Dahlhaus M, Almstadt E, Appel KE. The pentachlorophenol metabolite tetrachloro-p-hydroquinone induces the formation of 8-hydroxy-2-deoxyguanosine in liver DNA of male B6C3F1 mice. Toxicol Lett 1993; 74:265–274.
341. Dahlhaus M, Almstadt E, Henschke P, Luttgert S, Appel KE. Induction of 8-hydroxy-2-deoxyguanosine and single-strand breaks in DNA of V79 cells by tetrachloro-p-hydroquinone. Mutat Res 1995; 329:29–36.
342. Naito S, Ono Y, Somiya I, Inoue S, Ito K, Yamamoto K, Kawanishi S. Role of active oxygen species in DNA damage by pentachlorophenol metabolites. Mutat Res 1994; 310:79–88.
343. Sai-Kato K, Umemura T, Takagi A, Hasegawa R, Tanimura A, Kurokawa Y. Pentachlorophenol-induced oxidative DNA damage in mouse liver and protective effect of antioxidants. Food Chem Toxicol 1995; 33:877–882.
344. Bentley P, Calder I, Elcombe C, Grasso P, Stringer D, Wiegand HJ. Hepatic peroxisome proliferation in rodents and its significance for humans. Food Chem Toxicol 1993; 31:857–907.
345. Kluwe WM. Carcinogenic potential of phthalic acid esters and related compounds: structure-activity relationships. Environ Health Perspect 1986; 65:271–278.
346. Chevalier S, Roberts RA. Perturbation of rodent hepatocyte growth control by nongenotoxic hepatocarcinogens: Mechanisms and lack of relevance for human health. Oncol Rep 1998; 5:1319–1327.

347. Holden PR, Tugwood JD. Peroxisome proliferator-activated receptor alpha: Role in rodent liver cancer and species differences. J Mol Endocrinol 1999; 22:1–8.
348. Latruffe N, Vamecq J. Peroxisome proliferators and peroxisome proliferator activated receptors (PPARs) as regulators of lipid metabolism. Biochimie 1997; 79:81–94.
349. Johnson EF, Palmer CNA, Griffin KJ, Hsu MH. Role of the peroxisome proliferator-activated receptor in cytochrome P450 4A gene regulation. FASEB J 1996; 10:1241–1248.
350. Simpson AECM. The cytochrome P450 4 (CYP4) family. Gen Pharmacol 1997; 28:351–359.
351. Dogra SC, Whitelaw ML, May BK. Transcriptional activation of cytochrome P450 genes by different classes of chemical inducers. Clin Exp Pharmacol Physiol 1998; 25:1–9.
352. Clayson DE, Mehta R, Iverson F. Oxidative DNA damage—The effects of certain genotoxic and operationally non-genotoxic carcinogens. Mutat Res 1994; 317:25–42.
353. Kimie SK, Takagi A, Umemura T, Hasegawa R, Kurokawa Y. Role of oxidative stress in non-genotoxic carcinogenesis with special reference to liver tumors induced by peroxisome proliferators. Biomed Environ Sci 1995; 8:269–279.
354. Sausen PJ, Lee DC, Rose ML, Cattley RC. Elevated 8-hydroxydeoxyguanosine in hepatic DNA of rats following exposure to peroxisome proliferators: Relationship to mitochondrial alterations. Carcinogenesis 1995; 16:1795–1801.
355. Cattley RC, Deluca J, Elcombe C, Fenner-Crisp P, Lake BG, Marsman DS, Pastoor TA, Popp JA, Robinson DE, Schwetz B. Do peroxisome proliferating compounds pose a hepatocarcinogenic hazard to humans? Reg Toxicol Pharmacol 1998; 27:47–60.
356. Hanley K, Jiang Y, Crumrine D. Activation of the nuclear hormone receptor PPARa- and FXR accelerate the development of the fetal epidermal permeability barrier. J Clin Invest 1997; 100:705–712.
357. Fuchs C, Aneskievich BJ. Nuclear receptor regulation by peroxisome proliferators in epidermal keratinocytes. J Invest Dermatol 1997; 108:578.
358. Wechsler CJ, Shields HC. Potential reactions among indoor pollutants. Atmos Environ 1997; 31:3487–3495.
359. Mustafa MG. Health effects and toxicology of ozone and nitrogen dioxide. In: Nriagu JO, Simmons MS, eds. Environmental Oxidants. New York: John Wiley & Sons, Inc, 1994:351–404.
360. Menzel DB. Ozone: an overview of its toxicity in man and animals. J Toxicol Environ Health 1984; 13:183–204.
361. Mustafa MG. Biochemical basis of ozone toxicity. Free Radic Biol Med 1990; 9:245–265.
362. Pryor WA, Church DF. Aldehydes, hydrogen peroxide and organic radicals as mediators of ozone toxicity. Free Radic Biol Med 1993; 122:483–486.
363. Fukase O, Hashimoto K. The effects of exposure to ozone on collagen in lungs and skin. Nippon Eiseigaku Zasshi 1982; 37:694–700.
364. Podda M, Koh B, Cross CE, Packer L. Ozone exposure depletes lipophilic antioxidants in murine skin. J Invest Dermatol 1995; 104:639A.
365. Thiele JJ, Podda M, Packer L. Tropospheric ozone: an emerging environmental stress to skin. Biol Chem 1997; 378:1299–1305.
366. Thiele JJ, Traber MG, Tsang K, Cross CE, Packer L. In vivo exposure to ozone depletes vitamins C and E and induces lipid peroxidation in epidermal layers of murine skin. Free Radic Biol Med 1997; 23:385–391.
367. Thiele JJ, Traber MG, Polefka TG, Cross CE, Packer L. Ozone exposure depletes vitamin E and induces lipid peroxidation in murine stratum corneum. J Invest Dermatol 1997; 108: 753–757.
368. Thiele JJ, Traber MG, Re R, Espundo N, Yan LJ, Cross CE, Packer L. Macromolecular carbonyls in human stratum corneum: a biomarker for environmental oxidant exposure. FEBS Lett 1998; 422:403–406.
369. Thiele JJ, Traber MG, Packer L. Depletion of human stratum corneum vitamin E: an early

and sensitive in vivo marker of UV induced photo-oxidation. J Invest Dermatol 1998; 110: 756–761.
370. Weber SU, Thiele JJ, Cross CE, Packer L. Vitamin C, uric acid and glutathione gradients in murine stratum corneum and their susceptibility to ozone exposure. J Invest Dermatol 1999; 113:1128–1132.
371. Huie RE. The reaction kinetics of $NO_2$. Toxicology 1994; 89:193–216.
372. Beckman JS, Koppenol WH. Nitric oxide, superoxide, and peroxynitrite. The good, the bad, and the ugly. Am J Physiol 1996; 271:C1414–1437.
373. Beckman JS, Tsai JH. Reactions and diffusion of nitric oxide and peroxynitrite. Biochemist 1994; 16:8–10.
374. Eberlein-König B, Przybilla B, Kühnl P, Pechak J, Gebefügi I, Kleinschmidt J, Ring J. Influence of airborne nitrogen dioxide or formaldehyde on parameters of skin function and cellular activation in patients with atopic eczema and control subjects. J Allergy Clin Immunol 1998; 101:141–143.
375. Grosjean E, Grosjean D, Fraser MP, Cass GR. Air quality model evaluation data for organics. 3. Peroxyacetylnitrate and peroxylpropionyl nitrate in Los Angeles air. Environ Sci Technol 1996; 30:2704–2714.
376. Vyskocil A, Viau C, Lamy S. Peroxyacetyl nitrate: Review of toxicity. Human Exp Toxicol 1998; 17:212–220.
377. Neta P, Huie RE. Free radical chemistry of sulfite. Environ Health Perspect 1985; 64:209–217.
378. Constantin D, Mehrotra K, Jenstrom B, Tomasi A, Moldeus P. Alternative pathways of sulfite oxidation in human polymorphonuclear leukocytes. Pharmacol Toxicol 1994; 74:136–140.
379. Constantin D, Bini A, Meletti E, Moldeus P, Monti D, Tomasi A. Age-related differences in the metabolism of sulphite to sulphate and in the identification of sulphur trioxide radical in human polymorphonuclear leukocytes. Mech Ageing Devel 1996; 88:95–109.
380. Karoui H, Hogg N, Frejaville C, Tordo P, Kalyanaraman B. Characterization of sulfur-centered radical intermediates formed during the oxidation of thiols and sulfite by peroxynitrite. J Biol Chem 1996; 271:6000–6009.
381. Chamulitrat W. Activation of the superoxide generating NADPH oxidase of intestinal lymphocytes produces highly reactive free radicals from sulfite. Free Radic Biol Med 1999; 27: 411–421.
382. Jiang J, Liu KJ, Shi X, Swartz HM. Detection of short-lived free radicals by low-frequency electron paramagnetic resonance spin trapping in whole living animals. Arch Biochem Biophys 1995; 319:570–575.
383. Shi X. Generation of $SO_3^{\cdot -}$ and OH radicals in $SO_3^{2-}$ reactions with inorganic environmental pollutants and its implications to $SO_3^{2-}$ toxicity. J Inorg Biochem 1994; 56:155–165.
384. Shi X, Mao Y. 8-Hydroxy-2′-deoxyguanosine formation and DNA damage induced by sulfur trioxide anion radicals. Biochem Biophys Res Commun 1994; 205:141–147.
385. Langley-Evans SC, Phillips GJ, Jackson AA. Sulphur dioxide: A potent glutathione depleting agent. Comp Biochem Physiol C Pharmacol Toxicol Endocrinol 1996; 114:89–98.
386. Gunnison AF. Sulphite toxicity: a critical review of in vitro and in vivo data. Food Cosmet Toxicol 1981; 19:667–682.
387. Vena GA, Foti C, Angelini G. Sulfite contact allergy. Contact Derm 1994; 31:172–175.
388. Eedy DJ. Carbon-fibre-induced airborne irritant contact dermatitis. Contact Derm 1996; 35: 362–363.
389. Stam-Westerveld EB, Coenraads PJ, van der Valk PG, de Jong MC, Fidler V. Rubbing test responses of the skin to man-made mineral fibres of different diameters. Contact Derm 1994; 31:1–4.
390. Bjornberg A. Glass fiber dermatitis. Am J Indust Med 1985; 8:395–400.
391. Lockey JE, Ross CS. Radon and man-made vitreous fibers. J Allergy Clin Immunol 1994; 94:310–317.

392. Thriene B, Sobottka A, Willer H, Weidhase J. Man-made mineral fibre boards in buildings: Health risks caused by quality deficiencies. Toxicol Lett 1996; 88:299–303.
393. Pelin K, Husgafvel-Pursiainen K, Vallas M, Vanhala E, Linnainmaa K. Cytotoxicity and anaphase aberrations induced by mineral fibres in cultured human mesothelial cells. Toxicol in vitro 1992; 6:445–450.
394. Luoto K, Holopainen M, Sarataho M, Savolainen K. Comparison of cytotoxicity of man made vitreous fibers. Ann Occup Hyg 1997; 41:37–50.
395. Donaldson K, Gilmour PS, Beswick PH. Supercoiled plasmid DNA as a model target for assessing the generation of free radicals at the surface of fibers. Exp Toxicol Pathol 1995; 47:235–237.
396. Gilmour PS, Beswick PH, Brown DM, Donaldson K. Detection of surface free radical activity of respirable industrial fibres using supercoiled phi-X174 RF1 plasmid DNA. Carcinogenesis 1995; 16:2973–2979.
397. Donaldson K, Beswick PH, Gilmour PS. Free radical activity associated with the surface of particles: A unifying factor in determining biological activity? Toxicol Lett 1996; 88:293–298.
398. Gilmour PS, Brown DM, Beswick PH, Macnee W, Rahman I, Donaldson K. Free radical activity of industrial fibers: Role of iron in oxidative stress and activation of transcription factors. Environ Health Perspect 1997; 105 (suppl 5):1313–1317.
399. Erasmus LD. Scleroderma in gold miners on the Witwatersrand with particular reference to pulmonary manifestations. S Afr J Lab Clin Med 1957; 3:209–231.
400. Rodnan GP, Benedek TG, Medsger TA, Cammarata RJ. The association of progressive systemic sclerosis (scleroderma) with coal miners' pneumokoniosis an other forms of silicosis. Ann Intern Med 1967; 66:323–334.
401. Ziegler V, Pampel W, Zschunke E, Münzberger H, Mährlein W, Köpping H. Kristalliner Quarz (eine) Ursache der progressiven Sklerodermie? Dermatol Monschr 1982; 168:398–401.
402. Ziegler V, Haustein UF, Mehlhorn J, Münzberger H, Rennau H. Quarzinduzierte Sklerodermie. Sklerodermie-ähnliches Syndrom oder echte progressive Sklerodermie? Dermatol Monschr 1986; 172:86–90.
403. Haustein UF, Ziegler V. Sklerodermie und Sklerodermie ähnliche Erkrankungen durch Umweltsubstanzen. Dermatosen 1986; 34:61–67.
404. Haustein UF, Ziegler V, Herrmann K, Mehlhorn J, Schmidt C. Silica-induced scleroderma. J Am Acad Dermatol 1990; 22:444–448.
405. Kumagai Y, Abe C, Shiokawa Y. Scleroderma after cosmetic surgery. Arthr Rheumatol 1979; 22:532–537.
406. Kondo H, Kumagai Y, Shiokawa Y. Scleroderma following cosmetic surgery (''adjuvant disease''): A review of nine cases reported in Japan. In: Black CM, Myers AR, eds. Current Topics in Rheumatology: Systemic Sclerosis (scleroderma). New York: Grower Med Pub Ltd, 1985:135–137.
407. Spiera H. Scleroderma after silicone augmentation mammoplasty. J Am Med Assoc 1988; 260:236–238.
408. Varga J, Schumacher HR, Jimenez SA. Systemic sclerosis after augmentation mammoplasty with silicone implants. Ann Intern Med 1989; 111:377–383.
409. Gabriel SE, O'Fallon WM, Kurland LT, Beard CM, Woods JE, Melton LJ III. Risk of connective-tissue diseases and other disorders after breast implantation. N Engl J Med 1994; 330: 1697–1702.
410. Chow CK. Cigarette smoking and oxidative damage in the lung. Ann NY Acad Sci 1993; 686:289–298.
411. Leonard MB, Lawton K, Watson ID, Macfarlane I. Free radical activity in young adult cigarette smokers. J Clin Pathol 1995; 48:385–387.
412. Rahman I, MacNee W. Role of oxidants/antioxidants in smoking-induced lung diseases. Free Radic Biol Med 1996; 21:669–681.

413. Reilly M, Delanty N, Lawson JA, Fitzgerald GA. Modulation of oxidant stress in vivo in chronic cigarette smokers. Circulation 1996; 94:19–25.
414. Pryor WA. Cigarette smoke radicals and the role of free radicals in chemical carcinogenicity. Environ Health Perspect 1997; 105(suppl 4):875–882.
415. Pryor WA, Stone K, Zang LY, Bermudez E. Fractionation of aqueous cigarette tar extracts: Fractions that contain the tar radical cause DNA damage. Chem Res Toxicol 1998; 11:441–448.
416. Schaefer T, Dirschedl P, Kunz B, Ring J, Ueberla K. Maternal smoking during pregnancy and lactation increases the risk for atopic eczema in the offspring. J Am Acad Dermatol 1997; 36:550–556.
417. Smith JB, Fenske NA. Cutaneous manifestations and consequences of smoking. J Am Acad Dermatol 1996; 34:717–732.
418. O'Hare PM, Fleischer AB Jr, D'Agostino RB Jr, Feldman SR, Hinds MA, Rassette SA, McMichael AJ, Williford PM. Tobacco smoking contributes little to facial wrinkling. J Eur Acad Dermatol Venereol 1999; 12:133–139.
419. Chance B, Sies H, Boversi A. Hydroperoxide metabolism in mammalian organs. Physiol Rev 1979; 59:527–605.
420. Taffe BG, Kensler TW. Generation of free radicals from organic hydroperoxide tumor promoters by mouse epidermal cells. Pharmacologist 1986; 28:175–183.
421. Taffe BG, Takahashi N, Kensler TW, Mason RP. Generation of free radicals from organic hydroperoxide tumor promoters in isolated mouse keratinocytes. J Biol Chem 1987; 262: 12143–12149.
422. Athar M, Mukhtar H, Bickers DR, Khan IU, Kalyanaraman B. Evidence for the metabolism of tumor promoter organic hydroperoxides into free radicals by human carcinoma skin keratinocytes: an ESR-spin trapping study. Carcinogenesis 1989; 10:1499–503.
423. Vessey DA, Lee KW, Blacker KL. Characterization of the oxidative stress initiated in cultured human keratinocytes by treatment with peroxides. J Invest Dermatol 1992; 99:859–863.
424. Vessey DA, Lee KH. Inactivation of enzymes of the glutathione antioxidant system by treatment of cultured human keratinocytes with peroxides. J Invest Dermatol 1993; 100:829–833.
425. Iannone A, Marconi A, Zambruno G, Gianetti A, Vannini V, Tomasi A. Free radical production during metabolism of organic hydroperoxides by normal human keratinocytes. J Invest Dermatol 1993; 101:59–63.
426. Timmins GS, Davies MJ. Free radical formation in isolated murine keratinocytes treated with organic peroxides and its modulation by antioxidants. Carcinogenesis 1993; 14:1615–1620.
427. Kensler T, Guyton K, Egner P, McCarthy T, Lesko S, Akman S. Role of reactive intermediates in tumor promotion and progression. Prog Clin Biol Res 1995; 391:103–116.
428. Guyton KZ, Bhan P, Kuppusamy P, Zweier P, Trush MA, Kensler TW. Free radical-derived quinone methide mediates skin tumor promotion by butylated hydroxytoluene hydroperoxide: Expanded role for electrophiles in multistage carcinogenesis. Proc Natl Acad Sci USA 1991; 88:946–950.
429. Waravdekar VS, Saslaw LD, Jones WA, Kuhns JG. Skin changes induced by UV-irradiated linoleic acid extract. Arch Pathol 1965; 80:91–95.
430. Hayakawa R. Relation with facial skin diseases and lipids and lipoperoxide of serum and sebum, pH on the face and buffering power. Jpn J Dermatol 1971; 81:11–29.
431. Tanaka T. Skin damage and its prevention from lipoperoxide. Vitamins 1979; 53:577–586.
432. Ohsawa K, Watanabe T, Matsukawa R, Yoshimura Y, Imaeda K. The possible role of squalene and its peroxide of the sebum in the occurrence of sunburn and protection from the damage caused by UV irradiation. J Toxicol Sci 1984; 9:151–159.
433. Tanaka T, Hayakawa R. Lipid peroxides in cosmetic products and their effect to irritate the skin. J Clin Biochem Nutr 1986; 1:201–207.

434. Madzhunov N, Bajnova A, Khinkova L, Madzhunov M. Clinical and experimental studies on the irritating effect of dicumyl peroxide on the skin and upper respiratory tract. Dermatol Venereol 1990; 29:33–36.
435. Kitahara M, Ishiguro F, Takayama K, Isowa K, Nagai T. Evaluation of skin damage of cyclic monoterpenes, percutaneous absorption enhancers, by using cultured human skin cells. Biol Pharm Bull 1993; 16:912–916.
436. Koehler HBK, Knop J, Martin M, de Bruin A, Huckzermeyer B, Lehmann H, Kiezmann M, Meier B, Nolte I. Involvement of reactive oxygen species in TNF-α-mediated activation of the transcription factor NF-kappB in canine dermal fibroblasts. Vet Immunol Immunopathol 1999; 71:125–142.
437. Hellerström S, Thyresson N, Blohm SG, Widmark G. On the nature of the eczematogenic component of oxidized d3-carene. J Invest Dermatol 1955; 24:217–224.
438. Karlberg AT, Shao LP, Nilsson U, Gafvert E, Nilson JL. Hydroperoxides in oxidized d-limonene identified as potent contact allergens. Arch Derm Res 1994; 286:97–103.
439. Gäfvert E, Shao LP, Karlberg AT, Nilson U, Nilson JLG. Contact allergy to resin acid hydroperoxides. Hapten binding via free radicals and epoxides. Chem Res Toxicol 1994; 7:260–266.
440. Payne MP, Wals PT. Structure activity relationships for skin sensitization potential: development of structural alerts for use in knowledge-based toxicity prediction systems. J Chem Inf Com Sci 1994; 34:154–161.
441. Bezard M, Karlberg AT, Montelius J, Lepoittevin JP. Skin sensitization to linayl hydroperoxide: support for radical intermediates. Chem Res Toxicol 1997; 10:987–993.
442. Mannaioni PF, Masini E. The release of histamine by free radicals. Free Rad Biol Med 1988; 5:177–197.
443. Kotin P, Falk HL. Organic peroxides, hydrogen peroxide, epoxides and neoplasia. Rad Res 1963; 3(suppl):193–211.
444. Van Duuren BL, Nelson N, Orris L, Palmes ED, Schmitt FL. Carcinogenicity of epoxides, lactones, and peroxy compounds. J Natl Cancer Inst 1963; 31:41–55.
445. Pryor WA. The role of free radical reactions in biological systems. Free Rad Biol 1975; 1: 1–49.
446. Slaga TJ, Klein-Szanto AJP, Triplett LL, Yotti LP, Trosko JA. Skin tumor promoting activity of benzoyl peroxide, a widely used free radical generating compound. Science 1981; 213: 1023–1025.
447. Gimenez-Conti IB, Binder RL, Johnston D, Slaga TJ. Comparison of the skin tumor-promoting potential of different organic peroxides in SENCAR mice. Toxicol Appl Pharmacol 1998; 149:73–79.
448. Akman SA, Kensler TW, Doroshow JH, Dizdaroglu M. Copper ion-mediated modification of bases in DNA in vitro by benzoyl peroxide. Carcinogenesis 1993; 14:1971–1974.
449. King JK, Egner PA, Kensler TW. Generation of DNA base modification following treatment of cultured murine keratinocytes with benzoyl peroxide. Carcinogenesis 1996; 17:317–320.
450. Hazlewood C, Davies MJ. Benzoyl peroxide-induced damage to DNA and its components: Direct evidence for the generation of base adducts, sugar radicals, and strand breaks. Arch Biochem Biophys 1996; 332:79–91.
451. Kraus AL, Munro IC, Orr JC, Binder RL, Leboeuf RA, Williams GM. Benzoyl peroxide: An integrated human safety assessment for carcinogenicity. Reg Toxicol Pharmacol 1995; 21:87–107.
452. Barratt MD, Basketter DA. Structure activity relationships for skin sensitization: an expert system. In: Rougier A, Goldberg AM, Maibach HI, eds. In vitro toxicology: irritation, phototoxicity, sensitization. New York: Ann Liebert, 1994:293–301.
453. Hess JA, Molinari JA, Gleason MJ, Radecki C. Epidermal toxicity of disinfectants. Am J Dent 1991; 4:51–56.
454. Thompson DC, Perera K, London R. Quinone methide formation from para isomers of meth-

ylphenol (cresol), ethylphenol, and isopropylphenol: relationship to toxicity. Chem Res Toxicol 1995; 8:55–60.
455. Stoyanovsky DA, Goldman R, Claycamp HG, Kagan VE. Phenoxyl radical-induced thiol-dependent generation of reactive oxygen species: Implications for benzene toxicity. Arch Biochem Biophys 1995; 317:315–323.
456. BogadiSare A, Brumen V, Turk R, Karacic V, Zavalic M. Genotoxic effects in workers exposed to benzene: with special reference to exposure biomarkers and confounding factors. Ind Health 1997; 35:367–373.
457. Shedova AA, Kommineni C, Jeffries BA, Castranova V, Tyurina YY, Tyurin VA, Serbinova E, Fabisiak JE, Kagan V. Redox cycling of phenol induces oxidative stress in human epidermal keratinocytes. J Invest Dermatol 2000; 114:354–364.
458. Wilmer JL, Burleson FG, Kayama F, Kanno J, Luster MI. Cytokine induction in human epidermal keratinocytes exposed to contact irritants and its relation to chemical induced inflammation in mouse skin. J Invest Dermatol 1994; 102:915–922.
459. H Nohl. Quinones in biology: Functions in electron transfer and oxygen activation. Adv Free Rad Biol Med 1986; 2:211–279.
460. Powis G. Free radical formation by antitumor quinones. Free Radic Biol Med 1989; 6:63–101.
461. Monks TJ, Hanzlik RP, Cohen GM, Ross D, Graham DG. Quinone chemistry and toxicity. Toxicol Appl Pharmacol 1992; 112:2–16.
462. Lisi P, Hansel K. Is benzoquinone the prohapten in cross-sensitivity among aminobenzene compounds? Contact Derm 1998; 39:304–306.
463. Daugherty JP, Khurana A. Amelioration of doxorubicin induced skin necrosis in mice by butylated hydroxytoluene. Cancer Chemother Pharmacol 1985; 14:243–246.
464. Hajarizadeh H, Lebredo L, Barrie R, Woltering EA. Protective effect of doxorubicin in vitamin C or dimethyl sulfoxide against skin ulceration in the pig. Ann Surg Oncol 1994; 1: 411–414.
465. Bekerecioglu M, Kutluhan A, Demirtas I, Karaayvaz M. Prevention of adriamycin-induced skin necrosis with various free radical scavengers. J Surg Res 1998; 75:61–65.
466. Müller K, Eibler E, Mayer KK, Wiegrebe W. Dithranol, singlet oxygen and unsaturated fatty acids. Arch Pharm 1986; 319:2–9.
467. Müller K, Wiegrebe W, Younes M. Formation of active oxygen species by dithranol III. Dithranol, active oxygen species and lipid peroxidation in vivo. Arch Pharm 1987; 320:59–66.
468. Müller K, Kappus H. Hydroxyl radical formation by dithranol. Biochem Pharmacol 1988; 37:4277–4280.
469. Müller K. Antipsoriatic anthrones: Aspects of oxygen radical formation, challenges and prospects. Gen Pharmacol 1996; 27:1325–1335.
470. Müller K. Antipsoriatic and proinflammatory action of anthralin. Implications for the role of oxygen radicals. Biochem Pharmacol 1997; 53:1215–1221.
471. Schroot B, Brown C. Free radicals in skin exposed to dithranol and its derivatives. Arzneim Forsch/Drug Res 1986; 36:1253–1255.
472. Fuchs J, Packer L. Investigations on anthralin free radicals in model systems and in skin of hairless mice. J Invest Dermatol 1989; 92:677–682.
473. Hayden PJ, Free KE, Chignell CF. Structure-activity relationships for the formation of secondary radicals and inhibition of keratinocyte proliferation by 9-anthrones. Mol Pharmacol 1994; 46:186–198.
474. Mäder K, Baacic G, Swartz HM. In vivo detection of anthralin-derived free radicals in the skin of hairless mice by low-frequency electron paramagnetic resonance spectroscopy. J Invest Dermatol 1995; 104:514–517.
475. Schmidt KN, Podda M, Packer L, Baeuerle PA. Anti-psoriatic drug anthralin activates transcription factor NF-kappa-B in murine keratinocytes. J Immunol 1996; 156:4514–4519.

476. Lange RW, Hayden PJ, Chignell CF, Luster MI. Anthralin stimulates keratinocyte-derived proinflammatory cytokines via generation of reactive oxygen species. Inflamm Res 1998; 47:174–181.
477. Lange RW, Germolec DR, Foley JF, Luster MI. Antioxidants attenuate anthralin-induced skin inflammation in BALB-c mice: Role of specific proinflammatory cytokines. J Leukocyte Biol 1998; 64:170–176.
478. Peus D, Beyerle A, Rittner HL, Pott M, Meves A, Weyand C, Pittelkow MR. Anti-psoriatic drug anthralin activates JNK via lipid peroxidation: mononuclear cells are more sensitive than keratinocytes. J Invest Dermatol 2000; 114:688–692.
479. Fuchs J, Milbradt R. Antioxidant inhibition of skin inflammation induced by reactive oxidants: Evaluation of the redox couple dihydrolipoate/lipoate. Skin Pharmacol 1994; 7:278–284.
480. Viluksela M. Characteristics and modulation of dithranol (anthralin) induced skin irritation in the mouse ear model. Arch Dermatol Res 1991; 283:262–268.
481. Finnen MJ, Lawrence CM, Shuster S. Inhibition of dithranol inflammation by free radical scavengers. Lancet 1984; 17:1129–1130.
482. Parsle R, Rhodes LE, Friedmann PS. Anthralin induced erythema is mediated by nitric oxide. J Invest Dermatol 1996; 107:504.
483. Brennan RJ, Schiestl RH. Aniline and its metabolites generate free radicals in yeast. Mutagenesis 1997; 12:215–220.
484. Hlavica P, Golly I, Lehnerer M, Schulze J. Primary aromatic amines: their N-oxidative bioactivation. Hum Exp Toxicol 1997; 16:441–448.
485. Picardo M, Zompetta C, Marchese C, De Luca C, Faggioni A, Schmidt RJ, Santucci B. Paraphenylenediamine, a contact allergen, induces oxidative stress and ICAM-1 Expression in human keratinocyte. Br J Dermatol 1996; 126:450–455.
486. Rudzki E. Pattern of hypersensitivity to aromatic amines. Contact Derm 1975; 1:248–249.
487. Condae-Salazar L, Guimaraens D, Romero LV, Gonzalez MA. Unusual allergic contact dermatitis to aromatic amines. Contact Derm 1987; 17:42–44.
488. Santucci B, Cristaudo A, Cannistraci C, Amantea A, Picardo M. Hypertrophic allergic contact dermatitis from hair dye. Contact Derm 1994; 31:169–171.
489. Requena L, Requena C, Sanchez M, Jaqueti G, Aguilar A, Sanchez-Yus E, Hernandez-Moro B. Chemical warfare: Cutaneous lesions from mustard gas. J Am Acad Dermatol 1988; 19: 529–536.
490. Smith KJ, Hurst CG, Moeller RB, Skelton HG, Sidell FR. Sulfur mustard: its continuing threat as a chemical warfare agent, the cutaneous lesions induced, progress in understanding its mechanism of action, its long-term health effects, and new developments for protection and therapy. J Am Acad Dermatol 1995; 32:765–76.
491. Sidell FR, Smith WJ, Petrali JP, Hurst CG. Sulfur mustard: a chemical vesicant model. In: Marzulli FN, Maibach HI, eds. Dermatotoxicology. Washington DC: Taylor & Francis, 1996: 119–129.
492. Fox M, Scott D. The genetic toxicology of nitrogen and sulfur mustard. Mut Res 1980; 75: 131–168.
493. Weiss A, Weiss B. Karzinogenese durch Lost Exposition beim Menschen, ein wichtiger Hinweis für die Alkylantien Therapie. Deutsche Med Wochschr 1975; 100:9191–923.
494. Inada S, Hiragun K, Seo K, Ymura T. Multiple Bowen's disease observed in former workers of a poison gas factory in Japan, with special references to mustard gas exposure. J Dermatol 1978; 5:49–60.
495. Aasted A, Darre E, Wulf HC. Mustard gas: clinical toxicological, and mutagenic aspects based on modern experience. Ann Plast Surg 1984; 19:330–333.
496. Gentilhomme E, Neveux Y, Hua A, Thiriot C, Faure M, Thivolet J. Action of bis(betachloroethyl) sulphide (BCES) on human epidermis reconstituted in culture: Morphological alterations and biochemical depletion of glutathione. Toxicol In Vitro 1992; 6:139–147.

497. Gross CL, Meier HL, Papirmeister B, Brinkley FB, Johnson JB. Sulfur mustard lowers nicotinamide dinucleotide concentrations in human skin grafted to athymic nude mice. Toxicol Appl Pharmacol 1985; 81:85–90.
498. Millard CB, Bongiovanni R, Broomfield CA. Cutaneous exposure to bis-(2-chloroethyl)sulfide results in neutrophil infiltration and increased solubility of 180,000 M-r subepidermal collagens. Biochem Pharmacol 1997; 53:1405–1412.
499. Eldad A, Meir PB, Breiterman S, Chaouat M, Shafran A, Ben-Bassat H. Superoxide dismutase (SOD) for mustard gas burns. Burns 1998; 24:114–119.
500. Somani SM, Babu SR. Toxicodynamics of sulfur mustard. Int J Clin Pharmacol Ther Toxicol 1989; 27:419–435.
501. Vijayaraghavan R, Sugendran K, Pant SC, Husain K, Malhotra RC. Dermal intoxication of mice with bis (2-chloroethyl)sulfide and the protective effect of flavonoids. Toxicology 1991; 69:35–42.
502. Yourick JJ, Dawson JS, Benton CD, Craig ME, Mitcheltree LW. Pathogenesis of 2,2′-dichlorodiethyl sulfide in hairless guinea pigs. Toxicology 1993; 84:185–197.
503. Dacre JC, Goldman M. Toxicology and pharmacology of the chemical warfare agent sulfur mustard. Pharmacol Rev 1996; 48:289–326.
504. Arroyo CM, Carmichael AJ, Broomfield CA. Could nitrosyl chloride be produced by human skin keratinocytes and sulfur mustard? A magnetic resonance study. In Vitro Toxicol 1997; 10:253–261.
505. Arroyo CA. The role of cytokines in the inflammatory response. In: Rhodes CJ, ed. Toxicology of the Human Environment. London: Taylor and Francis, 2000:265–284.
506. Arroyo CM, von Tersch RL, Broomfield CA. Activation of alpha human tumor necrosis factor (TNFα) by human monocytes (THP-1) exposed to 2-chloroethyl-ethyl sulfide (H-MG). Hum Exp Toxicol 1995; 14:547–553.
507. Kurt EM, Schafer RJ, Arroyo CM. Effects of sulfur mustard on cytokines released from cultured human epidermal keratinocytes. Int J Toxicol 1998; 17:223–229.
508. Arroyo CM, Schafer RJ, Kurt EM, Broomfield CA, Carmichael AJ. Response of normal human keratinocytes to sulfur mustard (HD): Cytokine release using a non-enzymatic detachment procedure. Hum Exp Toxicol 1999; 18:1–11.
509. Cowan FM, Bongiovanni R, Broomfield CA, Yourick JJ, Smith WJ. Sulfur mustard increases elastase-like activity in homogenates of hairless guinea pig skin. J Toxicol Cut Ocul Toxicol 1994; 13:221–229.
510. Kam CM, Selzler J, Schulz SM, Bongiovanni R, Powers JC. Enhanced serine protease activities in the sulfur mustard-exposed homogenates of hairless guinea pig skin. Int J Toxicol 1997; 16:625–638.
511. Gray PJ. Sulphur mustards inhibit binding of transcription factor AP2 in vitro. Nucl Acids Res 1995; 23:4378–4382.
512. Colardyn F, De Keyser H, Ringoir S, De Bersaques J. Clinical observation and therapy of injuries with vesicants. J Toxicol Clin Exp 1986; 6:237–246.
513. Krüger M. Chemische Kampfstoffe (C-Waffen). Dermatosen/Occup Environ 1991; 39:179–193.
514. Lindsay CD, Rice P. Assessment of the biochemical effects of percutaneous exposure of sulphur mustard in an in vitro human skin system. Human Exp Toxicol 1996; 15:237–244.
515. Goldman M, Dacre JC. Lewisite: its chemistry, toxicology and biological effects. Rev Environ Contam Toxicol 1989; 110:76–115.
516. Ippen H, Mathies V. Die protrahierte Verätzung. Berufsdermatosen 1970; 18:144–165.
517. Fisher AA. Post-operative ethylene oxide dermatitis. Cutis 1973; 12:177–182.
518. Biro L, Fisher AA, Price E. Ethylene oxide burns, a hospital outbreak involving 19 women. Arch Dermatol 1974; 110:924–931.
519. Taylor JS. Dermatologic hazards from ethylene oxide. Cutis 1977; 19:189–192.
520. Bertolini JC. Hydrofluoric Acid: A Review of Toxicity. J Emerg Med 1992; 10:163–168.

521. Sebastian G. General practice-relevant therapeutic recommendations in hydrofluoric acid burns. Hautarzt 1994; 45:453–459.
522. Kirkpatrick JJ, Enion DS, Burd DA. Hydrofluoric acid burns: a review. Burns 1995; 21: 483–493.
523. Gallerani M, Bertoli V, Peron L, Manfredini R. Systemic and topical effects of intradermal hydrofluoric acid. Am J Emerg Med 1998; 16:521–522.
524. Della Bianca V, Grzeskowiak M, Dusi S, Rosi F. Fluoride can activate respiratory burst independently of $Ca^{2+}$ stimulation of phospholipid turnover, and protein kinase C translocation in primed human neutrophils. Biochem Biophys Res Commun 1988; 15:955–964.
525. Hong XJ, Francker A, Diamant B. Effects of N-acetylcysteine on histamine release by sodium fluoride and compound 48/80 from isolated rat mast cells. Int Arch All Appl Immunol 1991; 96:338–343.
526. Wang YY, Zhao BL, Li XJ. Spin trapping technique studies on active oxygen radicals from human polymorphonuclear leucocytes during fluoride stimulated respiratory burst. Fluoride 1997; 30:5–15.
527. Rzeuski R, Chlubek D, Machoy Z. Interactions between fluoride and biological free radical reactions. Fluoride 1998; 31:43–45.
528. Chinoy NJ, Sharma M, Michael M. Beneficial effects of ascorbic acid and calcium on reversal of fluoride toxicity in male rats. Fluoride 1993; 26:45–56.
529. Eldad A, Chaouat M, Weinberg A, Neuman A, Ben Meir P, Rotem M, Wexler MR. Phosphorus pentachloride chemical burn—a slowly healing injury. Burns 1992; 18:340–341.
530. Haycock JW, Ralston DR, Morris B, Freedlander E, Macneil S. Oxidative damage to protein and alterations to antioxidant levels in human cutaneous thermal injury. Burns 1997; 23: 533–540.
531. Eldad A, Wisoki M, Cohen H, Breiterman S, Chaouat M, Wexler MR, Ben-Bassat H. Phosphorous burns: evaluation of various modalities for primary treatment. J Burn Care Rehabil 1995; 16:49–55.

# 14

# Photooxidative Stress in Skin and Regulation of Gene Expression

**LEE ANN LAURENT-APPLEGATE and STEFAN SCHWARZKOPF**

*University Hospital, Lausanne, Switzerland*

## I. ULTRAVIOLET RADIATION AND ITS ROLE AS AN ENVIRONMENTAL OXIDANT STRESS

Solar ultraviolet radiation has long been known to have beneficial (vitamin D production as one of the most important) as well as deleterious effects on human skin. The roots of sunlight's beneficial effects have been known from stone-wall cave paintings and Egyptian temple hieroglyphics that illustrated that exposure to sunlight provided life-sustaining properties. Unfortunately, the beneficial aspect of the sun lasted a few thousand years and more emphasis has followed on the harmful effects of the sun on human skin. Hallmarks of therapeutic advances have been the introduction of ultraviolet (UV) light for the treatment of cutaneous tuberculosis at the beginning of 1900 (1) and of the photochemotherapy with oral psoralen and long-wave UV in the 1970s. However, by 1894, Unna (2) reported the association of sunlight and epidermal manifestations including skin cancer. Findley (3) experimentally produced the first UV-induced skin cancers in mice in 1928 and epidemiological studies in the 1940s to 1950s further associated the direct link between sunlight exposure and skin cancer induction (4,5). Since this time, it has become quite evident that human exposure to ultraviolet radiation from sunlight has many adverse effects on skin and eyes including sunburn, several types of photodermatoses, premature aging (as leathery texture, laxity, and mottled pigmentation), cancer, immune suppression, and cataract formation (6–9).

In this chapter, we would like to introduce ultraviolet radiation as an environmental oxidative stress describing its nature and mechanisms of action on cells and tissue on their molecular level. We will emphasize actions on human skin with gene induction and subsequent protein expression following ultraviolet radiation as an oxidative stress.

## II. SOLAR ULTRAVIOLET SPECTRUM

The solar UV spectrum has been divided into four different wavelength bands based on a variety of criteria including the human erythema action spectrum and the spectral energy distribution of sunlight at the earth's surface. The resulting wavebands include UVC (190–290 nm), UVB (290–320 nm), UVA II (320–340 nm), and UVA I (340–400 nm). The UVA waveband was divided because of similar biological effects encountered with UVA II and UVB radiation. Wavelengths in the UVC region of the sun's spectrum are filtered out by the stratospheric ozone layer, so these wavelengths do not reach the surface of the earth. The UVB region has generally been thought to be the most deleterious portion of the sun's spectrum, since it has been shown to be responsible for many damaging effects, including acute sunburn, mutation induction, cell mortality, and skin cancer (7,10–12). However, wavelengths in the UVA region are also potentially carcinogenic and can cause a wide variety of biological effects that have also been seen to be produced by the UVB spectrum (9,13–18). As the photons in the UVA waveband are less energetic, significantly more photons are needed to cause the same damage as that induced by the shorter wavelengths in the UVB region. This can be best illustrated by the difference in dose needed for a minimal erythema: ~.20 J/cm$^2$ is needed in the UVB region and ~100–200 J/cm$^2$ for UVA. Thus, while a dose of up to a 1000 times higher is needed for UVA than UVB in order to induce visible damage to the skin, it is important to remember that UVA photons are present in sunlight in much higher quantities than those of UVB and that these longer wavelengths have the potential to penetrate into the dermis to a far greater extent than UVB because of its less energetic potential. At this point, it is very important to emphasize that UVB and UVA differ from their mechanisms in reacting with cell substances, but indeed their action could lead to the same result. The pathway which photooxidated molecules undergo concerning UVA and UVB irradiation is described in the following section.

## III. BIOLOGICAL EFFECTS OF ULTRAVIOLET RADIATION

Exposure of human skin to solar radiation can result in short-term responses, such as erythema, pigmentation, and immune suppression and long-term responses, such as carcinogenesis and premature aging (7,11). These deleterious biological effects are mainly due to the UVB portion of the sun's spectrum but recently it was shown that wavelengths in the UVA range can produce similar biological effects by other mechanisms (19–22). These mechanisms involve endogenous chromophores such as quinones, steroids, flavins, free porphyrins, and heme-containing enzymes, which act as phtoosensitizers when induced by UVA irradiation. Reactive oxygen species (ROS), including singlet oxygen, superoxide radical anion, hydroxyl radical, and hydrogen peroxide are produced during these photosensitization reactions in such a manner that the proantioxidant balance can be disturbed (23). The resulting photooxidative stress due to these harmful ROS is considered to play a major role in causing modifications of DNA, lipids, proteins, and carbohydrates in human skin (17,22,24–25).

UVA radiation produces cellular modifications that considerably overlap those induced by oxidative damage (26,27). UVA radiation constitutes an oxidant stress that involves the generation of active species including singlet oxygen and hydroxyl radicals. Hydrogen peroxide can be generated by UVA irradiation of tryptophan (27), and superoxide can be produced by UVA irradiation of NADH and NADPH (28). Naturally occurring iron complexes can react in vivo with hydrogen peroxide in the presence of superoxide anion in the Haber–Weiss reaction, which produces the potentially lethal hydroxyl radical:

$$Fe^{2+} + H_2O_2 \rightarrow Fe^{3+} + OH^- + OH' \; Fe^{3+} + O_2^- \rightarrow Fe^{2+} + O_2$$

Tissue changes resulting from acute and chronic ultraviolet radiation exposure are themselves caused by many photobiological responses of cells. For any cellular or tissue alteration to occur, there is a common initiation point that is the absorption of the energy of ultraviolet and visible radiation by molecules in tissue called chromophores. These molecules absorb energy and become transformed into their excited state: they quickly undergo chemical changes (photochemistry), transfer their energy to other molecules, or give off the extra energy as light (fluorescence) or heat (Fig. 1). Indeed, it is mainly the UVA range with its less energetic potential that reacts with chromophores; UVB radiation directly penetrates the barrier (keratin layer) and causes mutations on DNA bases, lipids, proteins, and other cell molecules. The first steps in the chain of events leading to biological responses to UV and visible radiation are the photochemical reactions of chromophores and the reactions of molecules that accept energy from excited-state chromophores. Although many organic molecules and biological cells absorb radiation, only certain ones will either fluoresce or be damaged. Some can transfer the energy to another molecule that will then fluoresce or be damaged. In turn, the chemical reactions produce changes in the molecular structure of the chromophores as well as in the cellular molecules (i.e., DNA, proteins, lipids). Cellular activities can be disturbed by the photochemical products produced following UV irradiation and the new chemical structures may induce enzymatic processes, initiate repair processes, stimulate new gene expression, or alter metabolic activity.

The major cellular chromophores absorbing in the UVB (290–320 nm) wavelength range are nucleic acids and proteins (Table 1). There are many chromophores in the UVA (320–400 nm) region but they are usually present in low concentrations when compared to those absorbing in the shorter wavelengths. Because they have low extinction coefficients, the relationship for specific UVA-absorbing molecules and the effect of UVR on cells is more difficult to establish.

Up until now it has been illustrated that the majority of effects (such as cytotoxicity, mutagenesis, gene induction, and new protein synthesis) result following UVB radiation and are initiated by photochemistry of nucleic acids with particular emphasis on DNA. However, in the past years, some responses such as activation of membrane enzymes and induction of early response genes involve non-nuclear chromophores following absorption of both UVB and UVA radiation (8). There are a variety of biomolecules absorbing UVB radiation, including nicotinamide adenine dinucleotide (NADH), quinones, flavins, and other heterocyclic cofactors. In the UVA range, protein cofactors and soluble metabolites may be responsible for UVA-induced responses in cells. The actual amount of UVR absorbed by each tissue is proportional to its concentration and absorption coefficient, and each endogenous chromophore has its range of wavelengths absorbed and the maximum peak of absorption. There is evidence that a photobiological response may be initiated by a weakly absorbing chromophore and that all responses that arrive were not initiated by the dominant chromophore at the wavelength of interest. This is indeed represented by a recent study where the induction of pyrimidine dimers by UVA radiation was very similar to the level seen following UVB radiation for a given erythema dose to human skin (18).

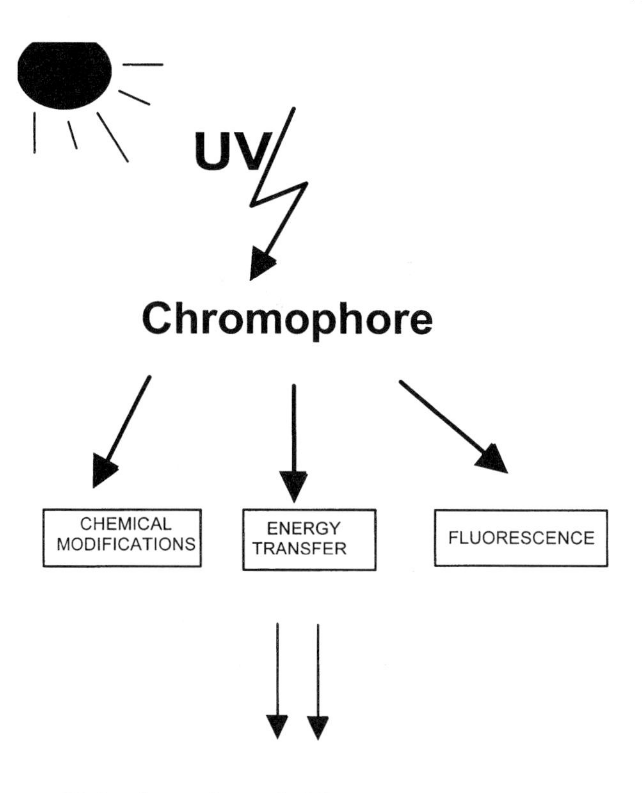


**Figure 1** Biological effects of sunlight. The absorption of sunlight by a chromophore can lead to fluorescence, energy transfer, or absorption of the radiation directly. The chemical reactions can alter the molecular structure of the chromophore or in cellular molecules including DNA, proteins, and lipids. Cellular activities can be altered by the photochemical products produced following exposure to UV radiation and the new chemical structures may induce enzymatic processes, initiate repair systems, stimulate new gene expression, and/or alter metabolic activity.

Protein and lipids are intimately organized in membranes and therefore both components are probably involved in photochemical processes. Since very few cell lipids absorb energy >290 nm, an oxidation of unsaturated lipids by photosensitized processes occurs following exposure to tissue or cells to UVA and UVB radiation. Absorption of UVB radiation can be seen in lipids containing two or more conjugated double bonds in their structure such as 7-dehydrocholesterol. Wavelengths <290 nm are absorbed by nonconjugated double bonds as those found in polyunsaturated fatty acids.

**Table 1** Major Chromophores of the Skin That Absorb Ultraviolet Radiation

| | Absorption Spectrum |
|---|---|
| DNA | 220–320 nm |
| Urocanic acid | 250–360 nm |
| Aromatic amino acids | 240–320 nm |
| Retinol esters | 260–380 nm |
| 7-Dehydrocholesterol | 270–315 nm |
| Flavins | 225–510 nm |
| Melanins | 250–700 nm |
| NADH, NADPH | 260–400 nm |
| Carotenes | 300–500 nm |
| Hemoglobin | 360–450 nm |
| Bilirubin | 300–530 nm |

Most likely candidates for chromophores in cell membranes are endogenous photosensitizing molecules since the reactive oxidizing species that they create diffuse small distances before reacting. Following a first photooxidation event, there are secondary oxidations of membrane components that can occur in the cytoplasm and nucleus when the antioxidant load is depleted by other factors. Secondary molecular alterations in cells following UVR can be the conversion of glutathione and ascorbic acid to oxidation products by reacting with oxidized lipids, for instance. Since the balance of antioxidant molecules in cells and tissue is altered following UVR, it may be this effect along with increased levels of lipid oxidation that acts as a signal for activation of other cellular processes. These include general signs of oxidative stress such as the alteration of heat shock proteins, heme oxygenase, ferritin, and other proteins that may help the cell to combat free radical presence (Table 2). This can also be illustrated by the global response of cell damage when photochemical events at or near the plasma membrane activate cytosolic transcription factors that control the activity of UV-responsive genes (Table 2). In the early stages, Src tyrosine kinase follows within minutes of exposure to UVR and is shortly followed by the activation of NF-κB. Although the chromophore has not been identified, the presence of membrane components is necessary and does not involve DNA photochemistry (29). Also shown in isolated membrane preparations was the induction of arachidonic acid (AA) from membrane phospholipids. The mechanism involves the stimulation of $PLA_2$ following UVB and both $PLA_2$ and phospholipase C following UVA radiation (30). The chromophores have not been identified for either of these UVA or UVB responses but the cellular response is very important since products of the cyclo-oxygenase pathway for AA metabolism are implicated in chronic UVR effects such as skin aging. The photooxidation of membranes is involved in chronic photodamage is supported by numerous studies where antioxidants are effective inhibitors of membrane lipid oxidation and since unsaturated lipids are readily oxidized, they are likely mediators of damage and also UVR-induced immunological effects. Other cellular responses of cells to UVR involving cell membrane photochemistry include UVA-induced inhibition of epidermal growth factor binding to its receptor and its subsequent complex procession (31), activation of protein kinase C (32) and alterations in the plasma membrane potential by UVB radiation (33).

**Table 2** Genes or Proteins Induced by UV Irradiation in Mammalian Cells

| *Stress Proteins* | *Proteases* |
|---|---|
| Heme-oxygenase | Collagenase |
| Ferritin | Gelatinase |
| Metallothioneine | Stromelysin |
| Hsp-70 | Plasminogin activator |
| Superoxide dismutase | *Signal and Regulator Molecules* |
| Catalase | |
| Nitric oxide | PKC-B |
| Cyclo-oxygenase | NF-κB |
| *Growth Factors* | Phospholipase $A_2$ |
| | Raf-1 |
| bFGF | Src tyrosine kinases |
| TGF-α, β | c-fos |
| NGF | SPR2 |
| ET-1 | c-jun |
| *Immunological Factors* | HIV-1 promoter |
| | bcl-1/bax |
| ELAM-1 | fas |
| IL-1α | fasL |
| Ornithine decarboxylase | Ras, N, K, H |
| IL-6, 8, 10, 12 | Map kinase |
| VCAM-1 | DAG |
| GM-CSF | Tyrosinase |
| TNF-α | |
| TNF-β | |
| ICAM-1 | |

## IV. BIOLOGICAL ACTIONS OF UVA RADIATION: CELLULAR LETHALITY

The lethal action of UVA radiation on mammalian cells is strongly dependent upon the presence of oxygen (34). There are two potential cellular targets which are probably involved in the cellular lethality induced by UVA radiation, namely, DNA and the membrane. Even though many studies have shown damage at the DNA level in mammalian cells, it is not certain that genetic damage is the ultimate factor in cellular inactivation (35,36). The specific lesions in DNA which can be induced by UVA radiation include pyrimidine dimers, single-strand breaks (both not thought to be the critical lesions in UVA radiation-induced cellular lethality), and, perhaps more importantly, DNA protein cross-links (37–40). UVA radiation generates more phosphodiester bond breaks in DNA than would be expected by the total amount of energy directly absorbed by the DNA. Therefore, most likely there is indirect damage to DNA accomplished by endogenous photosensitizers such as riboflavin, nicotinamide coenzymes, and rare RNA bases. These photosensitizers absorb photons of near-UV radiation and during deexcitation oxygen free radicals are formed that induce direct damage to a variety of molecular moieties including DNA.

Even though UVA radiation-induced cellular lethality cannot be associated with direct DNA damage, UVA radiation can cause structural changes in the DNA molecule.

However, these mutational changes seem to be related to cell type and assay conditions (41–46).

Membranes are another possible cellular target responsible for the UVA-induced cellular lethality due to the presence of free radicals that can produce lipid peroxidation and membrane damage (47). Peroxidized membranes become rigid and lose their selective permeability and integrity and a direct correlation between actual cell killing and general membrane damage in normal human skin fibroblasts from patients with the ''photosensitivity dermatitis/actinic reticuloid syndrome'' has been shown (48). In addition, there exists the possibility that lipid peroxidation products may form DNA adducts that give rise to mutations and altered patterns of gene expression (49). Therefore, it appears that both DNA and membranes could be equally involved in the lethal effects of UVA radiation on mammalian cells.

### A. Induction of Repair Systems: Lipids, Proteins, and DNA

There is evidence for oxidative damage to carbohydrates, lipids, proteins, and DNA even in healthy organisms (50–52). Damage to these cellular constituents necessitates recognition, removal, or repair to maintain the cells in functional form. Such repair systems would involve repair enzymes that directly restore biomolecules to their native form in addition to catabolic enzymes that function to degrade nonfunctional lipids, proteins, and nucleic acids and serve not only to remove oxidatively modified structures from the cytosol but also to replenish percursors for resynthesis.

As peroxidized membranes and lipid oxidation products are a constant threat to aerobic cells, a variety of mechanisms for maintaining membrane integrity and homeostasis by repairing oxidative-type damage have evolved. Compounds like vitamin E are capable of preventing initiation of peroxidation, phospholipase $A_2$ can preferentially hydrolyze fatty acids, which prevents further propagation reactions (53), and glutathione peroxidase can detoxify fatty acid hydroperoxides by reducing them to the corresponding hydroxy-fatty acids (53).

In addition to lipids, proteins are other cellular components that are susceptible to oxidant-type damage including UVA radiation. Following oxidizing damage, cells have been seen to undergo a proteolytic burst due to activation of latent proteolytic enzymes, and increase in the proteolytic susceptibility of oxidatively modified proteins or a combination of the two (54). Proteolytic complexes that have been identified seem to be responsible for proteolytic degradation of aberrant proteins that in turn prevent their accumulation and aggregation and provide amino acids for de novo synthesis of new proteins (55). Direct repair of altered proteins can prevent proteolytic degradation and direct enzymatic rereduction of sulfhydryl and heme groups can restore proteins to their native conformation (56).

Despite the necessity to guard the integrity of the genetic code, DNA is also a target for oxidative-type damages including UVA radiation. Even under normal physiological conditions, it has been reported that base modifications can exist as high as 1 in 130,000 bases in nuclear DNA and 1 in 8000 for mitochondrial DNA (51). Here again, it is in the interest of the organism to have efficient repair systems for the DNA. There are several repair mechanisms that include direct repair of altered DNA bases by GSH transferases and peroxidases, and DNA methylases (57,58). In addition, there is excision repair of damaged DNA where endonucleases and glycosolases have recently been identified in eukaryotic cells that have been previously described in prokaryotes with the induction of SOS genes (59–61).

The various DNA repair mechanisms are very important means for coping with UV-radiation-induced damage. A good illustration of the necessity of DNA repair is the rare autosomal recessive disease xeroderma pigmentosum where excision repair does not occur (62). These patients suffer from photosensitivity in early infancy followed by premature aging and multiple tumors of the skin very early on which can lead to the death.

## B. Gene Induction: Cellular Antioxidant Defense Mechanisms

As was discussed earlier, initial events that are required for UVB- and UVC-induced gene activation involve UV-induced DNA photodamage as well as membrane perturbations. Gene activation by UVA radiation has been less thoroughly studied to date, but the weaker absorption of UVA by biomolecules and the deeper penetration into skin would indicate that oxidative stress rather than DNA damage is responsible for UVA-related biological effects.

One of the first nonconstitutive defense mechanisms of the skin against oxidative stress including UVA radiation was discovered in the dermis of the skin: an inducible stress response of heme oxygenase-1 mRNA in human skin fibroblasts. Two specific heme oxygenase isozymes, HO-1 and HO-2, are known to catalyze heme degradation (63). HO-1 has been found to be inducible by a variety of oxidizing agents and glutathione-depleting agents, whereas HO-2 is not inducible (63). The observation that high levels of this enzyme are induced in cells from a tissue not involved in hemoglobin breakdown suggested that the breakdown of heme and heme-containing proteins could be involved in cellular protection against UVA radiation as well as other oxidant stresses (64–67). Interestingly, the end product of heme degradation, bilirubin, was also shown to be a powerful antioxidant in plasma (63). The inducible HO-1 response is characteristic of UVA radiation (rather than UVB or UVC) and high levels of expression are induced at sublethal fluencies (68). A further factor specific for the UVA radiation response is that compounds catabolized by HO are porphyrins which themselves can act as endogenous sensitizers leading to the generation of singlet oxygen. It has been shown that HO-1 is a crucial enzymatic intermediate in the UVA-inducible antioxidant defense involving ferritin in human skin fibroblasts (65). Ferritin constitutes the major storage site for nonmetabolized intracellular iron and therefore plays a critical role in regulating the availability of iron to catalyze certain harmful reactions such as the Fenton reaction and peroxidation of lipids. This inducible response, HO-1, is seen in human skin fibroblasts but not in epidermal keratinocytes (68). However, high HO activity is always notable in the keratinocytes (150–300% higher than in matching dermal fibroblasts) presumably due to the the constitutively high level of HO-2 mRNA present, which is also linked to ~3.0 times higher levels of the protective protein ferritin in the epidermis. As dermal fibroblasts are shielded from UV radiation to a considerable extent by the overlying epidermis with its stratum corneum, the adaptive response mediated by HO-1 is perhaps more appropriate and metabolically economic for fibroblasts, whereas keratinocytes require the constitutive pathway that appears to involve HO-2. In both epidermal and dermal skin cells, both the constitutive and inducible pathway of HO are closely linked to the intracellular levels of the iron storage protective protein, ferritin. The observation that levels of ferritin are higher in the epidermis than in the corresponding dermis is common to all other antioxidant molecules measured in human skin.

The response of ferritin following UV irradiation is complex. Previously, we reported that ferritin staining increased following UVA radiation (69). If we compare these data on the epidermal level only, video image analysis showed that the total epidermal ferritin content goes down, which is consistent with the behavior of the other antioxidants investigated. However, ferritin shows a more widespread staining pattern following UVA I exposure, characterized by scattered suprabasal keratinocytes containing ferritin. It still remains to be established whether ferritin is induced in suprabasal keratinocytes or whether it occurs there as a result of migration of basal keratinocytes. Furthermore, we have seen that skin cells derived from chronically sun-exposed areas have higher cellular levels of ferritin (70). In this respect we observed that the epidermal keratinocytes have up to sixfold higher levels of ferritin than the underlying dermal fibroblasts. Thus, the keratinocyte, as clearly the primary target for oxidative stress generated by solar UV irradiation, would benefit from the continuous protection provided by the high levels of ferritin. Antioxidant depletion following a first single exposure could therefore be considered as a single to induce the antioxidant defense system and that higher levels of protective antioxidants could suggest an adaptive response of cells that are exposed to chronic external insults. Superoxide dismutases (SOD) are metalloenzymes that catalyze the dismutation of superoxide anion into peroxide anion. These enzymes exist in two different types of copper–zinc SOD (Cu-Zn), one intracellular and one extracellular form. The intracellular form is a dimer, whereas the extracellular form represents a tetramer, binding to sulfated glycosaminoglycans on the cell surface. It has recently been shown that amyotrophic lateral sclerosis, a fatal neurodegenerative disease, is probably caused by mutations of the CuZn–SOD gene (71,72). Furthermore, it has been postulated that Alzheimer's disease could be due to the generation of superoxide radicals in endothelial cells, which is caused by lack or dysfunction of SOD, resulting in inflammatory stress and deposits of amyloid (73). In addition, the hyperproliferative epidermis of psoriasis was shown to express less Cu/Zn–SOD activity (74).

A third form of SOD, manganese SOD (Mn–SOD), is localized in mitochondria. It has been reported to be decreased in malignant cells, indicating that its gene could represent a tumor suppressor gene or, at least, a gene that is implicated in the regulation of p53 (75–77). Furthermore, it has been reported that prevention of apoptosis could be attributed to activation of Mn–SOD (78). It is therefore obvious that SODs play a major role in protecting cells from alterations caused by several kinds of oxidative stress factors (79).

All cells contain a set of highly conserved proteins in various intracellular compartments that are capable of increasing rapidly following environmental stress. These have been designated as heat shock proteins (Hsp) because the most studied stress is hyperthermia. They show increased synthesis when exposed to a variety of toxic agents, including heat, heavy metals, ethanol, anoxia, and UV light. Hsp provide a protective function to cells allowing them to recover from an inducing stress and survive subsequent stress that could otherwise be lethal. The high observation of the stress response and of genes involved suggests that this adaptive response is critical for the survival of both prokaryotic and eukaryotic organisms subjected to hostile environmental conditions. Hsp have been shown to play a protective role in ischemic diseases, carcinogenesis, and infection. Within these Hsp families, Hsp 70 is the best characterized. It exists in the cell in equilibrium between its free state in the cytoplasm and its bound state in the nucleolus. During recovery from heat shock when nucleoli begin to return to their normal activities, most of the Hsp 70 returns to the cytoplasm. Hsp 70 is linked with tumor suppressor genes such as p53 in pancreatic and skin cancer and the c-myc oncogene in human breast cancer (80,81). In a human in vivo study, immunohistochemical results have shown a dose- and wavelength-dependent perturbation of SOD and Hsp70 following a single irradiation with artificial

UV light sources of skin that is not sun exposed in normal life. The most prominent antioxidant depletion was seen resulting from irradiation with UVA I light. Biologically equivalent doses of UVA I + II (with regard to erythema induction) also lead to depletion of SOD and Hsp70, but less than UVA I, whereas solar UV-simulating light only provoked a weak antioxidant depletion.

The cause for the wavelength dependency for antioxidant depletion has not been identified to date. It could be related to the presence of a specific UVA receptor that occurs in the bilayer of the cell membrane. It has been shown in UV-induced apoptosis assays that activation of a receptor molecule could lead to the production of ceramides resulting in the arrest of cell cycle and protein synthesis (82). Immunohistochemical analysis revealed increased levels of the radical scavenging molecule SOD in the epidermis of chronically sun-exposed areas of the body. The presence of increased antioxidant levels in chronically sun-exposed skin is consistent with the recently published observation that sun-exposed skin is characterized by a higher erythema threshold dose for UVA radiation than non-exposed skin (17).

### C. Gene Induction: Cytokine Activation

The activation of the many cytokines in human skin keratinocytes following ultraviolet radiation can affect the immune response as well as the cell growth and differentiation of keratinocytes and other cell components in the skin. Skin exposed to UVA radiation has been shown to have an augmentative effect on the induction of erythema and carcinogenesis following UVB. In contrast, UVA may play a counterregulatory role in UV-induced immunosuppression. UVB radiation is capable of inducing mRNA and protein production for IL-10, GM-CSF, TNF-α, and TNF-β. UVA radiation can induce IL-1α, IL-6, IL-8, and IL-12, which would suggest that UVA augments UVB-induced inflammation through IL-1α but that it modulates skin immune function very distinctly from UVB by inducting a different series of immunomodulatory cytokines in keratinocytes (83). Therefore, although solar radiation containing both the UVA and UVB spectrum may be deleterious, it may also be beneficial for human health.

There are most likely other inducible antioxidant proteins that can provide cellular protection against subsequent oxidant (UVA and UVB radiation) challenges. Indeed, a genetic mammalian library that has known DNA damage sequences shows cross-hybridization following hydrogen peroxide treatment (66). Metallothionein has been shown to be induced by UVC and UVA radiation and the induction of this gene seems to correlate with a resistance to killing by several mutagenic agents and V79 hamster cells with elevated levels of metallothionein have been shown to have increased resistance to UVA radiation (67).

With the plethora of genes and subsequent proteins and molecules induced in human skin following photooxidative stress, there needs to be an equilibrium between hazardous products and defense systems for quotidian exposures.

''Benefit or harm,'' the sun's radiation reaching the earth's surface, depends mainly on the obtained dose. The generation of potentially harmful reactive oxygen species (ROS) takes place every time following an exposure to UV radiation. Knowing this aspect, it is obvious that in the case of sunburn, for example, too many ROS are produced and the ability of the tissue's antioxidant defense system is overpowered. As antioxidant molecules are considered of major importance in scavenging ROS, the overload of free ROS in human skin caused by UVA radiation is of particular interest due to the fact that the majority of sunscreens block UVB effectively but do not block the longer wavelengths efficiently.

Future studies will have to reveal which antioxidant molecules are the most important ones in protecting human skin against subsequent oxidative stress. Furthermore, it remains to investigate whether antioxidants added to sunscreens provide an efficient means in order to prevent harmful reactions and maintain the biological equilibrium of human skin.

## ACKNOWLEDGMENTS

The Laboratory of Oxidative Stress and Ageing is supported by grants from the Swiss National Fund for Scientific Research (31-49120.96), the Swiss League for Cancer Research (KFS 695-7-1998), the Braun Foundation, and the Heinz and Margot Kunz Foundation.

## REFERENCES

1. Finsen NR. Neue Untersuchungen über die Einwirkung des lichtes auf die Haut. Mitteilung Finsens Medizin Institut 1900; 1:8–34.
2. Unna PG. Histopathologie der Hautkrankheiten. Berlin: August Hirschwald, 1894.
3. Findlay GM. Ultraviolet light and skin cancer. Lancet 1928; 2:1070.
4. Blum HF. Sunlight as a causal factor in cancer of the skin of man. J Natl Cancer Inst 1948; 9:247–251.
5. Blum HF. On the mechanism of cancer induction by ultraviolet radiation. J Natl Cancer Inst 1950; 11:463–466.
6. Jagger J. Solar-UV actions on living cells. New York: Praeger Press, 1985.
7. Kripke ML, Applegate LA. Alteration in the immune response by ultraviolet radiation. In: LA Goldsmith LA, ed. Biochemistry and Physiology of the Skin, 2nd edition. City: Publisher, 1991:1304–1328.
8. Parrish JA, Anderson RR, Urbach F, Pitts D, eds. UV-A: Biological Effects of Ultraviolet Radiation with Emphasis on Human Responses to Longwave Ultraviolet. New York: Plenum Press, 1978.
9. Kripke ML. Immunologic mechanisms in UV radiation carcinogenesis. Adv Cancer Res 1981; 34:69–106.
10. Applegate LA, Lautier D, Frenk E, Tyrrell RM. Endogenous glutathione levels modulate the frequency of both spontaneous and long wavelength ultraviolet induced mutations in human cells. Carcinogenesis 1992; 13:1557–1560.
11. Soter NA. Acute effects of ultraviolet radiation on the skin. Semin Dermatol 1990; 9:11–15.
12. Gilchrest BA, Soter NA, Hawk JLM, Barr RM, Black AK, Hensby CN, Mallet AI, Greaves MW, Parrish JA. Histologic changes associated with ultraviolet A-induced erythema in normal human skin. J Am Acad Dermatol 1983; 9:213–219.
13. van Wheelden H, de Gruijl FR, van der Leun JC. Carcinogenesis by UVA, with an attempt to assess the carcinogenic risks of tanning with UVA and UVB. In: Urbach F, Gange RW. The Biological Effects of UVA Radiation. New York: Praeger Press, 1986:137–146.
14. Morrey JD, Bourn SM, Bunch TD, Jackson MK, Sidwell RW, Barrows LR, Daynes RA, Rosen CA. In vivo activation of human immunodeficiency virus type 1 long terminal repeat by UV type A (UV-A) light plus psoralen and UV-B light in the skin of transgenic mice. J Virol 1991; 65:5045–5051.
15. Freeman SE, Gange RW, Matzinger EA, Sutherland JC, Sutherland BM. Production of pyrimidine dimers in human skin exposed in situ to UVA irradiation. J Invest Dermatol 1987; 88: 430–433.

16. Rosario R, Mark GJ, Parrish JA. Histological changes produced in skin by equally erythemogenic doses of UV-A, UV-B, UV-C and UV-A with psoralens. Br J Dermatol 1979; 101:299–308.
17. Applegate LA, Scaletta C, Panizzon RG, Niggli H, Frenk E. In vivo induction of pyrimidine dimers in human skin by UVA radiation: Initiation of cell damage and/or intercellular communication? Int J Mol Med 1999; 3:467–472.
18. Burren R, Scaletta C, Frenk E, Panizzon RG, Applegate LA. Sunlight and carcinogenesis: Expression of p53 and pyrimidine dimers in human skin following UVA I, UVA I + II and solar simulating radiations. Int J Cancer 1998; 76:201–206.
19. Applegate LA, Noël A, Vile G, Frenk E, Tyrrell RM. Two genes contribute to different extents to the heme oxygenase enzyme activity measured in cultured human skin fibroblasts and keratinocytes: implication for protection against oxidant stress. Photochem Photobiol 1995; 61: 285–291.
20. Fuchs J, Huflejt ME, Rothfuss LM, Wilson DS, Carano G, Packer L. Acute effects on near ultraviolet and visible light on the cutaneous antioxidant defense system. Photochem Photobiol 1989; 50(6):739–744.
21. Moysan A, Marquis I, Gaboriau F, Santus R, Dubertret L, Morliere P. Ultraviolet A-induced lipid peroxidation and antioxidant defense systems in cultured human skin fibroblasts. J Invest Dermatol 1993; 100:692–698.
22. Moysan A, Clément-Lacroix P, Michel L, Dubertret L, Morliere P. Effects of ultraviolet A and antioxidant defense in cultured fibroblasts and keratinocytes. Photodermatol Photoimmunol Photomed 1995; 11:192–197.
23. Shindo Y, Witt E, Han D, Tzeng B, Aziz T, Nguyen L, Packer L. Recovery of antioxidants and reduction in lipid hydroperoxides in murine epidermis and dermis after acute ultraviolet radiation exposure. Photodermatol Photoimmunol Photomed 1994; 10:183–191.
24. Punnonen K, Jansen CT, Puntala A, Ahotupa M. Effects of in vitro UVA irradiation and PUVA treatment on membrane fatty acids and activities of antioxidant enzymes in human keratinocytes. J Invest Dermatol 1991; 96:255–259.
25. Treina G, Scaletta C, Fourtanier A, Seite S, Frenk E, Applegate LA. Expression of intracellular adhesion molecule-1 in UVA-irradiated human skin cells in vitro and in vivo. J Dermatol 1996; 135:241–247.
26. Tyrrell RM. UVA (320–380 nm) radiation as an oxidative stress. Oxidative Stress: Oxidants and Antioxidants. In: Sies H, ed. New York: Academic Press, 1991:57–78.
27. McCormick JP, Fisher JR, Pachlatko JP, Eisenstark A. Characterisation of a cell-lethal product from the photooxidation of tryptophan: hydrogen peroxide. Science 1976; 191:468–469.
28. Cunningham ML, Johnson JS, Giovanazzi SM, Peak MJ. Photosensitized production of superoxide anion by monochromatic (290–405 nm) ultraviolet irradiation of NADH and NADPH coenzymes. Photochem Photobiol 1985; 42:125–128.
29. Devary Y, Rosette C, DiDonato JA, Karin M. NFKB activation by ultraviolet light not dependent on a nuclear signal. Science 1993; 261:1442–1445.
30. Cohen D, DeLeo VA. Ultraviolet radiation-induced phospholipase $A_2$ occurs in mammalian cell membrane preparations. Photochem Photobiol 1993; 57:383–390.
31. Djauaher-Mergny M, Maziere C, Santus R, Dubertret L, Maziere JC. Ultraviolet A decreases epidermal growth factor (EGF) processing in cultured human fibroblast and keratinocytes: inhibition of EGF-induced diacylglycerol formation. J Invest Dermatol 1994; 102:192–196.
32. Matsui MS, Wang N, MacFarlane D, DeLeo VA. Long-wave ultraviolet radiation induces protein kinase C in normal human keratinocytes. Photochem Photobiol 1994; 59:53–57.
33. Gall RL, Kochevar IE, Granstein RD. Ultraviolet radiation induces a change in cell membrane potential in vitro: a possible signal for ultraviolet radiation induced alteration in cell activation. Photochem Photobiol 1989; 49:655–662.
34. Danpure HJ, Tyrrell RM. Oxygen dependence of near UV (365 nm) lethality and the interac-

tion of near-UV and X-rays in two mammalian cell lines. Photochem Photobiol 1976; 23: 171–177.
35. Smith PJ, Paterson MC. Lethality and the induction and repair of DNA damage in far, mid or near UV-irradiated human fibroblasts: comparison of effects in normal, xeroderma pigmentosum and Bloom's syndrome cells. Photochem Photobiol 1982; 36:333–343.
36. Roza L, van der Schans GP, Lohman PHM. The induction and repair of DNA damage and its influence on cell death in primary human fibroblasts exposed to UVA and UVC radiation. Mut Res 1982; 146:89–98.
37. Peak MJ, Peak JG, Carnes BA. Induction of direct and indirect single-strand breaks in human cell DNA by far- and near-ultraviolet radiations: action spectrum and mechanisms. Photochem Photobiol 1987; 45:381–387.
38. Rosenstein BS, Ducore JM. Induction of DNA strand breaks in normal human fibroblasts exposed to monochromatic ultraviolet and visible wavelengths in the 240–546 nm range. Photochem Photobiol 1983; 38:51–55.
39. Peak JG, Peak MJ, Sikorski RA, Jones RA. Induction of DNA-protein crosslinks in human cells by ultraviolet and visible radiations: action spectrum. Photochem Photobiol 1988; 47S: 26S.
40. Peak MJ, Peak JG, Jones CA. Different (direct and indirect) mechanisms for the induction of DNA-protein crosslinks in human cells by far- and near ultraviolet radiations (290 and 405 nm). Photochem Photobiol 1985; 42:141–146.
41. Applegate LA, Lautier D, Frenk E, Tyrrell RM. Endogenous glutathione levels modulate the frequency of both spontaneous and long wavelength ultraviolet induced mutations in human cells. Carcinogenesis 1992; 13:1557–1560.
42. Bradley MO, Sharkley NA. Mutagenicity and toxicity of visible fluorescent light to cultured mammalian cells. Nature (London) 1977; 266:724–726.
43. Hsie AW, Li AP, Machanoff R. A fluence response study of lethality and mutagenicity of white, black and blue fluorescent light, sunlamp and sunlight irradiation in Chinese hamster ovary cells. Mut Res 1977; 45:333–342.
44. Burki HJ, Sam CK. Comparison of the lethal and mutagenic effect of gold and white fluorescent lights on cultured mammalian cells. Mut Res 1978; 54:373–377.
45. Jones CA, Huberman ML, Cunningham ML, Peak MJ. Mutagenesis and cytotoxicity in human epithelial cells by far- and near-ultraviolet radiations: action spectra. Rad Res 1987; 110:244–254.
46. Tyrrell RM. Mutagenic action of monochromatic radiation in the solar range on human cells. Mut Res 1984; 129:103–110.
47. Morlière R, Moysan A, Santos R, Hüppe G, Mazière J-C, Dubertret L. UVA-induced lipid peroxidation in cultured human fibroblasts. Biochim Biophys Acta 1991; 1084:261–268.
48. Applegate LA, Frenk E, Gibbs N, Johnson B, Ferguson J, Tyrrell RM. Cellular sensitivity to oxidative stress in the photosensitivity dermatitis/actinic reticuloid syndrome. J Invest Dermatol 1994; 102:762–767.
49. Vaca CE, Wilhelm J, Harms-Ringdahl M. Interaction of lipid peroxidation products with DNA. A review. Mut Res 1988; 195:137–149.
50. Adelman R, Saul RL, Ames BN. Oxidative damage to DNA: Relation to species metabolic rate and life span. Proc Natl Acad Sci USA 1988; 85:2706–2708.
51. Ames BN. Dietary carcinogens and anticarcinogens. Science 1983; 221:1256–1264.
52. Richter C, Park JW, Ames BN. Normal oxidative damage to mitochondrial and nuclear DNA is extensive. Proc Natl Acad Sci (USA) 1988; 85:6465–6467.
53. Van Kuijk FJ, Sevanian GM, Handelmal GJ, Dratz EA. A new role for phospholipase $A_2$: protection of membranes from lipid peroxidation damage. Trends Biochem Sci 1987; 12:31–34.
54. Davies KJA, Goldberg AL. Oxygen radicals stimulate intracellular proteolysis and lipid peroxidation by independent mechanisms in erythrocytes. J Biol Chem 1987; 262:8220–8226.

55. Pacifici RE, Salo DC, Davies KJA. Macroxyproteinase (MOP): A 670 kDa proteinase complex that degrades oxidatively denatured proteins in red blood cells. Free Rad Biol Med 1989; 7: 521–536.
56. McFadden PN, Clarke S. Conversion of isoaspartyl peptides to normal peptides: implications for the cellular repair of damaged protein. Proc Natl Acad Sci USA 1987; 84:2595–2599.
57. Ketterer B, Meyer DJ. Glutathione transferases: a possible role in the detoxication and repair of DNA and lipid hydroperoxides. Mut Res 1989; 214:33–40.
58. Thomas JP, Maiorino M, Ursini F, Girotti AW. Protective action of phospholipid hydroperoxide glutathione peroxidase against membrane-damaging lipid peroxidation. J Biol Chem 1990; 265:454–461.
59. Wallace S. AP endonucleases and DNA glycosylases that recognize oxidative DNA damage. Environ Molec Mutagen 1988; 12:431–477.
60. Johnson AW, Demple B. Yeast DNA 3′-repair diesterase is the major cellular apurinic/apyrimidinic endonuclease: substrate specificity and kinetics. J Biol Chem 1988; 263:18017–18022.
61. Helland DE, Doetsch PW, Haseltine WA. Substrate specificity of a mammalian DNA repair endonuclease that recognizes oxidative base damage. Mol Cell Biol 1986; 6:1983–1990.
62. Cleaver JE. Defective repair of DNA in xeroderma pigmentosum. Nature (London) 1968; 218: 652–656.
63. Stocker R, Lai A, Peterhans E, Ames BN. Antioxidant activities of bilirubin and biliverdin. In: Hayaishi O, Niki E, Kondo M, Yoshikawa T. Medical, Biochemical and Chemical Aspects of Free Radicals. Amsterdam: Elsevier, 1989:465–468.
64. Vile GF, Basu-Modak S, Waltner C, Tyrrell RM. Haem oxygenase-1 mediates an adaptive response to oxidative stress in human skin fibroblasts. Proc Natl Acad Sci USA 1994; 91: 2607–2610.
65. Vile GF, Tyrrell RM. Oxidative stress resulting from ultraviolet A irradiation of human skin fibroblasts leads to a heme oxygenase-dependent increase in ferritin. J Biol Chem 1993; 268: 14678–14681.
66. Fornace AJ Jr, Alamo I Jr, Hollander MC. DNA damage-inducible transcripts in mammalian cells. Proc Natl Acad Sci USA 1988; 85:8800–8804.
67. Stein B, Rahmsdorf HJ, Steffen A, Litfin M, Herrlich P. UV-induced DNA damage is an intermediate step in UV-induced expression of human immunodeficiency virus type 1, collagenase, c-fos, and metallothionein. Mol Cell Biol 1989; 9:5169–5181.
68. Applegate LA, Luscher P, Tyrrell RM. Induction of heme oxygenase: a general response to oxidant stress in cultured mammalian cells. Cancer Res 1991; 51:974–978.
69. Applegate LA, Scaletta C, Panizzon R, Frenk E. Evidence that ferritin is UV inducible in human skin: part of a putative defense mechanism. J Invest Dermatal 1998; 111:159–163.
70. Applegate LA, Frenk E. Oxidative defense in cultured human skin fibroblasts and keratinocytes from sun-exposed and non-exposed skin. Photodermatol Photoimmunol Photomed 1995; 11: 95–101.
71. Rosen DR. Mutations in Cu/Zn superoxide dismutase gene are associated with familial amyotrophic lateral sclerosis. Nature 1993; 364:362.
72. Hugon J. Free radicals, SOD and neurodegeneration in sporadic Amyotrophic Lateral Sclerosis. Recent Advances and Clinical Applications. Abstract at the International Conference on Superoxide Dismutases, Institut Pasteur, Paris, 1998, p. 16.
73. Price JM, Rhodin JAG, Sutton ET. The protective effect of SOD on vascular endothelium dysfunction induced by Alzheimer's peptide beta-amyloid. Recent Advances and Clinical Applications. Abstract at the International Conference on Superoxide Dismutases. Paris: Institut Pasteur, 1998:15.
74. Iizuka H, Asaga H, Koike K. Decreased Cu, Zn-superoxide dismutase activity in psoriatic hyperproliferative epidermis. Eur J Dermatol 1993; 3:56–58.
75. Borrello S, Bedogni B, Pani G. Mn-SOD downregulation: A new potential mechanism for

the antitumoral activity of p53. Recent Advances and Clinical Applications. Abstract at the International Conference on Superoxide Dismutases. Paris Institut Pasteur, 1998:32.
76. Oberley LW. Overexpression of Mn-SOD inhibits cancer cell growth. Recent Advances and Clinical Applications. Abstract at the International Conference on Superoxide Dismutases. Paris: Institut Pasteur, p. 30, 1998.
77. Varachaud A, Berthier-Vergnes O, Rigaud M, Schmitt D, Bernard P. Expression and activity of Manganese Superoxide Dismutase in human melanoma cell lines with different tumorigenic potential. Eur J Dermatol 1998; 8:90–94.
78. Schranz N, Blanchard DA, Mitenne F. Manganese induces apoptosis of human B cells: activation of caspase-1 and caspase-3. Recent Advances and Clinical Applications. Abstract at the International Conference on Superoxide Dismutases. Paris: Institut Pasteur, 1998:12.
79. Montenegro L, Bonina F, Rigano L, Giogilli S. Protective effect evaluation of free radical scavengers on UVB induced human cutaneous erythema by skin reflectance spectrophotometry. Int J Cosmet Sci 1995; 17:91–93.
80. Ronai ZA, Okin E, Weinstein IB. Ultraviolet light induces the expression of oncogenes in rat fibroblast keratinocyte cells. Oncogene 1988; 2:201–204.
81. Poswig A, Wenk J, Brenneisen P, Wlaschek M, Hommel C, Quel G, Faisst K, Disemond J, Briviba K, Krieg T, Scharffetter-Kochanek K. Adaptive antioxidant response of manganese-superoxide dismutase following repetitive UVA irradiation. J Invest Dermatol 1999; 112:13–18.
82. Aragane Y, Kulms D, Metze D, Wilkes G, Pappelmann B, Luger TA, Schwarz T. Ultraviolet light induces apoptosis via direct activation of CD95 (FAS/APO-1) independently from its ligand CD95L. Abstr Int Invest Dermatol, Cologne, 1998.
83. Kondo S, Jimbow K. Induction of IL-12 but not IL-10 form human keratinocytes after exposure to UVA. J Cell Physiol 1998; 177:493–498.

# 15

# Ozone Stress in the Skin Barrier

**STEFAN U. WEBER**

*University of California, Berkeley, California*

**LESTER PACKER**

*University of Southern California School of Pharmacy, Los Angeles, California*

## I. INTRODUCTION

This chapter aims to provide a comprehensive view of the effects of ozone ($O_3$) exposure on the skin. Reports of hazardous effects induced by smog reach as far back as the 14th century when, during the reign of Richard III (1377–1399), human diseases were attributed to severe air pollution (1). Today it has been established that $O_3$ is one of the most prominent components of air pollution. Frequently $O_3$ concentrations exceed the federal safety limit of 0.12 ppm (1-h average), a concentration capable of decreasing the pulmonary function (1 $\mu g/m^{-3} = 0.501 \times 10^{-3}$ ppm $O_3$) (2). For example, in the south coast air basin of the U.S. in the greater Los Angeles region, this value was exceeded on 90 days in 1996, reaching maximal concentrations of 0.24 ppm (3). In heavily polluted metropoles, the maximal concentrations can reach up to 0.4 to 0.5 ppm (4).

The formation of $O_3$ in the troposphere, the ground level, is mediated by nitric oxides emitted as part of exhaust fumes from motor vehicles, industries, and private homes. $NO_2$ can be photolyzed by solar ultraviolet (UV) radiation ($\lambda < 370$ nm) resulting in NO and atomic oxygen. $O_2$ readily reacts with atomic oxygen to form $O_3$. $O_3$ can also be destroyed by nitric oxides. NO can react with $O_3$ to form $NO_2$ and $O_2$ (4). The emission of many species contributes to the generation of $O_3$. Sulfur dioxide and volatile organic compounds such as carbon monoxide and methane are involved as well. Other sources include photocopiers, electric arcs used in welding, and UV lamps (4).

The skin can be directly exposed to tropospheric $O_3$. The skin consists of two main layers, the outer layer called epidermis and the lower layers called dermis, which is placed on subcutaneous fat tissue (5). Dermal fibroblasts synthesize a complex extracellular matrix containing collagenous and elastic fibers. Blood capillaries reach the upper part of the dermis. The epidermis contains mostly keratinocytes that egress to the skin surface as they differentiate progressively to form enucleated corneocytes that comprise the outer part of the epidermis, the stratum corneum (SC). The SC contains neither viable cells nor nerve endings. Rather, it is composed of core and envelope proteins embedded in a lipophilic matrix (6). A total of 15 to 20 layers of so-called ''cornified envelopes'' turns over in approximately 2 weeks by desquamating the outer layer. The SC plays an important role as the skin barrier, limiting the penetration of exogenous compounds into the body. Moreover, it limits the evaporation of water across the skin, the so-called transepidermal water loss, thus preventing the body from drying out (7). Perturbation of the skin barrier is involved in several dermatological pathologies (6,8,9).

The skin is equipped with a network of protective antioxidants. They include enzymatic antioxidants such as glutathione peroxidase, superoxide dismutase and catalase, and nonenzymatic low-molecular-weight antioxidants such as vitamin E isoforms, vitamin C, glutathione (GSH), uric acid, and ubiquinol (10). In general, the outer part of the skin, the epidermis, contains higher concentrations of antioxidants than the dermis (11). In the lipophilic phase, $\alpha$-tocopherol is the most prominent antioxidant, while vitamin C and GSH have the highest abundance in the cytosol.

In the lung, $O_3$ has been shown to exhibit toxic effects. A drop in the forced expiratory volume ($FEV_1$) is caused by 0.12 ppm for 1 h or 0.08 ppm for 8 h (2). In asthmatic patients and also in healthy people, airway inflammation is triggered and a decrease in pulmonary function can be recorded. Recently, some reports indicated that $O_3$ may also directly affect the skin (12).

The following chapter will explain why the stratum corneum is the first target of environmental exposure to $O_3$. It will be summarized what types of biomolecules are oxidized by $O_3$ in the skin and how $O_3$ and other environmental oxidants can interact. The implications of oxidative stress by $O_3$ for epidermal function will be discussed.

## II. $O_3$ TARGETS STRATUM CORNEUM ANTIOXIDANTS

The SC was found to contain both hydrophilic and lipophilic antioxidants. Vitamins C and E (both $\alpha$- and $\gamma$-tocopherol) as well as GSH and uric acid were found to be present in the SC (13,14). Surprisingly, they were not distributed evenly, but in gradient fashion, with low concentrations on the outer layers and increasing concentrations toward the deeper layers of the SC. This phenomenon may be explained by the fact that $O_2$ partial pressure is higher in the upper SC, which already causes a mild oxidative stress resulting in the partial depletion of antioxidants. Moreover, the further up corneocytes are located, the longer they have been a part of the SC and the longer they have been exposed to the ambient environmental oxidative stress (13).

With a redox potential of +2.07, $O_3$ is one of the strongest known oxidants and readily oxidizes a wide range of compounds including rubber (4). For lung exposure, it has been calculated that $O_3$ already reacts within the lung lining fluid before it can reach the alveolar cells (15). In plant cells, the concentration of $O_3$ in the intercellular spaces was virtually zero, indicating that $O_3$ reacted within the first cell layers (16). Since the stratum corneum (SC) is designed to protect the skin from environmental stressors, it could be argued that it should be fairly resistant to $O_3$ exposure and that it should protect the viable part of the skin from $O_3$-induced damage.

In fact, similar findings as in the SC have been reported in the lung and in plants. When hairless mice were exposed to high concentrations of $O_3$ (10 ppm for 2 h), no

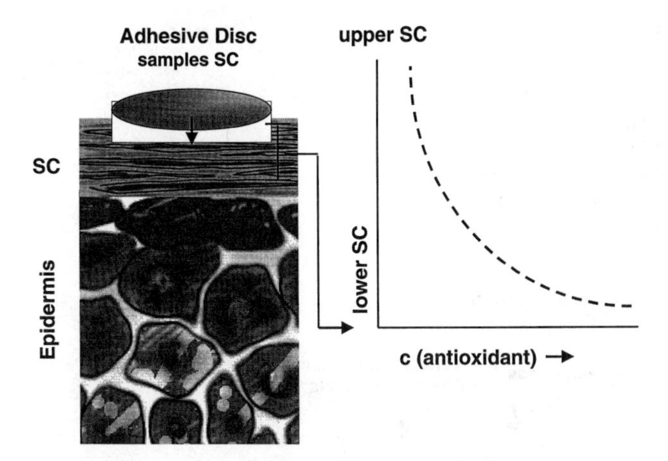


**Figure 1** Antioxidant gradients in stratum corneum (SC). The epidermis terminally differentiates into the SC, which forms the outermost layer of the skin. The SC can be sequentially removed by application of adehsive discs, and assayed for antioxidants. From the upper SC toward the lower SC, antioxidants form gradients with increasing concentrations toward the epidermis.

depletion of the sensitive antioxidant α-tocopherol occurred when full thickness skin was analyzed (17). Only when additional vitamin E was supplemented topically did approximately two-thirds of the applied vitamin E oxidize. To localize the $O_3$ effect, skin was sectioned into the upper epidermis, lower epidermis/papillary dermis, and dermis after $O_3$ exposure to the same dose as used in the previous experiment. Vitamins E and C were only found to be depleted in the upper epidermis while the antioxidant concentrations in the lower layers remained unaltered, indicating that the main interaction with $O_3$ occurs in the outer part of the epidermis (18). In an attempt to further localize the main area of interaction, new methods were developed to analyze antioxidants in the thin SC itself by sequential removal with adhesives (19). When the SC was sampled after $O_3$ exposure, tenfold lower doses effectively depleted all analyzed antioxidants. Exposure to 1 ppm for 2 h decreased α-tocopherol to approximately 80% (18), vitamin C to 80%, GSH to 40%, and urate to 45% (13). Lower $O_3$ doses were ineffective. These data indicate that the SC is the main target of environmental $O_3$ exposure within the skin. They demonstrate that, as in the lung tissue, $O_3$ mostly reacts in the uppermost layers without reaching the viable cells.

The depletion of antioxidants can be, in principle, due to either a direct reaction with $O_3$ or a reaction with a free radical induced by the reaction of $O_3$ with another biomolecule. In the case of vitamin C, it is assumed that it directly reacts with $O_3$ in a sacrificial process to protect the tissue from $O_3$, which is plausible since the rate constant of this reaction is high (20). Accordingly, it could be seen in other systems (e.g., plasma) that vitamin C and uric acid were both rapidly consumed after ozonization. Since the rate constant of vitamin E with $O_3$ is lower, it is hypothesized that vitamin E is depleted by reacting with lipid free radical intermediates arising from $O_3$-induced lipid peroxidation (20).

It has been demonstrated that vitamin C can regenerate α-tocopherol from its chromanoxyl radical (21), and the vitamin C radical may be recycled by GSH nonenzymati-

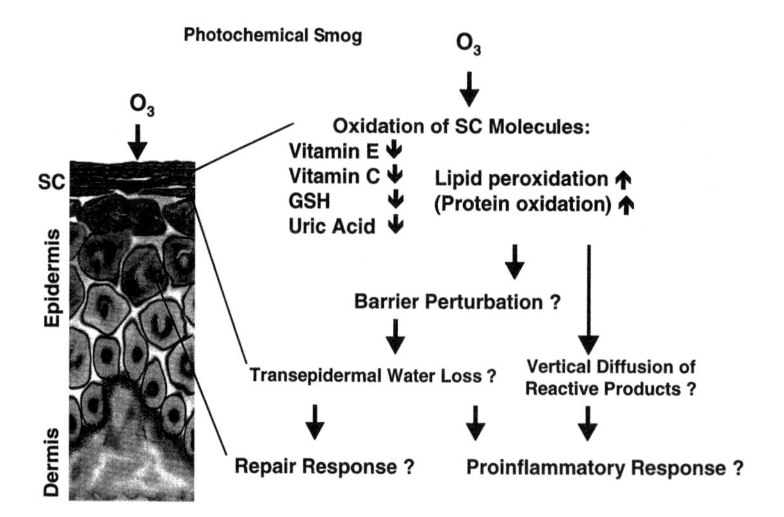


**Figure 2** Hypothetical mechanisms involved in topical $O_3$-exposure to the skin. $O_3$ oxidizes SC antioxidants and induces oxidation of structurally relevant SC molecules. Oxidative damage to the SC may result in a barrier perturbation, which, in turn, elicits repair responses and/or the release of proinflammatory cytokines. Alternatively, secondary products of lipid peroxidation may diffuse into the viable epidermis and cause a proinflammatory response there.

cally under slightly acidic conditions (22) that are present in the SC (23). If these recycling processes occur in the SC, GSH and vitamin C should be depleted prior to α-tocopherol. However, this was not the case in the experiments described above.

Speculations as to why this was not found are complicated by several factors. First, antioxidants are distributed in gradients in the SC, and the ratio between them varies from the upper to the lower SC. Second, antioxidants could react with each other or with $O_3$ directly. Third, it is not clear where in the ultrastructure of the SC antioxidants are located. The SC has a structure different from other tissues. The overall water content (15%) is very low (24). Hydrophilic antioxidants could be located either close to the core proteins or in so-called lacunar domains, which may serve as a hydrophilic pathway through the SC (25). It is not clear whether vitamin C and vitamin E are located in the SC in adjacent compartments, which is a conditio sine qua non for recycling. Therefore, it is not possible at this point to effectively model the possible interactions of low-molecular-weight antioxidants in the SC.

## III. $O_3$ DAMAGES LIPIDS AND PROTEINS

Apart from antioxidants, topical $O_3$ exposure to skin damages lipids and proteins in the SC. Both lipids and proteins are responsible for the structural integrity of the skin barrier (26,27). Therefore, it is important to assess damage to them induced by environmental oxidants. A HPLC-based method was developed for the SC to analyze malondialdehyde (MDA), a breakdown product of lipid peroxidation (18).

MDA was increased in full thickness skin after doses of 10 ppm $O_3 \times 2$ h and 5 ppm $\times$ 2 h (17). In the SC, however, a dose of 1 ppm $\times$ 2 h, which was capable of depleting vitamin E, failed to increase MDA, indicating that vitamin E may be able to partially block the $O_3$-induced peroxidation of lipids (18). In LDL particles, a total depletion of vitamin E preceded the peroxidation of lipids (28). This was not the case in the SC. Here, MDA was found in significant amounts when approximately 70% of the vitamin E was depleted.

These differences may be due to the fact that vitamin E is not able to inhibit lipid peroxidation in all cases. Vitamin E inhibits the lipid peroxidation of polyunsaturated fatty acids containing at least three double bonds, such as linolenic acid (29). However, $O_3$ can also react with polyunsaturated fatty acids containing two methylene-interrupted double bonds (30). This reaction is not blocked by α-tocopherol (31), which may explain why MDA increased while α-tocopherol was still present.

Since all evidence points to the fact that $O_3$ reacts completely within the SC, the question arises whether biochemical changes can be detected in the viable part of the epidermis. When murine skin was exposed to the high dose of 10 ppm × 2 h, MDA was also increased in the lower epidermis and the upper papillary dermis (32). The most probable explanation for this finding is that lipid peroxidation products diffuse from the SC where they are generated in high amounts to lower layers of the epidermis.

When murine skin was exposed chronically, cumulative damage could be observed. Exposure to 1 ppm $O_3$ × 2 h for six consecutive days resulted in a significant increase in MDA and caused cumulative depletion of vitamin E to less than 50% of the baseline (18).

As mentioned earlier, proteins, especially keratins and other structural proteins, are present in the SC. Protein oxidation frequently leads to the introduction of carbonyl groups. When SC was obtained by removal with adhesive disks, macromolecular carbonyl formation could be monitored in response to increasing $O_3$-doses of 1, 5, and 10 ppm × 2 h (33). Even though the assay used could not discriminate between carbonyl groups from proteins and lipids, it is reasonable to assume that a considerable portion of the increase in carbonyls was due to protein oxidation. These experiments need to be confirmed in in vivo experiments.

## IV. EXPOSURE TO MULTIPLE ENVIRONMENTAL OXIDANTS

As pointed out earlier, the lowest concentration to cause measurable oxidative stress in the SC was 1 ppm, a concentration that is about double the maxima encountered in heavily polluted metropoles. Frequently, at times when $O_3$ concentrations are high, the ground levels may also be exposed to increased UV irradiance (4), which is known to be an important oxidative stress in itself (34). It was therefore hypothesized that $O_3$ and solar UV radiation may synergize in terms of oxidative challenge to the skin, especially the SC.

A hairless mouse model was developed for exposure to UV and $O_3$ alone and in combination. α-Tocopherol was chosen as an indicator because it had been shown to be a sensitive and reproducible parameter of oxidative stress in the SC. A dose of 0.5 ppm $O_3$ for 2 h alone did not deplete vitamin E significantly. A small dose of solar-simulated UV radiation (0.3 minimal erythemal doses), however, destroyed a significant portion of SC vitamin E. Surprisingly, the combination of these two stressors resulted in a significant potentiation of the UV-induced α-tocopherol depletion (35). These findings indicate that UV radiation and $O_3$ can have additive oxidative effects.

This was true for low doses. When the UV dose was raised to 1 minimal erythemal dose, UV still potentiated the $O_3$ effect, but $O_3$ failed to add to the UV effect (35). This indicates that UV radiation in doses occurring in the environment is a stronger oxidant to the SC than $O_3$. One also needs to take into account that photochemical smog reduces the intensity of solar UV radiation on ground level, meaning it will take longer to reach 1 minimal erythemal dose. Under these circumstances, the relative importance of $O_3$ as an additional oxidative stressor may increase.

## V. IMPLICATIONS FOR EPIDERMAL FUNCTION

While several studies have characterized the biochemical effects of $O_3$ exposure in the skin, it is unclear whether the observed changes can be translated into pathophysiological events. Therefore, potential implications for epidermal function will be discussed.

As pointed out, the SC plays an important role as the skin barrier. Both proteins and lipids cooperate in maintaining the barrier function. Homeostasis of lipids has especially been shown to be crucial for functionality. SC lipids are mainly composed of ceramides, cholesterol, and free fatty acids, especially linoleic acid (7). Disturbance of the lipid ratios results in a reduction in barrier function and an increase in the water flux across the skin, the so-called transepidermal water loss (36). $O_3$ has been shown to oxidize lipids. It is certainly conceivable that $O_3$ may disturb the critical balance of lipids in the SC, which in turn could increase the transepidermal water loss.

A perturbed skin barrier is known to provoke several types of responses in the epidermis. Barrier disruption activates a repair response during which epidermal DNA synthesis is unregulated (37). The key enzymes for the production of SC lipids are turned on and the lipid classes are produced more rapidly (38). They are then packed into lamellar bodies and secreted into the perturbed skin barrier to repair the damage (39). Moreover, an increased transepidermal water loss elicits a proinflammatory response in the epidermal keratinocytes with an increased production of the proinflammatory cytokines IL-1$\alpha$, TNF, and GM-CSF, while manual occlusion of the perturbed barrier abrogated these events (40,41). In the lung, $O_3$ was shown to activate the transcription factor NF-$\kappa$B, which controls the expression of many important proinflammatory genes (42). Similar events may take place in the skin. For the lung it was also shown that $O_3$ treatment leads to the formation of aldehyde breakdown products of lipid peroxidation (15,43–47). These degradation products such as hexanal, heptanal, 4-hydroxy-nonenal and others may also occur in the SC after ozonization. They may diffuse from the SC into the viable layers of the epidermis and exhibit toxic effect there, as it is postulated for $O_3$-induced lung toxicity.

## VI. CONCLUSION AND PERSPECTIVE

Increasing evidence shows that topical exposure to tropospheric $O_3$ may induce oxidative stress in the skin. Most of the $O_3$ reacts within the SC. It depletes hydrophilic and lipophilic antioxidants dose dependently. Moreover, $O_3$ in concert with UV radiation may pose an additive oxidative stress close to doses occurring in polluted metropoles. Therefore, $O_3$-induced oxidative stress may be relevant in areas suffering from increased $O_3$ levels. $O_3$ damages SC lipids and presumably SC proteins, both of which are essential for the barrier properties of the skin. It is conceivable that $O_3$ exposure may compromise the structure and functionality of the skin barrier. A perturbed skin barrier is also seen in several skin pathologies such as psoriasis, atopic dermatitis, irritant contact dermatitis, and aging skin including pruritus. It can be speculated that in these diseases $O_3$ may have a potential to further compromise the barrier function. It may also be involved in a proinflammatory response of the epidermis. Thus far, no human studies have been carried out to verify these hypotheses. Further studies of the $O_3$-induced effects, preferably in humans, need to be carried out to elucidate the role of topical $O_3$ exposure on cutaneous tissues.

## ACKNOWLEDGMENTS

The authors would like to thank Nancy Han for editing the manuscript and Amir Tavakkol and Thomas G. Polefka for helpful discussion. This work was in part supported by the Colgate Palmolive Company.

## REFERENCES

1. Ayres SM, Evans RG, Buehler ME. Air pollution: a major public health problem. CRC Crit Rev Clin Lab Sci 1972; 3:1–40.
2. Goldring J, Morris RD. Ground-level ozone in Wisconsin: potential health effects. Wis Med J 1992; 91:240–242.
3. South-Coast-Air-Quality-Management-District. Annual Report. http://www.aqmd.gov, 1997.
4. Mustafa MG. Biochemical basis of ozone toxicity. Free Radic Biol Med 1990; 9:245–265.
5. Weber SU. Oxidants in skin pathophysiology. In: Sen CK, Packer L, Hanninen O, eds. Exercise and Oxygen Toxicity, 2nd ed. Amsterdam: Elsevier Science, 2000.
6. Elias PM, Feingold KR. Lipids and the epidermal water barrier: metabolism, regulation, and pathophysiology. Semin Dermatol 1992; 11:176–182.
7. Elias PM. Stratum corneum architecture, metabolic activity and interactivity with subjacent cell layers. Exp Dermatol 1996; 5:191–201.
8. Ghadially R, Brown BE, Sequeira-Martin SM, Feingold KR, Elias PM. The aged epidermal permeability barrier. Structural, functional, and lipid biochemical abnormalities in humans and a senescent murine model. J Clin Invest 1995; 95:2281–2290.
9. Mao-Qiang M, Feingold KR, Thornfeldt CR, Elias PM. Optimization of physiological lipid mixtures for barrier repair. J Invest Dermatol 1996; 106:1096–1101.
10. Shindo Y, Witt E, Packer L. Antioxidant defense mechanisms in murine epidermis and dermis and their responses to ultraviolet light. J Invest Dermatol 1993; 100:260–265.
11. Shindo Y, Witt E, Han D, Epstein W, Packer L. Enzymic and non-enzymic antioxidants in epidermis and dermis of human skin. J Invest Dermatol 1994; 102:122–124.
12. Thiele JJ, Podda M, Packer L. Tropospheric ozone: an emerging environmental stress to skin. Biol Chem 1997; 378:1299–1305.
13. Weber SU, Thiele JJ, Cross CE, Packer L. Vitamin C, uric acid, and glutathione gradients in murine stratum corneum and their susceptibility to ozone exposure. J Invest Dermatol 1999; 113:1128–1132.
14. Thiele JJ, Traber MG, Packer L. Depletion of human stratum corneum vitamin E: an early and sensitive in vivo marker of UV induced photo-oxidation. J Invest Dermatol 1998; 110: 756–761.
15. Frampton MW, Pryor WA, Cueto R, Cox C, Morrow PE, Utell MJ. Ozone exposure increases aldehydes in epithelial lining fluid in human lung. Am J Respir Crit Care Med 1999; 159: 1134–1137.
16. Moldau H, Laisk A. Ozone concentration in leaf intercellular air space is close to zero. Plant Physiol 1989; 90:1163–1167.
17. Thiele JJ, Traber MG, Podda M, Tsang K, Cross CE, Packer L. Ozone depletes tocopherols and tocotrienols topically applied to murine skin. FEBS Lett 1997; 401:167–170.

18. Thiele JJ, Traber MG, Polefka TG, Cross CE, Packer L. Ozone-exposure depletes vitamin E and induces lipid peroxidation in murine stratum corneum. J Invest Dermatol 1997; 108:753–757.
19. Weber SU, Jothi S, Thiele JJ. High pressure liquid chromatography analysis of ozone-induced depletion of hydrophilic and lipophilic antioxidants in murine skin. Meth Enzymol, in press.
20. Giamalva D, Church DF, Pryor WA. A comparison of the rates of ozonation of biological antioxidants and oleate and linoleate esters. Biochem Biophys Res Commun 1985; 133:773–779.
21. Packer JE, Slater TF, Willson RL. Direct observation of a free radical interaction between vitamin E and vitamin C. Nature 1979; 278:737–738.
22. Stocker R, Weidemann MJ, Hunt NH. Possible mechanisms responsible for the increased ascorbic acid content of Plasmodium vinckei-infected mouse erythrocytes. Biochim Biophys Acta 1986; 881:391–397.
23. Ohman H, Vahlquist A. In vivo studies concerning a pH gradient in human stratum corneum and upper epidermis. Acta Derm Venereol 1994; 74:375–379.
24. Mak VH, Potts RO, Guy RH. Does hydration affect intercellular lipid organization in the stratum corneum? Pharm Res 1991; 8:1064–1065.
25. Menon GK, Elias PM. Morphologic basis for a pore-pathway in mammalian stratum corneum. Skin Pharmacol 1997; 10:235–246.
26. Yang L, Mao-Qiang M, Taljebini M, Elias PM, Feingold KR. Topical stratum corneum lipids accelerate barrier repair after tape stripping, solvent treatment and some but not all types of detergent treatment. Br J Dermatol 1995; 133:679–685.
27. Thiele JJ, Hsieh SN, Briviba K, Sies H. Protein oxidation in human stratum corneum: susceptibility of keratins to oxidation in vitro and presence of a keratin oxidation gradient in vivo. J Invest Dermatol 1999; 113:335–339.
28. Esterbauer H, Striegl G, Puhl H, Rotheneder M. Continuous monitoring of in vitro oxidation of human low density lipoprotein. Free Rad Res Commun 1989; 6:67–75.
29. Pryor WA, Stanley JP, Blair E. Autoxidation of polyunsaturated fatty acids: II. A suggested mechanism for the formation of TBA-reactive materials from prostaglandin-like endoperoxides. Lipids 1976; 11:370–379.
30. Pryor WA. Can vitamin E protect humans against the pathological effects of ozone in smog? Am J Clin Nutr 1991; 53:702–722.
31. Roehm JN, Hadley JG, Menzel DB. Oxidation of unsaturated fatty acids by ozone and nitrogen dioxide. A common mechanism of action. Arch Environ Health 1971; 23:142–148.
32. Thiele JJ, Traber MG, Tsang K, Cross CE, Packer L. In vivo exposure to ozone depletes vitamins C and E and induces lipid peroxidation in epidermal layers of murine skin. Free Rad Biol Med 1997; 23:385–391.
33. Thiele JJ, Traber MG, Re R, Espuno N, Yan LJ, Cross CE, Packer L. Macromolecular carbonyls in human stratum corneum: a biomarker for environmental oxidant exposure? FEBS Lett 1998; 422:403–406.
34. Scharffetter-Kochanek K, Wlaschek M, Brenneisen P, Schauen M, Blaudschun R, Wenk J. UV-induced reactive oxygen species in photocarcinogenesis and photoaging. Biol Chem 1997; 378:1247–1257.
35. Valacchi G, Weber SU, Luu C, Cross CE, Packer L. Ozone potentiates vitamin E depletion by ultraviolet radiation in the murine stratum corneum [in process citation]. FEBS Lett 2000; 466:165–168.
36. Ghadially R, Brown BE, Hanley K, Reed JT, Feingold KR, Elias PM. Decreased epidermal lipid synthesis accounts for altered barrier function in aged mice. J Invest Dermatol 1996; 106:1064–1069.
37. Proksch E, Feingold KR, Man MQ, Elias PM. Barrier function regulates epidermal DNA synthesis. J Clin Invest 1991; 87:1668–1673.
38. Harris IR, Farrell AM, Grunfeld C, Holleran WM, Elias PM, Feingold KR. Permeability barrier

disruption coordinately regulates mRNA levels for key enzymes of cholesterol, fatty acid, and ceramide synthesis in the epidermis. J Invest Dermatol 1997; 109:783–787.
39. Menon GK, Feingold KR, Elias PM. Lamellar body secretory response to barrier disruption. J Invest Dermatol 1992; 98:279–289.
40. Wood LC, Feingold KR, Sequeira-Martin SM, Elias PM, Grunfeld C. Barrier function coordinately regulates epidermal IL-1 and IL-1 receptor antagonist mRNA levels. Exp Dermatol 1994; 3:56–60.
41. Wood LC, Jackson SM, Elias PM, Grunfeld C, Feingold KR. Cutaneous barrier perturbation stimulates cytokine production in the epidermis of mice. J Clin Invest 1992; 90:482–487.
42. Haddad EB, Salmon M, Koto H, Barnes PJ, Adcock I, Chung KF. Ozone induction of cytokine-induced neutrophil chemoattractant (CINC) and nuclear factor-kappa b in rat lung: inhibition by corticosteroids. FEBS Lett 1996; 379:265–268.
43. Cueto R, Squadrito GL, Bermudez E, Pryor WA. Identification of heptanal and nonanal in bronchoalveolar lavage from rats exposed to low levels of ozone. Biochem Biophys Res Commun 1992; 188:129–134.
44. Pryor WA. Mechanisms of radical formation from reactions of ozone with target molecules in the lung. Free Radic Biol Med 1994; 17:451–465.
45. Pryor WA, Squadrito GL, Friedman M. The cascade mechanism to explain ozone toxicity: the role of lipid ozonation products. Free Radic Biol Med 1995; 19:935–941.
46. Pryor WA, Bermudez E, Cueto R, Squadrito GL. Detection of aldehydes in bronchoalveolar lavage of rats exposed to ozone. Fundam Appl Toxicol 1996; 34:148–156.
47. Uppu RM, Cueto R, Squadrito GL, Pryor WA. What does ozone react with at the air/lung interface? Model studies using human red blood cell membranes. Arch Biochem Biophys 1995; 319:257–266.

# 16

# Immunotoxicity of Environmental Agents in the Skin

**SANTOSH K. KATIYAR and HASAN MUKHTAR**

*Case Western Reserve University, Cleveland, Ohio*

## I. INTRODUCTION

Skin is much more than a passive physical barrier between the external environment and internal tissues. A major role of the skin is to provide immune function at this crucial interface between inside and outside. Although it has been suggested that skin can function semiautonomously as an immunological organ (1,2), the ''skin immune system'' (3) functions most effectively when it draws upon the resources of the immune system from elsewhere in the body. Thus, recruitment of subsets of myeloid and lymphoid cells into the cutaneous microenvironment is essential for skin immune functions. For the proper homeostatic functioning of the skin immune system, the multiple proinflammatory signals that can be generated by skin cells must eventually be counterbalanced by mechanisms capable of promoting resolution of a cutaneous inflammatory process. Failure of these mechanisms may predispose skin to the development of chronic inflammatory process (4).

The cells that make up the skin immune system are distributed on both sides of the basement membrane separating epidermis and dermis and include several populations of nonleukocytes. Most of the attention has been focused on epidermal components (i.e., keratinocytes, Langerhans cells, and dendritic epidermal T-cells in mice). However, the dermal components of the skin immune system, including dermal fibroblasts, microvascular endothelial cells, dermal dendritic cells, mast cells, and resident perivascular T-cells, also participate in the cutaneous immune response (4). These epidermal and dermal components have evolved into a dynamic network of interacting cells capable of sensing a variety of perturbations including trauma, ultraviolet irradiation, toxic environmental chemicals, and pathogenic microorganisms in the cutaneous environment and rapidly sending appropriate signals that alert and recruit other branches of the immune system.

To better understand the immunological defense to the body by the skin and also immunotoxicity caused by environmental factors such as certain oxidizing agents in the skin, it will be important to take an overview of the major epidermal and dermal cells that are related with the immune functions of the skin. A brief summary of the cells responsible for immunological functions found in epidermis and dermis are described.

## II. CELLS OF THE SKIN IMMUNE SYSTEM

### A. Epidermal Cells of the Skin Immune System

#### 1. *Keratinocytes*

Keratinocytes constitute the major cellular part of the epidermis. The major mechanisms used by keratinocytes to participate in the immune system of the skin is the production of cytokines and responses to cytokines. The vast repertoire of cytokines produced by keratinocytes is perhaps the single most compelling reason to suggest that the skin functions as an immune organ. While resting keratinocytes produce some cytokines constitutively, keratinocyte cytokine production is markedly enhanced following activation. Keratinocytes can transduce multiple noxious stimuli into a signal, in the form of cytokines, that is transmitted to the other members of the skin immune system and the immune system at large. Keratinocyte activation may also be a response to release of proinflammatory cytokines from other cell types. Of all the cytokines produced by keratinocytes, only the ''primary'' cytokines, interleukin-1-alpha (IL-1$\alpha$), interleukin-1-$\beta$ (IL-1$\beta$), and tumor necrosis factor-alpha (TNF-$\alpha$) activate a sufficient number of effector mechanisms to trigger cutaneous inflammation independently (5). Upon stimulation, keratinocytes also produce at least two major immunoregulatory cytokines, IL-10 and IL-12, which play critical roles in regulating whether cutaneous immune responses become polarized toward cell-mediated or humoral responses. Keratinocytes produce low levels of IL-10 constitutively (6,7), and IL-10 production is enhanced following stimuli including UV light (7,8) and application of contact sensitizers (6,9). Keratinocyte-derived IL-10 can act to suppress immune responses locally and systemically (8,10). Various types of cytokines produced by keratinocytes are presented in Table 1 (4).

#### 2. *Langerhans Cells*

Langerhans cells (LC) are the professional antigen-presenting cells (APC) of the epidermis (11). LC cells possess specialized mechanisms permitting efficient capture and uptake of exogenous antigens, processing of these antigens into immunogenic peptides, and presentation of these peptides to T-cells in the context of MHC class II molecules. While they normally reside in the epidermis, LC can be mobilized by an inflammatory stimulus to leave the epidermis, enter dermal lymphatics, and travel to regional lymph nodes (12). Upon reaching the lymph node, LC have typically differentiated into cells that express higher levels of class II MHC molecules and costimulatory molecules like B7-1 (CD80) and B7-2 (CD86) (13,14), rendering them even more effective at stimulating the proliferation of naïve antigen-reactive T-cells they come in contact with in the lymph node.

LC are well suited for their role in antigen presentation to T cells; other epidermal cells can also become involved in antigen presentation. Keratinocytes are unable to migrate

**Table 1** Major Cytokines Produced by Keratinocytes

| |
|---|
| 1. Primary cytokines |
| IL-1α |
| IL-1β |
| TNF-α |
| 2. Cytokines regulating humoral vs. cellular immunity |
| IL-10 |
| IL-12 |
| 3. Cytokines that promote growth of T-lymphocytes |
| IL-7 |
| IL-15 |
| 4. Cytokines with colony-stimulating factor activity |
| Granulocyte-macrophage colony-stimulating factor |
| IL-6 |
| Stem cell factor |
| Macrophage colony-stimulating factor |
| IL-3 (mice) |
| 5. Cytokines with growth factor activity for nonleukocytes |
| Transforming growth factor-α |
| Platelet-derived growth factor |
| Vascular endothelial growth factor |
| Nerve growth factor |
| Amphiregulin |
| Heparin-binding epidermal growth factor |
| 6. Chemokines |
| MCP-1 |
| 7. Immunosuppressive cytokines |
| IL-10 |
| TGF-β |
| IL-1Ra |

to lymph nodes to sensitize naïve T cells, but they can present antigens to T-cells that reach the skin.

### *3. Dendritic Epidermal T-Cells*

Dendritic epidermal T-cells (DETC) are a population of γδ T-cells restricted to the epidermis in adult mice that are homogeneous for expression of an invariant T-cell receptor composed of Vγ5 (15) and Vγ1 chains without any junctional diversity (16). Although a homologous population of intraepidermal γδ T-cells does not exist in humans, the study of the biology of murine DETC has identified mechanisms for interaction between T-cells and keratinocytes that may be relevant to the analysis of human skin conditions associated with the development of epidermotropic T-cell infiltrates. Careful analysis of the pattern of cytokine production by DETC has provided an important clue to a possible physiological function in mouse skin. DETC produces significant quantities of keratinocyte growth factor after stimulation (17). Another function attributed to DETC in mice is suppression of autoreactivity.

#### 4. Cutaneous Nerves

Sensory nerve fibers are abundant in the epidermis and dermis (18). Many of these neuronal processes can elaborate members of a family of mediators termed neuropeptides (19,20). Some neuropeptides have shown to influence immune and inflammatory responses in skin via effects on endothelial cells, leukocytes, and keratinocytes. Neuropeptides that have received significant attention for their potential immunoregulatory effects include substance P and calcitonin gene-related peptide (CGRP). Substance P released from nerve fibers can act on vascular endothelial cells to increase permeability and vasodilation (20). Substance P and the related tachykinin neurokinin A have been shown to stimulate DNA synthesis by human keratinocytes (21). The release of neuropeptides within skin by cutaneous nerves allows an ongoing cross-talk to take place between different cell types featuring both neuropeptides and cytokines. Cutaneous nerves are not the exclusive source of neuropeptides in skin. Some observations indicate that neuropeptides may play significant roles in the homeostatic regulation of cutaneous inflammation (22).

### B. Dermal Cells of the Skin Immune System

#### 1. Dermal Fibroblasts

The mesenchyme-derived stromal cells of the dermis have not traditionally been considered a significant component of the immune system. However, several observations viewed together suggest that cytokine cross-talk between dermal fibroblasts and keratinocytes contributes significantly to maintaining homeostasis of the skin immune system. While both keratinocytes and fibroblasts can produce secondary cytokines in response to release of a primary cytokine such as keratinocyte IL-1$\alpha$, fibroblasts produce significantly greater quantities of secondary cytokines such as IL-6 (23,24). Dermal fibroblasts are also likely to be the major source of KGF used by epidermal cells, and production of KGF by dermal fibroblasts is substantially increased during wound healing (25) and in the presence of IL-1 (26). Most of the TNF-$\alpha$ produced in the skin following UVB irradiation is produced by dermal fibroblasts rather than keratinocytes (27). These observations suggest that fibroblasts play an important amplification role in responses to cytokines originating from keratinocytes.

#### 2. Dermal Dendritic Cells

Cells belonging to the dendritic cell lineage are present in almost parenchymal tissues including dermis (28–31). Both intraepidermal LC and dermal dendritic cells (DDC) seem to provide similar functions. Isolated DDC are comparable to cultured LC or dendritic cells isolated from lymphoid tissues at stimulating primary T-cell responses to mitogens (30,31). Immunostaining analysis of DDC have shown that most of their cell surface markers are shared with LC (30,31).

#### 3. Mast Cells

The dermis is a home to a large number of mast cells of the connective tissue type. Mast cells express high-affinity Fc receptors for IgE and are best known for their participation in IgE-mediated allergic responses initiated by allergen exposure. However, multiple other stimuli, including neuropeptides and complement components, can also provoke the release of mast-cell mediators (32). Activated mast cells release a variety of performed primary mediators upon degranulation and simultaneously initiate de novo synthesis of secondary mediators. The performed mediators present in mast-cell granules include vasoactive amines, neutrophil and eosinophil chemotactic factors, and multiple proteases. The cytokine TNF-α is also present preformed in mast-cell granules (33). The capacity of mast cells to produce cytokines has been studied most extensively in mice (34), and it remains to be seen if all of the same cytokines can be produced by human mast cells. Mast-cell production of IL-4 may be a significant factor in directing the orientation of T-cell responses to certain types of antigens. It has been proposed that some pathogenic microorganisms that efficiently trigger mast-cell release without significantly engaging macrophages will create a cutaneous milieu rich in IL-4 and relatively devoid of IL-12, a combination of cytokines that would distinctly favor evolution of a Th2-dominated immune response (35).

### 4. *Microvascular Endothelial Cells*

Endothelial cells play a critical role as gatekeeper that control the extravasation of leukocytes. Postcapillary venules are a specialized site within the microvasculature where many leukocytes exit. A major determinant of the behavior of leukocytes within vascular beds is the complement of adhesion molecules expressed by the leukocytes and by the endothelial cells. The major adhesion molecules are E-selectin, P-selectin, intercellular adhesion molecules (ICAM-1), (ICAM-2), and vascular adhesion molecules (VCAM-1), which participate in binding events between circulating leukocytes and endothelial cells in the skin. A critical conceptual advance in understanding leukocyte extravasation through the vasculature was the realization that it is a multistep process involving sequentially slowing of leukocytes, rolling on endothelial selectins, activation, firm adhesion, and finally diapedesis (36). Both of the endothelial selectins (E-selectin and P-selectin) can mediate rolling of neutrophils, monocytes, and T-cells. Endothelial cells also have the potential to function in antigen presentation. Like keratinocytes, endothelial cells expresses class II MHC molecules following stimulation with IFN-γ (37). Whether endothelial cells also express relevant costimulatory molecules that provide required second signals for full stimulation of T-cell activation has been a subject of controversy (38–40). While the antigen-presentation capabilities of endothelial cells are clearly relevant to immune responses against vascularized allografts, the contribution of endothelial antigen presentation to regulation of syngeneic T-cell responses to antigen remains to be established.

## III. RISK ASSESSMENT AND IMMUNOTOXICITY

A number of environmental contaminants can suppress immune responses and enhance susceptibility to infectious and/or neoplastic disease. Most of the evidence for immunotoxicity of such contaminants has been obtained from laboratory animal studies. In general toxicity testing, maximal acceptable concentrations are derived from "no observed adverse effect" levels in rodents. Risk assessment then considers safety factors for the interspecies difference and intraspecies variability. This approach can be used for assessing maximal acceptable concentrations for chemicals inducing direct immunotoxicity, resulting in reduced resistance to infections, for example. Generally, the assessment of risk for chemicals that induce contact sensitivity is limited to hazard identification, and risk management is restricted to labeling. An alternative type of evaluation of the risk of adverse effects due to exposure to immunotoxic chemicals may be the so-called parallelogram approach. In this kind of approach, there are four major points of consideration, one of which is the health effect of exposure to a chemical, assessed as an endpoint (e.g., infection model in experimental animals), and another the quantitative prediction of this endpoint in humans. The other main point of consideration is the assays of parameters that are relevant to the mechanism of the adverse effect in experimental animals and humans and are used for species comparison. Species comparison between the animal species used for hazard identification and humans is crucial for extrapolation of animal data to the human situation. This approach can be used to provide relevant information on the dose–response relationship in humans. Such approaches have been used for chemicals that exert direct immunotoxic activity and may hold promise for the risk evaluation of chemicals that exert skin-sensitizing properties (41,42).

## IV. UV RADIATION AND THE SKIN

The UV radiation present in sunlight is divided into three regions, short-wave UVC (200–280 nm), midrange UVB (280–320 nm), and long-wave UVA (320–400 nm). Although UVC is a potent mutagen that can induce immune suppression, all of the UVC radiation present in solar radiation is absorbed by the stratospheric ozone layer, so its role in human pathogenesis is minimal. UVB is also mutagenic and extensive epidemiological evidence has indicated that midrange UV radiation is responsible for inducing skin cancer (43). Moreover, wavelengths within the UVB portion of the spectrum are responsible for sunburn and immune suppression. UVA, the major component of the UV portion of the solar spectrum, does cause premature aging of the skin and can suppress some immunological functions (44).

Although many environmental and genetic factors contribute to the development of immunotoxicity in the skin, the most important and hazardous is chronic exposure to solar UV light, particularly UVB (290–320 nm) radiation. Each 1% decrease in stratospheric ozone allows an increase of 1.6% UVB to reach the earth's surface (45). Stratospheric ozone losses in northern midlatitudes have already occurred at a rate of 4 to 5% per decade over the past two decades (46), with measurable increases in UVB (47). The stratospheric chlorine measurements over eastern Canada and northern New England indicate that ozone depletion may reach 30 to 40% (48), with the expected result being very high peak UVB levels on clear days, a time when many people increase their outdoor activities. This UV exposure to skin is responsible for the development of oxidative stress and immunotoxicity in the open skin areas of the body. Following exposure to UV radiation, the ability to generate T-cell-mediated immune reactions such as delayed-type hypersensitivity is severely suppressed. Furthermore, the ability of UV radiation to induce suppression has been linked to its ability to induce skin cancer (49).

The target organ of UV radiation is the skin. UV exposure appears to promote the induction of skin cancer by two mechanisms. The first involves direct mutagenesis of epidermal DNA, which promotes the induction of neoplasia. The second is associated with immune suppression, which allows the developing tumor to escape immune surveillance and grow progressively. The suppression induced by UV radiation is unique. Rather than inhibiting all immune reactions, the suppression induced by UV radiation is highly selective. While cellular immune reactions such as contact and delayed-type hypersensitivity and natural killer cell function are suppressed following UV exposure, for the most part, antibody formation is not affected (44). UV alters the morphology and function of epidermal Langerhans cells, the antigen-presenting cell that resides in the skin. Sensitization with contact allergens through UV-irradiated skin results in immune suppression rather than immune stimulation. Studies have indicated that the modulation of antigen-presenting cell function by UV is one of the reasons for the induction of immune suppression.

Skin is a first defense barrier organ of the body from external environmental pollutants, including environmental chemicals and solar UV radiation. The coordinated functions of multiple epidermal and dermal cell population allows the skin immune system to respond rapidly and effectively to a wide variety of insults occurring at the interface of the organism and its environment. These environmental factors may stimulate the skin responsive system through their oxidizing properties. It is well documented that solar UV exposure to mammalian skin induces a number of pathological conditions, such as sunburn cell formation, hyperplastic response, and DNA damage that contributes to the development of several disease states including immunotoxicity. In addition, UV exposure to the skin results in the generation of reactive oxygen species (ROS), such as singlet oxygen, peroxy radicals, superoxide anion, and hydroxyl radicals, which damage cellular DNA and non-DNA cellular targets (50–54). UV-induced generation of ROS in the skin develops oxidative stress, when their formation exceeds the antioxidant defense ability of the target cell. The induction of oxidative stress and imbalance of the antioxidant defense system have been associated with the onset of several disease states including rheumatoid arthritis, inflammation, photoaging, immunotoxicity, and skin cancer. Thus UV radiation is the critical major oxidizing agent and hazardous to the immune system. Although the skin possesses an elaborate antioxidant defense system to deal with UV-induced oxidative stress and immunotoxicity, excessive and chronic exposure to UV light can overwhelm the cutaneous antioxidant and immune response capacity, leading to oxidative damage and immunotoxicity, premature skin aging, and skin cancer. A broad spectrum of observations are available on the effect of solar UV exposure on skin immunotoxicity, whereas very little is known about the immunotoxicity of other environmental pollutants or factors, such as ubiquitous chemical carcinogens, on the animal or human skin model system. In order to provide a broad view of the effect of oxidizing agents on the skin immune system or immunotoxicity, the effect of UV exposure on the skin is selected as a model to review the immunotoxicity caused by the oxidizing agents to the skin. Also, to provide an overview of other environmental factors responsible for adversely affecting the skin immune system, we will discuss the effect of chemical carcinogens, such as dimethylbenz(a)anthracene and benzo(a)pyrene, on the skin immune reactions.

## V. IMMUNOTOXICITY CAUSED BY UV RADIATION

### A. Immunotoxicity Caused by UV-Induced Damage in Epidermal APC

UV irradiation to skin can induce immunosuppression to a contact sensitizer through different mechanisms that operate primarily by influencing the type or function of the cell presenting the antigenic stimulus. UV irradiation may affect a contact sensitivity response in three ways: (1) immediate local with lack of induction of a contact sensitivity response (null or passive event); (2) delayed systemic induction of tolerance (active event); (3) delayed local induction of tolerance (active event). Whether a single high UV dose (55), a single low UV dose (56,57), or multiple low UV doses (55,58,59) are given, contact sensitivity induction is impaired if the sensitization is through the UV-irradiated site and not through a distant, non-UV-exposed site. This immediate local suppression of contact sensitization induction can be attributed to the effect of UV on the constitutive antigen-presenting cell population within the skin, the Langerhans cells, resulting in a lack of induction of a contact sensitivity response. UV irradiation drastically reduces the numbers of LC (60–62) either by induced migration (63,64) or by cell death (65). TNF-α plays an important role in this immediate effect of UV in mediating both UV induction of LC migration (66) and damage to LC morphology (67).

UV irradiation to skin can also result in a systemically active suppression of the ability to induce contact sensitivity responses (tolerance), as shown by the inability to immunize through sites distant to the UV exposure site, if the contact sensitizer is applied 24 to 72 h after the initial UV exposure, either with a single, high UV dose (55) or with multiple, low UV doses (59). This systemic suppression of contact sensitivity induction is not mediated by TNF-α (59) or IL-10 (68), but may be mediated by the systemic effects of $PGE_2$ (69) or *cis*-urocanic acid on splenic dendritic cells (70). The systemically altered APC not only fail to induce contact sensitivity, but also stimulate the development of Ag-specific T-cells (71) that result in tolerance (59,71). The initial effect of UV is on the constitutive APC within the skin, epidermal LC, and dermal LC-like APCs, such that cell death (65) or functional alteration (72) of these cells occurs. As a result of in vivo low-dose UV exposure, the constitutive skin APC are no longer capable of initiating in vivo contact sensitivity responses (56,57) or stimulating in vitro Th1-type cells (73).

The ability of UV irradiation to interfere with the induction of CHS responses is well documented, both in mice (74,75) and in humans (76,77). Application of a contact-sensitizing hapten to the UV-irradiated skin of certain strains of mice (75) leads to a decreased contact hypersensitivity response and the induction of specific immunological tolerance transferable by splenic T-cells (74,75). Numerous investigations have addressed the cellular mechanisms of this phenomenon. Although not all steps have been fully delineated, certain events have been shown to contribute to the modification of immune responsiveness. Epidermal LC, the primary APC of the epidermis, are altered morphologically and functionally after exposure to sublethal doses of UV radiation (60), and in vitro UV irradiation of these cells impairs their ability to activate Th1 cells (78). Hapten-bearing LC and macrophages collected from the draining lymph nodes (DLN) of mice sensitized through UV-irradiated skin are deficient in their ability to induce contact hypersensitivity responses in vivo (79). Macrophages that infiltrate the skin several days after UV irradiation differ from normal LC in their antigen-presenting activity and may be responsible for T suppressor cell induction (56). TNF-α may play a central role in modifying this cellular response because administration of antibody against this cytokine permits the development of contact hypersensitivity in UV-irradiated mice (66,67).

## B. Immunotoxicity Caused by UV-Induced Infiltrating Leukocytes

It has been demonstrated that a single UV-dose-induced immunosuppression and tolerance is due to the appearance within the UV-irradiated skin of tolerance-inducing infiltrating class II $MHC^+$ $CD11b^+$ cell monocyte/macrophages in both mice (56,57) and humans (76,80,81). Thus, UV exposure, in addition to affecting constitutive LC, must generate the appropriate conditions, such as expression of ICAM-1 and VCAM-1 on endothelial cells (82) and production of inflammatory cytokines (IL-1 and TNF-α) (83–86) necessary for diapedesis of tolerance-inducing class II $MHC^+$ $CD11b^+$ monocyte/macrophages into the UV-exposed skin. It has been found that in vivo anti-CD11b treatment does reverse tolerance induction, and this reversal is associated with a 50% reduction in the infiltration of a class II $MHC^+$ $CD11b^+$ $Gr\text{-}1^+$ monocyte/macrophage population into both the dermis and epidermis of UV-irradiated skin. In addition to the reversal of tolerance and partial blockade of class II MHC$^+$ CD11b$^+$ Gr-1$^+$ monocyte/macrophage infiltration, anti-CD11b treatment completely restored the initial induction of a contact sensitivity response (87).

The above-mentioned changes in class II MHC$^+$ cells within UV-exposed skin of anti-CD11b-treated mice could have resulted from the blockade of infiltrating CD11b$^+$ leukocytes (neutrophils and macrophages), which may cause additional damage to the LC of dermal LC-like cells by their production of ROS-producing CD11b$^+$ leukocyte infiltration may also explain the reduced damage to the UV-irradiated epidermis that was observed in anti-CD11b-treated mice (88,89). This important observation in vivo, using anti-CD11b treatment, affects UV-irradiated mice in two ways: (1) tolerance induction is blocked; and (2) protection is provided to the UV-irradiated epidermis. Both effects may be attributed to the blockade of the infiltrating CD11b$^+$ monocyte/macrophages and neutrophils. The infiltrating leukocytes could provide a second hit, via reactive oxygen species, to the UV-damaged epidermis that results in keratinocyte disassociation and further breakdown in the structure of the UV-exposed epidermis. A schematic diagram, as shown in Figure 1, depicts the mechanism of UV-induced effects and infiltration of CD11b$^+$ cells in the skin.

To demonstrate further that UV-induced leukocyte infiltration and the ROS derived from them are responsible for UV-induced immunosuppression or immunotoxicity, Katiyar et al. (90) recently showed that skin treatment with (−)-epigallocatechin-3-gallate (EGCG), a strong antioxidant from green tea, to C3H/HeN mouse skin before UVB exposure reverses UVB-induced immunosuppression concomitant with the reduction in number of UVB-induced infiltration of CD11b$^+$ (monocyte/macrophages and neutrophils) cells in the UVB-irradiated skin site. This study also demonstrated that these infiltrating leukocytes are the major source of ROS when measured in the form of hydrogen peroxide

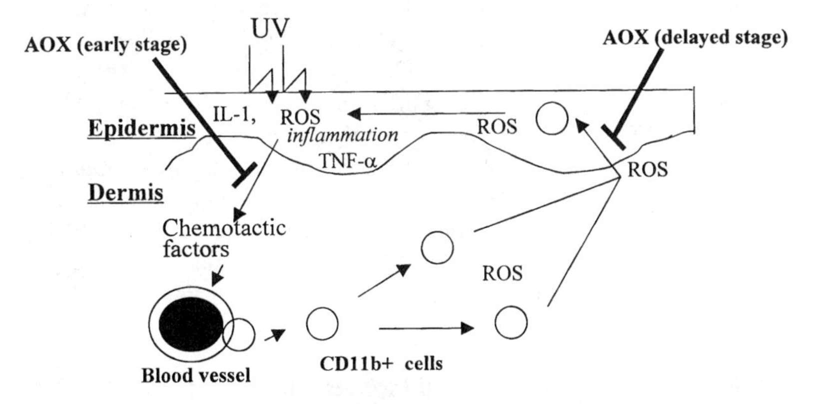


**Figure 1** Schematic diagram depicts UVB-induced inflammatory and oxidizing effects in the skin. UVB exposure induces inflammatory and oxidative effects at early stage and infiltration of CD11b$^+$ cells (monocyte/macrophages and neutrophils) into the skin at delayed stage. Infiltration of CD11b$^+$ cells is one of the possible causes for UVB-induced immune suppression, and considered tolerogenic in nature. Infiltrating CD11b$^+$ cells are also the major source of reactive oxygen species (ROS). Treatment with antioxidant (AOX) before UVB exposure may inhibit UVB-induced ROS generation (early stage) and infiltration of CD11b$^+$ cells (delayed stage). Inhibitory effect of AOX against UVB-induced adverse effects may inhibit or reverse UVB-induced immune suppression/immunotoxicity.

production by these infiltrating cells (91). Antioxidant EGCG treatment to mouse skin as well as to human skin before UV exposure inhibited hydrogen peroxide/oxidative stress induction through blockade of CD11b$^+$ cells (91,92). Thus it can be suggested that environmental solar UV radiation is a strong oxidizing agent, and UV-induced immunotoxicity in the skin, in part, may be caused by the UV-induced reactive oxygen species.

## C. Immunotoxicity Caused by UVB-Radiation-Induced Epidermal Cytokines

Because it is assumed that most of the UVB is absorbed within the epidermis and cannot directly affect cells outside the irradiated layer, one potential mechanism for its action involves the release of immunomodulatory cytokines by epidermal cells (93). Thus chronic exposure to UVB radiation induces immune suppression through the induction of certain cytokines in the skin. Keratinocytes in humans and mice are known to produce a wide variety of cytokines, including interleukin (IL)-1α, IL-1β, IL-3, IL-6, IL-7, IL-8, IL-10, IL-12, IL-15, colony-stimulating factor-1 (CSF-1), granulocyte-macrophage colony-stimulating factor (GM-CSF), TNF-α, transforming growth factor-α (TGF-α), nerve growth factor, and platelet-derived growth factor-α (PDGF-α). LC, which are resident APC in epidermis, have been shown to produce IL-1β, IL-6, TNF-α, and macrophage inflammatory protein (MIP)-α. These cytokines, produced locally within the epidermal microenvironment and termed ''epidermal cytokines,'' play critical roles during the induction, amplification, and resolution of immune responses in skin (94). A primary role for the cytokine IL-10, produced by keratinocytes, in mediating UV-induced systemic immunosuppression has been suggested by Rivas and Ullrich (8). Intraperitoneally injected IL-10 suppresses the effector phase, but not the induction phase, of contact hypersensitivity, and also the induction phase of delayed-type hypersensitivity (95). However, others have suggested that this systemic suppression of CHS induction is not mediated by IL-10 (68). Recently, Katiyar et al. (90) have demonstrated that UVB exposure to mouse skin induces IL-10 production in skin by infiltrating leukocytes and in draining lymph nodes after 48 h. Thus, the increased production of IL-10, an immunoregulatory cytokine, may play a crucial role in UVB-induced immune responses. Furthermore, topical treatment with an antioxidant from green tea, EGCG, before UVB exposure to mouse skin reverses the UVB-induced immunosuppression. The reversal of UVB-induced immune suppression may be because of alterations in IL-10 and IL-12 production in skin and draining lymph nodes caused by EGCG treatment. A schematic diagram (Fig. 2) depicting the migration of APC from skin to lymph nodes under the influence of UV light, and the treatment with antioxidant, modulated the effect of UVB-induced production of immunoregulatory cytokines IL-10 and IL-12.

## D. Immunotoxicity Caused by UV-Induced Isomerism in Urocanic Acid

Along with the progress in our understanding of the cellular events leading to immune suppression, there has been increasing interest in delineating the precise molecular pathways that initiate these events. A theory that has received much attention involves the photo-isomerization of urocanic acid (UCA), which is a metabolite of the histidine synthesis pathway (96). This metabolite is abundant in the stratum corneum and is isomerized from the *trans* to the *cis* configuration upon exposure to UV radiation. *Cis*-UCA can be found in the circulation after the skin is exposed to UV radiation (97), and it has

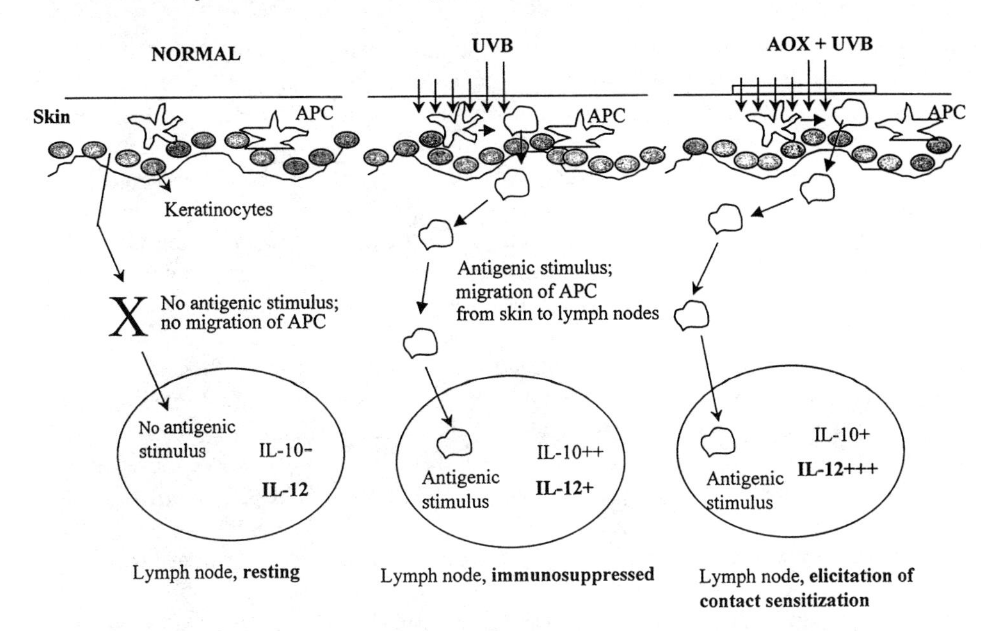


**Figure 2** A working model depicting the migration of antigen-presenting cells (APC) from skin to regional lymph nodes under the influence of UVB light. Migrating APC activates lymph node cells and stimulates production of immunoregulatory cytokines IL-10 and IL-12, which contribute in the development of immune responses. Treatment with antioxidant (AOX) before UVB exposure may elicit contact sensitization reactions to contact sensitizers against UVB-induced immunosuppressed state.

immunosuppressive activity, most notably, the ability to interfere with the induction of delayed-type hypersensitivity to herpes simplex virus (98). *Cis*-UCA also causes morphological alterations in LC that resemble those induced by UV radiation (99). These and other observations support the hypothesis that UCA is a photoreceptor in the skin that transforms radiant energy into an immunosuppressive signal, ultimately impairing the induction of certain immune responses (100). The cellular target of *cis*-UCA and its relationship to cytokine mediators of immune suppression, such as TNF-α, remain undefined.

## E. Immunotoxicity Caused by UVB-Induced DNA damage

An equally important hypothesis is that DNA damage is the initiating event for immune suppression. Support for this idea comes from studies of the South American opossum *Monodelphis domestica*, whose cells contain a photoreactivating enzyme that absorbs visible light to repair UV-induced cyclobutane pyrimidine dimers (CPD) in DNA. In these marsupials, exposure of the skin to visible light after UV irradiation prevents UV-induced local suppression of contact hypersensitivity (101) and altered morphology of LC (102), presumably by reversing CPD.

Another possibility was suggested by recent studies demonstrating that exposing cells in vitro to UV radiation can, by a DNA-independent mechanism, activate transcription factors such as AP-1 and nuclear factor-κB (NF-κB), which in turn increase production of TNF-α and other immunoregulatory and proinflammatory cytokines (103,104). Thus, it is not clear which molecular events trigger the various forms of UV-induced immune suppression, such as local or systemic suppression of the induction, elicitation, of contact hypersensitivity and delayed-type hypersensitivity responses, inhibition of tumor rejection. Even within a single model of immune suppression, such as that described above, different steps in the pathway suppressor T-cell leading to induction may be initiated by different molecular events. Vink et al. (105) examined the role of DNA damage in a particular step of the pathway of UV-induced local suppression of contact hypersensitivity, namely, the alteration in the function of cutaneous APC that migrate to the DLN after epicutaneous sensitization. In these studies, they have attempted to delineate the molecular event that triggers one step in the cascade of cutaneous responses initiated by UV irradiation that culminates in impaired induction of contact hypersensitivity. DNA-damaged cells were found not only in the epidermis of UV-irradiated mice but also in the dermis, which suggests that dermal macrophages, as well as epidermal LC, are potential immunological targets for the DNA-damaging effects of UV radiation, and complements the study of Kurimoto et al. (106) indicating the importance of these cells in UV-induced local suppression of contact hypersensitivity. UV-damaged and CPD$^+$ cells were also detected in DLN from 1 to 4 days after UV exposure of the skin to UV radiation. When fluorescein isothiocyanate was applied to UV-exposed skin, the DLN contained cells that were Ia$^+$, FITC$^+$, and CPD$^+$; such cells from mice sensitized 3 days after UV irradiation exhibited reduced antigen-presenting function in vivo. This study supported the hypothesis that DNA damage is an essential initiator of one or more of the steps involved in impaired APC function after UV irradiation.

### F. Immunotoxicity Caused by UVB-Induced Activation of Complement Component 3

Complement components, especially complement component 3 (C3), have been implicated in the regulation of T-cell-mediated responses (107,108). C3, a critical regulator of innate immunity, may also play a role in the regulation of cognate immunity, such as contact sensitivity responses. Hammerberg et al. (109) showed that activated C3 is required for UV induction of immunosuppression and antigenic tolerance. This was addressed through the use of C3-deficient mice, blockade of C3 cleavage to C3b, and accelerated degradation of iC3b by soluble complement receptor type 1. Mice deficient in C3 when exposed to a single UV dose were able to show the contact sensitization response to contact sensitizer applied through the UV-irradiated skin (109,110).

## VI. IMMUNOTOXICITY CAUSED BY CHEMICAL CARCINOGENS

Several studies demonstrate the interplay between the immune system and the chemical carcinogens most notably with polyaromatic hydrocarbons, such as dimethylbenz(a)anthracene (DMBA) and benzo(a)pyrene (BaP). These aromatic hydrocarbons, particularly BaP, are found ubiquitously as environmental pollutants and are potent carcinogens, mutagens, and teratogens (111). Studies have shown that both aromatic hydrocarbons DMBA and BaP are immunogens, producing contact or delayed-type hypersensitivity reactions when applied topically or injected into the skin (112–115). In addition to their capacity to act as immunogens, the polyaromatic hydrocarbons are also known to suppress the cutaneous cell-mediated immune response to other chemicals. Inhibition of cell-mediated immunity by intraperitoneally administered 3-methylcholanthrene does not occur in Ah-receptor-negative strains that metabolize 3-methylcholanthrene poorly (116), implying that a metabolite is responsible for the immune suppression. Topically applied DMBA and BaP both inhibit the development of DNFB-induced contact hypersensitivity in C3H mice (117,118). However, unlike those intraperitoneally administered, topically applied polyaromatic hydrocarbons act to inhibit cutaneous immunological function only at their site of application (117,119) (e.g., when DMBA is applied at one skin site and animals are immunized to the contact allergen at another site, a normal contact hypersensitivity response is observed).

At least three genes were identified that influence the development of immune response (120–122). The reaction does not occur in strains of mice that have a genetic deficiency in Ah receptor expression, the intracellular receptor to which polyaromatic hydrocarbons bind. Ah receptor expression is governed by a single gene, the Ah receptor gene. Because the Ah receptor gene is required for metabolism of polyaromatic hydrocarbons, this implies that a metabolite rather than the parent compound is the actual immunogenic moiety in these reactions (120,121). The contact hypersensitivity response is also governed by genes present within the major histocompatibility complex. In comparative studies with strains of mice congenic at the murine MHC H-$2^a$ and H-$2^k$, haplotype strains were shown to develop a much stronger immune response to this class of chemicals than H-$2^b$ or H-$2^d$ haplotype strains. Third, the *Lps* locus, a genetic locus that controls macrophage activation and synthesis and release of several cytokines in response to lipopolysaccharide, also appears to influence the cell-mediated immune response to polyaromatic hydrocarbons. C3H/HeJ mice, which have a mutation at that locus and, as a result, are deficient in the immunological capabilities referred to above, develop a smaller contact hypersensitivity response to DMBA than do nonmutant C3H/HeN mice that respond in an immunologically normal fashion to insults that depend on that locus.

When the inbred strains of mice that differ in their immunological responsiveness to polyaromatic hydrocarbons were subjected to chemical carcinogenesis protocols utilizing these agents as initiators, a close correlation was observed between the magnitude of the immunological response to these agents and their susceptibility to skin cancer (121,123). For example, in C3H.SW mice, a strain that fails to mount a contact hypersensitivity response to DMBA, relatively large numbers of tumors develop in response to the DMBA initiation-12-O-tetradecanoylphorbol-13-acetate (TPA) promotion protocol. In contrast, in C3H/HeN mice, a strain in which vigorous contact hypersensitivity response is observed, relatively few tumors develop. C3H.SW mice differ from C3H/HeN mice only in the MHC, and the level of DMBA-DNA adducts that are present in C3H.SW mice substantially less than in C3H/HeN mice. These results are compatible with the hypothesis that the immune response to polyaromatic hydrocarbons serves to protect animals from the carcinogenic activities of these compounds by eradicating cells that are destined to develop into malignant neoplasms through immunological means.

DMBA-induced immunosuppression is associated with the development of antigen-specific suppressor T-lymphocytes that inhibit the induction of both cellular and humoral immune responses when adoptively transferred to naïve syngeneic recipients (117,124). Their activation in this system appears to be caused by the direct effects of DMBA on the skin: when the reactive hapten, trinitrophenol, is conjugated to skin cells previously treated with DMBA in vivo and those cells are administered to naïve syngeneic recipients, suppressor cell induction occurs. This does not occur when regional lymph node cells from the skin draining the site of DMBA application are substituted for epidermal cells in these experiments (125). Based on the mechanism by which UV radiation mediates its immunosuppressive activities, it was postulated that polyaromatic hydrocarbons may exert their immunosuppressive effects by reducing Langerhans cell numbers and/or function. This does not appear to be the case. Although chronic application of DMBA to the skin of mice (126) and to the hamster cheek pouch (127) results in a marked reduction in Langerhans cell densities, topical application of BaP and catechol produces an increase in the number of identifiable Langerhans cells (118). In spite of the variation in the response of Langerhans cell densities to these three compounds, all suppress cutaneous cell-mediated immune responses (117,118), indicating Langerhans cell depletion is not sufficient to explain polyaromatic-hydrocarbon-induced inhibition of the contact hypersensitivity response.

## ACKNOWLEDGMENTS

Studies in the authors' laboratory are supported by USPHS Grant CA 78809 (HM), and by Grants from Ohio Cancer Research Associates and Cancer Research Foundation of America (SKK).

## REFERENCES

1. Streilein JW. Skin-associated lymphoid tissues (SALT): origins and functions. J Invest Dermatol 1983; 80:12S–16S.
2. Edelson RL, Fink JM. The immunologic function of skin. Sci Am 1985; 252:46–53.
3. Bos JD, Kapsenberg ML. The skin immune system: progress in cutaneous biology. Immunol Today 1993; 14:75–78.
4. Williams IR, Kupper TS. Immunity at the surface: Homeostatic mechanisms of the skin immune system. Life Sci 1996; 58:1485–1507.
5. Kupper TS. Immune and inflammatory processes in cutaneous tissues. Mechanisms and speculations. J Clin Invest 1990; 86:1783–1789.
6. Enk AH, Katz SI. Identification and induction of keratinocyte-derived IL-10. J Immunol 1992; 149:92–95.
7. Enk CD, Sredni D, Blauvelt A, Katz SI. Induction of IL-10 gene expression in human keratinocytes by UVB exposure in vivo and in vitro. J Immunol 1995; 154:4851–4856.
8. Rivas JM, Ullrich SE. Systemic suppression of delayed-type hypersensitivity by supernatants from UV-irradiated keratinocytes. An essential role for keratinocyte-derived IL-1. J Immunol 1992; 149:3865–3871.
9. Ferguson TA, Dube P, Griffith TS. Regulation of contact hypersensitivity by interleukin 10. J Exp Med 1994; 179:1597–1604.
10. Schwarz A, Grabbe S, Riemann H, Aragane Y, Simon M, Manon S, Andrade S, Luger TA, Zlotnik A, Schwarz T. In vivo effects of interleukin-10 on contact hypersensitivity and delayed-type hypersensitivity reactions. J Invest Dermatol 1994; 103:211–216.
11. Streilein JW, Bergstresser PR. Langerhans cells: antigen presenting cells of the epidermis. Immunobiology 1984; 168:285–300.
12. Kripke MI, Munn CG, Jeevan A, Tang JM, Bucana C. Evidence that cutaneous antigen-presenting cells migrate to regional lymph nodes during contact sensitization. J Immunol 1990; 145:2833–2838.
13. Larsen CP, Ritchie SC, Hendrix R, Linsley PS, Hathcock KS, Hodes RJ, Lowry RP, Pearson TC. Regulation of immunostimulatory function and costimulatory molecule (B7-1,B7-2) expression on murine dendritic cells. J Immunol 1994; 152:5208–5219.
14. Kawamura T, Furue M. Comparative analysis of B7-1 and B7-2 expression in Langerhans cells: differential regulation by T helper type 1 and T helper type 2 cytokines. Eur J Immunol 1995; 25:1913–1917.

15. Takagaki Y, Nakanishi N, Ishida I, Kanagawa O, Tonegawa S. T cell receptor-gamma and -delta genes preferentially utilized by adult thymocytes for the surface expression. J Immunol 1989; 142:2112–2121.
16. Asarnow DM, Kuziel WA, Bonyhadi M, Tigelaar RE, Tucker PW, Allison JP. Limited diversity of gamma delta antigen receptor genes of Thy-1+ dendritic epidermal cells. Cell 1988; 55:837–847.
17. Boismenu R, Havran WI. Modulation of epithelial cell growth by intraepithelial gamma delta T cells. Science 1994; 266:1253–1255.
18. Hilliges M, Wang L, Johansson O. Ultrastructural evidence for nerve fibers within all vital layers of the human epidermis. J Invest Dermatol 1995; 104:134–137.
19. Eedy DJ. Neuropeptides in skin. Br J Dermatol 1993; 128:597–605.
20. Lotti T, Hautmann G, Panconesi E. Neuropeptides in skin. J Am Acad Dermatol 1995; 33: 482–496.
21. Tanaka T, Danno K, Ikai K, Imamura S. Effects of substance P and substance K on the growth of cultured keratinocytes. J Invest Dermatol 1988; 90:399–401.
22. Schauer E, Trautinger F, Kock A, Schwarz A, Bhardwaj R, Simon M, Ansel JO, Schwarz T, Luger TA. Propiomelanocortin-derived peptides are synthesized and released by human keratinocytes. J Clin Invest 1994; 93:2258–2262.
23. Boxman I, Lowik C, Aarden L, Ponec M. Modulation of IL-6 production and IL-1 activity by keratinocyte-fibroblast interaction. J Invest Dermatol 1993; 101:316–324.
24. Waelti ER, Inaebnit SP, Rast HP, Hunziker T, Limat A, Braathen LR, Wiesmann U. Co-culture of human keratinocytes on post-mitotic human dermal fibroblast feeder cells: production of large amount of interleukin 6. J Invest Dermatol 1995; 98:805–808.
25. Werner S, Peters KG, Longaker MT, Fuller-Pace R, Banda MJ, Williams LT. Large induction of keratinocyte growth factor expression in the dermis during wound healing. Proc Natl Acad Sci USA 1992; 89:6896–6900.
26. Chedid M, Rubin JS, Csaky KG, Aaronson SA. Regulation of keratinocyte growth factor gene expression by interleukin 1. J Biol Chem 1994; 269:10753–10757.
27. de Kossodo S, Cruz PD Jr, Dougherty I, Thompson P, Silva-Valdez M, Beutler B. Expression of the tumor necrosis factor gene by dermal fibroblasts in response to ultraviolet irradiation or lipopolysaccharide. J Invest Dermatol 1995; 104:318–322.
28. Headington JT, Cerio R. Dendritic cells and the dermis: 1990. Am J Dermatopathol 1990; 12:217–220.
29. Meunier L, Gonzalez-Ramos A, Cooper KD. Heterogeneous populations of class II MHC+ cells in human dermal cell suspensions. Identification of a small subset responsible for potent dermal antigen-presenting cell activity with features analogous to Langerhans cells. J Immunol 1993; 151:4067–4080.
30. Lenz A, Heine M, Schuler G, Romani N. Human and murine dermis contain dendritic cells. Isolation by means of a novel method and phenotypical and functional characterization. J Clin Invest 1993; 92:2587–2596.
31. Nestle FO, Zheng X-G, Thompson CB, Turka LA, Nickoloff BJ. Characterization of dermal dendritic cells obtained from normal human skin reveals phenotypic and functionally distinctive subsets. J Immunol 1993; 151:6535–6545.
32. Marshall JS, Bienenstock J. The role of mast cells in inflammatory reactions of the airways, skin and intestine. Curr Opin Immunol 1994; 6:853–859.
33. Gordon JR, Galli SJ. Mast cells as a source of both preformed and immunologically inducible TNF-alpha/cachectin. Nature 1990; 346:274–276.
34. Galli SJ. New concepts about the mast cell. N Engl J Med 1993; 328:257–265.
35. Garside P, Mowat AM. Polarization of Th-cell responses: a phylogenetic consequence of nonspecific immune defence? Immunol Today 1995; 16:220–223.
36. Springer TA. Traffic signals for lymphocyte recirculation and leukocyte emigration: the multistep paradigm. Cell 1994; 76:301–314.

37. Pober JS, Collins T, Gimbrone MA Jr, Libby P, Reiss CS. Inducible expression of class II major histocompatibility complex antigens and the immunogenicity of vascular endothelium. Transplantation 1986; 41:141–146.
38. Savage CO, Hughes CC, Pepinsky RB, Wallner BP, Freedman AS, Pober JS. Endothelial cell lymphocyte function-associated antigen-3 and an unidentified ligand act in concert to provide costimulation to human peripheral blood CD4+ T cells. Cell Immunol 1991; 137: 150–163.
39. St Louis JD, Lederer JA, Lichtman AH. Costimulator deficient antigen presentation by an endothelial cell line induces a nonproliferative T cell activation response without anergy. J Exp Med 1993; 178:1597–1605.
40. Murray AG, Khodadoust MM, Pober JS, Bothwell AL. Porcine aortic endothelial cells activate human T cells: direct presentation of MHC antigens and costimulation by ligands for human CD2 and CD28. Immunity 1994; 1:57–63.
41. Van Loveren H, De jong WH, Vandebriel RJ, Vos JG, Garssen J. Risk assessment and immunotoxicology. Toxicol Lett 1998; 102-103:261–265.
42. Selgrade MK. Use of immunotoxicity data in health risk assessments: uncertainties and research to improve the process. Toxicology 1999; 133:59–72.
43. Urbach F. Evidence of epidemiology of UV-induced carcinogenesis in man. Natl Cancer Inst Monogr 1978; 50:5–10.
44. Ullrich SE. Potential for immunotoxicity due to environmental exposure to ultraviolet radiation. Hum Exp Toxicol 1995; 14:89–91.
45. Van Der Leun JC. Human Health. In: Van Der Leun JC, Tevini M, eds. United Nations Environmental Program Report on the Environmental Effects of Ozone Depletion Washington, DC: EPA, 1989.
46. Kerr RA. Ozone destruction worsens. Science 1991; 252:204.
47. Blumthaler M, Ambach W. Indication of increasing solar ultraviolet-B radiation flux in Alpine Regions. Science 1990; 248:206–208.
48. Kerr RA. New assaults seen on earth's ozone shield. Science 1992; 255:797–798.
49. ML Kripke. Photoimmunology. Photochem Photobiol 1990; 52:919–924.
50. Cadet J, Berger M, Decarroz C, Wagner JR, Van Liet JE, Ginot YM, Vigny P. Photosensitized reactions of nucleic acids. Biochimie 1986; 68:813–834.
51. Peak MJ, Ito A, Foote CS, Peak JG. Photosensitized inactivation of DNA by monochromatic 334-nm radiation in the presence of 2-thiouracil: genetic activity and backbone breaks. Photochem Photobiol 1988; 47:809–813.
52. Beehler BC, Przybyszewski J, Box HB, Kulesz-Martin MF. Formation of 8-hydroxydeoxyguanosine within DNA of mouse keratinocytes exposed in culture to UVB and $H_2O_2$. Carcinogenesis (Lond) 1992; 13:2003–2007.
53. Berton TR, Mitchell DL, Fischer SM, Locniskar MF. Epidermal proliferation but not the quantity of DNA photodamage is correlated with UV-induced mouse skin carcinogenesis. J Invest Dermatol 1997; 109:340–347.
54. Li G, Mitchell DL, Ho VC, Reed JC, Tron VA. Decreased DNA repair but normal apoptosis in ultraviolet-irradiated skin of p53-transgenic mice. Am J Pathol 1996; 148:1113–1123.
55. Noonan FP, DeFabo EC. Ultraviolet-B dose-response curves for local and systemic immunosuppression are identical. Photochem Photobiol 1990; 52:801–810.
56. Cooper KD, Duraiswamy N, Hammerberg C, Allen E, Kimbrough-Green C, Dillon W, Thomas D. Neutrophils, differentiated macrophages, and monocyte/macrophage antigen presenting cells infiltrate murine epidermis after UV injury. J Invest Dermatol 1993; 101:155–163.
57. Hammerberg C, Duraiswamy N, Cooper KD. Active induction of unresponsiveness (tolerance) to DNFB by in vivo ultraviolet-exposed epidermal cells is dependent upon infiltrating class II MHC+ CD11b (bright) monocytic/macrophagic cells. J Immunol 1994; 153:4915–4924.

58. Cruz PD Jr, Nixon-Fulton J, Tigelaar RE, Bergstresser PR. Local effects of UV radiation on immunization with contact sensitizers. I. Down-regulation of contact hypersensitivity by application of TNCB to UV-irradiated skin. Photodermatology 1988; 5:126–132.
59. Shimizu T, Streilein JW. Local and systemic consequences of acute, low-dose ultraviolet B radiation are mediated by different immune regulatory mechanisms. Eur J Immunol 1994; 24:1765–1770.
60. Toews GB, Bergstresser PR, Streilein JW. Epidermal Langerhans cell density determines whether contact hypersensitivity or unresponsiveness follows skin painting with DNFB. J Immunol 1980; 124:445–453.
61. Aberer G, Schuler G, Stingl G, Honigsmann H, Wolff K. Ultraviolet light depletes surface markers of Langerhans cells. J Invest Dermatol 1981; 76:202–210.
62. Noonan FP, Bucana C, Sauder DN, DeFabo EC. Mechanism of systemic immune suppression by UV irradiation in vivo II. The UV effects on number and morphology of epidermal Langerhans cells and the UV-induced suppression of contact hypersensitivity have different wavelength dependencies. J Immunol 1984; 132:2408–2416.
63. Moodycliffe AM, Kimber I, Norval M. The effect of ultraviolet B irradiation and urocanic acid isomers on dendritic cell migration. Immunology 1992; 77:394–399.
64. Sontag Y, Guikers CLH, Vink AA, De Gruijl FR, Van Loveren H, Garssen J, Roza L, Kripke ML, Van Der Leun JC, van Vloten WA. Cells with UV-specific DNA damage are present in murine lymph nodes after in vivo UV irradiation. J Invest Dermatol 1995; 104:734–738.
65. Tang A, Udey MC. Effects of ultraviolet radiation on murine epidermal Langerhans cells: doses of ultraviolet radiation that modulate ICAM-1 (CD54) expression and inhibit Langerhans cell function cause delayed cytotoxicity in vitro. J Invest Dermatol 1992; 99:83–89.
66. Moodycliffe AM, Kimber I, Norval M. Role of tumor necrosis factor-α in ultraviolet B light-induced dendritic cell migration and suppression of contact hypersensitivity. Immunology 1994; 81:79–84.
67. Vermeer M, Streilein JW. Ultraviolet B light-induced alterations in epidermal Langerhans cells are mediated in part by tumor necrosis factor-alpha. Photodermatol Photoimmunol Photomed 1990; 7:258–265.
68. Rivas JM, Ullrich SE. The role of Il-4, IL-10 and TNF-alpha in the immune suppression induced by ultraviolet radiation. J Leukoc Biol 1994; 56:769–775.
69. Chung H-T, Burnham DK, Robertson B, Roberts LK, Daynes RA. Involvement of prostaglandins in the immune alterations caused by the exposure of mice to ultraviolet radiation. J Immunol 1986; 137:2478–2484.
70. Noonan FP, DeFabo EC, Morrison H. Cis-urocanic acid, a product formed by ultraviolet B irradiation of the skin, initiates an antigen presentation defect in splenic dendritic cells in vivo. J Invest Dermatol 1988; 90:92–99.
71. Greene MI, Sy MS, Kripke ML, Benacerraf B. Impairment of antigen presenting cell function by ultraviolet radiation. Proc Natl Acad Sci USA 1979; 76:6591–6595.
72. Tang A, Udey MC. Inhibition of epidermal Langerhans cell function by low dose ultraviolet B radiation: ultraviolet B radiation selectively modulates ICAM-1 (CD54) expression by murine Langerhans cells. J Immunol 1991; 146:3347–3355.
73. Simon JC, Cruz PD Jr, Bergstresser PR, Tigelaar RE. Low dose ultraviolet B-irradiated Langerhans cells preferentially activate CD4+ cells of the T helper 2 subset. J Immunol 1990; 145:2087–2091.
74. Kripke ML. Immunological unresponsiveness induced by ultraviolet radiation. Immunol Rev 1984; 80:87–102.
75. Cruz PD, Bergstresser PR. The low-dose model of UVB-induced immunosuppression. Photodermatol Photoimmunol Photomed 1988; 5:151–161.
76. Cooper KD, Oberhelman L, Hamilton TA, Baadsgaard O, Terhuna M, LeVee G, Anderson T, Koren H. UV exposure reduces immunization rates and promotes to epicutaneous antigens

in humans: relationship to dose, CD1a-DR+ epidermal macrophage induction, and Langerhans cell depletion. Proc Natl Acad Sci USA 1992; 89:8497–8501.
77. Yoshikawa T, Rae V, Bruins-Slot W, van den Berg JW, Taylor JR, Streilein JW. Susceptibility to effects of UVB radiation on induction of contact hypersensitivity as a risk factor for skin cancer in humans. J Invest Dermatol 1990; 95:530–536.
78. Simon JC, Tigelaar RE, Bergstresser PR, Edelbaum D, Cruz PD. Ultraviolet B radiation converts Langerhans cells from immunogenic to tolerogenic antigen-presenting cells. J Immunol 1991; 146:485–491.
79. Okamoto H, Kripke ML. Effector and suppressor circuits of the immune response are activated in vivo by different mechanisms. Proc Natl Acad Sci USA 1987; 84:3841–3845.
80. Baadsgaard O, Fox DA, Cooper KD. Human epidermal cells from ultraviolet light-exposed skin preferentially activate autoreactive CD4+2H4+ suppressor-inducer lymphocytes and CD8+ suppressor/cytotoxic lymphocytes. J Immunol 1988; 140:1738–1744.
81. Cooper KD, Fox P, Neises G, Katz SI. Effects of ultraviolet radiation on human epidermal cell alloantigen presentation: initial depression of Langerhans cell-dependent function is followed by the appearance of T6-Dr+ cells that enhance epidermal alloantigen presentation. J Immunol 1985; 134:129–137.
82. Norris P, Poston RN, Thomas DS, Thornhill M, Hawk J, Haskard DO. The expression of endothelial leukocyte adhesion molecule-1 (ELAM-1), intercellular adhesion molecule-1 (ICAM-1), and vascular cell adhesion molecule-1 (VCAM-1) in experimental cutaneous inflammation: a comparison of ultraviolet B erythema and delayed hypersensitivity. J Invest Dermatol 1991; 96:763–770.
83. Kock A, Schwarz T, Kirnbauer R, Urbanski A, Perry P, Ansel JC, Luger TA. Human keratinocytes are a source for tumor necrosis factor alpha:evidence for synthesis and release upon stimulation with endotoxin or ultraviolet light. J Exp Med 1990; 172:1609–1614.
84. Murphy GM, Dowed PM, Hudspith BN, Brostoff J, Greaves MW. Local increase in interleukin-1-like activity following UVB irradiation of human skin in vivo. Photodermatology 1989; 6:268–274.
85. Oxholm A, Oxholm P, Staberg B, Bendtzen K. Immunohistological detection of interleukin-1-like molecules and tumor necrosis factor in human epidermis before and after UVB-irradiation in vivo. Br J Dermatol 1988; 118:369–376.
86. Kupper TS, Chua AO, Flood P, McGuire J, Gubler U. Interleukin 1 gene expression in cultured human keratinocytes is augmented by ultraviolet irradiation. J Clin Invest 1987; 80: 430–436.
87. Hammerberg C, Duraiswamy N, Cooper KD. Reversal of immunosuppression inducible through ultraviolet-exposed skin by in vivo anti-CD11b treatment. J Immunol 1996; 157: 5254–5261.
88. Dizdaroglu M, Olinski R, Doroshow JH, Akman SA. Modification of DNA bases in chromatin of intact target human cells by activated human polymorphonuclear leukocytes. Cancer Res 1993; 53:1269–1272.
89. Rosin MP, Anwar WA, Ward AJ. Inflammation, chromosomal instability, and cancer: the schistosomiasis model. Cancer Res 1994; 54:1929s–1933s.
90. Katiyar SK, Challa A, McCormick TS, Cooper KD, Mukhtar H. Prevention of UVB-induced immunosuppression in mice by the green tea polyphenol (−)-epigallocatechin-3-gallate may be associated with alterations in IL-10 and IL-12 production. Carcinogenesis 1999; 20:2117–2124.
91. Challa A, Katiyar SK, Cooper KD, Mukhtar H. Inhibition of UV radiation-caused induction of oxidative stress and immunosuppression in C3H/HeN mice by polyphenols from green tea (abstr). J Invest Dermatol 1998; 110:695.
92. Katiyar SK, Matsui MS, Mukhtar H. Protection against ultraviolet (UV) light-induced thymine dimers, oxidative stress and depletion of antioxidant enzymes in human skin by polyphenols from green tea. Proc Am Assoc Cancer Res 2000; 41:532.

93. Schwarz T, Urbanska A, Gschnait F, Luger TA. Inhibition of contact hypersensitivity by a UV-mediated epidermal cutokine. J Invest Dermatol 1986; 87:289–291.
94. Takashima A, Bergstresser PR. Impact of UVB radiation on the epidermal cytokine network. Photochem Photobiol 1996; 63:397–400.
95. Schwarz A, Grabbe S, Riemann H, Aragane Y, Simon M, Manon S, Andrade S, Luger TA, Zlotnik A, Schwarz T. In vivo effects of interleukin-10 on contact hypersensitivity and delayed-type hypersensitivity reactions. J Invest Dermatol 1994; 103:211–216.
96. DeFabo EC, Noonan FP. Mechanism of immune suppression by ultraviolet irradiation in vivo. I. Evidence for the existence of a unique photoreceptor in skin and its role in photoimmunology. J Exp Med 1983; 157:84–98.
97. Moodycliffe AM, Norval M, Kimber I, Simpson TJ. Characterization of a monoclonal antibody to *cis*-urocanic acid: Detection of *cis*-urocanic acid in the serum of irradiated mice by immunoassay. Immunology 1993; 79:667–672.
98. Norval M, Simpson TJ, Bardshiri E, Howie SEM. Urocanic acid analogues and the suppression of the delayed-type hypersensitivity response to herpes simplex virus. Photochem Photobiol 1989; 49:633–638.
99. Kurimoto I, Streilein JW. cis-Urocanic acid suppression of contact hypersensitivity induction is mediated via tumor necrosis factor-α. J Immunol 1992; 148:3072–3078.
100. Noonan FP, DeFabo EC. Immunosuppression by ultraviolet B radiation: initiation by urocanic acid. Immunol Today 1992; 13:250–254.
101. Applegate LA, Ley RD, Kripke ML. Identification of the molecular target for the suppression of contact hypersensitivity by ultraviolet radiation. J Exp Med 1989; 170:1117–1131.
102. LeVee GJ, Applegate LA, Ley RD. Photoreversal of the ultraviolet radiation-induced disappearance of ATPase-positive Langerhans cells in the epidermis of Monodelphis domestica. J Leukocyte Biol 1988; 44:508–513.
103. Devary Y, Rosette C, DiDonato JA, Karin M. NF-κB activation by ultraviolet light is not dependent on a nuclear signal. Science 1993; 261:1442–1445.
104. Simon MM, Aragane Y, Schwarz A, Luger TA, Schwarz T. UVB light induces nuclear factor κB (NFκB) activity independently from chromosomal DNA damage in cell-free cytosolic extracts. J Invest Dermatol 1994; 102:422–427.
105. Vink AA, Strickland FM, Bucana C, Cox PA, Roza L, Yarosh DB, Kripke ML. Localization of DNA damage and its role in altered antigen-presenting cell function in ultraviolet-irradiated mice. J Exp Med 1996; 183:1491–1500.
106. Kurimoto I, Arana M, Streilein JW. Role of dermal cells from normal and ultraviolet B-damaged skin in induction of contact hypersensitivity and tolerance. J Immunol 1994; 152: 3317–3323.
107. Feldbush TL, Hobbs MV, Severson CD, Ballas ZK, Weiler JM. Role of complement in the immune response. Fed Proc 1984; 43:2548–2552.
108. Erdei A, Fust G, Gergely J. The role of C3 in the immune response. Immunol Today 1991; 12:332–337.
109. Hammerberg C, Katiyar SK, Carroll MC, Cooper KD. Activated complement component 3 (C3) is required for ultraviolet induction of immunosuppression and antigenic tolerance. J Exp Med 1998; 187:1133–1138.
110. Katiyar SK, Cooper KD, Hammerberg C. Complement activation is critical for local UV immunosuppression (abstr). J Invest Dermatol 1996; 106:824.
111. Conney AH. Induction of microsomal enzymes by foreign chemicals and carcinogenesis by polyaromatic hydrocarbons. Cancer Res 42:4875–4917.
112. Old LJ, Benacerraf B, Carswell E. Contact reactivity to carcinogenic polycyclic hydrocarbons. Nature 1963; 198:1215–1216.
113. Pomeranz JR. Preliminary studies of tolerance to contact sensitization in carcinogen-fed guinea pigs. J Natl Cancer Inst 1972; 48:1513–1517.

114. Pomeranz JR, Carney JF, Alarif A. The induction of immunologic tolerance in guinea pigs infused with dimethylbenz(a)anthracene. J Invest Dermatol 1980; 75:488–490.
115. Klemme JC, Mukhtar H, Elmets CA. Induction of contact hypersensitivity to dimethylbenz (a)anthracene and benzo(a)pyrene in C3H/HeN mice. Cancer Res 1987; 47:6074–6078.
116. Frank DM, Yamashita TS, Blumer JL. Genetic differences in methylcholanthrene-mediated suppression of cutaneous delayed hypersensitivity in mice. Toxicol Appl Pharmacol 1982; 64:31–41.
117. Halliday GM, Muller HK. Induction of tolerance via skin depleted of Langerhans cells by a chemical carcinogen. Cell Immunol 1986; 99:220–227.
118. Ruby JC, Halliday GM, Muller HK. Differential effects of benzo(a)pyrene and dimethylbenz(a)anthracene on Langerhans cell distribution and contact sensitization in murine epidermis. J Invest Dermatol 1989; 92:150–155.
119. Anderson C, Hehr A, Robbins R, Hasan R, Athar M, Mukhtar H, Elmets CA. Metabolic requirement for induction of contact hypersensitivity to immunotoxic polyaromatic hydrocarbons. J Immunol 1995; 155:3530–3537.
120. Elmets CA, Athar M, Zaidi SIA, Mukhtar H. Contact hypersensitivity of dimethylbenz (a)anthracene is influenced by genes within the major histocompatibility complex and at the Ah receptor locus. Clin Res 1992; 40:309A.
121. Elmets CA, Athar M, Tubesing KA, Rothaupt D, Xu H, Mukhtar H. Susceptibility to the biological effects of polyaromatic hydrocarbons is influenced by genes of the major histocompatibility complex. Proc Natl Acad Sci USA 1998; 95:14915–14919.
122. Elmets CA, Zaidi SIA, Bickers DR, Mukhtar H. Immunogenetic influences on the initiation stage of the cutaneous chemical carcinogenesis pathway. Cancer Res 1992; 52:6106–6109.
123. Halliday GM, Muller HK. Sensitization through carcinogen-induced Langerhans cell-deficient skin activates specific long-lived suppressor cells for both cellular and humoral immunity. Cell Immunol 1987; 109:206–221.
124. Halliday GM, Cavanaugh LL, Muller HK. Antigen presented in the local lymph node by cells from dimethylbenzanthracene-treated murine epidermis activates suppressor cells. Cell Immunol 1988; 117:289–302.
125. Muller HK, Halliday GM, Knight BA. Carcinogen-induced depletion of cutaneous Langerhans cells. Br J Cancer 1985; 52:81–85.
126. Schwartz J, Solt DB, Pappo J, Weichselbaum R. Distribution of Langerhans cells in normal and carcinogen-treated mucosa of bucchal pouches of hamster. J Dermatol Surg Oncol 1981; 7:1005–1010.

# 17

# Environmental Stressors and the Eye

**JOHN R. TREVITHICK**

*University of Western Ontario, London, Ontario, Canada*

**KENNETH P. MITTON**

*Kellogg Eye Center, University of Michigan, Ann Arbor, Michigan*

## I. INTRODUCTION

Of all our senses, the most valued is sight. The eye, by virtue of its function in vision, must be in contact with the external environment. The cornea admits light, which is fine-focused by the lens into images and passes through aqueous and vitreous humors before reaching the retina.

All of the component parts of the eye are susceptible to environmental stresses, depending on whether direct or indirect effects are involved, because of its external exposure. Indirect effects can arise when, for instance, one component of the eye influences another, either by (1) damaging lipid peroxides originating in the degenerating retina, causing damage to the lens posterior (1–4); or (2) immune damage to the lens via the aqueous humor from plasma antibodies to lens proteins (5). Direct effects can occur when toxic substances or physical trauma contact the eye surface, damaging the cornea. Penetrating wounds can damage the lens epithelium, and chemicals in the environment can diffuse across the cornea and aqueous humors to damage the lens epithelium or trabecular meshwork. Unfortunately, medications necessary for serious medical conditions, such as steroidal anti-inflammatory drugs, can have damaging side effects on the lens or cause a harmful elevation in intraocular pressure. Nutritional influences can also be very important, potentiating in severe deficiencies causing damage to the cornea, lens, or retina.

## A. The Functional Components of the Eye: Targets of Stressors

The eye is referred to as an ''organ,'' but it is one of the more complex organs of the human body. In fact, it is best to think of the eye as a compact collection of organs itself, which together form the functioning eye. These various components of the eye provide many different functions, all of which are required to create the final sensation of vision processed in the visual cortex of our brains. Any stressors that damage or compromise any of the component functions become potential threats to vision. Embryonically, the eye is elegantly formed and molded from external ectodermal tissue that meets migrating neural tissue. The final result, our eyes, are complex organs that are both external and yet contain the only part of our brains outside the cranium—the neural retina. Light must enter the eye through the optical elements formed by the cornea, the liquid aqueous humor, and the lens. Most of the refracting power actually occurs at the air–cornea interface, while the lens can be placed under varied tension by the cablelike zonules around its periphery to focus on far or near objects. This lens shaping brings the desired image into focus on the neural retina at the back of the eye. Special photoreceptor cells in the neural retina have the ability to detect photons of light through the use of photosensitive compounds synthesized from beta-carotene from our diets. The rod photoreceptor cells contain the rhodopsin photopigment and these cells detect light at very low intensities, providing night vision. In sunlight, our color perception is more active and depends on three different types of cone photoreceptor cells that are sensitive to the red, green, or blue parts of the light spectrum. A healthy set of photoreceptor cells is also very dependent on the retinal pigment epithelium to which they are adjacent. The RPE provides some indispensable metabolic support to the photoreceptor cells.

Turning our attention to the front of the eye, we require a clear lens and cornea that are metabolically active and need nutrient supplies without a direct blood supply to deliver them. This supply depends on the production of aqueous humor (similar to blood but devoid of red blood cells) by the ciliary epithelial cells. This supply flows from behind the iris and over the front of the lens to nourish the lens and the cornea. The lens epithelium and the corneal endothelium must pump and exchange nutrients and products to keep these organs healthy. The eye is a closed system that maintains an internal pressure; the production of aqueous humor (inflow) must be balanced by the outflow of aqueous humor from the eye to prevent elevated intraocular pressure. This flow exits via the trabecular meshwork system, which is in the angle formed by the cornea and the iris. The trabecular meshwork is maintained by the trabecular meshwork cells and the meshwork is both a kind of filtering system and a one-way valve. Outflow passes this system and is collected in a canal that eventually leads back into the venous bloodflow. This is the eye, in a nutshell, and all of these very different functioning suborgans must work to maintain vision. Therefore, they are all potential targets of stressors that threaten vision.

## B. How Light Reaches the Retina

The primary refraction of light occurs at the tear film covering the cornea, at the air–liquid interface (Fig. 1). The corneal surface curvature is critical for focusing the image, and its correction to improve the sharpness of the retinal image has traditionally been accomplished by glasses or contact lenses; more recently, surgical alteration of the corneal by photo refractive keratotomy (PRK) procedures in which excimer lasers are used to alter the corneal surface curvature (after temporarily removing a flap of overlying corneal epithelium and replacing it after the excimer laser ablation of the underlying corneal stro-

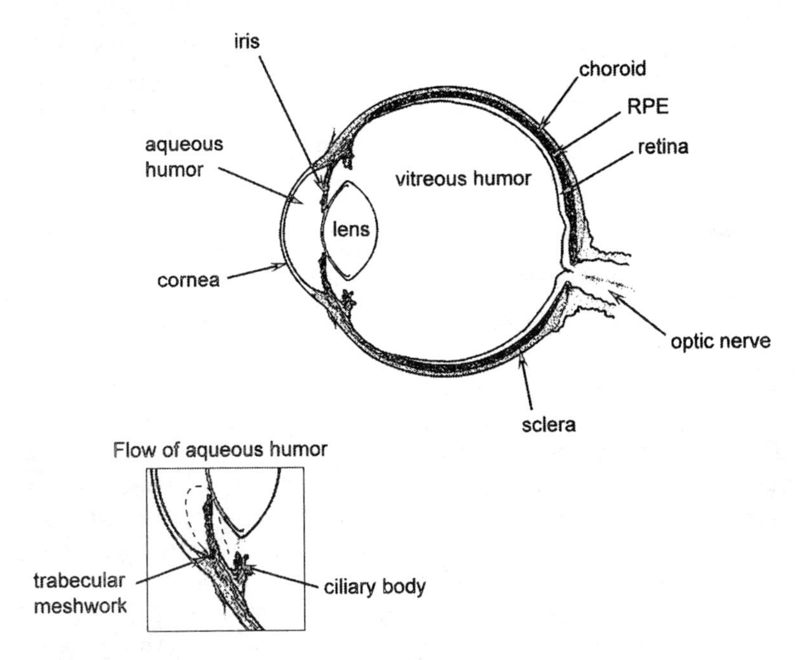


**Figure 1** Diagram of vertical meridional section of the human eye, side view, showing aqueous humor, choroid, cornea, lens, iris, optic nerve, retina, and retina pigmented epithelium (RPE), sclera, vitreous humor, and optic nerve. (Inset) Detail of flow of aqueous humor, showing entry by ultrafiltration through ciliary body, passage through space between lens and cornea, and exit through trabecular meshwork.

mal surface). The pressure wave generated by excimer laser pulses has been reported to generate superoxide free radical, which may damage the corneal endothelium and lens. In Korea, a group (6,7) has shown that PRK treatment induced peryoxynitrite (identified in the aqueous humor), which is a risk factor for cataract.

The lens functions in fine-focusing of the light; any opacity in the lens scatters the light, particularly if it is on the visual axis (8). Random scattering in the forward direction, toward the retina, results in overall increase in background light level and decreases the contrast of the retinal image. Many elderly people have cortical opacities that do not obstruct the path of light along the optical axis but cause such forward scattering that obscures their retinal images when they are facing oncoming headlights while driving at night, or when viewing a distant scene in bright sunlight (8). Cataract surgery, to remove the opaque lens and replace it by a plastic or silicone lens, has a very high success rate (better than 99%) (9,10), but regrowth of remaining cells from the lens epithelium can repopulate the surface of the posterior capsule, causing a cloudy secondary cataract in approximately half of these cases (11). Treatment with a YAG laser is necessary to break open the cell-coated capsule at the visual axis, to permit the light rays to pass through to form the retinal image.

The vitreous humor serves as a conduit via the passage of Cloquet from the retina to the area at the lens posterior. The incidence of cataracts rises with age from age 55 so that by age 75 approximately half of the population has a cataract (12), while the incidence of macular degeneration rises proportionally to that of cataract (13). This suggests that macular degeneration may be a contributory cause of cataract. As suggested by Zigler and Hess (1,2) and Clarke et al. (3,4), the polyunsaturated fatty acids may form lipoperoxides that can diffuse from the degenerating retina to the posterior of the lens. In the retina, light enters the photo receptors after passing through the neural retina, and any remaining light photons are finally absorbed in the retinal pigment epithelium (14,15).

## II. PATHOGENESIS

### A. Pathogenesis of Corneal Damage

Direct corneal damage can occur since the cornea is most directly exposed to noxious influences of the environment (14,15). It is well documented, however, that other indirect influences such as nutritional deficiency of vitamin A can result in corneal damage.

Chemical burns can be caused by organic solvents, acids, and alkali (14,15). Of these, the most difficult to heal are alkali burns.

Oxidizing agents are also able to damage the corneal epithelium: chlorine or hypochlorite, hydrogen peroxide, etc. Iodine is used clinically to chemically cauterize corneal scars to permit normal healing of corneal epithelium. Swimming pool overchlorination can result in corneal irritation if people do not wear goggles. Diffusion of chlorine across the cornea to the lens may also increase the risk of cataracts in swimmers.

Corneal damage (xerophthalmia) is caused by vitamin A deficiency, which can be remedied by vitamin A treatment (16). Until very recently, in third world countries this was a serious problem and still is during famines and in refugee camps. The associated diarrhea disease exacerbates the problem, interfering with absorption of whatever vitamin A is present in the diet. This seems to be worse during the rainy season. Since vitamin A is essential for maintenance of the lens epithelium (17), it is quite likely that vitamin A deficiency in early life might predispose to cataract in adults.

Amelioration of keratoconus can be achieved by adding vitamin E to the diet (18).

### B. Labrador Keratitis

Snow blindness, or Labrador keratitis, long known by Arctic explorers, Labrador fishermen, shipwrecked sailors, etc., is associated with corneal epithelial inflammation, which results in corneal swelling and corneal cloudiness (14,15). Although it results in temporary blindness, in the absence of ultraviolet light, healing leads to dehydration of the cornea and clearing to regain normal corneal clarity.

### C. Cataract

Cataracts are opacities of the eye lens that interfere with vision (8,19,20). They are characteristically associated with aging, but are considered to be multifactorial, so that the age-related cataract may have several contributing interacting causes. Within the lens, the opaque areas may be cortical (in the lens cortex or outer layers), nuclear (in the lens nucleus), or mixed cortical and nuclear. Nuclear cataracts appear to be associated with aggregation of lens proteins to large oligomeric aggregates that scatter light (21). Recent work has also shown cell membrane damage in the nucleus, which may also result in light scattering. In contrast, in cortical cataracts oxidation permits intracellular ionic calcium to increase so that cortical cataracts appear to involve cellular processes; cellular surface globules form (22) (Fig. 2) by cytoskeletal digestion by calpain, and further changes lead to formation of water clefts and vacuoles (20). An additional factor increasing light scatter-

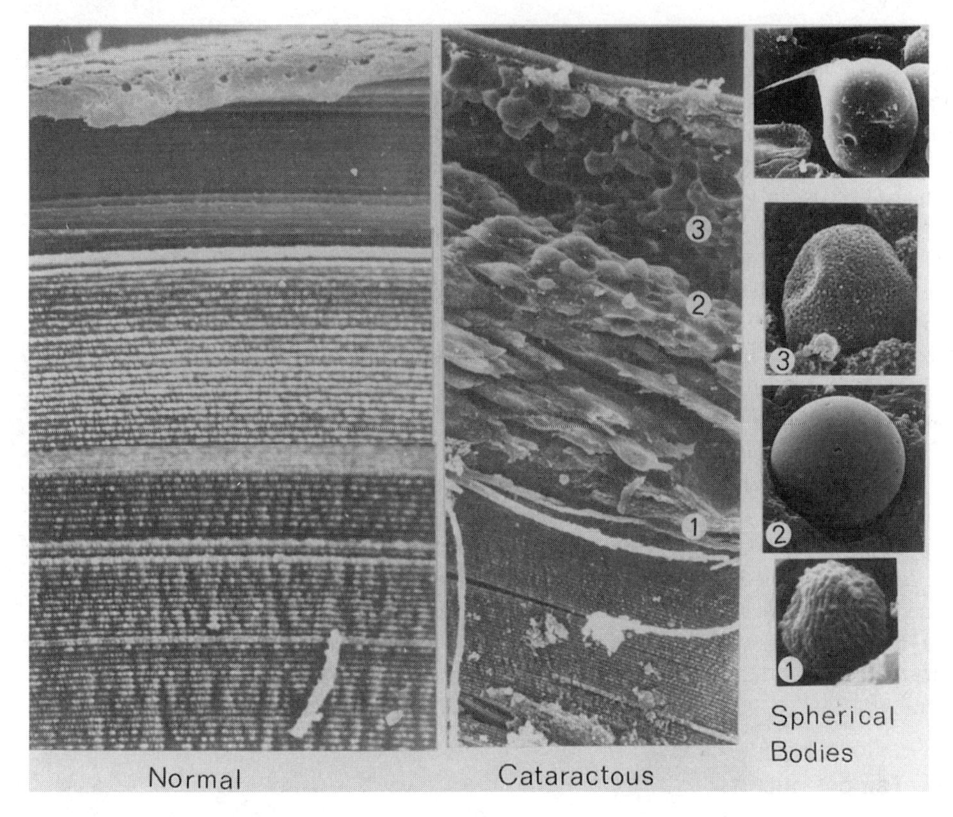


**Figure 2** Normal morphology by scanning microscopy of human lens compared to posterior cortical cataract, with globules typically located at locations numbered as shown enlarged at right. Note the thread of adhering cytoplasm still attached to globule located at 1.

ing is the phase change because of cooling the concentrated solution of lens proteins to the critical point (when phase separation occurs). Light scattering and reflection occurs at the boundaries of the individual phase particles (23). The critical temperature rises as the protein concentration of the lens increases, so that for the young lens it is lower. In addition, autoimmune attack on the lens may occur (5).

A number of environmental stress factors can result in increased risk of cataracts. This may involve free radical damage. We showed that in the rat eye (24), superoxidelike radical concentrations per unit protein increased with age. Other major risk factors may include ultraviolet light or diarrhea (25) and radiation; gamma radiation (26,27), protons, and neutrons (28,29), and nonionizing radiation; microwaves and millimeter waves (30–34), all of which are associated with free radical damage. The Chernobyl accident resulted in increased risk of cataracts among the workers involved in the cleanup (27). The potential risk of cataract in long space voyages should not be minimized (28,29). In the diabetic lens, stoichiometric loss of antioxidant osmolytes taurine, glutathione, ascorbate, and amino acids, occurs in response to the increased osmolar level of sorbitol (35–37). Additional generation of free radicals occurs during nonenzymatic glycation of proteins (38), and during normal metabolism during oxidative phosphorylation in mitochondria (39). This oxidative stress seems to potentiate release of calcium from degenerating mitochondria (40). The involvement of mitochondria in cataract was suggested previously by the observation that a drug used during the decade of the 1930s for weight reduction, dinitrophenol, caused an increased risk of cataract in humans and animal models (41). Naphthalene also causes cataract by an oxidative mechanism; vitamin E is able to decrease damage in naphthalene cataractogenesis (41). The damage reduction is consistent with oxidative damage because of vitamin E free radical scavenging. Hyperbaric oxygen, which increases the partial pressure of oxygen in tissues, may also increase risk of cataracts if data from animal models are an indication (42).

Perhaps the recent observation that socioeconomic status and mortality are related to increased cataract risk (44) may be associated with increased free radical damage from tobacco smoke (45). Smoke from tobacco or cooking fires (46,47) carries increased free radicals (45) that can potentiate the development of cataracts. In addition, tobacco smoke increases the concentration of metals such as zinc and lead in the lens, increasing the risk of cataract as demonstrated in rat models (47).

Although alcoholic beverages consumed in excess cause increased risk of cataracts, one drink per day appears to reduce the risk of cataract by 50% in a J-shaped curve, increasing to approximately 90% of normal risk for 2 drinks per day (49). Antioxidants are contained in the beverages; catechol and epicatechin and related polyphenols in beer, wine, and whiskies, etc., appear to be able to scavenge free radicals that potentiate cataract (50–53). Higher intakes of alcoholic beverages probably result in increased acetaldehyde which, like malondialdehyde from lipid peroxidation, potentiates development of the cataract.

## D. Temperature

Although the Singapore–Japan Cataract Study (54) suggests that increased ultraviolet radiation is associated with increased cataract risk, the other factor that may have a bearing on the increased cataract risk is the ambient temperature. Ambient temperature and cataract risk rise going from northern Japan to Kanazawa to Okinawa to Singapore. Consistent with this possibility, Lydahl showed that cataracts associated with glass blowing were found in the eye that was exposed to the molten glass and glass furnace (55). Iron workers show a similar increased risk, associated with exposure to molten steel. In vitro experiments on isolated rat lenses (56) indicated that such heat stress appears to involve an oxidative stage, since vitamin E can reduce the damage of the lens cortex caused by elevated temperature.

## E. Other Sources of Antioxidants in the Environment That May Reduce Cataract Risk

Many plant materials contain polyphenols and flavonoids that are precursors of lignin in plant cell walls. Because they are water soluble, compounds such as catechin and epicatechin and other polyphenols and flavonoids can serve as free radical scavengers by virtue of their phenolic hydroxyls. For this reason, any beverages that are obtained from leaves or fruit contain these antioxidants. Consumption of tea (5 cups/day) in our study (57) resulted in approximately half the risk of cataract found for normal people. The similarity to the 50% risk reduction for one alcoholic drink per day is perhaps due to the polyphenols in alcoholic beverages (50–53)—from red and white wine, beer which is fermented from aqueous extract of barley malt, wheat, and hops (53), matured spirits that extract polyphenols from the oak casks (whiskies, bourbon, rum, etc.) (52). In the case of dark rum, additional polyphenols may be extracted from the lilycane leaves used as caulking between the barrel staves.

## F. Glaucoma and Environmental Factors

Glaucoma is a family of conditions that ultimately lead to destruction of the optic nerve of the eye, which is composed of over 1,000,000 neurons that are required to transmit electrochemical signals from the photoreceptors to the brain. The most common form is open-angle glaucoma, where degeneration of the peripheral retina occurs because of increased intraocular pressure. Aqueous humor is generated by ultrafiltration of blood plasma in the ciliary body, which has a structure similar to the kidney nephron.

The aqueous humor contains small proteins only, such as ApoB lipoprotein, which have molecular weights not much larger than 100,000 Da. ApoB can bind vitamin E, but it has not been shown to be responsible for transporting it into the eye. Exiting the eye through the trabecular meshwork (TM), the aqueous pressure can build up if the egress is inhibited. The antioxidant level seems to be important in glaucoma, since several articles, most from Russian workers, seem to indicate that antioxidants can maintain the permeability of the TM (58–67).

Since the outflow facility of the eye is mostly via the TM (80%), and the TM cells are very active metabolically as they migrate on and maintain the meshwork, they themselves could be compromised by elevated oxidation stress over a lifetime. The TM cells are also finite in number; therefore, damage to these cells cannot be repaired by cell turnover. One can also hypothesize that any stresses from various insults to tissues in the front of the eye could indirectly affect the TM, since the aqueous outflow leads to and through the TM. This flow could carry damaging substances from the cornea, lens, and iris to the TM itself.

The meshwork maintained by the TM cells is composed of trabecular beams that are basically an extracellular matrix constructed from structural proteins such as elastin, laminin, collagens, and other extracellular glycoproteins. TM cells cover the trabecular beams in a monolayer and migrate upon the network as well. They also seem to be very active in phagocytosis, having some form of cleanup role that is presumably to prevent larger insoluble material from flowing into the meshwork and blocking it physically. As these cells internalize such debris, they have the potential not only to accumulate their own lifetime store of oxidation-damaged lipoprotein that cannot be broken down, but also to engulf and become burdened with the indigestible protein oxidation products that flow into the meshwork from other anterior tissues. TM cells are known to have a large amount of the stress protein αB-crystallin (also a lens crystallin) and can vary its levels with mechanical stress (68) and steroidal drug exposure. Steroidal drugs used for treating ocular inflammation (uveitis) and as part of treatment to resist organ graft rejection are known to pose a risk of causing elevation of intraocular pressure. Steroid drugs and mechanical stretching both trigger similar rearrangements of the TM cell cytoskeleton (69,70). Therefore, the trabecular meshwork is another potential target of environmental stress, that could result in some percentage of glaucoma incidence.

## G. Retina

Although retinal degeneration is a general accompanying disease of aging, called age-related macular degeneration (AMD), environmental nutritional factors seem to play a role in this disease. In the normal population, almost 20% of the people will have macular degeneration by age 75. Sheila West's group (71) has shown that vitamin E supplementation appears to reduce this by about 57%, indicating that oxidative damage seems to be one component of the pathogenesis process involved. Risk of diabetic retinopathy, too, appears to be reduced by vitamin E (72). Hypoxia associated with diabetic retinopathy may paradoxically initiate oxidative damage by damaging retinal and kidney capillaries (73). Prevention of such damage by addition of antioxidants appears also to involve prevention of protein kinase C (PKC) activation (72). In addition to increasing cataract risk, hyperbaric oxygen also appears to cause retinal damage (74,75). Increasing the partial pressure of oxygen also increases the risk of retinal damage in premature infants. We have previously examined the effect of vitamin E on retinopathy of prematurity (ROP) (19). This proliferation of retinal vascular capillaries resulting in frequent blindness in low-birth-weight (21–23 week gestation <1500 g) newborn infants can cause permanent blindness. Although vitamin E treatment was tested in the 1980s, toxic side effects of one parenteral preparation forced an end to this line of research. The medical establishment concluded that use of vitamin E to treat this condition was not justified, but it did recognize that use of vitamin E to treat hemolytic disease of the newborn was. Because of this, the manufacturers of parenteral baby nutrition and baby foods adopted a level of supplementation equivalent to 400 IU (mg) vitamin E in a normal adult, a supplemental dose (18,74–79). Although the question of efficacy of vitamin E for ROP treatment is still controversial, our meta-analysis of the data from trials published until 1990 clearly indicated that the vitamin-E-treated group received significant protection (19). Delaying treatment of ROP until stage 3, when high doses could be used with minimum risk, appears to have an excellent outcome (80).

### H. UV Light (Longer Wavelength)

In the aphakic eye, the lens, which normally acts as a powerful filter of UV light, is replaced during a cataract operation by a plastic intraocular lens (IOL). Because of this, it has been felt that the retina is at increased risk of UV-induced damage leading to earlier onset of macular degeneration as a result of the increased free radical levels caused by the UV. For this reason, if their IOLs do not prevent UV from reaching the retina, elderly aphakics are at particular risk of macular degeneration. Taken with the predisposition of the elderly to live or at least spend winters in a state with copious amounts of sunshine, playing golf or spending time at the beach, it is obviously important for these people to wear UV-absorbing sunglasses. People who work in bright sun should take particular care to ensure that their IOLs absorb UV light and, after cataract surgery, minimize exposure to UV on a continuing basis.

### I. Nutritional Factors

Deficiencies of selenium and vitamins A and E also can result in retinal capillary and/or rod and cone photoreceptor cell damage (81–88).

### J. Toxic Drugs

Chlorpromazine, though a powerful tranquilizer, has prooxidative effects that may potentiate light damage to the retina pigment epithelial cells (89–91) and cornea. Fluorescent dyes can also act as photosensitizers, leading to retinal damage (92).

## K. Nature and Nurture: Genetics and Environment in Multifactorial Diseases of the Eye

Almost all of the ocular disease mentioned in this chapter are conditions that are closely associated with aging as a major risk factor. While cataract and glaucoma risk is slightly greater for individuals with affected ancestors, most of these conditions do not appear to be overtly due to genetics. A small percentage of cataract and glaucoma is congenital genetic disease, showing obvious inheritance within a pedigree. Some have been mapped to specific mutations in certain genes. Retinal degenerations are frequently genetic in origin, while age-related macular degeneration does not display such a simple genetic component.

Therefore, ocular pathologies display the entire spectrum of nature versus nurture, or genetic predisposition versus environmental effects. Some are fully genetic, showing simple Mendelian inheritance, while others show no known or obvious genetic predisposition. Some are in between these two extremes.

Congenital or infantile cataracts occur in 1 to 6 out of every 10,000 births (93). While it has been thought that most cataract occurrence is only influenced by external stressors, there is some evidence that both nuclear and cortical cataracts may have a larger genetic component than previously suspected (94,95).

Diabetic retinopathy, glaucoma, and age-related macular degeneration are major causes of blindness that have nongenetic and genetic components. Mendelian retinal diseases affect a much smaller portion of the population, but still a significant portion of about 1/2000 (96). As for congenital (Mendelian) cataracts, the Mendelian retinal diseases tend to display an earlier age of onset and a more severe progression than their age-related counterparts.

The Mendelian diseases of the eye are being mapped and characterized at an ever-increasing rate. For cataracts, mutations have been mapped to the genes for connexins, and α-, β-, and γ-crystallins (97). For retinal diseases, mutations have been mapped to many genes of the phototransduction cascade (rhodopsin, transducin, arrestin, rhodopsin kinase, cGMP phosphodiesterase, cyclic nucleotide gated channel, guanylate cyclase, guanylate cyclase activating protein); structural and metabolic/transport proteins [peripherin, rod outer segment membrane protein-1 (ROM-1), Rab escort protein-1 (REP-1), myosin VIIA, USH2A]; visual cycle proteins [cellular retinaldehyde binding protein (CRALBP), 11-*cis*-retinol dehydrogenase, RPE65, ABCR, bestrophin, and tissue inhibitor of metalloproteinases-3 (TIMP-3) (96). More recently, mutations in retinal-specific developmental proteins have been discovered in the transcription factors NRL (neural retina leucine zipper) and CRX (cone rod homeobox). These mutations presumably alter gene expression of such retinal proteins as rhodopsin (98–100).

While most cataract and retinal diseases do not show a Mendelian pattern of inheritance, genetics is still influential on how the effects of various environmental stresses are tolerated. Likewise, even in animal models of Mendelian retinal dystrophies that affect known genes, diet and pharmacological interventions have been used to alter the rate of progression of photoreceptor loss. Therefore, we must understand that no matter how large or small the genetic component, there is an interaction of environment and genetic predisposition that results in a multifactorial condition. The situation is like that of the tires on a car. Some tires are made to last 40,000 miles and some 70,000. Both types are viable tires, both do their job, but one may last longer due to their different construction. We are only at the very start of attempting to grasp and understand all of these interactions, with the hope of using this knowledge in the future design of therapeutic strategies that may involve combinations of genetic, nutritional, and pharmacological components.

## REFERENCES


1. Zigler JS, Hess HH. Cataracts in the Royal College of Surgeons rat: evidence for initiation by peroxidation products. Exp Eye Res 1985; 41:67–76.
2. Hess HH, Kuwabara T, Zigler S, Westney IV. Posterior subcapsular cataracts in the Royal College of Surgeons (RCS) rat: light as a factor in development and maturation. Invest Ophthalmol Visual Sci 1986; 27(suppl):203.
3. Clarke IS, Dzialoszynski T, Sanford SE, Chevendra V, Trevithick JR. Dietary prevention of damage to retinal proteins in RCS rats. In: Lerman S, Tripathi R, eds. Ocular Toxicity. New York: Marcel Dekker, 1989:253–272.
4. Clarke IS, Dzialoszynski T, Sanford SE, Trevithick JR. A possible relationship between increased levels of the major heat shock protein HSP 70 and decreased levels of S-antigen in the retina of the RCS rat. Exp Eye Res 1991; 53:545–548.
5. Ibaraki N, Lin LR, Dang L, Reddy VN, Singh DP, Sueno T, Chylack LT, Jr., Shinohara T. Anti-beta-crystallin antibodies (mouse) or sera from humans with age-related cataract are cytotoxic for lens epithelial cells in culture. Exp Eye Res 1997; 64:229–38.
6. Hayashi S, Ishimoto S, Wu GS, Wee WR, Rao NA, Mcdonnell PJ. Oxygen free radical damage in the cornea after excimer laser therapy. Br J Ophthalmol 1997; 81:141–144.
7. Ryu JH, Bae NY, Kim JC, Shyn KH. The change of nitric oxide concentration in aqueous humor of rabbit eye after excimer laser PRK: role of NO in cataractogenesis. Abstracts, Second Asian Cataract Research Conference, June 4–6, 1998, available from Dept. Ophthalmology, Yongsan Hospital, Chung Ang University, Soeul, Korea, 140-757. 1998:71.
8. Creighton MO, Trevithick JR, Mousa GY, Percy DH, McKinna AJ, Dyson C, Maisel H, Bradley R. Globular bodies: A primary cause of the opacity in senile and diabetic posterior cortical subcapsular cataracts? Can J Ophthalmol 1978; 13:166–181.
9. Steinberg EP, et al. National Study of Cataract Surgery Outcomes: Variation in 4-month postoperative outcomes as reflected in multiple outcome measures. Ophthalmology 1994; 101:1131–1141.
10. Javitt JC, Tielsch JM, Canner JK, Kolb MM, Sommer A, Steinberg EP. National outcomes of cataract extraction; increased risk of retinal complications associated with Nd: YAG laser capsulotomy. Ophthalmology 1992; 99:1487–1498.
11. Davidson MG, Harned J, Grimes AM, Duncan G, Wormstone IM, McGahan MC. Transferrin in after-cataract and as a survival factor for lens epithelium. Exp Eye Res 1998; 66(2):207–15.
12. Kahn HA, Leibowitz HM, Ganley JP, et al. The Framingham eye study I. Outline and major prevalence findings. Am J Epidemiol 1977; 106:17–32.
13. Liebowitz HM, Krueger DE, Maunder LR, Milton RC, Kini MM, Kahn HA, et al. Framingham eye study monograph. Surv Ophthalmol 1980; 24(suppl):335–610.
14. Garner A, Klintworth GC, eds. The Pathobiology of Ocular Disease. New York: Marcel Dekker, 1982.
15. Yanoff M, Fine BS, eds. Ocular Pathology. New York: Harper & Row, 1975.
16. Sommer A. Xerophthalmia and vitamin A status. Prog Retin Eye Res 1998; 17(1):9–31.
17. Linklater HA, Dzialoszynski T, McLeod HL, Sanford SE, Trevithick JR. Modelling cortical cataractogenesis XII supplemented vitamin A treatment reduces gamma-crystallin leakage from lenses in diabetic rats. Lens Eye Toxicity Res 1992; 9:115–126.
18. Puchkowskaya NA, Titarenko ZD. Neues in der bahandlung des keratoconus. Klin Mbl Augenheilk 1986; 189:11–14.

18a. Titarenko ZD. The action of vitamin E on the cornea in keratoconus. Oftalmologicheskii Zhurnal 1985; 35:163–165.

19. Trevithick JR, McD Robertson J, Mitton KP. Vitamin E and the eye. In: Packer L, Fuchs J, eds. Vitamin E in Health and Disease. New York: Marcel Dekker, 1993:873–926.
20. Bellows JG, Bellows RJ. Presenile and senile cataract. In: Bellows JG, ed. Cataract and Abnormalities of the Lens. New York: Grune & Stratton, 1975:303–313.
21. Benedek GB. Theory of transparency of the eye. Appl Optics 1971; 10:459–473.
22. Kilic F, Trevithick JR. Modeling cortical cataractogenesis XXXIX. Calpain proteolysis of lens fodrin in cataract. Biochem Molec Biol Int 1998; 45:963–978.
23. Eccarius S, Clark JI. Effect of aspirin and vitamin E on phase separation in calf lens homogenate. Ophthalm Res 1987; 19:65–71.
24. Trevithick JR, Dzialoszynski T. Endogenous superoxide-like species and antioxidant activity in ocular tissues detected by luminol luminescence. Biochem Molec Biol Int 1997; 41:695–705.
25. Harding JJ. Cataract: sanitation or sunglasses? Lancet 1982; 1(8262):39.
26. Ross WM, Creighton MO, Trevithick JR. Radiation cataractogenesis induced by neutron or gamma irradiation in the rat lens is reduced by vitamin E. Scanning Microsc 1990; 4:641–650.
27. Buzunov V, Fedirko P. Ophthalmo-pathology in victims of the Chernobyl catastrophe—Results of a clinical epidemiological study. In: Junk AK, Kndiev Y, Vitte P, Worgul BV, eds. Ocular Radiation Risk Assessment in Populations Exposed to Environmental Radiation Contamination, NATO ASI series 2. Dordrecht: Kluwer Academic Publishers, 1998; 50:57–67.
28. Worgul BV, Medvedovsky C, Huang Y, Marino SA, Randers-Pehrson G, Brenner DJ. Quantitative assessment of the cataractogenic potential of very low doses of neutrons. Radiat Res 1996; 145:343–349.
29. Cox AB, Ainsworth EJ, Jose JG, Lee AC, Lett JT. Cataractogenesis from high-LET radiation and the Casarett model. Adv Space Res 1983; 3:211–219.
30. van Ummerson CA, Cogan FC. Effects of microwave irradiation on the lens epithelium in the rabbit eye. AMA Arch Ophthalmol 1976; 94:828–834.
31. Carpenter RL, van Ummersen CA. The action of microwave irradiation on the eye. J Microwave Power 1968; 3:3–19.
32. Creighton MO, Larsen LE, Stewart-Dehaan PJ, Jacobi JH, Sanwal M, Baskerville JL, Bassen HI, Brown DO, Trevithick JR. In vitro studies of microwave-induced cataract II comparison of damage observed for continuous wave and pulsed microwaves. Exp Eye Res 1987; 45: 357–373.
33. Stewart-DeHann PJ, Creighton MO, Larsen LE, Jacobi JH, Ross WM, Sanwal M, Guo TC, Guo WW, Trevithick JR. In vitro studies of microwave-induced cataract: separation of field and heating effects. Exp Eye Res 1983; 36:75–90.
34. Stewart-DeHann PJ, Creighton MO, Larsen LE, Jacobi JH, Ross WM, Sanwal M, Baskerville J, Trevithick JR. In vitro studies of microwave-induced cataract: reciprocity between exposure duration and dose rate for pulsed microwaves. Exp Eye Res 1985; 40:1–13.
35. Mitton KP, Dean PAW, Dzialoszynski T, Xiong H, Sanford SE, Trevithick JR. Modeling cortical cataractogenesis 13: Early effects on lens ATP/ADP and glutathione in the streptozotocin rat model of diabetic cataract. Exp Eye Res 1993; 56:187–198.
36. Mitton KP, Linklater HA, Dzialoszynski T, Sanford SE, Starkey K, Trevithick JR. Modeling cortical cataractogenesis 21: in diabetic rat lenses taurine supplementation partially reduces damage resulting from osmotic compensation leading to osmolyte loss and antioxidant depletion. Exp Eye Res 1999; 69:279–89.
37. Kilic F, Bhardwaj R, Caulfeild J, Trevithick JR. Modeling cortical cataractogenesis 22: is in vitro reduction of damage in model diabetic rat cataract by taurine due to its antioxidant activity? Exp Eye Res 1999; 69:291–300.
38. Wolff SP, Bascal ZA, Hunt JV. Autoxidative glycosylation: free radicals and glycation theory. Prog Clin Biol Res 1989; 304:259–75.

39. Turrens JF, Alexandre A, Lehninger AL. Ubisemiquinone is the electron donor for superoxide formation by complex III of heart mitochondria. Arch Biochem Biophys 1985; 237:408–14.
40. Trevithick JR, Bantseev V, Sivak JG. Model diabetic cataract opacification depends on mitochondrial calcium release. Ophthal Res 1999; 31(suppl 1):35.
41. Holmen JB, Ekesten B, Lundgren B. Inhibition of naphthalene-induced cataract by pantethine and vitamin E in brown Norway rats. Invest Ophthalmol Vis Sci 1999; 40:S526.
42. Hollwich F, Boateng A, Kolck B. Toxic cataract. In: Bellows JG, ed. Cataract and Abnormalities of the Lens. New York: Grune and Stratton, 1975:230–243.
43. Padgaonkar VA, Lin LR, Leverenz VR, Rinke A, Reddy VN, Giblin FJ. Hyperbaric oxygen in vivo accelerates the loss of cytoskeletal proteins and MIP26 in guinea pig lens nucleus. Exp Eye Res 1999; 68:493–504.
44. Meddings DR, Hertzman C, Barer ML, Evans RG, Kazanjian A, McGrail K, Sheps SB. Socioeconomic status, mortality, and development of early cataract. Soc Sci Med 1997; 46: 1451–1457.
45. Pryor WA, Prier DG, Church DF. Electron-spin resonance study of mainstream and sidestream cigarette smoke: Nature of the free radicals in gas-phase smoke and in cigarette tar. Environ Health Perspect 1983; 47:345–355.
46. Shalini VK, Luthra M, Srinivas L, Rao SH, Basti S, Reddy M, Balasubramanian D. Oxidative damage to the eye lens caused by cigarette smoke and fuel smoke condensates. Ind J Biochem Biophys 1994; 31:261–266.
47. Dilsiz N, Olcucu A, Cay M, Naziroglu M, Cobanoglu D. Protective effects of selenium, vitamin C and vitamin E against oxidative stress of cigarette smoke in rats. Cell Biochem Funct 1999; 17:1–7.
48. Avunduk AM, Yardimci S, Avunduk MC, Kurnaz L, Kockar MC. Determinations of some trace and heavy metals in rat lenses after tobacco smoke exposure and their relationships to lens injury. Exp Eye Res 1997; 65:417–23.
49. Clayton RM, Cuthbert J, Duffy J, Seth J, Phillips CI, Bartholomew RS, Reid JMK. Some risk factors associated with cataract in S.E. Scotland: a pilot study. Trans Ophthalmol Soc UK 1982; 102:331–336.
50. Trevithick CC, Vinson JA, Caulfeild J, Rahman F, Derksen T, Bocksch L, Hong S, Stefan A, Teufel K, Wu N, Hirst M, Trevithick JR. Is ethanol an important antioxidant in alcoholic beverages associated with risk reduction of cataract and atherosclerosis? Redox Rep 1999; 4:89–93.
51. Duthie GG, Pedersen MW, Gardner PT, Morrice PC, Jenkinson AM, McPhail DB, Steele GM. The effect of whisky and wine consumption on total phenol content and antioxidant capacity of plasma from healthy volunteers. Eur J Clin Nutr 1998; 52:733–736.
52. Goldberg DM, Hoffman B, Yang J, Soleas GJ. Phenolic constituents, furans, and total antioxidant status of distilled spirits. J Agric Food Chem 1999; 47:3978–3985.
53. Madigan D, McMurrough I, Smyth MR. Determination of proanthocyanidins and catechins in beer and barley by high-performance liquid chromatography. Analyst 1993; 119:863–868.
54. Sasaki H, Kojima M, Shui YB, Chen HM, Nagai K, Kasuga T, Katoh N, Jin CS, Sasaki K. The Singapore-Japan Cooperative Eye Study. Abstracts, US-Japan Cooperative Cataract Research Group Meeting, Kona HI, November 16–19, 1997:66.
55. Lydahl E, Glansholm A. Infrared radiation and cataract. III. Differences between the two eyes of glass workers. Acta Ophthalmol (Copenh) 1985; 63:39–44.
56. Stewart-DeHaan PJ, Creighton MO, Sanwal M, Ross WM, Trevithick JR. Effects of vitamin E on cortical cataractogenesis induced by elevated temperature in intact rat lenses in medium 199. Exp Eye Res 1981; 32:54–60.
57. McD Robertson J, Donner AP, Trevithick JR. Vitamin E intake and risk of cataracts in humans. Ann NY Acad Sci 1989; 570:372–382.

58. Alekseev VN, Ketlinskii SA, Sharonov BP, Martynova EB, Lauta VF. [Lipid peroxidation in experimental glaucoma and the possibilities for its correction (preliminary report)]. [Russian] Original Title Perekisnoe okislenie lipidov pri eksperimental'noi glaukome i vozmozhnosti ego korrektsii (predvaritel'noe soobshchenie). Vestn Oftalmol 1993; 109:10–12.
59. Kurysheva NI, Vinetskaia MI, Erichev VP, Demchuk ML, Kuryshev SI. [Contribution of free-radical reactions of chamber humor to the development of primary open-angle glaucoma]. [Russian] Original Title Rol'svobodnoradikal'nykh reaktsii kamernoi vlagi v razvitii pervichnoi otkrytougol'noi glaukomy. Vestn Oftalmol 1996; 112:3–5.
60. Lebedev OI, Dumenova NV, Kovalevskii VV. [Laser irradiation: study of total effect of irradiation of the eyeball. A clinical study]. [Russian] Lazernoe izluchenie: issledovanie obshchego deistviia pri obluchenii glaznogo bloka. Klinicheskoe issledovanie. Vestn Oftalmol 1995; 111:17–19.
61. Popova ZS, Kuz'minov OD. [Treatment of primary open-angle glaucoma by the method of combined use of hyperbaric oxygenation and antioxidants]. [Russian] Original Title Lechenie pervichnoi otkrytougol'noi glaukomy metodom sochetannogo primeneniia giperbaricheskoi oksigenatsii i antioksidantov. Vestn Oftalmol 1996; 112:4–6.
62. Filina AA, Davydova NG, Endrikhovskii SN, Shamshinova AM. Lipoic acid as a means of metabolic therapy of open-angle glaucoma. Vestn Oftalmol 1995; 111:6–8.
63. Bunin AI, Ermakov VN, Filina AA. [New directions in the hypotensive therapy of open-angle glaucoma]. [Russian] Original Title: Novye napravleniia gipotenzivnoi terapii otkrytougol'noi glaukomy (eksperimental'no-klinicheskie issledovaniia). Vestn Oftalmol 1993; 109:3–6.
64. Schumer RA, Podos SM. The nerve of glaucoma. Arch Ophthalmol 1994; 112:37–44.
65. Bonne C. New perspectives in the pharmacological treatment of glaucoma. Therapie 1993; 48:559–65.
66. Osborne NN. Serotonin and melatonin in the iris/ciliary processes and their involvement in intraocular pressure. [Review] Acta Neurobiol Exp (Warsz) 1994; 54(suppl):57–64.
67. Russell P, Johnson DH. Enzymes protective of oxidative damage present in all decades of life in the trabecular meshwork, as detected by two-dimensional gel electrophoresis protein maps. J Glaucoma 1996; 5:317–312.
68. Mitton KP, Tumminia SJ, Arora J, Zelenka P, Epstein DL, Russell P. Transient loss of alphaB-crystallin: an early cellular response to mechanical stretch. Biochem Biophys Res Commun 1997; 9:235(1):69–73.
69. Clark AF, Wilson K, McCartney MD, Miggans ST, Kunkle M, Howe W. Glucocorticoid-induced formation of cross-linked actin networks in cultured human trabecular meshwork cells. Invest Ophthalmol Vis Sci 1994; 35(1):281–294.
70. Tumminia SJ, Mitton KP, Arora J, Zelenka P, Epstein DL, Russell P. Mechanical stretch alters the actin cytoskeletal network and signal transduction in human trabecular meshwork cells. Invest Ophthalmol Vis Sci 1998; 39(8):1361–71.
71. West S, Vitale S, Hallfrisch J, Munoz B, Muller D, Bressler S, Bressler NM. Are antioxidants or supplements protective for age-related macular degeneration? Arch Ophthalmol 1994; 112: 222–227.
72. Bursell SE, Clermont AC, Aiello LP, Aiello LM, Schlossman DK, Feener EP, Laffel L, King GL. High-dose vitamin E supplementation normalizes retinal blood flow and creatinine clearance in patients with type I diabetes. Diabetes Care 1999; 22:1245–1251.
73. Takagi H, King GL, Aiello LP. Hypoxia upregulates glucose transport activity through an adenosine-mediated increase of GLUT1 expression in retinal capillary endothelia cells. Diabetes 1998; 47:1480–1488.
74. Stone WL, Henderson RA, Howard GH, Jr, Hollis AL, Payne PH, Scott RL. The role of antioxidant nutrients in preventing hyperbaric oxygen damage to the retina. Free Rad Biol Med 1989; 6:505–512.
75. Bresnick GH. Oxygen induced visual cell degeneration in the rabbit. Invest Ophthalmol Visual Sci 1970; 9:372–387.

76. Shvedova AA, Kagan VE, Kuliev IY, Dobrina SK, Prilipko LL, Meerson FZ, Koslov UP. Lipid peroxidation and retinal injury in stress. Bull Exp Biol Med 1982; 93:408–411.
77. National Research Council, Food and Nutrition Board. Recommended Dietary Allowances, 9th ed. Washington, DC: National Academy of Sciences, 1980.
78. Nutritional Advisory Group: Guidelines for multivitamin preparation for parenteral use. Chicago: American Medical Association, 1975.
79. American Medical Association Department of Foods and Nutrition, 1975: Multivitamin preparations for parenteral use. A statement by the nutrition advisory group. J Paren Enteral Nutr 1979; 3:258.
80. Johnson L, Quinn GE, Abassi S, Delevoria-Papadopoulos M, Peckham G, Bowen FW, Jr. Bilateral stage 3-plus retinopathy of prematurity (ROP). Effect of treatment (Rx) with high-dose vitamin E. Ann NY Acad Sci 1989; 570:464–466.
81. Ameniya T. Effects of vitamin E and selenium deficiencies on rat capillaries. Int J Vit Nutr Res 1989; 59:122–126.
82. Ostrovsky MA, Sakina NL, Dontsov AE. An antioxidative role of ocular screening pigments. Vision Res 1987; 27:893–899.
83. Kager S. Langzertbehandlung arteriosklerotisch bedingter durchblutungsstorungen der netz und aderhaut mit den vitaminen A and E sowie einem nikotinsaurederivat. Klin Monatsbl Augenheilk 1968; 153:571–577.
84. Satya-Murti S, Howard L, Krohel G, Wolfe B. The spectrum of neurologic disorder from vitamin E deficiency. Neurology 1986; 36:917–921.
85. Kagan VE, Barybina GV, Novikov KN. Peroxidation of lipids and degeneration of photoreceptors in the retina of rats with avitaminosis E. Zh Exp Biol Med 1977; 83:473–476.
86. Hermann RK, Robison WG, Bieri JG, Spitznas M. Lipofuscin accumulation in extra ocular muscle of rats deficient in vitamins E and A. Graefe's Arch Clin Exp Ophthalmol 1985; 223: 272–277.
87. Hermann RK, Robison WG, Bieri JG. Deficiencies of vitamins E and A in the rat: lipofuscin accumulation in the choroid. Invest Ophthalmol Vis Sci 1984; 25:429–433.
88. Runge P, Muller DRP, McAlister J, Calver D, Lloyd JK, Taylor D. Oral vitamin E supplements can prevent the retinopathy of abetalipoproteinemia. Br J Ophthalmol 1986; 70:166–173.
89. Greener AC, Berry K. Skin pigmentation and corneal and lens opacities with prolonged chlorpromazine therapy. Can Med Assoc J 1964; 90:663–665.
90. Silverman HI. The adverse effects of commonly used systemic drugs on the human eye II. Am J Optometry 1972; 49:335–362.
91. Persad S, Menon IA, Basu PK, Carre F. Phototoxicity of chlorpromazine on retinal pigment epithelial cells. Curr Eye Res 1988; 7:1–9.
92. Shvedova AA, Kagan VE, Kuliev IY, Vekshina OM. Mechanisms of the harmful action of fluorescent dyes on the retina. Bull Exp Biol Med 1983; 96:1085–1089.
93. Lambert SR, Drack AV. Infantile cataracts. Surv Ophthalmol 1996; 40:427–458.
94. Heiba IM, Elston RC, Klein BE, Klein R. Genetic etiology of nuclear cataract: evidence for a major gene. Am J Med Genet 1993; 47:1208–1214.
95. Heiba IM, Elston RC, Klein BE, Klein R. Evidence for a major gene for cortical cataract. Invest Ophthalmol Vis Sci 1995; 36:227–235.
96. Rattner A, Sun H, Nathans J. Molecular genetics of human retinal disease. Ann Rev Genet 1999; 33:89–131.
97. Ionides A, Francis P, Berry V, Mackay D, Bhattacharya S, Shiels A, Moore A. Clinical and genetic heterogeneity in autosomal dominant cataract. Br J Ophthalmol 1999; 83(7):802–808.
98. Bessant DAR, Payne AM, Mitton KP, Wang Q-L, Swain PK, Plant C, Bird AC, Zack DJ, Swaroop A, Bhattacharya SS. A mutation in NRL is associated with autosomal dominant retinitis pigmentosa. Nat Gen 1999; 21:355–356.

99. Swain PK, Chen S, Wang Q-L, Affatigato LM, Coats CL, Brady KD, Fishman GA, Jacobson SG, Swaroop A, Stone E, Sieving PA, Zack DJ. Mutations in the cone-rod homeobox gene are associated with the cone-rod dystrophy photoreceptor degeneration. Neuron 1997; 19: 1329–1336.
100. Swaroop A, Wang QL, Wu W, Cook J, Coats C, Xu S, Chen S, Zack DJ, Sieving PA. Leber congenital amaurosis caused by a homozygous mutation (R90W) in the homeodomain of the retinal transcription factor CRX: direct evidence for the involvement of CRX in the development of photoreceptor function. Hum Mol Gen 1999; 8:299–305.

# 18

# Free Radicals and Aging of Anterior Segment Tissues of the Eye

**KEITH GREEN**

*Medical College of Georgia, Augusta, Georgia*

## I. OCULAR STRUCTURE

The anterior segment of the eye consists of the cornea, anterior chamber, trabecular meshwork, iris, and ciliary body (Fig. 1). The cornea usually contains five layers. From anterior to posterior they are: the 50-μm-thick epithelium; Bowman's membrane (10 μm); stroma (500 μm); Descemet's membrane (5 μm); and endothelium (5 μm). The values given for the thickness of each component of the cornea reflect those of humans, where the total corneal thickness is approximately 550 μm. Other species have proportional thicknesses of the component layer of stroma relative to their total thickness (bovine ≃0.8 mm; rabbit ≃0.4 mm; mouse ≃0.1 mm).

The epithelium is a five- or six-cell layered membrane that is covered by the 7- to 10-μm tear film and thus represents the first barrier against invasion from the environment or the first line of defense (including the tear film) against environmental stress. The cells are of the squamous type at the surface, while the most posterior layer of the epithelium consists of columnar basal cells: wing cells fill the space between the other cell layers. The surface cells are desquamated into the tear film (1,2) on a regular basis as a result of the shear forces of lid movements, with replacement centripetally from both conjunctiva at the corneal periphery and underlying basal cells (3–5). Replacement of epithelial cells occurs on a 5- to 6-day cycle such that total epithelial replacement occurs over about a 2-week period (2,6–9).

The epithelium forms a relatively tight barrier to the movement of substances in either direction (i.e., from the tear film into the stroma or vice versa) (10–13). The barrier characteristics of the epithelium are largely dependent upon the lipophilicity of the com-

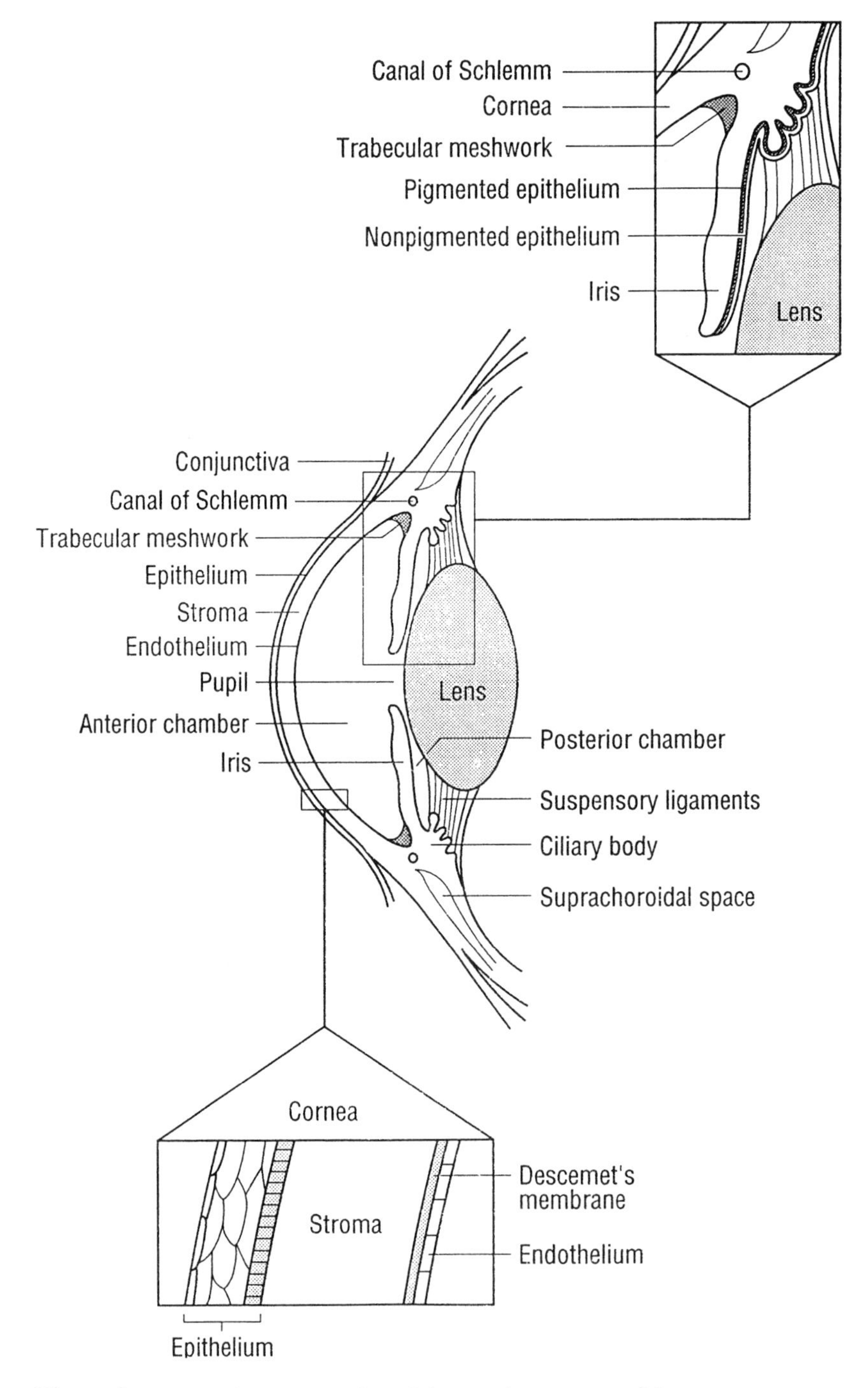


**Figure 1** Schematic representation of the anterior segment of the eye.

pound under consideration (13) due to the highly lipid nature of this membrane as well as the presence of many desmosomal junctions between the cells. Bowman's membrane is the anterior portion of the stroma and acts as a basement membrane for the epithelium to which the cells are attached by hemidesmosomes and other adhesion complexes (14, 15).

The bulk of the cornea consists of the stroma, which contains collagen fibrils of a consistent diameter embedded in a glycosaminoglycan matrix with a structural quasiordered arrangement such that the cornea is transparent at normal thickness (≃400 μm for rabbit) and hydration (75% to 78%) (16–19). The stroma is capable of imbibing large volumes of fluid and increasing thickness by at least two- or threefold (depending upon the species) (20), with a concomitant increase in hydration and loss of transparency due to fluid imbibition and subsequent swelling; the collagen fibril diameter remains constant during stromal expansion. Swelling occurs rapidly in the antero-posterior direction, reaching near equilibrium in about 1 or 2 h after removal of the two bounding membranes with free exposure to solution (21). Experimentally induced stromal swelling was created in human or rabbit tissue from either the anterior or posterior face of the cornea (22). Swelling occurred more rapidly from the posterior surface in both species. Corneal swelling normally only occurs when either of the bounding membranes are compromised by increased permeability or attenuation of ion transport mechanisms.

Descemet's membrane is the basement membrane of the single cell layer of endothelium. The latter monolayer membrane, facing the aqueous humor fluid in the anterior chamber, is responsible for maintaining corneal thickness constant in face of the large imbibition or swelling pressure (70 to 90 mmHg) exerted by the stroma at normal hydration (23). The cells are arranged with their apical surfaces facing the aqueous humor and with an apical flap over the terminal cell junction. Endothelial cells have a characteristic regular hexagonal shape when viewed by specular microscopy or scanning electron microscopy.

The relatively leaky endothelium is the site of a fluid pump that moves fluid back into the aqueous humor from the stroma to maintain an equilibrium in the pump-leak system (24), which is operative in this membrane in order to keep corneal thickness constant. Failure to maintain the balance between fluid absorption by the stroma and extrusion of an equal volume of fluid by the pump leads to corneal swelling and the development of opacity. The endothelium is also the location of bicarbonate and sodium ion transport systems, but the relationship of these ion pumps and coexistent ion channels to the fluid transfer mechanism remains to be determined (25–39).

Aqueous humor fills the anterior chamber (≃250 μL) and is constantly being exchanged at about 2 μL/min (40) by a fluid flow from the ciliary epithelium through the posterior and anterior chambers and out of the eye via one of two pathways. The ciliary epithelium consists of a bilayered membrane that covers the aqueous-facing surface of the ciliary body and its processes. This epithelium consists of a nonpigmented epithelium that abuts the aqueous, and a pigmented cell layer that faces the blood vessels and ground substance of the ciliary body. The arrangement of the cells is unique in that the cell layers join at their apical membranes; the basolateral surfaces of the nonpigmented cells thus face the aqueous humor while the basolateral surfaces of the pigmented cells face the ciliary body matrix. Despite this unusual physical arrangement, the cells act as a syncytium through cell–cell connections via gap junctions as revealed by electron microscopy (41) and intracellular dye and electrical signal transfer in all directions among the cell bilayers (42). The ciliary body is located at the iris root at the posterior surface of the iris (Fig.

1), but exceptions, such as the rabbit, exist where extended processes occur on the posterior iris face. The ciliary body extends from the ora serrata (the anterior-most extension of the retina) to the scleral spur (located at the posterior-most part of the cornea).

The outflow paths for fluid loss from the anterior chamber are the conventional outflow via the trabecular meshwork and the unconventional route that involves fluid escape through the root of the iris and ciliary body to the suprachoroidal space. From this latter site, aqueous humor exits the eye either by diffusion across the sclera (43) or it is absorbed by choroidal vessels with loss via the vortex veins (44–47).

The trabecular meshwork consists of a matrix of glycosaminoglycans and cells that secrete these components. The cells are located on the trabecular beams. These connective tissue beams consist of collagen and elastic fibrils enveloped with endothelial cells that secrete the glycosaminoglycans. This structure is located at the angle between Schwalbe's line, the most posterior region of the cornea, and the scleral spur with many labyrinthine passages between the trabeculae leading from the anterior chamber to the aqueous-facing epithelial lining of Schlemm's canal. The meshwork is organized such that it becomes more compacted the closer it is to Schlemm's canal. Fluid then passes from this circumferential channel to episcleral and aqueous veins before entering the venous circulation. The major site of resistance to fluid outflow lies in the juxtacanalicular tissue and the adjacent inner wall of Schlemm's canal (48). The rate of fluid exit from the eye is termed outflow facility, which is the reciprocal of the resistance to fluid flow via this pathway. It is reported in units of μL/min/mmHg.

The iris consists primarily of connective tissue, radially oriented dilator muscle, and circular constrictor muscle that regulate pupil diameter in response to different stimuli (light, drugs, etc.). This tissue is well vascularized and contains a large number of relatively leaky vessels.

## II. CHANGES WITH AGING

### A. Cornea

Several changes occur in the anterior eye segment in parallel with aging. There are alterations in the physical dimensions of Bowman's and Descemet's membranes in humans, both increasing in thickness up to at least double the thickness that is seen at early maturity (49,50). Correspondingly, there is a loss of human corneal endothelial cells with age (51–55). A similar cell loss also occurs in several species including dog, rabbit, cat, cow, pig, deer, monkey, and rat (56–62).

The rate of cell loss in humans occurs at a steady rate, with a decrease in cell count from about 3500 cells/mm$^2$ at 10 years of age to about 2300 cells/mm$^2$ at 90 years of age (approximately a total 32% decrease at 4% per decade). There is a rapid decline in cell count from birth through 10 years (63) and, even before birth, the endothelial cell density declines from about 10,000 cells/mm$^2$ to 6000 cells/mm$^2$. After birth the number of cells shows a 13% loss from 5 through 7 years of age and a 12% loss from 7 to 10 years of age (64). The early rapid decrease in cell density is caused by the growth of the cornea and the change in size, whereas later in life (beyond 10 years of age) the cell count decrease occurs due to cell loss. The endothelium does not regenerate in humans, hence when cells are lost they are not replaced and nearby cells expand to cover the area. Accelerated cell loss occurs over long time periods following either anterior segment surgery such as cataract extraction (65,66) and corneal transplantation (67,68), or long-term contact lens wear (69,70).

Together with this change in cell population is an alteration in endothelial cell morphology, becoming less regular with increased polymegethism (variations in cell area) and pleomorphism (greater variation in cell shape) and an increased permeability to fluorescein (63,71,72). Carlson et al. (71) found a 28% decrease in human endothelial cell density from 5 to 80 years with a simultaneous increase in cell area from 280 $\mu m^2$ to 360 $\mu m^2$ and a 23% increase in fluorescein permeability. Hexagonal-shaped cells, which constitute the majority of cell shapes (32), decreased by 14% (from 74% to 60% of the cell total), while pentagonal and heptagonal cell shapes increased by 50% and 40%, respectively. The adjustment in cell shape and area is not surprising given that there is spreading and expansion of cells to replace those that are lost.

The endothelial fluid pump also showed decreased function with aging in humans as determined from deswelling rates after corneal (stromal) swelling was induced under a hard contact lens of low oxygen diffusibility (73,74). The ion pump site density, on the other hand, measured by ouabain binding, showed a constant value with age (75). The recovery to preswollen corneal thickness was much faster in young, 24-year-old persons than in 72-year-old persons (207 $\pm$ 42 min vs. 452 $\pm$ 117 min) (72). Recovery rates showed a similar tendency in other groups of subjects (73), 35.6 μm/h for a young age group (27 years) and 26.5 μm/h for an older age group (67 years). This functional change appears to parallel the morphological changes in hexagonal-shaped cells. While the functional changes may not result in steady-state corneal thickness alterations, physiological changes are revealed when the corneal endothelium is placed under stress such as hypoxia. Therefore, older endothelia appear to be more subject to decompensation than younger endothelia.

Donor human corneas were studied in vitro for endothelial viability with permeability measured by electrical conductivity and ion pump activity measured by short-circuit current (76). Donor age was correlated with some loss of ion pump activity but endothelial permeability remained constant. These results differ somewhat from in vivo studies in humans and may reflect the difficulties associated with the use of stored corneas for studies of physiological processes.

Corneal epithelial permeability studies using fluorophotometry indicate greater fluorescein penetration from the tear film into the stroma with age. The increase in epithelial permeability possibly represents a subclinical increase in barrier function (77), although others (78) attribute the increased penetration of dye with age to greater contact time of the eye drop with the cornea.

## B. Aqueous Humor Dynamics

Other less discernible changes in humans include an increase in intraocular pressure at about 0.1 mmHg/year, or about 0.6% year from a baseline value between 15 and 20 mmHg (40,79–81). Together with this change there are alterations in other parameters of aqueous humor dynamics as a function of age (82,83). These alterations include a reduction in true outflow facility via the trabecular meshwork that is a measure of the ease with which fluid exits the eye (82,84–86) and pseudofacility (82). [Pseudofacility is a measure of the effect of changes in intraocular pressure upon the flow of aqueous humor into the eye (87–91).] The volume (1 to 2 μL/year or 0.6%/year) and depth (0.01 mm/year) of the anterior chamber decrease with age (due to changes in the lens surface), as does the aqueous humor flow rate (0.006 μL/min/year or about 0.4%/year) (40,92–94). Of these alterations, probably the most important is the decrease in outflow facility since this will lead to an increased intraocular pressure as long as aqueous humor formation remains constant.

Baseline outflow facility declined with age in humans (95), but the decrease in intraocular pressure and the response of outflow facility to pilocarpine did not. The facility increase to pilocarpine was independent of age. There is probably an age-related loss of ciliary muscle mobility that compromises the baseline function of the trabecular meshwork. Pilocarpine increases outflow facility through contraction of the ciliary muscle that, in turn, leads to an alteration in the configuration of the trabecular meshwork to allow greater fluid outflow (96). Baseline outflow facility also declined with age in rhesus monkeys (97) with the average facility in young animals being 50% greater than in older monkeys. After atropine treatment, outflow facility decreased by 25% in young animals but no change occurred in older animals. Presumably there are changes in the ciliary muscle with age that prevent the normal muscle movement and its subsequent effects on the trabecular meshwork.

### C. Trabecular Meshwork

The structural changes that occur in the trabecular meshwork with aging show a continuum with glaucomatous changes. This is not surprising given the substantially greater occurrence of primary open-angle glaucoma with increasing age (79,98–101). The disease-induced pathological changes seen in glaucomatous eyes result in dimensions that deviate from eyes in which aging occurs as a single event (102–104). The anterior chamber depth is significantly shallower than in age- and sex-matched normal aging subjects. It would seem, therefore, that eyes with glaucoma have different dimensions than those changes caused by aging alone (103).

Coupled with the functional change, there are structural and biochemical alterations in the human trabecular meshwork that include cell loss and a redistribution and a different mix of glycosaminoglycans (105–119). The cell population of the trabecular meshwork decreases with normal aging and the loss is more severe in primary open-angle glaucoma (110,120,121). Human trabecular meshwork cells can be stimulated to undergo apoptosis through the Fas/FasL pathway, providing evidence that cell death in these cells may be under some regulation (122). These cells also have functional growth factor receptors that may play a role in governing biochemical changes in glycosaminoglycan production (123,124). The level of TGF-$\beta_2$ is increased in the aqueous humor of open-angle glaucoma patients and enhances the synthesis and secretion of extracellular matrix molecules by trabecular cells as well as decreasing the proliferation of these cells (125). The receptors may also be altered by the microenvironment surrounding the cells, such as aqueous humor containing free radicals or free radical products. This influence could change the cellular responsiveness to growth factors and lead to the type of changes seen in glaucoma.

Examination of the glycosaminoglycans of normal and primary open-angle glaucoma trabecular meshwork indicated that there was a depletion of hyaluronic acid and an accumulation of chondroitin sulfates, especially in the juxtacanalicular region immediately adjacent to Schlemm's canal, which may account for the increase in aqueous humor outflow resistance in glaucoma (116,117). A similar pattern of increased deposits was found for fibronectin in the trabecular meshwork of glaucomatous eyes (126). Ascorbate was found to increase the production of fibronectin by cells from bovine trabecular meshwork, thus suggesting that it may be a factor in regulating the fluid outflow pathway of the eye (127).

Experimentally, artificially increasing the intraocular pressure in monkey eyes even for the relatively short time of 1 h has been associated with changes in the distribution of carbohydrate-rich materials in the trabecular meshwork (128). This result supports the findings in eyes with open-angle glaucoma that also reveal a redistribution of materials in the trabecular meshwork. Studies such as those described below, but on normotensive glaucoma eyes where no elevation of intraocular pressure occurs in spite of the other glaucomatous changes in the eye, might reveal the role of intraocular pressure alone on the trabecular meshwork.

As outflow facility declined with age, several morphological changes occurred as revealed by electron microscopy (129). The thickness of the cornea-scleral meshwork membrane, as well as the fiber sheath of the corneo-scleral meshwork increased with age. Age also revealed an accumulation of amorphous material underneath and within 10 μm of the trabecular wall endothelium. On the other hand, McMenamin and Lee (107) found in an electron microscopic study that the amount of amorphous ground substance decreased with age while the percentage area occupied by electron dense material increased. Fine et al. (130) identified some basic changes between glaucomatous and aging eyes. Two major differences were found in glaucoma; first, there was collapse and accretion of uveal meshwork, and second, proliferation of endothelial connective tissues into Schlemm's canal (131).

Rohen et al. (132) examined specimens of trabecular meshwork obtained from cases of untreated primary open-angle glaucoma. They found a decrease in uveal meshwork cellularity, thickened trabeculae, and increased amounts of plaque material in the cribriform (or juxtacanalicular) layer. The amount of sheath-derived plaque material in the inner and outer walls of Schlemm's canal was greater than in normal controls of a similar age. When the aging changes become excessive, glaucomatous changes occur in the eye that both morphologically and biochemically may reflect changes in both the quantity and distribution of glycosaminoglycans. These changes correlate with the decrease in outflow facility thus linking morphological, physiological, and biochemical responses.

Recently, evidence has been presented that indicates that the tight junctions of the endothelial cells lining Schlemm's canal may play a significant role in governing the flow of fluid from the eye (133). These studies showed that the cells had altered tight junctions when exposed to dexamethasone that correlated with increased intraocular pressure changes (steroid glaucoma) seen after in vivo dexamethasone treatment. The increased resistance to flow across the endothelial cells in culture paralleled a severalfold increase in the expression of the junction associated protein Z0-1. Whether the tight junctions will show a relationship to aging or glaucoma remains to be determined.

## D. Glaucoma

Glaucoma, the most common optic neuropathy, defines a group of conditions that produce characteristic optic nerve and visual field changes. Axonal damage at the optic nerve head cribriform plate results in death of retinal ganglion cells and cupping of the optic disk. A major risk factor for the development of glaucoma is an elevated intraocular pressure. In primary open-angle glaucoma, the most common form of this condition, changes in the trabecular meshwork produce a reduction in aqueous humor outflow facility and an increase in intraocular pressure with subsequent optic nerve damage. The severity of optic nerve damage is correlated with changes in the trabecular meshwork that involve an increase in the amount of plaque material in the cribriform or juxtacanalicular region of the trabecular meshwork (134). Increased intraocular pressure is thought to cause ischemia which, in turn, leads to glutamate release that is responsible for neuronal death in ganglion cells containing N-methyl-D-aspartate receptors. Cell death is accompanied by calcium influx into the cells, activation of nitric oxide synthases, and free radical generation (135–138). It is entirely possible that antioxidants, acting against reactive oxidant species and/or nitric oxide, may participate in a neuroprotective role in retinal ganglion cells.

The incidence of glaucoma increases markedly with age from a value of 1% of the population over the age of 60 years to 9% over 80 years of age. Glaucoma is more prevalent in African-Americans and African-Caribbeans than in Caucasians by severalfold (139–141), thus race plays a significant role in determining the occurrence of this disease. The three major risk factors for glaucoma, therefore, are age, race, and elevated intraocular pressure.

### E. Blood-Aqueous Barrier

The blood-aqueous barrier permeability is apparently unaffected by aging when determined by fluorescein entry into the eye after intravenous administration (142). Nevertheless, there are some structural changes in the ciliary body that occur with age. Alterations occur in the nonpigmented ciliary epithelium and the inner limiting membrane over the entire ciliary body (143). There are minor cellular changes in the side walls of the ciliary processes that develop irregularities of the basal cell surface. This surface faces the posterior chamber. Flocculent deposits have also been noted both in intercellular spaces and the basement membrane of the ciliary epithelium (144,145).

Increasing age has been associated in other studies with increased fluorescein concentrations in the aqueous humor following intravenous dye injection (146). The time needed for dye to reach the pupillary margin of the iris from the limbal vessels increased linearly with age at 0.15 s/year. The increase in aqueous humor dye concentration with age reflects an increase in permeability of both iris and ciliary body blood vessels to fluorescein. In addition, aqueous humor protein concentration increases with age (147), indicating either an increased leakiness of the ciliary body or a decrease in aqueous flow rate. The latter is known to occur but the change appears too small to account for the amount of increase in protein concentration.

### F. Aging/Glaucoma

It is apparent, therefore, that aging of the eye is accomplished by both morphological, biochemical, and functional changes in several tissues. These alterations occur more quickly and to a greater extent in glaucomatous eyes, indicating that glaucoma, perhaps, is an acceleration of the aging process. It is of interest that all tissues perfused by aqueous humor show some form of aging change, implicating the latter fluid as being the carrier of a compound, or compounds, that contributes to the aging process.

## III. AQUEOUS HUMOR COMPOSITION

Aqueous humor composition in adult humans has been well documented (148), although variation with age has not been studied in detail. Aqueous humor composition is influenced by the lens and vice versa (149,150). Aqueous humor glucose concentration has been found to decrease with age (151) coupled with a decrease in the plasma:aqueous ratio. Several specific chemicals (urea, calcium, phosphorus) that show age-related changes have been noted (152). Bruun Laursen and Lorentzen (153) found no correlation between age and pyruvate or citrate concentrations or aqueous:plasma ratios of glucose, pyruvate, or citrate in humans; however, the number of samples was small making it difficult to assess the data with confidence.

Glutathione is present in the aqueous humor at concentrations ranging from 2 to 17 µmol/L with most being in the reduced state (154). This antioxidant is also present in the corneal endothelium where about 10 to 15% is in the oxidized form (155,156). Glutathione is taken up by the endothelium by a saturable process that is inhibited in galactose-fed guinea pigs. The uptake was similar in control corneas of 10-day- and 4-week-old animals but a significant decrease in uptake (about 80%) occurred in normal 9-month-old guinea pigs. The latter aging effect may play a role in the development of corneal pathology associated with aging that could be caused by decreased intracellular glutathione (157).

Ascorbate is present in the aqueous humor of many mammals at a concentration of about 1 mM (158–160). The ascorbic acid concentration of the aqueous humor is about 25 to 40 times greater than plasma levels in those species in which aqueous ascorbate is elevated, and is sustained by an active transport process located in the ciliary epithelium (161–164). The aqueous humor ascorbate concentration is higher in animals considered to be diurnal compared to nocturnal animals. Nocturnal animals, however, have higher concentrations of the antioxidants glutathione and cysteine than diurnal species (165). This would suggest that ascorbate may play a protective role in those species most exposed to light (166) while other protective mechanisms have evolved in nocturnal animals. Ascorbate is regenerated by bovine iris-ciliary body from its oxidative byproduct, dehydroascorbic acid. This conversion is enzymatically driven and is dependent upon reduced glutathione concentration (167). Ascorbate serves two roles in the anterior chamber of the eye—first as an antioxidant that can offer protection against light-induced damage, and second as a source of oxidants.

Coincident with ascorbic acid is hydrogen peroxide, which has been reported to occur at concentrations of either 0.5 to 5 µM (168–170) or between 20 and 50 µM (171–176) with even higher concentrations in cataractous eyes (171). The variation in the reported hydrogen peroxide concentration is dependent upon the method employed, with the higher range being detected with the dichlorophenol-indophenol method or by a radioisotopic determination. It has been reported that the value of 0.5 µM is the lower detection limit of the dichlorophenol-indophenol method which generates peroxide during the analysis procedure (177). Another report (170) indicated that a method based on the decarboxylation of α-[1-$^{14}$C-]ketoglutarate also gives a value in the higher range, whereas methods based on phenol red (178), a titanium porphyrin complex, or on the oxidation of ferroammonium sulfate (170) gave values in the range of 1 µM for bovine and human aqueous humor hydrogen peroxide concentration. These findings necessitate a reevaluation of the hydrogen peroxide concentration in the aqueous humor of normal and diseased eyes.

Nevertheless, a linear relationship exists between the concentrations of ascorbate and hydrogen peroxide in the aqueous humor (174,175,179,180). Ascorbate directly modulates the hydrogen peroxide concentration by acting as a source of this chemical. A constant source of hydrogen peroxide is thus available in the anterior eye segment, regardless of either other peroxides or free radical species generated by inflammatory cells or by photochemical reactions.

Investigations of ascorbate levels in primary open-angle glaucoma have revealed that the concentration was decreased in this disease (181). In addition, products of lipid peroxidation were increased. It is suggested that increased lipid peroxidation is linked to the decrease in the level of the antioxidant ascorbate and to the pathophysiological changes in glaucoma. In hereditary buphthalmic rabbits that have an elevated intraocular pressure and low outflow facility, there is a low ascorbate concentration in the aqueous humor (182), suggesting that ascorbate is related to an attenuation in outflow and movement of fluid in the anterior chamber. This may be similar to the increased fibronectin caused by ascorbate in experimental animal tissue (127) or in trabecular meshwork specimens of open-angle glaucoma patients (126). The link, if any, between ascorbate and trabecular meshwork constituents is worthy of further study.

Hydrogen peroxide is one of the products of free radical metabolism with superoxide dismutase catalyzing the dismutation of superoxide anion to hydrogen peroxide, and catalase and peroxidases such as the selenoenzyme glutathione peroxidase breaking down hydrogen peroxide into water (183,184). Any accumulation of hydrogen peroxide would result in the formation of hydroxyl radicals that can lead to lethal injuries (185). The continuous presence of hydrogen peroxide in the anterior segment offers one of the products of free radical metabolism bathing the corneal endothelium, iris, ciliary epithelium, and trabecular meshwork (and the lens). The aqueous humor is a source of nutrients and a means of disposing of metabolic waste products, especially for the avascular cornea and lens. It is apparent that it also provides a source of oxidants that can contribute to the aging process.

## IV. EXPERIMENTAL STUDIES

### A. Cornea

Corneal endothelial cells are known to contain catalase and the constituents of the glutathione cycle (172,174,175). These enzyme systems are capable of degrading hydrogen peroxide to water and oxygen. In endothelia of young rabbits (4 to 6 weeks old) there is 50% more catalase activity than in the endothelial of adult rabbits (3 to 6 months old) (173,175). Normal young animals receiving hydrogen peroxide injected into the anterior chamber to an initial concentration of 3.6 mM revealed only minor alterations in the morphology of the corneal endothelium. Adult animals receiving hydrogen peroxide into their anterior chamber at the same concentration as in young animals revealed substantial changes in the structure of the corneal endothelium. Cells became disrupted, contained many vacuoles, and showed a posterior surface that was uneven in contrast to the normally flat posterior endothelial surface (186). Concurrent with the morphological changes was a marked increase in ionic and nonionic permeability of the endothelium (187). A functional change thus paralleled the morphological changes.

The half-life of hydrogen peroxide in the anterior chamber of young rabbits was found to be 94 s while that of adult rabbits was 109 s (188) with baseline values reached by about 10 min. There is, therefore, a rapid degradation of exogenous hydrogen peroxide to the normal aqueous humor concentration of 25 to 50 μM in this series of experiments. The difference in half-life offers a comparison between the experimental and clinical findings. Clinically, endothelial cell loss occurs with age and experimentally the tissues increase in sensitivity to hydrogen peroxide due to a decrease in catalase activity. If the same process exists in human as in animal tissue, then a ready explanation is evident for the loss of corneal endothelial cells and the increase in sensitivity to corneal decompensation.

Photochemical damage to the corneal endothelium was demonstrated by perfusing the posterior surface of the isolated rabbit cornea with rose bengal at $10^{-6}$ M with exposure to light. The corneal endothelium became impaired, either through increased leakiness or decreased pump activity, as judged by an increased corneal thickness (189–192). Phototoxicity was also observed following corneal endothelial perfusion with chlorpromazine (193), trifluoperazine (194), and hematoporphyrin derivative (195) when simultaneously exposed to light. Potassium superoxide (196) at concentrations of 0.5 mM and above also resulted in corneal swelling caused by severe anatomical and physiological alteration of endothelial cells. Catalase offered protection but the toxic effect was unaltered by superoxide dismutase. Through parallel studies using coperfusion of the endothelial surface of isolated corneas with scavengers such as catalase, superoxide dismutase, ascorbate, EDTA, and mannitol in the presence of photoxicants and potassium superoxide, hydrogen peroxide was identified as the agent causing corneal swelling (184,197). The threshold for rabbit corneal endothelial toxicity to hydrogen peroxide is 50 μM in the absence of glucose (198) and between 300 and 400 μM in the presence of glucose (190,191). Active transport processes are affected first (197), while passive permeability is affected at higher concentrations (187,191). Following a series of experiments, it was concluded that photodynamically induced corneal swelling was not the result of failure of the glutathione redox system (199).

The corneal epithelium also has been shown to contain catalase (200,201) and superoxide dismutase (202–204). These enzymes, located in the outermost layer of the cornea, offer protection against oxidative insults reaching the tissue through the environment. Antioxidant enzymes are also present in the tear film (205) that forms a thin layer (7 to 10 μm) on the outer surface of the corneal epithelium and thus faces the environment. In addition, antioxidants such as ascorbic acid, glutathione, and cysteine are present in the tear film of humans (206). The presence of these compounds offers protection against potential damage from radiation, oxygen toxicity, and environmental chemicals. It has been shown that patients with dry eye show a significantly higher than normal level of lipid peroxides as well as a greater inflammatory activity in their tears compared with healthy control subjects. The authors concluded that both oxidative tissue damage and polymorphonuclear leukocytes occurred in the tear film or corneal epithelium indicating an oxidative potential in the tear film of dry eye patients (207).

### B. Aqueous Humor Dynamics

Experimental studies on intraocular pressure after anterior chamber injection of hydrogen peroxide into young and adult rabbits indicate that the aqueous dynamics response is very similar. No significant difference occurred in the intraocular pressure response between the two age groups. The recovery of normal intraocular pressure, however, was slower in adult rabbits (173). The reduced recovery rate is presumably the result of both the greater amount of damage to all tissues in the anterior segment caused by anterior chamber hydrogen peroxide and the decrease in healing rate with increasing age (208).

### C. Trabecular Meshwork

Exposure of tissue-cultured human trabecular meshwork cells to hydrogen peroxide caused substantial changes in the cellular morphology (209,210) and in the adhesion of trabecular cells to extracellular adhesion proteins (211). These changes have not yet been associated with alterations in the secretory pattern of the glycosaminoglycans that constitute portions of the trabecular meshwork or the tight junctions of the endothelial cells of Schlemm's canal (133), but are highly suggestive of accompanying functional change. The loss of cellular adhesion may relate to the greater cell loss with increasing time as the effects of prolonged exposure to hydrogen peroxide become manifest. Some index of functional change in outflow facility through the trabecular meshwork has been found in enucleated pig eyes (212), where oxidative damage caused by hydrogen peroxide resulted in an increased outflow facility at normal intraocular pressures, and decreased outflow at higher than normal pressures.

Studies have indicated that glutathione may not participate directly in regulating aqueous humor outflow but can protect the trabecular meshwork against hydrogen peroxide–induced oxidative damage. The enzymes glutathione peroxidase and glutathione reductase have both been shown to be present in calf trabecular meshwork (213,214). The trabecular meshwork, therefore, possesses mechanisms to detoxify hydrogen peroxide in order to protect against oxidative damage (215,216). Similarly, catalase and superoxide dismutase exist in cell-free extracts of calf trabecular meshwork (217). That these enzymes play a role in removing hydrogen peroxide from perfused solutions has been demonstrated experimentally.

The rate of removal of hydrogen peroxide was determined by harvesting the outflow fluid from the cut ends of aqueous veins following perfusion of hydrogen peroxide–containing solutions through the anterior chamber of enucleated calf eyes. At 1 mM hydrogen peroxide in the anterior chamber, the amount of oxidant in the emergent fluid was undetectable. At 5 mM, the concentration of hydrogen peroxide in the emerging fluid was 150 to 1000-fold less than the anterior chamber concentration. 3-Amino-triazole (3AT), an inhibitor of catalase, reduced the rate of peroxide removal, with a maximal inhibition of about 50%. Depletion of glutathione with diamide and inhibition of glutathione reductase with 1,3-bis-(2-chlorethyl)-1-nitrosourea (BCNU) caused a marked increase in outflow facility upon perfusion of the anterior chamber with 10 mM cysteamine (218), again indicating a role of trabecular meshwork glutathione in modulating oxidants.

Since oxidative stress to the trabecular meshwork may contribute to changes seen in aging and primary open-angle glaucoma, studies were made to determine if human or monkey trabecular meshwork cells expressed αβ-crystallin. This small heat-shock protein might be overexpressed under stress conditions such as exposure to hydrogen peroxide (219,220). Cell cultures were exposed to 200 μmol hydrogen peroxide for 1 h; at 4 or 7 days after the oxidative stress, the cells showed an increase in β-crystallin mRNA. The authors concluded that overexpression of αβ-crystallin might be an important mechanism for trabecular meshwork to prevent cellular damage associated with oxidative stress. It is of interest that this stress protein has been found to occur in greater amounts in trabecular meshwork specimens from primary open-angle glaucoma patients (221). This finding suggests that oxidative stress may be a contributory factor in the development of open-angle glaucoma.

Oxidative stress, in the form of hydrogen peroxide exposure of tissue-cultured human trabecular meshwork cells, has been shown to cause production of a trabecular meshwork-induced glucocorticoid response protein that is the product of a newly identified gene, trabecular meshwork inducible glucocorticoid response (TIGR). The very large, progressive induction of intracellular protein related to TIGR suggested that it should be considered a candidate gene for outflow obstruction in glaucoma (222). The TIGR gene mapped to the center of the GLC1A locus on chromosome 1 (223), which was the location of the myocilin gene and the two genes have been found to be identical, producing the same intracellular protein (224,225). Mutations of this gene have been identified as being involved with primary open-angle glaucoma (223,226). The TIGR stress protein, found in trabecular meshwork cells, is more pronounced in samples from primary open-angle glaucoma patients (221), indicating a possible connection between the manufacture of the protein, as a cellular protection, and environmental insult (elevated intraocular pressure; hydrogen peroxide; free radicals). Modulation of gene expression by reactive oxygen species (including superoxide anion, hydrogen peroxide, hydroxyl radical, or singlet oxygen) has been identified in a number of studies, usually as a downregulation (227). Other reactions are quite possible, however, and these may be the prevailing action with the interaction between oxidative stress and TIGR.

Clinical studies have shown that the level of lipid peroxidation products, specifically malonic dialdehyde, is more than twice as large in aqueous humor of patients with advanced glaucoma compared with earlier stages of the disease (228). The total antioxidant activity of aqueous humor was significantly lowered in glaucoma patients compared to normals. Other studies have found a relationship between vitamin E and open-angle glaucoma (229). While serum vitamin concentrations were the same in the two study groups, the aqueous humor concentration was decreased in glaucomatous eyes, suggesting that there is a lower scavenger activity in glaucoma.

After the oral administration of vitamin E to patients and animals, enhanced levels of reduced glutathione were found in the aqueous humor of rabbits and humans (230). The percentage of reduced glutathione converted to the oxidized form was the same in vitamin E–treated and control groups. In another series with open-angle glaucoma patients, the level of sulfhydryl groups was significantly lower in the aqueous humor compared to nonglaucomatous cataract patients. Tissue samples of the trabecular area of glaucoma patients also showed a lower content of sulfhydryl groups (231). These studies indicate that changes in glutathione concentration may play a role in the pathogenesis of glaucoma.

Given the relatively limited information on the trabecular meshwork, it is possible that damage due to oxidative agents and other free radicals is responsible for either the loss of, or change in function of, the meshwork cells. Not only might the physical pathway for fluid outflow from the eye be altered but the secretory pattern of glycosaminoglycans that constitutes the ground substance of the meshwork matrix may also be altered. These changes could lead to those seen in primary open-angle glaucoma and aging (106), which include a decrease in outflow facility or increase in resistance to fluid outflow from the eye.

## D. Iris/Ciliary Body

The iris/ciliary body of the rabbit also demonstrated the presence of catalase (232), which decreased with increasing age while superoxide dismutase activity was unchanged (175). Under these conditions, the production of hydrogen peroxide would be sustained while the capacity to detoxify the peroxide would be decreased; an increased peroxide concentration could result. After anterior chamber hydrogen peroxide injection, the ciliary epithelium of young rabbits showed little difference from control, water-injected animals. After pretreatment with 3AT, however, young animals responded to hydrogen peroxide in a manner similar to adult animals in that the nonpigmented epithelium became disrupted and contained many vacuoles. In addition, the ciliary processes were swollen. Adult animals treated with 3AT and hydrogen peroxide demonstrated a severe breakdown of the ciliary epithelium, including additional changes to the pigmented epithelium and cyst formation between the two cell layers that was more disruptive than in the absence of 3AT (186).

The ciliary body and iris have been shown to contain glutathione peroxidase, catalase, and superoxide dismutase and thus have the protective antioxidants necessary to detoxify reactive oxygen species (233). In the ciliary body, the nonpigmented epithelium contained significantly higher activities of catalase, superoxide dismutase, and selenium-dependent glutathione peroxidase than did the pigmented epithelium (234). Bovine and human ciliary epithelium were also found to contain glutathione peroxidase (235). The detoxification of hydrogen peroxide and organic hydroperoxides by the bovine ciliary body was dependent upon the ambient glutathione concentration (236).

### E. Catalase/Glutathione

The interplay between catalase and glutathione redox systems in the regulation of rabbit anterior chamber hydrogen peroxide has been explored. It was determined that the glutathione system was more involved at low hydrogen peroxide levels, whereas catalase became of greater importance at higher peroxide levels (174,175). Similar observations have been made in lens epithelium where catalase and glutathione both play a role in hydrogen peroxide detoxification (237,238).

Manipulations of enzyme activity of the anterior segment tissues were made using intravenous 3AT to decrease catalase activity, and intravitreal buthionine sulfoximine to inhibit glutathione synthesis. The 3AT doses used were 2, 4, or 6 mL of a 3M solution per kilogram body weight given intravenously. Changes in total glutathione did not appear to occur with aging, but the level of oxidized glutathione was significantly altered after anterior chamber hydrogen peroxide (172). This change indicates a role of the glutathione redox system in the detoxification of hydrogen peroxide. Suppression of glutathione systems with intravitreal buthionine sulfoximine at 4 mg also increased the sensitivity of the corneal endothelium to hydrogen peroxide (187,239). The passive permeability of the endothelium was increased after such inhibition. Similarly, the rabbit endothelium became more responsive to hydrogen peroxide as the tissue increased in age (184,190). Presumably this is a reflection of the natural decrease in catalase activity of the endothelium that occurs with increasing age (173,175).

Pretreatment of rabbits with 3AT resulted in substantial inhibition of catalase activity (172–175,188,240). Young animals treated with 3AT that also received anterior chamber hydrogen peroxide showed morphological changes in the corneal endothelium and ciliary body epithelium that closely resembled the tissues of an adult non-3AT-treated rabbit after hydrogen peroxide. In the presence of 3AT and hydrogen peroxide, adult tissues became worse than when receiving only hydrogen peroxide (186). The data indicate that catalase offers a protective role against hydrogen peroxide in the tissues of the anterior segment. Treatment with 3AT extends the half-life of anterior chamber hydrogen peroxide from 109 s to 147 s in adult animals treated with 2 mL intravenous 3AT, to 161 s after 4 mL of 3AT, and to 184 s after 6 mL of 3AT (188). The decrease in catalase activity with age allows hydrogen peroxide to duplicate the effects seen in adult tissues where the catalase activity is about half that of young eye tissues (172,175,241).

### F. Hydrogen Peroxide

Hydrogen peroxide decreases adenine uptake in tissues of the rabbit anterior segment (corneal endothelium, iris, ciliary body) (242) and also can cause cellular injury to tissue cultured human trabecular meshwork at high concentrations (210). A mixture of hypoxanthine and xanthine oxidase, which generates free radicals when injected into the rabbit anterior chamber, has a marked effect on iris vascular permeability to fluorescein that is significantly increased (243). Similarly, anterior chamber injections of hydrogen peroxide increased iris vascular permeability to a greater extent in adult compared with young rabbits (173). It is evident, therefore, that the presence of hydrogen peroxide in the aqueous humor on a continuous basis and at a relatively low concentration offers the opportunity for long-term effects to occur on the corneal endothelium, ciliary epithelium, iris, and trabecular meshwork.

### G. Aging

From the preceding it is evident that experimental manipulations of either oxidative compounds or the protective enzymes leads to changes in various anterior segment tissues that resemble those seen in the aging or glaucomatous human eye. This relationship supports the concept that naturally occurring hydrogen peroxide and free radicals contribute significantly to the aging process. The ability to experimentally duplicate effects in young animals that normally only occur in aged animals indicates that there is a causal relationship between the events. Much remains to be determined regarding the interaction between reactive oxygen species and aging of the anterior segment of the eye but the relationship, thus far, seems quite strong.

## V. ENVIRONMENTAL STRESSORS

### A. Ultraviolet Radiation

The cornea, together with areas of skin, is constantly exposed to ultraviolet radiation when outdoors. Reflections of light off water, snow, or sand and rocks enhances the radiation reaching the eye (244,245), indeed the incidence of cataract is increased in persons who work outdoors and in situations where light reflection is pronounced (246,247). Radiation at 400 nm can pass through the cornea, albeit at a transmission level about 60% of visible light (17), and below a wavelength of 310 to 320 nm corneal transmission decreases precipitously (248,249). Transmission through the whole cornea is at or near zero below 290 nm. Ultraviolet radiation is often considered to have two components, ultraviolet B which ranges from 290 to 320 nm, and ultraviolet A, which covers wavelengths from 320 to 400 nm. For a cornea swollen by 50%, light transmission at all wavelengths is decreased but the effects are more pronounced at shorter wavelengths. A net reduction in light transmission of about 20% occurs at 450 nm in normal thickness corneas compared to the transmission of longer wavelengths, but in corneas swollen by 50% the reduction is 40%.

Exposure to ultraviolet radiation causes a photokeratitis that is characterized by photophobia and a painful keratitis accompanied by increased lacrimation and conjunctival hyperemia. Depending upon the source of the radiation, the resulting response may be called snow blindness, welder's eye, or actinic keratitis (250,251). The effects are seen after a period of a few hours and are generally resolved within 3 to 5 days. The progressive change over a period of hours differs markedly from the effects of infrared where the response, which is thermal, occurs very rapidly (252). The slow rate of onset of changes after ultraviolet radiation is indicative of a response that reflects the involvement of systems that require time for the fulfillment of an effect (e.g., the induction of an inflammatory response).

Experimentally, many studies have been made on the effects of ultraviolet radiation on the eye, with a focus on the cornea. While the bulk of the studies have been made in the rabbit, this species has by no means been used exclusively (251). Low-to-moderate levels of ultraviolet radiation exposure produce a granular appearance of the corneal epithelium. This normally resolves in a few days. Higher levels of exposure cause residual haze of the epithelium and anterior stroma (253). These results have most often been obtained with single exposures to ultraviolet radiation, but the usual pattern of exposure in humans is of repeated exposure over many years and that has only rarely been attempted to be duplicated in an animal model. Corneal degenerations have been linked to this long-term exposure and response to ultraviolet radiation. Both acute and chronic exposure of the cornea to ultraviolet radiation produces corneal endothelial polymegethism (254). The regularity of hexagonal cells was compared between workers chronically exposed to solar radiation and an age-matched control group. Polymegethism was significantly greater in corneas that had received chronic ultraviolet exposure.

Exposure of the rabbit eye to radiation between 0.1 and 0.5 J/cm$^2$ at 300 nm caused a variety of changes in the cornea and surrounding tissues, including an increase in corneal thickness. Irradiation at less than 0.1 J/cm$^2$ and 0.16 J/cm$^2$ caused minor in vivo swelling of 30 and 70 μm, respectively. These corneas returned to normal thickness by 4 h upon in vitro perfusion in a specular microscope (255). Irradiation, however, at 0.24 J/cm$^2$ caused an in vivo swelling of 100 μm that did not reverse upon in vitro perfusion. It was concluded that the response to ultraviolet radiation reflected an increase in the leak component of the pump–leak balance that is normally responsible for the regulation of normal corneal thickness and hydration (256).

Rabbit eyes were exposed to 300-nm radiation at 0.125 J/cm$^2$ total dose and evaluated regularly over the following 336 days (251,257). At 48 h, the cornea showed an increased thickness of 20% that returned to near normal thickness by 6 days. This edema was accompanied by keratitis. Granular opacities appeared in the corneal epithelium that peaked at 72 to 96 h and resolved by 28 days. The corneal stroma, particularly the anterior stroma, also showed some disruption that began at 72 h and persisted throughout the evaluation period. Corneas were also taken for in vitro assessment of function. Large reductions were found in the rate and magnitude of deswelling; this was accompanied by large reductions in endothelial fluid pump activity, with both occurring by 36 h after irradiation. Both functions recovered to preirradiation levels by 112 days after irradiation. Scanning electron microscopy revealed corneal epithelial cellular exfoliation that peaked at 48 h and underwent considerable recovery by 96 h (258,259).

The exposure of the eye to ultraviolet radiation in experimental studies (251, 255,260) approximates those levels encountered by humans exposed to sunlight. Solar radiation delivers about 10 kJ/cm$^2$ annually at 300 nm (32,261,262), thus exposure to sunlight for 6 to 8 h a day gives a dose of 0.5 J/cm$^2$/h. With humans being in an erect posture, the value is reduced by about 80%, and this is about the threshold determined experimentally. These values are underestimates for persons who work on water or snow where reflected radiation will increase the exposure by as much as 16-fold (263). The amount of ultraviolet radiation reaching the eye is also dependent on the maintenance of the atmospheric ozone layer that offers protection from ultraviolet radiation at wavelengths below 300 nm (264,265). Exposure to ultraviolet radiation has been suggested to be the cause of the pleomorphism of the corneal endothelium (266,267). Ultraviolet B radiation

(290 to 320 nm) can, therefore, cause damage to the endothelium at a sufficiently high level of irradiance.

Ultraviolet radiation effects on the cornea have been strongly linked to reactive oxygen species. In studies using rabbits, exposure to ultraviolet radiation of 254- or 312-nm wavelength damaged the anterior segment, whereas 365 nm had no effect. Irradiation for 2 min at 254 nm caused a reduction in corneal epithelial catalase and 5 min of irradiation caused a total depletion of catalase. Catalase activity was also reduced in the corneal endothelium. The catalase changes were accompanied by decreased activities of $Na^+$-$K^+$-dependent adenosine triphosphatase and gamma-glutamyl transpeptidase. All these events occurred as the cornea increased in thickness as judged by a decrease in transparency. Irradiation with 312 nm radiation caused ocular changes, although a longer exposure time was needed to induce effects. The corneal disturbances could be prevented by dropping a catalase solution on the eye surface either during irradiation or immediately afterward. The changes in native catalase and the prevention of irradiation effects by the use of a catalase solution suggest a key role of oxyradicals in the damage to the eye by ultraviolet radiation (268,269).

Xanthine oxidase normally exists in the corneal epithelium and endothelium (270) and repeated irradiation with ultraviolet B for 5 min once daily for 1 to 4 days increased the activity of this enzyme. The application of catalase to the eye surface during irradiation attenuated the increase in xanthine oxidase activity. Pretreatment with 3AT enhanced the increase in xanthine oxidase activity. The authors suggest that xanthine oxidase is involved in the generation of reactive oxygen species in the anterior eye segment during early irradiation of rabbit eyes with ultraviolet B radiation and participates in the resultant tissue damage. Other studies have supported the involvement of free radicals in ultraviolet-induced corneal changes (271). The action spectrum for ultraviolet-induced free radical formation in human cornea indicated that opacification in actinic keratopathy proceeded through free radicals. Shimmura et al. (272) identified that hydrogen peroxide was formed by cultured corneal epithelial cells after exposure to subthreshold ultraviolet radiation. Experiments in mice exposed to ultraviolet B radiation revealed an increase in the level of hydrogen peroxide in the aqueous humor together with increases in lipid peroxides and depletion of glutathione (273).

## B. Lasers

Lasers are widely used in ophthalmology for a variety of uses ranging from the correction of mild myopia to eradication of new vessel growth in the retina. Their employment in many procedures means that the anterior segment of the eye, especially the cornea and aqueous humor, is exposed to different wavelengths of visible and nonvisible light.

Radiation from lasers at 193 nm (excimer; argon-fluoride) and 213 nm [q-switched neodymium:yttrium-aluminum-garnet, Nd:YAG; these lasers normally emit in the infrared range at about 1050 nm, the 213 nm wavelength is that of the fifth harmonic (274)] causes damage to the cornea that can include the endothelium (275–282). Considerable evidence has been accumulated to indicate the involvement of free radicals in the modulation of the tissue damage seen after laser use. This has occurred through the use of antioxidants or the measurement of lipid peroxidation products (280,282–284). Direct determinations have been made of hydroxyl radicals but the quantities were very small (285).

Jain et al. (283) performed phototherapeutic keratectomy in rabbit eyes. This procedure is designed to change the shape of the anterior cornea in order to correct myopia.

After completing a 40-μm epithelial ablation, eyes received a 1-min application of either dimethyl sulfoxide or superoxide dismutase. A 6-mm diam, 100-μm-deep anterior corneal ablation followed and corneal light scattering was measured over the course of 18 weeks. Corneal haze, indicating increased light scattering, increased to a peak at 2 to 3 weeks, with a gradual regression thereafter. There was a lesser light scattering at 3 weeks, which assumed high significance at 9 weeks. The conclusion was reached that ultraviolet-induced free radicals play a role in the pathogenesis of corneal light scattering following excimer laser keratectomy since treatment with the antioxidants dimethyl sulfoxide and superoxide dismutase decreased the appearance of corneal light scattering. Presumably the antioxidants scavenged the generated free radicals with less damage induced.

Bilgihan et al. (284) measured aqueous humor levels of lipid peroxide and superoxide dismutase in the aqueous humor of rabbits after laser photoablation of anterior stroma. Aqueous humor lipid peroxide levels were unchanged after laser treatment, but the superoxide dismutase levels decreased. These authors also suggest that free radicals formed in the cornea during laser keratectomy may be responsible for some of the complications of excimer laser surgery such as corneal opacities. The decrease in superoxide dismutase levels would allow the continuance of free radicals generated by laser treatment. The measurement of oxidized lipids in the cornea was made after excimer laser ablation of corneal epithelium and anterior stroma (280). Significant increases in both conjugated dienes and ketodienes were found, indicating the presence of lipid peroxidation in the superficial cornea. The authors concluded that the lipid peroxidation could arise from oxygen radicals generated by infiltrating polymorphonuclear leukocytes at the site of tissue damage.

Topical ascorbic acid at 10% concentration was applied to rabbit corneas following 193-nm argon excimer laser phototherapeutic keratectomy (282). Application of the drops was made every 3 h for 24 h. Comparisons were made with control laser-treated eyes that did not receive ascorbate. Lipid peroxidation and polymorphonuclear cell counts were significantly reduced in the superficial cornea of ascorbate-treated eyes compared with controls. The use of an antioxidant to reduce tissue damage could aid in minimizing postoperative stromal opacification following the clinical use of lasers to modify corneal shape. The occurrence of stromal haze due to light scattering after corneal swelling remains one of the major side effects of this ablative laser surgery.

Shimmura et al. (286) investigated the effect of excimer laser upon cultured stromal keratocytes. They found that hydroxyl radicals were formed (identified using spin-trapping) and that the number of dead cells increased after laser treatment. The number of dead cells was decreased with cotreatment with the hydroxyl radical scavenger dimethylsulfoxide or an L-ascorbic acid analog. It appears that hydroxyl radicals may, in part, be responsible for stromal keratocyte apoptosis after excimer photoablation.

The increasing use of excimer laser in the treatment of myopia through photoablation of epithelium and anterior stroma will lead to a greater occurrence of corneal changes as more surgery is performed. Because recent studies have increasingly revealed the involvement of free radicals in the generation of the tissue responses to laser radiation, there is a recognized need for antioxidant therapy immediately after treatment. Whether the source of reactive oxygen species is the residual cells of the anterior segment tissues or recruited cells such as polymorphonuclear leukocytes, antioxidant therapy should reduce the sequelae of laser treatment. These phenomena need further investigation to verify the detailed mechanisms.

## C. Uveitis

Oxidants are present in all inflamed states. Activated polymorphonuclear leukocytes undergo a respiratory burst during which molecular oxygen is reduced to the superoxide anion radical by NADPH-dependent reductase. The superoxide anion radical is dismutated either enzymatically or spontaneously to hydrogen peroxide. The peroxide in the presence of chloride and myeloperoxidase produces hypochlorous acid, a very potent bactericidal and cytotoxic agent. This response, in all likelihood, evolved for bactericidal purposes, but produces host tissue damage regardless of the stimulating source.

Uveitis, an inflammation in the eye, is characterized by an initial large influx of polymorphonuclear leukocytes and protein into the ocular fluids in response to different stimuli ranging from trauma to chemical injury. Tissue disruption frequently occurs with alterations in function. The experimental generation of superoxide in the anterior chamber of the rabbit eye, initiated by xanthine oxidase injection, caused an infiltration of leukocytes as a primary response (287,288). The oxidants produced during the respiratory burst of these cells can damage anterior segment tissues. The invasion of leukocytes usually does not occur until about 12 to 24 h after initiation of the insult (186,239,289). A decrease in ascorbate concentration in rabbit aqueous humor occurred during an experimentally induced inflammatory response that corresponded to the infiltration of leukocytes (290). The normally high ascorbate levels in the aqueous humor may afford protection against oxygen radicals released by invading leukocytes during inflammatory episodes.

The role of oxygen radicals in rat endophthalmitis was studied by adding catalase treatment to the experimental antigenic material (retinal S-antigen) injection that was used to induce uveitis (291,292). The protective effect of catalase in this experiment indicates that hydrogen peroxide or a product of peroxide may play an important role in the development of the inflammatory-induced aspects of experimental endophthalmitis. A marked reduction in experimentally induced inflammation was also found in guinea pigs after treatment with either catalase or superoxide dismutase (293). This result supports the notion that reactive oxygen species play a role in the destruction of ocular tissues and amplification of the inflammatory processes (294,295). This conclusion is supported by other studies on the protective effect of superoxide dismutase against leukocyte invasion in several forms of uveitis generated by different stimuli (296). Augustin et al. (297) created a lens-induced uveitis and challenged it with allopurinol at high concentrations. Allopurinol, acting as a free radical scavenger and antioxidant, reduced the oxidative damage as assessed by determinations of lipid peroxides.

Experimental studies were made utilizing acute corneal inflammation coupled with the topical addition of the antioxidants 0.2% superoxide dismutase and 0.5% dimethylthiourea. Corneal inflammation was induced using a 60-s application of 1N sodium hydroxide given to the cornea using a piece of soaked circular filter paper. Corneal ulcers occurred as well as an increased corneal thickness and a simultaneous loss of corneal transparency. Treatment with dimethylthiourea improved all parameters that were measured while superoxide dismutase reduced only the size of the corneal ulcers. It was suggested that antioxidant therapy be considered as complementary to the other treatment modalities for corneal ulcers (298).

Both sodium hyaluronate and hydroxypropylmethylcellulose have been shown to offer protection to the iris (299) and corneal endothelium (300) from free radical damage. Hydrogen peroxide was injected into rabbit anterior chamber and various sequelae followed, including lipid peroxidation of the iris or corneal edema. These studies, and others using hyaluronate, seem to suggest that the viscous solutions have an inherent antioxidant capability.

Experimental autoimmune uveitis was induced in rats using human S-antigen. Conjugated dienes and ketodienes were extracted from the cornea, iris-ciliary body, and lens of the uveitic eyes. Both conjugated dienes and ketodienes were increased in the cornea and ciliary body of uveitic eyes compared to controls. In eyes with uveitis, portions of the iris, trabecular meshwork, and corneal endothelium revealed peroxidized carboxyl products (301). The conclusion was reached that free radicals and lipid peroxidation products were generated in the anterior segment in uveitis. The extent of tissue damage from lipid peroxidation may represent a balance between fatty acid composition and tissue antioxidant distribution.

Evidence has been obtained by ultrastructural localization of the involvement of hydrogen peroxide in uveitis. The injection of S-antigen creates reactive oxygen metabolites that enter the surrounding tissues. Subcellular hydrogen peroxide was localized using cerium perhydroxide, and electron-dense granules were seen in the plasma membranes of leukocytes that invaded the uvea including extravascular spaces (302). This direct demonstration of the occurrence of hydrogen peroxide suggests the possibility that it is an inflammatory mediator in uveitis. Superoxide anion has also been invoked as a mediator of ocular inflammatory disease (303). Vitamin E has been shown to have a greater effect than catalase or superoxide dismutase on lens-induced uveitis in rats (304). Recent studies (305) have shown that a positively charged submicron emulsion has intrinsic free radical scavenger ability that enhanced corneal re-epithelialization after an alkali burn.

One of the sequelae of inflammation in the anterior segment is the concurrent release of eicosanoids from cell membranes (306). It has been shown that the anterior chamber injection of hydrogen peroxide caused an increase in eicosanoid concentrations in tissues and fluids of the anterior segment of the rabbit eye (307). Studies have been made on the influence of ascorbic acid on the generation of lipoxygenase products by the rabbit cornea where ascorbate inhibited corneal lipoxygenase. This result suggested that the response could be due to the antioxidant properties of ascorbate (308). The synthesis of 12S-hydroxyeicosatetraenoic acid (12-HETE), a major metabolite formed by the cornea (309–311), was significantly inhibited by physiological concentrations of ascorbic acid. Prostaglandin $E_2$ formation by the cornea was unaffected by ascorbate. The authors conclude that ascorbic acid in the aqueous humor might modulate corneal lipoxygenase activity. A similar conclusion was reached following studies using tissue-cultured corneal cells (312).

Another study of oxidants or antioxidants on arachidonate metabolism was made that evaluated the effects of hydrogen peroxide or ascorbic acid and glutathione on archidonic acid metabolism (313). Hydrogen peroxide effectively inhibited corneal, but not ciliary body, arachidonate metabolism, but this might be caused by a more efficient detoxification of peroxide by the iris-ciliary body. Ascorbate stimulated arachidonate metabolism in the iris-ciliary body whereas it reduced metabolism to HETE in the cornea while stimulating cyclooxygenase pathways. In both the cornea and iris-ciliary body, reduced glutathione inhibited arachidonate metabolism. Arachidonate metabolism in ocular tissues appears to be sensitive to the oxidative environment.

These observations have been confirmed by later studies (314), where attention was drawn to the multiplicity of responses induced by inflammation. The authors indicated that the generation of free radicals is a major factor in the impairment of function during inflammation. Whether the free radicals are generated by cells that constitute the ocular tissues or are formed by invading cells, such as polymorphonuclear leukocytes, they result in lipid peroxidation and cellular damage. These results have important implications for the treatment of uveitis.

Uveitis can have many etiologies yet, despite this, has common features such as cellular and protein invasion into ocular tissues and fluids. The cells, at least in the initial phases of the inflammation, are primarily polymorphonuclear leukocytes that are capable of a respiratory burst that is accompanied by the release of superoxide anion and other free radicals that can be converted into hydrogen peroxide and other chemicals that have a longer half-life than superoxide in the ocular tissues and fluids. Our studies (184) have shown that the corneal endothelium is particularly sensitive to hydrogen peroxide rather than superoxide anion. As such, suppression of cellular production of superoxide anions and/or hydrogen peroxide would seem to be a priority in the treatment of uveitis.

Glucose oxidase can show prolonged retention in tissues, which results in the production of hydrogen peroxide from the glucose/glucose oxidase reaction. Amidated glucose oxidase was injected into rabbit corneas and an opacification resulted in 3 to 4 days with severe corneal damage by 7 days (315). Heat-inactivated amidated glucose oxidase caused a gradual opacification after 4 days. The results are consistent with the production of oxidative metabolites, such as hydrogen peroxide, that produce stromal changes.

### D. Contact Lens Care Solutions

One of the major problems facing wearers of contact lenses is the maintenance of adequate disinfection of the lenses and neutralization of the disinfection system before reinsertion of the lens onto the corneal surface (316). Hydrogen peroxide has been shown to provide adequate disinfection of lenses when used for an appropriate number of hours as an overnight disinfectant and can be equivalent to heat disinfection. The disinfection problem is true for both hard and soft lenses but the neutralization or desorption of the disinfectant is more severe for soft lenses because of the penetration of solutes into the lens matrix during immersion of the lens in the solution. Despite lens uptake of the immersion solution, hydrogen peroxide joins the parabens and sorbic acid as candidates for an ideal disinfectant since effects of these compounds on the ocular surface are minimal or even nonexistent (317). These same disinfectants, plus low concentration benzalkonium chloride, are also used as preservative agents in topical ophthalmic drops.

Several studies have been made of hydrogen peroxide effects on the cornea when placed on the anterior corneal or ocular surface. In vitro studies using cultured epithelial cells have been found by some to show that the epithelium has only a low level of catalase and glutathione peroxidase. These results indicate that catalase prevents the lysis of cells caused by hydrogen peroxide and that the latter is the primary damaging chemical (318). In a related study, Hayden et al. (319) used a xanthine/xanthine oxidase system to generate reactive oxygen species. A chromium 51 release assay was used to determine the lysis of either rat or human cultured epithelial cells. In addition to inducing cell lysis, cell proliferation was decreased as was DNA. Partial protection was afforded by superoxide dismutase and complete protection was provided by catalase.

Other studies (320,321) have examined the toxicity of hydrogen peroxide to primary cell cultures of human corneal epithelial cells. The concentrations employed ranged from 30 to 100 ppm (approximately 1 to 3 mM). The lowest concentration of 30 ppm caused rapid cell retraction, together with cessation of mitotic activity. The formation of membranous vesicles preceded cell death which occurred at 7 to 8 h of exposure. Cell activity ceased immediately at 50 ppm with cell death at 4 to 5 h. Cell death occurred within a few minutes at hydrogen peroxide concentrations above 70 ppm.

Other in vitro studies have used perfused whole rabbit corneas where corneal thickness was measured in order to assess the impact of epithelial-side placement of hydrogen peroxide. In one study, the relative effects of a 10-min exposure to peroxide was compared to a 150-min sustained exposure (322). No swelling occurred with a 10-min pulse of up to 235 ppm hydrogen peroxide. Using the sustained application of peroxide, swelling was found between 72 and 150 ppm. The results indicate that hydrogen peroxide normally present at the epithelial surface after the placement of a peroxide soaked contact lens on the corneal surface does not represent a concentration sufficient to alter corneal function.

It has been stressed that the corneal endothelium is sensitive to hydrogen peroxide (184,323) and should peroxide pass through the cornea to reach the endothelium, corneal swelling might ensue. Contact lenses were soaked in hydrogen peroxide solutions of 1 to 20 mM (34 to 680 ppm) and placed on isolated rabbit corneas with the latter in a perfusion chamber. No hydrogen peroxide entered the artificial aqueous humor compartment up to the highest concentration used at 20 mM. In the absence of an epithelium, a transient pulse of peroxide appeared in the endothelial perfusion fluid, indicating that the cornea can rapidly metabolize hydrogen peroxide. Replacement of the contact lens with a solution containing 3 to 4 mM hydrogen peroxide led to peroxide occurring in the endothelial perfusion fluid (324), indicating that only a constant source of a high peroxide concentration could cause penetration through the cornea.

Obviously, the concentration, volume applied, and the duration of exposure will determine if hydrogen peroxide reaches the anterior chamber when placed on the epithelial surface. Given the kinetics of drop application to the ocular surface (325), it is highly unlikely that a hydrogen peroxide–containing solution would have sufficient retention time, or reach a concentration in the tear film to allow penetration into the anterior chamber. The rate of desorption of hydrogen peroxide from a soft contact lens would be slow enough that the tears would be able to dilute and remove the peroxide from the ocular surface in order to sustain a nonharmful concentration. The presence of detoxifying enzymes in the cornea and conjunctiva also make a major contribution to the maintenance of a peroxide-free solution on the ocular surface. The experiments described above indicate that with the proper use of peroxide-containing contact lens solutions, rinsing of the lens surface before reapplication to the eye and an eye with an intact epithelium would not allow the corneal endothelium or other intraocular tissues to be damaged by hydrogen peroxide.

In vivo experiments where up to 600 μL of 60-mM hydrogen peroxide was applied to the entire rabbit ocular surface, cornea, and conjunctiva, revealed that no change occurred in aqueous humor peroxide concentration. When applied only to the cornea, however, penetration occurred at 18 mM (326,327). Peroxide seems to be rapidly eliminated at the ocular surface and even if some penetration could occur into the anterior chamber there is a sufficiently high catalase activity that peroxide would be rapidly eliminated (175). The cornea and conjunctiva form a highly effective barrier that prevents both extraocular and intraocular damage from hydrogen peroxide, whether employed in contact lens solutions or as a preservative in ophthalmic medications (326). If the contact lens is rinsed correctly after disinfection, slow diffusion of peroxide from the lens into the tear film would provide only a small load for the enzymes of the corneal epithelium and conjunctiva to metabolize.

Some in vivo studies have found a change in central corneal thickness of the rabbit cornea whether applied as drops or from soft contact lenses soaked in peroxide-containing solutions (328). These authors found a small (up to 10%) increase in central corneal thickness after instillation of drops into the cul-de-sac containing either 100 ppm or 300 ppm hydrogen peroxide. The hydrogen peroxide concentration in the aqueous humor was increased by 160% after 300 ppm ($\approx$10 mM) peroxide was topically applied in drop form. Wearing of soft contact lenses that had been presoaked in peroxide caused an increase of corneal thickness of between 4.6 and 5.8%. The increase in corneal thickness and interference with the deswelling property of the cornea was interpreted as a toxic effect on the endothelial fluid pump.

A study was made in contact lens wearers, who used peroxide as a lens disinfectant, of the effect of hydrogen peroxide on the corneal barrier function. The latter was determined by fluorophotometric measurement of corneal epithelial fluorescein permeability in 30 subjects after use of the solution to disinfect contact lenses on a regular basis for 1 month. All participants had corneal epithelial permeability values in the range of non-contact lens–wearing healthy subjects. It was considered that hydrogen peroxide had no harmful effect on corneal epithelial barrier function under conditions where the solutions were handled as instructed (329).

A randomized study in human subjects determined the toxic level of hydrogen peroxide when presented to the eye from a 75% water, hydrogel contact lens. Even 800 ppm of peroxide failed to induce corneal or conjunctival staining, although the higher levels of peroxide caused conjunctival hyperemia. Residual maximal concentrations of hydrogen peroxide in contact lenses were recommended not to exceed 100 ppm (330).

In this instance of an environmental stress being deliberately applied to the ocular surface, there is more than adequate protection of the eye from oxidative damage that is afforded by the protective enzymes in the tissues. The amount of hydrogen peroxide reaching the ocular surface would be small, even from soft contact lenses that offer a source of peroxide to the tear film due to absorption of the solution into the lens matrix and its subsequent elution. The experimental data provide substantial support to the notion that the protective enzyme systems are most capable of detoxifying peroxide before it has a chance to pass through the cornea, especially when the disinfecting peroxide solution is used correctly.

## E. Miscellaneous

Studies have shown that, even in situations where free radicals would not seem to be present, free radical scavengers can provide tissue protection. Isolated perfused rabbit corneas receiving superoxide dismutase, catalase, or lipid-soluble antioxidants such as vitamin E on their endothelial surface show a doubling of survival time relative to corneas perfused in the absence of antioxidants (331,332). Vitamin E was also found to enhance corneal preservation even at low temperatures (333). Perhaps the free radical scavengers in solution are acting to replace native compounds that may be eluted from the cornea.

Corneal lesions heal at a slower than normal rate in diabetes. Corneal epithelial lesions were created in tissues isolated from normal or diabetic rats; diabetic rats were either untreated or treated with Trolox, a water-soluble vitamin E analog. Trolox-treated corneas produced a significantly smaller lesion than found in untreated diabetic rat corneas (334). This result suggests that free radicals are involved in the delay of corneal epithelial wound healing in diabetes.

Few studies have been made on the effect of environmental pollutants on the cornea or conjunctiva. The air pollutant nitric oxide produces nitrite when in aqueous solution (335). Nitrites have been shown to induce oxidation of corneal epithelial thiols and reduced glutathione especially in the presence of 365 nm ultraviolet radiation (336). Nitrite is considered to be an effective phototoxicant which has possible pathophysiological implications to external eye tissues since nitric oxide could dissolve in the watery tear film and thereby reach the epithelium. Ascorbate was effective in preventing thiol oxidation indicating that the nitrite was inducing free radical formation.

Nitric oxide (NO) is a known free radical that is formed by the action of NO synthases; at least one of these enzymes is inducible while others are constitutive. The role of NO in normal conditions and during inflammation is beginning to be unraveled in ocular tissues, but many aspects remain unidentified. Nitric oxide has been shown to play a role in the regulation of corneal thickness (337,338), possibly in the regulation of fluid flow through the trabecular meshwork especially related to open-angle glaucoma (339, 340), and anterior segment inflammation (341–347). The interaction between NO and superoxide produces the highly reactive peroxynitrite, which may also be involved in the effects of NO on ocular tissues. Much remains to be learned of the role of NO in the normal and pathophysiological functions of the anterior segment.

Cigarette smoke condensates contain polycyclic aromatics (348) that can photodynamically generate reactive oxygen species. Effects of the condensates on isolated lenses were partially inhibited by antioxidants. The effects of cigarette smoke on the cornea have been studied in vitro (349) and in vivo (350), but no positive connection was made to the role of reactive oxygen species. Another study, however, assessed cigarette smoke damage to rabbit corneal cells (351). Cells exposed to *N*-acetyl-L-cysteine showed a significant reduction in cellular damage. Buthionine sulfoximine, a glutathione synthesis inhibitor, reduced the effect of the cysteine. The authors concluded that *N*-acetyl-L-cysteine prevented damage in two ways; as a glutathione precursor and directly as an antioxidant capable of scavenging nonperoxide radicals.

The eye is somewhat unique in that drugs are frequently applied to the ocular surface in order to treat disease processes of both an extra- and intraocular nature. Frequently, the drug penetrates the cornea and enters the eye where it distributes according to its solubility in lipid membranes and binding to melanin. This presents a situation where the drug, or some metabolite, is subjected to light (including ultraviolet) and can create a phototoxic response when it is present unbound in the aqueous humor. Systemically administered drugs, such as chlorpromazine, may suffer a similar fate and subject the anterior segment tissues to free radicals when the drug enters the aqueous humor.

Examinations of environmental pollutant effects on the eye have been limited to ocular irritation effects. The involvement of free radicals in the generation of eye responses remains to be studied. The conjunctiva offers a large area for interaction with the environment (352) and is known to contain antioxidant enzymes including glutathione peroxidase, glutathione S-transferase, and superoxide dismutase (353). Taurine has been shown to protect the ocular surface from damage induced by hypochlorous acid and might be clinically useful in the treatment of oxidant-induced damage (354).

While not truly being an environmental stress, cataract removal using phacoemulsification is an intrusion upon the eye. Lens extraction by this process involves the use of a probe that vibrates at ultrasonic frequencies. During this procedure, cavitation bubbles are formed that subsequently implode. Simultaneously with the pressure changes that accompany bubble implosion, free radicals form that possibly involve the presence of singlet oxygen (355–357). The addition of superoxide dismutase offers protection against surgically induced damage (356). Free radicals, therefore, could have a significant role in the postoperative complications that occur clinically following phacoemulsification.

## VI. SUMMARY

Numerous changes occur in the anterior segment of the eye with increasing age. Many of these alterations are related to the increasing prevalence of primary open-angle glaucoma, particularly in the trabecular meshwork, where readily distinguishable morphological changes occur. Metabolic differences also occur with increasing age as antioxidant enzyme activities in several tissues of the anterior segment decrease and secretory products from cells undergo change.

Oxidative damage and antioxidant protection in ocular tissues is of increasing interest because the eye is exposed to, and susceptible to, sunlight, pollutants, oxygen, and various chemicals (358). The eye is especially vulnerable to oxidative damage due to the high level of activated oxygen species and a high concentration of unsaturated fatty acids in the ocular tissues (359). Hydrogen peroxide appears to play a central role in the aging process and its regulation is also of vital importance in pathophysiological conditions. Ascorbic acid plays a dual role since it is not only an antioxidant but also a source of hydrogen peroxide. The high concentration of ascorbate in the aqueous humor of many mammals surely plays a significant role in sustaining the peroxide levels. This is illustrated by the systemic administration of ascorbate, which not only increases aqueous humor ascorbate concentration but also hydrogen peroxide concentration. It appears that the levels of oxidants do not vary themselves, but rather that the protective mechanisms decrease with increasing age. This allows the oxidants to reveal more effects on tissues and result in pathophysiological changes that lead to events such as increased intraocular pressure, decreased outflow facility, and loss of corneal endothelial cells. Decreases in catalase and glutathione peroxidase would undoubtedly lead to increased hydrogen peroxide in the aqueous humor, as long as superoxide dismutase was not inhibited by the greater peroxide concentrations (360). The changes in the dynamics of aqueous humor flow through the anterior chamber can, if progressing quickly, lead to glaucoma. Loss of corneal endothelial cells renders the corneal endothelium more responsive to changes in the local environment, which place stress on its ability to maintain a constant corneal thickness and may lead to corneal decompensation with resultant swelling.

The relationship between antioxidant enzyme activity and the effects of hydrogen peroxide to anterior segment tissues is well illustrated in young animals that do not normally respond to exogenous aqueous humor peroxide. When catalase activity is inhibited such that its activity reflects that of an adult rabbit, the injection of hydrogen peroxide into the anterior chamber of a young rabbit induces tissue changes similar to those seen in a normal adult. This demonstrates the strong connection between enzyme activity, peroxide, and functional and morphological changes. The common factor in these effects is free radicals and their products, which appear to play a significant role in the development of aging processes. The ubiquitous presence of reactive oxygen species or their end-products in the anterior chamber ensures that all the tissues perfused with aqueous humor are exposed to these chemicals.

Oxidant by-products of normal metabolism cause damage to DNA, proteins, and lipids (361) and this is a major contributor to aging and diseases of aging. Increasing evidence from both clinical and experimental approaches supports the notion that anterior eye segment aging and some disease processes are products of reactive oxygen species and their metabolites, with an emphasis on hydrogen peroxide. The demonstrated involvement of free radicals in the generation of ultraviolet radiation damage and uveitis, together with the suggested participation in other effects in the anterior segment, is highly indicative of a strong relationship between the various events. The ability to duplicate clinical findings in experimental studies, albeit on a different time scale, and often with higher concentrations of oxidants, provides support to the concept that oxidants are involved in the pathophysiological changes in aging and glaucoma. Considerably more work is required, however, to fully explore the connections between aging and reactive radicals in the anterior segment of the eye.

## ACKNOWLEDGMENTS

The author is the recipient of a Senior Scientific Investigator Award from Research to Prevent Blindness, Inc., New York, New York. Supported in part by an Unrestricted Departmental Award from Research to Prevent Blindness, Inc. I thank Brenda Sheppard for typing the manuscript and providing valuable secretarial help, Laura McKie for drawing Figure 1, and Anastasios P. Costarides, M.D., Ph.D. and Malcolm N. Luxenberg, M.D. for their constructive input.

## REFERENCES

1. Hazlett LD, Spann B, Wells P, Berk RS. Desquamation of the corneal epithelium in the immature mouse: a scanning and transmission microscopy study. Exp Eye Res 1980; 31: 21–30.
2. Lemp MA, Mathers WD. Corneal epithelial cell movement in humans. Eye 1989; 3:438–445.
3. Shapiro MS, Friend J, Thoft RA. Corneal re-epithelialization from the conjunctiva. Invest Ophthalmol Vis Sci 1981; 21:135–142.
4. Thoft RA, Friend J. The X, Y, Z, hypothesis of corneal epithelium maintenance. Invest Ophthalmol Vis Sci 1983; 24:1442–1443.
5. Danjo S, Friend J, Thoft RA. Conjunctival epithelium in healing of corneal epithelial wounds. Invest Ophthalmol Vis Sci 1987; 28:1445–1449.
6. Fogle JA, Yoza BK, Neufeld AH. Diurnal rhythm of mitosis in rabbit corneal epithelium. Graefes Arch Clin Exp Ophthalmol 1980; 213:143–148.
7. Gloor BP, Rokos L, Kaldarar-Pedotti S. Cell cycle time and life-span of cells in the mouse eye. Measurements during the postfetal period using repeated 3H-thymidine injections. Dev Ophthalmol 1985; 12:70–129.
8. Haskjold E, Bjerknes R, Bjerknes E. Migration of cells in the rat corneal epithelium. Acta Ophthalmol (Copenh) 1989; 67:91–96.
9. Cenedella RJ, Fleschner CR. Kinetics of corneal epithelium turnover in vivo. Studies of lovastatin. Invest Ophthalmol Vis Sci 1990; 31:1957–1962.
10. Mishima S, Hedbys BO. The permeability of the corneal epithelium and endothelium to water. Exp Eye Res 1967; 6:10–32.
11. Green K. Anatomic study of water movement through rabbit corneal epithelium. Am J Ophthalmol 1969; 67:110–116.
12. Marshall WS, Klyce SD. Cellular and paracellular pathway resistances in the ''tight'' $Cl^-$-secreting epithelium of rabbit cornea. J Membr Biol 1983; 73:275–282.
13. Prausnitz MR, Noonan JS. Permeability of cornea, sclera, and conjunctiva: a literature analysis for drug delivery to the eye. J Pharm Sci 1998; 87:1479–1488.

14. Gipson IK, Spurr-Michaud S, Tisdale A, Keough M. Reassembly of the anchoring structures of the corneal epithelium during wound repair in the rabbit. Invest Ophthalmol Vis Sci 1989; 30:425–434.
15. Gipson IK. Adhesive mechanisms of the corneal epithelium. Acta Ophthalmol (Copenh) 1992; 202(suppl):13–17.
16. Cox JL, Farrell RA, Hart RW, Langham ME. The transparency of the mammalian cornea. J Physiol (Lond) 1970; 210:601–616.
17. Farrell RA, McCally RL, Tatham PER. Wave-length dependencies of light scattering in normal and cold swollen rabbit corneas and their structural implications. J Physiol (Lond) 1973; 233:589–612.
18. Farrell RA, McCally RL. On corneal transparency and its loss with swelling. J Opt Soc Am 1976; 66:342–345.
19. McCally RL, Farrell RA. Light scattering from cornea and corneal transparency. In: Masters BR, ed. Noninvasive Diagnostic Techniques in Ophthalmology. New York: Springer-Verlag, 1990:189–210.
20. Ehlers N. Variations in hydration properties of the cornea. Acta Ophthalmol (Copenh) 1966; 44:461–471.
21. Friedman MH, Green K. Swelling rate of corneal stroma. Exp Eye Res 1971; 12:239–250.
22. Cristol SM, Edelhauser HF, Lynn MJ. A comparison of corneal stromal edema induced from the anterior or the posterior surface. Refract Corneal Surg 1992; 8:224–229.
23. Friedman MH, Kearns JP, Michenfelder CJ, Green K. Contribution of the Donnan osmotic pressure to the swelling pressure of corneal stroma. Am J Physiol 1972; 222:1565–1570.
24. Riley MV. Pump and leak in regulation of fluid transport in rabbit cornea. Curr Eye Res 1985; 4:371–376.
25. Huff JW, Green K. Demonstration of active sodium transport across the isolated rabbit corneal endothelium. Curr Eye Res 1981; 1:113–114.
26. Huff JW, Green K. Characteristics of bicarbonate, sodium, and chloride fluxes in the rabbit corneal endothelium. Exp Eye Res 1983; 36:607–615.
27. Green K, Simon S, Kelly GM Jr, Bowman KA. Effects of $[Na^+]$, $[Cl^-]$, carbonic anhydrase, and intracellular pH on corneal endothelial bicarbonate transport. Invest Ophthalmol Vis Sci 1981; 21:586–591.
28. Fischbarg J, Lim JJ. Fluid and electrolyte transports across corneal endothelium. Curr Top Eye Res 1984; 4:201–223.
29. Geroski DH, Edelhauser HF. Quantitation of Na/K ATPase pump sites in the rabbit corneal endothelium. Invest Ophthalmol Vis Sci 1984; 25:1056–1060.
30. Fischbarg J, Hernandez J, Liebovitch LS, Koniarek JP. The mechanism of fluid and electrolyte transport across corneal endothelium: critical revision and update of a model. Curr Eye Res 1985; 4:351–360.
31. Kuang KY, Xu M, Koniarek JP, Fischbarg J. Effects of ambient bicarbonate, phosphate and carbonic anhydrase inhibitors on fluid transport across rabbit corneal endothelium. Exp Eye Res 1990; 50:487–493.
32. Green K. Corneal endothelial structure and function under normal and toxic conditions. Cell Biol Rev 1991; 25:169–207.
33. Bonanno JA, Giasson C. Intracellular pH regulation in fresh and cultured bovine corneal endothelium. II. $Na^+:HCO_3^-$ cotransport and $Cl^-/HCO_3^-$ exchange. Invest Ophthalmol Vis Sci 1992; 33:3068–3079.
34. Riley MV, Winkler BS, Peters MI, Czajkowski CA. Relationship between fluid transport and in situ inhibition of Na(+)-K+ adenosine triphosphatase in corneal endothelium. Invest Ophthalmol Vis Sci 1994; 35:560–567.
35. Wigham CG, Turner HC, Ogbuehi KC, Hodson SA. Two pathways for electrogenic bicarbonate ion movement across the rabbit corneal endothelium. Biochim Biophys Acta 1996; 1279: 104–110.

36. Riley MV, Winkler BS, Starnes CA, Peters MI. Fluid and ion transport in corneal endothelium: insensitivity to modulators of $Na(^{+})$-$K(^{+})$-$2Cl^{-}$ cotransport. Am J Physiol 1997; 273: C1480–C1486.
37. Bourne WM. Clinical estimation of corneal endothelial pump function. Trans Am Ophthalmol Soc 1998; 96:229–239.
38. Diecke FP, Zhu Z, Kang F, Kuang K, Fischbarg J. Sodium, potassium, two chloride cotransport in corneal endothelium: characterization and possible role in volume regulation and fluid transport. Invest Ophthalmol Vis Sci 1998; 39:104–110.
39. Lane JR, Wigham CG, Hodson SA. Sodium ion uptake into isolated plasma membrane vesicles: indirect effects of other ions. Biophys J 1999; 76:1452–1456.
40. Brubaker RF, Nagataki S, Townsend DJ, Burns RR, Higgins RG, Wentworth W. The effect of age on aqueous humor formation in man. Ophthalmology 1981; 88:283–288.
41. Raviola G, Raviola E. Intercellular junctions in the ciliary epithelium. Invest Ophthalmol Vis Sci 1978; 17:958–981.
42. Green K, Bountra C, Georgiou P, House CR. An electrophysiologic study of rabbit ciliary epithelium. Invest Ophthalmol Vis Sci 1985; 26:371–381.
43. Bill A. Uveoscleral drainage of aqueous humor: physiology and pharmacology. Prog Clin Biol Res 1989; 312:417–427.
44. Green K, Sherman SH, Laties AM, Pederson JE, Gaasterland DE, MacLellan HM. Fate of anterior chamber tracers in the living rhesus monkey eye with evidence for uveo-vortex outflow. Trans Ophthalmol Soc UK 1977; 97:731–739.
45. Sherman SH, Green K, Laties AM. The fate of anterior chamber fluorescein in the monkey eye. 1. The anterior chamber outflow pathways. Exp Eye Res 1978; 27:159–173.
46. Pederson JE, Gaasterland DE, MacLellan HM. Experimental ciliochoroidal detachment. Effect on intraocular pressure and aqueous humor flow. Arch Ophthalmol 1979; 97:536–541.
47. Toris CB, Gregerson DS, Pederson JE. Uveoscleral outflow using different-sized fluorescent tracers in normal and inflamed eyes. Exp Eye Res 1987; 45:525–532.
48. Mäepea O, Bill A. The pressures in the episcleral veins, Schlemm's canal, and the trabecular meshwork in monkeys: effects of changes in intraocular pressure. Exp Eye Res 1989; 49: 645–663.
49. Alvarado J, Murphy C, Juster R. Age-related changes in the basement membrane of the human corneal epithelium. Invest Ophthalmol Vis Sci 1983; 24:1015–1028.
50. Murphy C, Alvarado J, Juster R. Prenatal and postnatal growth of the human Descemet's membrane. Invest Ophthalmol Vis Sci 1984; 25:1402–1415.
51. Bourne WM, Kaufman HE. Specular microscopy of human corneal endothelium. Am J Ophthalmol 1976; 81:319–323.
52. Laing RA, Sanstrom MM, Berrospi AR, Leibowitz HM. Changes in the corneal endothelium as a function of age. Exp Eye Res 1976; 22:587–594.
53. Yee RW, Matsuda M, Schultz RO, Edelhauser HF. Changes in the normal corneal endothelial cellular pattern as a function of age. Curr Eye Res 1985; 4:671–678.
54. Sherrard ES, Novakavic P, Speedwell L. Age-related changes of the corneal endothelium and stroma as seen in vivo by specular microscopy. Eye 1987; 1:197–203.
55. Daus W, Volcker HE, Meysen H. Klinische Bedeutung der altersabhangigen, regional unterschiedlichen Verteilung des menschlichen Hornhautendothels. Klin Monatsbl Augenheilkd 1990; 196:449–455.
56. Bahn CF, MacCallum DK, Pachtman MA, Meyer RF, Martonyi CL, Lillie JH, Robinson BJ. Effect of age and keratoplasty on the postnatal development of feline corneal endothelium. Cornea 1982; 1:233–240.
57. Bahn CF, Glassman RM, MacCallum DK, Lillie JH, Meyer RF, Robinson BJ, Rich NM. Postnatal development of corneal endothelium. Invest Ophthalmol Vis Sci 1986; 27:44–51.
58. Fitch KL, Nadakavukaren MJ, Richardson A. Age-related changes in the corneal endothelium of the rat. Exp Gerontol 1982; 17:179–183.

59. Gwin RM, Lerner I, Warren JK, Gum G. Decrease in canine corneal endothelial cell density and increase in corneal thickness as functions of age. Invest Ophthalmol Vis Sci 1982; 22: 267–271.
60. Baroody RA, Bito LZ, DeRousseau CJ, Kaufman PL. Ocular development and aging. 1. Corneal endothelial changes in cats and in free-ranging and caged rhesus monkeys. Exp Eye Res 1987; 45:607–622.
61. Tailoi C, Curmi J. Changes in corneal endothelial morphology in cats as a function of age. Curr Eye Res 1988; 7:387–392.
62. Doughty MJ. The cornea and corneal endothelium in the aged rabbit. Optom Vis Sci 1994; 71:809–818.
63. Murphy C, Alvarado J, Juster R, Maglio M. Prenatal and postnatal cellularity of the human corneal endothelium. A quantitative histologic study. Invest Ophthalmol Vis Sci 1984; 25: 312–322.
64. Nucci P, Brancato R, Mets MB, Shevell SK. Normal endothelial cell density range in childhood. Arch Ophthalmol 1990; 108:247–248.
65. Binkhorst CD, Nygaard P, Loones LH. Specular microscopy of the corneal endothelium and lens implant surgery. Am J Ophthalmol 1978; 85:597–605.
66. Price N, Jacobs P, Cheng H. Rate of endothelial cell loss in the early postoperative period after cataract surgery. Br J Ophthalmol 1982; 66:709–713.
67. Bourne WM, Hodge DO, Nelson LR. Corneal endothelium five years after transplantation. Am J Ophthalmol 1994; 118:185–196.
68. Bourne WM, Shearer DR. Effects of long-term rigid contact lens wear on the endothelium of corneal transplants for keratoconus 10 years after penetrating keratoplasty. CLAO J 1995; 21:265–267.
69. Setala K, Vasara K, Vesti E, Ruusuvaara P. Effects of long-term contact lens wear on the corneal endothelium. Acta Ophthalmol Scand 1998; 76:299–303.
70. Bourne WM, Holtan SB, Hodge DO. Morphologic changes in corneal endothelial cells during 3 years of fluorocarbon contact lens wear. Cornea 1999; 18:29–33.
71. Carlson KH, Bourne WM, McLaren JW, Brubaker RF. Variations in human corneal endothelial cell morphology and permeability to fluorescein with age. Exp Eye Res 1988; 47:27–41.
72. Polse KA, Brand R, Mandell R, Vastine D, Demartini D, Flom R. Age differences in corneal hydration control. Invest Ophthalmol Vis Sci 1989; 30:392–399.
73. O'Neal MR, Polse KA. Decreased endothelial pump function with aging. Invest Ophthalmol Vis Sci 1986; 27:457–463.
74. Siu AW, Herse PR. The effect of age on the edema response of the central and mid-peripheral cornea. Acta Ophthalmol (Copenh) 1993; 71:57–61.
75. Geroski DH, Matsuda M, Yee RW, Edelhauser HF. Pump function of the human corneal endothelium. Effects of age and cornea guttata. Ophthalmology 1985; 92:759–763.
76. Wigham CG, Hodson SA. Physiological changes in the cornea of the ageing eye. Eye 1987; 1:190–196.
77. Chang SW, Hu FR. Changes in corneal autofluorescence and corneal epithelial barrier function with aging. Cornea 1993; 12:493–499.
78. Nzekwe EU, Maurice DM. The effect of age on the penetration of fluorescein into the human eye. J Ocul Pharmacol 1994; 10:521–523.
79. Armaly MF. On the distribution of applanation pressure. I. Statistical features and the effect of age, sex and family history of glaucoma. Arch Ophthalmol 1965; 73:11–18.
80. Hollows FC, Graham PA. Intraocular pressure, glaucoma and glaucoma suspects in a defined population. Br J Ophthalmol 1966; 50:570–586.
81. Green K, Balazs EA, Denlinger JL. Aqueous humor and vitreous production. In: Platt D, ed. Gerontology, Vol. II. Geriatrics 3. Berlin: Springer-Verlag, 1984:352–372.
82. Gaasterland DE, Kupfer C, Milton R, Ross K, McCain L, MacLellan H. Studies of aqueous

humor dynamics in man. VI. Effect of age upon parameters of intraocular pressure in normal human eyes. Exp Eye Res 1978; 26:651–656.
83. Green K, Gaasterland DE, Milton R, Bowman K. Influence of aging on aqueous humor production. Interdiscipl Top Gerontol 1978; 13:14–20.
84. Goldmann H. Abflussdruck, Minutenvolumen und Widerstand der Kammerwasser-Strömung des Menschen. Doc Ophthalmol 1951; 5:278–356.
85. Becker B. The decline in aqueous secretion and outflow facility with age. Am J Ophthalmol 1958; 46:731–736.
86. Armaly MF, Sayegh RD. Water drinking test. II. The effect of age on tonometric and tonographic measures. Arch Ophthalmol 1970; 83:176–181.
87. Bill A, Barany EH. Gross facility, facility of conventional routes, and pseudofacility of aqueous humor outflow in the cynomolgus monkey. The reduction in aqueous humor formation rate caused by moderate increments in intraocular pressure. Arch Ophthalmol 1966; 75:665–673.
88. Brubaker RF, Kupfer C. Determination of pseudofacility in the eye of the rhesus monkey. Arch Ophthalmol 1966; 75:693–697.
89. Kupfer C. Clinical significance of pseudofacility. Sanford R Gifford Memorial Lecture. Am J Ophthalmol 1973; 75:193–204.
90. Gaasterland D, Kupfer C, Ross K. Studies of aqueous humor dynamics in man. IV. Effects of pilocarpine upon measurements in young normal volunteers. Invest Ophthalmol 1975; 14:848–853.
91. Green K, Elijah D. Drug effects on aqueous humor formation and pseudofacility in normal rabbit eyes. Exp Eye Res 1981; 33:239–245.
92. Weekers R, Luyckx-Bacus J, Weekers JF. Etude ultrasonique des dimensions respectives des segments anterieur et posterieur du globe oculaire dans diverses affections genetiques. In: Oksala A, Gennet H, eds. Ultrasonics in Ophthalmology: Proceedings on the Munster Symposium. Basel: Karger, 1966:215.
93. Fontana S, Brubaker RF. Volume and depth of the anterior chamber in the normal aging human eye. Arch Ophthalmol 1980; 98:1803–1808.
94. Toris CB, Yablonski ME, Wang YL, Camras CB. Aqueous humor dynamics in the aging human eye. Am J Ophthalmol 1999; 127:407–412.
95. Croft MA, Oyen MJ, Gange SJ, Fisher MR, Kaufman PL. Aging effects on accommodation and outflow facility responses to pilocarpine in humans. Arch Ophthalmol 1996; 114:586–592.
96. Kaufman PL, Barany EH. Loss of acute pilocarpine effect on outflow facility following surgical disinsertion and retrodisplacement of the ciliary muscle from the scleral spur in the cynomolgus monkey. Invest Ophthalmol 1976; 15:793–807.
97. Kiland JA, Croft MA, Gabelt BT, Kaufman P. Atropine reduces but does not eliminate the age-related decline in perfusion outflow facility in monkeys. Exp Eye Res 1997; 64:831–835.
98. Leydhecker W, Akiyama K, Neumann HG. Der Interokulare Druck gesunder menschlicher Augen. Klin Monatsbl Augenheilkd 1958; 133:662–670.
99. Bankes JLK, Perkins ES, Tsolakis S, Wright JE. Bedford glaucoma survey. Br Med J 1968; 1:791–796.
100. Kahn HA, Liebowitz HM, Ganley JP. The Framingham eye study. I. Outline and major prevalence findings. Am J Epidemiol 1977; 106:17–32.
101. Colton T, Ederer F. The distribution of intraocular pressures in the general population. Surv Ophthalmol 1980; 25:123–129.
102. Tomlinson A, Leighton DA. Ocular dimensions in the heredity of angle-closure glaucoma. Br J Ophthalmol 1973; 57:475–486.
103. Tomlinson A, Leighton DA. Ocular dimensions and the heredity of open-angle glaucoma. Br J Ophthalmol 1974; 58:68–74.

104. Lowe RF, Clark BAJ. Posterior corneal curvature. Correlations in normal eyes and in eyes involved with primary angle-closure glaucoma. Br J Ophthalmol 1973; 57:464–470.
105. Rohen JW, Lutjen-Drecoll E. Age changes of the trabecular meshwork in human and monkey eyes. A light and electron microscopic study. Altern Entwickl Aging Dev 1971; 1:1–36.
106. Rohen JW, Lutjen-Drecoll E. Age-related changes in the anterior segment of the eye. In: Platt D, ed. Gerontology, Vol II, Geriatrics 3. Berlin: Springer-Verlag, 1984:326–351.
107. McMenamin PG, Lee WR. Age related changes in extracellular materials in the inner wall of Schlemm's canal. Graefes Arch Clin Exp Ophthalmol 1980; 212:159–172.
108. Alvarado J, Murphy C, Polansky J, Juster R. Age-related changes in trabecular meshwork cellularity. Invest Ophthalmol Vis Sci 1981; 21:714–727.
109. Horstmann HJ, Rohen JW, Sames K. Age-related changes in the composition of proteins in the trabecular meshwork of the human eye. Mech Ageing Dev 1983; 21:121–136.
110. Alvarado J, Murphy C, Juster R. Trabecular meshwork cellularity in primary open-angle glaucoma and nonglaucomatous normals. Ophthalmology 1984; 91:564–579.
111. McMenamin PG, Lee WR, Aitken DA. Age-related changes in the human outflow apparatus. Ophthalmology 1986; 93:194–209.
112. Miyazaki M, Segawa K, Urakawa Y. Age-related changes in the trabecular meshwork of the normal human eye. Jpn J Ophthalmol 1987; 31:558–569.
113. Grierson I, Howes RC. Age-related depletion of the cell population in the human trabecular meshwork. Eye 1987; 1:204–210.
114. Millard CB, Tripathi BJ, Tripathi RC. Age-related changes in protein profiles of the normal human trabecular meshwork. Exp Eye Res 1987; 45:623–631.
115. Gong H, Freddo TF, Johnson M. Age-related changes of sulfated proteoglycans in the normal human trabecular meshwork. Exp Eye Res 1992; 55:691–709.
116. Knepper PA, Goossens W, Hvizd M, Palmberg PF. Glycosaminoglycans of the human trabecular meshwork in primary open-angle glaucoma. Invest Ophthalmol Vis Sci 1996; 37:1360–1367.
117. Knepper PA, Goossens W, Palmberg PF. Glycosaminoglycan stratification of the juxtacanalicular tissue in normal and primary open-angle glaucoma. Invest Ophthalmol Vis Sci 1996; 37:2414–2425.
118. Lutjen-Drecoll E, Rohen JW. Morphology of aqueous outflow pathways in normal and glaucomatous eyes. In: Rich R, Shields MB, Krupin T, eds. The Glaucomas, 2nd ed. St. Louis: Mosby, 1996:89–123.
119. Lutjen-Drecoll E. Functional morphology of the trabecular meshwork in primate eyes. Prog Retin Eye Res 1999; 18:91–119.
120. Grierson I, Wang Q, McMenamin PG, Lee WR. The effects of age and antiglaucoma drugs on the meshwork cell population. Res Clin Forums 1982; 4:69–92.
121. Grierson I, Hogg P. The proliferative and migratory activities of trabecular meshwork cells. Prog Retin Eye Res 1995; 15:33–67.
122. Agarwal R, Talati M, Lambert W, Clark AF, Wilson SE, Agarwal N, Wordinger RJ. Fasactivated apoptosis and apoptosis mediators in human trabecular meshwork cells. Exp Eye Res 1999; 68:583–590.
123. Tripathi RC, Borisuth NSC, Tripathi BJ. Growth factors in the aqueous humor and their therapeutic implications in glaucoma and anterior segment disorders of the human eye. Drug Devel Res 1991; 22:1–23.
124. Wordinger RJ, Clark AF, Agarwal R, Lambert W, McNatt L, Wilson SE, Qu Z, Fung BK-K. Cultured human trabecular meshwork cells express functional growth factor receptors. Invest Ophthalmol Vis Sci 1998; 39:1575–1589.
125. Li J, Tripathi BJ, Tripathi RC. Increased level of TGF-β2 in patients with primary open-angle glaucoma is implicated in the pathogenesis of the disease. J Toxicol Cutan Ocul Toxicol 1999; 18:266–267.
126. Babizhayev MA, Brodskaya MW. Fibronectin detection in drainage outflow system of human

eyes in ageing and progression of open-angle glaucoma. Mech Ageing Dev 1989; 47:145–157.
127. Yue BY, Higginbotham EJ, Chang IL. Ascorbic acid modulates the production of fibronectin and laminin by cells from an eye tissue-trabecular meshwork. Exp Cell Res 1990; 187:65–68.
128. Grierson I, Lee WR. Pressure effects on the distribution of extracellular materials in the rhesus monkey outflow apparatus. Graefes Arch Clin Exp Ophthalmol 1977; 203:155–168.
129. Segawa K. Electron microscopic changes of the trabecular tissue in primary open-angle glaucoma. Ann Ophthalmol 1979; 11:49–54.
130. Fine BS, Yanoff M, Stone RA. A clinicopathological study of four cases of primary open-angle glaucoma compared to normal eyes. Am J Ophthalmol 1981; 91:88–105.
131. Yanoff M, Fine BS. Ocular Pathology. A Text and Atlas. Hagerstown: Harper & Row, 1975: 594–598.
132. Rohen JW, Lutjen-Drecoll E, Flugel C, Meyer M, Grierson I. Ultrastructure of the trabecular meshwork in untreated cases of primary open-angle glaucoma (POAG). Exp Eye Res 1993; 56:683–692.
133. Underwood JL, Murphy CG, Chen J, Franse-Carman L, Wood I, Epstein DL, Alvarado JA. Glucocorticoids regulate transendothelial fluid flow resistance and formation of intercellular junctions. Am J Physiol 1999; 277:C330–C342.
134. Gottanka J, Johnson DH, Martus P, Lutjen-Drecoll E. Severity of optic nerve damage in eyes with POAG is correlated with changes in the trabecular meshwork. J Glaucoma 1997; 6:123–132.
135. Neufeld AH. New conceptual approaches for pharmacological neuroprotection in glaucomatous neuronal degeneration. J Glaucoma 1998; 7:434–438.
136. Neufeld AH, Sewada A, Becker B. Inhibition of nitric oxide synthase 2 by aminoguanidine provides neuroprotection of retinal ganglion cells in a rat model of chronic glaucoma. Proc Natl Acad Sci USA 1999; 96:9944–9948.
137. Osborne NN, Cazevieille C, Carvalho AL, Larsen AK, DeSantis L. In vivo and in vitro experiments show that betaxalol is a retinal neuroprotective agent. Brain Res 1997; 751: 113–123.
138. Osborne NN, Ugarte M, Chao M, Chidlow G, Bae JH, Wood JP, Nash MS. Neuroprotection in relation to retinal ischemia and relevance to glaucoma. Surv Ophthalmol 1999; 43(suppl 1):S102–S128.
139. Sommer A, Tielsch JM, Katz J, Quigley HA, Gottsch JD, Javitt J, Singh K. Relationship between intraocular pressure and primary open angle glaucoma among white and black Americans. The Baltimore Eye Survey. Arch Ophthalmol 1991; 109:1090–1095.
140. Leske MC, Connell AM, Schachat AP, Hyman L. The Barbados Eye Study: Prevalence of open angle glaucoma. Arch Ophthalmol 1994; 112:821–829.
141. Quigley HA. Number of people with glaucoma worldwide. Br J Ophthalmol 1996; 80:389–393.
142. van Best JA, Kappelhof JP, Laterveer L, Oosterhuis JA. Blood aqueous barrier permeability versus age by fluorophotometry. Curr Eye Res 1987; 6:855–863.
143. Hogan MJ, Alvarado JA, Weddell JE. Histology of the Human Eye. Philadelphia: Saunders, 1971.
144. Fine BS, Zimmerman LE. Light and electron microscopic observations on the ciliary epithelium in man and rhesus monkey. Invest Ophthalmol Vis Sci 1963; 2:105–137.
145. Missoten L. L'ultrastructure des tissus occulaire. Bull Soc Belge Ophthalmol 1964; 136:3–204.
146. Van Nerom PR, Rosenthal AR, Jacobson DR, Pieper I, Schwartz H, Grieder BW. Iris angiography and aqueous photofluorometry in normal subjects. Arch Ophthalmol 1981; 99:489–493.

147. Inada K, Murata T, Baba H, Murata Y, Ozaki M. Increase of aqueous humor proteins with aging. Jpn J Ophthalmol 1988; 32:126–131.
148. PL Altman, DS Dittmer, eds. Blood and Other Body Fluids. Bethesda, MD: Federation of American Societies for Experimental Biology, 1974:478–483.
149. Kuck JFR, Jr, Croswell HH, Jr. Lens as the chief source of fructose in rabbit aqueous humor. Ophthal Res 1974; 6:189–196.
150. Hockwin O, Koch HR. Lens. In: Heilmann K, Richardson KT, eds. Glaucoma. Philadephia: Saunders, 1978:67.
151. Pohjola S. The glucose content of the aqueous humor in man. Acta Ophthalmol (Copenh) 1966; 88(suppl):11–80.
152. LeRebeller MJ, Crockett R, Maurain C, Sirieix F. Nouvelle contribution à l'étude des composants de l'humeur aqueuse chez le sujet normal par utilisation des techniques d'analyses automatiques. Arch Ophthalmol (Paris) 1976; 36:749–765.
153. Bruun Laursen A, Lorentzen SE. Glucose, pyruvate and citrate concentrations in the aqueous humor of human cataractous eyes. Acta Ophthalmol (Copenh) 1974; 52:477–489.
154. Riley MV, Meyer RF, Yates EM. Glutathione in the aqueous humor of human and other species. Invest Ophthalmol Vis Sci 1980; 19:94–96.
155. Riley MV, Yates EM. Glutathione in the epithelium and endothelium of bovine and rabbit cornea. Exp Eye Res 1977; 25:385–389.
156. Riley MV, Ng MC, Yates EM, Soppet DR, Whikehart DR. Oxidized glutathione in the corneal endothelium. Exp Eye Res 1980; 30:607–609.
157. Kannan R, Mackic JB, Zlokovic BV. Corneal transport of circulating glutathione in normal and galactosemic guinea pigs. Cornea 1999; 18:321–327.
158. Taylor A, Jacques PF, Nadler D, Morrow F, Sulsky SI, Shepard D. Relationship in humans between ascorbic acid consumption and levels of total and reduced ascorbic acid in lens, aqueous humor, and plasma. Curr Eye Res 1991; 10:751–759.
159. Shichi H, Page T, Sahouri MJ, Shin DH. Microplate assay of ascorbic acid in aqueous humor with bicinchoninic acid. J Ocul Pharmacol Ther 1997; 13:201–206.
160. Jampel HD, Moon JI, Quigley HA, Barron Y, Lam KW. Aqueous humor uric acid and ascorbic acid concentrations and outcome of trabeculectomy. Arch Ophthalmol 1998; 116: 281–285.
161. Chu TC, Candia OA. Active transport of ascorbate across the isolated rabbit ciliary epithelium. Invest Ophthalmol Vis Sci 1988; 29:594–599.
162. DiMattio J. A comparative study of ascorbic acid entry into aqueous and vitreous humors of the rat and guinea pig. Invest Ophthalmol Vis Sci 1989; 30:2320–2331.
163. Rose RC, Bode AM. Ocular ascorbate transport and metabolism. Comp Biochem Physiol A 1991; 100:273–285.
164. Mead A, Sears J, Sears M. Transepithelial transport of ascorbic acid by the isolated intact ciliary epithelial bilayer of the rabbit eye. J Ocul Pharmacol Ther 1996; 12:253–258.
165. Richer SP, Rose RC. Water soluble antioxidants in mammalian aqueous humor: interaction with UV B and hydrogen peroxide. Vision Res 1998; 38:2881–2888.
166. Reiss GR, Werness PG, Zollman PE, Brubaker RF. Ascorbic acid levels in the aqueous humor of nocturnal and diurnal mammals. Arch Ophthalmol 1986; 104:753–755.
167. Bode AM, Green E, Yavarow CR, Wheeldon SL, Bolken S, Gomez Y, Rose RC. Ascorbic acid regeneration by bovine iris-ciliary body. Curr Eye Res 1993; 12:593–601.
168. Garcia-Castineiras S, Velazquez D, Martinez P, Torres N. Aqueous humor hydrogen peroxide analysis with dichlorophenol-indophenol. Exp Eye Res 1992; 55:9–19.
169. Bleau G, Giasson C, Brunette I. Measurement of hydrogen peroxide in biological samples containing high levels of ascorbic acid. Anal Biochem 1998; 263:13–17.
170. Spector A, Ma W, Wang RR. The aqueous humor is capable of generating and degrading $H_2O_2$. Invest Ophthalmol Vis Sci 1998; 39:1188–1197.

171. Spector A, Garner WH. Hydrogen peroxide and human cataracts. Exp Eye Res 1981; 33: 673–681.
172. Csukas S, Costarides A, Riley MV, Green K. Hydrogen peroxide in the rabbit anterior chamber: Effects on glutathione, and catalase effects on peroxide kinetics. Curr Eye Res 1987; 6:1395–1402.
173. Csukas S, Green K. Effects of intracameral hydrogen peroxide in the rabbit anterior chamber. Invest Ophthalmol Vis Sci 1988; 29:335–339.
174. Costarides A, Recasens JF, Riley MV, Green K. The effects of ascorbate, 3-aminotriazole, and 1,3-bis-(2-chloroethyl)-1-nitrosourea on hydrogen peroxide levels in the rabbit aqueous humor. Lens Eye Toxic Res 1989; 6:167–173.
175. Costarides A, Riley MV, Green K. Roles of catalase and the glutathione redox cycle in the regulation of anterior-chamber hydrogen peroxide. Ophthalmic Res 1991; 23:284–294.
176. Ramachandran S, Morris SM, Devamanoharan P, Henein M, Varma SD. Radio-isotopic determination of hydrogen peroxide in aqueous humor and urine. Exp Eye Res 1991; 53:503–506.
177. Garcia-Castineiras S. Peroxido de hidrogeno en el humor acuoso: 1992–1997. P R Health Sci J 1998; 17:335–343.
178. Sharma Y, Druger R, Mataic D, Basnett S, Beebe DC. Aqueous humor hydrogen peroxide and cataract. Invest Ophthalmol Vis Sci (ARVO abstract) 1997; 38(suppl):1149.
179. Giblin FJ, McCready JP, Kodama T, Reddy VN. A direct correlation between the levels of ascorbic acid and $H_2O_2$ in aqueous humor. Exp Eye Res 1984; 38:87–93.
180. Riley MV, Schwartz CA, Peters MI. Interactions of ascorbate and $H_2O_2$: Implications of in vitro studies of lens and cornea. Curr Eye Res 1986; 5:207–216.
181. Aleksidze AT, Beradze IN, Golovachev OG. Vliianie askorbinovoi kisloty vodianistoi vlagi na protsess perekisnogo okisleniia lipidov v glazu pri pervichnoi otkrytougol'noi glaukome. Oftalmol Zh 1989; 2:114–116.
182. Lee PF, Fox R, Henrick I, Lam WK. Correlation of aqueous humor ascorbate with intraocular pressure and outflow facility in hereditary buphthalmic rabbits. Invest Ophthalmol Vis Sci 1978; 17:799–802.
183. Freeman BA, Crapo JD. Biology of disease: Free radicals and tissue injury. Lab Invest 1982; 47:412–426.
184. Hull DS, Green K. Oxygen free radicals and corneal endothelium. Lens Eye Toxic Res 1989; 6:87–91.
185. Fridovitch I. Biological effects of the superoxide radical. Arch Biochem Biophys 1986; 247: 1–11.
186. Costarides A, Birnbaum D, Csukas S, Forbes E, Green K. Morphological sequelae of anterior segment hydrogen peroxide in young and adult rabbits with or without 3-aminotriazole treatment. In: Simic MG, Taylor KA, Ward JF, Von Sonntag C, eds. Oxygen Radicals in Biology and Medicine. New York: Plenum Press, 1988:1039–1042.
187. Hull DS, Pendarvis RW, Cheeks L, Green K. Hydrogen peroxide effects on ionic and nonionic permeability of the rabbit corneal endothelium. Lens Eye Toxic Res 1991; 8:9–25.
188. Csukas S, Costarides A, Riley MV, Green K. Anterior chamber hydrogen peroxide: Effects of 3-aminotriazole on peroxide kinetics and on the status of glutathione. In: Simic MG, Taylor KA, Ward JF, Von Sonntag C, eds. Oxygen Radicals in Biology and Medicine. New York: Plenum Press, 1988:1035–1038.
189. Hull DS, Strickland EC, Green K. Photodynamically induced alteration of cornea endothelial cell function. Invest Ophthalmol Vis Sci 1979; 18:1226–1231.
190. Hull DS, Csukas S, Green K, Livingston V. Hydrogen peroxide and corneal endothelium. Acta Ophthalmol (Copenh) 1981; 59:409–421.
191. Hull DS, Green K, Csukas S, Livingston V. Photodynamic alteration of cornea endothelium. Relation to bicarbonate fluxes and oxygen concentration. Biochim Biophys Acta 1981; 640: 231–239.

192. Hull DS, Csukas S, Green K. Rose bengal induced corneal swelling: relation to inciting wavelength. Curr Eye Res 1981; 1:487–490.
193. Hull DS, Csukas S, Green K. Chlorpromazine-induced corneal endothelial phototoxicity. Invest Ophthalmol Vis Sci 1982; 22:502–508.
194. Hull DS, Csukas S, Green K. Trifluoperazine: corneal endothelial phototoxicity. Photochem Photobiol 1983; 38:425–428.
195. Hull DS, Green K, Hampstead D. Effect of hematoporphyrin derivative on rabbit corneal endothelial cell function and ultrastructure. Invest Ophthalmol Vis Sci 1985; 26:1465–1474.
196. Hull DS, Green K, Hampstead D. Potassium superoxide induction of rabbit corneal endothelial cell damage. Curr Eye Res 1984; 3:1321–1328.
197. Hull DS, Green K, Thomas L, Alderman N. Hydrogen peroxide-mediated corneal endothelial damage: induction by oxygen free radical. Invest Ophthalmol Vis Sci 1984; 25:1246–1253.
198. Riley MV, Giblin FJ. Toxic effects of hydrogen peroxide on corneal endothelium. Curr Eye Res 1982/83; 2:451–458.
199. Hull DS, Riley MV, Csukas S, Green K. Corneal endothelial glutathione after photodynamic change. Invest Ophthalmol Vis Sci 1982; 22:405–408.
200. Bhuyan KC, Bhuyan DK. Catalase in ocular tissue and its intracellular distribution in corneal epithelium. Am J Ophthalmol 1970; 69:147–153.
201. Yuen VH, Zeng LH, Wu TW, Rootman DS. Comparative antioxidant protection of cultured rabbit corneal epithelium. Curr Eye Res 1994; 13:815–818.
202. Redmond TN, Duke EJ, Coles WH, Simson JA, Crouch RK. Localization of corneal superoxide dismutase by biochemical and histocytochemical techniques. Exp Eye Res 1984; 38:369–378.
203. Rao NA, Thaete LG, Delmage JM, Sevanian A. Superoxide dismutase in ocular structures. Invest Ophthalmol Vis Sci 1985; 26:1778–1781.
204. Behndig A, Svensson B, Marklund SL, Karlsson K. Superoxide dismutase isoenzymes in the human eye. Invest Ophthalmol Vis Sci 1998; 39:471–475.
205. Crouch RK, Goletz P, Snyder A, Coles WH. Antioxidant enzymes in human tears. J Ocul Pharmacol 1991; 7:253–258.
206. Gogia R, Richer SP, Rose RC. Tear fluid content of electrochemically active components including water soluble antioxidants. Curr Eye Res 1998; 17:257–263.
207. Augustin AJ, Spitznas M, Kaviani N, Meller D, Koch FH, Grus F, Gobbels MJ. Oxidative reactions in the tear fluid of patients suffering from dry eyes. Graefes Arch Clin Exp Ophthalmol 1995; 233:694–698.
208. Green K. Free radicals and aging of anterior segment tissues of the eye: A hypothesis. Ophthalmic Res 1995; 27:143–149.
209. Polansky JR, Wood IS, Maglio MT, Alvarado JT. Trabecular meshwork cell culture in glaucoma research: Evaluation of biological activity and structural properties of human trabecular cells in vitro. Ophthalmology 1984; 91:580–585.
210. Polansky JR, Fauss DJ, Hydom T, Bloom E. Cellular injury from sustained vs acute hydrogen peroxide exposure in cultured human corneal endothelium and human lens epithelium. CLAO J 1990; 16(suppl):S23–S28.
211. Zhou L, Li Y, Yue BY. Oxidative stress affects cytoskeletal structure and cell-matrix interactions in cells from an ocular tissue: the trabecular meshwork. J Cell Physiol 1999; 180:182–189.
212. Yan DB, Trope GE, Ethier CR, Menon IA, Wakeham A. Effects of hydrogen peroxide-induced oxidative damage on outflow facility and washout in pig eyes. Invest Ophthalmol Vis Sci 1991; 32:2515–2520.
213. Scott DR, Karageuzian LN, Anderson PJ, Epstein DL. Glutathione peroxidase of calf trabecular meshwork. Invest Ophthalmol Vis Sci 1984; 25:599–602.
214. Nguyen KP, Weiss H, Karageuzian LN, Anderson PJ, Epstein DL. Glutathione reductase of calf trabecular meshwork. Invest Ophthalmol Vis Sci 1985; 26:887–890.

215. Kahn MG, Giblin FJ, Epstein DL. Glutathione in calf trabecular meshwork and its relation to aqueous humor outflow facility. Invest Ophthalmol Vis Sci 1983; 24:1283–1287.
216. Nguyen KP, Chung ML, Anderson PJ, Johnson M, Epstein DL. Hydrogen peroxide removal by the calf aqueous outflow pathway. Invest Ophthalmol Vis Sci 1988; 29:976–981.
217. Freedman SF, Anderson PJ, Epstein DL. Superoxide dismutase and catalase of calf trabecular meshwork. Invest Ophthalmol Vis Sci 1985; 26:1330–1335.
218. Epstein DL, DeKater AW, Lou M, Patel J. Influences of glutathione and sulfhydryl containing compounds on aqueous humor outflow function. Exp Eye Res 1990; 50:785–793.
219. Tamm ER, Russell P, Johnson DH, Piatigorsky J. Human and monkey trabecular meshwork accumulate alpha β-crystallin in response to heat shock and oxidative stress. Invest Ophthalmol Vis Sci 1996; 37:2402–2413.
220. Tamm ER, Russell P, Piatigorsky J. Development and characterization of an immortal and differentiated murine trabecular meshwork cell line. Invest Ophthalmol Vis Sci 1999; 40: 1392–1403.
221. Lutjen-Drecoll E, May CA, Polansky JR, Johnson DH, Bloemendal H, Nguyen TD. Localization of the stress proteins alpha B-crystallin and trabecular meshwork inducible glucocorticoid response protein in normal and glaucomatous trabecular meshwork. Invest Ophthalmol Vis Sci 1998; 39:517–525.
222. Polansky JR, Fauss DJ, Chen P, Chen H, Lutjen-Drecoll E, Johnson D, Kurtz RM, Ma ZD, Bloom E, Nguyen TD. Cellular pharmacology and molecular biology of the trabecular meshwork inducible glucocorticoid response gene product. Ophthalmologica 1997; 211:126–139.
223. Stone EM, Fingert JH, Alward WLM, Nguyen TD, Polansky JR, Sunden SLF, Nishimura D, Clark AF, Nystuen A, Nichols BE, Mackey DA, Ritch R, Kalenak JW, Craven ER, Sheffield VC. Identification of a gene that causes primary open angle glaucoma. Science 1997; 275:668–670.
224. Fingert JH, Ying L, Swiderski RE, Nystuen AM, Arbour NC, Alward WL, Sheffield VC, Stone EM. Characterization and comparison of the human and mouse GLC1A glaucoma genes. Genome Res 1998; 8:377–384.
225. Nguyen TD, Chen P, Huang WD, Chen H, Johnson AH, Polansky JR. Gene structure and properties of TIGR, an olfactomedin-related glycoprotein cloned from glucocorticoid-induced trabecular meshwork cells. J Biol Chem 1998; 273:6341–6350.
226. Alward WL, Fingert JH, Johnson AT, Lerner SF, Junqua D, Durcan FJ, McCartney PJ, Mackey DA, Sheffield VC, Stone EM. Clinical features associated with mutations in the chromosome 1 open angle glaucoma gene (GLC1A). N Engl J Med 1998; 338:1022–1027.
227. Morel Y, Barouki R. Repression of gene expression by oxidative stress. Biochem J 1999; 342:481–496.
228. Kurysheva NI, Vinetskaia MI, Erichev VP, Demchuk ML, Kuryshev SI. Rol' svobodnoradikal'nykh reakstii kamernoi vlagi v razvitii pervichnoi otkrytougol'noi glaukomy. Vestn Oftalmol 1996; 112:3–5.
229. Scorolli L, Grossi G, Scorolli L, Bozza D, Savini G, Preda P, Scalinci SZ, Sprovieri C, Meduri R. Antioxidants and glaucoma: blood and aqueous vitamin E dosage in primary open angle glaucoma. Invest Ophthalmol Vis Sci 1997; 38(suppl):1053.
230. Costagliola C, Iuliano G, Menzione M, Rinaldi E, Vito P, Auricchio G. Effect of vitamin E on glutathione content in red blood cells, aqueous humor and lens of humans and other species. Exp Eye Res 1986; 43:905–914.
231. Bunin AI, Filina AA, Erichev VP. Defitsit glutationa pri otkrytougol'noi glaukome I podkhody k ego korrektsii. Vestn Oftalmol 1992; 108:13–15.
232. Delamere NA, Williams RN. Detoxification of hydrogen peroxide by the rabbit iris-ciliary body. Exp Eye Res 1985; 40:805–811.
233. Armstrong D, Santangelo G, Connole E. The distribution of peroxide regulating enzymes in the canine eye. Curr Eye Res 1981; 1:225–242.
234. Ng MC, Susan SR, Shichi H. Bovine non-pigmented and pigmented ciliary epithelial cells

in culture: comparison of catalase, superoxide dismutase and glutathione peroxidase activities. Exp Eye Res 1988; 46:919–928.
235. Martin-Alonso JM, Ghosh S, Coca-Prados M. Cloning of the bovine plasma selenium-dependent glutathione peroxidase (GP) cDNA from the ocular ciliary epithelium: expression of the plasma and cellular forms within the mammalian eye. J Biochem (Tokyo) 1993; 114: 284–291.
236. Shichi H. Glutathione-dependent detoxification of peroxide in bovine ciliary body. Exp Eye Res 1990; 50:813–818.
237. Reddan JR, Giblin FJ, Dziedzic DC, McCready JP, Schrimscher L, Reddy VN. Influence of the activity of glutathione reductase on the response of cultured lens epithelial cells from young and old rabbits to hydrogen peroxide. Exp Eye Res 1988; 46:209–221.
238. Giblin FJ, Reddan JR, Schrimscher L, Dziedzic DC, Reddy VN. The relative roles of the glutathione redox cycle and catalase in the detoxification of $H_2O_2$ by cultured rabbit lens epithelial cells. Exp Eye Res 1990; 50:795–804.
239. Costarides AP, Nelson E, Green K. Morphological sequelae of intracameral hydrogen peroxide after inhibition of glutathione synthesis. Lens Eye Toxic Res 1991; 8:441–447.
240. Bhuyan KC, Bhuyan DK. Regulation of hydrogen peroxide in eye humors: Effect of 3-amino-1H,1,2,4-triazole on catalase and glutathione peroxidase of the rabbit eye. Biochim Biophys Acta 1977; 497:641–651.
241. Birnbaum D, Csukas S, Costarides A, Forbes E, Green K. 3-Amino-triazole effects on the eye of young and adult rabbits in the presence and absence of hydrogen peroxide. Curr Eye Res 1987; 6:1403–1414.
242. Green K, Greiff M, Cheeks L. Influence of hydrogen peroxide on adenine uptake by iris, ciliary epithelium and corneal endothelium. Lens Eye Toxic Res 1989; 6:405–414.
243. Hull DS, Green K, Elijah RD. Effect of oxygen free radical products on rabbit iris vascular permeability. Acta Ophthalmol (Copenh) 1985; 63:513–518.
244. Taylor HR. The environment and the lens. Br J Ophthalmol 1980; 64:303–310.
245. Taylor HR. Ultraviolet radiation and the eye: an epidemiologic study. Trans Am Ophthalmol Soc 1989; 87:802–853.
246. Bochow TW, West SK, Azar A, Munoz B, Sommer A, Taylor HR. Ultraviolet light exposure and risk of posterior subcapsular cataracts. Arch Ophthalmol 1989; 107:369–372.
247. Taylor HR, West S, Munoz B, Rosenthal FS, Bressler SB, Bressler NM. The long-term effects of visible light on the eye. Arch Ophthalmol 1992; 110:99–104.
248. McLaren JW, Brubaker RF. Measurement of transmission of ultraviolet and visible light in the living rabbit cornea. Curr Eye Res 1996; 15:411–421.
249. Dillon J, Zheng L, Merriam JC, Gaillard ER. The optical properties of the anterior segment of the eye: implications for cortical cataract. Exp Eye Res 1999; 68:785–795.
250. Pitts DG, Cullen AP, Hacker PD. Ocular effects of ultraviolet radiation from 295 to 365 nm. Invest Ophthalmol Vis Sci 1977; 16:932–939.
251. Doughty MJ, Cullen AP. Long-term effects of a single dose of ultraviolet-B on albino rabbit cornea. I. In vivo analyses. Photochem Photobiol 1989; 49:185–196.
252. Bargeron CB, Deters OJ, Farrell RA, McCally RL. Epithelial damage in rabbit corneas exposed to $CO_2$ laser radiation. Health Phys 1989; 56:85–95.
253. Cogan DG, Kinsey VE. Action spectrum of keratitis produced by ultraviolet radiation. Arch Ophthalmol 1946; 35:670–677.
254. Good GW, Schoessler JP. Chronic solar radiation exposure and endothelial polymegethism. Curr Eye Res 1988; 7:157–162.
255. Riley MV, Susan S, Peters MI, Schwartz CA. The effects of UV-B irradiation on the corneal endothelium. Curr Eye Res 1987; 6:1021–1033.
256. Maurice DM, Riley MV. The cornea. In: Graymore CN, ed. Biochemistry of the Eye. London: Academic Press, 1970:1–103.
257. Doughty MJ, Cullen AP. Long-term effects of a single dose of ultraviolet-B on albino rabbit

cornea. II. Deturgescence and fluid pump assessed. Photochem Photobiol 1990; 51:439–449.

258. Ringvold A. Damage of the cornea epithelium caused by ultraviolet radiation. A scanning electron microscopic study in rabbit. Acta Ophthalmol (Copenh) 1983; 61:898–907.
259. Clarke SM, Doughty MJ, Cullen AP. Acute effects of ultraviolet-B irradiation on the corneal surface of the pigmented rabbit studied by quantitative scanning electron microscopy. Acta Ophthalmol (Copenh) 1990; 68:639–650.
260. Cullen AP, Chou BR, Hall MG, Jany SE. Ultraviolet-B damages corneal endothelium. Am J Optom Physiol Optics 1984; 61:473–478.
261. Zigman S, Datiles M, Torczynski E. Sunlight and human cataracts. Invest Ophthalmol Vis Sci 1979; 18:462–467.
262. Rosenthal FS, Safran M, Taylor HR. The ocular dose of ultraviolet radiation from sunlight exposure. Photochem Photobiol 1985; 42:163–171.
263. Blumthaler M, Ambach W, Daxecker F. On the threshold radiant exposure for keratitis solaris. Invest Ophthalmol Vis Sci 1987; 28:1713–1716.
264. Charman WN. Ocular hazards arising from depletion of the natural atmospheric ozone layer: a review. Ophthalmic Physiol Opt 1990; 10:333–341.
265. Longstreth J, de Gruijl FR, Kripke ML, Abseck S, Arnold F, Slaper HI, Velders G, Takizawa Y, van der Leun JC. Health risks. J Photochem Photobiol B 1998; 46:20–39.
266. Ringvold A, Davanger M, Olsen EG. Changes in the corneal endothelium after ultraviolet radiation. Acta Ophthalmol (Copenh) 1982; 60:41–53.
267. Karai I, Matsumura S, Takise S, Horiguchi S, Matsuda M. Morphological change in the corneal endothelium due to ultraviolet radiation in welders. Br J Ophthalmol 1984; 68:544–548.
268. Cejkova J, Lojda Z. The damaging effect of UV rays (with the wavelength shorter than 320 nm) on the rabbit anterior eye segment. I. Early changes and their prevention by catalase-aprotinin application. Acta Histochem 1994; 96:281–286.
269. Cejkova J, Lojda Z. The damaging effect of UV rays below 320 nm on the rabbit anterior eye segment. II. Enzyme histochemical changes and plasmin activity after prolonged irradiation. Acta Histochem 1995; 97:183–188.
270. Cejkova J, Lojda Z. Histochemical study on xanthine oxidase activity in the normal rabbit cornea and lens and after repeated irradiation of the eye with UVB rays. Acta Histochem 1996; 98:47–52.
271. Yamanashi BS, Hacker H, Klintworth GK. Wavelength dependence and kinetics of UV-induced free radical formation in the human cornea and lens. Photochem Photobiol 1979; 30:391–395.
272. Shimmura S, Suematsu M, Shimoyama M, Tsubota K, Oguchi Y, Ishimura Y. Subthreshold UV radiation-induced peroxide formation in cultured corneal epithelial cells: the protective effects of lactoferrin. Exp Eye Res 1996; 63:519–526.
273. Babu V, Misra RB, Joshi PC. Ultraviolet-B effects on ocular tissues. Biochem Biophys Res Commun 1995; 210:417–423.
274. Trokel SL. History of the excimer and other ophthalmic refractive lasers. In: McGhee CNJ, Taylor HR, Gartry DS, Trokel SL, eds. Excimer Lasers in Ophthalmology. Principles and Practice. Boston: Butterworth-Heinemann, 1997:1–13.
275. Kerr Muir MG, Sherrard ES. Damage to the corneal endothelium during Nd/YAG photodisruption. Br J Ophthalmol 1985; 69:77–85.
276. Martin NF, Gaasterland DE, Rodrigues MM, Thomas G, Cummins CE, III. Endothelial damage thresholds for retrocorneal Q-switched neodymium: YAG laser pulses in monkeys. Ophthalmology 1985; 92:1382–1386.
277. Dehm EJ, Puliafito CA, Adler CM, Steinert RF. Corneal endothelial injury in rabbits following excimer laser ablation at 193 and 248 nm. Arch Ophthalmol 1986; 104:1364–1368.

278. Thoming C, Van Buskirk EM, Samples JR. The corneal endothelium after laser therapy for glaucoma. Am J Ophthalmol 1987; 103:518–522.
279. Ediger MN, Pettit GH, Matchette LS. In vitro measurements of cytotoxic effects of 193 nm and 213 nm laser pulses at subablative fluences. Lasers Surg Med 1997; 21:88–93.
280. Hayashi S, Ishimoto S, Wu GS, Wee WR, Rao NA, McDonnell PJ. Oxygen free radical damage in the cornea after excimer laser therapy. Br J Ophthalmol 1997; 81:141–144.
281. Isager P, Guo S, Hjortdal JO, Ehlers N. Endothelial cell loss after photorefractive keratectomy for myopia. Acta Ophthalmol Scand 1998; 76:304–307.
282. Kasetsuwan N, Wu FM, Hsieh F, Sanchez D, McDonnell PJ. Effect of topical ascorbic acid on free radical tissue damage and inflammatory cell influx in the cornea after excimer laser corneal surgery. Arch Ophthalmol 1999; 117:649–652.
283. Jain S, Hahn TW, McCally RL, Azar DT. Antioxidants reduce corneal light scattering after excimer keratectomy in rabbits. Lasers Surg Med 1995; 17:160–165.
284. Bilgihan K, Bilgihan A, Akata F, Hasanreisoglu B, Turkozkan N. Excimer laser corneal surgery and free oxygen radicals. Jpn J Ophthalmol 1996; 40:154–157.
285. Timberlake GT, Gemperli AW, Larive CK, Warren KA, Mainster MA. Free radical production by Nd:YAG laser photodisruption. Ophthalmic Surg Lasers 1997; 28:582–589.
286. Shimmura S, Masumizu T, Nakai Y, Urayama K, Shimazaki J, Bissen-Miyajima H, Kohno M, Tsubota K. Excimer laser-induced hydroxyl radical formation and keratocyte death in vitro. Invest Ophthalmol Vis Sci 1999; 40:1245–1249.
287. Mittag T. Role of oxygen radicals in ocular inflammation and cellular damage. Exp Eye Res 1984; 39:759–769.
288. Mittag TW, Hammond BR, Eakins KE, Bhattacherjee P. Ocular responses to superoxide generated by intraocular injection of xanthine oxidase. Exp Eye Res 1985; 40:411–419.
289. Williams RN, Paterson CA, Eakins KE, Bhattacherjee P. Quantification of ocular inflammation: evaluation of polymorphonuclear leucocyte infiltration by measuring myeloperoxidase activity. Curr Eye Res 1982/83; 2:465–470.
290. Williams RN, Paterson CA. A protective role for ascorbic acid during inflammatory episodes in the eye. Exp Eye Res 1986; 42:211–218.
291. Rao NA, Fernandez MA, Sevanian A, Till GO, Marak GE, Jr. Antiphlogistic effect of catalase on experimental phacoanaphylactic endophthalmitis. Ophthalmic Res 1986; 18:185–191.
292. de Kozak Y, Nordman JP, Faure JP, Rao NA, Marak GE, Jr. Effect of antioxidant enzymes on experimental uveitis in rats. Ophthalmic Res 1989; 21:230–234.
293. Rao NA, Romero JL, Fernandez MA, Sevanian A, Marak GE, Jr. Role of free radicals in uveitis. Surv Ophthalmol 1987; 32:209–213.
294. Rao NA, Sevanian A, Fernandez MA, Romero JL, Faure JP, de Kozak Y, Till GO, Marak GE, Jr. Role of oxygen radicals in experimental allergic uveitis. Invest Ophthalmol Vis Sci 1987; 28:886–892.
295. Rao NA. Role of oxygen free radicals in retinal damage associated with experimental uveitis. Trans Am Ophthalmol Soc 1990; 88:797–850.
296. Yamada M, Shichi H, Yuasa T, Tanouchi Y, Mimura Y. Superoxide in ocular inflammation: human and experimental uveitis. Free Radical Biol Med 1986; 2:111–117.
297. Augustin AJ, Boker T, Blumenroder SH, Lutz J, Spitznas M. Free radical scavenging and antioxidant activity of allopurinol and oxypurinol in experimental lens-induced uveitis. Invest Ophthalmol Vis Sci 1994; 35:3897–3904.
298. Alio JL, Ayala MJ, Mulet ME, Artola A, Ruiz JM, Bellot J. Antioxidant therapy in the treatment of experimental acute corneal inflammation. Ophthalmic Res 1995; 27:136–143.
299. Artola A, Alio JL, Bellot JL, Ruiz JM. Lipid peroxidation in the iris and its protection by means of viscoelastic substances (sodium hyaluronate and hydroxypropylmethylcellulose). Ophthalmic Res 1993; 25:172–176.
300. Artola A, Alio JL, Bellot JL, Ruiz JM. Protective properties of viscoelastic substances (so-

dium hyaluronate and 2% hydroxymethylcellulose) against experimental free radical damage to the corneal endothelium. Cornea 1993; 12:109–114.
301. Ishimoto S, Wu GS, Hayashi S, Zhang J, Rao NA. Free radical tissue damages in the anterior segment of the eye in experimental autoimmune uveitis. Invest Ophthalmol Vis Sci 1996; 37:630–636.
302. Wu GS, Gritz DC, Atalla LR, Stanforth DA, Sevanian A, Rao NA. Ultrastructural localization of hydrogen peroxide in experimental autoimmune uveitis. Curr Eye Res 1992; 11:955–961.
303. Sery TW, Petrillo R. Superoxide anion radical as an indirect mediator in ocular inflammatory disease. Curr Eye Res 1984; 3:243–252.
304. Koch FH, Augustin AJ, Grus FH, Spitznas M. Effects of different antioxidants on lens-induced uveitis. Ger J Ophthalmol 1996; 5:185–188.
305. Benita S. Prevention of topical and ocular oxidative stress by positively charged submicron emulsion. Biomed Pharmacother 1999; 53:193–206.
306. Jaramillo A, Bhattacherjee P, Sonnenfeld G, Paterson CA. Modulation of immune responses by cyclo-oxygenase inhibitors during intraocular inflammation. Curr Eye Res 1992; 11:571–579.
307. Green K, Nye RA, Nelson E, Cheeks L. Role of eicosanoids in the ocular response to intracameral hydrogen peroxide. Lens Eye Toxic Res 1990; 7:419–426.
308. Williams RN, Paterson CA. Modulation of corneal lipoxygenase by ascorbic acid. Exp Eye Res 1986; 43:7–13.
309. Schwartzman ML, Balazy M, Masferrer J, Abraham NG, McGiff JC, Murphy RC. 12(R)-hydroxyeicosatetraenoic acid: a cytochrome-P450-dependent arachidonate metabolite that inhibits $Na^+$, $K^+$-ATPase in the cornea. Proc Natl Acad Sci USA 1987; 84:8125–8129.
310. Murphy RC, Falck JR, Lumin S, Yadagiri P, Zirrolli JA, Balazy M, Masferrer JL, Abraham NG, Schwartzman ML. 12(R)-hydroxyeicosatrienoic acid: a vasodilator cytochrome P-450-dependent arachidonate metabolite from the bovine corneal epithelium. J Biol Chem 1988; 263:17197–17202.
311. Hurst JS, Balazy M, Bazan HE, Bazan NG. The epithelium, endothelium, and stroma of the rabbit cornea generate (12S)-hydroxyeicosatetraenoic acid as the main lipoxygenase metabolite in response to injury. J Biol Chem 1991; 266:6726–6730.
312. Taylor L, Menconi M, Leibowitz MH, Polgar P. The effect of ascorbate, hydroperoxides, and bradykinin on prostaglandin production by corneal and lens cells. Invest Ophthalmol Vis Sci 1982; 23:378–382.
313. Hurst JS, Paterson CA, Short CS. Oxidant and anti-oxidant effects on arachidonate metabolism by rabbit ocular tissues. J Ocul Pharmacol 1989; 5:51–64.
314. Bazan NG, de Abreu MT, Bazan HE, Belfort R, Jr. Arachidonic acid cascade and platelet-activating factor in the network of eye inflammatory mediators: therapeutic implications in uveitis. Int Ophthalmol 1990; 14:335–344.
315. Carubelli R, Nordquist RE, Rowsey JJ. Role of active oxygen species in corneal ulceration. Effect of hydrogen peroxide generated in situ. Cornea 1990; 9:161–169.
316. Chandler JW. Biocompatibility of hydrogen peroxide in soft contact lens disinfection: antimicrobial activity vs. biocompatibility—the balance. CLAO J 1990; 16:S43–S45.
317. Green K. The effect of preservatives on corneal permeability of drugs. In: Edman P, ed. Biopharmaceutics of Ocular Drug Delivery. Boca Raton, FL: CRC Press, 1991:43–60.
318. Crouch R, Ling Z, Hayden BJ. Corneal oxygen scavenging systems: lysis of corneal epithelial cells by superoxide anions. In: Simic MG, Taylor KA, Ward JF, Von Sonntag C, eds. Oxygen Radicals in Biology and Medicine. New York: Plenum Press, 1988:1043–1046.
319. Hayden BJ, Zhu L, Sens D, Tapert MJ, Crouch RK. Cytolysis of corneal epithelial cells by hydrogen peroxide. Exp Eye Res 1990; 50:11–16.
320. Tripathi BJ, Tripathi RC. Hydrogen peroxide damage to human corneal epithelial cells in vitro. Implications for contact lens disinfection systems. Arch Ophthalmol 1989; 107:1516–1519.

321. Tripathi BJ, Tripathi RC, Millard CB, Borisuth NS. Cytotoxicity of hydrogen peroxide to human corneal epithelium in vitro and its clinical implications. Lens Eye Toxic Res 1990; 7:385–401.
322. Wilson GS, Chalmers RL. Effect of $H_2O_2$ concentration and exposure time on stromal swelling: an epithelial perfusion model. Optom Vis Sci 1990; 67:252–255.
323. Riley MV. Physiologic neutralization mechanisms and the response of the corneal endothelium to hydrogen peroxide. CLAO J 1990; 16:S16–S21.
324. Riley MV, Kast M. Penetration of hydrogen peroxide from contact lenses or tear-side solutions into the aqueous humor. Optom Vis Sci 1991; 68:546–551.
325. Green K. Principles and methods of ocular pharmacokinetic evaluation. In: Hobson DW, ed. Dermal and Ocular Toxicology: Fundamentals and Methods. Boca Raton, FL: CRC Press, 1991:541–584.
326. Riley MV, Wilson G. Topical hydrogen peroxide and the safety of ocular tissues. CLAO J 1993; 19:186–190.
327. Wilson G, Riley MV. Does topical hydrogen peroxide penetrate the cornea? Invest Ophthalmol Vis Sci 1993; 34:2752–2760.
328. Yuan N, Pitts DG. Residual $H_2O_2$ compromises deswelling function of in vivo rabbit cornea. Acta Ophthalmol (Copenh) 1991; 69:241–246.
329. Boets EP, Kerkmeer MJ, van Best JA. Contact lens care solutions and corneal epithelial barrier function: a fluorophotometric study. Ophthalmic Res 1994; 26:129–136.
330. Paugh JR, Brennan NA, Efron N. Ocular response to hydrogen peroxide. Am J Optom Physiol Opt 1988; 65:91–98.
331. Neuwirth Lux O, Dikstein S. Survival of isolated rabbit cornea and free radical scavengers. Curr Eye Res 1985; 4:153–154.
332. Lux-Neuwirth O, Millar TJ. Lipid soluble antioxidants preserve rabbit corneal cell function. Curr Eye Res 1990; 9:103–109.
333. Travkin AG, Vinetskaia MI, Slepukhina LV. Vliianie antioksidantov obladaiushchikh razlichnoi antiradikal'noi aktivnost'iu na konservatsiiu rogovitsy. Biofizika 1976; 21:1064–1066.
334. Hallberg CK, Trocme SD, Ansari NH. Acceleration of corneal wound healing in diabetic rats by the antioxidant trolox. Res Commun Mol Pathol Pharmacol 1996; 93:3–12.
335. Snell JC, Colton CA, Chernyshev ON, Gilbert DL. Location-dependent artifact for NO measurement using multiwell plates. Free Radical Biol Med 1996; 20:361–363.
336. Varma SD, Ali AH, Devamanoharan PS, Morris SM. Nitrite-induced photo-oxidation of thiol and its implications in smog toxicity to the eye: prevention by ascorbate. J Ocul Pharmacol Ther 1997; 13:179–187.
337. Yanagiya N, Akiba J, Kado M, Yoshida A, Kono T, Iwamoto J. Transient corneal edema induced by nitric oxide synthase inhibition. Nitric Oxide 1997; 1:397–403.
338. Behar-Cohen FF, Savoldelli M, Parel JM, Goureau O, Thillaye-Goldenberg B, Courtois Y, Pouliquen Y, de Kozak Y. Reduction of corneal edema in endotoxin-induced uveitis after application of L-NAME as nitric oxide synthase inhibitor in rats by iontophoresis. Invest Ophthalmol Vis Sci 1998; 39:897–904.
339. Nathanson JA, McKee M. Identification of an extensive system of nitric oxide-producing cells in the ciliary muscle and outflow pathway of the human eye. Invest Ophthalmol Vis Sci 1995; 36:1765–1773.
340. Nathanson JA, McKee M. Alterations of ocular nitric oxide synthase in human glaucoma. Invest Ophthalmol Vis Sci 1995; 36:1774–1784.
341. Tilton RG, Chang K, Corbett JA, Misko TP, Currie MG, Bora NS, Kaplan HJ, Williamson JR. Endotoxin-induced uveitis in the rat is attenuated by inhibition of nitric oxide production. Invest Ophthalmol Vis Sci 1994; 35:3278–3288.
342. Goureau O, Bellot J, Thillaye B, Courtois Y, de Kozak Y. Increased nitric oxide production

in endotoxin-induced uveitis. Reduction of uveitis by an inhibitor of nitric oxide synthase. J Immunol 1995; 154:6518–6523.
343. Wang ZY, Hakanson R. Role of nitric oxide (NO) in ocular inflammation. Br J Pharmacol 1995; 116:2447–2450.
344. McMenamin PG, Crewe JM. Cellular localisation and dynamics of nitric oxide synthase expression in the rat anterior segment during endotoxin-induced uveitis. Exp Eye Res 1997; 65:157–164.
345. Allen JB, McGahan MC, Fleisher LN, Jaffe GJ, Keng T, Privalle CT. Nitric oxide in ocular inflammation. In: Green K, Edelhauser HF, Hackett RB, Hull DS, Potter DE, Tripathi RC, eds. Advances in Ocular Toxicology. New York: Plenum Press, 1997:121–131.
346. Bellot JL, Alcoriza N, Palmero M, Blanco A, Espi R, Hariton C, Orts A. Effects of the inhibition of nitric oxide synthase and lipoxygenase on the development of endotoxin-induced uveitis. In: Green K, Edelhauser HF, Hackett RB, Hull DS, Potter DE, Tripathi RC, eds. Advances in Ocular Toxicology. New York: Plenum Press, 1997:151–158.
347. Taniguchi T, Kawakami H, Sawada A, Iwaki M, Tsuji A, Sugiyama K, Kitazawa Y. Effects of nitric oxide synthase inhibitor on intraocular pressure and ocular inflammation following laser irradiation in rabbits. Curr Eye Res 1998; 17:308–315.
348. Shalini VK, Luthra M, Srinivas L, Rao SH, Basti S, Reddy M, Balasubramanian D. Oxidative damage to the eye lens caused by cigarette smoke and fuel smoke condensates. Indian J Biochem Biophys 1994; 31:261–266.
349. Adachi M, Kasai T, Adachi T, Budna JN, Pollak OJ. Effect of cigarette smoke on rabbit corneal cells in vitro. Tohoku J Exp Med 1966; 90:41–51.
350. Adachi M, Kasai T, Adachi T, Budna JN, Pollak OJ. Effect of cigarette smoke on rabbit cornea in vivo. Tohoku J Exp Med 1966; 90:169–173.
351. Pelle E, Ingrassia M, Mammone T, Marenus K, Maes D. Protection against cigarette smoke-induced damage to intact transformed rabbit corneal cells by N-acetyl-L-cysteine. Cell Biol Toxicol 1998; 14:253–259.
352. Watsky MA, Jablonski MM, Edelhauser HF. Comparison of conjunctival and corneal surface areas in rabbit and human. Curr Eye Res 1988; 7:483–486.
353. Modis L, Jr, Marshall GE, Lee WR. Distribution of antioxidant enzymes in the normal aged human conjunctiva: an immunocytochemical study. Graefes Arch Clin Exp Ophthalmol 1998; 236:86–90.
354. Koyama I, Nakamori K, Nagahama T, Ogasawara M, Nemoto M. The reactivity of taurine with hypochlorous acid and its application for eye drops. Adv Exp Med Biol 1996; 403:9–18.
355. Shimmura S, Tsubota K, Oguchi Y, Fukumura D, Suematsu M, Tsuchiya M. Oxiradical-dependent photoemission induced by a phacoemulsification probe. Invest Ophthalmol Vis Sci 1992; 33:2904–2907.
356. Holst A, Rolfsen W, Svensson B, Ollinger K, Lundgren B. Formation of free radicals during phacoemulsification. Curr Eye Res 1993; 12:359–365.
357. Svensson B, Mellerio J. Phaco-emulsification causes the formation of cavitation bubbles. Curr Eye Res 1994; 13:649–653.
358. Rose RC, Richer SP, Bode AM. Ocular oxidants and antioxidant protection. Proc Soc Exp Biol Med 1998; 217:397–407.
359. Florence TM. The role of free radicals in disease. Aus NZ J Ophthalmol 1995; 23:3–7.
360. Bray RC, Cockle SA, Fielden EM, Roberts PB, Rotilio G, Calabrese L. Reduction and inactivation of superoxide dismutase by hydrogen peroxide. Biochem J 1974; 139:43–48.
361. Ames BN, Shigenaga MK, Hagen TM. Oxidants, antioxidants, and the degenerative diseases of aging. Proc Natl Acad Sci USA 1993; 90:7915–7922.

# 19

# The Role of Ocular Free Radicals in Age-Related Macular Degeneration

**PAUL S. BERNSTEIN**

*Moran Eye Center, University of Utah School of Medicine, Salt Lake City, Utah*

**NIKITA B. KATZ**

*Institute for Natural Resources, Berkeley, California*

## I. INTRODUCTION

Age-related macular degeneration (AMD), the leading cause of irreversible blindness in the western world, is a major ocular public health problem of increasing concern, especially in light of the world's rapidly growing elderly population. It is a particularly frustrating disease for ophthalmologists and their patients due to its relentless progressive course, which can ultimately lead to legal blindness. For the vast majority of patients, there are no proven interventions that can halt or reverse the damage wrought by the disease. Since treatment options in the late stages of AMD are so limited, there is considerable interest in identifying modifiable environmental risk factors that in turn could be used to guide early intervention strategies to lessen the risk of visual loss from AMD in susceptible individuals. As will be discussed below, there is substantial evidence that retinal pathology due to AMD is in part mediated by oxidative damage to photoreceptors and other ocular cells. The possibility that individuals can modulate their risk of visual loss from AMD by decreasing their exposure to environmental oxidants or by increasing their dietary consumption of antioxidants has been enthusiastically embraced by many members of the ophthalmic community, by the nutraceutical industry, and by the general public. The scientific basis to support these interventions, however, often lags behind the popular wisdom. In this chapter, the pathogenesis of AMD will be reviewed, the possible mechanisms for

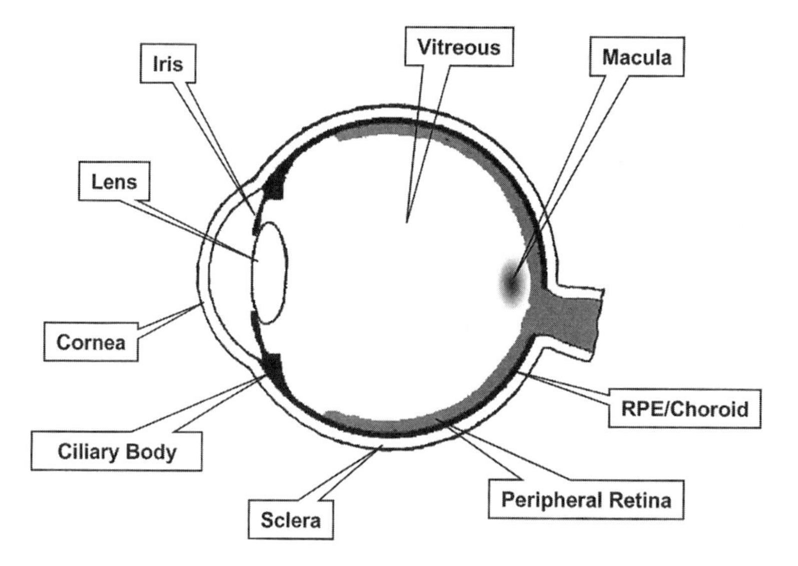


**Figure 1** Cross-sectional anatomy of the human eye.

free radical damage to the retina will be explored, and the evidence to support antioxidant interventions against AMD will be examined.

## II. MACULAR ANATOMY AND BIOLOGY

The macula, an anatomical region unique to the primate retina, is defined as that part of the retina lying within the major vascular arcades where the ganglion cell layer is two or more layers thick and where xanthophyll pigment is deposited (Figs. 1 and 2) (1). It is the area of the human retina responsible for high-resolution visual acuity required for

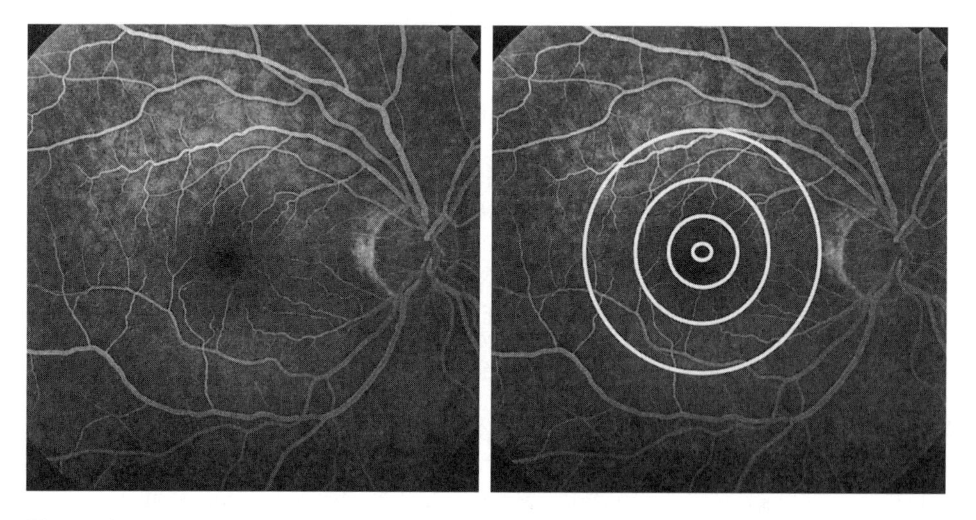

**Figure 2** Macula of the human eye. The concentric circles mark the boundaries of the regions of the human macula: foveola, fovea, parafovea, and perifovea (smallest to largest).

reading, driving, and recognizing faces. The macular region is approximately 6 mm in diameter, less than 5% of the total retinal area. Its center (the umbo) lies approximately 4 mm temporal to and 0.8 mm inferior to the middle of the optic disc. The macula is subdivided anatomically into concentric circular regions with progressively decreasing densities of cone photoreceptors: the foveola (diameter ~ 0.35 mm); the fovea (diameter ~ 2 mm); the parafovea (diameter ~ 3 mm); and the perifovea (diameter ~ 6 mm) (2). The fovea is recognized clinically by its increased pigmentation. At the fovea's outer edge, the retina is relatively thick (0.55 mm) because of multiple layers of ganglion cell nuclei. It thins progressively toward the center, such that at the foveola, the retina is only 0.13 mm thick since no ganglion cell nuclei are present. At the foveola, the photoreceptors are exclusively red and green cones in the central 100 μm (3). Blue cones are most dense between 100 and 300 μm from the center of the fovea. Rod cell density rises rapidly in an inverse manner to cone density, such that peak retinal rod density lies in an annulus just outside the border of the macula (4). Light focused by the eye's lens must traverse the vitreous cavity before reaching the surface of the retina. It must then pass through several layers of neural cells, including the nerve fiber layer, the ganglion cell layer, the inner plexiform layer, the inner nuclear layer, and the outer plexiform layer before it finally reaches the photoreceptor cell layer (Fig. 3). Specialized glial cells known as Müller cells span these cell layers and provide structural and organizational support. There are four different photoreceptor cells of the human retina: the rod cells and the red, green, and blue cone cells. Rods are predominant in the peripheral retina while cones predominate in the foveal area. All of these photoreceptors consist of an inner segment containing nuclei, mitochondria, Golgi apparatus, and other cellular organelles connected by a modified cilium to the outer segment, a specialized structure packed with photoreceptor pig-

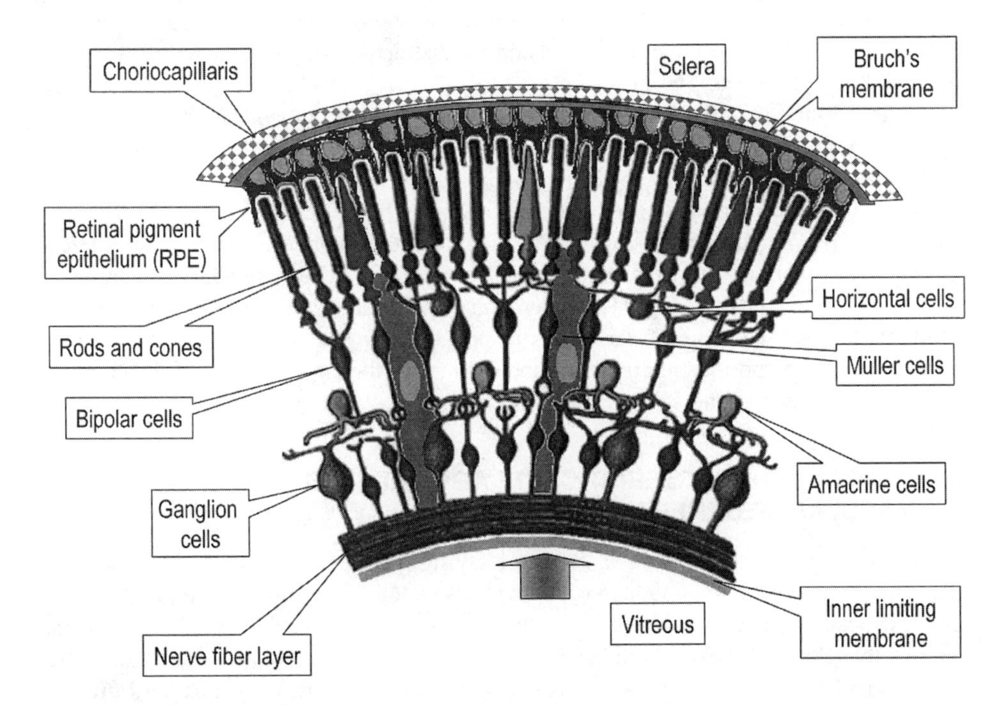


**Figure 3** Cross-sectional anatomy of the human retina. (Courtesy of Helga Kolb, Ph.D.)

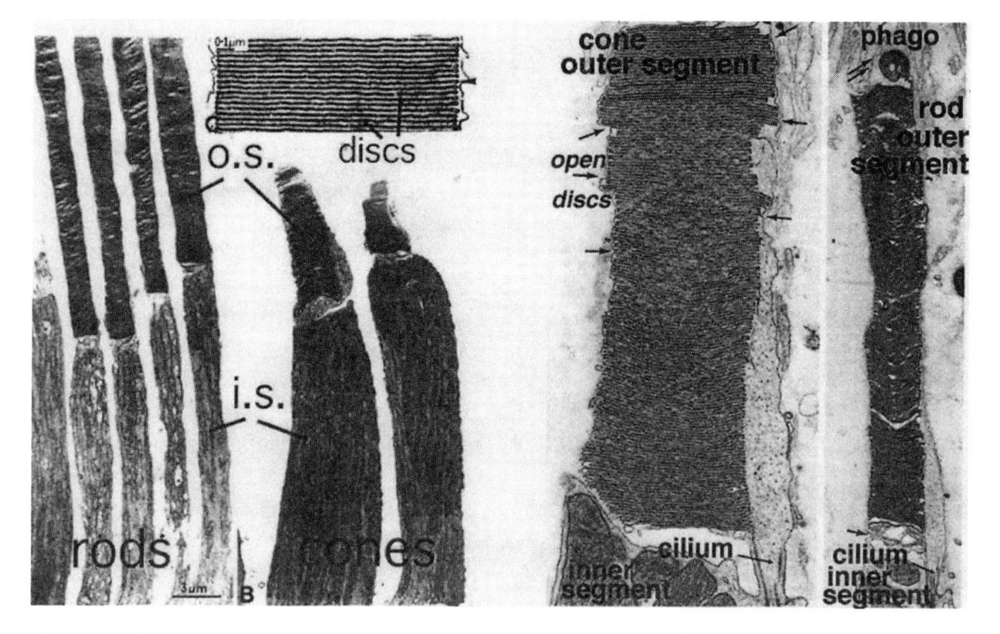


**Figure 4** Electron micrographs of primate photoreceptors. (Courtesy of Helga Kolb, Ph.D.)

ments, and the various enzymes and channels responsible for visual transduction (5). Rod outer segments contain membrane discs distinct from the plasma membrane, while cone cells have deep invaginations of their plasma membranes (Fig. 4). The predominant protein in the rod outer segment disc is the visual pigment rhodopsin with a peak absorption at ~500 nm, while the three cone types each contain distinct visual pigments with absorption maxima at 437, 533, and 564 nm (6).

Distal to the photoreceptor layer is the retinal pigment epithelium (RPE), a monolayer of hexagonal cells (7). These cells are essential to the maintenance of photoreceptor health since they phagocytose shed tips of rod and cone photoreceptors (8); pump ions and fluid out of the subretinal space to ensure adhesion of the retina (9); contain enzymes essential for regeneration of rhodopsin's 11-*cis*-retinal chromophore (10); and their intercellular tight junctions form the blood-retinal barrier (11). The basement membrane of the RPE is densely adherent to Bruch's membrane, a sheetlike condensation of the choriocapillaris, the dense plexus of capillaries responsible for the nourishment of the outer retinal layers. The choriocapillaris is fed and drained by the larger vessels of the choroid, the layer juxtaposed to the scleral coat of the eye.

## III. CLINICAL CHARACTERISTICS OF AGE-RELATED MACULAR DEGENERATION

AMD follows a clinical course that varies widely from patient to patient (Figs. 5 and 6). The initial signs visible on ophthalmoscopic exam include formation of drusen (12), yellowish deposits of material located between the RPE and a thickened Bruch's membrane that are thought to consist of oxidized lipids (13) and proteins (14). At first, they are usually small ($\leq$63 μm) with discrete borders, and are therefore known as hard drusen. Nonextensive macular hard drusen are so commonly observed in individuals over age 50

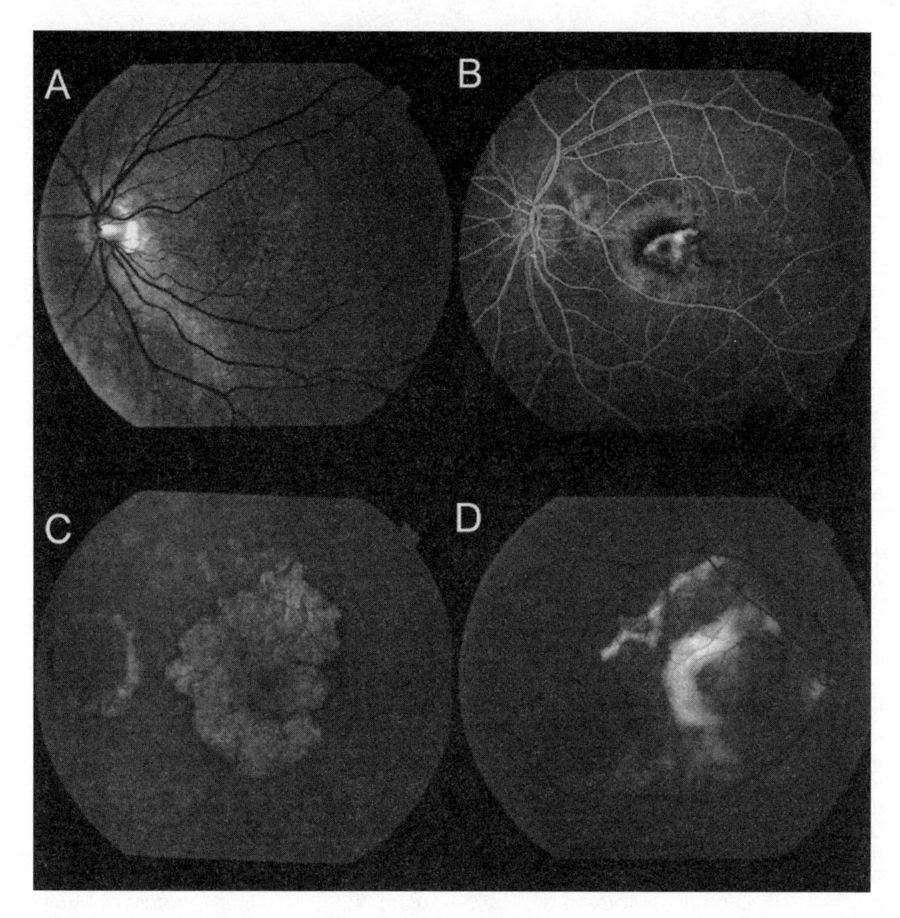


**Figure 5** Examples of age-related macular degeneration. (A) Extensive drusen. (B) Choroidal neovascularization. (C) Geographic atrophy. (D) Disciform (fibrotic) scarring.

that they are often considered normal. It is only when they are extensive and associated with pigmentary abnormalities of the RPE that they are considered part of the early stages of AMD known as age-related maculopathy (ARM). In intermediate ARM, the drusen become larger with ill-defined borders, and they are referred to as soft drusen. At this point, some patients begin to notice disturbances of central vision and loss of central acuity, although many are still asymptomatic. Eventually, the disruption of retinal and RPE structure and function can become so severe that substantial portions of the macula become atrophic. This advanced form of dry AMD takes many years to evolve and is known as geographic atrophy. Ultimately, these patients may become legally blind with visual acuities of 20/200 or worse. In other patients, the dry form of AMD progresses into the exudative or wet form of AMD. The integrity of Bruch's membrane is disrupted, and blood vessels derived from the choroid invade the subretinal and sub-RPE space where they bleed and leak fluid and lipid causing severe disruption of photoreceptor function. The progression of wet AMD may be so rapid that many lines of vision can be permanently lost within a matter of weeks. If a patient has developed wet AMD in one eye, there is a 5 to 10% risk per year that the fellow eye will also develop exudative changes (15).

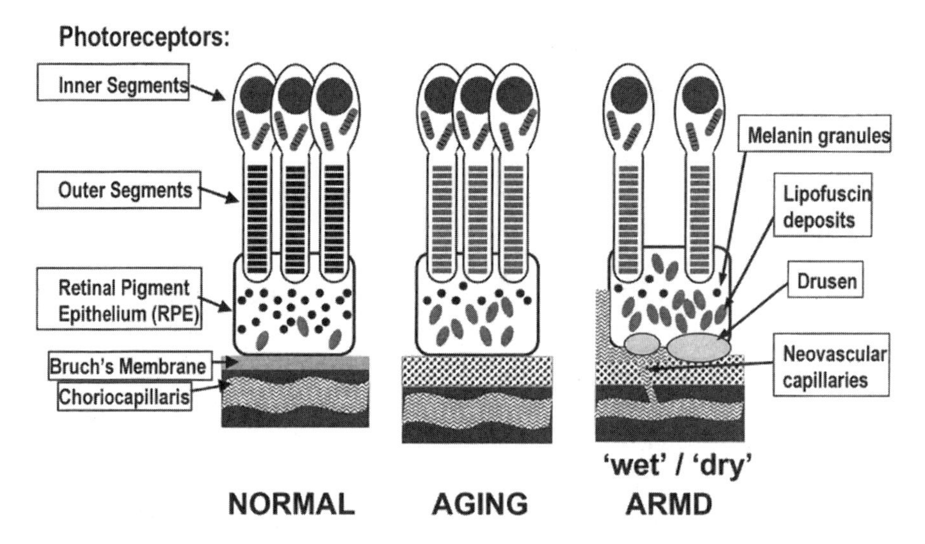


**Figure 6** Aging of the retina/RPE organocomplex during the course of AMD. With increasing age, photoreceptors are lost, disc membrane lipids are oxidized, lipofuscin accumulates in the RPE, Bruch's membrane thickens, drusen are deposited between the RPE and Bruch's membrane, and choroidal neovascularization may occur.

Treatment options for advanced AMD are limited. If the choroidal neovascular membrane is small, well-defined, and outside of the fovea, it can be treated with laser ablation (16), although the persistence and recurrence rate after treatment can be as high as 50% (17). If the lesion extends under the fovea, laser ablation can still be performed, but the permanent laser-induced tissue damage is often unacceptably high for the physician and the patient (18). Recently, photodynamic therapy using light-activated photosensitizing dyes has been introduced as a less damaging alternative for the treatment of subfoveal neovascularization (19). The majority of patients, however, are ineligible for any laser treatments due to the size or other angiographic characteristics of the lesion. Heroic surgical interventions such as subfoveal surgery or macular translocation benefit only a few patients, and other interventions such as external radiation and antiangiogenic compounds are still in clinical trials. Ultimately, the vast majority of patients with wet AMD will end up legally blind in the affected eye. Treatment options for patients with geographic atrophy are even more limited, since there are no proven interventions that can halt the progression of the disease. Restorative treatments that could benefit these patients such as retinal and RPE transplantation, gene therapy, and various prosthetic devices are still in the very early phases of research.

## IV. RISK FACTORS FOR AMD

AMD is a disease that results from a complex interaction of a variety of environmental and acquired risk factors, some more established scientifically than others. Identification of many of these risk factors originates in large part from the study of the epidemiology of AMD (20). These studies of large populations are designed to determine whether or not associations exist between the disease and an inherited or acquired factor and whether or not there is a cause and effect (21). They can have a variety of designs, including cross-sectional to look for the prevalence of the disease in a particular population at a particular time; case-control, whereby affected patients are matched to unaffected individuals with similar demographic characteristics; or prospective cohort studies to look for the incidence of new disease or significant progression in well-characterized study participants. These studies are all subject to the usual pitfalls of epidemiological research, including chance associations, inadequate sample sizes, selection bias, investigator bias, recall bias, misclassification, loss to follow-up, and various other confounding factors. As a result of these pitfalls and of the fact that independent studies often have overt or subtle differences in design and execution, they frequently come to opposite conclusions as to whether a particular risk factor is associated with AMD. Thus, the best evidence to support a risk factor's true association with AMD occurs when multiple independent studies reach the same conclusion.

Increasing age is a definite risk factor for AMD. Below age 50, virtually all patients with significant macular pathologies resembling AMD actually have inherited macular dystrophies such as Stargardt's disease, Best's disease, Sorsby's fundus dystrophy, dominant drusen, or the pattern dystrophies (22). Within families, these diseases are highly concordant in clinical presentation, and it is often possible to establish dominant or recessive inheritance patterns. The prevalence of AMD progressively rises with age, such that approximately 30% of the United States population over age 75 exhibits at least the early signs of AMD, and 7% have advanced stages of the disease including geographic atrophy or exudative disease (23). As will be discussed in detail later, this increased prevalence with age is thought to be due in part to cumulative oxidative damage to ocular structures from environmental insults.

Since it is commonly noted by patients and their clinicians that AMD seems to ''run in families,'' hereditary predisposition is thought to play an important role. Various epidemiological studies have confirmed this notion. Siblings of AMD patients are more likely to exhibit signs and symptoms of AMD relative to the general population (24), and monozygotic twins are more likely to be concordant for AMD than dizygotic twins (25). Actually proving there is a genetic risk factor for AMD has been quite difficult, however. The large kindreds (typically >10 definitely affected individuals and a similar number of age-matched definitely unaffected members) required for genetic linkage analysis are relatively hard to ascertain due to the late onset of the disease and its variable clinical presentation. Moreover, linkage analysis is further complicated by the high prevalence of the disease and the high likelihood that multiple genes are responsible for genetic susceptibility to visual loss from AMD. To date, genetic linkage for typical AMD has been established for only one large family with dominant inheritance and linkage to a locus on chromosome 1q (26). Alternatively, some investigators have used the candidate disease approach whereby the genetic defects responsible for inherited early-onset macular dystrophies that resemble AMD are screened for in AMD patients. So far, this approach has failed for all of the dominant macular dystrophies examined. Recessively inherited Stargardt's disease (27), on the other hand, does seem to be a risk factor for the development of AMD. Initially, it was reported that up to 16% of AMD patients have heterozygous mutations in ABCR (28), the gene that encodes for rim protein, a rod-cell-specific protein that appears to be involved in vitamin A transport out of the outer segment (29). This report has been disputed by other groups who found much lower mutation rates (30). Recently, the association of ABCR mutations and risk of AMD has been confirmed by a large international consortium of researchers who demonstrated that that the two most common ABCR mutations, G1961E and D2177N, are significantly more prevalent in

AMD patients relative to control populations (31). They found that 53/1385 affected patients had G1961E or D2177N mutations versus 13/1478 control individuals ($p < 0.0001$) (22).

Although there is a general perception by many ophthalmologists that AMD is more common in women and in lightly pigmented people, the epidemiological evidence is surprisingly weak (20). The perception that AMD is more common in women may be clouded by the fact that there are substantially more elderly women than elderly men. Studies of pigmentation have come to mixed conclusions, but in general, it seems that darkly pigmented individuals have somewhat lower rates of advanced AMD (32). Some studies have also found that patients with light colored irises or a history of an age-related decrease in iris pigmentation have a higher rate of AMD (33). The possible mechanistic basis for the protective effects of pigmentation will be explored in later sections of this chapter.

The identification of possible environmental risk factors for AMD is an area of very active research interest because these factors could potentially be modified by individuals to reduce their risk of visual loss. There is also a substantial amount of public misinformation in this area since the scientific evidence is still incomplete, yet many members of the public and their clinicians are eager to embrace any possible approach to lessen risk of blindness when AMD is still in its early stages. As will be discussed extensively in the following sections, many of these environmental risk factors appear to share the common mechanism of promoting oxidative damage to sensitive ocular structures such as the highly unsaturated lipids of the rod outer segments, while others may alter blood circulation patterns to the retina and the choroid.

Cigarette smoking is the best supported environmental risk factor for AMD. Multiple epidemiological studies have determined that smokers have higher rates of AMD, especially the exudative form (34,35). The underlying mechanisms are likely to be complex due to the myriad effects of smoking on the body. Smokers are subject to significant whole-body oxidative stress from ingested toxins that can dramatically reduce the body's levels of circulating and tissue-based antioxidants (36). Heavy smokers also have substantially impaired small vessel circulation, which may contribute to alterations in retinal and choroidal blood flow that may predispose to choroidal neovascularization (37).

Cardiovascular associations with AMD have been examined in several studies. Poorly controlled hypertension appears to increase one's risk of exudative AMD (38), and some epidemiological investigations have suggested that there is a significant association between coronary artery disease and AMD (20). Some of this association may derive from alterations in choroidal blood flow induced by cardiovascular disease or from common predisposing risk factors such as smoking and diet.

Excessive light exposure has often been implicated as a risk factor for AMD, but this has been quite challenging to prove epidemiologically (39). Since it is difficult to quantify an individual's long-term light exposure with any degree of accuracy, published studies supporting a positive link between light exposure and risk of AMD have generally had to focus on populations with extremely high daily light exposures such as fishermen on the Chesapeake Bay (40). It is not clear that studies such as this are relevant to the moderate daily light levels typically encountered by the general population. It is certainly well known that light, especially at the shorter wavelengths of the visible spectrum, is an efficient initiator of free radical reactions. The macula is clearly at substantial risk for light-initiated damage since it is an oxygen-rich environment with highly unsaturated lipid membranes where environmental light is actively focused by the eye's cornea and lens. Cell culture and laboratory animal models for ocular light damage have been developed to explore underlying mechanisms and to provide a means to test possible protective strategies. Although these models have some inherent limitations that may prevent extrapolation to the human system, recent studies to be discussed in the following sections have yielded intriguing results that may help to guide rational interventions against AMD.

The general category of nutritional risk factors and interventions for AMD is a very active area of AMD research. The American public is quick to embrace nutritional interventions, since most supplements have relatively low risks of toxicity, are available without a physician's prescription, and there is abundant advisory information available in the lay press and on the Internet. From an epidemiological standpoint, nutritional studies are complex and fraught with numerous pitfalls. Cross-sectional and case-control nutritional studies are usually single ''snapshots in time'' that may not reflect dietary patterns earlier in life. These studies often use food frequency questionnaires that have substantial limitations and are subject to recall bias and other artifacts. Prospective cohort studies, whether observational or interventional, may provide better quality data, but they may require very large populations followed for long periods of time to yield statistically significant data. Some studies try to assess nutritional status through measurement of blood levels of relevant nutrients. There are severe limitations with this approach also, since blood levels for many nutrients can vary widely over brief spans of time, and blood levels of a particular nutrient may be a poor indicator of levels of the nutrient in the relevant target tissue. Myriad studies have tried to address the role of nutrition in AMD, but no consensus recommendations have emerged (41). Many, but not all, of these studies have focused on specific antioxidant nutrients such as vitamins A, C, and E and the carotenoids or on minerals such as zinc and selenium that may act as cofactors for antioxidant enzymes. Other studies have focused on more general aspects of nutrition, such as fat and alcohol intake, which can have important influences on cardiovascular status. Currently, the best hope for any definitive answers as to whether supplementation with some of these antioxidant nutrients can play a role in preventing progression of AMD is expected in the near future from the age-related eye disease study (AREDS) (42), a large multicenter randomized placebo-controlled study designed to assess the effect of vitamins C, E, β-carotene, and zinc in more than 5000 subjects followed for approximately 5 years, but this study's impact will be limited by the fact that two key carotenoid nutrients, lutein and zeaxanthin, were not included.

## V. MECHANISMS OF FREE RADICAL PRODUCTION IN THE RETINA

### A. Metabolic Free Radicals Produced in the Retina

The retina is characterized by a high rate of metabolism and the highest rate of oxygen utilization in the body (direct oxygen tension measurements in guinea pig retinae indicate an average consumption of $1.1 \pm 0.09$ mL $O_2$/min per 100 g of tissue) (43). Presumably, this physiological phenomenon is required to facilitate the high turnover of vision-specific compounds (e.g., retinoids and opsins) in the photoreceptor cells and by the high intensity of electrical and chemical transmission that is observed in the neural retina. As a result, the human retina is well vascularized, consumes high amounts of nutrients, has a large number of mitochondria, and is very sensitive to hypoxia. Clinically, even brief ischemia may lead to irreversible vision changes (both in the absence or in the presence of reperfusion injury).

Metabolic free radicals, the majority of which are reactive oxygen species (ROS), arise in the retina as a result of several mechanisms (44):

1. Electron leakage from the normal pathway of reduction of oxygen via cytochrome oxidase.
2. The ubiquinone mechanism in the intact mitochondria ($UQ\bullet^- + O_2 \rightarrow UQ + O_2\bullet^-$).
3. Oxidation of various biomolecules such as glucose by glucose oxidase and autooxidation of hemoglobin, catecholamines, and other compounds.
4. Normal enzymatic action of flavin dehydrogenases and xanthine oxidase (e.g., xanthine + $H_2O + O_2 \rightarrow$ uric acid + $O_2\bullet^- + H^+$).
5. The Fenton–Haber–Weiss reaction [approximated as $Fe^{2+}$ (or $Cu^+$) + ROS ($H_2O_2$, $O_2\bullet^-$) $\rightarrow Fe^{3+}(Cu^{2+}) + HO\bullet$].

It is worth noting that xanthine oxidase has recently been positively identified in human cone photoreceptors (45), a finding that emphasizes the importance of these metabolic free radicals in the general picture of free radical generation in the intact retina. Once produced, these metabolic free radicals enter the general pool, where they react with each other and with various cellular and extracellular compounds.

## B. Free Radical Production via the Lipid Peroxidation Pathway

Retinal photoreceptors are characterized by the presence of sophisticated membrane constructs consisting of stacks of closed discs in the rods and open discs or invaginations in the color-sensitive cones. As a result, the retina contains higher than usual quantities of lipid bilayers characterized by low rigidity and extraordinarily high relative concentrations of polyunsaturated fatty acids. For instance, docosahexaenoic acid (22:6) forms 19.5% of the retinal phosphatidylcholine, 34.2% of the retinal phosphatidylethanolamine, and 34.1% of phosphatidylserine, while corresponding numbers for saturated linoleic acid (18:0) are 14.9%, 36.2%, and 28.1% (46). Retinal phospholipids (PL), thus, are prone to extensive lipid peroxidation that, in the absence of scavenging compounds, may proceed indefinitely as a chain of the following reactions:

$$PL\text{-}H + HO\bullet \rightarrow PL\bullet + H_2O$$
$$PL\bullet + O_2 \rightarrow PLO_2\bullet$$
$$PLO_2\bullet + PL\text{-}H \rightarrow PLOOH + PL\bullet$$

Empirically, certain general rules appear to hold. Of all the reactive oxygen species, lipid peroxidation is thought to be most commonly initiated by either singlet oxygen ($^1O_2$) or hydroxyl radical ($HO\bullet$) attack, with the rest of the common ROS such as superoxide radical ($O_2\bullet^-$) and hydrogen peroxide ($H_2O_2$) playing a somewhat secondary role (47). Also, the first step, abstraction of hydrogen, occurs more readily in phospholipids formed from fatty acids with high degrees of unsaturation.

Organic equivalents of the Fenton reaction may also contribute to generation of peroxyl ($RO_2\bullet$) and alkoxyl ($RO\bullet$) radicals via the following mechanisms (48):

$$Fe^{2+}(Cu^+) + ROOH \rightarrow Fe^{3+}(Cu^{2+}) + RO\bullet + OH^-$$
$$Fe^{3+}(Cu^{2+}) + ROOH \rightarrow Fe^{2+}(Cu^+) + RO_2\bullet + H^+$$

These reactions are deleterious to the function of the lipid bilayer and, hence, the cell per se. One of the most important mechanisms that mediates the loss of function and structural damage to the outer segments of photoreceptors appears to be loss of membrane fluidity. Lipid peroxidation accounts for the major share of such loss observed in older subjects, more so than age-related accumulation of another rigidifying agent, cholesterol. Similar changes occur in experimental models after dietary restriction of polyunsaturated fatty acids, especially 22:6, 22:5, and 20:4 varieties (49).

An interesting attempt to measure the ''normal'' extent of peroxidation of retinal lipids involved chromatographic separation of lipid peroxides and nonoxidized lipids from Bruch's membrane, a nonvascular acellular layer of the retina where artifactual postmortem oxidation would be expected to be minimal, making these data more representative of the living state. The ratio was found to be high (about 1:200), which, as the authors surmise, indicated that the high concentration of polyunsaturated fatty acids in the retina makes it more susceptible to peroxidation (50).

## C. Free Radicals Arising from Neurotransmitters

In addition to the metabolic and lipid peroxidation pathways, the retina, as a neural tissue, is prone to the production of free radicals that arise from auto-oxidation and enzymatic oxidation of neurotransmitters. Although more than 40 chemicals have been implicated as retinal neurotransmitters, only a few of them appear to be established as sources of free radicals. Most interesting are dopamine, which is present in the CA-1 and CA-2 amacrine cells, and nitric oxide (NO), which serves as a signaling molecule that may be involved in the inhibitory modulation of color and contrast signals by the spiny amacrine cells. In addition, glial cells of the retina, the Müller cells, possess the inducible NO-synthase and have been shown to produce significant amounts of NO after stimulation with IFN-$\gamma$ and lipopolysaccharide both in vitro and in retinal pathology. This process has been implicated in development of the RCS retinal degeneration in rats and, thus, may be of significance in the case of AMD (51,52).

Dopamine is well established as a source of $H_2O_2$ and $O_2{\bullet}^-$, both from its auto-oxidation and its oxidation by monoamine oxidase (53). Accumulation of dopamine is observed in the aging brain, and it is positively correlated with increased rates of lipid peroxidation, a finding that might play an important role in the age-related diseases of the retina whose lipids are highly polyunsaturated (54). In addition, a reactive semiquinone product of dopamine auto-oxidation has been reported, which may serve an important role in generation of superoxide radical via the following mechanism:

$$DSq{\bullet}^- + O_2 \rightarrow Dq + O_2{\bullet}^-$$

In this reaction $DSq{\bullet}^-$ is the free radical anion of the dopamine semiquinone, and Dq is the dopamine oxidation product (54).

Nitric oxide is a highly reactive free radical that is believed to have a role in many physiological processes. By itself, it is not a pure prooxidant; however, its reaction with the superoxide radical $O_2{\bullet}^-$ produces a potent oxidant, peroxynitrite ($ONOO^-$), known to cause extensive cellular damage via reaction with thiols, lipids, and selenocysteine. Since the enzyme superoxide dismutase competes with NO for the superoxide radical, it has been theorized that formation of $ONOO^-$ may occur in the vicinity of cells that produce NO only if local concentrations of NO are in the 1 to 10 $\mu$M range (55). The retinal photoreceptors and neural cells appear to fit the latter criterion, since peroxynitrite-induced plaques have been recently observed in vivo in autoimmune uveitis (56). The authors concluded that the presence of $ONOO^-$ correlated with the pathological oxidation demonstrated in this area, especially lipid peroxidation in the photoreceptors (56).

Yet another important question concerns the balance of positive and potentially detrimental activities of NO and $ONOO^-$ in the retina. Some light on this matter was shed by coculturing of retinal neuronal cells (NO-synthase negative) with retinal glial cells (Müller cells) that may produce NO upon activation. NO appears not to be toxic to Müller cells themselves, but causes death of neuronal cells. Peroxynitrite or superoxide scavengers, but not selected antiapoptotic agents, appear to prevent such lethality (52).

Modulation of NO production, especially by the endothelial cells, may be afforded by glutamate, the most abundant neurotransmitter in the retina. It is believed to convey signals from rods and cones to the bipolar cells and to the next order neurons. It is also known to freely diffuse from these cells, thus affecting the vascular structures of the retina (57). In addition, in ischemic injury, glutamate causes production of ROS via overstimulation of the glutamate ionotropic receptors, which in turn leads to calcium-dependent activation of enzymes such as NO-synthase and phospholipase A2 (58).

An in vivo experiment appears to further the claim that neurotransmitters might be major sources of free radicals in the retina. According to Bush and Williams, optic nerve section a week or more before illumination affords protection to retinas exposed to toxic amounts of light (59). This might be a consequence of the known phenomenon of cessation of production of neurotransmitters in the sensory organ after the connection with the brain has been severed.

## D. Lipofuscin and Retinoids as Ocular Photosensitizers

Photochemical damage has been implicated as both an etiologic and pathogenic factor in the evolution of AMD. Short-wavelength (blue) light is thought to be particularly important because of its ability to induce free radicals from lipofuscin (LF), a pigment produced in ocular cells during aging. Ocular lipofuscin's major constituent is A2E, a pyridinium bis-retinoid produced by the condensation of two molecules of retinal with one molecule of ethanolamine (60,61). The initial step in this process, the Schiff base reaction of two retinal molecules with one molecule of phosphatidylethanolamine, is thought to occur in the outer segments of photoreceptors during light-induced activation of the visual cycle of rhodopsin regeneration, due to the large amounts of all-*trans*-retinal released during bleaching of rhodopsin (Fig. 7). The ABCR/rim protein of the outer segments plays an important role in facilitating the ATP-dependent transport of retinal out of the discs so that it can be reduced to retinol by an alcohol dehydrogenase enzyme in preparation for the retinoid's transport back to the RPE, where it will be esterified to long-chain fatty acids and then isomerized to the visually active 11-*cis*-conformation by an isomerohydrolase enzyme (29). Under high light conditions, ABCR/rim may not be able to keep up with the massive amounts of cycling retinoids, allowing substantial amounts of retinal to accumulate in the outer segment membranes where it can react with phosphatidylethanolamine to begin the process of lipofuscin generation. This pathological process of lipofuscin accumulation is amplified in Stargardt's macular dystrophy, a recessively inherited disease in which both copies of the ABCR gene are defective (27). Up to 16% of AMD patients may have dominant mutations in ABCR, indicating that enhanced lipofuscin production may play a significant role in the progression of some cases of AMD (28). A transgenic ABCR knockout mouse model has been shown to have a dramatic enhancement of lipofuscin production when raised in high light conditions relative to dark-reared litter

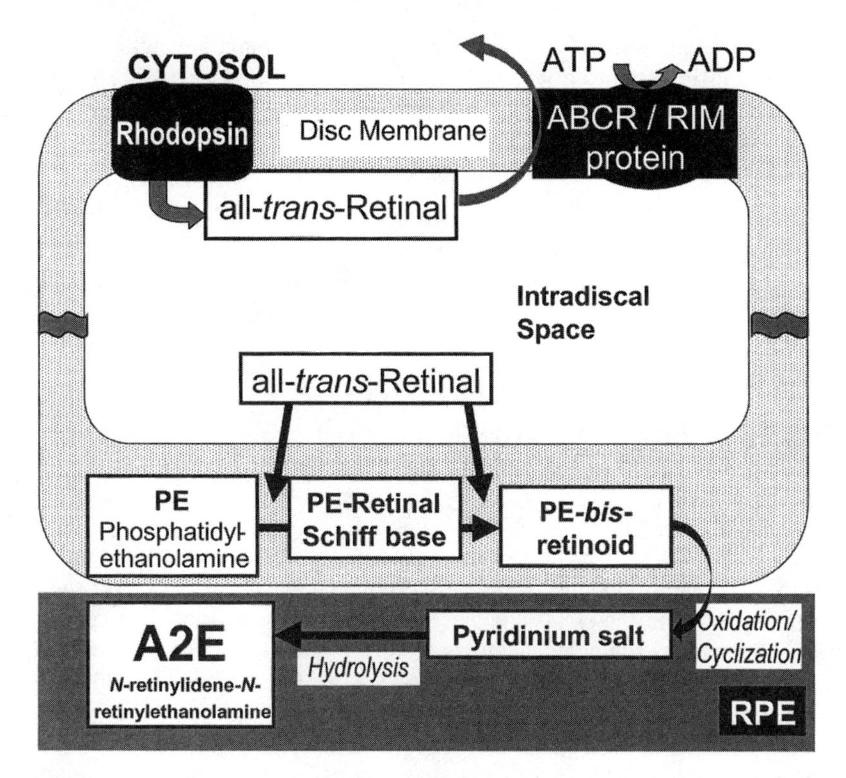


**Figure 7** Schematic representation of the formation of A2E, the major component of lipofuscin. Ordinarily, all-*trans*-retinal released from bleached rhodopsin is pumped out of the disc membrane by ABCR/rim. In high light conditions or in the presence of a dysfunctional ABCR/rim protein, excessive amounts of all-*trans*-retinal can accumulate in the disc membrane. Two molecules of retinal can then react with phosphatidylethanolamine (PE) to form a PE-bis-retinoid adduct. After disc phagocytosis by the RPE, the final reactions required for A2E formation, oxidation/cyclization and fatty acid ester hydrolysis, take place in the lysosomes.

mates with the same mutation (62). This phenotype, as expected, is more severe in homozygous ABCR knockout mice relative to heterozygous ABCR knockout mice.

Lipofuscin precursors generated in the photoreceptor outer segments eventually end up in the RPE cells as photoreceptor disk membranes are shed and phagocytosed by the RPE (63). Within the RPE lysosomes, the final steps of lipofuscin production, cyclization, and hydrolysis of the fatty acid esters are thought to occur. Lipofuscin is indigestible by RPE, rendering accumulation of lipofuscin granules an excellent marker of the aging process in these postmitotic cells (64). Since RPE cell processes are intimately apposed to the photoreceptor cells, free radicals produced by the granules of lipofuscin can easily cross back into the outer layers of the retina, where they can attack the polyunsaturated fatty-acid-enriched outer segments of photoreceptors.

Lipofuscin is a known phototoxic agent that has been extensively characterized. Higher levels of LF are observed in cells with lower levels of vitamin E and other antioxidants (65). Its presence in RPE and its unique photochemical properties might account for the unusual ability of blue (as opposed to more energetic purple and minimally less energetic blue–green) light to cause extensive and irreversible damage to the RPE and retina (66). Isolated lipofuscin granules are known to fluoresce and generate free radicals upon illumination in various media (67). In nonpolar media, generation of singlet oxygen by extracted lipofuscin drops tenfold when the incident light's wavelength is changed from 420 to 520 nm (68). The corresponding drop in energy of the light is less than 20%, suggesting that a wavelength-specific mechanism is in place.

The following features of extractable lipofuscin appear to be agreed upon (68).

1. Illumination of lipofuscin leads to formation of a transient species with a broad absorption peak at approximately 440 nm.
2. The lifetime of these species is approximately 10.3 μs, a relatively long time in free radical chemistry.
3. Illuminated lipofuscin enters the triplet state:

   $^0$LF(ground state) + hν → $^3$LF

4. Triplet lipofuscin reacts with oxygen (whose ground state is a triplet), producing singlet oxygen and ground-state lipofuscin:

   $^3$LF + $^3O_2$ → $^0$LF + $^1O_2$ (singlet oxygen species)

Given that ocular lipofuscin consists largely of condensation products of amines and free retinoids, it is also important to look at the photochemistry of the latter, since free retinoids may be transiently present in ocular membranes at high concentrations during intense light illumination. The photochemistry of all-*trans*-retinal and all-*trans*-retinol closely resembles that of lipofuscin (69) with the triplet state of both retinoids exhibiting broad visible light absorption with extinction coefficients around $10^5$, lifetimes of around 10 μs, and absorption maxima in the range of 440 to 470 nm (70,71). Of course, the most common fate of excited retinal is photoisomerization, a process that is employed in light perception in virtually all living organisms. This extensively studied reaction could be called a ''mixed blessing,'' since it might have distracted researchers from addressing what we propose as the ''Multi-Photon Blue Light Toxicity Hypothesis.''

Like lipofuscin, free retinoids may enter the triplet state via a variety of mechanisms including reaction with another excited molecule, especially if it is in its triplet state:

$^0$RET + Sensitizer* → $^3$RET + Sensitizer

or

$^0$RET + $^3$Sensitizer → $^3$RET + Sensitizer

(other mechanisms are possible).

Triplet retinoids can serve as a sensitizer in reaction with ground-state oxygen, generating singlet oxygen and ground-state retinoids according to the following scheme, which does not involve absorption of additional photons.

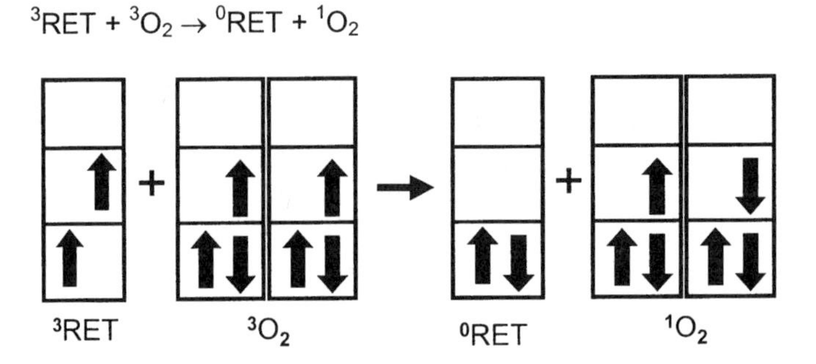

Triplet states of retinoids, however, are relatively long-lived species that absorb light of a broad spectrum with maximum absorption in the blue range around 440 to 450 nm. The energy of this light (somewhat less than 3 eV) may be sufficient to regenerate the triplet retinoids in the presence of the ground-state oxygen and/or other photosensitizers. The essence of this hypothesis is that by absorbing blue light, certain triplet-state compounds (retinoids and lipofuscin) could act as photosensitizers to generate substantial amounts of singlet oxygen from ground-state oxygen in the lipophilic environments of the lipofuscin deposits in the RPE and in the membranes of the photoreceptors according to the following scheme:

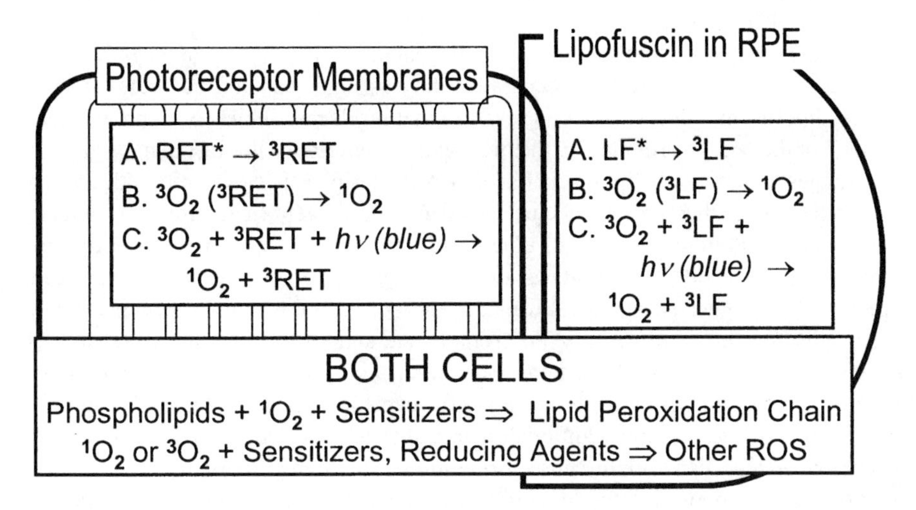


This reaction might occur in several stages:

First, a higher energy triplet state ($^3$RET*) could be generated from the initial triplet state by absorption of blue light:

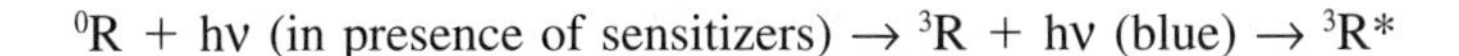

$$^0R + h\nu \text{ (in presence of sensitizers)} \rightarrow {}^3R + h\nu \text{ (blue)} \rightarrow {}^3R^*$$

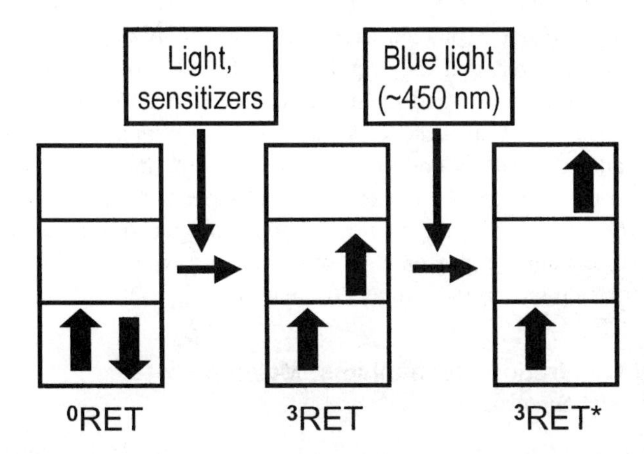


$^3$R* can then react with triplet oxygen to generate $^3$R and singlet oxygen:

$$^3R^* + {}^3O_2 \rightarrow {}^3R + {}^1O_2$$

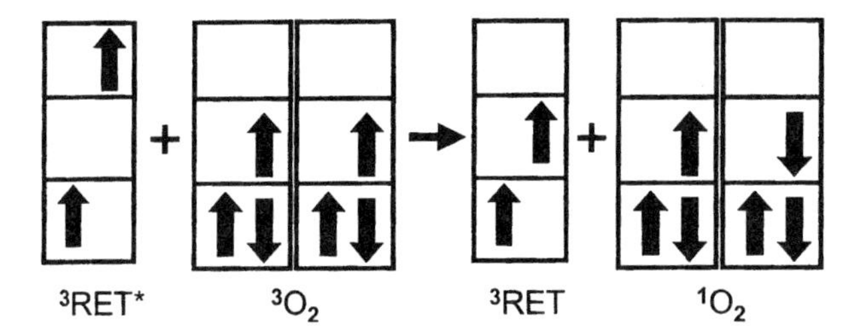


The triplet retinoid of lipofuscin is then available to absorb more blue light and generate even more singlet oxygen.

This theory departs from the free radical generation hypothesis as suggested by Rozanowska and colleagues (68). Specifically, they assume that one quantum of blue light excites lipofuscin, which leads to a reaction that produces singlet oxygen (as a sole ''possessor'' of the energy provided by the blue photon) and leaves lipofuscin in its ground state. We suggest that not only lipofuscin but also excited triplet species of retinoids absorb blue light in the retina, thus serving as blue sensitizers for oxygen. In addition, we suggest that absorption of additional photons of blue light (with approximate energy of 3 eV) by triplet-state retinoids inside a lipid bilayer or similar hydrophobic environment may lead to the formation of higher-energy-level triplet species of retinoids with enough energy to generate singlet oxygen from the oxygen ground state and still remain in the sensitized triplet state, thus permitting a self-sustaining chain reaction. The same consideration may be applied to lipofuscin, whose triplet state is also capable of absorption of blue light.

It is also possible that the additional energy that is provided by absorption of the second photon of blue light might cause formation of the highly energetic $^1\Sigma^+_g$ species of singlet oxygen as opposed to the less energetic $^1\Delta_g$ singlet species. The former species is known to be highly unstable (life time $< 200$ ns), however, and its chemistry is unknown, thus precluding us from further speculation (72).

The most severe criticism of the hypothesis proposed above might be offered on the basis of the empirical claim that the singlet-generation mechanism is not traditionally considered to be a major free radical producer in the retina due to its limited effectiveness and the claim that electron paramagnetic resonance (EPR) spectra of ex vivo products of acute light damage to retina are better explained by an electron (or hydrogen) process (73). It must be remembered, however, that the specifics of retinoid photochemistry in polar systems (such as dimethylformamide and methanol) cannot be readily assumed to be representative of the photochemistry of these substances in the hydrophobic environment of the cellular membranes or inside a grain of lipofuscin. Furthermore, blue-light-induced singlet-oxygen generation has certainly been noted in model systems that utilize actual lipofuscin granules (68,74). Finally, the singlet-oxygen chain reaction hypothesis proposed here is designed to explain the phenomenon of excessive toxicity of blue light and not other kinds of light (wide spectrum or monochromatic).

### E. Other Free Radical Generation Mechanisms: Melanin and Prostaglandin G/H Synthase

Melanin, the common dark brown (eu-melanin) or reddish brown (pheo-melanin) pigment of RPE has been variously implicated as a prooxidant and an antioxidant. Its prooxidative function in the RPE has recently been investigated. Melanin granules isolated from cultured RPE cells appear to have a direct catalytic effect on photoxidation of docosahexaenoic acid (22:6) when illuminated with argon laser light (a mixture of 488- and 514-nm blue–green light). The relation appears to be near linear in the physiologically relevant dosages of light (0 to 300 mW/cm$^2$) (74).

Proposed mechanisms focus on melanin semiquinone's role in abstraction of hydrogen from polyunsaturated fatty acids:

$$\text{MSq}\bullet \text{ (semiquinone radical)} + \text{Lipid-H} \rightarrow \text{Lipid}\bullet + \text{MHQ (hydroquinone of melanin)}$$

This might happen in conjunction with or in competition with the following superoxide-generating reaction:

$$\text{MSq}\bullet^- + O_2 \rightarrow O_2\bullet^- + \text{MQ (quinone of melanin)}$$

The fact that retina is more sensitive to shorter wavelength (blue) light has also prompted a more detailed investigation of light-activated enzymatic mechanisms of such toxicity. In vivo experiments indicate that blue light activates prostaglandin G/H synthase in the retina, leading to an increased rate of lipid peroxidation, especially in the photoreceptors (75).

## VI. MECHANISMS OF FREE RADICAL PROTECTION IN THE RETINA

### A. Enzymatic Scavenging: The Role of SOD, GPx, Catalase, and GR

It has been hypothesized that susceptibility to visual loss from AMD may stem from age-related deficiencies of free-radical-scavenging enzymes and cofactors found in the retina and/or RPE. These antioxidant enzymes include superoxide dismutase (SOD) (76), glutathione peroxidase (GPx) (77), catalase, and glutathione reductase (GR) (78). Relevant cofactors include zinc, copper, and manganese for various species of SOD, heme for catalase, and selenium for GPx and GR. Of these, zinc and selenium are most commonly promoted by the nutraceutical industry as ''eye friendly'' antioxidants, possibly because of the concerns with copper and iron prooxidant activity in Fenton-style reactions. Although no scientific data have been published on the frequency of consumption of these mineral supplements for the specific reason of ''eye health promotion,'' our informal data indicate that a majority of ophthalmic patients exhibit at least superficial knowledge of the claim that ''zinc and selenium are good for your eyes.'' A review of questionnaires filled by attendees of a continuing medical education seminar shows that approximately 20% (75 out of 400) have encountered use of zinc and/or selenium by their patients with the intent of ''other antiaging benefits, such as preservation of senses, like sight, hearing, balance, taste, etc.'' (NB Katz, unpublished data).

The effect of age on antioxidant enzymes in the retina has been studied extensively. In one of the most authoritative studies, SOD and GPx activities were found to decline in the peripheral retina but not in the macula, while the other antioxidant enzymes remain unaffected. At the same time, interindividual variability of these enzymes appears to be high and unrelated to age (78).

The link between expression of individual enzymes and photic damage appears to be more reliable. For instance, manganese-SOD, a naturally occurring enzyme in the retina, can be induced by bright light exposure and has been shown to be deactivated by phototoxic doses of light (79).

The role of retinal catalase has been elucidated in an animal model in which administration of the specific inhibitor of this enzyme, 3-aminotriazole, led to a measurable increase of retinal concentrations of lipid peroxides, measured as thiobarbituric acid-reactive substances (TBARS), suggesting a role for catalase in the termination of the chain reaction of lipid peroxidation (80).

Moderate zinc deficiency leads to higher oxidative stress in experimental animals, a finding that supports the claim that, under normal conditions, it is zinc and not copper that limits the activity of Cu,Zn-SOD. In addition, retinal and RPE metallothionein levels appear to be important for the overall antioxidant protection of the retina (81). Zinc is widely promoted in supplements targeted toward the population at risk for AMD based largely on a clinical study by Newsome and colleagues, which reported significant protection against the progression of AMD in individuals randomized to zinc supplementation (82). Subsequent studies by other researchers have failed to confirm these findings. It has been noted that the Newsome study did not measure visual acuity in a standardized manner and that the placebo group had unusually rapid progression of AMD (41). It is hoped that the upcoming results from the AREDS study will provide more definitive guidance on the role of zinc supplementation for AMD.

### B. Vitamins C and E

Vitamins C and E are often considered ''gold standard'' antioxidant vitamins since their function and importance in maintenance of pro- and antioxidative homeostasis is well characterized in a wide variety of biological systems. Histological studies show that vitamin E predominates in the lipophilic structures of the retina and the RPE, while water-soluble vitamin C appears to be predominantly localized in the neural retina, away from the photoreceptors (42,83,84). In numerous animal models, vitamin E has been shown to afford protection to photoreceptors, especially with respect to photic damage (85), and it appears that its concentration is higher in photoreceptors in animals that are subjected to photic shock, which might suggest a specialized mechanism of uptake and storage of this vitamin (86). Vitamin E levels are high in the fovea, and they have a relative minimum in the perifoveal area, a region that often exhibits the first signs of AMD-related ocular pathology (87).

It has been established that the primary effect of experimental vitamin E deficiency is on the photoreceptors. Ultrastructural studies, however, do not show significant morphological changes in the photoreceptors that could explain receptor dysfunction. Better clues are provided by chemical analysis of the phospholipids in the cellular membranes of the photoreceptors. Such analysis shows an irreversible loss of the long-chain polyunsaturated fatty acids from the retina, increased lipid peroxidation, and alterations in membrane fluidity that might contribute to vision loss in these animals and, theoretically, might be important for the pathogenesis of human AMD (88).

The findings discussed above have prompted a number of epidemiological studies of the role of vitamin E in prevention of AMD. Unfortunately, their results are contradictory, with some studies establishing even an adverse association of vitamin E intake with the incidence of the wet form of AMD, although such relationships did not appear to be statistically significant (89,90). Other studies have found that vitamin E is a protective factor against AMD (91), and others have failed to see any effect (92). Currently, there is insufficient evidence to assign clinically significant ''retina-protective'' qualities to vitamin E. A similar mixture of contradictory and insignificant epidemiological results has been found in studies addressing the role of vitamin C. Definitive recommendations for vitamin E and C supplementation for AMD will need to come from large-scale randomized clinical trials such as AREDS.

### C. Antioxidant Properties of Melanin

Ocular melanin in the highly pigmented RPE, in addition to acting as a prooxidant, could serve as an antioxidant, either via direct optical filtering of the potentially damaging light or via interception of free radicals, leading to the formation of relatively nonreactive free radicals of melanin. Interestingly, phototoxicity of blue and ultraviolet-A light appears to be no higher in albino rats relative to pigmented rats (93,94), arguing against an important photoprotective or photosensitizing role of melanin in vivo. On the other hand, melanin appeared to be a protective antioxidant in a laboratory model of lens-induced uveitis in which TBARS concentrations in albino guinea pigs increased by more than 200%, while in pigmented animals TBARS increased by just 40% (95).

Synthetic melaninlike substances have been shown to exhibit free radical scavenging ability with both the eu-melanin model (DOPA melanin) and pheo-melanin model (Cys-DOPA melanin). They are active in the interception of perchlorate radical with kinetic rates of $1.2 \times 10^8$ and $1.3 \times 10^8$ $M^{-1}s^{-1}$, respectively, suggesting that both molecules may scavenge $O_2\bullet^-$, HO•, and $H_2O_2$ in a site-specific manner (96).

### D. Flavonoids and Experimental Antioxidants

While herbal products such as bilberry and gingko biloba extracts have long been purported to improve vision and combat ocular degenerative diseases, presumably due to their high content of antioxidant flavonoids and related compounds, relatively little is known about their uptake and concentrations in the eye. Quercetin is the only flavonoid that has been definitively detected in the mammalian eye (97). It is found in low amounts in bovine retina, but no similar human studies have been performed. Quercetin exhibits strong antioxidant properties comparable to both vitamin E and vitamin C (rate constant for quercetin reaction with trichloromethylperoxy radical is estimated to be $3 \times 10^8$ versus $1.2 \times 10^8$ for ascorbate and $4.9 \times 10^8$ for dl-$\alpha$-tocopherol) (98). Quercetin has also been observed to afford in vitro protection from oxidative stress provoked by glutathione depletion in sensory neurons at effective concentrations of approximately 200 $\mu$M, reducing the death rate by 50% (99).

Other antioxidants that appear to afford protection to neural and retinal cells include $\alpha$-lipoic acid and its water-soluble amides (100), lazaroids (a class of synthetic steroids with no hormonal activity and purported strong antioxidant features) (101), idebenone, a novel pharmaceutical antioxidant that appears to extend some protection even if administered after the oxidative insult (102), and ebselen, a metallo-organic compound of selenium that mimics the antioxidant enzyme glutathione peroxidase and may scavenge peroxynitrite directly, leading to decreased lipid peroxidation in endotoxin-induced uveitis (103). Other promising compounds are epigallocatechin gallate, an active ingredient in green tea (104), and *N*-acetylserotonin, a dose-dependent inhibitor of lipid peroxidation, that is present in the retina as a physiological byproduct of the neurotransmitter serotonin (105).

Considerable enthusiasm has surrounded the use of the popular over-the-counter antioxidant supplement, EGb (standardized extract of gingko biloba), which contains quercetin glycosides and other polyphenolic compounds. In one study, the effects of EGb in experimental uveitis were compared to the effects of intramuscular injections of purified

Cu,Zn-SOD, with no appreciable difference noted in either group. The issue of biological availability of intramuscular SOD to the ocular tissues was not addressed, however (106).

In general, most of the evidence to support the efficacy of herbal supplements for AMD has been anecdotal and uncontrolled. Much work remains to be done to demonstrate that sufficient quantities of the active compounds are accumulated by ocular structures and that they can function as physiologically relevant antioxidants to prevent the progression of AMD.

## E. Carotenoids in the Retina

In recent years, the macular pigment carotenoids lutein and zeaxanthin have emerged as important protective compounds for the human macula. While the existence of the ''yellow spot'' or macula lutea has been well known for over 100 years, their definitive identification as the xanthophyll carotenoids comprising the macular pigment is relatively recent (107,108). They are members of the large class of compounds known as the carotenoids, most of which share a common $C_{40}H_{56}$ isoprenoid chemical structure (Fig. 8). Since they are hydroxylated, they are part of the carotenoid subclass known as the xanthophylls. β-Carotene, the most widely known carotenoid, has long been known to serve an important role in the eye because it is a dietary precursor for the vitamin A essential for the formation of visually active rod and cone rhodopsin, yet little intact β-carotene is found in the retina (109). Lutein and zeaxanthin, on the other hand, are found in abundant quantities in the retina and in other ocular tissues such as lens, iris, ciliary body, and RPE/choroid (110), yet they have no vitamin A activity due to the hydroxylation of both of their ionone rings. Since lutein and zeaxanthin cannot be synthesized by higher animals, they must be ingested preformed in the diet from sources such as green leafy vegetables (spinach, kale,

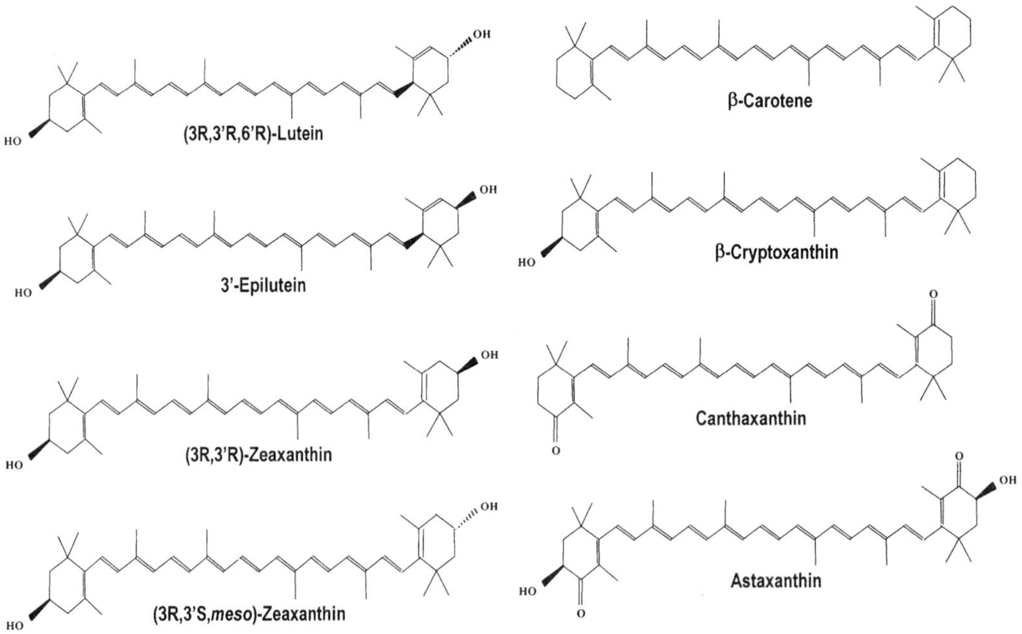

**Figure 8** Carotenoid structures.

collard greens, broccoli) and orange-yellow fruits and vegetables (corn, peaches, oranges, persimmons) (111).

At the fovea, the concentration of carotenoids is exceedingly high (approximately 1 mM), over 100 times as high as peripheral retina and more than 1000 times the concentration in the serum and other body tissues (112). Psychophysical, spectroscopic, and biochemical studies all agree that that there is a sharp peak of carotenoids centered on the fovea, which drops rapidly with increasing eccentricity (113–115); measurements in our laboratory indicate that 25 to 50% of all of the retina's carotenoids are localized to the macula, even though it accounts for less than 5% of the total retinal area. At the fovea, the concentration of zeaxanthin exceeds lutein by a factor of 2:1 (116). This ratio is inverted in the peripheral retina, which has a lutein-to-zeaxanthin ratio between 2:1 and 3:1, similar to that of the serum and most other tissues in the body (116). Cross-sectionally, the macular carotenoids have been localized to the Henle fiber layer of retina, which corresponds to the axons of the foveal cone cells (114). This anatomical localization has been challenged recently by Gass, who feels that the macular carotenoids are actually localized to the Müller cells of the fovea (117). The highly specific uptake of lutein and zeaxanthin into the macula implies the existence of specific binding proteins similar to known invertebrate xanthophyll binding proteins such a crustacyanin from lobster and a lutein binding protein isolated from the silkworm (118,119). Our laboratory has been actively pursuing this avenue of research (120). We have recently solubilized and partially purified and characterized a membrane-associated xanthophyll binding protein (XBP) from human retina and macula. Preliminary binding studies indicate that it is specific for monohydroxy- and dihydroxy-carotenols with relative affinities of lutein > zeaxanthin > cryptoxanthin > isozeaxanthin > astaxanthin. Canthaxanthin and β-carotene do not bind to XBP. Further purification and characterization of XBP is in progress.

High dietary intakes of lutein and zeaxanthin, as well as high blood levels of these same carotenoids, were found to be strongly associated with decreased risk of exudative AMD in a large case-control study—patients with the highest quintile of lutein and zeaxanthin intake had a 43% lower risk of exudative AMD relative to subjects in the lowest quintile (89,90). Several other studies have failed to see this protective effect, but they had much lower numbers of patients with exudative disease, so their statistical power was low (121–123). These various studies are summarized in Table 1. Animal studies using a primate model have demonstrated that monkeys fed a carotenoid-deficient diet for several years have depleted macular carotenoid levels and develop pathology reminiscent of early AMD (124). Several small clinical studies utilizing psychophysical techniques have demonstrated that macular pigment levels can be raised in many individuals by long-term dietary supplementation with lutein and/or zeaxanthin (115,125), and a biochemical study has suggested that eyes from donors diagnosed with AMD have lower levels of lutein and zeaxanthin in the macula and peripheral retina relative to age-matched control eyes without a known diagnosis of AMD (115). In the near future, we plan to perform expanded large-scale clinical studies on the protective role of macular carotenoids using a newly developed noninvasive objective instrument designed to measure ocular carotenoids by Raman spectroscopy (126).

Two major theories attempt to account for lutein and zeaxanthin's protective effects against macular degeneration (127,128). The blue light filtration hypothesis focuses on the extraordinarily high extinction coefficients of the carotenoids (typically ~140,000 $cm \cdot M^{-1}$ at ~450 nm) and their broad visible absorption band between 400 and 500 nm. These absorption properties are ideally suited to filter out the wavelengths of light thought to

**Table 1** Summary of Selected Epidemiology Studies Investigating the Role of Carotenoids in AMD

| Principal author(s), year of publication (Ref.) | Sample | Method of measurement | Relative risk of AMD (95% confidence intervals) for carotenoids and selected other antioxidants | Notes |
|---|---|---|---|---|
| EDCC Study Group, 1992 (89) | 421 cases, 615 controls, ages 45–74 | Biochemistry of plasma | Lutein + zeaxanthin: 0.3 (0.2–0.6)<br>β-carotene: 0.3 (0.2–0.6)<br>Vitamin C: 0.7 (0.5–1.2)<br>Vitamin E: 0.6 (0.4–1.04)<br>Lycopene: 0.8 (0.5–1.3) | Highest vs. lowest quintile analyzed, multiple factor adjustment |
| Sanders, 1993 (121) | 65 cases, 65 controls, ages 66–87 | Biochemistry of plasma | Lutein: 1.4 (0.6–3.4)<br>β-carotene: 0.5 (0.2–1.2)<br>Retinol (vitamin A): 1.2 (0.7–2.1)<br>Lycopene: 1.0 (0.4–2.6) | Highest vs. lowest tertile analyzed, age and sex matched |
| Seddon, 1994 (90) | 356 cases, 520 controls, ages 55–80 | Analysis of dietary data | Lutein + zeaxanthin: 0.4 (0.2–0.7)<br>β-carotene: 0.6 (0.4–0.96)<br>Vitamin C: 0.8 (0.5–1.3)<br>Vitamin E: 1.5 (0.9–2.4)<br>Lycopene: 1.2 (0.7–1.8)<br>Retinol (vitamin A): 0.6 (0.4–0.96) | Highest vs. lowest quintile analyzed, multiple factor matching, data for vitamins E and C excludes supplements |
| Mares-Perlman, 1995 (Beaver Dam nested case-control study) (122) | 167 cases, 167 controls, ages 43–86 | Biochemistry of plasma | Lutein + zeaxanthin: 0.7 (0.4–1.4)<br>β-carotene: 0.8 (0.4–1.5)<br>Vitamin E: 0.8 (0.4–1.5)<br>Lycopene: 1.2 (0.8–1.8) | Lowest quintile vs all others analyzed, multiple factors matched, vitamin E measured as α-tocopherol |
| Mares-Perlman, 1996 (Beaver Dam retrospective cohort study) (123) | 314 cases, 1968 subjects, ages 43–86 | Analysis of dietary data | Lutein + zeaxanthin: 1.0 (0.7–1.5)<br>β-carotene: 1.1 (0.8–1.7)<br>Vitamin C: 0.8 (0.5–1.2)<br>Vitamin E: 0.7 (0.5–1.1)<br>Lycopene: 1.2 (0.8–1.8) | Highest vs. lowest quintile analyzed, adjusted for sex and age, data for vitamins C, E includes supplements |

be responsible for the blue light hazard discussed in previous sections. Typical optical densities of foveal carotenoids range from 0.1 to 1.0, corresponding to blue light transmittance ranges from 79% to 10%, respectively (112). The anatomical localization of the macular pigment is ideal to shield the foveal photoreceptors, but the rapid fall in pigment concentration with increasing eccentricity means that large portions of the retina are relatively unprotected.

The other hypothesis proposes that the macular carotenoids function as local antioxidants in the macula. This hypothesis is problematic when one considers that the highest concentration of macular pigment is in the Henle fiber layer, well separated (~ 50 μm) from the photoreceptors they are supposed to protect. More recently, preliminary reports have indicated that a portion of the retina's carotenoids may be associated with the photoreceptor outer segments where they could more reasonably function as local antioxidants to prevent oxidative damage to the polyunsaturated membrane phospholipids (129). Biochemically, there is substantial circumstantial evidence that the macular carotenoids are subject to oxidation in vivo. Approximately 5 to 10% of the lutein and zeaxanthin in retinal tissues is present in an oxidized form, 3-hydroxy-β,ε-caroten-3′-one, a much higher percentage than what is found in the blood and nonocular tissues (130). The retina also contains a number of other nondietary lutein and zeaxanthin diastereomers, such as meso-zeaxanthin, and 3′-epilutein, compounds that may be part of metabolic pathways for the ocular interconversion of lutein and zeaxanthin and for the regeneration of these compounds from oxidation products generated in situ during light exposure (Fig. 9) (130).

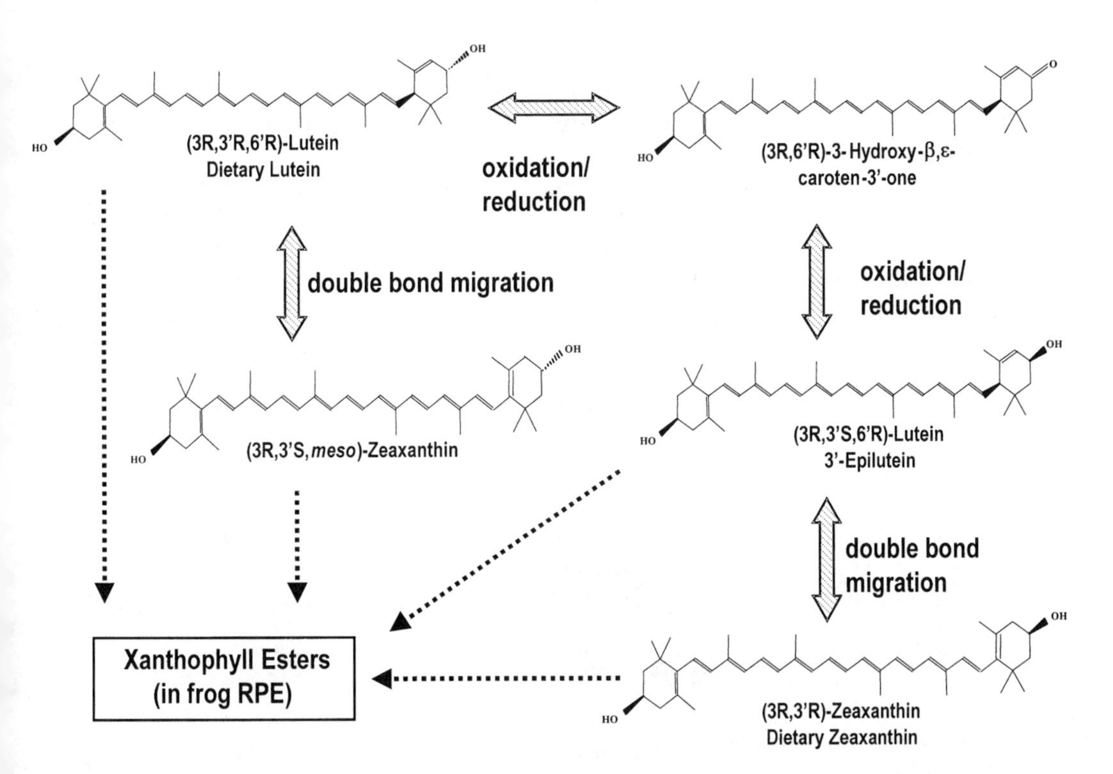


**Figure 9** Proposed metabolic conversions of carotenoids in the retina. Note that xanthophyll ester formation is not known to occur in the human eye.

We have recently obtained preliminary data that similar carotenoid metabolism probably occurs in the frog eye, making this a viable animal model to explore the ocular carotenoid metabolic pathways, although it must be remembered that the frog eye differs from the human eye in that approximately 95% of its ocular lutein and zeaxanthin is stored as esters to fatty acid in the oil droplets of the RPE (126).

Traditionally, several types of reactions of xanthophylls with free radicals have been proposed:

1. Formation of the adduct between xanthophyll and the free radical:

   $X + R\bullet \rightarrow \{X \ldots R\}\bullet$ (adduct)

   or

   $X + R\bullet \rightarrow X\bullet^{+} + R\bullet^{-}$(ion pair)

2. Oxidation (electron transfer) of the xanthophyll:

   $X \rightarrow X\bullet^{+} + e^{-}$

3. Dismutation

   $2\ X\bullet^{+} \leftrightarrow X + X^{2+}$(not a radical)

According to Rice-Evans and colleagues, the reaction immediately above is an equilibrium reaction that leads to partial regeneration of the intact xanthophyll (131).

Another group compared various xanthophylls to find that they exhibit a net antioxidant (free radical scavenging) activity, while a related compound, β-carotene appears to have more prominent prooxidative features. Among various components of the retinal xanthophyll, pure lutein appears to be a more efficient scavenger than pure zeaxanthin (rate constant of formation of lutein$\bullet^{+}$ in chloroform is $7.8 \times 10^{4}\ s^{-1}$, compared to zeaxanthin$\bullet^{+}$ formation rate of $5.7 \times 10^{4}\ s^{-1}$) (132).

Most, if not all, carotenoids are believed to exhibit the ability to terminate chain reactions of lipid peroxidation. Not surprisingly, the relative efficiencies of members of this superfamily of chemicals vary depending on the solvent and the chemical composition of the experimental phase. The keto-xanthophylls (astaxanthin, canthaxanthin) appear to be more effective than either carotenes (β-carotene) or hydroxy-xanthophylls (lutein and zeaxanthin) if the reaction involves the methyllinoleate/2,2′-azobis(2,4-dimethylvaleronitrile) (AVMN) system (133). On the other hand, in phosphatidylcholine liposomes, the same chemicals line up as astaxanthin > zeaxanthin > canthaxanthin ≫ β-carotene if a lipid-soluble inducer such as AMVN is used, and as zeaxanthin ≫ β-carotene > canthaxanthin if a water-soluble inducer such as 2,2′-azobis-(2-amidinopropane) dihydrochloride is used (134,135).

Rice-Evans et al. cite the following TROLOX–equivalent antioxidant capacity for various xanthophylls (based on oxidation of 2,2′-azobis(3-ethylbenzothiazoline-6-sulfonic acid) in medium of unspecified polarity) (131).

Lutein—1.5 ± 0.1
Zeaxanthin—1.4 ± 0.04
Astaxanthin—0.03 ± 0.03
Canthaxanthin—0.02 ± 0.02

The TROLOX equivalent, although a more ''physiological'' measurement, does not escape the criticism of the limited applicability of conclusions derived from model systems to clinical and pathological situations. For instance, the TROLOX equivalent of β-carotene is 1.9, which is higher than either vitamin E or vitamin C, both assumed to be 1.0; however, β-carotene is known to be a potential prooxidant, as indicated by the results of the CARET and ATBC clinical trials (136,137).

Shalch et al. propose to measure the effectiveness of various carotenoids by either their ability to quench singlet oxygen (zeaxanthin > meso-zeaxanthin > lutein) or their ability to regenerate the α-tocopheryl radical cation (zeaxanthin > meso-zeaxanthin > lutein), which might indeed be the most clinically relevant system of ranking (138). Of course, rate constants obtained in organic solvents (benzene, hexane, etc.) may change as the experiment is moved into more physiological conditions such as lipid bilayers.

The fact that the rate of antioxidant action of closely related chemicals depends so dramatically on minimal changes of their structure (after all, zeaxanthin differs from astaxanthin only by one carbonyl moiety on each ionone ring) prompted the investigation of the geometry of the distribution of these compounds in lipid bilayers. While β-carotene and other nonpolar carotenes are distributed randomly with local concentrations governed by the thermal equilibrium, both lutein and zeaxanthin adopt a rigid, membrane-reinforcing orientation and have been compared with ''molecular rivets'' (139,140).

The importance of xanthophylls as optical filters and local antioxidants has led to the investigation of their possible metabolism in the tissues of the eye. Xanthophyll metabolic pathways share some similarities with the metabolic pathways of retinoid compounds and, at the same time, exhibit significant differences. In frog eyes (126), xanthophylls are predominantly located in the RPE and are present mostly as xanthophyll di- and monoesters of fatty acids, which resembles the mammalian metabolism of vitamin A in which isomerization of all-*trans*-retinal is achieved via esterification in the RPE (10). In primate eyes, however, xanthophyll esters are undetectable, but they do appear to undergo oxidation, reduction, and isomerization reactions (Fig. 9), again exhibiting resemblance to the retinoid metabolic pathways (130). The xanthophyll metabolic pathways are the focus of active investigation in our laboratory, with research efforts concentrated on isolation of potential redox enzymes and hypothetical double bond isomerases. We theorize that these enzymatic actions might be attributed to members of the XBP family of proteins and may require absorption of photic energy and/or formation of xanthophyll free radicals.

Considering the unique importance and critical functions of lutein and zeaxanthin as the macular pigment, the question arises, should they be considered ocular vitamins? We offer the following arguments for and against their ocular vitamin status as ''Vitamin X.''

### 1. Vitamin X: Principles

1. Xanthophylls in the form of two chemical isomers, lutein and zeaxanthin, play critical roles in the maintenance of ocular health that may qualify their designation as human ''vitamin X.''
2. Vitamin X cannot be synthesized by humans, and its metabolism is limited to interconversion of the two components (probably via intermediates such as 3′-epilutein and meso-zeaxanthin) and limited regeneration of oxidized vitamin X by antioxidant factors.
3. Vitamin X is accumulated in high quantities in various tissues of the human body and in exceedingly high quantities in several crucial tissues of the visual system: the macula lutea and the rest of the retina, the ciliary body, and the rest of the uvea. Accumulation in the macula cannot be accounted for by passive lipophilic absorption and storage, and specific uptake likely requires specialized

xanthophyll binding proteins (XBP) that we have partially purified and characterized (141).

4. Vitamin X is a potent antioxidant with TROLOX equivalent of 1.4 to 1.5, which has the ability to intercept free radicals (oxygen and nitrogen reactive species) and to prevent their formation in tissues exposed to light by optically filtering highly damaging photons with wavelengths of 400 to 500 nm.
5. Deficiency in vitamin X is highly likely to contribute to development of a degenerative disease, age-related macular degeneration, the leading cause of new blindness in the Western countries. Deficiency states may also be associated other ocular pathological conditions.
6. High dietary consumption of vitamin X–enriched compounds leads to a statistically significant decline of relative risk for development of AMD.
7. Vitamin X levels in the circulating blood and in the ocular tissues of humans can be increased by dietary modification and supplementation, and ocular carotenoid levels have been shown to decrease in monkeys following initiation of a vitamin X-restricted diet. This warrants establishment of the recommended daily intake and recommended daily allowance for this biophysically and biochemically active substance.

### 2. *Vitamin X: Problems*

1. Little is known about which of the various forms of vitamin X (lutein, 3′-epilutein, meso-zeaxanthin, or 3R,3′R-zeaxanthin) are relatively more active or more efficient as anti-AMD factors.
2. Due to the ongoing process of characterization of XBP and the XBP-encoding gene(s), it is unknown whether an XBP-related mutation could be the genetic cause of a form of AMD.
3. Vitamin X interconverting enzymes have not been characterized yet, and the evidence for vitamin X metabolism is indirectly obtained from animal models and/or postmortem tissues.
4. Relatively little effort has gone into the establishment of the exact mechanisms for vitamin X antioxidant activity in physiologically relevant environments (lipid bilayers, cellular organelles), while there have been more studies utilizing nonbiological environments such as methanol and hexane.
5. The exact pathogenesis of AMD remains unclear, and it appears that AMD might be a mere common moniker for a variety of clinically related, yet pathogenically distinct, diseases. Vitamin X might play a significant role in some and a secondary or even an insignificant role in others.
6. The role of vitamin X as a preventative factor for AMD needs to be clarified in an extensive controlled clinical trial on a prospective basis.
7. The activity of vitamin X needs to be measured, and a system of assigning units of activity and recommended daily requirements needs to be developed for both the photochemical and the antioxidant activities of the substance.

Although we admit the existence of many open questions, we firmly believe that the preponderance of evidence supports our proposition that lutein and zeaxanthin may function as an ocular vitamin (vitamin X). Potential benefits (such as a possible decline in incidence of a sight-robbing diseases) outweigh the doubts and, thus, we feel that it is warranted to consider classifying ocular xanthophylls as a human vitamin and that a recommended daily allowance be established for this substance.

## VII. CONCLUSIONS

Since age-related macular degeneration will remain a major ocular public health problem for the foreseeable future, what advice should be given to individuals at risk for visual loss from the disease? Although there is abundant evidence suggesting that free radical–mediated oxidative damage is likely to be a significant factor contributing to the progression of AMD, definitive results supporting the consumption of antioxidant supplements is still lacking. Due to AMD's slow progression and its variable clinical course, long-term, large-scale randomized prospective clinical trials will be required. Until such trials are complete, the best recommendations for patients wishing to modify their risk of AMD is to quit smoking, avoid excessive sunlight, and to consume a healthy diet rich in fruits and vegetables. Since many elderly are unable to maintain such a healthy diet, it is not unreasonable to consider supplementation with a multivitamin containing recommended daily allowances (RDA) of the antioxidant nutrients with at least some supporting data for a beneficial effect against macular degeneration such as vitamin E, C, and zinc. Although no RDA is yet established for lutein or zeaxanthin, we support supplementation with these carotenoids for those patients with inadequate fruit and vegetable intake, since these nutritional factors probably have the best supported scientific rationale for anti-AMD activity. Excessive doses of any of these antioxidants should not be encouraged unless convincing prospective clinical trial results are published in the peer-reviewed literature. Patients who wish to pursue the use of gingko biloba extract (Egb), bilberry fruit and extract, and other herbal/nutraceutical products need to understand that, so far, there is little evidence beyond uncontrolled anecdotal studies to support their efficacy.

The pace of progress in the field of antioxidants for AMD is certainly quickening, but it is still not fast enough to satisfy the majority of clinicians and patients. We believe that a sufficient body of scientific evidence exists to emphasize the importance of minimizing prooxidative influences such as smoking and excessive exposure to short-wavelength blue light and of maximizing the antioxidative protection afforded by antioxidant enzymes and by exogenous diet-derived factors such as lutein and zeaxanthin. Hopefully, in the near future, a clear consensus on rational, scientifically based lifestyle, photoprotective, nutritional, and pharmacological interventions will emerge so that AMD can be prevented and excellent vision can be preserved for the vast majority of our elderly population.

## REFERENCES


1. Gass JDM. Stereoscopic Atlas of Macular Diseases, 4th ed. St. Louis: Mosby, 1997:1–18.
2. Hogan MJ, Alvarado JA, Weddell JE. Histology of the human eye: an atlas and textbook. Philadelphia: WB Saunders, 1971:508–519.
3. Curcio CA, Allen KA, Sloan KR, Lerea CL, Hurley JB, Klock IB, Milam AB. Distribution and morphology of human photoreceptors stained with anti-blue opsin. J Comp Neurol 1990; 292:497–523.
4. Curcio CA, Sloan KR, Packer O, Hendrickson AE, Kalina RE. Distribution of cones in human and monkey retina: individual variability and radial asymmetry. Science 1987; 236:579–582.
5. Bemstein PS. Macular biology. In: Berger JW, Fine SL, Maguire MG, eds. Age-Related Macular Degeneration. St. Louis: Mosby, 1999:1–16.
6. Neitz M, Neitz J, Jacobs GH. Spectral tuning of pigments underlying red-green color vision. Science 1991; 252:971–974.
7. Rapaport DH, Rakic P, Yasamura D, LaVail MM. Genesis of the retinal pigment epithelium in the macaque monkey. J Comp Neurol 1995; 363:359–376.

8. Young RW. Visual cells and the concept of renewal. Invest Ophthalmol 1976; 15:700–725.
9. Marmor MF. Control of subretinal fluid: experimental and clinical studies. Eye 1990; 4:340–344.
10. Rando RR, Bernstein PS, Barry RJ. New insights into the visual cycle. Prog Retinal Res 1990; 10:161–178.
11. Suleiman J, Abdal-Monaim MM, Ashraf M. Morphological features of the retinal pigment epithelium in the toad, Bufo marinus. Anat Rec 1997; 249:128–134.
12. Green WR, Harlan JB Jr. Histopathologic features. In: Berger JW, Fine SL, Maguire MG, eds. Age-Related Macular Degeneration. St. Louis: Mosby, 1999:81–154.
13. Schraemeyer U, Kayatz P, Heimann K. Ultrastructural localization of lipid peroxides in the eye: presentation of a new method. Ophthalmologie 1995; 95:291–295.
14. Mullins RF, Johnson LV, Anderson DH, Hageman GS. Characterization of drusen-associated glycoconjugates. Ophthalmology 1997; 104:288–294.
15. Chang B, Yannuzzi LA, Ladas ID, Guyer DR, Slakter JS, Sorenson JA. Choroidal neovascularization in second eyes of patients with unilateral exudative age-related macular degeneration. Ophthalmology 1995; 102:1380–1386.
16. Macular Photocoagulation Study Group. Argon laser photocoagulation for senile macular degeneration: results of a randomized clinical trial. Arch Ophthalmol 1982; 100:912–918.
17. Macular Photocoagulation Study Group. Argon laser photocoagulation for neovascular maculopathy: five year results from a randomized clinical trial. Arch Ophthalmol 1991; 109:1109–1114.
18. Macular Photocoagulation Study Group. Visual outcome after laser photocoagulation of subfoveal choroidal neovascularization secondary to age-related macular degeneration: the influence of initial lesion size and initial visual acuity. Arch Ophthalmol 1994; 112:480–488.
19. Miller JW, Schmidt-Erfurth U, Sickenberg M, Pournaras CJ, Laqua H, Barbazetto I, Zografos L, Piguet B, Donati G, Lane AM, Birngruber R, van-den-Berg H, Strong A, Manjuris U, Gray T, Fsadni M, Bressler NM, Gragoudas ES. Photodynamic therapy with verteporfin for choroidal neovascularization caused by age-related macular degeneration: results of a single treatment in a phase 1 and 2 study. Arch Ophthalmol 1999; 117:1177–1187.
20. Klein R. Epidemiology. In: Berger JW, Fine SL, Maguire MG, eds. Age-Related Macular Degeneration. St. Louis: Mosby, 1999:31–55.
21. Christen WG. Evaluation of epidemiologic studies of nutrition and cataract. In: Taylor A, ed. Nutritional and Environmental Influences on the Eye. Boca Raton: CRC Press, 1999:95–104.
22. Allikmets R. Molecular genetics of age-related macular degeneration: current status. Eur J Ophthalmol 1999; 9:255–265.
23. Klein R, Klein BE, Linton KL. Prevalence of age-related maculopathy: the Beaver Dam eye study. Ophthalmology 1992; 99:933–943.
24. Seddon JM, Ajani UA, Mitchell BD. Familial aggregation of age-related maculopathy. Am J Ophthalmol 1997; 123:199–206.
25. Meyers SM, Greene T, Gutman F. A twin study of age-related macular degeneration. Am J Ophthalmol 1995; 120:757–766.
26. Klein ML, Schultz DW, Edwards A, Matise TC, Rust K, Berselli CB, Trzupek K, Weleber RG, Ott J, Wirtz MK, Ascott TS. Age-related macular degeneration: clinical features in a large family and linkage to chromosome 1q. Arch Ophthalmol 1998; 116:1082–1088.
27. Allikmets R, Singh N, Sun H, Shroyer NF, Hutchinson A, Chidambaram A, Gerrard B, Baird L, Stauffer D, Peiffer A, Rattner A, Smallwood P, Li Y, Anderson KL, Lewis RL, Nathans J, Leppert M, Dean M, Lupski JR. A photoreceptor cell-specific ATP-binding transporter gene (ABCR) is mutated in Stargardt macular dystrophy. Nat Gen 1997; 15:236–246.
28. Allikmets R, Shroyer NF, Singh N, Seddon JM, Lewis RA, Bernstein PS, Peiffer A, Zabriskie NA, Li Y, Hutchinson A, Dean M, Lupski JR, Leppert M. Mutation of the Stargardt disease gene (ABCR) in age-related macular degeneration. Science 1997; 277:1805–1807.

29. Sun H, Molday RS, Nathans J. Retinal stimulates ATP hydrolysis by purified and reconstituted ABCR, the photoreceptor-specific ATP-binding cassette transported responsible for Stargardt disease. J Biol Chem 1999; 274:8269–8281.
30. Stone EM, Webster AR, Vandeburgh K, Streb LM, Hockey RR, Lotery AJ, Sheffield VC. Allelic variation in ABCR associated with Stargardt disease but not age-related macular degeneration. Nat Genet 1998; 20:328–329.
31. Allikmets R and the International ABCR Screening Consortium. Further evidence for an association of ABCR alleles with age-related macular degeneration. Am J Hum Genet 2000; 67:487–491.
32. Jampol LM, Tielsch J. Race, macular degeneration, and the Macular Photocoagulation Study. Arch Ophthalmol 1992; 110:1699–1700.
33. Holz FG, Piguet B, Minassian DC, Bird AC, Weale RA. Decreasing stromal iris pigmentation as a risk factor for age-related macular degeneration. Am J Ophthalmol 1994; 117:19–23.
34. West SK. Smoking and the risk of eye diseases. In: Taylor A, ed. Nutritional and Environmental Influences on the Eye. Boca Raton: CRC Press, 1999:151–164.
35. Klein R, Klein BE. Smoke gets in your eye too. JAMA 1996; 276:1178–1179.
36. Taylor A, Jacques PP, Epstein EM. Relations among aging, antioxidant status, and cataract. Am J Clin Nutr 1995; 62:1439S–1447S.
37. Morgado PB, Chen HC, Patel V, Herbert L, Kohner EM. The acute effect of smoking in retinal blood flow in subjects with and without diabetes. Ophthalmology 1994; 101:1220–1226.
38. Macular Photocoagulation Study Group. Risk factors for choroidal neovascularization in the second eye of patients with juxtafoveal or subfoveal choroidal neovascularization secondary to age-related macular degeneration. Arch Ophthalmol 1997; 115:741–747.
39. McCarty S, Taylor HR. Light and risk for age-related eye diseases. In: Taylor A, ed. Nutritional and Environmental Influences on the Eye. Boca Raton: CRC Press, 1999:135–150.
40. West SK, Rosenthal FS, Bressler NM, Bressler SB, Munoz B, Fine SL, Taylor HR. Exposure to sunlight and other risk factors for age-related macular degeneration. Arch Ophthalmol 1989; 107:875–879.
41. Cho E, Hung S, Seddon JM. Nutrition. In: Berger JW, Fine SL, Maguire MG, eds. Age-Related Macular Degeneration. St. Louis: Mosby, 1999:57–67.
42. Seddon JM. Vitamins, minerals, and macular degeneration, promising, but unproven hypotheses. Arch Ophthalmol 1994; 112:176–179.
43. Cringle S, Yu D-Y, Alder V, Su E-N, Yu P. Oxygen consumption in the avascular guinea pig retina. APStracts (American Physiological Society Abstracts) 1996; 3:0129H.
44. Packer L. Oxidative stress, antioxidants, aging and disease. In: Gutler RG, Packer L, Bertram J, Mori A, eds. Oxidative Stress and Aging. Basel: Birkhauser, 1995:1–15.
45. Fox NE, van Kuijk FJ. Immunohistochemical localization of xanthine oxidase in human retina. Free Radic Biol Med 1998; 24:900–905.
46. Anderson RE, Andrews LM. Biochemistry of photoreceptor membranes in vertebrates and invertebrates. In: Westfall J, ed. Visual Cells in Evolution. New York: Raven Press, 1982: 1–22.
47. Krinsky NI. Mechanism of action of biological antioxidants. Proc Soc Exp Biol Med 1992; 200:248–254.
48. Garnier-Suillerot A, Gattegno L. Interaction of adriamycin with human erythrocyte membranes. Biochim Biophys Acta 1988; 936:50–60.
49. Choe M, Jackson C, Yu BP. Lipid peroxidation contributes to age-related membrane rigidity. Free Radic Biol Med 1995; 18:977–984.
50. Spaide RF, Ho-Spaide WC, Browne RW, Armstrong D. Characterization of peroxidized lipids in Bruch's membrane. Retina 1999; 19:141–147.
51. Cotinet A, Goureau O, Hicks D, Thillaye-Goldenberg B, de Kozak Y. Tumor necrosis factor

and nitric oxide production by retinal Müller glial cells from rats exhibiting inherited retinal dystrophy. Glia 1997; 20:59–69.

52. Goureau O, Regnier-Ricard F, Courtois Y. Requirement for nitric oxide in retinal neuronal cell death induced by activated Müller glial cells. J Neurochem 1999; 72:2506–2515.
53. Cohen G. Monoamine oxidase, hydrogen peroxide, and Parkinson's disease. Adv Neurol 1987; 45:119–125.
54. Spencer JP, Jenner A, Butler J, Aruoma OI, Dexter DT, Jenner P, Halliwell B. Evaluation of the pro-oxidant and antioxidant actions of L-dopa and dopamine in vitro: implications for Parkinson's disease. Free Radic Res 1996; 24:95–105.
55. Koppenol WH. The basic chemistry of nitrogen monoxide and peroxynitrite. Free Radic Biol Med 1998; 25:385–391.
56. Wu GS, Zhang J, Rao NA. Peroxynitrite and oxidative damage in experimental autoimmune uveitis. Invest Ophthalmol Vis Sci 1997; 38:1333–1339.
57. Meringer CJ, Wu GJ. L-glutamate inhibits nitric oxide synthesis in bovine venular endothelial cells. Pharmacol Exp Ther 1997; 281:448–453.
58. Bonne C, Muller A, Villain M. Free radicals in retinal ischemia. Gen Pharmacol 1998; 30: 275–280.
59. Bush RA, Williams TP. The effect of unilateral optic nerve section on retinal light damage in rats. Exp Eye Res 1991; 52:139–153.
60. Eldred GE, Lasky MR. Retinal age pigments generated by self-assembling lysmotropic detergents. Nature 1993; 361:724–726.
61. Parrish CA, Hashimoto M, Nakanishi K, Dillon J, Sparrow J. Isolation and one-step preparation of A2E and iso-A2E, fluorophores from human retinal pigment epithelium. Proc Natl Acad Sci 1993; 95:14609–14613.
62. Weng J, Mata NL, Azarian SM, Tzekov RT, Birch DG, Travis G. Insights into the function of rim protein in photoreceptors and etiology of Stargardt's disease from the phenotype in ABCR knockout mice. Cell 1999; 98:13–23.
63. Sparrow JR, Parrish CA, Hashimoto M, Nakanishi K. A2E, a lipofuscin fluorophore, in human retinal pigmented epithelial cells in culture. Inv Ophthalmol Vis Sci 1999; 40:2988–2995.
64. Holz FG, Schutt F, Kopitz J, Eldred GE, Kruse FE, Volcker HE, Cantz M. Inhibition of lysosomal degredative functions in RPE cells by a retinoid component of lipofuscin. Invest Ophthalmol Vis Sci 1999; 40:737–743.
65. Katz ML, Robinson WG, Dratz EA. Potential role of autooxidation in age changes of the retina and retinal pigment epithelium of the eye. In: Armstrong D, Sohal RS, Cutler RG, Slater TF, eds. Free Radicals in Molecular Biology, Aging and Disease. New York: Raven Press, 1984:163–180.
66. Wassell J, Davies S, Bardsley W, Boulton M. The photoreactivity of the retinal age pigment lipofuscin. J Biol Chem 1999; 274:23828–23832.
67. Rozanowska M, Jarvis Evans J, Korytowski W, Boulton ME, Burke JM, Sarna T. Blue light-induced reactivity of retinal age pigment. In vitro generation of oxygen-reactive species. J Biol Chem 1995; 270:18825–18830.
68. Rozanowska M, Wassell J, Boulton M, Burke JM, Rodgers MA, Truscott TG, Sarna T. Blue light-induced singlet oxygen generation by retinal lipofuscin in non-polar media. Free Radic Biol Med 1998; 24:1107–1112.
69. Bensasson RV, Land EJ, Truscott TG. Excited States and Free Radicals in Biology and Medicine. Oxford: Oxford University Press, 1993:201–228.
70. Bensasson RV, Dawe EA, Long DA, Land EJ. Singlet-triplet intersystem crossing quantum yields of photosynthetic and related polyenes. J Chem Soc Faraday Trans 1, 1977; 73:1319–1325.
71. Chattopadhyay SK, Kumar CV, Das PK. Quantum yields of triplet states of polyenes. J Photochem 1984; 24:1–9.

72. Hild M, Schmidt R. The mechanism of the collision-induced enhancement of the a1(Delta)g → X3(Sigma)g- and b1(Sigma)g+ → a1(Delta)g radiative transitions of $O_2$. J Phys Chem A 1999; 103:6091–6096.
73. Dillon J, Gallard ER, Bilski P, Chignell CF, Reszka KJ. The photochemistry of the retinoids as studied by steady-state and pulsed methods. Photochem Photobiol 1996; 63:680–685.
74. Dontsov AE, Glickman RD, Ostrovsky MA. Retinal pigment epithelium pigment granules stimulate the photo-oxidation of unsaturated fatty acids. Free Radic Biol Med 1999; 26:1436–1446.
75. Hanna N, Peri KG, Abran D, Hardy P, Doke A, Lachapelle P, Roy MS, Orquin J, Varma DR, Chemtlob S. Light induces peroxidation in retina by activating prostaglandin G/H synthase. Free Radic Biol Med 1997; 23:885–897.
76. Behndig A, Svensson B, Marklund SL, Karisson K. Superoxide dismutase isoenzymes in the human eye. Invest Ophthalmol Vis Sci 1998; 39:471–475.
77. Wang L, Lam TT, Lam KW, Tso MOM. Correlation of phospholipid hydroperoxide glutathione peroxidase activity to the sensitivity of rat retinas to photic injury. Ophthalmic Res 1994; 26:60–64.
78. De La Paz MA, Zhang J, Fridovich I. Antioxidant enzymes of the human retina: effect of age on enzyme activity of macula and periphery. Curr Eye Res 1996; 15:273–278.
79. Yamamoto M, Lidia K, Gong H, Onitsuka S, Kotani T, Ohira A. Changes in manganese superoxide dismutase expression after exposure of the retina to intense light. Histochem J 1999; 31:81–87.
80. Ohta Y, Yamasaki T, Niwa T, Niimi K, Majima Y, Ishiguro I. Role of catalase in retinal antioxidant defence system: its comparative study among rabbits, guinea pigs, and rats. Ophthalmic Res 1996; 28:336–342.
81. Miceli MV, Tate DJ, Alcock NW, Newsome DA. Zinc deficiency and oxidative stress in the retina of pigmented rats. Invest Ophthalmol Vis Sci 1999; 40:1238–1244.
82. Newsome DA, Swartz M, Leone NC, Elston RC, Miller E. Oral zinc in macular degeneration. Arch Ophthalmol 1988; 106:192–198.
83. Friedrichson T, Kalbach HL, Buck P, van Kuijk FJ. Vitamin E in macular and peripheral tissues of the human eye. Curr Eye Res 1995; 14:693–701.
84. Tso MO, Woodford BJ, Lam KW. Distribution of ascorbate in normal primate retina and after photic injury: a biochemical, morphological correlated study. Curr Eye Res 1984; 3: 181–191.
85. Katz ML, Parker KR, Handelman GJ, Bramel TL, Dratz EA. Effects of antioxidant nutrient deficiency on the retina and retinal pigment epithelium of albino rats: a light and electron microscopic study. Exp Eye Res 1992; 34:339–369.
86. Wiegand RD, Joel CD, Rapp LM, Nielsen JC, Maude MB. Polyunsaturated fatty acids and vitamin E in rat rod outer segments during light damage. Invest Ophthalmol Vis Sci 1986; 27:727–733.
87. Crabtree DV, Adler AJ, Snodderly DM. Radial distribution of tocopherols in rhesus monkey retina and retinal pigment epithelium. Invest Ophthalmol Vis Sci 1996; 37:61–76.
88. Goss-Sampson MA, Kriss T, Muller DP. Retinal abnormalities in experimental vitamin E deficiency. Free Radic Biol Med 1998; 25:457–462.
89. Eye Disease Case-Control Study Group. Risk factors for advanced age-related macular degeneration. Arch Ophthalmol 1992; 110:1701–1708.
90. Seddon JM, Sperduto RD, Hiller R, Blair N, Burton TC, Farber MD, Gragoudas ES, Haller J, Miller DT, Yannuzzi LA, Willett W. Dietary carotenoids, vitamins A, C, and E, and advanced age-related macular degeneration. Eye Disease Case–Control Study Group. JAMA 1994; 272:1413–1420.
91. Mares-Perlman JA, Klein R, Klein BE, Greger JL, Brady WE, Palta M, Ritter LL. Association of zinc and antioxidant nutrients with age-related maculopathy. Arch Ophthalmol 1996; 114: 991–997.

92. Sanders TA, Haines AP, Wromald R, Wright LA, Obeid O. Essential fatty acids, plasma cholesterol, and fat-soluble vitamins in subjects with age-related maculopathy and matched control subjects. Am J Clin Nutr 1993; 57:428–433.
93. Rapp LM, Smith SC. Evidence against melanin as the mediator of retinal phytotoxicity by short-wavelength light. Exp Eye Res 1992; 54:55–62.
94. Gergels TG, Van Norren D. Two spectral types of retinal light damage occur in albino as well as in pigmented rat: no essential role for melanin. Exp Eye Res 1998; 66:155–162.
95. Bilgihan A, Bilgihan MK, Akata RF, Aricioglu A, Hasanreisoglu B. Antioxidative role of ocular melanin pigment in the model of lens induced uveitis. Free Radic Biol Med 1995; 19:883–885.
96. Rozanowska M, Sarna T, Land EJ, Truscott TG. Free radical scavenging properties of melanin interaction of eu- and pheo-melanin models with reducing and oxidising radicals. Free Radic Biol Med 1999; 26:518–525.
97. Paulter EL, Maga JA, Tengerdy C. A pharmacologically potent natural product in the bovine retina. Exp Eye Res 1986; 42:285–288.
98. Aruoma OI, Spencer JP, Butler J, Halliwell B. Reaction of plant-derived and synthetic antioxidants with trichloromethylperoxyl radicals. Free Radic Res 1995; 22:187–190.
99. Skaper SD, Fabris M, Ferrari V, Dalle-Carbonar M, Leon A. Quercetin protects cutaneous tissue-associated cell types including sensory neurons from oxidative stress induced by glutathione depletion: cooperative effects of ascorbic acid. Free Radic Biol Med 1997; 22:669–678.
100. Tirosh O, Sen CK, Roy S, Kobayashi MS, Packer L. Neuroprotective effects of alpha-lipoic acid and its positively charged amide analogue. Free Radic Biol Med 1999; 26:1418–1426.
101. Muller A, Villain M, Favreau B, Sandillion F. Differential effect of ischemia/reperfusion on pigmented and albino rabbit retina. J Ocul Pharmacol Ther 1996; 12:337–342.
102. Rego AC, Santos MS, Oliveira CR. Influence of the antioxidants vitamin E and idebenone on retinal cell injury mediated by chemical ischemia, hypoglycemia, or oxidative stress. Free Radic Biol Med 1999; 26:1405–1417.
103. Bosch-Morell F, Roma J, Puertas FJ, Marin N, Diaz-Llopis M. Efficacy of the antioxidant ebselen in experimental uveitis. Free Radic Biol Med 1999; 27:388–391.
104. Ueda T, Ueda T, Armstrong D. Preventive effect of natural and synthetic antioxidants on lipid peroxidation in the mammalian eye. Ophthalmic Res 1996; 28:184–192.
105. Longoni B, Pryor WA, Marchiafava P. Inhibition of lipid peroxidation by N-acetylserotonin and its role in retinal physiology. Biochem Biophys Res Commun 1997; 233:778–780.
106. Baudouin C, Pisella PJ, Ettaiche M, Goldschild M, Becquet F. Effects of EGb761 and superoxide dismutase in an experimental model of retinopathy generated by intravitreal production of superoxide anion radical. Graefes Arch Clin Exp Ophthalmol 1999; 237:58–66.
107. Bone RA, Landrum JT, Tarsis SL. Preliminary identification of the human macular pigment. Vision Res 1985; 25:1531–1535.
108. Bone RA, Landrum JT, Hime GW, Cains A. Stereochemistry of the human macular carotenoids. Invest Ophthalmol Vis Sci 1993; 34:2033–2040.
109. Handelman GJ, Snodderly DM, Adler AJ, Russett MD, Dratz EA. Measurement of carotenoids in human and monkey retinas. Methods Enzymol 1992; 213:220–230.
110. Bernstein PS, Khachik F, Carvalho LS, Muir GJ, Zhao D-Y, Katz NB. Identification and quantitation of carotenoids and their metabolites in the tissues of the human eye. Exp Eye Res 2001; 72:215–223.
111. United States Department of Agriculture (USDA) Carotenoid Database. {http://www.nal.usda.gov/fnic/foodcomp/Data/car98/car98.html}.
112. Landrum JT, Bone RA, Moore LL, Gomez CM. Analysis of zeaxanthin distribution within individual human retinas. Methods Enzymol 1999; 299:457–467.
113. Hammond BR, Wooten BR, Snodderly DM. Individual variations in the spatial profile of human macular pigment. J Opt Soc Am A 1997; 14:1187–1196.

114. Snodderly DM, Auran JD, Delori FC. The macular pigment, I: absorbance spectra, localization, and discrimination from other yellow pigments in primate retinas. Invest Ophthalmol Vis Sci 1984; 25:660–673.
115. Landrum JT, Bone RA, Kilburn MD. The macular pigment: a possible role in protection from age-related macular degeneration. Adv Pharmacol 1997; 38:537–556.
116. Bone RA, Landrum JT, Fernandez L, Tarsis SL. Analysis of the macular pigment by HPLC: retinal distribution and age study. Invest Ophthalmol Vis Sci 1988; 29:843–849.
117. Gass JDM. Müller cell cone, an overlooked part of the anatomy of the fovea centralis. Arch Ophthalmol 1999; 117:821–823.
118. Keen JN, Caceres I, Eliopoulos EE, Zagalsky PF, Findlay JBC. Complete sequence and model for the A2 subunit of the carotenoid pigment complex, crustacyanin. Eur J Biochem 1991; 197:407–414.
119. Jouni ZE, Wells MA. Purification and partial characterization of lutein-binding protein from the midgut of the silkworm Bombyx mori. J Biol Chem 1996; 271:14722–14726.
120. Bernstein PS, Balashov NA, Tsong ED, Rando RR. Retinal tubulin binds macular carotenoids. Inv Ophthalmol Vis Sci 1997; 38:167–175.
121. Sanders TAB, Haines AP, Wormland R, Wright LA, Obeid O. Essential fatty acids, plasma cholesterol, and fat-soluble vitamins in subjects with age-related maculopathy and matched control subjects. Am J Clin Nutr 1993; 57:428–433.
122. Mares-Periman JA, Brady WE, Klein R, Klein BEK, Bowen P, Stacewicz-Sapuntzakis M, Palta M. Serum antioxidants and age-related macular degeneration in a population-based case-control study. Arch Ophthalmol 1995; 113:1518–1523.
123. Mares-Perlman JA, Klein R, Klein BEK, Greger JL, Brady WE, Palto M, Ritter LL. Association of zinc and antioxidant nutrients with age-related maculopathy. Arch Ophthalmol 1996; 114:991–997.
124. Malinow MR, Feeney-Burns L, Peterson LH, Klein ML, Neuringer M. Diet related macular anomalies in monkeys. Invest Ophthalmol Vis Sci 1980; 19:857–863.
125. Hammond BR, Johnson EJ, Russell RM, Krinsky NI, Yeum K-J, Edwards RB, Snodderly DM. Dietary modification of human macular pigment density. Inv Ophthalmol Vis Sci 1997; 38:1795–1801.
126. Bernstein PS, Yoshida MD, Katz NB, McClane RW, Gellermann W. Raman detection of macular carotenoid pigments in intact human retina. Inv Ophthalmol Vis Sci 1998; 39:2003–2011.
127. Schalch W. Carotenoids in the retina: a review of their possible role in preventing or limiting damage caused by light of oxygen. In: Emerit I, Chance B, eds. Free Radicals and Aging. Basel, Switzerland: Birkhauser-Verlag, 1992:280–298.
128. Snodderly DM. Evidence for protection against age-related macular degeneration by carotenoids and antioxidant vitamins. Am J Clin Nutr 1995; 62(suppl):1448S–1461S.
129. Van Kuijk FJGM, Seims WG, Sommerberg O. Carotenoid localization in human eye tissues. Inv Ophthalmol Vis Sci 1997; 39:S1030.
130. Khachik F, Bernstein PS, Garland DL. Identification of lutein and zeaxanthin oxidation products in human and monkey retinas. Inv Ophthalmol Vis Sci 1997; 38:1802–1811.
131. Rice-Evans CA, Sampson J, Bramley PM, Holloway DE. Why do we expect carotenoids to be antioxidants in vivo? Free Radic Res 1997; 26:381–398.
132. Mortensen A, Skibsted LH. Free radical transients in photobleaching of xanthophylls and carotenes. Free Rad Res 1997; 26:549–563.
133. Terao J. Antioxidant activity of beta-carotene-related carotenoids in solution. Lipids 1989; 24:659–661.
134. Lim BP, Koga T. Antioxidant action of xanthophylls. Biochim Biophys Acta 1989; 112: 178–184.
135. Jorgensen K, Skibsted LH. Carotenoid scavenging of radicals. Lebensmit Unt Forsch 1998; 196:423–429.

136. Omenn GS, Goodman GE, Thornquist MD, Balmes J, Cullen MR, Glass A, Keogh JP. Effects of a combination of beta carotene and vitamin A on lung cancer and cardiovascular disease. N Engl J Med 1996; 334:1150–1155.
137. ATBC Study Group. Report of the ATBC Study Group. N Engl J Med 1994; 330:1029–1035.
138. Schalch W, Dayhaw-Barker P, Barker FM. The carotenoids of the human retina. In: Taylor A, ed. Nutritional and Environmental Influences on the Eye. Boca Raton: CRC Press, 1999: 215–250.
139. Subczynski WK, Markowska E, Gruszecki WI, Sielewiesiuk J. Effects of polar carotenoids on dimyristoylphosphatidylcholine membranes: a spin-label study. Biochim Biophys Acta 1992; 1105:97–108.
140. Gruszecki WI, Sielewiesiuk J. Orientation of xanthophylls in phosphatidylcholine multibilayers. Biochim Biophys Acta 1990; 1023:405–412.
141. Yemelyanov AY, Katz NB, Bernstein PS. Ligand-binding characterization of xanthophyll carotenoids to solubilized membrane proteins derived from human retina. Exp Eye Res 2001; 72; in press.

# 20

# Nutritional Influences on Risk for Cataract

**ALLEN TAYLOR**

*Jean Mayer USDA Human Nutrition Research Center on Aging, Tufts University, Boston, Massachusetts*

## I. INTRODUCTION

It is clear that oxidative stress is associated with compromises to the lens. Recent literature indicates that antioxidants may ameliorate that risk and may actually decrease risk for cataract. This chapter will briefly review etiology of cataract and then review the epidemiological information.

## II. CATARACT AS A PUBLIC HEALTH ISSUE

### A. Cataract: Age-Related Prevalence and Annual Cost

Cataract is one of the major causes of preventable blindness throughout the world (1–3). In the United States, the prevalence of visually significant cataract increases from approximately 5% at age 65 to about 50% for persons older than 75 years (4–6). In the U.S. and much of the developed world, cataract surgery, albeit costly, is readily available and routinely successful in restoring sight. Nevertheless, given the negative impact on quality of life, productivity, and the Medicare budget, it would be best to diminish risk for this debility. In less developed countries, such as India (7), China (8), and Kenya (9), cataracts are more common and develop earlier in life than in more developed countries (6,7). The impact of cataract on impaired vision is much greater in less developed countries, where more than 90% of the cases of blindness and visual impairment are found (10,11), and where there is a dearth of ophthalmologists to perform lens extractions.

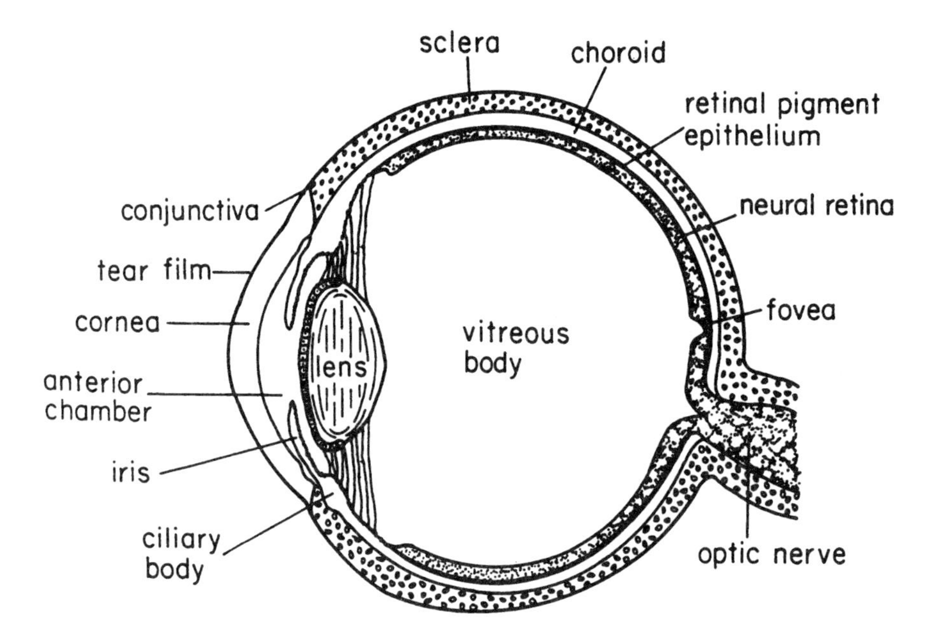


**Figure 1** Cross-section of the eye. (From Ref. 11, with permission.)

Given both the extent of disability caused by age-related cataract and its costs ($5 to 6 billion per year* (12) in the U.S.), it is urgent that we elucidate causes of cataract and identify strategies to slow the development of this disorder. It is estimated that a delay in cataract formation of about 10 years would reduce the prevalence of visually disabling cataract by approximately 45% (1). Such a delay would enhance the quality of life for much of the world's older population and substantially reduce the economic burden due to cataract-related disability and cataract surgery. Such data provide the impetus for this research.

## III. AGE-RELATED CHANGES IN LENS FUNCTION

The primary function of the eye lens is to collect and focus light on the retina (Figs. 1, 2a, 2d). To do so, it must remain clear throughout life. The lens is located posterior to the cornea and iris and receives nutriture from the aqueous humor. The lens is exquisitely organized. A single layer of epithelial cells is found directly under the anterior surface of the collagenous membrane in which it is encapsulated (Fig. 2b). The epithelial cells at the germinative region divide, migrate posteriorly, and differentiate into lens fibers. As their primary gene products, the fibers elaborate the predominant proteins of the lens, called crystallins. They also lose their organelles. New cells are formed throughout life, but older cells are usually not lost. Instead, they are compressed into the center, or nucleus, of the lens. As the lens ages, or upon stress due to light exposure (13) or smoking (14), the proteins are photooxidatively damaged and aggregate. Among the oxidative insults

* Congressional Testimony of S.J. Ryan, May 5, 1993.

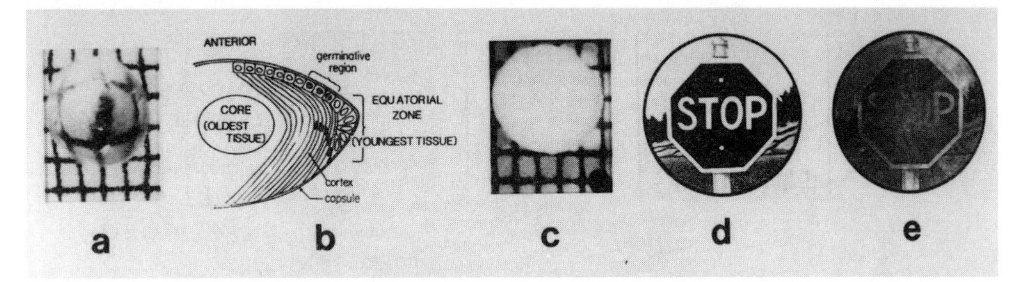


**Figure 2** Clear and cataractous lens. (a) Clear lens allows an unobstructed view of the wire grid placed behind it. (b) Cartoon of the structure of the lens. The anterior surface of the lens has a unicellular layer of epithelial cells (youngest tissue). Cells at the anterior equatorial region divide and migrate to the cortex as they are overlaid by less mature cells. These cells produce a majority of thc crystallins. As development and maturation proceed, the cells denucleate and elongate. Tissue originally found in the embryonic lens is found in the core or nucleus (oldest tissue). (c) The cataractous lens prohibits viewing the wire grid behind it. (d) Artist's view through a clear, uncolored young lens. The image is clear and crisp. (e) Artist's view through a lens with developing cataract. The image is partially obscured and the field is darkened due to browning of the lens that accompanies aging. (From Ref. 10, with permission.)

are high-energy radiation, reactive oxygen species, sunlight exposure, as well as failure of secondary defense systems (Fig. 3). Such systems involve proteolytic enzymes that selectively recognize and remove damaged and cytotoxic proteins. There is a coincident dehydration of the proteins and of the lens itself. Consequently, protein concentrations rise to hundreds of mg/mL (15). Together with other age-related modifications of the protein and other constituents, these changes result in a less flexible lens with limited accommodative capability. Eventually, the large aggregates of protein precipitate in lens opacities or cataract.

The term ''age-related cataract'' is used to distinguish lens opacification associated with old age from opacification associated with other causes, such as congenital and metabolic disorders or trauma. There are several systems for evaluating and grading cataracts. Most of these employ an assessment of extent, or density and location of the opacity (16,17). Usually evaluated are opacities in the posterior subcapsular, nuclear, cortical, and multiple (mixed) locations. However, it is not established that cataract at each location has completely different etiology. Coloration or brunescence is also quantified, since these diminish visual function (18,19).

## IV. PRIMARY AND SECONDARY DEFENSE SYSTEMS

Only a brief introduction to some of these topics is offered here.

### A. Cellular Antioxidants as Primary Defenses Against Lens Damage

Protection against photooxidative insult can be conceived of as due to two interrelated processes. Primary defenses offer protection of proteins and other constituents by lens antioxidants and antioxidant enzymes (Fig. 3). Secondary defenses include proteolytic and repair processes, which degrade and eliminate damaged proteins and other biomolecules in a timely fashion (20).

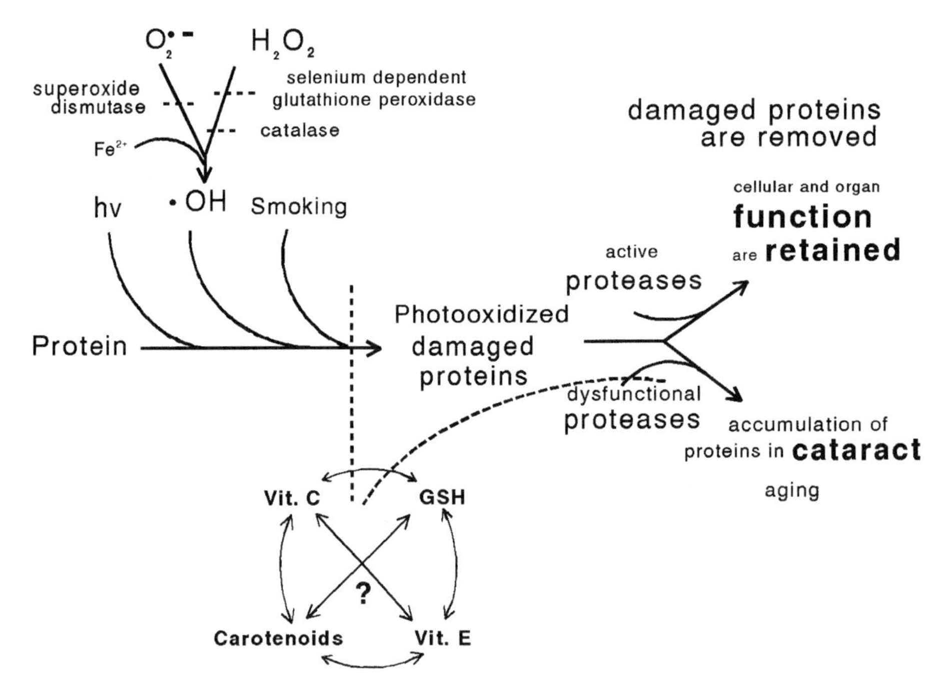


**Figure 3** Proposed interaction between lens proteins, oxidants, light, smoking, antioxidants, antioxidant enzymes, and proteases. Lens proteins are extremely long-lived and are subject to alteration by light and various forms of oxygen. They are protected indirectly by antioxidant enzymes: superoxide dismutase, catalase, and glutathione reductase/peroxidase. These enzymes convert active oxygen to less damaging species. Direct protection is offered by antioxidants: glutathione (GSH), ascorbate (vitamin C), tocopherol (vitamin E), and carotenoids. Levels of reduced and oxidized forms of some, but perhaps not all (?), of these molecules are determined by interaction between the four and with the environment (131–136). In many systems, GSH and ascorbate levels are related, but we did not find this to be the case in ascorbate-sufficient, ascorbate-requiring rats (26). When the proteolytic capability is sufficient, obsolete and damaged proteins may be reduced to their constituent amino acids. Upon aging, some of the eye antioxidant supplies are diminished, antioxidant enzymes inactivated, and proteases less active. This appears to be related to the accumulation, aggregation, and eventual precipitation in cataractous opacities of damaged proteins. (Adapted from Ref. 10, with permission.)

The major aqueous antioxidants in the lens are ascorbate (21) and GSH (22–28). Both are present in the lens at millimolar concentrations.

Ascorbate is probably the most effective, least toxic, antioxidant identified in mammalian systems (29,30). Interest in the function of ascorbate in the lens was prompted by teleological arguments, which suggested age-related compromises in ascorbate and compromises in lens function might be related. Thus, it was observed that: (1) the lens and aqueous humor have concentrations of ascorbate greater than tenfold the level found in guinea pig and human plasma (21,23,31,32); (2) in the lens core (Fig, 2b), the oldest part of the lens and the region involved in much senile cataract, the concentration of ascorbate is only 25% of the surrounding cortex (33); (3) lens ascorbate concentrations are lower in cataract than in the normal lens (34); and (4) ascorbate levels in the lens are significantly lower in old guinea pigs than in young animals with the same dietary intake of ascorbate (21,31). The same pertains in Emory mice (28). These data suggest either that there is age-related depletion of ascorbate in the lens or that the bioavailability of this compound changes with age. Enthusiasm for nutrient antioxidants has been fueled by observations that ocular levels of ascorbate are related to dietary intake in humans and animals that require ascorbate (21,23,31,32). Thus, the concentration of vitamin C in the lens was increased with dietary supplements beyond levels achieved in persons who already consumed more than two times the RDA (60 mg/day) for vitamin C (21,23).

Feeding elevated ascorbate delayed progress of, or prevented, galactose cataract in guinea pigs (35,36) and rats (37), selenite-induced cataracts in rats (38), lens opacification in GSH-depleted chick embryos (39), and delayed UV-induced protein and protease damage in guinea pig lenses (40–43). Increasing lens ascorbate concentrations by only twofold is associated with protection against cataract-like damage (41).

Since ascorbate is a carbohydrate, it is biochemically plausible that vitamin C induces damage in the lens in vivo (44,45). However, presently there are no data to support this as a medical concern. While we did find that glycohemoglobin levels increased with increasing dietary ascorbate in ascorbate-requiring rats (26), mice fed 8% of the weight of their diet as ascorbate did not develop cataract (46). In the event that glycation induces pathological lens damage, antiglycating agents such as phenacylthiazoliums may be useful (47). It is interesting that comparable compounds have been tried as anticataractogens and were assumed to act as reducing cysteine prodrug agents (48).

Glutathione (GSH) levels are severalfold the levels found in whole blood and orders of magnitude greater than the concentration observed in the plasma. GSH levels also diminish in the older and cataractous lens (24). There have been several attempts to exploit the reducing capabilities of GSH. Injection of GSH-OMe was associated with delayed buthionine sulfoxamine–induced (49) and naphthalene cataract (24,50,51). Preliminary evidence from studies with galactose-induced cataract also indicates some advantage of maintaining elevated GSH status in rats (27). However, it is not clear that feeding GSH is associated with higher ocular levels of this antioxidant (27). Other compounds, such as pantetheine, which also include sulfhydryls, are under investigation as anticataractogenic agents (52). However, the efficacy of this compound in later-life cataract remains to be established (53). In many systems, GSH and ascorbate levels are related. We did not find this to be the case in ascorbate-sufficient, ascorbate-requiring rats (26).

Pharmacological opportunities are suggested by observations that incorporating the industrial antioxidant 0.4% butylated hydroxy toluene in diets of galactose-fed (50% of diet) rats diminished prevalence of cataract (54).

Tocopherols and carotenoids are lipid-soluble antioxidants (55–57) with probable roles in maintaining membrane integrity (58) and GSH recycling (59). Concentrations of tocopherol in the whole lens are in the micromolar range (Table 1), and levels are elevated in younger tissues (60), but it appears that lens and dietary levels of tocopherol are unrelated (61). Since most of the compound is found in the membranes, particularly in the younger tissues (55), the concentrations can be orders of magnitude higher. Age-related changes in levels of tocopherol and carotenoids have not been documented. Tocopherol is reported to be effective in delaying a variety of induced cataracts in animals, including galactose- (62–64) and aminotriazole-induced cataracts in rabbits (65).

Elevated carotenoid intake is frequently associated with health benefits. However, little experimental work has been done regarding lens changes in response to variations in levels of this nutrient. It is intriguing that β-carotene levels in human lenses are limited (55,60) (Table 1). Instead, major lens carotenoids are lutein/zeaxanthin. These are also

**Table 1** Endogenous Concentrations of Ascorbate, Carotenoids, Retinoids, and Tocopherols in Human Lenses

| | Level in whole lens[a] |
|---|---|
| Ascorbate (mM) | 3.6[b] |
| Lutein/zeaxanthin ng/g wet weight | 11–44 |
| α-Tocopherol ng/g wet weight | 1232–2550 |
| β-Carotene ng/g wet weight | ND[c] |
| Retinol ng/g wet weight | 21–50 |
| Lycopene ng/g wet weight | ND |

[a] Generally, levels of these antioxidants are higher in the lens epithelium as opposed to lens nuclear tissue (55,60).
[b] Average of value for women and men (23).
[c] None detected.

the major carotenoids in the macula (66–68). Also present are retinol and retinol ester, and tocopherols. In beef, β-carotene was occasionally observed in lenses. This apparent quixotic appearance of β-carotene appears to be due to seasonal and dietary availability.

The lens also contains antioxidant enzymes: glutathione peroxidase/reductase, catalase, and superoxidase dismutase and enzymes of the glutathione redox cycle (50,69–72). These interact via the forms of oxygen, as well as with the antioxidants (i.e., GSH is a substrate for glutathione peroxidase). The activities of many antioxidant enzymes are compromised upon development, aging, and cataract formation (73).

## B. Proteolytic Enzymes Provide Secondary Defenses Against Lens Damage

Proteolytic systems can be considered secondary defense capabilities that remove cytotoxic damaged or obsolete proteins from lenses and other tissues (20,36,74–88). Such proteolytic systems exist in young lens tissue, and damaged proteins are usually maintained at harmless levels by primary and secondary defense systems in younger lenses and in younger lens tissues within older lenses.

Several studies indicate interactions between primary and secondary defense systems. A direct sparing effect of ascorbate on a photooxidatively induced compromise of proteolytic function has been demonstrated (40). GSH also spares activity of enzymes involved in the conjugation of ubiquitin to substrates (77,86). Ubiquitin conjugation is required for selective targeting of substrates for degradation. However, upon aging or oxidative stress, most of these enzymatic capabilities are found in a state of reduced activity (20) (Table 2). The observed accumulation of oxidized (and/or otherwise modified) proteins in older lenses is consistent with the failure of these protective systems to keep pace with the insults that damage lens proteins. This occurs in part because, like bulk proteins, enzymes that comprise some of the protective systems are damaged by photooxidation (40,74,77,86,89). From these data it is clear that the young lens has significant primary and secondary protection. However, age-related compromises in the activity of antioxidant enzymes, concentrations of the antioxidants, and activities of secondary defenses may lead to diminished protection against oxidative insults. This diminished protection leaves the long-lived proteins and other constituents vulnerable. Lens opacities develop as the damaged proteins aggregate and precipitate. Current data predict that elevated

**Table 2** Effect of Aging and Development on Proteolytic Activities

| Activity | Source | Activity upon aging |
|---|---|---|
| Intracellular proteolysis (137) | BLEC[a,b] | ↓ |
| Ubiquitin-dependent proteolysis in response to stress (88) | Tissue | ↓ |
| Neutral proteinase/proteasome/high molecular weight protease (138–141) | Tissue | ↓ |
| Endopeptidase (87,88) | BLEC | ↓ |
| LAP (75,141) | Tissue | ↓ |
| | BLEC | ↓ |
| Cathepsins (75) | BLEC | ↓ |
| Calpain (142,143) | Tissue | ↑ |
| | BLEC | ND[c] |

[a,b] In cultured cells, aging was simulated by progressive passage of cells in culture. BLEC = beef lens epithelial cells.
[c] Not determined.

antioxidant intake can be exploited to extend the function of some of these proteolytic capabilities (see below).

## V. EPIDEMIOLOGICAL STUDIES REGARDING ASSOCIATIONS BETWEEN ANTIOXIDANTS AND CATARACT

Since cataract is due in part to oxidative stress on lens constituents and the enzymes that might normally remove these damaging moieties (reviewed in Fig. 3), a considerable effort is being dedicated to determine if antioxidants can be used to diminish risk for cataract. I believe that the overall impression created by the data indicates that nutrient intake is related to risk for cataract and that nutrition may be exploited to diminish risk for this debility. As noted above, cataracts are classified into subcapsular, cortical, nuclear, mixed, or cataract extraction (the endpoint we want to avoid). Cataracts in the different lens zones are distinguished because of the (largely unproven) assumption that the opacities in these metabolically and developmentally distinguishable areas of the lens have different etiologies. Nutritional status has been assessed using various questionnaires and/or by measuring blood nutrient levels.

Since studies varied in design, comparisons are not always straightforward. Some of these design issues are discussed below and at the end of the chapter.

To date most of the studies were retrospective case-control or cross-sectional studies in which levels of cataract patients were compared with levels of individuals with clear lenses (56,90–99). Our ability to interpret data from retrospective studies such as these is limited by the concurrent assessment of lens status and levels. Prior diagnosis of cataract might influence behavior of cases including diet, and it might also bias reporting of usual diet.

Other studies (100–111) assessed levels and/or supplement use, and then followed individuals with intact lenses for up to 8 years. Such prospective studies are less prone to bias because assessment of exposure is performed before the outcome is present. Some of these studies (102–107) used cataract extraction or reported diagnosis of cataract as a measure of cataract risk. Extraction may not be a good measure of cataract incidence (development of new cataract), because it incorporates components of both incidence and progression in severity of existing cataract. However, extraction is the result of visually disabling cataract and is the endpoint that we wish to prevent. Whereas risk ratios that are presented in some studies are adjusted for many potentially confounding variables, others are not.

Duration of measurement of dietary intake of nutrients and frequency of assessment may also affect the accuracy of these analyses. This is because cataract develops over many years and frequency of one measure may not provide as accurate an assessment of usual intake as multiple measures over time.

In addition to the different study designs noted above, various studies used different lens classification schemes, different definitions of high and low levels of nutrients, and different age groups of subjects.

This description of the epidemiological data is generally organized by source of nutrient: supplement or dietary intake from foods, or blood levels. This is done to get the clearest impression about the effect of a nutrient on risk for cataracts. Supplements frequently provide more of a nutrient than can be achieved in a normal diet and frequently provide levels of nutrients that come closer to saturating tissues. Data from retrospective and prospective studies are separated.

## A. Ascorbate

The available epidemiological data regarding ascorbate intake and risk for cataract is particularly intriguing, since dietary ascorbate intake is related to eye tissue ascorbate levels and because there are reports that indicate potential anticataractogenic and procataractogenic roles for the vitamins.

Most studies investigated relations between vitamin C intake and risk for cataract (90–96,98,100,102,106,110,112–114). In many of these studies, vitamin C—particularly vitamin C supplement use—was inversely associated with at least one type of cataract (Fig. 4).

In our Vitamin C and Cataract Study, age-adjusted analyses based on 165 women with high vitamin C intake (mean = 294 mg/day) and 136 women with low vitamin C intake (mean = 77 mg/day) indicated that the women who took vitamin C supplements ≥10 years had >70% lower prevalence of early opacities (RR: 0.23; CI: 0.09–0.60) (Fig. 4a) and >80% lower risk of moderate opacities (RR: 0.17; CI: 0.03–0.87) at any site compared with women who did not use vitamin C supplements (112). Data from the baseline phase of the Nutrition and Vision Project corroborate the prior observation. The RR was 0.36 (CI: 0.18–0.72) for early nuclear opacities as assessed using LOCS III for persons who took vitamin C supplements for >10 years (98) (Fig. 4c). Blood level measures also indicated inverse relations between vitamin C status and risk for nuclear opacity (98). These studies were consistent with work by Hankinson et al. (106), who noted that women who consumed vitamin C supplements for >10 years had a 45% reduction in rate

**Figure 4** Cataract risk ratio, high-versus-low intake (with or without supplements), or plasma levels of vitamin C. Types of cataract are any, moderate/advanced, nuclear, cortical, posterior subcapsular, mixed, or cataract extraction. Data for retrospective and prospective studies are presented independently. (Adapted from Ref. 11, with permission.)

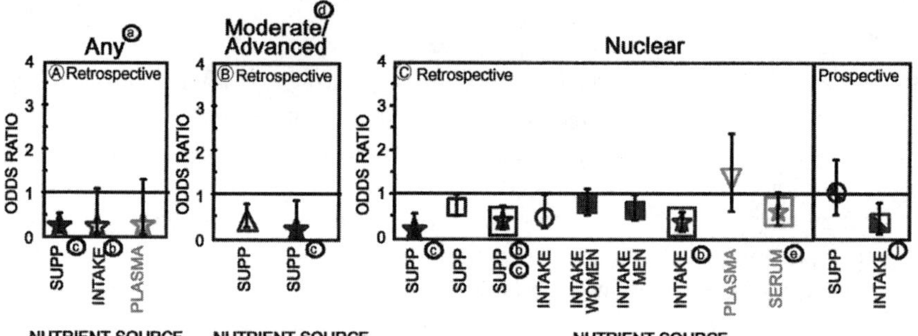


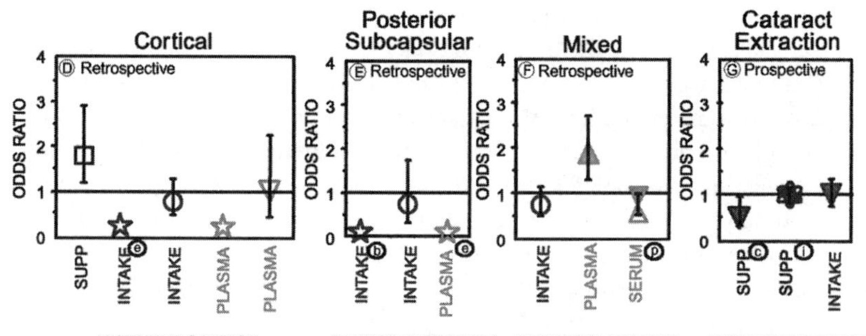


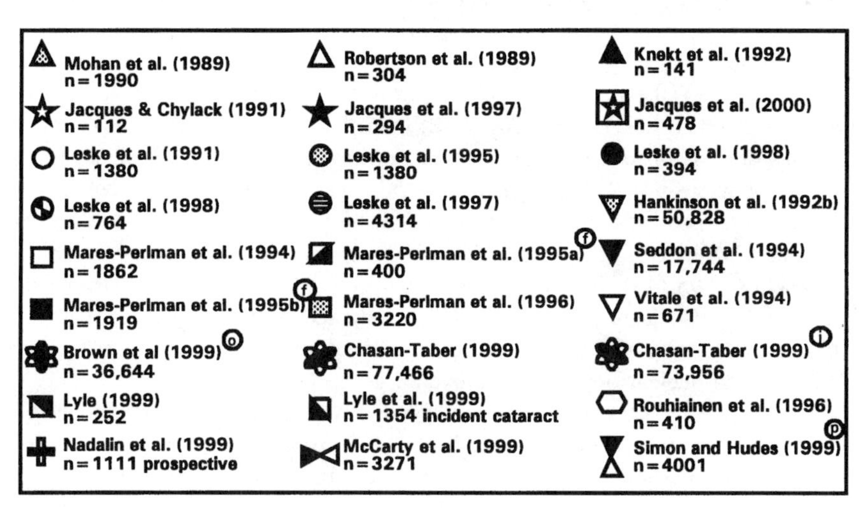


(a) Any = cataract with LOCS II grade ≥ 1 (b) $p \leq 0.1$ (c) supp use > 10 years.
(d) Adv = cataract with LOCS II grade ≥ 2 (e) $p \leq 0.05$ (f) when data were available for comparable groups of subjects, the most recent data from the largest sample were used for these displays. (g) applicable for persons <70 years of age (h) lutein/zeaxanthin, particularly broccoli and kale. (i) vitamins E, C, A and multivitamin use > 10 years. (j) risk for incident opacities for persons with ≥25 pack-years smoking (n=80). Same OR for persons with hypertension (n=90). For overall group there was no significant relation between vitamin C intake and risk for nuclear opacity. (k) risk for incident opacities for persons <65 years of age (n=102) for diet assessed 1978-1980. Not significant for diet assessed 1988-1990. Decreased OR was also observed for persons with high egg and spinach consumption. (l) risk for incident opacities for persons ≥65 years of age (n=143). For overall group there was no significant relation between vitamin E intake and risk for nuclear opacities. (m) sum of $\alpha$ and $\tau$-tocopherol. Risk for incident opacities over 5 years. (n) risk for incident opacities in 5 year follow-up. (o) particularly broccoli and spinach. (p) self-reported cataract in 60-74 year-olds. Each mg/dl increase in serum ascorbate is associated with a 26% decrease in prevalence of cataract (CI 0.56-0.97). (q) multivariate risk factor including interaction with ultraviolet B exposure. (r) for 5-10 mg/day intake vs. intake of <5mg/day.
(s) notes from Jacques.

of cataract surgery (RR: 0.55; CI: 0.32–0.96) (Fig. 4g). However, after controlling for nine potential confounders including age, diabetes, smoking, and energy intake, they did not observe an association between vitamin C intake and rate of cataract surgery (see below). Moreover, looking at the same groups Chasen-Taber (104) also found no benefit of prolonged vitamin C supplement use with respect to risk for cataract extraction (Fig. 4g).

In comparison to the data noted above, Mares-Perlman and coworkers (91) report that past use of supplements containing vitamin C was associated with a reduced prevalence of nuclear cataract (RR: 0.7; CI: 0.5–1.0) (Fig. 4c), but an increased prevalence of cortical cataract (adjusted RR: 1.8; CI: 1.2–2.9) after controlling for age, sex, smoking, and history of heavy alcohol consumption (Fig. 4d).

The inverse relationship is corroborated by data from other studies. Robertson and coworkers (92) compared cases (with cataracts that impaired vision) to age- and sex-matched controls who were either free of cataract or had minimal opacities that did not impair vision. The prevalence of cataract in consumers of daily vitamin C supplements of >300 mg/day was approximately one-third the prevalence in persons who did not consume vitamin C supplements (RR: 0.30; CI: 0.24–0.77 (Fig. 4b). Elevated dietary ascorbate was also related to benefit with respect to cataract in some studies. Leske and coworkers (93) observed that persons with vitamin C intake in the highest 20% of their population group had a 52% lower prevalence for nuclear cataract (RR: 0.48; CI: 0.24–0.99) compared with persons who had intakes among the lowest 20% after controlling for age and sex (Fig. 4c). Weaker inverse associations were noted for other types of cataract (Fig. 4f). Jacques and Chylack observed that among persons with higher vitamin C intakes (>490 mg/day), the prevalence of cataract was 25% of the prevalence among persons with lower intakes (<125 mg/day) (RR: 0.25; CI: 0.06–1.09) (Fig. 4a) (94). In our assessments of early opacities in the baseline Nutrition and Vision Project, we found decreased risk for nuclear (RR: 0.31; CI: 0.16–0.58) opacities in the entire cohort and decreased risk for cortical opacities in persons under 60 years of age who had elevated vitamin C intake (99). Intakes of less than 150 to 300 mg/day vitamin C appear to provide maximum benefit (23,98,115). In a more recent prospective study from the Beaver Dam group, Lyle found a 70% decreased risk for nuclear cataract for heavy smokers (RR: 0.3; CI: 0.1–0.8) and persons with hypertension, but they found no significant relation between vitamin C intake and nuclear opacity in the overall group (101) (Fig. 4c).

However, Vitale and coworkers observed no differences in cataract prevalence between persons with high (>261 mg/day) and low (<115 mg/day) vitamin C intakes (96). The Italian-American Studies Group (90) also failed to observe any association between prevalence of cataract and vitamin C intake. In addition, in a large prospective study, comparison of women with high intakes (median = 705 mg/day) to women with low intakes (median = 70 mg/day) failed to reveal any significant correlation with risk for cataract extraction (RR: 0.98; CI: 0.72–1.32) (106).

Attempts to corroborate the above inverse associations between cataract risk and intake using plasma vitamin C levels have yielded mixed results. Jacques and Chylack (94) observed that persons with high plasma vitamin C levels (> 90 μM) had less than one-third the prevalence of early cataract as persons with low-plasma vitamin C (<40 μM), although this difference was not statistically significant (risk ratio [RR]: 0.29; 95% CI: 0.06–1.32) after adjustment for age, sex, race, and history of diabetes (Fig. 4a). As noted above, we corroborated these results in our baseline Nutrition and Vision Project (98). (Fig. 4c) Mohan (95) noted an 87% (RR: 1.87; CI: 1.29–2.69) increased prevalence of mixed cataract (posterior subcapsular and nuclear involvement) for each standard deviation increase in plasma vitamin C levels. Vitale and coworkers (96) observed that persons with plasma levels greater than 80 μM and below 60 μM had similar prevalences of both nuclear (RR: 1.31; CI: 0.61–2.39) and cortical (RR: 1.01; CI: 0.45–2.26) cataract after controlling for age, sex, and diabetes.

Results from one 5-year intervention trial showed no benefits of vitamin C supplementation to the population studied (116) (Fig. 4).

## B. Vitamin E

A variety of studies examined relations between vitamin E supplement use, dietary intake, and plasma levels and risk for various forms of cataract. The results are mixed. Consumption of vitamin E supplements was inversely (significantly) correlated with cataract risk in two retrospective studies (Fig. 5). Robertson and coworkers (92) found among age- and sex-matched cases and controls that the prevalence of advanced cataract was 56% lower (RR: 0.44; CI: 0.24–0.77) (Fig. 5) in persons who consumed vitamin E supplements (>400 I.U./day) than in persons not consuming supplements. Jacques and Chylack (unpublished) observed a 67% (RR: 0.33; CI: 0.12–0.96) reduction in prevalence of cataract for vitamin E supplement users after adjusting for age, sex, race, and diabetes. In addition, in a prospective study, Lyle et al. found a nonsignificant inverse relationship between that vitamin intake and nuclear cataract (RR: 0.5; CI: 0.3–1.1). In the baseline Nutrition and Vision Project, we found that risk for nuclear cataract was decreased by 51% (RR = 0.49: CI: 0.22–1.09) in supplement users (98). By comparison, Mares-Perlman and coworkers (91) observed only weak, nonsignificant associations between vitamin E supplement use and nuclear cataract (RR: 0.9; CI: 0.6–1.5, Fig. 5c). However, cortical cataract (RR: 1.2; CI: 0.6–2.3) (Fig. 5d) was positively (although nonsignificantly) related to vitamin E supplement intake by Mares-Perlman (91). McCarty et al. (117) also noted no effect of supplemental intake of vitamin E on risk for nuclear cataract (Fig. 5c). These investigators also found a positive relationship between vitamin E supplementation and posterior subcapsular cataract (RR: 1.47; CI: 1.04–2.098) (Fig. 5e). Nadalin (118) found supplement use to be associated with a 53% decreased risk for incident cortical cataract in a prospective study (RR: 0.47; CI: 0.28–0.83) (Fig. 5d), but not to be related to reduced risk for nuclear cataract (Fig. 5c). In partial contrast, vitamin E supplementation was related to a lower risk for progress of nuclear opacity (RR: 0.43; CI: 0.19–0.99) (110).

Nonetheless, Simon and Hudes found no significant relationship between vitamin E supplement use and self-reported cataract (RR: 0.93; CI: 0.52–1.67) (Fig. 5b) (114), and Chasen-Taber did not find cataract extraction to be related to intake of vitamin E supplements (RR: 0.99; CI: 0.74–1.32), even when intake was for 10 years (Fig. 5g) (104).

Leske et al. observed that, after controlling for age and sex, persons with vitamin E intakes among the highest 20% had an approximately 40% lower prevalence of cortical (RR: 0.59; CI: 0.36–0.97) (Fig. 5d) and mixed (RR: 0.58; CI: 0.37–0.93, Fig. 5f) cataract relative to persons with intakes among the lowest 20% (93). Jacques and Chylack observed a nonsignificant inverse association when they related total vitamin E intake (combined dietary and supplemental intake) to cataract prevalence (94). Persons with vitamin E intake greater than 35.7 mg/day had a 55% lower prevalence of early cataract (RR: 0.45; CI: 0.12–1.79) than did persons with intakes less than 8.4 mg/day (94). They also showed a significant reduction in risk for cortical cataract (Fig. 5d) (RR: 0.37; $p < 0.1$) for persons with higher vitamin E intakes.

However, Hankinson et al. (106) found no association between vitamin E intake

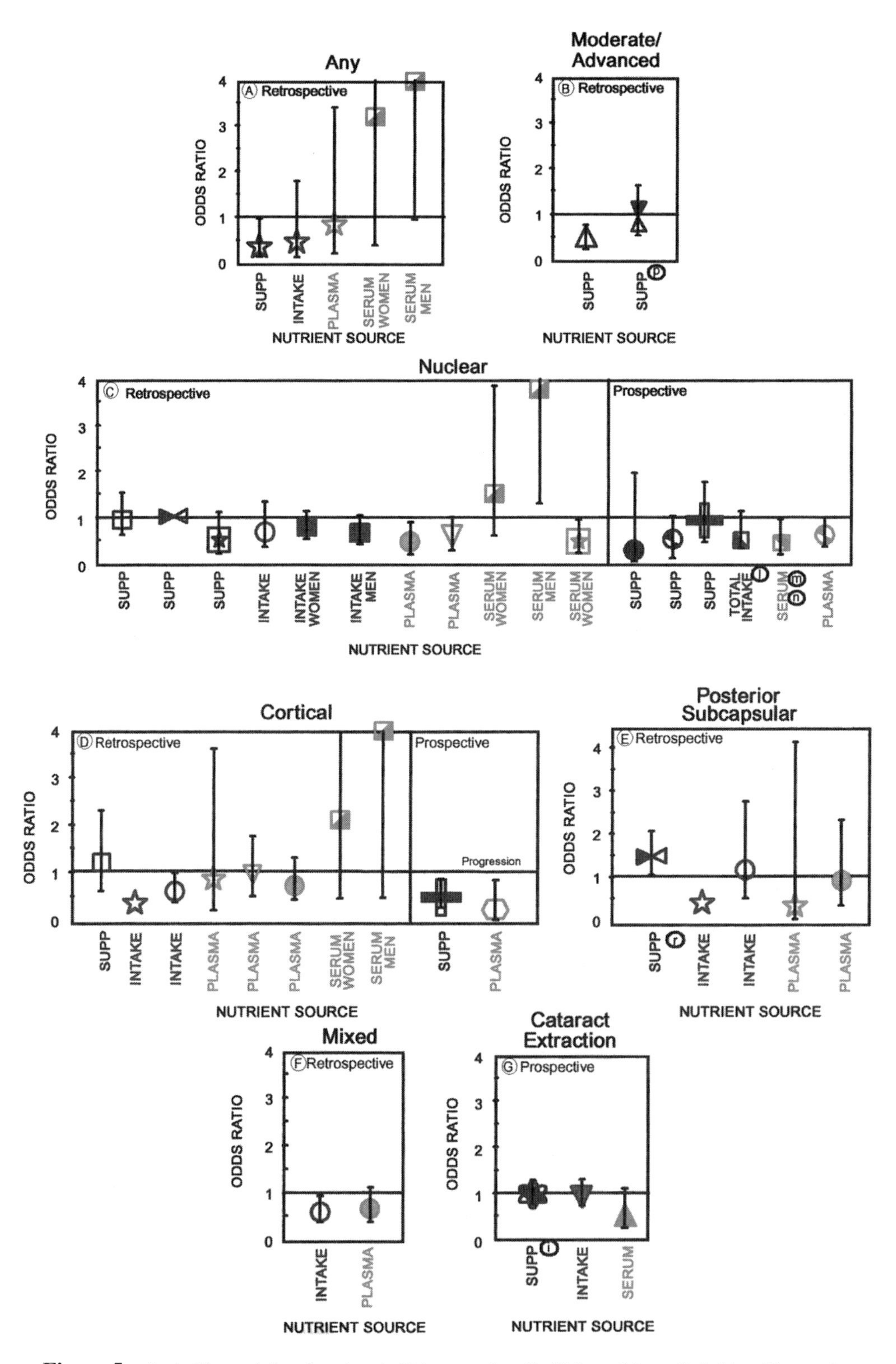


**Figure 5** As in Figure 4, but for vitamin E (α-tocopherol). (Adapted from Ref. 11, with permission.)

and cataract surgery (Fig. 5g). Women with high vitamin E intakes (median = 210 mg/day) had a similar rate of cataract surgery (RR: 0.96; CI: 0.72–1.29) as women with low intakes (median = 3.3 mg/day). This is consistent with data regarding supplement intake in the same cohort (104). Mares-Perlman et al. found that dietary vitamin E was associated (nonsignificantly) with diminished risk for nuclear cataract in men, but not in women (Fig. 5c), although this information contrasts with their positive correlations between serum α-tocopherol levels and cataract (Figs. 5a, 5c, and 5d) (see below) (113).

Working later in the same community, Lyle et al. (101) in prospective studies showed that total intake was inversely, but nonsignificantly, related to incident cataract in Beaver Dam (RR: 0.5; CI: 0.3–1.1), a result which is corroborated by serum vitamin E determinations (Fig. 5c) (RR: 0.4; CI: 0.2–0.9)(100). Several other studies assessing plasma vitamin E levels also reported significant inverse associations with cataract (Fig. 5). We observed a 52% decreased risk for nuclear cataract (RR: 0.48; CI: 0.25–0.95) for women in the baseline phase of the Nutrition and Vision Project with serum vitamin E levels of 0.41 μmol/L versus women with 23 μmol/L (98). Knekt and coworkers (107) followed a cohort of 1419 Finns for 15 years and identified 47 patients admitted to ophthalmological wards for mature cataract. They selected two controls per patient matched for age, sex, and municipality. These investigators reported that persons with serum vitamin E concentrations above approximately 20 μM had about one-half the rate of subsequent cataract surgery (RR: 0.53; CI: 0.24–1.1) (Fig. 5g) compared with persons with vitamin E concentrations below this concentration. In another retrospective study, Vitale and coworkers (96) observed the age-, sex-, and diabetes-adjusted prevalence of nuclear cataract to be about 50% less (RR: 0.52; CI: 0.27–0.99) (Fig. 5c) among persons with plasma vitamin E concentrations greater than 29.7 μM compared to persons with levels below 18.6 μM. A similar comparison showed that the prevalence of cortical cataract did not differ between those with high and low plasma vitamin E levels (RR: 0.96; CI: 0.52–0.178) (Fig. 5d). Jacques and Chylack (94) observed the prevalence of posterior subcapsular cataract to be 67% (RR: 0.33; CI: 0.03–4.13) (Fig. 5e) lower among persons with plasma vitamin E levels above 35 μM relative to persons with levels below 21 μM after adjustment for age, sex, race, and diabetes; however, the effect was not statistically significant. Prevalence of any early cataract (RR: 0.83; CI: 0.20–3.40) (Fig. 5a) or cortical cataract (RR: 0.84; CI: 0.20–3.60) (Fig. 5d) did not differ between those with high and low plasma levels. In a prospective study that examined risk for cortical cataract progress among individuals with higher plasma vitamin E, Rouhiainen et al. found a 73% reduction in risk (RR: 0.27; CI: 0.08–0.83) (Fig. 5d) (109). Plasma vitamin E was also inversely associated with prevalence of cataract in a large Italian study after adjusting for age and sex, but the relationship was no longer statistically significant after adjusting for other factors such as education, sunlight exposure, and family history of cataract (90). Leske et al. (56) also demonstrated that individuals with high plasma vitamin E levels had significantly lower prevalence of nuclear cataract (RR: 0.44; CI: 0.21–0.90), but vitamin E was not associated with cataracts at other lens sites. In a prospective study, Leske et al. reported a 42% reduction in risk for nuclear cataract progression among supplement users (RR: 0.58; CI: 0.36–0.94) (Fig. 6c) (110).

Mares-Perlman and coworkers, working with similar cohorts as Lyle (100), noted a significant elevated prevalence of nuclear cataract in men with high serum vitamin E (RR: 3.74; CI: 1.25–11.2) and in women (RR: 1.47; CI: 0.57–3.82) (Fig. 5c) (119). One other study failed to observe any association between cataract and plasma vitamin E levels (95).

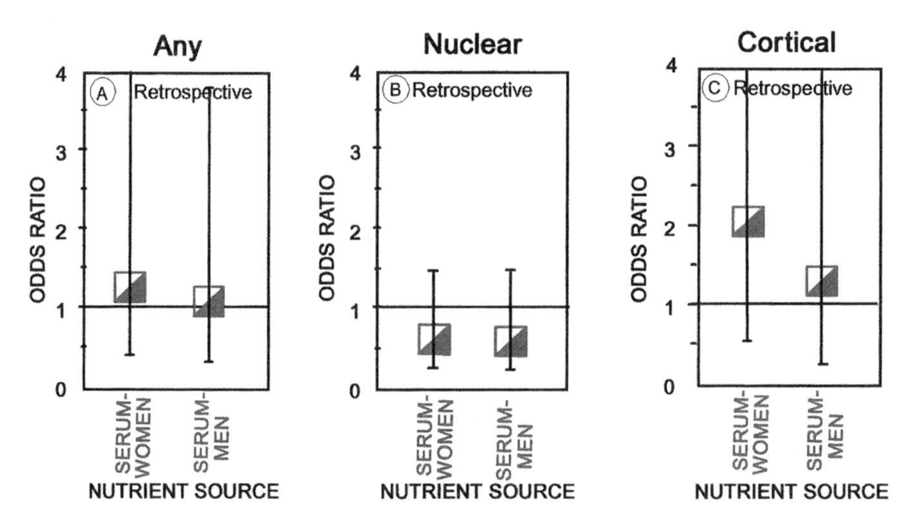


**Figure 6** As in Figure 4, but for γ-tocopherol. (Adapted from Ref. 11, with permission.)

Since the recognition of antioxidant potential of γ-tocopherol, studies have begun to ask if this form of tocopherol is associated with altered risk for cataract. γ-Tocopherol has lower biological vitamin E activity compared to α-tocopherol. Mares-Perlman et al. (119) observed an inverse (nonsignificant) relationship (RR: 0.61; CI: 0.32–1.19) between serum γ-tocopherol (Fig. 6b) and severity of nuclear sclerosis, but a positive, significant relationship between elevated serum γ-tocopherol levels and severity of cortical cataract for men and women (Fig. 6c).

## C. Carotenoids

The carotenoids, like vitamin E, are also natural lipid-soluble antioxidants (58). β-carotene is the best known carotenoid because of its importance as a vitamin A precursor. It exhibits particularly strong antioxidant activity at low partial pressures of oxygen (15 Torr) (120). This is similar to the ≦20 Torr partial pressure of oxygen in the core of the lens (121). However, it is only one of ~400 naturally occurring carotenoids (122), and other carotenoids may have similar or greater antioxidant potential (58,68,123,124). In addition to β-carotene, α-carotene, lutein, zeaxanthin, and lycopene are important carotenoid components of the human diet (125). Carotenoids have been identified in the lens in ≈10 nG/g wet weight concentrations (Table 1) (60,126). There is a dearth of laboratory data that relate carotenoids to cataract formation, but data that relate carotenoid supplement use to risk for cataracts are beginning to appear and carotenoid supplements are already being included in several ''eye vitamin'' preparations.

Jacques and Chylack (94) were the first to observe that persons with carotene intakes above 18,700 IU/day had the same prevalence of cataract as those with intakes below 5677 IU/day (RR: 0.91; CI: 0.23–3.78) (Fig. 7a). Hankinson et al. (106) followed this report with a study that reported that the multivariate-adjusted rate of cataract surgery was about 30% lower (RR: 0.73; CI: 0.55–0.97) for women with high carotene intakes (median = 14,558 IU/day) compared with women with low intakes of this nutrient (median = 2935 IU/day) (Fig. 7e). In another group of women, Chasen-Taber showed that risk for cataract extraction was slightly reduced in users of supplements (Fig. 7e) (104)

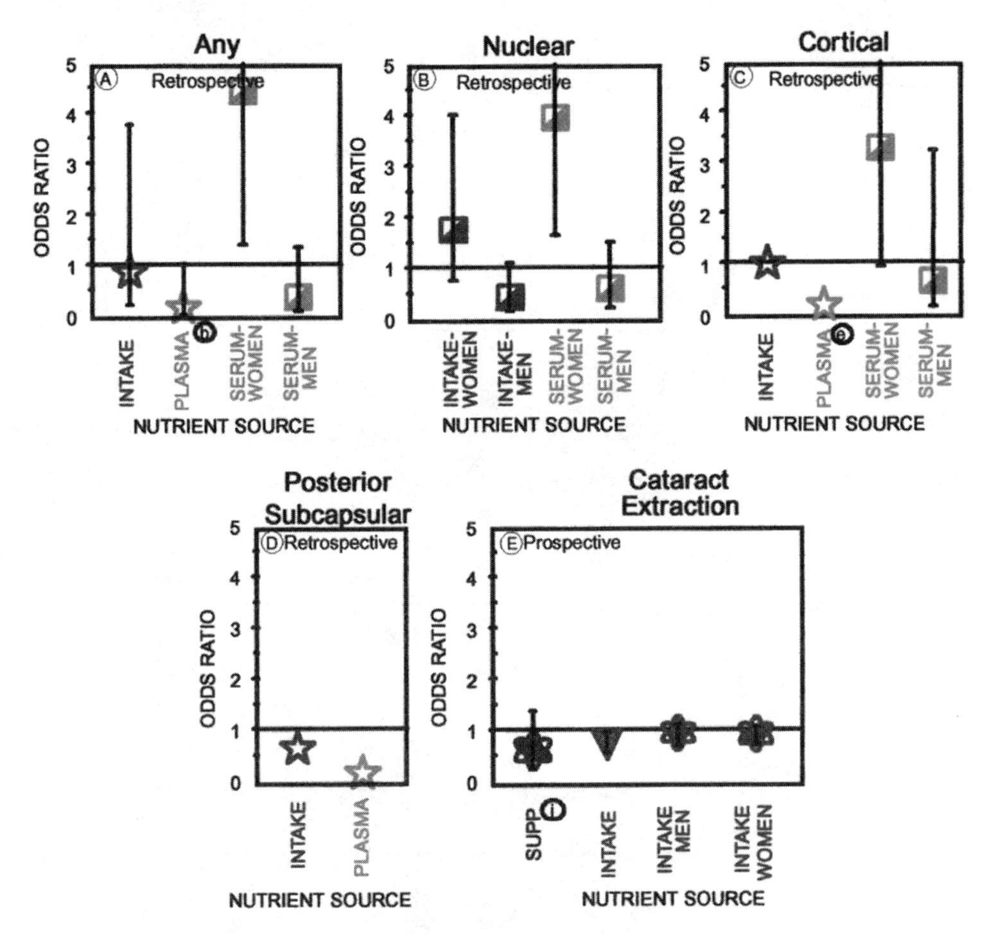


**Figure 7** As in Figure 4, but for carotenoids (generally measured as β-carotene). (Adapted from Ref. 11, with permission.)

and in women with higher total carotenoid intake (RR: 0.85; CI: 0.7–1.03) (103). Similarly, Brown showed that men with higher carotenoid intakes had reduced risk for cataract extraction (RR:0.85; CI: 0.68–1.07) (102). However, while cataract surgery was inversely associated with total carotene intake in these studies, it was not strongly associated with consumption of carotene-rich foods, such as carrots. Rather, cataract surgery was associated with lower intakes of foods such as spinach that are rich in lutein and xanthin carotenoids, rather than β-carotene. This would appear to be consistent with our observation that the human lens contains lutein and zeaxanthin but no β-carotene (Table 1). Unfortunately, cataract surgery was not an endpoint in other studies that considered xanthaphylls (91,119). Mares-Perlman found no significant associations between specific carotenoid intake and risk for cataract (113).

In another prospective study, Lyle showed that intakes of α-carotene (RR: 0.95; CI: 0.7–2.1) (Fig. 8b) and β-carotene (RR: 0.9; CI: 0.5–1.4) (Fig. 9b) were not significantly related to incident nuclear cataract. Neither were intakes of α-carotene, β-carotene, lycopene, and β-cryptoxanthin related to risk for cataract extraction (103). For men, intakes of α-carotene (RR: 0.89; CI: 0.72–1.1), β-carotene (RR: 0.92; CI:0.73–1.16), lycopene

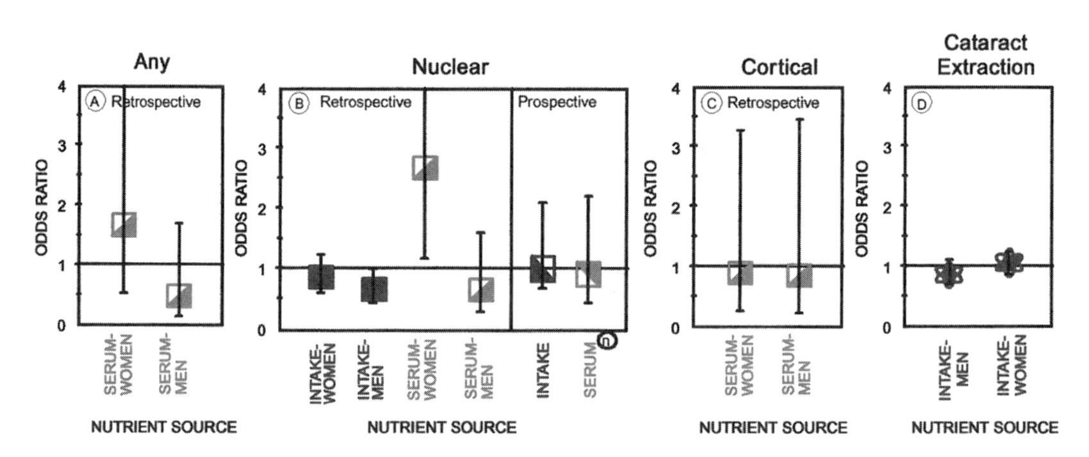


**Figure 8** As in Figure 4, but for α-carotene. (Adapted from Ref. 11, with permission.)

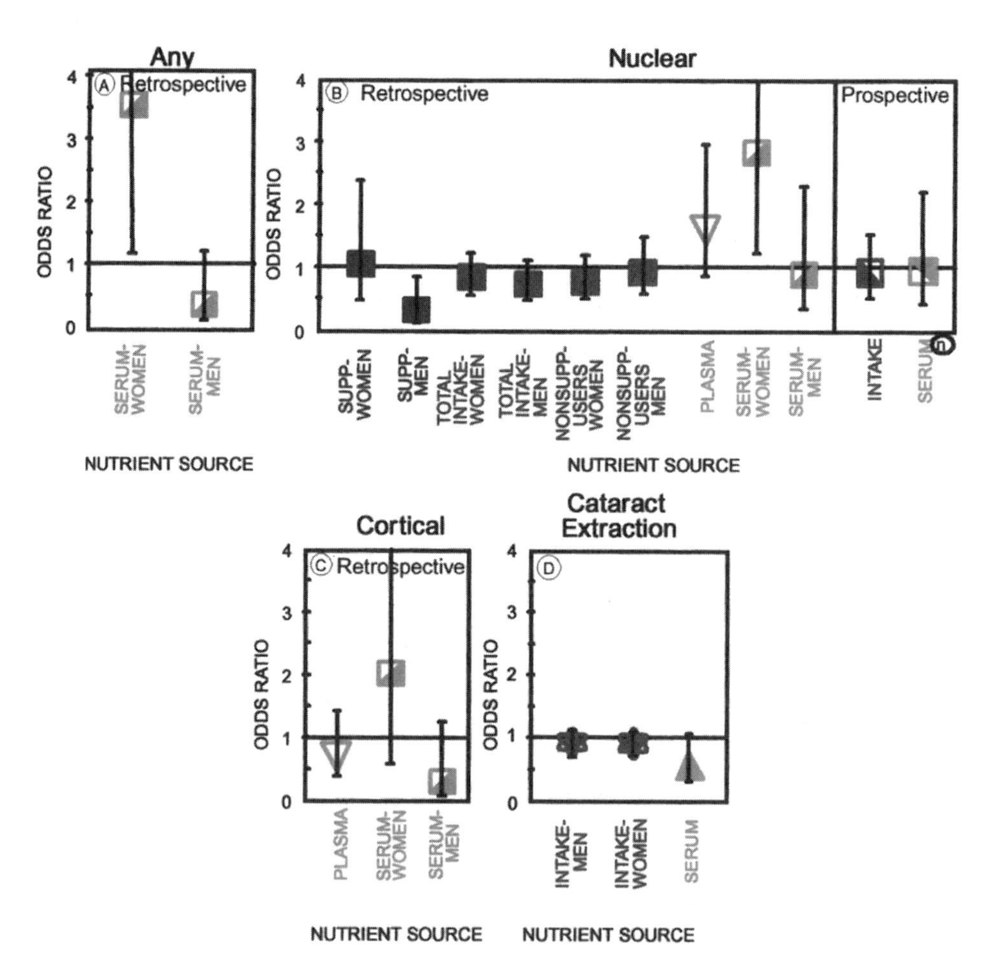


**Figure 9** As in Figure 4, but for β-carotene (see also data in Fig. 8). (Adapted from Ref. 11, with permission.)

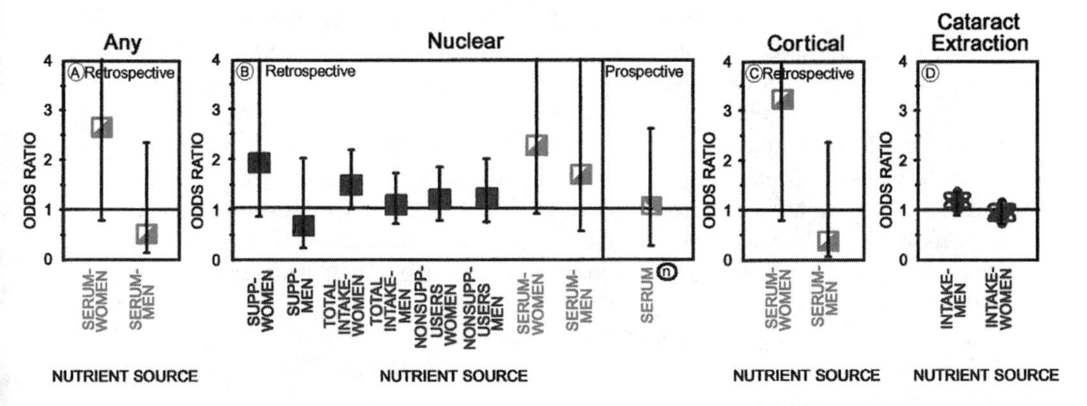


**Figure 10** As in Figure 4, but for lycopene. (Adapted from Ref. 11, with permission.)

(RR: 1.10; CI: 0.88–1.36) or β-cryptoxanthin (R: 1.09; CI: 0.87–1.37) were not significantly related to risk for cataract extraction (Figs. 8–11) (102), and Mares-Perlman also did not detect significantly altered risk for cataract among consumers of these nutrients (Figs. 8–11) (91,113).

Plasma levels of carotenoids were also related to risk for cataract. Jacques and Chylack noted that persons with high plasma total carotenoid concentrations (>3.3 μM) had less than one-fifth the prevalence of cataract compared to persons with low plasma carotenoid levels (<1.7 μM) (RR: 0.18; CI: 0.03–1.03) after adjustment for age, sex, race, and diabetes (Fig. 7a). However, they were unable to observe an association between carotene intake and cataract prevalence (94). Knekt and coworkers (107) reported that among age- and sex-matched cases and controls, persons with serum β-carotene concentrations above approximately 0.1 μM had a 40% reduction in the rate of cataract surgery compared with persons with concentrations below this level (RR: 0.59; CI: 0.26–1.25) (Fig. 9d).

Mares-Perlman correlated serum carotenoids and severity of nuclear and cortical opacities (119), finding that higher levels of individual or total carotenoids in the serum were not associated with less severe nuclear or cortical cataract overall (Figs. 7–9). Asso-

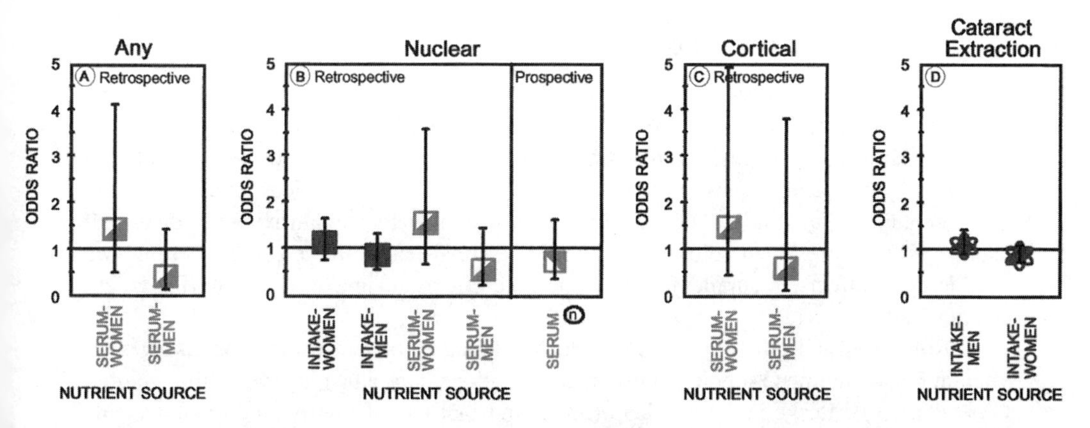


**Figure 11** As in Figure 4, but for β-cryptoxanthin. (Adapted from Ref. 11, with permission.)

ciations between risk for some forms of cataract and nutriture differed between men and women (e.g., nuclear cataract and α-carotene intake) (Fig. 8b) (113). Other nutrients for which cataract risk in women versus men showed opposing relationships include serum β-carotene (Fig. 9a and b) and serum lycopene (Fig. 10c). A marginally significant trend for lower risk ratio for cortical opacity with increasing serum levels of β-carotene was observed in men, but not women. Higher serum levels of α-carotene, β-cryptoxanthin, and lutein were significantly related to lower risk for nuclear sclerosis only in men who smoked. In contrast, higher levels of some carotenoids were often directly associated with elevated risk for nuclear sclerosis and cortical cataract (Figs. 7–9), particularly in women. Following up on these relations Lyle (100) found little effect of serum levels of α-carotene (RR: 0.9; CI: 0.4–2.2), β-carotene (RR: 0.9; CI: 0.4–2.2), lycopene (RR: 1.1; CI: 0.5–2.6), β-cryptoxanthin (RR: 0.7; CI: 0.3–1.6), or lutein (RR: 0.7; CI: 0.3–1.6) and incident cataract (Figs. 8–10) (98,99).

Vitale and colleagues (96) also examined the relationships between plasma β-carotene levels and age-, sex-, and diabetes-adjusted prevalence of cortical and nuclear cataract (Fig. 9b and c). Although the data suggested a weak inverse association between plasma β-carotene and cortical cataract and a weak positive association between this and nuclear cataract, neither association was statistically significant. Persons with plasma β-carotene concentrations above 0.88 μM had a 28% lower prevalence of cortical cataract (RR: 0.72; CI: 0.37–1.42) and a 57% (RR: 1.57; CI: 0.84–2.93) higher prevalence of nuclear cataract compared to persons with levels below 0.33 μM.

Lutein intake has been related to risk for age-related maculopathy, and it has recently been given more careful examination with respect to risk for cataract. As indicated above, several studies failed to find significant inverse relations with respect to lutein intake and risk for cataract. However, other studies found indications that elevated intake of foods rich in lutein is related to decreased risk for cataract. In keeping with food analyses, intake of lutein/zeaxanthin was inversely related to risk for incident nuclear cataract (RR: 0.4; CI: 0.2–0.8) (Fig. 12b) (101) and to cataract extraction in women (Fig. 12d) (103) (RR: 0.78; CI: 0.63–0.95) and in men (RR: 0.81; CI: 0.65–1.01) (102). The foods most clearly associated with decreased risk for cataract extraction are broccoli and cooked spinach. These are foods that are rich in lutein and zeaxanthin (102). Recently, we found that risk for nuclear cataract was 51% lower (RR: 0.49; CI: 0.25–0.94) in persons who had intake of ≥7.4 versus ≤5.7 mg/day (98).

## D. Antioxidant Combinations

In order to more closely approximate combined effects on cataract risk of the multiple antioxidants that are contained in food, this group was the first to adopt ''antioxidant indices.'' However, it is possible that single nutrients appear to have strong influences on the indices, and we now question the utility of the indices. Nevertheless, in an attempt to offer a complete summary, data regarding relationships between antioxidant indices and cataract risk are presented below (Fig. 13). To be consistent with the earlier sections, we first describe effects of supplement use; then the effects of intake, including foods, are summarized.

Robertson and coworkers (92) found no enhanced benefit to persons taking both vitamin E and vitamin C supplements compared with persons who only took either vitamin C or vitamin E. Leske and coworkers (93) found that use of multivitamin supplements was associated with decreased prevalence for each type of cataract: 60%, 48%, 45%, and

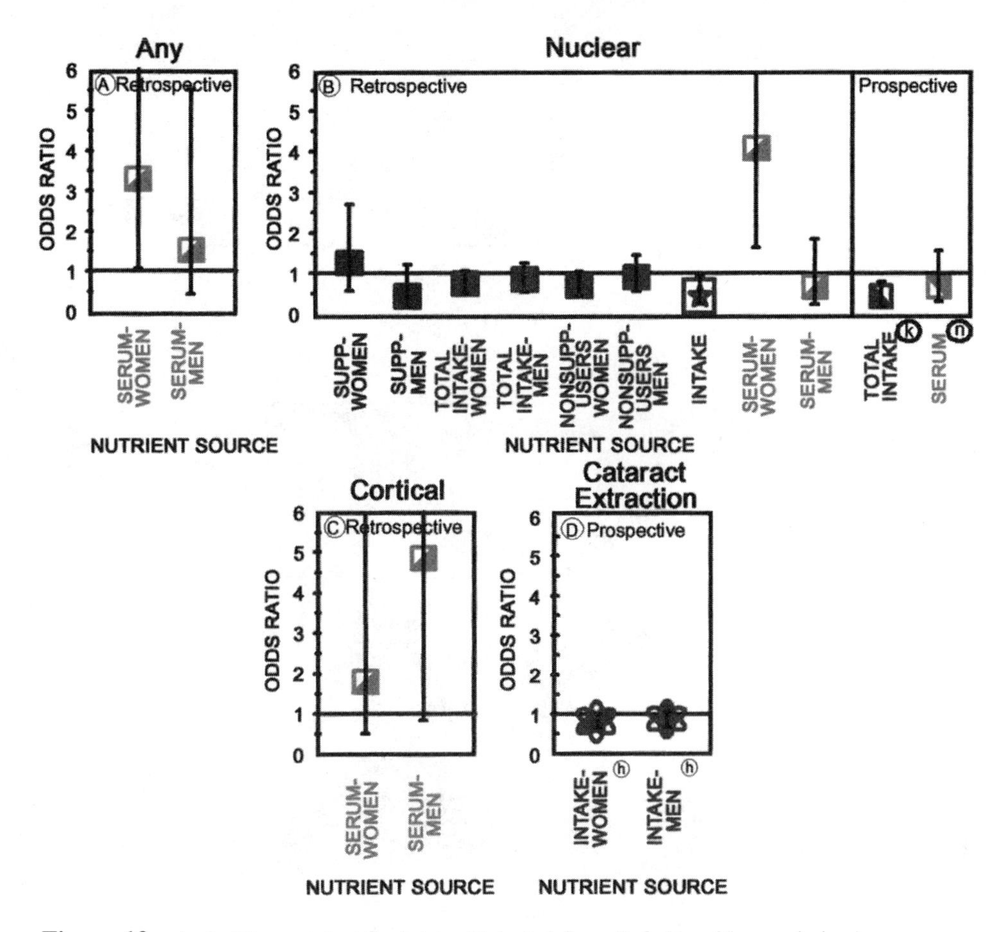


**Figure 12** As in Figure 4, but for lutein. (Adapted from Ref. 11, with permission.)

30%, respectively, for posterior subcapsular (RR: 0.40; CI: 0.21–0.77), cortical (RR: 0.52; CI: 0.36–0.72), nuclear (RR: 0.55; CI: 0.33–0.92), and mixed (RR: 0.70; CI: 0.51–0.97) cataracts (Fig. 13b–13e). Luthra et al. (97) and Leske et al. (127) reported that supplement consumption was associated with a slight reduction in risk for cortical cataract in a black population (RR: 0.77; CI: 0.61–0.98) in those <70 years (Fig. 13c).

Multivitamins were also reported to reduce the risk of incident cataracts, as well as progression of existing cataracts in several studies (Fig. 13). Seddon and coworkers (105) observed a reduced risk for incident cataract for users of multivitamins (RR: 0.73; CI: 0.54–0.99). Mares-Perlman et al. (108) reported that mutivitamin users had significant 20% (RR: 0.8; CI: 0.6–1.0) and 30% (RR: 0.7; CI: 0.5–1.0) reduction of cortical cataract progression and incidence, respectively (Fig. 13c). This report corroborates earlier data regarding nuclear cataract (Fig. 13b), but it is at odds with their data regarding cortical cataract (Fig. 13c) (91). Leske et al. reported a 31% reduced risk for progression of nuclear cataract among users of multivitamins (RR: 0.69; CI: 0.48–0.99) (110) (Fig. 13c). Hankinson found no relationship between multivitamin use and risk for cataract extraction (106). Although we noted diminished risk for nuclear opacities among users of multivitamins in the baseline Nutrition and Vision Project (RR: 0.73; CI: 0.44–1.23), the results were

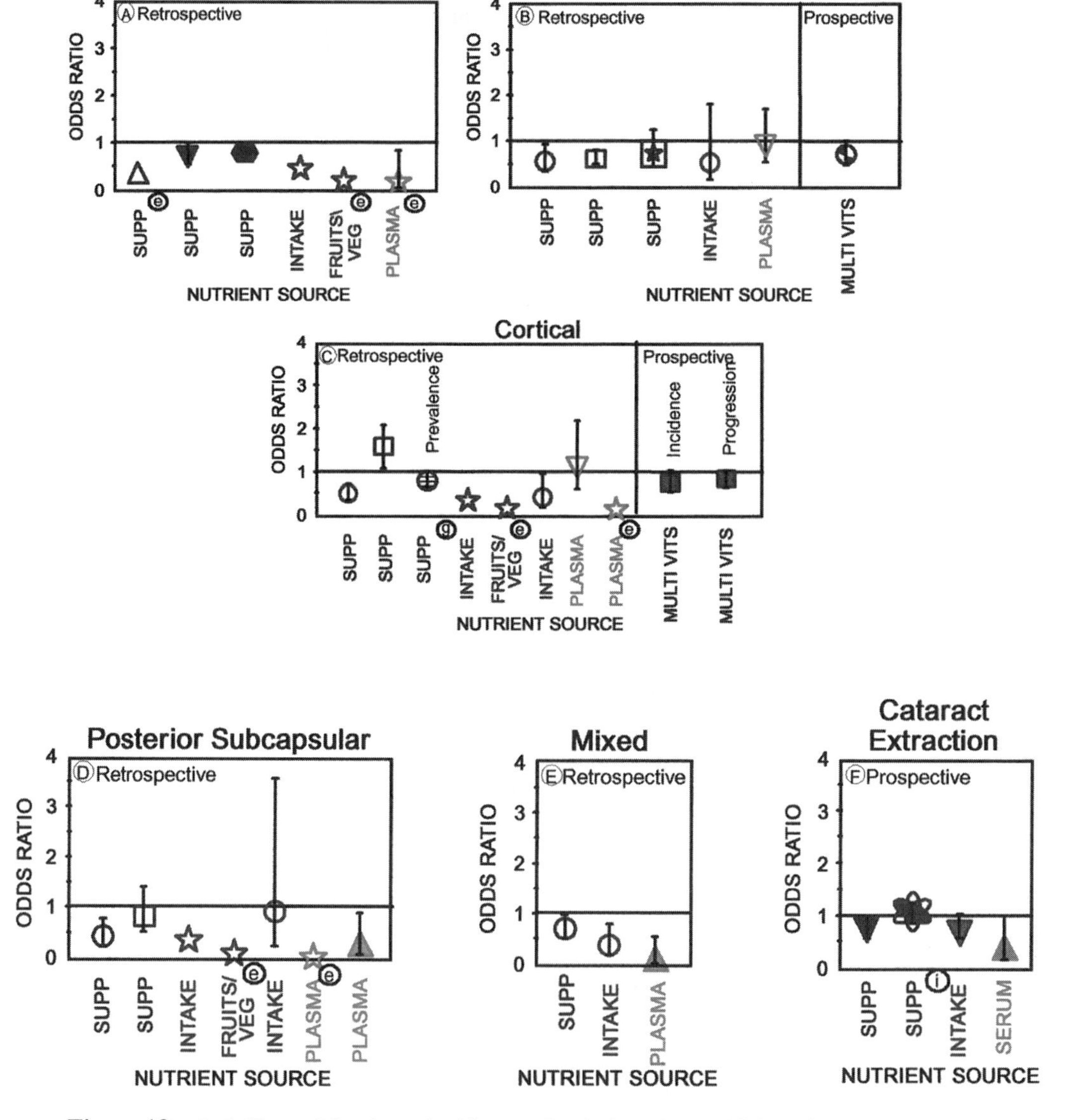


**Figure 13** As in Figure 4, but for antioxidant nutrient index using multiple antioxidants. (Adapted from Ref. 11, with permission.)

not statistically significant after correcting for intake of the supplements (98). In these studies it is not clear that synergy between nutrients is indicated with respect to conferring diminished risk for cataract.

Foods contain multiple nutrients and, as such, intake of foods might be related to risk for cataract in ways that are different from relations that are observed for specific nutrients. The first, and perhaps most important, study in terms of revealing the utility of foods indicates that persons who consumed $\geq$ 1.5 servings of fruits and/or vegetables had only ≃20% the risk of developing cataract as those who did not (Fig. 13a) (94). These workers also found that intakes of fruits and vegetables that provide high levels of lutein and zeaxanthin are associated with decreased risk for cataract. This observation has found support in several of the recent studies noted above. Hankinson and coworkers (106) calculated an antioxidant score based on intakes of carotene, vitamin C, vitamin E, and riboflavin and observed a 24% reduction in the adjusted rate of cataract surgery among women with high antioxidant scores relative to women with low scores (RR: 0.76; CI: 0.57–1.03) (Fig. 13f). Using a similar index based on combined antioxidant intakes (vitamin C, vitamin E, and carotene, as well as riboflavin), Leske and coworkers (93) found that persons with high scores had 60% lower adjusted prevalence of cortical (RR: 0.42; CI: 0.18–0.97) (Fig. 13c) and mixed (RR: 0.39; CI: 0.19–0.80) (Fig. 13e) cataract compared to those who had low scores. Whereas risk for nuclear cataract was also inversely related in this index, the data were not significant. Risk for posterior subcapsular cataract was not related on this index (Fig. 13d).

Jacques and Chylack (94) also found that the adjusted prevalence of all types of cataract was 40% (RR: 0.62; CI: 0.12–1.77) and 80% (RR: 0.16; CI: 0.04–0.82) lower for persons with moderate and high antioxidant index scores (based on combined plasma vitamin C, vitamin E, and carotenoid levels), as compared with persons with low scores (Fig. 13a). Risk for cortical and posterior subcapsular cataract was also diminished in the study (Figs. 13c and 13d, respectively). Mohan and coworkers (95) constructed a somewhat more complex antioxidant scale that included red blood cell levels of glutathione peroxidase, glucose-6-phosphate dehydrogenase, and plasma levels of vitamin C and vitamin E. Even though they failed to see any protective associations with any of these individual factors, and even reported a positive association between plasma vitamin C and prevalence of cataract, they found that persons with high antioxidant index scores had a substantially lower prevalence of cataracts involving the posterior subcapsular region (RR: 0.23; CI: 0.06–0.88) (Fig. 13d) or mixed cataract with posterior subcapsular and nuclear components (RR: 0.12; CI: 0.03–0.56) after multivariate adjustment (Fig. 13e). Knekt and coworkers (107) observed that the rate of cataract surgery for persons with high levels of both serum vitamin E and β-carotene concentrations appeared lower than the rate for persons with either high vitamin E or high β-carotene levels (Fig. 13f). Persons with high serum levels of either had a rate of cataract surgery that was 40% less than persons with low levels of both nutrients (RR: 0.38; CI: 0.15–1.0).

Vitale and coworkers (96) also examined the relationship between antioxidant scores (based on plasma concentrations of vitamin C, vitamin E, and β-carotene) and prevalence of cataract, but did not see evidence of any association. The age-, sex-, and diabetes-adjusted risk ratios were close to 1 for both nuclear (RR: 0.96; CI: 0.54–1.70) and cortical (RR: 1.17; CI: 0.62–2.20) cataract (Fig. 13b and 13c).

## E. Intervention Studies

To date only one intervention trial designed to assess the effect of vitamin supplements on cataract risk has been completed. Sperduto and coworkers (116) took advantage of two ongoing, randomized, double-blinded vitamin and cancer trials to assess the impact of vitamin supplements on cataract prevalence. The trials were conducted among almost 4000 participants aged 45 to 74 years from rural communes in Linxian, China. Participants in one trial received either a multisupplement or placebo. In the second trial, a more complex factorial design was used to evaluate the effects of four different vitamin/mineral combinations: retinol (5000 IU) and zinc (22 mg); riboflavin (3 mg) and niacin (40 mg); vitamin C (120 mg) and molybdenum (30 μg); and vitamin E (30 mg), β-carotene (15 mg), and selenium (50 μg). At the end of the 5- to 6-year follow-up, the investigators conducted eye examinations to determine the prevalence of cataract (Fig. 14).

In the first trial there was a significant 43% reduction in the prevalence of nuclear cataract for persons aged 65 to 74 years receiving the multisupplement (RR: 0.57; CI: 0.36–0.90) (Fig. 14a). The second trial demonstrated a significantly reduced prevalence of nuclear cataract in persons receiving the riboflavin/niacin supplement relative to those persons not receiving this supplement (RR: 0.59; CI: 0.45–0.79). The effect was strongest in those aged 65 to 74 years (RR: 0.45; CI: 0.31–0.64). However, the riboflavin/niacin supplement appeared to increase the risk of posterior subcapsular cataract (RR: 2.64; CI: 1.31–5.35) (Fig. 14c). The results further suggested a protective effect of the retinol/zinc supplement (RR: 0.77; CI: 0.58–1.02) and the vitamin C/molybdenum supplement (RR: 0.78; CI: 0.59–1.04) on prevalence of nuclear cataract.

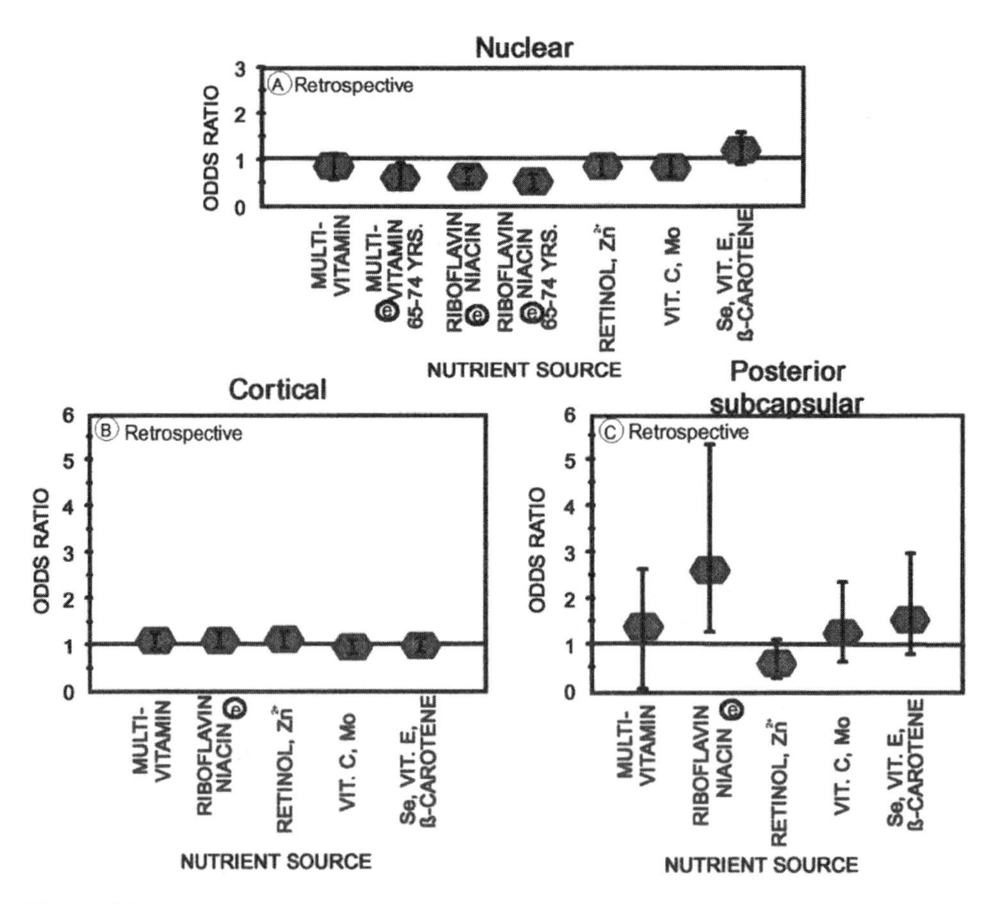


**Figure 14** As in Figure 4, but for intervention trials. (Adapted from Ref. 11, with permission.)

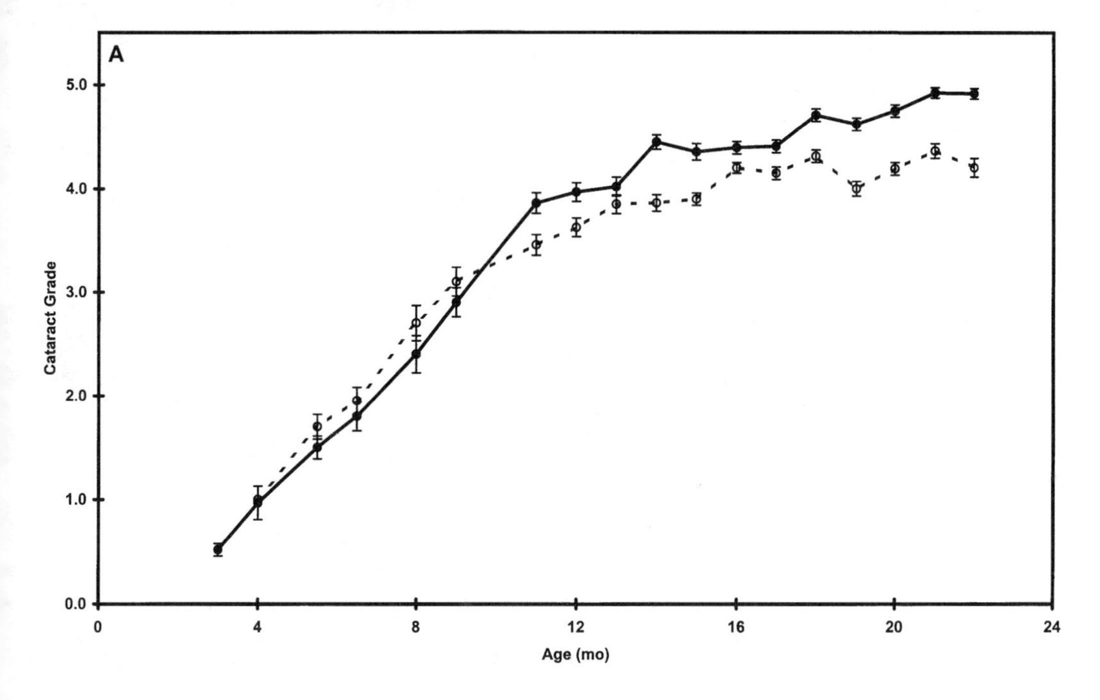


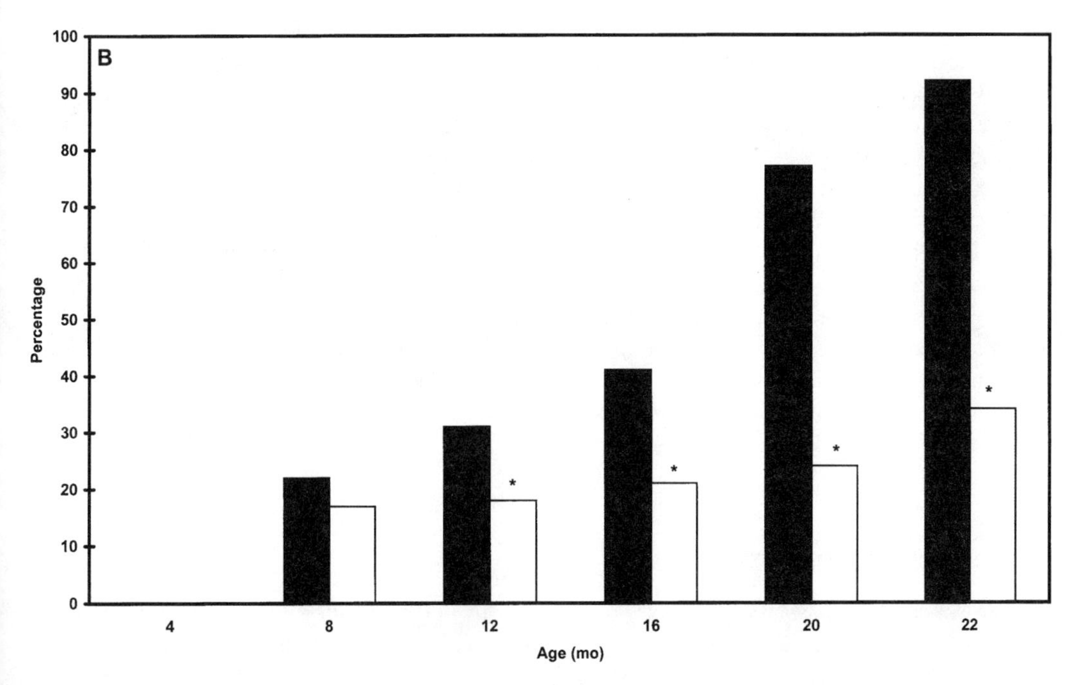


**Figure 15** (A) Mean cataract grade in calorie-restricted and control Emory mice at 3 to 22 months of age. ● = control mice; ○ = restricted mice; values are means ± SEM. At older ages error bars are approximately equal in size to the symbol for control animals. (B) Percent of lenses with grade 5 cataract. $*p = 0.05$. (Adapted from Ref. 11, with permission.)

## VI. CONCLUSION

Light and oxygen appear to be both a boon and a bane. While necessary for physiological function, when present in excess or in uncontrolled circumstances they appear to be causally-related to cataractogenesis. Upon aging, compromised function of the lens is exacerbated by depleted or diminished primary antioxidant reserves, antioxidant enzyme capabilities, and diminished secondary defenses such as proteases. Smoking (14) and light exposure (13) appear to provide oxidative challenge and are also associated with an elevated risk of cataract. Poor education and lower socioeconomic status also markedly increased risk for cataract (93,95,128,129). These are related to poor nutrition. The impression created by the literature is that there is some benefit to enhanced antioxidant intake with respect to diminished risk for cataract. Optimal levels of ascorbate would appear to be ≥250 mg/day (23,98). It is possible to adjust normal dietary practice to obtain close-to-saturating levels of plasma ascorbate (21,106,112,130). Because the bioavailability of ascorbate may decrease with age, slightly higher intakes may be required in the elderly. Intake of other antioxidant nutrients such as vitamin E and lutein also appear to be inversely related to risk for cataract, but more work is needed to confirm these results and define optimal levels of these nutrients. It is difficult to compare the various studies. In addition to the conclusions reached, that the correlations were not always with the same form of cataract may indicate that the cataracts were graded differently and/or that there are common etiological features of each of the forms of cataract described. Most of the studies noted above utilized case-control designs, and most assessed status only once. Since intake or status measures are highly variable and the effects of diet are likely to be cumulative, studies should be performed on populations for which long-term dietary records are available. It appears that intake studies are preferable to use of plasma measures, if a single measure of status must be chosen. Longitudinal studies and more intervention studies are certainly essential in order to truly establish the value of antioxidants and to determine the extent to which cataract progress is affected by nutriture. More uniform methods of lens evaluation, diet recording, and blood testing, etc., would facilitate conclusions regarding the merits of antioxidants. Optimization of nutriture can be achieved through better diets and supplement use, especially among the poor and at-risk population. In addition to quantifying optimal intake, it is essential to know when and for how long intake of the nutrients would be useful with respect to delaying cataract. Thus, the overall impression created by these data suggests that nutrition, possibly including antioxidant vitamin supplement use, may provide the least costly and most practicable means to delay cataract.

## ACKNOWLEDGMENTS

We acknowledge the assistance of Tom Nowell in the preparation of figures and Paul Jacques for invaluable assistance in evaluating the epidemiological data.

## REFERENCES

1. Kupfer C. The conquest of cataract: a global challenge. Trans Ophthal Soc UK 1984; 104: 1–10.
2. Schwab L. Cataract blindness in developing nations. Int Ophthalmol Clin 1990; 30:16–18.

3. World Health Organization. Use of intraocular lenses in cataract surgery in developing countries. Bull World Health Org 1991; 69:657–666.
4. Klein BEK, Klein R, Linton KLP. Prevalence of age-related lens opacities in a population: The Beaver Dam Eye Study. Ophthalmology 1992; 99:546–552.
5. Klein R, Klein BE, Linton KL, DeMets DL. The Beaver Dam Eye Study: the relation of age-related maculopathy to smoking. Am J Epidemiol 1993; 137:190–200.
6. Leibowitz H, Krueger D, Maunder C, Milton RC, Mohandas MK, Kahn HA, Nickerson RJ, Pool J, Colton TL, Ganley JP, Loewenstein JI, Dawber TR. The Framingham Eye Study Monograph. Surv Ophthalmol 1980; 24(suppl):335–610.
7. Chatterjee A, Milton RC, Thyle S. Prevalence and etiology of cataract in Punjab. Br J Ophthalmol 1982; 66:35–42.
8. Wang GM, Spector A, Luo CQ, Tang LQ, Xu LH, Guo WY, Huang YQ. Prevalence of age-related cataract in Ganzi and Shanghai. The Epidemiological Study Group. Chinese Med J 1990; 103:945–951.
9. Whitfield R, Schwab L, Ross-Degnan D, Steinkuller P, Swartwood J. Blindness and eye disease in Kenya: ocular status survey results from the Kenya Rural Blindness Prevention Project. Br J Ophthalmol 1990; 74:333–340.
10. Taylor A. Lens and retina function: Introduction and challenge. In: Taylor A, ed. Nutrition and Environmental Influences on the Eye. Boca Raton, FL: CRC Press, 1999:1–4.
11. Taylor A. Nutritional and environmental influences on risk for cataract. In: Taylor A, ed. Nutritional and Environmental Influences on Vision. Boca Raton, FL: CRC Press, 1999:53–93.
12. Young R. Optometry and the preservation of visual health. Optom Vis Sci 1992; 70:255–262.
13. McCarty C, Taylor HR. Light and risk for age-related diseases. In: Taylor A, ed. Nutritional and Environmental Influences on the Eye. Boca Raton, FL: CRC Press, 1999:135–150.
14. West SK. Smoking and the risk of eye disease. In: Taylor A, ed. Nutritional and Environmental Influences on the Eye. Boca Raton, FL: CRC Press, 1999:151–164.
15. Taylor A, Tisdell F, Carpenter FH. Leucine aminopeptidase (bovine lens): synthesis and kinetic properties of ortho, meta, and para substituted leucyl-analides. Arch Biochem Biophys 1981; 210:90–97.
16. Chylack LTJ, Wolfe JK, Singer DM, Leske MC, Bullimore MA, Bailey IL, Friend J, McCarthy D, Wu S. The lens opacities classification system III. Arch Ophthalmol 1993; 111:831–836.
17. Chylack LT, Jr. Function of the lens and methods of quantifying cataract. In: Taylor A, ed. Nutritional and Environmental Influences on the Eye. Boca Raton, FL: CRC Press, 1999: 25–52.
18. Chylack LTJ, Wolfe JK, Singer DM, Wu SY, Leske MC. Nuclear cataract: relative contributions to vision loss of opalescence and brunescence. Invest Ophthalmol Vis Sci 1994; 35: 42632 (abstr).
19. Wolfe JK, Chylack LT, Leske MC, Wu SY, Group L. Lens nuclear color and visual function. Invest Ophthalmol Vis Sci 1993; 34 (suppl):2550.
20. Taylor A, Davies KJA. Protein oxidation and loss of protease activity may lead to cataract formation in the aged lens. Free Radic Biol Med 1987; 3:371–377.
21. Taylor A, Jacques PF, Nadler D, Morrow F, Sulsky SI, Shepard D. Relationship in humans between ascorbic acid consumption and levels of total and reduced ascorbic acid in lens. aqueous humor, and plasma. Curr Eye Res 1991; 10:751–759.
22. Bunce GE, Kinoshita J, Horwitz J. Nutritional factors in cataract. Annu Rev Nutr 1990; 10: 233–254.
23. Taylor A, Jacques PF, Nowell T, Jr, Perrone G, Nadler D, Joswiak B. Vitamin C in human and guinea pig aqueous, lens, and plasma in relation to intake. Exp Eye Res 1997; 16:857–864.

24. Reddy V. Glutathione and its function in the lens—an overview. Exp Eye Res 1990; 150: 771–778.
25. Mune M, Meydani M, Jahngen-Hodge J, Martin A, Smith D, Palmer V, Blumberg JB, Taylor A. Effect of calorie restriction on liver and kidney glutathione in aging Emory mice. Age 1995; 18:43–49.
26. Smith D, Shang F, Nowell TR, Asmundsson G, Perrone G, Dallal G, Scott L, Kelliher M, Gindelsky B, Taylor A. Decreasing ascorbate intake does not affect the levels of glutathione, tocopherol or retinol in the ascorbate-requiring osteogenic disorder Shionogi rats. J Nutr 1999; 129:1229–32.
27. Sastre J, Meydani M, Martin A, Biddle L, Blumberg J. Effect of glutathione monoethyl ester administration on galactose-induced cataract in the rat. Life Chem Rep 1994; 12:89–95.
28. Taylor A, Jahngen-Hodge J, Smith D, Palmer V, Dallal G, Lipman R, Padhye N, Frei B. Dietary restriction delays cataract and reduces ascorbate levels in Emory mice. Exp Eye Res 1995; 61:55–62.
29. Levine M. New concepts in the biology and biochemistry of ascorbic acid. N Engl J Med 1986; 314:892–902.
30. Frei B, Stocker R, Ames BN. Antioxidant defenses and lipid peroxidation in human blood plasma. Proc Natl Acad Sci USA 1988; 85:9748–9752.
31. Berger J, Shepard D, Morrow F, Taylor A. Relationship between dietary intake and tissue levels of reduced and total vitamin C in the guinea pig. J Nutr 1989; 119:734–740.
32. Berger J, Shepard D, Morrow F, Sadowski J, Haire T, Taylor A. Reduced and total ascorbate in guinea pig eye tissues in response to dietary intake. Curr Eye Res 1988; 7:681–686.
33. Nakamura B, Nakamura O. Ufer das vitamin C in der linse und dem Kammerwasser der menschlichen katarakte. Graefes Arch Clin Exp Ophthalmol 1935; 134:197–200.
34. Wilczek M, Zygulska-Machowa H. Zawartosc witaminy C W. roznych typackzaem. J Klin Oczna 1968; 38:477–480.
35. Yokoyama T, Sasaki H, Giblin FJ, Reddy VN. A physiological level of ascorbate inhibits galactose cataract in guinea pigs by decreasing polyol accumulation in the lens epithelium: a dehydroascorbate-linked mechanism. Exp Eye Res 1994; 58:207–18.
36. Kosegarten DC, Maher TJ. Use of guinea pigs as model to study galactose-induced cataract formation. J Pharm Sci 1978; 67:1478–1479.
37. Vinson JA, Possanza CJ, Drack AV. The effect of ascorbic acid ongalactose-induced cataracts. Nutr Rep Int 1986; 33:665–668.
38. Devamanoharan PS, Ramachandran S, Varma SD. hydrogen peroxide in the eye lens: radioisotopic determination. Curr Eye Res 1991; 10:831–838.
39. Nishigori H, Lee JW, Yamauchi Y, Iwatsuru M. The alteration of lipid peroxide in glucocorticoid-induced cataract of developing chick embryos and the effect of ascorbic acid. Curr Eye Res 1986; 5:37–40.
40. Blondin J, Taylor A. Measures of leucine aminopeptidase can be used to anticipate UV-induced age-related damage to lens proteins: ascorbate can delay this damage. Mech Aging Dev 1987; 41:39–46.
41. Blondin J, Baragi V, Schwartz E, Sadowski J, Taylor A. Delay of UV-induced eye lens protein damage in guinea pigs by dietary ascorbate. J Free Radic Biol Med 1986; 2:275–281.
42. Taylor A, Jacques PF. Relationships between aging, antioxidant status, and cataract. Am J Clin Nutr 1995; 62:1439S–1447S.
43. Blondin J, Baragi VJ, Schwartz E, Sadowski J, Taylor A. Dietary vitamin C delays UV-induced age-related eye lens protein damage. Ann NY Acad Sci 1987; 498:460–463.
44. Garland DL. Ascorbic acid and the eye. Am J Clin Nutr 1991; 54:1198S.
45. Naraj RM, Monnier VM. Isolation and characterization of a blue fluorophore from human eye lens crystallins: in vitro formation from Maillard action with ascorbate and ribose. Biochim Biophys Acta 1992; 1116:34–42.

46. Bensch KG, Fleming JE, Lohmann W. The role of ascorbic acid in senile cataract. Proc Natl Acad Sci USA 1985; 82:7193–7196.
47. Vasan S, Zhang X, Zhang X, Kapurniotu A, Bernhagen J, Teichburg S, Basgen J, Wagle D, Shih D, Terlecky I, Bucala R, Cerami A, Egan J, Ulrich P. An agent cleaving glucose-derived protein crosslinks in vitro and in vivo. Nature 1996; 382:275.
48. Rathbun WB, Holleschau AM, Cohen JF, Nagasawa HT. Prevention of acetaminophen- and naphthalene-induced cataract and glutathione loss by CySSME. Invest Ophthalmol Vis Sci 1996; 37:923–929.
49. Martenssen J, Steinhertz R, Jain A, Meister A. Glutathione ester prevents buthionine sulfoximine-induced cataracts and lens epithelial cell damage. Biochemistry 1989; 86:8727–8731.
50. Rathbun WB, Killen CE, Holleschau AM, Nagasawa HT. Maintenance of hepatic glutathione homeostasis and prevention of acetaminophen-induced cataract in mice by L-cysteine prodrugs. Biochem Pharmacol 1996; 51:1111–1116.
51. Vina J, Perez C, Furukawa T, Palacin M, Vina JR. Effect of oral glutathione on hepatic glutathione levels on rats and mice. Br J Nutr 1989; 62:683–691.
52. Clark JI, Livesey JC, Steele JE. Delay or inhibition of rat lens opacification using pantethine and WR-77913. Exp Eye Res 1996; 62:75–84.
53. Congdon NG, Duncan DD, Fisher D, Rieger K, Urist J, Sanchez AM, Vitale S, West SK, Pham T, Cole L, McNaughton C. UV light and lenticular opacities in the Emory Mouse. Invest Ophthalmol Vis Sci 1997; 38:S1020.
54. Srivastava SK, Ansari NH. Prevention of sugar-induced cataractogenesis in rats by butylated hydroxytoluene. Diabetes 1988; 37:1505–1508.
55. Yeum K-J, Shang F, Schalch W, Russell RM, Taylor A. Fat-soluble nutrient concentrations in different layers of human cataractous lens. Curr Eye Res 1999; 19:5021–505.
56. Leske MC, Wu SY, Sperduto R, Underwood B, Chylack LT Jr, Milton RC, Srivastava S, Ansari N. Biochemical factors in the lens opacities. Case-control study. The lens opacities case-control study group. Arch Ophthalmol 1995; 113:1113–1119.
57. Schalch W, Weber P. Vitamins and carotenoids—a promising approach to reducing the risk of coronary heart disease, cancer and eye diseases. Adv Exp Med Biol 1994; 366:335–3350.
58. Machlin LJ, Bendich A. Free radical tissue damage: protective role of antioxidant nutrients. FASEB J 1987; 1:441–445.
59. Costagliola C, Iuliano G, Menzione M, Rinaldi E, Vito P, Auricchio G. Effect of vitamin E on glutathione content in red blood cells, aqueous humor and lens of humans and other species. Exp Eye Res 1986; 43:905–914.
60. Yeum K-J, Taylor A, Tang G, Russell RM. Measurement of carotenoids, retinoids, and tocopherols in human lenses. Invest Ophthalmol Vis Sci 1995; 36:2756–2761.
61. Stevens RJ, Negi DS, Short SM, van Kuijk FJGM, Dratz EA, Thomas DW. Vitamin E distribution in ocular tissues following long-term dietary depletion and supplementation as determined by microdissection and gas chromatography-mass spectrometry. Exp Eye Res 1988; 47:237–245.
62. Jacques PF, Taylor A. Micronutrients and age-related cataracts. In: Bendich A, Butterworth CE, eds. Micronutrients in Health and in Disease Prevention. New York: Marcel Dekker, 1991:359–379.
63. Creighton MO, Ross WM, Stewart-DeHaan PJ, Sanwai M, Trevithick JR. Modeling cortical cataractogenesis. VII: Effects of vitamin E treatment on galactose induced cataracts. Exp Eye Res 1985; 40:213–222.
64. Bhuyan DK, Podos SM, Machlin LT, Bhagavan HN, Chondhury DN, Soja WS, Bhuyan KC. Antioxidant in therapy of cataract II: Effect of all-roc-alpha-tocopherol (vitamin E) in sugar-induced cataract in rabbits. Invest Ophthalmol Vis Sci 1983; 24:74.
65. Bhuyan KC, Bhuyan DK. Molecular mechanism of cataractogenesis: III. Toxic metabolites of oxygen as initiators of lipid peroxidation and cataract. Curr Eye Res 1984; 3:67–81.
66. Hammond BR Jr, Wooten BR, Snodderly DM. The density of the human crystalline lens is

related to the macular pigment carotenoids, lutein and zeaxanthin. Optom Vis Sci 1997; 74: 499–504.

67. Snodderly DM, Hammond BR Jr. In vivo psychophysical assessment of nutritional and environmental influences on human ocular tissues: lens and macular pigment. In: Taylor A, ed. Nutritional and Environmental Influences on the Eye. Boca Raton, FL: CRC Press, 1999: 251–285.
68. Schalch W, Dayhaw-Barker P, Barker FM II. The carotenoids of the human retina. In: Taylor A, ed. Nutritional and Environmental Influences on the Eye. Boca Raton, FL: CRC Press, 1999:215–250.
69. Zigler JS, Goosey JD. Singlet oxygen as a possible factor in human senile nuclear cataract development. Curr Eye Res 1984; 3:59–65.
70. Varma SD, Chand D, Sharma YR, Kuck JF Jr, Richards RD. Oxidative stress on lens and cataract formation: role of light and oxygen. Curr Eye Res 1984; 3:35–57.
71. Fridovich I. Oxygen: Aspects of its toxicity and elements of defense. Curr Eye Res 1984; 3:1–2.
72. Giblin FJ, McReady JP, Reddy VN. The role of glutathione metabolism in detoxification of $H_2O_2$ in rabbit lens. Invest Ophthalmol Vis Sci 1992; 22:330–335.
73. Berman ER. Biochemistry of the Eye. New York: Plenum Press, 1991:210–308.
74. Taylor A, Jacques PF, Dorey CK. Oxidation and aging: impact on vision. J Toxicol Indust Health 1993; 9:349–371.
75. Eisenhauer DA, Berger JJ, Peltier CZ, Taylor A. Protease activities in cultured beef lens epithelial cells peak and then decline upon progressive passage. Exp Eye Res 1988; 46:579–590.
76. Jahngen-Hodge J, Laxman E, Zuliani A, Taylor A. Evidence for ATP and ubiquitin degradation of proteins in cultured bovine lens epithelial cells. Exp Eye Res 1991; 52:41–47.
77. Shang F, Taylor A. Oxidative stress and recovery from oxidative stress are associated with altered ubiquitin conjugating and proteolytic activities in bovine lens epithelial cells. Biochem J 1995; 307:297–303.
78. Huang LL, Jahngen-Hodge JJ, Taylor A. Bovine lens epithelial cells have a ubiquitin-dependent proteolysis system. Biochim Biophys Acta 1993; 1175:181–187.
79. Jahngen JH, Lipman RD, Eisenhauer DA, Jahngen EGE Jr, Taylor A. Aging and cellular maturation cause changes in ubiquitin-eye lens protein conjugates. Arch Biochem Biophys 1990; 276:32–37.
80. Jahngen-Hodge J, Cyr D, Laxman E, Taylor A. Ubiquitin and ubiquitin conjugates in human lens. Exp Eye Res 1992; 55:897–902.
81. Obin M, Nowell T, Taylor A. The photoreceptor G-protein transducin ($G_t$) is a substrate for ubiquitin-dependent proteolysis. Biochem Biophys Res Commun 1994; 200:1169–1176.
82. Jahngen JH, Haas AL, Ciechanover A, Blondin J, Eisenhauer D, Taylor A. The eye lens has an active ubiquitin-protein conjugation system. J Biol Chem 1986; 261:13760–13767.
83. Obin M, Jahngen-Hodge J, Nowell T, Taylor A. Ubiquitinylation and ubiquitin-dependent proteolysis in vertebrate photoreceptors (rod outer segments): evidence for ubiquitinylation of Gt and rhodopsin. J Biol Chem 1996; 271:14473–14484.
84. Obin M, Shang F, Gong X, Handelman G, Blumberg J, Taylor A. Redox regulation of ubiquitin-conjugating enzymes: mechanistic insights using the thiol-specific oxidant diamide. FASEB J 1998; 12:561–569.
85. Obin M, Mesco E, Gong X, Haas A, Joseph J, Taylor A. Neurite Outgrowth in PC12 Cells (Distinguishing the roles of ubiquitinylation and ubiquitin-dependent proteolysis). J Biol Chem 1999; 27:11789–11795.
86. Jahngen-Hodge J, Obin M, Gong X, Shang F, Nowell T, Gong J, Abasi H, Blumberg J, Taylor A. Regulation of ubiquitin conjugating enzymes by glutathione following oxidative stress. J Biol Chem 1997; 272:28218–28226.

87. Shang F, Gong X, Palmer HJ, Nowell TR, Taylor A. Age-related decline in ubiquitin conjugation in response to oxidative stress in the lens. Exp Eye Res 1997; 64:21–30.
88. Shang F, Gong X, Taylor A. Activity of ubiquitin-dependent pathway in response to oxidative stress: Ubiquitin-activating enzye (E1) is transiently upregulated. J Biol Chem 1997; 272: 23086–23093.
89. Shang F, Gong X, Taylor A. Changes in ubiquitin conjugation activities in young and old lenses in response to oxidative stress. Invest Ophthalmol Vis Sci 1995; 36:S528.
90. The Italian-American Cataract Study Group. Risk factors for age-related cortical, nuclear, and posterior subcapsular cataracts. Am J Epidemiol 1991; 133:541–553.
91. Mares-Perlman JA, Klein BEK, Klein R, Ritter LL. Relationship between lens opacities and vitamin and mineral supplement use. Ophthalmology 1994; 101:315–355.
92. Robertson J McD, Donner AP, Trevithick JR. Vitamin E intake and risk for cataracts in humans. Ann NY Acad Sci 1989; 570:372–382.
93. Leske MC, Chylack LT, Wu S-Y, Group LOC-C. The lens opacities case-control study: risk factors for cataract. Arch Ophthalmol 1991; 109:244–251.
94. Jacques PF, Chylack LT. Epidemiologic evidence of a role for the antioxidant vitamins and carotenoids in cataract prevention. Am J Clin Nutr 1991; 53:352S–3525S.
95. Mohan M, Sperduto RD, Angra SK, Milton RC, Mathur RL, Underwood B, Jafery N, Pandya CB. India-US case-control study of age-related cataract. Arch Ophthalmol 1989; 107:670–676.
96. Vitale S, West S, Hallfrisch J, Alston C, Wang F, Moorman C, Muller D, Singh V, Taylor HR. Plasma antioxidants and risk of cortical and nuclear cataract. Epidemiology 1993; 4: 195–203.
97. Luthra R, Wa S-Y, Leske MC. Lens opacities and use of nutritional supplements: The Barbados Study. Invest Ophthalmol Vis Sci 1997; 8:S450.
98. Jacques PF, Chylack LT Jr, Hankinson SE, Khu PM, Rogers G, Friend J, Tung W, Wolfe JK, Padhye N, Willett WC, Taylor A. Long-term nutrient intake and early age-related nuclear lens opacities. In press.
99. Taylor A, Jacques PF, Chylack LT Jr, Hankinson SE, Khu P, Rogers G, Friend J, Tung W, Wolfe JK, Padhye N, Willett WC. Long-term nutrient intake and early age-related cortical and posterior subcapsular cataracts. In press.
100. Lyle BJ, Mares-Perlman J, Klein BEK, Klein R, Palta M, Bowen PE, Greger JL. Serum carotenoids and tocopherols and incidence of age-related nuclear cataract. Am J Clin Nutr 1999; 69:272–277.
101. Lyle BJ, Mares-Perlman JA, Klein BE, Klein R, Greger JL. Antioxidant intake and risk of incident age-related nuclear cataracts in the Beaver Dam Eye Study. Am J Epidemiol 1999; 149:801–809.
102. Brown L, Rimm EB, Seddon JM, Giovannucci EL, Chasan-Taber L, Spiegelman D, Willett WC, Hankinson SE. A prospective study of carotenoid intake and risk of cataract extraction in US men. Am J Clin Nutr 1999; 70:517–524.
103. Chasan-Taber L, Willett WC, Seddon JM, Stampfer MJ, Rosner B, Colditz GA, Speizer FE, Hankinson SE. A prospective study of carotenoid and vitamin A intakes and risk of cataract extraction in US women. Am J Clin Nutr 1999; 70:509–516.
104. Chasan-Taber L, Willett WC, Seddon JM, Stampfer MJ, Rosner B, Colditz GA, Hankinson SE. A prospective study of vitamin supplement intake and cataract extraction among US women. Epidemiology 1999; 10:679–684.
105. Seddon JM, Christen WG, Manson JE, LaMotte FS, Glynn RJ, Buring JE, Hennekens CH. The use of vitamin supplements and the risk of cataract among US male physicians. Am J Publ Health 1994; 84:788–792.
106. Hankinson SE, Stampfer MJ, Seddon JM, Colditz GA, Rosner B, Speizer FE, Willett WC. Intake and cataract extraction in women: a prospective study. Br Med J 1992; 305:335–339.

107. Knekt P, Heliovaara M, Rissanen A, Aromaa A, Aaran R. Serum antioxidant vitamins and risk of cataract. Br Med J 1992; 305:1392–1394.
108. Mares-Perlman JA, Brady WE, Klein BEK, Klein R, Palta M. Supplement use and 5-year progression of cortical opacities. Invest Ophthalmol Vis Sci 1996; 37:S237.
109. Rouhiainen P, Rouhiainen H, Salonen JT. Association between low plasma vitamin E concentration and progression of early cortical lens opacities. Am J Epidemiol 1996; 144:496–500.
110. Leske MC, Chylack LT Jr, He Q, Wu SY, Schoenfeld E, Friend J, Wolfe J, LSC Group. Antioxidant vitamins and nuclear opacities—The Longitudinal Study of Cataract. Ophthalmology 1998; 105:831–836.
111. Leske MC, Chylack LT Jr, He Q, Wu SY, Schoenfeld E, Friend J, Wolfe J, LSC Group. Risk factors for nuclear opalescence in a longitudinal study. Am J Epidemiol 1998; 147:36–41.
112. Jacques PF, Taylor A, Hankinson SE, Lahav M, Mahnken B, Lee Y, Vaid K, Willett WC. Long-term vitamin C supplement use and prevalence of early age-related lens opacities. Am J Clin Nutr 1997; 66:911–916.
113. Mares-Perlman JA, Brady WE, Klein BEK, Klein R, Haus GJ, Palta M, Ritter LL, Shoff SM. Diet and nuclear lens opacities. Am J Epidemiol 1995; 141:322–334.
114. Simon JA, Hudes ES. Serum ascorbic acid and other correlates of self-reported cataract among older Americans. Clin Epidemiol 1999; 52:1207–1211.
115. Taylor A, Jacques PF, Nadler D, Morrow F, Sulsky SI, Shephard D. Relationship in humans between ascorbic acid consumption and levels of total and reduced ascorbic acid in lens, aqueous and plasma. Curr Eye Res 1991; 10:751–759.
116. Sperduto RD, Hu T-S, Milton RC, Zhao J, Everett DF, Cheng Q, Blot WJ, Bing L, Taylor PR, Jun-Yao L, Dawsey S, Guo W. The Linxian cataract studies: Two nutrition intervention trials. Arch Ophthalmol 1993; 111:1246–1253.
117. McCarty CA, Mukesh BN, Fu CL, Taylor HR. The epidemiology of cataract in Australia. Am J Ophthalmol 1999; 446–465.
118. Nadalin G, Robman LD, McCarty CA, Garrett SK, McNeil JJ, Taylor HR. The role of past intake of vitamin E in early cataract changes. Ophthalm Epidemiol 1999; 6:105–12.
119. Mares-Perlman JA, Brady WE, Klein BEK, Klein R, Palta M, Bowen P, Stacewicz-Sapuntzakis M. Serum carotenoids and tocopherols and severity of nuclear and cortical opacities. Invest Ophthalmol Vis Sci 1995; 36:276–288.
120. Burton W, Ingold KU. Beta-carotene: an unusual type of lipid antioxidant. Science 1984; 224:569–573.
121. Kwan M, Niinikoski J, Hunt TK. In vivo measurement of oxygen tension in the cornea, aqueous humor, and the anterior lens of the open eye. Invest Ophthalmol Vis Sci 1972; 11: 108–114.
122. Erdman J. The physiologic chemistry of carotenes in man. Clin Nutr 1988; 7:101–106.
123. Di Mascio P, Murphy ME, Sies H. Antioxidant defense systems: The role of carotenoids, tocopherols and thiols. Am J Clin Nutr 1991; 53:194S–200S.
124. Krinsky NI, Deneke SS. Interaction of oxygen and oxy-radicals with carotenoids. J Natl Cancer Inst 1982; 69:205–210.
125. Micozzi MS, Beecher GR, Taylor HR, Khachik F. Carotenoid analyses of selected raw and cooked foods associated with a lower risk for cancer. J Natl Cancer Inst 1990; 82:282–285.
126. Daicker B, Schiedt K, Adnet JJ, Bermond P. Canthaxanthin retinopathy. An investigation by light and electron microscopy and physiological analyses. Graefe's Arch Clin Exp Ophthalmol 1987; 225:189–197.
127. Leske MC, Wu SY, Connell AMS, Hyman L, et al. Lens opacities, demographic factors and nutritional supplements in the Barbados Eye Study. Int J Epidemiol 1997; 26:1314–1322.
128. Harding JJ, van Heyningen R. Epidemiology and risk factors for cataract. Eye 1987; 1:537–541.

129. McLaren DS. Nutritional Ophthalmology, 2nd ed. London: Academic Press, 1980.
130. Jacob RA, Otradovec C, Russell RM, Munro HN, Hartz SC, McGandy RB, Morrow FD, Sadowski JA. Vitamin C status and nutrient interactionsin a healthy elderly population. Am J Clin Nutr 1988; 48:1436–1442.
131. Wefers H, Sies H. The protection by ascorbate and glutathione against microsomal lipid peroxidation is dependent on vitamin E. FEBS Lett 1988; 174:353–357.
132. Burton GW, Wronska U, Stone L, Foster DO, Ingold KU. Biokinetics of dietary RRR-alpha-tocopherol in the male guinea pig at three dietary levels of vitamin C and two levels of vitamin E: Evidence that vitamin C does not ''spare'' vitamin E in vivo. Lipids 1990; 25: 199–210.
133. Chen S. A protective role for glutathione-dependent reduction of dehydroascorbic acid in lens epithelium. Invest Ophthalmol Vis Sci 1995; 36:1805.
134. Sasaki H, Giblin FJ, Winkler BS, Chakrapani B, Leverenz V, Ch-Johnston CS, Meyer CG, Srilakshmi JC. Vitamin C elevates red blood cell glutathione in healthy adults. Am J Clin Nutr 1993; 58:103–105.
135. Bohm F, Edge R, Land EJ, McGarvey DJ, Truscott TG. Carotenoids enhance vitamin E antioxidant efficiency. Am Chem Soc 1997; 119:621–622.
136. Valgimigli L, Lucarini M, Pedulli GF, Ingold KU. Does β-carotene really protect vitamin E from oxidation. J Am Chem Soc 1997; 119:8095–8096.
137. Taylor A, Berger J, Reddan J, Zuliani A. Effects of aging in vitro on intracellular proteolysis in cultured rabbit lens epithelial cells in the presence and absence of serum. In Vitro Cell Devel Biol 1991; 27A:287–292.
138. Fleshman KR, Wagner BJ. Changes during aging in rats lens endopeptidase activity. Exp Eye Res 1984; 39:543–551.
139. Ray K, Harris H. Purification of neutral lens endopeptidase: close similarity to a neutral proteinase in pituatory. Proc Natl Acad Sci USA 1985; 82:7545–7549.
140. Murakami K, Jahngen JH, Lin S, Davies KJA, Taylor A. Lens proteasome shows enhanced rates of degradation of hydroxyl radical modified alpha-crystallin. Free Rad Biol Med 1990; 8:217–222.
141. Taylor A, Brown MJ, Daims MA, Cohen J. Localization of leucine aminopeptidase in hog lenses using immunofluorescence and activity assays. Invest Ophthalmol Vis Sci 1983; 24: 1172–1181.
142. Varnum MD, David LL, Shearer TR. Age-related changes in calpain II and calpastatin in rat lens. Exp Eye Res 1989; 49:1053–1065.
143. Yoshida H, Yumoto N, Tsukahara I, Murachi T. The degradation of alpha-crystallin at its carboxyl-terminal portion by calpain in bovine lens. Invest Ophthalmol Vis Sci 1986; 27: 1269–1273.

# Index